INTERMEDIATE ALGEBRA

AN INTERACTIVE APPROACH

INTERMEDIATE ALGEBRA

AN INTERACTIVE APPROACH

Third Edition

Linda Pulsinelli   Patricia Hooper

Department of Mathematics, Western Kentucky University

MACMILLAN PUBLISHING COMPANY
NEW YORK

COLLIER MACMILLAN CANADA
TORONTO

MAXWELL MACMILLAN INTERNATIONAL

NEW YORK OXFORD SINGAPORE SYDNEY

Printed in the United States of America

Macmillan Publishing Company
866 Third Avenue, New York, New York 10022

Collier Macmillan Canada, Inc.
1200 Eglinton Avenue, E.
Suite 200
Don Mills, Ontario, M3C 3N1

Library of Congress Cataloging-in-Publication Data

Pulsinelli, Linda Ritter.
Intermediate algebra : an interactive approach / Linda Pulsinelli, Patricia Hooper. — 3rd ed.
p. cm.
Includes index.
ISBN 0-02-396994-6
1. Algebra. I. Hooper, Patricia (Patricia L.) II. Title.
QA154.2.P85 1991 90-35859
512.9—dc20 CIP

Printing: 2 3 4 5 6 7 8 Year: 3 4 5

Preface

Intermediate Algebra: An Interactive Approach is designed to be useful to students in bridging the gap between high school and college algebra. It is generally assumed that a student using this textbook has been exposed to an introductory algebra course but is not ready to tackle traditional college algebra.

In order to make this mathematics textbook readable, we have worded explanations in clear, concise language understandable to students at this level. Our general approach is to start from a fundamental idea with which students are familiar and proceed to a related concept in the most straightforward, intuitive way possible.

Our experience with intermediate algebra students leads us to believe that they must practice each new skill as soon as it has been presented. For this reason the book includes several unique features that are designed to provide maximum reinforcement. Each chapter in the book follows the same basic structure.

Motivational Applied Problem

At the beginning of each chapter we have presented an applied problem that can be solved after the student has mastered the skills in that chapter. Its solution appears within the chapter.

Explanations

We have tried to avoid the "cookbook" approach to algebra by including a straightforward and readable explanation of each new concept. Realizing that students at the intermediate level become easily bogged down in reading lengthy explanations, we have attempted to make our explanations as brief as possible without sacrificing rigor.

Highlighting

Definitions, properties, theorems, and formulas are highlighted in boxes throughout the book for easy student reference. In most cases a rephrasing of a generalization in words accompanies the symbolic statement, and it is also highlighted in a box.

Examples

Immediately following the presentation of a new idea, several completely worked-out examples appear along with several partially worked-out examples with blanks to be filled in by the student. These examples are completed correctly at the end of each section and the student is advised to check his or her work immediately.

Trial Runs

Sprinkled throughout each section are several Trial Runs, a short list of problems to check the student's grasp of a new skill. The answers appear at the end of the section.

Exercise Sets

Each section concludes with an extensive Exercise Set in which each odd-numbered problem corresponds closely to the following even-numbered problem.

Stretching the Topics

At the end of each Exercise Set there are several problems designed to challenge students by extending to the next level of difficulty the skills learned in that section.

Checkups

Following each Exercise Set, a list of about 10 problems checks the student's mastery of the most important concepts in the section. Each Checkup problem is keyed to comparable examples in the section for restudy if necessary.

Problem Solving

One section of almost every chapter involves switching from words to algebra. By including such a section in each chapter, we are attempting to treat problem solving as a natural outgrowth of acquiring algebraic skills.

Chapter Summaries

Each chapter concludes with a summary in which the important ideas are again highlighted, in tables when possible. New concepts are presented in symbolic form and verbal form, accompanied by a typical example.

Speaking the Language of Algebra and Writing about Mathematics

Following the summary, we have included a group of sentences to be completed *with words* and questions to be answered *in sentences* by the student. Algebra students (especially those in self-paced programs) often lack the opportunity to "speak mathematics." We hope that these short sections will help them develop a better mathematics vocabulary and become more comfortable expressing themselves in mathematical terms.

Review Exercises

A list of exercises reviewing all the chapter's important concepts serves to give the student an overview of the content. Each problem is keyed to the appropriate section and examples.

Practice Test

A Practice Test is included to help the student prepare for a test over the material in the chapter. Once again, each problem is keyed to the appropriate chapter section and examples.

Sharpening Your Skills

Finally, we have included a short list of exercises that will provide a cumulative review of concepts and skills from earlier chapters. Retention seems to be a very real problem with students at this level, and we hope that these exercises will serve to minimize that problem. Each cumulative review exercise is keyed to the appropriate chapter and section.

Throughout the book we have adhered to a rather standard order of topics, making an attempt to connect new concepts to old ones whenever appropriate. This modified spiraling technique is designed to help students maintain an overview of the content. Success in future courses seems to us to hinge on students' seeing that algebra is a logical progression of ideas rather than a set of unrelated skills to be memorized and forgotten.

In preparing the third edition, we wished to make the text even more readable for students. We have eliminated wasted space, added more annotation to many examples, and increased the number of applied problems. We have also revised our treatment of the parabola and added interval notation to our discussion of inequalities.

The answers to the odd-numbered exercises in the Exercise Sets appear in the back of the book together with answers for *all* items in Stretching the Topics, Checkups, Speaking the Language of Algebra, Review Exercises, Practice Tests, and Sharpening Your Skills.

More assistance for students and instructors can be found among the supplementary materials that accompany this book.

Instructor's Manual with Test Bank

The Instructor's Manual contains the answers for all exercises in the Exercise Sets and Stretching the Topics. In addition there are eight Chapter Tests (four open-ended and four multiple choice) for each chapter and four Final Examinations (two open-ended and two multiple choice). Answers to these tests and examinations also appear in the Instructor's Manual. The chapter tests are also available on computer disks.

Student's Solutions Manual

The Student's Solutions Manual, written by Rebecca Stamper, contains step-by-step solutions for the even-numbered exercises in the Exercise Sets and for *all* items in the Review Exercises, Practice Tests, Sharpening Your Skills, and exercises involving word problems. Using the same style as appears in the text, these solutions emphasize the procedure as well as the answer.

Video Tapes

A series of video tapes (each 20 to 30 minutes in length) provides explanations for many topics in the course.

Computerized Test Generator

A set of computer-generated tests is available for producing tests of any length for each chapter. Cumulative tests and final exams may also be constructed.

Interactive Software Program

This computer-assisted program is available for Apple and IBM systems and is a series of lessons including problems at differing levels of difficulty.

Geometry Supplement

A supplementary geometry chapter is available upon request to adopters of the third edition. The chapter contains an overview of geometric measurement and a detailed discussion of the triangle, including the basic concepts of similarity and congruence.

Acknowledgments

The writing of this book would not have been possible without the assistance of many people. We express our appreciation to our typist Maxine Worthington, to Rebecca Stamper of Western Kentucky University and Kathy Rodgers of the University of Southern Indiana for carefully working all our problems, to our families for tolerating our obsessive work schedules, to our editors Gary Ostedt and Bob Pirtle for their enthusiastic support, and to Ed Burke and the production staff at Hudson River Studio for their expert guidance and efficient supervision.

We also thank our reviewers for their careful scrutiny and helpful comments.

Marshall Cates, California State University, Los Angeles
Robert Kaiden, Lorain County Community College
Mark Serebransky, Camden County College
Howard Anderson, Skagit Valley College
Bruce F. Sloan, Simpson College
E. James Peake, Iowa State University
Robert A. Davies, Cuyahoga Community College
Jack R. Barone, Baruch College
Edward Hancock, DeVry Institute of Technology, Phoenix
Barbara Duch, University of Delaware
John R. Martin, Tarrant County Junior College, Northeast Campus
Nancy S. Heyl, University of Georgia
Linda Mudge, Illinois Valley Community College
Cheryl I. Hobneck, Illinois Valley Community College
Gerald J. LePage, Bristol Community College
David Price, Tarrant County Junior College, South Campus
Dan Streeter, Portland State University
William C. Dunning, Simpson College
Diane Geerken, SUNY, Cobleskill

The cover painting, *Elegant Journey*, was executed in acrylic, gauze, and dyes by Frances Wells, a Kentucky artist and teacher. Her prints, oils, and watercolors appear in numerous galleries and in corporate and private collections throughout the United States.

L.R.P.
P.I.H.

Contents

INTERMEDIATE ALGEBRA

AN INTERACTIVE APPROACH

C H A P T E R

1

Working with Numbers and Variables

Charlie is a part-time college student. He must pay \$50 per course credit hour. He has saved \$800. Write an algebraic expression for the amount Charlie will have left after enrolling for *h* credit hours.

We review operations with real numbers here at the start, so that we can use them with confidence throughout the remainder of the text. In this chapter we recall how to

1. Recognize different kinds of numbers.
2. Perform operations with real numbers.
3. Use the properties of the real numbers.
4. Work with variables and constants in algebraic expressions.
5. Find the absolute value of a quantity.
6. Switch from word expressions to algebraic expressions.

1.1 Working with Sets of Numbers

In the same way that your knowledge of different kinds of numbers has changed throughout your life, mathematicians introduced new sets of numbers as they were needed. We shall use standard set notation to represent our number sets, naming each with a capital letter and listing the members of each set with commas between them and enclosed in braces.

Recognizing Sets of Numbers

The first set of numbers ever used for measuring and counting was the set of **natural numbers**, represented by N.

Natural Numbers

$N = \{1, 2, 3, 4, \dots\}$

The dots following the number 4 indicate that this set continues forever; it is an **infinite set** containing no last element. However, N is a **well-defined set**, which means that we can determine exactly what numbers belong to the set without listing all its members.

The need for a symbol to represent *none* of some measure led to the introduction of **zero**, and the set of **whole numbers** was invented. We shall use W to represent this set.

Whole Numbers

$W = \{0, 1, 2, 3, 4, \dots\}$

Notice that the set of whole numbers is also a well-defined, infinite set. Notice, too, that every natural number is a whole number.

In order to measure losses, mathematicians expanded their set of numbers to include the opposites (or **negatives**) of the natural numbers, and the set of **integers** came into existence. We represent the set of integers with the letter J.

Integers

$J = \{\dots, -3, -2, -1, 0, 1, 2, 3, \dots\}$

This set is also well defined and infinite, and it contains all the whole numbers.

Let's take a moment here to introduce some mathematical shorthand. To say that a number belongs to a set (or is an **element** of a set) we use the symbol "ϵ."

Symbol	Meaning	Examples
$\in$	Is an element of	$5 \in N$ $-6 \in J$
$\notin$	Is not an element of	$-2 \notin W$ $\frac{1}{2} \notin J$
$=$	Is equal to	$2 + 3 = 5$ $7 \cdot 4 = 28$
$\neq$	Is not equal to	$-6 \neq 0$ $\frac{2}{3} \neq \frac{3}{2}$

A need to represent *parts* of integers led mathematicians to define the set of **rational numbers** containing every number that can be expressed as a fraction with an integer as the numerator and a nonzero integer as the denominator. Using a to represent the numerator, b to represent the denominator, and Q to represent the set of rational numbers, we define this new set.

Rational Numbers

$$Q = \left\{\frac{a}{b} : a \in J,\ b \in J,\ b \neq 0\right\}$$

To be a member of Q a number *must* be expressible as a quotient of integers. You should agree that these numbers are all rational numbers.

$$\frac{2}{3},\quad \frac{1}{2},\quad \frac{7}{93},\quad -\frac{5}{8},\quad \frac{17}{11},\quad -\frac{1}{6},\quad -\frac{9}{5}$$

Since every integer can be expressed as a quotient of integers, every integer is also a rational number. Some decimal numbers are also rational numbers; in particular, any decimal number that *terminates* or *repeats* in a fixed block of digits is a rational number.

$6 \in Q$	because	$6 = \frac{6}{1}$
$-3 \in Q$	because	$-3 = \frac{-3}{1}$
$0 \in Q$	because	$0 = \frac{0}{1}$
$0.37 \in Q$	because	$0.37 = \frac{37}{100}$
$0.\overline{3} \in Q$	because	$0.\overline{3} = \frac{1}{3}$
$-0.\overline{27} \in Q$	because	$-0.\overline{27} = \frac{-3}{11}$

To recognize members of the set of rational numbers, we must look for numbers that are integers or common fractions or repeating decimals or terminating decimals.

Although you may think that we have considered all the possible sets of numbers, there are many numbers, called **irrational numbers**, which are *not* expressible as quotients of integers. Some examples might jog your memory:

$$\sqrt{2}, \quad \sqrt{3}, \quad -\sqrt{5}, \quad \pi, \quad 3 + \sqrt{7}$$

are members of the set of irrational numbers, H.

If we consider all the rational numbers together with all the irrational numbers, we can form the very important set of **real numbers**, represented by R.

Real Numbers

R = {numbers that are rational *or* irrational}

Every number that we have discussed so far belongs to the set of real numbers. Until further notice, all our work in algebra will deal with real numbers.

Example 1. To which of the number sets discussed does 3 belong?

Solution

$3 \in N$

$3 \in W$

$3 \in J$

$3 \in Q$

$3 \in R$

Now try Example 2.

Example 2. To which of the number sets discussed does -0.5 belong?

Solution

$-0.5 \in$ _____

$-0.5 \in$ _____

Check your work on page 9. ▶

You complete Example 3.

Example 3. List the members of the set $\{-5, 0, 0.1, \sqrt{5}, \pi, 4, \frac{24}{5}\}$ that belong to each set.

Natural numbers: 4

Whole numbers: 0, 4

Integers: _____, _____, _____

Rational numbers: _____, _____, _____, _____, _____

Irrational numbers: _____, _____

Check your work on page 9. ▶

Using the Number Line

To better understand the different sets of numbers, we can use a **real number line**. To construct a number line, we draw a straight line and then choose a zero point and a length to represent 1 unit. All points spaced 1 unit apart are labeled to correspond to the integers in order, with *positive* integers to the *right* of zero and *negative* integers to the *left* of zero.

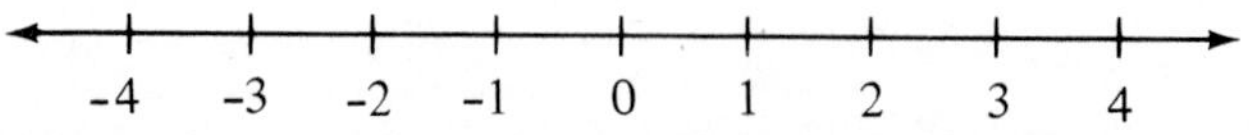

The arrows at either end of the number line show that the line extends indefinitely in both directions, so **we can locate a point corresponding to any rational or irrational number**

on this number line. Although the location of irrational numbers will not be completely accurate, we shall be satisfied with a good approximation. On the number line, we locate a point corresponding to a real number by placing a solid dot in the appropriate position. This is called **graphing a number**.

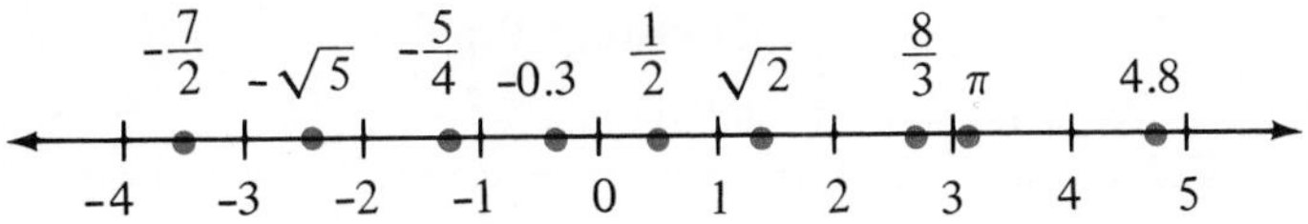

Always be sure to *label* your points so that a reader will not be confused about your meaning. See if you can graph the numbers in Example 4.

Example 4. On a real number line, graph the numbers 3, 1.1, $-\sqrt{2}$, $\frac{9}{2}$, -2.5, and 0.

Solution

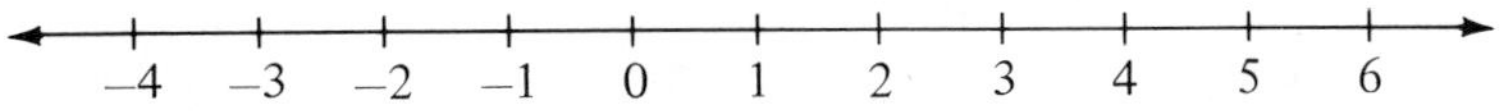

Check your work on page 9. ▶

The number line provides us with a handy means of visualizing the set of real numbers and allows us to make observations about **order** within that set. Suppose that we let a and b represent any real numbers.

If a lies to the *right* of b on the number line, a is **greater than** b.
If a lies to the *left* of b on the number line, a is **less than** b.
If a and b occupy the *same* position on the number line, a is **equal to** b.

Mathematicians have invented symbols to represent these three situations.

Symbols	Meaning	Number Line
$a > b$	a is greater than b	
$a < b$	a is less than b	
$a = b$	a is equal to b	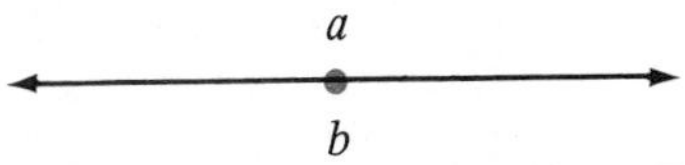

Because of the orderly nature of the real numbers, it is always possible to compare two real numbers using exactly one of these three statements.

Trichotomy Principle. Given any two real numbers, a and b, exactly *one* of the following statements must be true:

$$a > b \quad \text{or} \quad a < b \quad \text{or} \quad a = b$$

The statements $a > b$ and $a < b$ are called **inequalities**, and the number line helps us see that the following property for inequalities makes sense.

> **Transitive Property.** Let a, b, and c be real numbers. If $a < b$ and $b < c$, then $a < c$.

For instance, if we know that $\sqrt{2} < 1.5$ and $1.5 < \sqrt{3}$, we may safely conclude that $\sqrt{2} < \sqrt{3}$.

Example 5. Compare the numbers using the symbols $<$, $>$, or $=$.

$$-2 < 4$$

$$0 > -1$$

$$3 < \pi$$

$$\sqrt{25} = 5$$

Now you try Example 6.

Example 6. Compare the numbers using the symbols $<$, $>$, or $=$.

$$-6 \_\_\_\_ -5.5$$

$$2 \_\_\_\_ \sqrt{3}$$

$$\sqrt{9} \_\_\_\_ 3$$

Check your work on page 9. ▶

If a real number lies to the *right* of zero on the number line, we know that it is *positive*. If a real number lies to the *left* of zero on the number line, we know that it is *negative*. Therefore, we can state whether a number is positive or negative using our symbols of inequality.

$a > 0$	means	"a is a positive real number."
$a < 0$	means	"a is a negative real number."

Example 7. Use an inequality to state whether each number is positive or negative.

$$5, \quad 0.01, \quad -\pi$$

Solution

$$5 > 0$$

$$0.01 > 0$$

$$-\pi < 0$$

Now you try Example 8.

Example 8. Use an inequality to state whether each number is positive or negative.

$$-\tfrac{9}{2}, \quad \sqrt{3}, \quad -\sqrt{3}$$

Solution

$$-\tfrac{9}{2} \_\_\_\_ 0$$

$$\sqrt{3} \_\_\_\_ 0$$

$$-\sqrt{3} \_\_\_\_ 0$$

Check your work on page 9. ▶

Suppose that we look at the real numbers 5 and -5. On the number line, each of these numbers corresponds to a point 5 units from zero. The point corresponding to 5 lies 5 units to the *right* of zero; the point corresponding to -5 lies 5 units to the *left* of zero. Such numbers are called **opposites** of each other and we say that

$$-5 \text{ is the opposite of } 5$$

$$5 \text{ is the opposite of } -5$$

Every real number has an opposite. The opposite of 0 is 0.

Let's complete Example 9.

Example 9. Complete each statement.

-2 is the opposite of _____ .　　_____ is the opposite of π.

1.6 is the opposite of _____ .　　_____ is the opposite of $-\sqrt{2}$.

_____ is the opposite of $\frac{2}{3}$.　　_____ is the opposite of 0.

Check your work on page 9. ▶

Trial Run

1. On the number line graph the numbers $\frac{7}{2}$, $-\sqrt{3}$, 2.2, -3.1, 0, and 1.

Compare the numbers using $<$, $>$, or $=$.

2. 4 _____ $\sqrt{10}$　　**3.** $\frac{5}{2}$ _____ 3　　**4.** -2 _____ -2.5

Use an inequality to state whether each number is positive or negative.

_____ **5.** 0.53　　_____ **6.** $-\sqrt{5}$　　_____ **7.** $\frac{2}{3}$

Complete each statement.

8. 3 is the opposite of _____ .

9. _____ is the opposite of $-\frac{4}{5}$.

10. -0.25 is the opposite of _____ .

Answers are on page 9.

Finding Absolute Values of Numbers

The number line is a useful tool for understanding the important mathematical concept of **absolute value**. Stated very simply, the **absolute value of a real number is its distance from zero on the number line**.

The absolute value of 3 is 3 because its distance from zero is 3 units. The absolute value of -3 is also 3 because its distance from zero is 3 units.

> To find the absolute value of a real number, count the units between that number and zero on the number line.

To denote the absolute value of a number, we enclose it within a pair of vertical bars.

Symbols	Meaning	Evaluation
$\lvert 3\rvert$	Absolute value of 3	$\lvert 3\rvert = 3$
$\lvert -3\rvert$	Absolute value of -3	$\lvert -3\rvert = 3$

Example 10. Find $\lvert 2\rvert$, $\lvert 0\rvert$, and $\lvert -\sqrt{2}\rvert$.

Solution

$\lvert 2\rvert = 2$ because there are 2 units between 0 and 2.

$\lvert 0\rvert = 0$ because there are 0 units between 0 and 0.

$\lvert -\sqrt{2}\rvert = \sqrt{2}$ because there are $\sqrt{2}$ units between 0 and $-\sqrt{2}$.

You complete Example 11.

Example 11. Find $\lvert -16\rvert$, $\lvert 0.25\rvert$, and $\lvert -\frac{2}{3}\rvert$.

Solution

$\lvert -16\rvert =$ ______ because there are ______ units between 0 and ______ .

$\lvert 0.25\rvert =$ ______ because there are ______ units between 0 and ______ .

$\lvert -\frac{2}{3}\rvert =$ ______ because there are ______ units between 0 and ______ .

Check your work on page 9. ▶

From these examples, we make an important observation about absolute value.

> The absolute value of a real number is never negative.

Trial Run

Evaluate.

______ 1. $\lvert 5\rvert$

______ 2. $\lvert -3.2\rvert$

______ 3. $\lvert 0\rvert$

______ 4. $\lvert -\sqrt{3}\rvert$

______ 5. $\lvert 7\rvert + \lvert -2\rvert$

______ 6. $\lvert -8\rvert - \lvert 3\rvert$

______ 7. $\lvert 7 - 2\rvert$

______ 8. $\lvert -8\rvert + \lvert -3\rvert$

Answers are on page 9.

▶ Examples You Completed

Example 2. To which of the number sets discussed does -0.5 belong?

Solution

$$-0.5 \in Q$$

$$-0.5 \in R$$

Example 3 (*Solution*)

Natural numbers: 4

Whole numbers: 0, 4

Integers: $-5, 0, 4$

Rational numbers: $-5, 0, 4, 0.1, \frac{24}{5}$

Irrational numbers: $\sqrt{5}, \pi$

Example 4. On a real number line, graph the numbers 3, 1.1, $-\sqrt{2}$, $\frac{9}{2}$, -2.5, and 0.

Solution

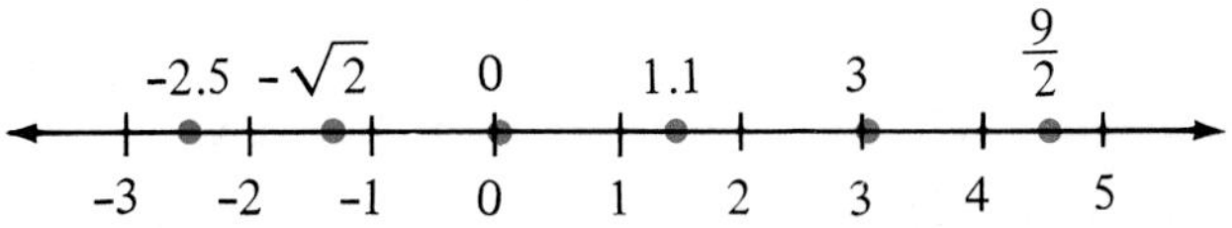

Example 6. Compare the numbers using the symbols $<$, $>$, or $=$.

$$-6 < -5.5$$

$$2 > \sqrt{3}$$

$$\sqrt{9} = 3$$

Example 8. Use an inequality to state whether each number is positive or negative.

$$-\tfrac{9}{2}, \quad \sqrt{3}, \quad -\sqrt{3}$$

Solution

$$-\tfrac{9}{2} < 0$$

$$\sqrt{3} > 0$$

$$-\sqrt{3} < 0$$

Example 9 (*Solution*)

-2 is the opposite of 2.

1.6 is the opposite of -1.6.

$-\frac{2}{3}$ is the opposite of $\frac{2}{3}$.

$-\pi$ is the opposite of π.

$\sqrt{2}$ is the opposite of $-\sqrt{2}$.

0 is the opposite of 0.

Example 11 (*Solution*)

$|-16| = 16$ because there are 16 units between 0 and -16.

$|0.25| = 0.25$ because there are 0.25 units between 0 and 0.25.

$|-\frac{2}{3}| = \frac{2}{3}$ because there are $\frac{2}{3}$ units between 0 and $-\frac{2}{3}$.

Answers to Trial Runs

page 7 1.

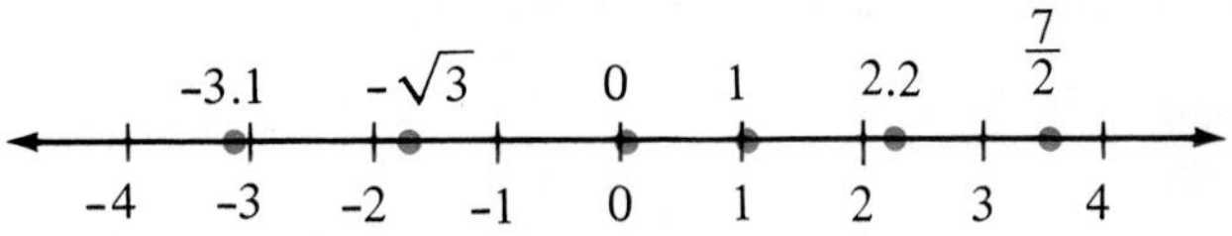

2. $4 > \sqrt{10}$ 3. $\frac{5}{2} < 3$ 4. $-2 > -2.5$ 5. $0.53 > 0$ 6. $-\sqrt{5} < 0$ 7. $\frac{2}{3} > 0$
8. -3 9. $\frac{4}{5}$ 10. 0.25

page 8 1. 5 2. 3.2 3. 0 4. $\sqrt{3}$ 5. 9 6. 5 7. 5 8. 11

Name ________________________________ Date ____________

EXERCISE SET 1.1

Name the number sets to which each of the following numbers belongs.

______ 1. $-\frac{2}{3}$ ______ 2. $\frac{4}{5}$

______ 3. 0 ______ 4. 1

______ 5. $\sqrt{2}$ ______ 6. $\sqrt{5}$

______ 7. 3.52 ______ 8. -1.39

List the members of the set $\{-8, -5.2\ -\sqrt{3}, -\frac{1}{2}, 0, \sqrt{2}, \frac{7}{2}, \pi, 5.\overline{6}, 9\}$ *that belong to the following sets.*

______ 9. Whole numbers ______ 10. Natural numbers

______ 11. Rational numbers ______ 12. Irrational numbers

______ 13. Negative integers ______ 14. Nonpositive integers

Decide whether each statement is true or false.

______ 15. Every natural number is a whole number.

______ 16. Every real number is a rational number.

______ 17. Every rational number can be written as a quotient of integers.

______ 18. Every integer is a rational number.

______ 19. Every irrational number is a real number.

______ 20. Every real number is rational or irrational.

21. Use a real number line to graph the numbers $\frac{8}{3}$, $-\sqrt{7}$, 3.4, -5.12, 0, -2, $0.\overline{3}$.

22. Use a real number line to graph the numbers $\frac{9}{2}$, 5.2, -3.6, 1, -3, $1.\overline{6}$, $-\sqrt{17}$.

Compare the following numbers using $<$, $>$, *or* $=$.

______ 23. 3 and $\sqrt{7}$ ______ 24. $\sqrt{10}$ and 4

______ 25. $-\frac{9}{2}$ and -3 ______ 26. $-\frac{7}{3}$ and -5

______ 27. $\frac{10}{2}$ and $\sqrt{25}$ ______ 28. $\frac{18}{3}$ and $\sqrt{36}$

______ 29. -1.6 and -0.7 ______ 30. -3.9 and -2.5

______ 31. $-\frac{8}{3}$ and $-\sqrt{16}$ ______ 32. $-\frac{12}{5}$ and $-\sqrt{4}$

Use an inequality to state whether each of the following numbers is positive or negative.

_______ 33. 0.29

_______ 34. 0.74

_______ 35. $-\sqrt{6}$

_______ 36. $-\sqrt{11}$

_______ 37. $\frac{3}{4}$

_______ 38. $\frac{2}{3}$

_______ 39. -9

_______ 40. 7

Find the missing number in each statement.

41. 5 is the opposite of _____ .

42. -4 is the opposite of _____ .

43. $-\frac{7}{8}$ is the opposite of _____ .

44. $\frac{4}{5}$ is the opposite of _____ .

45. -3.2 is the opposite of _____ .

46. 5.6 is the opposite of _____ .

47. $\sqrt{5}$ is the opposite of _____ .

48. $-\sqrt{6}$ is the opposite of _____ .

Evaluate each expression.

_______ 49. $|6|$

_______ 50. $|13|$

_______ 51. $|-6.8|$

_______ 52. $|-1.3|$

_______ 53. $|-\sqrt{3}|$

_______ 54. $|-\sqrt{5}|$

_______ 55. $|3| + |-2|$

_______ 56. $|7| + |-8|$

_______ 57. $|-10| - |-2|$

_______ 58. $|-6| - |-4|$

☆ Stretching the Topics

_______ 1. Evaluate $|a + 8|$ for $a = -15$.

_______ 2. Complete $-4(7 + 3)$ _____ $- |-41|$ using $<$, $>$, or $=$.

_______ 3. Evaluate $|-40.39| - (|-3.72| + |18.02|) - |-18.65|$.

Check your answers in the back of your book.

If you can complete **Checkup 1.1**, you are ready to go on to Section 1.2.

Name ______________________________ **Date** ______________

✓ CHECKUP 1.1

________ 1. To which number sets does -9 belong?

________ 2. List the members of the set $\{-3, -1.2, -\frac{3}{4}, 0, \sqrt{2}, \frac{5}{4}, 4\}$ that belong to the set of integers.

Compare the following numbers using $<$, $>$, or $=$.

________ 3. $\frac{8}{2}$ and $\sqrt{16}$

________ 4. $-\frac{7}{2}$ and -2.5

Use an inequality to state whether each number is positive or negative.

________ 5. $-\frac{9}{5}$

________ 6. $\sqrt{36}$

Find the missing number in each statement.

7. $-\frac{5}{12}$ is the opposite of ____ .

8. 9 is the opposite of ____ .

Evaluate each of the following.

________ 9. $|-3|$

________ 10. $|7| + |-2|$

Check your answers in the back of your book.

If You Missed Problems:	You Should Review Examples:
1, 2	1–3
3, 4	5, 6
5, 6	7, 8
7, 8	9
9, 10	10, 11

1.2 Operating with Real Numbers

Although you learned to add, subtract, multiply, and divide real numbers in earlier algebra courses, a brief review is provided here to refresh your memory.

Adding and Subtracting Real Numbers

Recall that the result of performing **addition** is called a **sum** and the numbers being added are called **terms**. Some students find the number line helpful in finding sums of real numbers.

Most of us would agree that $6 + 7 = 13$. Let's see how we could find that sum using the number line. We start at the point corresponding to the number 6 and then move 7 more units to the *right*.

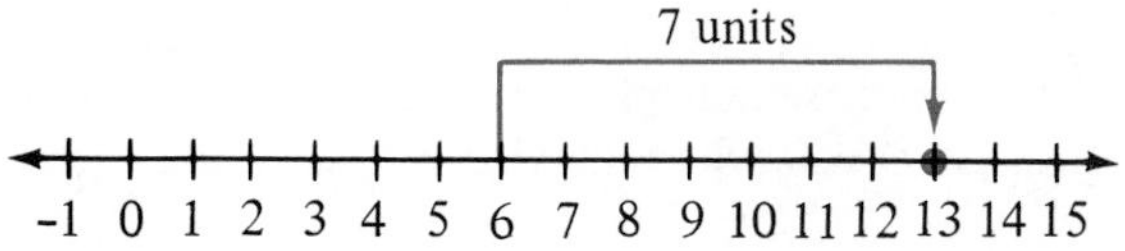

Indeed, we end up at the point corresponding to positive 13.

Let's try the sum $-3 + (-5)$. Here we must start at the point corresponding to -3. Then we must move 5 more units to the *left*, since we are adding -5.

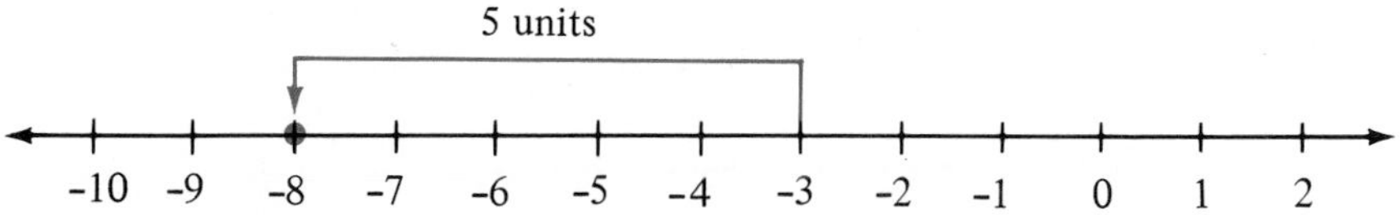

Our final point is located 8 units to the *left* of zero, so our sum is negative 8.

$$-3 + (-5) = -8$$

When we added two positive numbers, our sum was positive. When we added two negative numbers, our sum was negative. The following rule describes the results of our additions.

> To add two numbers having the *same* sign, add the units (absolute values) and give your answer that sign.

Example 1. Find $0.5 + 0.1$.

Solution. Both numbers are *positive* and the sum of the units is 0.6, so

$$0.5 + 0.1 = 0.6$$

You complete Example 2.

Example 2. Find $-1 + (-3)$.

Solution. Both numbers are ________ and the sum of the units is _____ , so

$$-1 + (-3) = \_\_\_\_$$

Check your work on page 20. ▶

Now let's consider the sum $-2 + 8$. On the number line we start at the point corresponding to -2. Since we are adding a *positive* 8, we now move 8 units to the *right*.

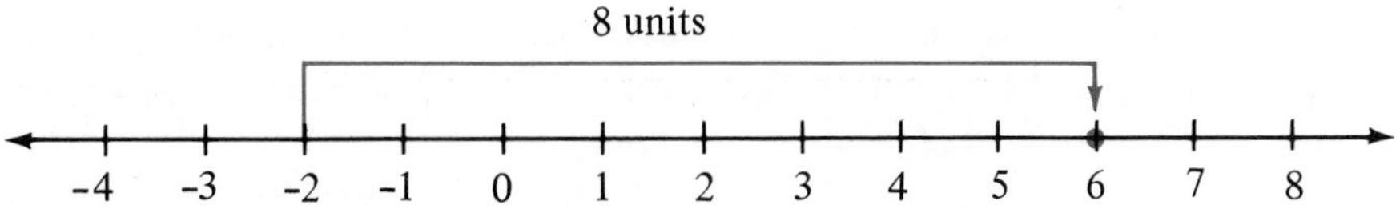

Our final point lies 6 units to the right of zero, so

$$-2 + 8 = 6$$

Example 3. Use the number line to find the sum $10 + (-14)$.

Solution. We start at positive 10 and move 14 units to the *left*.

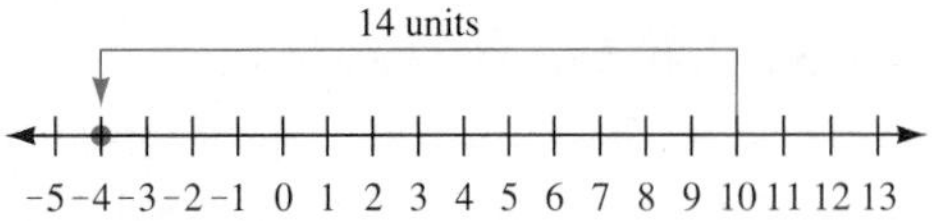

So $10 + (-14) = -4$.

You complete Example 4.

Example 4. Use the number line to find the sum $-7 + 6$.

Solution. We start at ________ 7 and move 6 units to the ________ .

So $-7 + 6 =$ _____ .

Check your work on page 20. ▶

From these examples, can you see a pattern for finding the sum of two numbers with *different* signs? Doesn't the following rule seem reasonable?

> To add two numbers with different signs, find the *difference* between the units (absolute values) and give your answer the sign of the original number having greater absolute value.

Example 5. Find the sum $5 + (-17)$.

Solution. The signs are different. The answer will have a *negative* sign. The difference between 5 units and 17 units is 12 units.

$$5 + (-17) = -12$$

You complete Example 6.

Example 6. Find the sum $-3 + 9$.

Solution. The signs are different. Here the answer will have a ________ sign. The difference between 3 units and 9 units is _____ units.

$$-3 + 9 = \_\_\_\_$$

Check your work on page 20. ▶

In performing these addition problems, it was assumed that you could find the difference between numbers of units. Now, however, we must consider a more detailed discussion of **subtraction**. Remember that the result of performing subtraction is called a **difference**.

Some subtraction problems cause us no difficulty. We know that

$$27 - 16 = 11 \quad \text{because} \quad 11 + 16 = 27$$
$$8 - 3 = 5 \quad \text{because} \quad 5 + 3 = 8$$

> If a, b, and c are real numbers, then
>
> $a - b = c$ provided that $c + b = a$

Unfortunately, this fact does not help us to see quickly answers to more complicated subtraction problems. Let's try an approach that will change each subtraction problem into an *addition* problem.

Subtraction Statement	*Addition Statement*
$27 - 16 = 11$	$27 + (-16) = 11$
$8 - 3 = 5$	$8 + (-3) = 5$
$22 - 7 = 15$	$22 + (-7) = 15$

In fact, every subtraction statement can be written as an addition statement.

To subtract two real numbers, *add* the *opposite* of the number being subtracted.

This is the mathematical definition for subtraction, and letting a and b represent any real numbers, we may write

Definition of Subtraction

$$a - b = a + (-b)$$

Subtraction Statement	Addition Statement	Answer
$3 - 10$	$3 + (-10)$	-7
$-3 - 4$	$-3 + (-4)$	-7
$5 - (-6)$	$5 + 6$	11
$-8 - (-2)$	$-8 + 2$	-6

Example 7. Find $-22 - 14$.

Solution

$$-22 - 14 = -22 + (-14)$$
$$= -36$$

You try Example 8.

Example 8. Find $13 - (-7)$.

Solution

$$13 - (-7) = 13 + \_\_\_\_$$
$$= \_\_\_\_$$

Check your work on page 20. ▶

Example 9. Find $-\frac{2}{3} - (-\frac{1}{5})$.

Solution

$$-\frac{2}{3} - (-\frac{1}{5}) = -\frac{2}{3} + \frac{1}{5} \qquad \text{LCD: 15}$$
$$= -\frac{10}{15} + \frac{3}{15}$$
$$= -\frac{7}{15}$$

Now complete Example 10.

Example 10. Find $-18 - (-18)$.

Solution

$$-18 - (-18)$$
$$= -18 + \_\_\_\_$$
$$= \_\_\_\_$$

Check your work on page 20. ▶

Trial Run

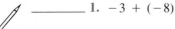

Perform the indicated operation.

______ 1. $-3 + (-8)$

______ 2. $-11 + 23$

______ 3. $\frac{15}{4} + (-\frac{9}{2})$

______ 4. $-8 + 8$

______ 5. $23 - 50$

______ 6. $-9 - 13$

______ 7. $-25 - (-38)$

______ 8. $3.5 - 5.0$

Answers are on page 21.

Multiplying and Dividing Real Numbers

To review the operation of **multiplication**, we should recall that quantities being multiplied are called **factors** and the result of multiplication is called a **product**.

It is assumed that you know your basic multiplication tables, but you may need a short refresher on the rules for *signs* of products. Recall that the sign of a product is determined by the signs of the factors being multiplied. Let's review the basic rules for multiplying real numbers.

> The product of two positive numbers is a *positive* number.
>
> *Example:* $5 \cdot 7 = 35$
>
> To multiply two negative numbers, multiply the units (absolute values) and give your answer a *positive* sign.
>
> *Example:* $(-2)(-6) = 12$
>
> To multiply a positive number and a negative number, multiply the units (absolute values) and give your answer a *negative* sign.
>
> *Examples:* $-3(6) = -18$
>
> $2(-7) = -14$

Example 11. Find the product $6(-1.2)$.

Solution

$$6(-1.2) = -7.2$$

Complete Example 12.

Example 12. Find the product $-11(-3)$.

Solution

$$-11(-3) = \_\_\_\_$$

Check your work on page 20. ▶

In a **division** problem, the number being divided is called the **dividend** and the number doing the dividing is called the **divisor**. The result of division is called the **quotient**, and the quotient of two real numbers, a and b, can be written as $\frac{a}{b}$ or $a \div b$. To investigate the rules for the sign of a quotient, recall that there is a multiplication statement that corresponds to every division statement.

Division Statement		*Multiplication Statement*
$\frac{15}{3} = 5$	because	$15 = 3 \cdot 5$
$28 \div 7 = 4$	because	$28 = 7 \cdot 4$
$\frac{-15}{3} = -5$	because	$-15 = 3(-5)$
$\frac{28}{-7} = -4$	because	$28 = -7(-4)$
$\frac{-18}{-6} = 3$	because	$-18 = -6(3)$

From these examples, we may conclude that

> The quotient of two positive numbers is positive.
> The quotient of two negative numbers is positive.
> The quotient of a positive number and a negative number is negative.

Example 13. Find the quotient $\frac{49}{-7}$.

Solution

$$\frac{49}{-7} = -7$$

You complete Example 14.

Example 14. Find the quotient $\frac{-125}{-5}$.

Solution

$$\frac{-125}{-5} = \_\_\_\_$$

Check your work on page 20. ▶

Example 15. Find the quotient $-\frac{3}{4} \div \frac{3}{5}$.

Solution

$$-\frac{3}{4} \div \frac{3}{5} = -\frac{3}{4} \cdot \frac{5}{3}$$

$$= -\frac{5}{4}$$

Now complete Example 16.

Example 16. Find the quotient $\frac{17}{-1}$.

Solution

$$\frac{17}{-1} = \_\_\_\_$$

Check your work on page 20. ▶

Trial Run

Perform the indicated operation.

_______ 1. $-11(5)$

_______ 2. $(-8)(-7)$

_______ 3. $9(-6)$

_______ 4. $-1(-1.9)$

_______ 5. $\frac{72}{-8}$

_______ 6. $\frac{-110}{-10}$

______ 7. $\frac{5}{8} \div \left(-\frac{3}{4}\right)$

______ 8. $\frac{-8.3}{-1}$

Answers are on page 21.

Working with Zero

Having discussed addition, subtraction, multiplication, and division of real numbers, we turn our attention to performing those operations with the real number **zero**. Let's review the properties of zero. We shall let a represent any real number.

Property Name	Statement	Examples
Addition property of zero	$a + 0 = a$	$-6 + 0 = -6$
	$0 + a = a$	$0 + (-\frac{1}{2}) = -\frac{1}{2}$
		$\pi + 0 = \pi$
Subtraction property of zero	$a - 0 = a$	$5.3 - 0 = 5.3$
	$0 - a = -a$	$0 - 2 = -2$
		$0 - (-1) = 1$
Multiplication property of zero	$a \cdot 0 = 0$	$0(-2.5) = 0$
	$0 \cdot a = 0$	$3 \cdot 0 = 0$

You complete Examples 17, 18, and 19.

Example 17. Find each sum.

$-11 + 0 =$ ____

$0 + 8 =$ ____

$0 + (-1) =$ ____

$0 + 0 =$ ____

Check your work on page 21. ▶

Example 18. Find each difference.

$20 - 0 =$ ____

$-1 - 0 =$ ____

$0 - \frac{2}{3} =$ ____

$0 - 0 =$ ____

Check your work on page 21. ▶

Example 19. Find each product.

$10 \cdot 0 =$ ____

$0(-1) =$ ____

$\frac{1}{2}(0) =$ ____

$0 \cdot 0 =$ ____

Check your work on page 21. ▶

Let's investigate the use of zero in *division* problems. What happens when we divide zero by a real number other than zero?

$$\frac{0}{6} = 0 \quad \text{because} \quad 0 = 6 \cdot 0$$

$$\frac{0}{-5} = 0 \quad \text{because} \quad 0 = -5 \cdot 0$$

If we divide zero by a nonzero number, the quotient is zero.

What about dividing a nonzero number by zero?

$$\frac{13}{0} = \underline{\qquad} \quad \text{because} \quad 13 = 0(\underline{\qquad})$$

$$\frac{-25}{0} = \underline{\qquad} \quad \text{because} \quad -25 = 0(\underline{\qquad})$$

We can find *no* number to use in these blanks. We conclude that division *by* zero is impossible or **undefined**.

Suppose that we try to divide zero by zero. Will we get an answer this time?

$$\frac{0}{0} = \underline{\qquad} \quad \text{because} \quad 0 = 0(\underline{\qquad})$$

As a matter of fact, we can use *any* number in the blanks.

$$\frac{0}{0} = 0 \quad \text{because} \quad 0 = 0(0)$$

$$\frac{0}{0} = 1 \quad \text{because} \quad 0 = 0(1)$$

$$\frac{0}{0} = 837 \quad \text{because} \quad 0 = 0(837)$$

In this case there are *too many* answers! We conclude that division of zero by zero is **not uniquely defined**. Let's summarize our observations about zero in division.

Property Name	Statement	Examples
Division of zero	$\frac{0}{a} = 0$ $(a \neq 0)$	$\frac{0}{11} = 0$ $\frac{0}{-1} = 0$
Division by zero	$\frac{a}{0}$ is undefined $(a \neq 0)$	$\frac{3 \cdot 9}{0}$ is undefined
Division of zero by zero	$\frac{0}{0}$ is not uniquely defined	$\frac{5-5}{0} = \frac{0}{0}$ is not uniquely defined

Now complete Examples 20 and 21.

Example 20. Complete each statement.

$$\frac{0}{11} = \underline{\qquad} \quad \text{because} \quad 0 = 11(\underline{\qquad})$$

$$\frac{0}{-1} = \underline{\qquad} \quad \text{because} \quad 0 = -1(\underline{\qquad})$$

Check your work on page 21. ▶

Example 21. Find each quotient.

$$\frac{0}{12} = \underline{\qquad}$$

$$\frac{-1}{0} = \underline{\qquad}$$

$$\frac{0}{-1} = \underline{\qquad}$$

$$\frac{0}{0} = \underline{\qquad}$$

Check your work on page 21. ▶

Trial Run

Perform the indicated operation.

_______ 1. $7 + 0$

_______ 2. $-9 + 0$

_______ 3. $0 - 3.2$

_______ 4. $16 \cdot 0$

_______ 5. $0(-4)$

_______ 6. $\frac{0}{-2}$

_______ 7. $0\left(\frac{2}{3}\right)$

_______ 8. $\frac{8}{0}$

_______ 9. $\frac{0}{0}$

_______ 10. $0 - (-1)$

Answers are on page 21.

Examples You Completed

Example 2. Find $-1 + (-3)$.

Solution. Both numbers are negative and the sum of the units is 4, so

$$-1 + (-3) = -4$$

Example 4. Use the number line to find the sum $-7 + 6$.

Solution. We start at negative 7 and move 6 units to the right.

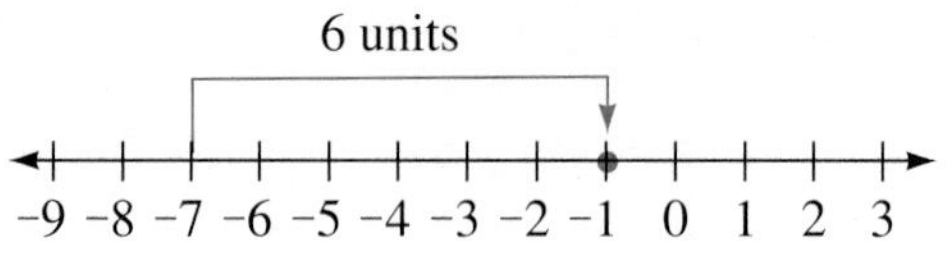

So $-7 + 6 = -1$.

Example 6. Find the sum $-3 + 9$.

Solution. The signs are different. Here the answer will have a positive sign. The difference between 3 units and 9 units is 6 units.

$$-3 + 9 = 6$$

Example 8. Find $13 - (-7)$.

Solution

$$\begin{aligned} 13 - (-7) &= 13 + 7 \\ &= 20 \end{aligned}$$

Example 10. Find $-18 - (-18)$.

Solution

$$\begin{aligned} -18 - (-18) &= -18 + 18 \\ &= 0 \end{aligned}$$

Example 12. Find the product $-11(-3)$.

Solution

$$-11(-3) = 33$$

Example 14. Find the quotient $\frac{-125}{-5}$.

Solution

$$\frac{-125}{-5} = 25$$

Example 16. Find the quotient $\frac{17}{-1}$.

Solution

$$\frac{17}{-1} = -17$$

Example 17. Find each sum.

$$-11 + 0 = -11$$
$$0 + 8 = 8$$
$$0 + (-1) = -1$$
$$0 + 0 = 0$$

Example 18. Find each difference.

$$20 - 0 = 20$$
$$-1 - 0 = -1$$
$$0 - \tfrac{2}{3} = -\tfrac{2}{3}$$
$$0 - 0 = 0$$

Example 19. Find each product.

$$10 \cdot 0 = 0$$
$$0(-1) = 0$$
$$\tfrac{1}{2}(0) = 0$$
$$0(0) = 0$$

Example 20. Complete each statement.

$$\frac{0}{11} = 0 \quad \text{because} \quad 0 = 11(0)$$
$$\frac{0}{-1} = 0 \quad \text{because} \quad 0 = -1(0)$$

Example 21. Find each quotient.

$$\frac{0}{12} = 0 \qquad \frac{0}{-1} = 0$$
$$\frac{-1}{0} \text{ is undefined} \qquad \frac{0}{0} \text{ is not uniquely defined}$$

Answers to Trial Runs

page 16 1. -11 2. 12 3. $-\frac{3}{4}$ 4. 0 5. -27 6. -22 7. 13 8. -1.5

page 17 1. -55 2. 56 3. -54 4. 1.9 5. -9 6. 11 7. $-\frac{5}{6}$ 8. 8.3

page 20 1. 7 2. -9 3. -3.2 4. 0 5. 0 6. 0 7. 0 8. Undefined
9. Not uniquely defined 10. 1

Name ____________________ Date ____________

EXERCISE SET 1.2

Perform the indicated operation.

____ **1.** $8 + 6$

____ **2.** $9 + 7$

____ **3.** $(-3) + (-7)$

____ **4.** $(-9) + (-3)$

____ **5.** $(-4) + 8$

____ **6.** $(-5) + 9$

____ **7.** $4 + (-13)$

____ **8.** $7 + (-15)$

____ **9.** $-\frac{2}{9} + (-\frac{5}{9})$

____ **10.** $-\frac{3}{20} + (-\frac{11}{20})$

____ **11.** $5 - 12$

____ **12.** $3 - 8$

____ **13.** $(-6) - (-4)$

____ **14.** $(-8) - (-3)$

____ **15.** $(-1.7) - (-3.2)$

____ **16.** $(-1.9) - (-4.1)$

____ **17.** $(-27) - 13$

____ **18.** $(-25) - 19$

____ **19.** $32 - (-42)$

____ **20.** $47 - (-22)$

____ **21.** $(-8)(0.5)$

____ **22.** $(-6)(1.5)$

____ **23.** $(-9)(-8)$

____ **24.** $(-3)(-21)$

____ **25.** $7(-15)$

____ **26.** $5(-13)$

____ **27.** $(-9)(0)$

____ **28.** $15(0)$

____ **29.** $(-2)(-8)$

____ **30.** $(5)(-2)$

____ **31.** $\dfrac{-24}{6}$

____ **32.** $\dfrac{-42}{7}$

____ **33.** $\dfrac{-21}{-3}$

____ **34.** $\dfrac{-25}{-5}$

____ **35.** $-\dfrac{5}{6} \div \dfrac{35}{36}$

____ **36.** $-\dfrac{6}{7} \div \dfrac{12}{49}$

____ **37.** $\dfrac{-144}{-12}$

____ **38.** $\dfrac{-169}{-13}$

____ **39.** $\dfrac{0}{-3}$

____ **40.** $\dfrac{0}{8}$

____ **41.** $0 - (-23)$

____ **42.** $0 - (-37)$

____ **43.** $\frac{4}{5}(0)$

____ **44.** $\frac{7}{8}(0)$

____ **45.** $\dfrac{0}{-3 - (-3)}$

____ **46.** $\dfrac{0}{8 + (-8)}$

____ **47.** $0 + (-13)$

____ **48.** $0 + (-51)$

____ **49.** $\dfrac{-9}{0}$

____ **50.** $\dfrac{12}{0}$

☆ Stretching the Topics

________ 1. Evaluate $-4a + (-7b) + 6c$ when $a = -1$, $b = -3$, and $c = -5$.

________ 2. Simplify $\dfrac{-9 - 5 + 6 - 2}{-5 - 7 + 8 + 4}$.

________ 3. The temperature is dropping 3° per hour. The temperature right now is 10°. Find the temperature 3 hours from now and 4 hours ago.

Check your answers in the back of your book.

If you can complete **Checkup 1.2**, you are ready to go to Section 1.3.

Name ____________________ Date ____________

✓ CHECKUP 1.2

Perform the indicated operation.

______ 1. $(-31) + (-21)$

______ 2. $14 + (-52)$

______ 3. $24 - (-9)$

______ 4. $(-19) - (-26)$

______ 5. $0 - (-7)$

______ 6. $(-4)(-7)$

______ 7. $(-3)(9)$

______ 8. $\frac{-81}{9}$

______ 9. $\frac{-75}{-5}$

______ 10. $\frac{7 - (-2)}{4 + (-4)}$

Check your answers in the back of your book.

If You Missed Problems:	You Should Review Examples:
1, 2	1–6
3, 4	7–10
5	17, 18
6, 7	11, 12
8, 9	13–16
10	21

1.3 Working with Real Number Properties

Number properties are the rules involved in working in algebra. They tell us what we can do and cannot do in performing operations with real numbers.

We have observed that sums and products of real numbers are always themselves real numbers. Add two real numbers and the sum is a real number; multiply two real numbers and the product is a real number. Moreover, there is one and only one answer when a given operation is performed with two numbers. Every time we add $2 + 3$, the answer is 5; every time we multiply $-6 \cdot 7$, the answer is -42. Think how confusing mathematics would be if this were not so; we depend on our answers' being **unique**.

We summarize these first facts as the **closure properties** for addition and multiplication of real numbers.

Closure Properties. If a and b are real numbers, then	*Illustrations*
(1) $a + b$ is a unique real number	(1) $3 + 2 = 5$ and $5 \in R$
(2) $a \cdot b$ is a unique real number	(2) $5 \cdot 2 = 10$ and $10 \in R$

The order in which we add or multiply two real numbers does not change the sum or product. We can summarize these facts as the **commutative properties** for addition and multiplication of real numbers.

Commutative Properties. If a and b are real numbers, then	*Illustrations*
(1) $a + b = b + a$	(1) $-5 + 6 = 6 + (-5)$
(2) $a \cdot b = b \cdot a$	(2) $(-7)(-2) = (-2)(-7)$

How do we add or multiply *three* numbers? To find

$$2 + 6 + 11 \quad \text{or} \quad 7 \cdot 2 \cdot 3$$

requires some mental grouping, since our minds can handle only two numbers at once. We shall use **parentheses** to indicate such grouping; parentheses always tell us to "do this first."

There are two approaches to grouping the terms in our sum.

Approach 1	*Approach 2*
$2 + 6 + 11$	$2 + 6 + 11$
$= (2 + 6) + 11$	$= 2 + (6 + 11)$
$= 8 + 11$	$= 2 + 17$
$= 19$	$= 19$

With either approach, the sum is 19. The way in which we group the terms does not change the sum.

Once again, there are two approaches to grouping the factors in our product.

Approach 1	*Approach 2*
$7 \cdot 2 \cdot 3$	$7 \cdot 2 \cdot 3$
$= (7 \cdot 2) \cdot 3$	$= 7 \cdot (2 \cdot 3)$
$= 14 \cdot 3$	$= 7 \cdot 6$
$= 42$	$= 42$

Notice again that the answer is the same in each case.

These examples lead us to the **associative properties** for addition and multiplication of real numbers.

Associative Properties. If a, b, and c are real numbers, then

(1) $a + (b + c) = (a + b) + c$

(2) $a(bc) = (ab)c$

Illustrations

(1) $10 + (7 + 1) = (10 + 7) + 1$

(2) $-1(3 \cdot 5) = (-1 \cdot 3) \cdot 5$

Notice how the associative properties can help you cut down on complicated arithmetic in the following examples.

Example 1. Find $-99 + (99 + 76)$.

Solution

$$\begin{aligned} &-99 + (99 + 76) \\ &= (-99 + 99) + 76 \\ &= 0 + 76 \\ &= 76 \end{aligned}$$

You try Example 2.

Example 2. Find $(37 \cdot \frac{1}{2})20$.

Solution

$$\begin{aligned} (37 \cdot \tfrac{1}{2})20 &= 37(\tfrac{1}{2} \cdot 20) \\ &= 37(\underline{\qquad}) \\ &= \underline{\qquad} \end{aligned}$$

Check your work on page 33. ▶

The commutative and associative properties allow us to change the positions of numbers in a sum or product to make our computation more convenient. In an addition problem, you may wish to group all the positive numbers and all the negative numbers before adding.

Example 3. Find

$$-13 + 17 + 4 - 18 - 1$$

Solution

$$\begin{aligned} &-13 + 17 + 4 - 18 - 1 \\ &= -13 - 18 - 1 + 17 + 4 \\ &= -32 + 21 \\ &= -11 \end{aligned}$$

You try Example 4.

Example 4. Find

$$2 - 5 - 7 + 3 - (-9)$$

Solution

$$\begin{aligned} &2 - 5 - 7 + 3 - (-9) \\ &= 2 + 3 + 9 - 5 - 7 \end{aligned}$$

Check your work on page 33. ▶

Using the commutative and associative properties in a multiplication problem, the "easy" products can be found first.

Example 5. Find $6(-3)(2)(5)(-1)$.

Solution

$$\begin{aligned} & 6(-3)(2)(5)(-1) \\ &= 6(-6)(-5) \\ &= 6(30) \\ &= 180 \end{aligned}$$

You try Example 6.

Example 6. Find $-2(3)(8)(10)$.

Solution

$$\begin{aligned} & -2(3)(8)(10) \\ &= \end{aligned}$$

Check your work on page 33. ▶

When a problem contains several different operations, some guidelines are needed. Recall that grouping symbols (parentheses, brackets, and braces) say "do this first" and should be dealt with by working from the innermost pairs outward. Fraction bars also act as symbols of grouping, separating the numerator and the denominator.

Order of Operations

1. Deal with any symbols of grouping.
2. Perform multiplications and/or divisions from left to right.
3. Perform additions and/or subtractions from left to right.

Try completing Example 7.

Example 7. Find

$$5[3 - 2(7 - 10)] - 2$$

Solution

$$\begin{aligned} & 5[3 - 2(7 - 10)] - 2 \\ &= 5[3 - 2(-3)] - 2 \\ &= 5[3 + \underline{\qquad}] - 2 \end{aligned}$$

Check your work on page 33. ▶

Example 8. Find

$$\frac{5(9 - 4) - 2(7 - 3)}{-2(\frac{1}{2})}$$

Solution

$$\begin{aligned} & \frac{5(9 - 4) - 2(7 - 3)}{-2(\frac{1}{2})} \\ &= \frac{5(5) - 2(4)}{-2(\frac{1}{2})} \\ &= \frac{25 - 8}{-1} \\ &= \frac{17}{-1} \\ &= -17 \end{aligned}$$

To be sure that we understand how to deal with expressions involving several operations, let's try a few more examples.

Example 9. Find

$$3 - 2\left[\frac{5(-4) - 2(-1)}{10(3) - 6(4)}\right]$$

Solution

$$\begin{aligned} & 3 - 2\left[\frac{5(-4) - 2(-1)}{10(3) - 6(4)}\right] \\ &= 3 - 2\left[\frac{-20 + 2}{30 - 24}\right] \\ &= 3 - 2\left[\frac{-18}{6}\right] \\ &= 3 - 2(-3) \\ &= 3 + 6 \\ &= 9 \end{aligned}$$

Now carefully complete Example 10.

Example 10. Find

$$-7\left(8 + \frac{4 - 1}{2 - 3}\right) - 10$$

Solution

$$\begin{aligned} & -7\left(8 + \frac{4 - 1}{2 - 3}\right) - 10 \\ &= -7\left(8 + \frac{3}{-1}\right) - 10 \\ &= -7(8 - \underline{\qquad}) - 10 \end{aligned}$$

Check your work on page 33. ▶

Suppose that we try two different approaches to find the product $-2(5 + 1)$.

Approach 1	*Approach 2*
$-2(5 + 1)$	$-2(5 + 1)$
$= -2(6)$	$= -2(5) + (-2)(1)$
$= -12$	$= -10 + (-2)$
	$= -12$

The second approach illustrates the **distributive property** for multiplication over addition of real numbers.

> **Distributive Property.** If a, b, and c are real numbers, then
>
> $$a(b + c) = a \cdot b + a \cdot c$$

Illustration

$$3(2 + 5) = 3 \cdot 2 + 3 \cdot 5$$

Example 11. Use the distributive property to find $-6(-2 - 8)$.

Solution

$$\begin{aligned} & -6(-2 - 8) \\ &= -6(-2) + (-6)(-8) \\ &= 12 + 48 \\ &= 60 \end{aligned}$$

You try Example 12.

Example 12. Use the distributive property to find $7(-4 + 3)$.

Solution

$$\begin{aligned} & 7(-4 + 3) \\ &= 7(-4) + 7(3) \end{aligned}$$

Check your work on page 33. ▶

You may be thinking that the distributive property is more complicated than our earlier method for finding these products, and for problems involving only rational numbers, you are probably right. However, we shall soon encounter algebraic expressions that can be simplified only by using the distributive property. As a matter of fact, *the distributive property is one of the most useful of all the properties of the real numbers.*

Trial Run

Perform the indicated operations.

_______ 1. $(19 + 15) - 15$

_______ 2. $\frac{1}{3}(12 \cdot 11)$

_______ 3. $4(5 + 2) - 3 \cdot 10$

_______ 4. $9 - 2[7(3 - 9) + 2]$

_______ 5. $\dfrac{7(10 - 5) - 3(8 - 7)}{-16(\frac{1}{2})}$

_______ 6. $5 + 2\left[\dfrac{7 + (-2)(-10)}{6 - 3(5)}\right]$

Simplify using the distributive property for real numbers.

_______ 7. $-3(6 + 2)$

_______ 8. $-7(3 - 11)$

_______ 9. $2(\pi - 9)$

Answers are on page 33.

We have already discussed the very special nature of the real number zero. We should emphasize that zero is unique in that it is the *only* real number that leaves every real number unchanged under the operation of addition. Because zero added to any real number gives an answer that is *identical* to that original number, we call zero the **identity for addition.**

Do you suppose that the set of real numbers also contains an identity for multiplication? Can you think of a real number that leaves every real number unchanged under multiplication? You should realize that 1 is the only such number. We conclude that 1 is the unique **identity for multiplication**.

Identity for Addition Is 0	Identity for Multiplication Is 1
$3 + 0 = 3$	$3 \cdot 1 = 3$
$-7 + 0 = -7$	$-7 \cdot 1 = -7$
$\frac{1}{3} + 0 = \frac{1}{3}$	$\frac{1}{3} \cdot 1 = \frac{1}{3}$
$\pi + 0 = \pi$	$\pi \cdot 1 = \pi$

We summarize the identity properties for the real numbers as follows.

Identity Properties

1. The unique *identity for addition* is 0. For every $a \in R$, $a + 0 = a$.
2. The unique *identity for multiplication* is 1. For every $a \in R$, $a \cdot 1 = a$.

In our discussion of the number line, we discovered that every real number has an *opposite* that is also a real number. The opposite of a number is also called its **additive inverse**. Every time we add a number to its additive inverse (or opposite), the sum is 0 (the additive identity).

Number	Additive Inverse	Reason
5	-5	$5 + (-5) = 0$
0.6	-0.6	$0.6 + (-0.6) = 0$
-8	$-(-8) = 8$	$-8 + 8 = 0$
a	$-a$	$a + (-a) = 0$

We may conclude that *the additive inverse (or negative) of a number can be found by multiplying that number by* -1.

Example 13. Find the additive inverse of $\pi + 2$.

Solution. The additive inverse of $\pi + 2$ is $-(\pi + 2)$.

$$-(\pi + 2) = -1(\pi + 2)$$
$$= -1 \cdot \pi + (-1)(2)$$
$$= -\pi - 2$$

Notice that we used the distributive property to multiply our quantity by -1.

You try completing Example 14.

Example 14. Find the additive inverse of $5 - \sqrt{3}$.

Solution. The additive inverse of $5 - \sqrt{3}$ is $-(5 - \sqrt{3})$.

$$-(5 - \sqrt{3})$$
$$= -1(5 - \sqrt{3})$$
$$= (-1)(\_\_\_\_) + (-1)(\_\_\_\_)$$
$$=$$

Check your work on page 33. ▶

Now that we have discussed additive inverses, we consider the existence of **multiplicative inverses**. The unique multiplicative inverse for every real number (except 0) is its **reciprocal**. When we multiply a number by its multiplicative inverse, the product is 1 (the multiplicative identity).

Number	Multiplicative Inverse	Reason
5	$\frac{1}{5}$	$5 \cdot \frac{1}{5} = 1$
-7	$-\frac{1}{7}$	$-7\left(-\frac{1}{7}\right) = 1$
$\frac{5}{8}$	$\frac{8}{5}$	$\frac{5}{8}\left(\frac{8}{5}\right) = 1$
-0.3	$-\frac{1}{0.3}$	$-0.3\left(-\frac{1}{0.3}\right) = 1$
π	$\frac{1}{\pi}$	$\pi\left(\frac{1}{\pi}\right) = 1$

As you can see, we find the multiplicative inverse of a real number by inverting it, or turning it upside down.

Inverse Properties

1. Every real number a has a unique *additive inverse* (or opposite) denoted by $-a$, and

$$a + (-a) = 0$$

2. Every real number a (except 0) has a unique *multiplicative inverse* (or reciprocal) denoted by $\frac{1}{a}$, and

$$a \cdot \frac{1}{a} = 1$$

Example 15. Find the additive inverse and the multiplicative inverse of $\frac{1}{7}$.

Solution. The additive inverse of $\frac{1}{7}$ is $-\frac{1}{7}$ because

$$\frac{1}{7} + \left(-\frac{1}{7}\right) = 0$$

The multiplicative inverse of $\frac{1}{7}$ is 7 because

$$\frac{1}{7} \cdot 7 = 1$$

You complete Example 16.

Example 16. Find the additive inverse and the multiplicative inverse of -4.

Solution. The additive inverse of -4 is _____ because

$$-4 + \_\_\_\_\_ = 0$$

The multiplicative inverse of -4 is _____ because

$$-4(\_\_\_\_\_) = 1$$

Check your work on page 33. ▶

The existence of a multiplicative inverse corresponding to every nonzero real number allows us to write every division expression as a multiplication expression.

Division Expression	Multiplication Expression	Answer
$\frac{6}{3}$	$6 \cdot \frac{1}{3}$	2
$\frac{-10}{5}$	$-10 \cdot \frac{1}{5}$	-2
$\frac{24}{-8}$	$24\left(-\frac{1}{8}\right)$	-3

To divide two numbers, we may find the reciprocal of the divisor and multiply.

Indeed, this is the formal definition of division in terms of multiplication. Letting a and b represent real numbers, we may write

Definition of Division

$$\frac{a}{b} = a \cdot \frac{1}{b} \quad \text{where } b \neq 0$$

Trial Run

Name the property of real numbers illustrated by each statement.

_______ 1. $-6 + 0 = -6$

_______ 2. $8\left(\frac{1}{8}\right) = 1$

_______ 3. $-3(1) = -3$

_______ 4. $5 + (-5) = 0$

5. The additive inverse of -10 is _____ .

6. The additive inverse of $-3 + \sqrt{2}$ is _____ .

7. The multiplicative inverse of -5 is _____ .

Rewrite each division expression as a multiplication expression.

_______ 8. $\frac{47}{6}$

_______ 9. $\frac{52}{-3}$

Answers are on page 33.

We mention here that although both addition and multiplication are commutative and associative, subtraction and division are *not*. For example,

$$3 - 5 \neq 5 - 3 \quad \text{and} \quad 8 \div 2 \neq 2 \div 8$$

$$9 - (7 - 2) \neq (9 - 7) - 2 \quad \text{and} \quad 24 \div (6 \div 2) \neq (24 \div 6) \div 2$$

▶ Examples You Completed

Example 2. Find $(37 \cdot \frac{1}{2})\ 20$.

Solution

$$\begin{aligned}(37 \cdot \tfrac{1}{2})20 &= 37(\tfrac{1}{2} \cdot 20)\\ &= 37(10)\\ &= 370\end{aligned}$$

Example 4. Find $2 - 5 - 7 + 3 - (-9)$.

Solution

$$\begin{aligned}&2 - 5 - 7 + 3 - (-9)\\ &= 2 + 3 + 9 - 5 - 7\\ &= 14 - 12\\ &= 2\end{aligned}$$

Example 6. Find $-2(3)(8)(10)$.

Solution

$$\begin{aligned}&-2(3)(8)(10)\\ &= -6(80)\\ &= -480\end{aligned}$$

Example 7. Find $5[3 - 2(7 - 10)] - 2$.

Solution

$$\begin{aligned}&5[3 - 2(7 - 10)] - 2\\ &= 5[3 - 2(-3)] - 2\\ &= 5[3 + 6] - 2\\ &= 5 \cdot 9 - 2\\ &= 45 - 2\\ &= 43\end{aligned}$$

Example 10. Find

$$-7\left(8 + \frac{4 - 1}{2 - 3}\right) - 10$$

Solution

$$\begin{aligned}&-7\left(8 + \frac{4 - 1}{2 - 3}\right) - 10\\ &= -7\left(8 + \frac{3}{-1}\right) - 10\\ &= -7(8 - 3) - 10\\ &= -7(5) - 10\\ &= -35 - 10\\ &= -45\end{aligned}$$

Example 12. Use the distributive property to find $7(-4 + 3)$.

Solution

$$\begin{aligned}&7(-4 + 3)\\ &= 7(-4) + 7(3)\\ &= -28 + 21\\ &= -7\end{aligned}$$

Example 14. Find the additive inverse of $5 - \sqrt{3}$.

Solution. The additive inverse of $5 - \sqrt{3}$ is $-(5 - \sqrt{3})$.

$$\begin{aligned}&-(5 - \sqrt{3})\\ &= -1(5 - \sqrt{3})\\ &= (-1)(5) + (-1)(-\sqrt{3})\\ &= -5 + \sqrt{3}\end{aligned}$$

Example 16. Find the additive inverse and the multiplicative inverse of -4.

Solution. The additive inverse of -4 is 4 because

$$-4 + 4 = 0$$

The multiplicative inverse of -4 is $-\frac{1}{4}$ because

$$-4\left(-\frac{1}{4}\right) = 1$$

Answers to Trial Runs

page 29 1. 19 2. 44 3. -2 4. 89 5. -4 6. -1 7. -24 8. 56 9. $2\pi - 18$

page 32 1. Additive identity 2. Multiplicative inverse 3. Multiplicative identity 4. Additive inverse 5. 10 6. $3 - \sqrt{2}$ 7. $-\frac{1}{5}$ 8. $47 \cdot \frac{1}{6}$ 9. $52(-\frac{1}{3})$

Name ______________________ Date __________

EXERCISE SET 1.3

Name the property of real numbers that is illustrated by each of the following statements.

_______ 1. $5(3 \cdot 9) = (5 \cdot 3)9$

_______ 2. $2 + 0 = 2$

_______ 3. $\frac{4}{5} + (-5)$ is a real number

_______ 4. $(-3 + 5) + 10 = -3 + (5 + 10)$

_______ 5. $(-12)(1) = -12$

_______ 6. $23 + (-23) = 0$

_______ 7. $-8(7 + 5) = (-8)(7) + (-8)(5)$

_______ 8. $9 + (-13) = (-13) + 9$

_______ 9. $9\left(\frac{1}{9}\right) = 1$

_______ 10. $5 \cdot 2$ is a real number

Rewrite each statement using the indicated property.

11. $-4 + 8 =$ _____ (commutative)

12. $-2(3 \cdot 5) =$ _____ (associative)

13. $(-9 + 3) + 4 =$ _____ (associative)

14. $(-0.9)(1) =$ _____ (identity)

15. $(-15) + 15 =$ _____ (inverse)

16. $-7(6 + 4) =$ _____ (distributive)

17. $(-19) + 0 =$ _____ (identity)

18. $\frac{1}{4}(4) =$ _____ (inverse)

19. $2(3) + 2(-5) =$ _____ (distributive)

20. $5(7) + 5(8) =$ _____ (distributive)

Perform the indicated operations.

_______ 21. $-55 + (55 + 23)$

_______ 22. $15 + (-15 + 6)$

_______ 23. $(29 \cdot \frac{1}{3})30$

_______ 24. $15(\frac{1}{5} \cdot 17)$

_______ 25. $-8 + 12 + 7 - 13 - 1$

_______ 26. $-9 + 15 - 18 + 3 - (-6)$

_______ 27. $4(-3)(2)(-7)$

_______ 28. $-3(6)(2)(-5)$

_______ 29. $6(-1)(9)(0)(21)$

_______ 30. $9(-13)(7)(0)(15)$

_______ 31. $-9(5 - 7)$

_______ 32. $-10(5 - 12)$

_______ 33. $(-6)(11) - 21$

_______ 34. $9(-5) - 15$

_______ 35. $5(\frac{1}{5}) - 3(7)$

_______ 36. $9(\frac{1}{9}) - 4(5)$

_______ 37. $7 - 9[3 - 5(-1)]$

_______ 38. $6 - 5[8 - 3(-2)]$

_______ 39. $(7 - 9)[4 - 7(0)]$

_______ 40. $(6 - 5)[13 - 9(0)]$

_______ 41. $\dfrac{5(-2) - \frac{1}{2}(-20)}{3(8 - 6)}$

_______ 42. $\dfrac{6(-3) - \frac{1}{2}(-36)}{7(9 - 5)}$

_______ 43. $\dfrac{20(-4)}{10} + \dfrac{21 - 6}{-3}$

_______ 44. $\dfrac{36(-2)}{12} + \dfrac{27 - 9}{-6}$

_______ 45. $\dfrac{-9(7 - 1) - 8(-3)}{-5[(-8) + 6]}$

_______ 46. $\dfrac{-8(10 - 3) - 9(-4)}{-2[(-9) + 4]}$

_______ 47. $\dfrac{-6(\frac{1}{2}) + 3[2 - (-3)]}{9[12 - 2(6)]}$

_______ 48. $\dfrac{-8(\frac{1}{4}) + 6[8 - (-1)]}{11[15 - 3(5)]}$

_______ 49. $\dfrac{\frac{-15}{5} + \frac{0}{7}}{\frac{35}{-5} + 11}$

_______ 50. $\dfrac{\frac{18}{6} + \frac{0}{9}}{\frac{-30}{5} - 9}$

_______ 51. $8 - (-6)\left[\dfrac{2(-3) - 5(4)}{-8(6) - 4}\right]$

_______ 52. $-9 - 6\left[\dfrac{11(-1) + 9}{4(-3) - 2(-5)}\right]$

_______ 53. $8 - (-2)\left[\dfrac{3 - 5(7 - 1)}{6 - 1(-3)}\right]$

_______ 54. $9 - (-5)\left[\dfrac{8 - 3(6 - 12)}{7 - 3(-2)}\right]$

_______ 55. $4\left(-9 + \dfrac{7 - 3}{6 - 8}\right) - 3$

_______ 56. $5\left(-10 + \dfrac{8 - 3}{15 - 20}\right) - 6$

_______ 57. $\dfrac{8(-5 + 7) - (-9)}{(-2)4 + 13(-4)}$

_______ 58. $\dfrac{12(-6 + 4) - (-4)}{-3(4) + 7(-4)}$

☆ Stretching the Topics

_______ 1. Evaluate $a + bc - d$ when $a = 2.1$, $b = -3.2$, $c = 4.5$, and $d = -5.6$.

_______ 2. Evaluate $\dfrac{m + (2n - 3) - (n - 4)}{3(n - 1) - 2n}$ when $n = 2$ and $m = -3$.

_______ 3. The population of a city was 495,000. Over the next 5 years the urban planning chairman recorded these changes: a gain of 2365, a loss of 3729, a gain of 5728, a loss of 1796, and a loss of 768. Find the population of the city after 5 years. Find the average gain or loss.

Check your answers in the back of your book.

If you can complete **Checkup 1.3**, you are ready to go on to Section 1.4.

Name ______________________ Date ____________

✓ CHECKUP 1.3

Name the property of real numbers that is illustrated by each statement.

________ 1. $5(-2) = -2(5)$

________ 2. $-\frac{1}{3}(-3) = 1$

Perform the indicated operations.

________ 3. $-23 + (23 + 17)$

________ 4. $-4(5 - 12)$

________ 5. $-5 + 7 - 10 - (-13)$

________ 6. $5(-3)(6)(2)$

________ 7. $4 - 8[6 - 2(-3)]$

________ 8. $(5 - 10)[7 - 2(5)]$

________ 9. $\dfrac{5(-6) - \frac{1}{5}(15)}{-6(8 - 19)}$

________ 10. $\dfrac{-12(5) - 3[8 - 2(4)]}{-9[17 - 3(5)]}$

Check your answers in the back of your book.

If You Missed Problems:	You Should Review Examples:
1, 2	(Real number properties)
3	1
4	11, 12
5	3, 4
6	5, 6
7, 8	7
9, 10	8–10

1.4 Working with Variables

Our work with the properties of the real numbers is intended to lay the groundwork for the study of algebra, which requires the use of constants and variables. A **constant** is a number such as 3 or -4 whose value is fixed and never changes. A **variable** is a symbol that stands for a number. We generally use letters to represent variables. For example,

17 is a constant	y is a variable
x is a variable	0 is a constant

Combinations of constants and variables involving the operations of addition, subtraction, multiplication, and division are called **algebraic expressions**. We continue to use familiar symbols to show what operation is to be performed.

Expression	Meaning
$10 + x$	The sum of 10 and x
$y - 3$	The difference when 3 is subtracted from y
$\frac{a}{2}$	The quotient when a is divided by 2
$9x$	The product of 9 and x
$-3y$	The product of -3 and y

Evaluating Algebraic Expressions

An algebraic expression has different numerical values depending on what values the variables represent. Finding the numerical value of an algebraic expression for particular numerical values of the variables is called **evaluating** an expression. We evaluate an expression by substituting the given numbers wherever the variables appear in the expression and then performing the necessary arithmetic.

Example 1. Evaluate $3(x + 6) - 7$ when $x = -1$.

Solution. We must substitute -1 for x in the expression.

$$3(x + 6) - 7$$
$$= 3(-1 + 6) - 7$$
$$= 3(5) - 7$$
$$= 15 - 7$$
$$= 8$$

You complete Example 2.

Example 2. Evaluate $9y - 37z$ when $y = 0$ and $z = 1$.

Solution. We must substitute ______ for y and 1 for ______ in the expression.

$$9y - 37z$$
$$= 9(\_\_\_\_) - 37(\_\_\_\_)$$

Check your work on page 45. ▶

Example 3. Evaluate $x(x + 3) - 4$ when $x = -6$.

Solution

$$\begin{aligned} & x(x + 3) - 4 \\ &= -6(-6 + 3) - 4 \\ &= -6(-3) - 4 \\ &= 18 - 4 \\ &= 14 \end{aligned}$$

Now complete Example 4.

Example 4. Evaluate $(x + 1)(1 - x)$ when $x = -2$.

Solution

$$\begin{aligned} & (x + 1)(1 - x) \\ &= (-2 + 1)[1 - (-2)] \end{aligned}$$

Check your work on page 45. ▶

Algebraic expressions often contain quantities with **exponents**. In general, if a is a real number (called the **base**) and n is a natural number (called the exponent), then a^n is called the nth power of a.

Definition of a Power

$$a^n = \underbrace{a \cdot a \cdot a \cdot \ldots \cdot a}_{n \text{ factors}}$$

This definition allows us to give meaning to more complicated algebraic expressions.

Exponent Form	Factor Form
5^2	$5 \cdot 5$
x^2	$x \cdot x$
$(-2)^4$	$(-2)(-2)(-2)(-2)$
-2^4	$-1 \cdot 2 \cdot 2 \cdot 2 \cdot 2$
$-2y^3$	$-2 \cdot y \cdot y \cdot y$
$(x + 4)^2$	$(x + 4)(x + 4)$

Use the definition of a power to evaluate the expression in Example 5.

Example 5. Evaluate $-x^2 - 3x + 2$ when $x = -4$.

Solution

$$\begin{aligned} & -x^2 - 3x + 2 \\ &= -(-4)^2 - 3(-4) + 2 \\ &= -1(-4)(-4) - 3(-4) + 2 \end{aligned}$$

Check your work on page 45. ▶

Example 6. Evaluate $4(x - 3)^2 - 2xy - y^3$ when $x = 3$ and $y = -1$.

Solution

$$\begin{aligned} & 4(3 - 3)^2 - 2(3)(-1) - (-1)^3 \\ &= 4(0)^2 - 6(-1) - (-1)(-1)(-1) \\ &= 4(0)(0) + 6 - (-1) \\ &= 0 + 6 + 1 \\ &= 7 \end{aligned}$$

Example 7. The surface area of a box with a square base and no top is found by the formula $S = 4sh + s^2$, where s is the length of one side of the base and h is the height. Evaluate this formula when $s = 12$ inches and $h = 20$ inches.

Solution

$$S = 4sh + s^2$$

$= 4(12)(20) + (12)^2$ Substitute 12 for s, 20 for h.

$= 960 + 144$ Remove parentheses.

$= 1104$ Find the sum.

The surface area is 1104 square inches.

Combining Like Terms

In an algebraic expression the parts being added are called the **terms** of that expression. Each term consists of a constant alone *or* the product of a constant and one or more variables. The constant part of a variable term is called the **numerical coefficient** for that term.

Expression	Constant Term	Variable Terms	Numerical Coefficients
$x^2 - 7x + 2$	2	x^2 $-7x$	1 -7
$3 - x$	3	$-x$	-1
$5x^2 - xy + y^2$	None	$5x^2$ $-xy$ y^2	5 -1 1

In addition and subtraction of algebraic expressions, we may combine only **like terms**, which are terms whose variable parts are *exactly* the same. The expression

$$3x + 5x - 2x$$

contains three like terms and their sum can be found using the *distributive property*.

$$3x + 5x - 2x = (3 + 5 - 2)x$$
$$= 6x$$

This illustration should help you see that

> To combine like terms, we combine the numerical coefficients and keep the same variable part.

The commutative and associative properties of real numbers allow us to rearrange the terms in an expression with the like terms together. We then use our addition and subtraction rules to combine the coefficients of the like terms.

Example 8. Simplify

$$3x + 2yx - 4x - xy + 3$$

Solution

$$\begin{aligned} &3x + 2yx - 4x - xy + 3 \\ &= 3x - 4x + 2yx - 1xy + 3 \\ &= -1x + 1xy + 3 \\ &= -x + xy + 3 \end{aligned}$$

You complete Example 9.

Example 9. Simplify

$$x^2 - 6x + 3x^2 - x$$

Solution

$$\begin{aligned} &x^2 - 6x + 3x^2 - x \\ &= \end{aligned}$$

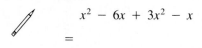

Check your work on page 45. ▶

From Example 8, notice that $-1x$ is written as $-x$ and $1xy$ is written as xy. Also note that xy and yx are like terms.

Trial Run

Evaluate each expression.

_______ 1. $(x - 6)(x + 7)$ when $x = -7$

_______ 2. $x^2 - 5x + 6$ when $x = -1$

_______ 3. $4(x - 2)^2 - 12x$ when $x = 0$

_______ 4. $9x^2 - xy + y^2$ when $x = 1$ and $y = -2$

Simplify by combining like terms.

_______ 5. $3x - 5 - 7x + 10$

_______ 6. $x^2 + 3x - 5 - 2x^2 + 7x - 1$

_______ 7. $2x^2 - 5xy + 3y^2 - 4x^2 + 7yx - y^2$

_______ 8. $a^2 - 2a + 3 - a^2 - 2a - 5$

Answers are on page 46.

Multiplying Variable Terms by Constants

To multiply a constant times a variable term, we recall the associative property for multiplication.

$$\begin{aligned} 2(5x) &= (2 \cdot 5)x &&= 10x \\ -2(-\tfrac{1}{3}x) &= [(-2)(-\tfrac{1}{3})]x &&= \tfrac{2}{3}x \\ 7(-x^2) &= [7(-1)]x^2 &&= -7x^2 \\ -(3x) &= (-1 \cdot 3)x &&= -3x \\ -(-4y^2) &= [-1(-4)]y^2 &&= 4y^2 \end{aligned}$$

To multiply a constant times a variable term, we multiply the constant times the numerical coefficient and keep the same variable part.

Example 10. Simplify $3(2x) - 7(4x)$.

Solution

$$\begin{aligned} & 3(2x) - 7(4x) \\ &= 6x - 28x \\ &= -22x \end{aligned}$$

You complete Example 11.

Example 11. Simplify $-3(2y) - 5(-y)$.

Solution

$$\begin{aligned} & -3(2y) - 5(-y) \\ &= \end{aligned}$$

Check your work on page 46. ▶

Since we agreed earlier that division can be expressed in terms of multiplication, the following example of dividing a variable term by a constant should make sense.

$$\frac{12x}{3} = (12x)\left(\frac{1}{3}\right) \qquad \text{Use the definition of division.}$$

$$= \left(\frac{1}{3}\right)(12x) \qquad \text{Use the commutative property for multiplication.}$$

$$= 4x \qquad \text{Multiply the constant times the variable term.}$$

We conclude that this quotient could be found more easily as

$$\begin{aligned} \frac{12x}{3} &= \frac{12}{3} \cdot x \\ &= 4x \end{aligned}$$

To divide a variable term by a constant, we divide the numerical coefficient by the constant and keep the same variable part.

$$\frac{10y}{2} = \frac{10}{2} \cdot y = 5y$$

$$\frac{-2x^2}{2} = \frac{-2}{2} \cdot x^2 = -x^2$$

$$\frac{16ab}{-4} = \frac{16}{-4} \cdot ab = -4ab$$

Simplify the expression in Example 12.

Example 12. Simplify $\frac{10y^2}{5} - \frac{14x^2}{7}$.

Solution

$$\frac{10y^2}{5} - \frac{14x^2}{7}$$

$$=$$

Check your work on page 46. ▶

Example 13. Simplify

$$\frac{27x}{3} - (-2x) + \frac{5x}{-5}$$

Solution

$$\frac{27x}{3} - (-2x) + \frac{5x}{-5}$$

$$= 9x + 2x + (-1x)$$

$$= 11x - x$$

$$= 10x$$

A product such as $3(x + 1)$ should remind you of the **distributive property** that we used earlier. According to that property,

$$3(x + 1)$$

$$= 3 \cdot x + 3 \cdot 1$$

$$= 3x + 3$$

Example 14. Simplify $-5(x^2 + x - 2)$.

Solution

$$-5(x^2 + x - 2)$$

$$= -5 \cdot x^2 + (-5 \cdot x) + (-5)(-2)$$

$$= -5x^2 - 5x + 10$$

You complete Example 15.

Example 15. Simplify $0.2(4x - 11y)$.

Solution

$$0.2(4x - 11y)$$

$$= 0.2(4x) + 0.2(-11y)$$

$$=$$

Check your work on page 46. ▶

As soon as you feel more comfortable with the distributive property, you will need fewer steps to reach your answer. Be careful, however, to avoid careless errors.

$$-3(x - 1) = -3x + 3 \quad \text{but} \quad -3(x - 1) \neq -3x - 1$$

$$-3(x - 1) \neq -3x - 3$$

Let's practice simplifying some expressions involving several operations. Recall the rules for the order in which we perform operations.

Order of Operations

1. Simplify powers.
2. Remove symbols of grouping (from the inside out).
3. Perform multiplications and/or divisions from left to right.
4. Perform additions and/or subtractions from left to right.

You try simplifying the expression in Example 16.

Example 16. Simplify

$$\tfrac{1}{2}(4a - 10) - 3(4 - a)$$

Solution

$$\frac{1}{2}(4a - 10) - 3(4 - a)$$

$$= \frac{1}{2} \cdot 4a - \frac{1}{2} \cdot 10 - 3 \cdot 4 - 3(-a)$$

Check your work on page 46. ▶

Example 17. Simplify

$$7^2 + 2[x - 8(1 + x)]$$

Solution

$$7^2 + 2[x - 8(1 + x)]$$
$$= 49 + 2[x - 8 - 8x]$$
$$= 49 + 2[-8 - 7x]$$
$$= 49 + 2(-8) + 2(-7x)$$
$$= 49 - 16 - 14x$$
$$= 33 - 14x$$

In Example 17 you may have been tempted to combine 49 + 2. This would be incorrect because the rules for the order of operations require that we multiply the 2 times the quantity within the brackets *before* adding the result to 49.

Trial Run

Perform the indicated operations and simplify.

_______ 1. $5(3x) - 4(-x)$

_______ 2. $-(2x) + 3(4y) + \frac{16x}{2}$

_______ 3. $\frac{45x^2}{-9} - (-3x^2) + 5(3x) + \frac{28x}{-14}$

_______ 4. $3(a - 1) - 2(5 + a)$

_______ 5. $8^2 + 3[x - 7(1 - x)]$

_______ 6. $-\frac{1}{3}(3x - 6y) + 5[x - (x - 2y)]$

Answers are on page 46.

Finding the Absolute Value of an Expression

In Section 1.1 we learned that the absolute value of a constant is its distance from zero on the number line. This definition works very well as long as we are dealing with constants because the numerical value of a constant is *fixed*; it never changes.

At this time, however, we are ready to consider a definition for absolute value that will work for constants *and* for variables. We must find a definition for the absolute value of any real number x, which we write as $|x|$.

Suppose that x represents a positive number or zero ($x \geq 0$):

if $x = 3$ then $|x| = |3| = 3$

if $x = 0$ then $|x| = |0| = 0$

if $x = \frac{2}{3}$ then $|x| = |\frac{2}{3}| = \frac{2}{3}$

If x is a *positive* number or *zero*, the absolute value of x is x itself.

Suppose that x represents a negative number ($x < 0$):

$$\text{if } x = -2 \quad \text{then } |x| = |-2| = 2$$

$$\text{if } x = -\tfrac{1}{5} \quad \text{then } |x| = |-\tfrac{1}{5}| = \tfrac{1}{5}$$

$$\text{if } x = -183 \quad \text{then } |x| = |-183| = 183$$

If x is a *negative* number, the absolute value of x is the *opposite* of x, which is written $-x$.

The definition for the absolute value of x is a *two-part* definition. If we *know* that x represents a *positive number or zero*, we use the first part of the definition. If we *know* that x represents a *negative number*, we use the second part of the definition. However, if we *don't know* whether x represents a positive number or a negative number or zero, we must consider both parts of the definition.

Definition of Absolute Value. If x represents a real number, then

$$|x| = \begin{cases} x & \text{if } x \geq 0 \\ -x & \text{if } x < 0 \end{cases}$$

Use the definition of absolute value to complete Example 18.

Example 18. Find $|\pi|$ and $|-\sqrt{2}|$.

Solution. Since $\pi > 0$,

$$|\pi| = \_\_\_\_$$

Check your work on page 46. ▶

Solution. Since $-\sqrt{2} < 0$,

$$|-\sqrt{2}| = -(-\sqrt{2})$$
$$= \_\_\_\_$$

Example 19. Find $|x + 2|$.

Solution. Since we do not know the nature of $x + 2$, we must say that

$$|x + 2| = \begin{cases} x + 2 & \text{if } x + 2 \geq 0 \\ -(x + 2) & \text{if } x + 2 < 0 \end{cases}$$

Example 20. Find $|1 - \sqrt{2}|$.

Solution. Since $1 < \sqrt{2}$, we know that $1 - \sqrt{2}$ is *negative*. Therefore,

$$|1 - \sqrt{2}| = -(1 - \sqrt{2})$$
$$= -1 + \sqrt{2}$$

Example 21. Find $-|3 - 7|$.

Solution. We must find the absolute value first.

$$-|3 - 7| = -|-4|$$
$$= -(4)$$
$$= -4$$

Complete Example 22.

Example 22. Find $-|3| - |-7|$.

Solution. We first find the absolute values.

$$-|3| - |-7|$$
$$= -(3) - (\_\_\_\_)$$
$$= \_\_\_\_$$

Check your work on page 46. ▶

Trial Run

Write each expression without absolute value bars.

______ 1. $|-3|$

______ 2. $|-\sqrt{5}|$

______ 3. $|x - 3|$

______ 4. $|\pi - 2|$

______ 5. $|2 - \sqrt{11}|$

______ 6. $|-3| - |5|$

Answers are on page 46.

▶ Examples You Completed

Example 2. Evaluate $9y - 37z$ when $y = 0$ and $z = 1$.

Solution. We must substitute 0 for y and 1 for z in the expression.

$$9y - 37z$$
$$= 9(0) - 37(1)$$
$$= 0 - 37$$
$$= -37$$

Example 4. Evaluate $(x + 1)(1 - x)$ when $x = -2$.

Solution

$$(x + 1)(1 - x)$$
$$= (-2 + 1)[1 - (-2)]$$
$$= (-1)(1 + 2)$$
$$= (-1)(3)$$
$$= -3$$

Example 5. Evaluate $-x^2 - 3x + 2$ when $x = -4$.

Solution

$$-x^2 - 3x + 2$$
$$= -(-4)^2 - 3(-4) + 2$$
$$= -1(-4)(-4) - 3(-4) + 2$$
$$= -16 + 12 + 2$$
$$= -2$$

Example 9. Simplify $x^2 - 6x + 3x^2 - x$.

Solution

$$x^2 - 6x + 3x^2 - x$$
$$= x^2 + 3x^2 - 6x - x$$
$$= 4x^2 - 7x$$

Example 11. Simplify $-3(2y) - 5(-y)$.

Solution

$$\begin{aligned} &-3(2y) - 5(-y) \\ &= -6y + 5y \\ &= -y \end{aligned}$$

Example 12. Simplify $\frac{10y^2}{5} - \frac{14x^2}{7}$.

Solution

$$\begin{aligned} &\frac{10y^2}{5} - \frac{14x^2}{7} \\ &= 2y^2 - 2x^2 \end{aligned}$$

Example 15. Simplify $0.2(4x - 11y)$.

Solution

$$\begin{aligned} &0.2(4x - 11y) \\ &= 0.2(4x) + 0.2(-11y) \\ &= 0.8x + (-2.2y) \\ &= 0.8x - 2.2y \end{aligned}$$

Example 16. Simplify

$$\tfrac{1}{2}(4a - 10) - 3(4 - a)$$

Solution

$$\begin{aligned} &\frac{1}{2}(4a - 10) - 3(4 - a) \\ &= \frac{1}{2} \cdot 4a - \frac{1}{2} \cdot 10 - 3 \cdot 4 - 3(-a) \\ &= 2a - 5 - 12 + 3a \\ &= 2a + 3a - 5 - 12 \\ &= 5a - 17 \end{aligned}$$

Example 18. Find $|\pi|$ and $|-\sqrt{2}|$.

Solution. Since $\pi > 0$, $|\pi| = \pi$.

Since $-\sqrt{2} < 0$,

$$\begin{aligned} |-\sqrt{2}| &= -(-\sqrt{2}) \\ &= \sqrt{2} \end{aligned}$$

Example 22. Find $-|3| - |-7|$.

Solution

$$\begin{aligned} &-|3| - |-7| \\ &= -(3) - (7) \\ &= -10 \end{aligned}$$

Answers to Trial Runs

page 40 1. 0 2. 12 3. 16 4. 15 5. $-4x + 5$ 6. $-x^2 + 10x - 6$
7. $-2x^2 + 2xy + 2y^2$ 8. $-4a - 2$

page 43 1. $19x$ 2. $6x + 12y$ 3. $-2x^2 + 13x$ 4. $a - 13$ 5. $24x + 43$ 6. $-x + 12y$

page 45 1. 3 2. $\sqrt{5}$ 3. $x - 3$ if $x - 3 \geq 0$; $-(x - 3)$ if $x - 3 < 0$ 4. $\pi - 2$
5. $-2 + \sqrt{11}$ 6. -2

Name ________________ Date ________

EXERCISE SET 1.4

Evaluate each expression.

______ 1. $x(2x - 3)$ when $x = 3$

______ 2. $2x(x - 9)$ when $x = 5$

______ 3. $(x - 8)(x + 1)$ when $x = -2$

______ 4. $(x - 1)(x + 5)$ when $x = -9$

______ 5. $-x^2 - 4x - 3$ when $x = -3$

______ 6. $-x^2 - 7x - 9$ when $x = -1$

______ 7. $5(x - 3)^2 - 9x$ when $x = 0$

______ 8. $9(x - 2)^2 - 8x$ when $x = 2$

______ 9. $7x^2 - 3xy - 4y^2$ when $x = 2$ and $y = -1$

______ 10. $10x^2 - 3xy + y^2$ when $x = 1$ and $y = -4$

______ 11. $P = 2l + 2w$ when $l = 12$ and $w = 7$.

______ 12. $K = \frac{1}{2}h(a + b)$ when $h = 16$, $a = 21$, and $b = 13$.

______ 13. $V = lwh$ when $l = 8.2$, $w = 5.6$, and $h = 10$.

______ 14. $i = prt$ when $p = \$4000$, $r = 0.09$, and $t = 4$.

______ 15. $M = \dfrac{c(1 + p)}{1 - d}$ when $c = 7.5$, $p = 0.30$, and $d = 0.25$.

______ 16. $s = \dfrac{rl - a}{r - 1}$ when $r = 5$, $l = 18.4$, and $a = 0.4$.

______ 17. $S = 100t - 16t^2$ when $t = 5$.

______ 18. $A = p(1 + r)^2$ when $p = \$1000$ and $r = 0.05$.

Perform the indicated operations and simplify.

______ 19. $3a - 9 - 7a + 2$

______ 20. $5a - 6 - 9a + 3$

______ 21. $9x^2 - 4x - 7x^2 - 7x$

______ 22. $13x^2 - 5x - 9x^2 - 4x$

______ 23. $-7y^2 + 3y - 8 - 2y^2 - 3y + 7$

______ 24. $-15y^2 + 9y - 1 - 5y + 4y^2 - 6$

______ 25. $17x - 8y + 3 - 6x + 2y - 5$

______ 26. $21x - 9y + 5 - 8x + 3y - 9$

______ 27. $8x^2 - 3xy - 7y^2 - 5x^2 - 2xy + 8y^2$

______ 28. $21x^2 - 7xy - 9y^2 - 3xy + 5y^2$

______ 29. $-17a^2 - 9b^2 + c^2 + 5a^2 - 2b^2 - c^2$

______ 30. $8a^2 - 3b^2 - 2c^2 - 11a^2 + 3b^2 - 2c^2$

______ 31. $3(7a) - 5(4a)$

______ 32. $-3(2a) + 7(-a)$

_______ 33. $-(3x) + 5(-2x) + \frac{18x}{3}$

_______ 34. $-(4x) + 6(-3x) + \frac{21x}{3}$

_______ 35. $\frac{28y^2}{4} + 3(-y^2) - 2(-4y^2)$

_______ 36. $\frac{32y^2}{8} + 5(-y^2) - 3(-5y^2)$

_______ 37. $\frac{63}{-9} + 4(2x) - 2(-3x^2) - \frac{72x}{9}$

_______ 38. $\frac{56x^2}{-8} + 3(5x) + \frac{14x^2}{2} - 2(-10x)$

_______ 39. $-8(2y) + \frac{91y^2}{7} - 2(-7)$

_______ 40. $5(-2y) + \frac{38y^2}{2} - 5(-3)$

_______ 41. $\frac{1}{2}(16x) - 5(-3y) - 4(13)$

_______ 42. $\frac{1}{3}(-15x) - 7(-2y) - 2(15)$

_______ 43. $(-9x - 2y) - (5x - 2y)$

_______ 44. $(-5m - 2n) - (m - 2n)$

_______ 45. $3(5m - 4) + 4(m + 2)$

_______ 46. $5(2m - 1) - 2(m + 5)$

_______ 47. $-3(x - 2y) + 4(-x + y)$

_______ 48. $-2(2x - y) + 3(-x + 2y)$

_______ 49. $-4[3(2x - 1) + 10x] - \frac{1}{5}(15x)$

_______ 50. $-8[2(3x + 2) + 7x] - \frac{1}{8}(16x)$

_______ 51. $7^2 - 5[x - 3(2 - x)]$

_______ 52. $8^2 - 3[x - 6(5 - x)]$

_______ 53. $2x - 4[x - 6(3x + 5)]$

_______ 54. $7x - 3[x - 8(5x - 7)]$

_______ 55. $-\frac{1}{2}(4x - 6y) + 7[x - (x - 5y)]$

_______ 56. $-\frac{1}{3}(9x - 12y) + 8[x - (x - y)]$

_______ 57. $-2[5(a - 2b) + 7(2a + b)]$

_______ 58. $-3[4(2a - b) + 5(a + 3b)]$

_______ 59. $7[3x - (x + 2y)] - [5(x - 2y) - 9y]$

_______ 60. $3[4x - (2x + y)] - [5(x - 2y) - 7y]$

Write each expression without absolute value bars.

_______ 61. $|-11|$

_______ 62. $|-8|$

_______ 63. $|3x|$

_______ 64. $|5x|$

_______ 65. $|5 - \pi|$

_______ 66. $|9 - \pi|$

_______ 67. $|x - 2|$

_______ 68. $|x + 1|$

_______ 69. $|\sqrt{7} - 3|$

_______ 70. $|\sqrt{6} - 1|$

_______ 71. $|-4| - |9|$

_______ 72. $|-5| - |11|$

_______ 73. $|-4 - 9|$

_______ 74. $|-5 - 11|$

☆ Stretching the Topics

________ 1. Simplify $-2[5x - 2(2x - 5y)] + 7[4(x - 2y) - (3x + y)]$. Then evaluate the expression when $x = -8$ and $y = -1$.

________ 2. Subtract the opposite of $[3 - (4 - x)]$ from $8x - (4 - 2x)$.

________ 3. Evaluate $2|x + 3| - |2x + 5| - 4(3x - 1)$ when $x = -2$.

Check your answers in the back of your book.

If you can complete **Checkup 1.4**, you are ready to do Section 1.5.

Name ______________________ Date __________

✓ CHECKUP 1.4

Evaluate each expression.

______ 1. $(x - 3)(2x - 1)$ when $x = 5$

______ 2. $s = \frac{1}{2}gt^2$ when $g = 32$ and $t = 7$

Perform the indicated operations and simplify.

______ 3. $-5x^2 + 3x - 7 + x^2 - 2x + 3$

______ 4. $18x^2 - 4xy - 5y^2 - 15x^2 + 4xy - 4y^2$

______ 5. $7(-2a) - 3(4a)$

______ 6. $-2(-3y) + \dfrac{34y^2}{2} - 4(2)$

______ 7. $-3(2n - 7) + 8[n - (2n - 3)]$

______ 8. $9y - 7[x - 3(2y - x)]$

Write each expression without absolute value bars.

______ 9. $|4x|$

______ 10. $|\sqrt{2} - 9|$

Check your answers in the back of your book.

If You Missed Problems:	You Should Review Examples:
1	4
2	7
3, 4	8, 9
5	10, 11
6	12, 13
7, 8	16, 17
9, 10	18–21

1.5 Switching from Words to Algebraic Expressions

Now that we have learned how to deal with simple algebraic expressions, we must try our hand at switching some everyday word expressions into expressions involving variables and constants.

To switch a word expression to a variable expression, we must first decide what the variable is to represent. Then we translate the words into a variable expression. The key to success here is in writing down each word phrase as a variable phrase. (Skipping steps will only confuse the matter.) Finally, we write a sentence that summarizes our findings.

Example 1. Martina has a part-time job paying \$3.50 per hour. Write an algebraic expression for her total earnings after she has worked for x hours.

Solution

We Think	***We Write***
The variable x represents the number of hours worked.	Let x = number of hours
Earnings are found by *multiplying* rate per hour times number of hours worked.	$\$3.50x$ = earnings

The expression $\$3.50x$ represents Martina's earnings.

Example 2. If a rectangular garden is 15 feet longer than it is wide, write an algebraic expression for the number of feet of fencing needed to enclose this garden.

Solution

We Think	***We Write***
The number of feet of fencing needed is just the *perimeter* of the rectangle.	Perimeter = 2 lengths + 2 widths
We need expressions for length and width.	Let w = width (feet)
We know less about the width than we know about the length.	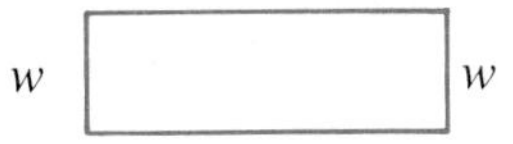
We know that the length is 15 feet *more than* the width.	Let $w + 15$ = length (feet) 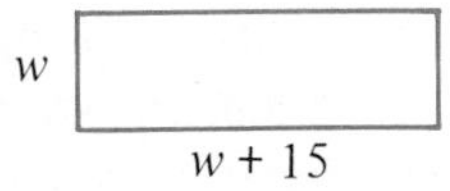
We know that the perimeter is the *sum* of 2 lengths and 2 widths.	$2(w + 15) + 2w$ = perimeter (feet)
We can simplify this expression.	$2w + 30 + 2w$ $= 4w + 30$

The expression $4w + 30$ represents the number of feet of fencing needed to enclose this garden.

You try completing the missing parts of Example 3.

Example 3. Charlie is a part-time college student. He has saved \$800 for his tuition. He must pay \$50 per course credit hour. Write an algebraic expression for the amount Charlie will have left after enrolling for h credit hours.

Solution

We Think	***We Write***
The variable h represents the number of credit hours.	Let ____ = number of credit hours
The total cost is found by *multiplying* the rate per credit hour times the number of credit hours.	____ = total cost (dollars)
The money left after enrolling is found by *subtracting* the total cost from \$800.	800 − ____ = dollars left

After enrolling, Charlie will have ____ dollars left. What happens if $h > 16$? ____________ What does that mean to Charlie? ____________

Check your work on page 56. ▶

Example 4. On a recent trip, Francine drove for 8 hours, part through towns and part on the highway. Her rates were 35 miles per hour in town and 55 miles per hour on the highway. Write an expression for the total distance Francine drove.

Solution

We Think	***We Write***
Francine's total distance will be the *sum* of her distance through towns and her distance on the highway.	Distance = town distance + highway distance
Each distance is found by *multiplying* rate times time.	Distance = rate × time
Town distance is found by *multiplying* rate in town times hours in town.	Town distance = 35 · hours in town
Highway distance is found by *multiplying* rate on highway times hours on highway.	Highway distance = 55 · hours on highway
All we know about the two times is that they *sum* to 8 hours. We can choose the variable t to represent the time spent driving in town.	Let t = hours in town
The time remaining for highway driving is the total time *minus* the time spent in town.	$8 - t$ = hours on the highway
Now we can write expressions for town distance and highway distance.	Town distance = $35t$ miles Highway distance = $55(8 - t)$ miles
Finally, we can *sum* the two distances to write an expression for total distance.	Distance = $35t + 55(8 - t)$ miles
We can simplify this expression.	$35t + 55(8 - t)$ $= 35t + 440 - 55t$ $= 440 - 20t$

The expression $440 - 20t$ represents Francine's total distance in miles.

To switch word expressions to variable expressions, it is helpful to

1. Read the problem carefully and sort out the given information.
2. Identify the variable.
3. Switch each word expression in the problem to a variable expression.
4. Be sure that your expression represents the quantity you are asked to find.
5. Simplify the variable expression if possible.

Suppose that you try completing Example 5 using these steps.

Example 5. Each month Mark's mother earns twice as much as he does, and his father earns $900 more than Mark. Write an algebraic expression for the total monthly earnings of Mark's family.

Solution

We Think	***We Write***
The family's total earnings will be the *sum* of the members' separate earnings.	Total earnings = Mark's earnings + mother's earnings + father's earnings
We know the least about Mark's earnings.	Let x = Mark's earnings
Mark's mother earns *twice* as much as Mark earns.	____ = mother's earnings
Mark's father earns $900 *more than* Mark earns.	x + ____ = father's earnings
Now we find the *sum* of all the earnings.	Total earnings = x + ____ + (x + ____)
We can simplify this expression.	x + ____ + x + ____ = ____ + ____

The expression ____ + ____ represents the total monthly earnings for Mark's family.

Check your work on page 56. ▶

When a company manufactures and sells a product, the money received from selling the product is called the company's **revenue**. The dollars spent in manufacturing the product are called the company's **costs**. The *difference* between the company's revenue and costs represents the company's **profit**.

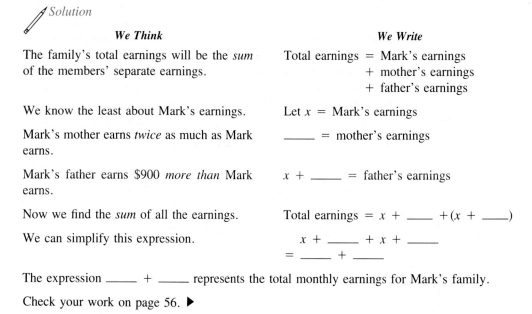

$$\text{Revenue} = \text{dollars received}$$
$$\text{Costs} = \text{dollars spent}$$
$$\text{Profit} = \text{revenue} - \text{costs}$$

Usually, the *revenue* from the sale of a product is found by multiplying the selling price times the number of units sold. For instance, if a shoe company sells 500 pairs of shoes at $20 a pair, the company's revenue is

$$\begin{aligned}\text{Revenue} &= (\text{selling price})(\text{number of units}) \\ &= (\$20)(500) \\ \text{Revenue} &= \$10{,}000\end{aligned}$$

Example 6. If Supercal Calculator Company sells x calculators at \$23.95 each, write an algebraic expression for the company's revenue.

Solution

We Think	***We Write***
We know the selling price.	\$23.95 = selling price
The number of units (calculators) is x.	x = number of units
Revenue is found by multiplying the selling price times the number of units sold.	Revenue = $23.95x$

The company's revenue is represented by the expression $23.95x$ dollars.

In manufacturing a product (say, shoes), a company must consider two kinds of *costs*. **Variable costs** are the costs that depend on the number of pairs of shoes produced. Variable costs include wages, raw materials, and utilities used in the manufacturing process. **Fixed costs**, on the other hand, are expenses that the company must pay even if it does not manufacture any shoes for a period of time. Fixed costs include rent, depreciation of equipment, and payments on company loans.

Total costs are the *sum* of variable costs and fixed costs.

Costs = variable costs + fixed costs

For instance, suppose the shoe company knows that its *fixed costs* are \$1000 per week and its *variable costs* are \$9 for each pair of shoes. If the company produces 500 pairs of shoes in a week, what are the company's costs?

$$\begin{aligned}
\text{Fixed costs} &= \$1000 \\
\text{Variable costs} &= \$9(500) \\
&= \$4500 \\
\text{Costs} &= \text{variable costs} + \text{fixed costs} \\
&= \$4500 + \$1000 \\
&= \$5500
\end{aligned}$$

Example 7. The Supercal Calculator Company knows that its fixed costs are \$7300 per month and its variable costs are \$8.45 for each calculator. Write an algebraic expression for the company's costs when x calculators are produced in a month.

Solution

We Think	***We Write***
We know that fixed costs are \$7300.	7300 = fixed costs
The variable costs are found by *multiplying* the number of calculators times the cost of producing each calculator.	8.45 = cost to produce each calculator x = number of calculators produced $8.45x$ = variable costs
Total costs are found by *adding* variable costs and fixed costs.	Costs = $8.45x + 7300$

The expression $8.45x + 7300$ dollars represents the company's cost for producing x calculators in one month.

To calculate *profit*, a company must *subtract* its costs from its revenue. For instance, when the shoe company produces and sells 500 pairs of shoes in one week, we found

$$\text{Revenue} = \$20(500) = \$10{,}000 \qquad \text{and} \qquad \text{Costs} = \$9(500) + \$1000 = \$5500$$

Now we can find the shoe company's profit.

$$\begin{aligned}\text{Profit} &= \text{revenue} - \text{costs}\\ &= \$10{,}000 - \$5500\\ &= \$4500\end{aligned}$$

Example 8. The Supercal Company manufactures calculators at a cost of \$8.45 each and sells them for \$23.95 each. The company's fixed costs are \$7300 per month. Write an expression for the company's *profit* if x calculators are produced and sold in one month.

Solution

We Think	***We Write***
Profit is found by *subtracting* costs from revenue.	Profit = revenue − costs
Revenue is found by *multiplying* selling price times number of units sold.	Revenue = $23.95x$
Costs are found by *adding* variable costs and fixed costs.	Variable costs = $8.45x$ Fixed costs = 7300 Costs = $8.45x + 7300$
Profit = revenue − costs.	Profit = $23.95x - (8.45x + 7300)$ $= 23.95x - 8.45x - 7300$ $= 15.5x - 7300$

The expression $15.5x - 7300$ dollars represents the Supercal Company's profit for one month.

Now you try Example 9.

Example 9. The Hamburger Heaven makes hamburgers for \$0.81 each and sells them for \$1.59 each. The restaurant's fixed costs are \$2100 per week. Write an expression for the restaurant's *profit* if x hamburgers are made and sold in one week.

Solution

We Think	***We Write***
Profit is found by *subtracting* costs from revenue.	Profit = ____ − ____
Revenue is found by *multiplying* selling price times the number of units sold.	Revenue = ____ x
Costs are found by *adding* variable costs and fixed costs.	Variable costs = ____ x Fixed costs = ____ Costs = ____ x + ____
Profit = revenue − costs.	Profit = $1.59x - (0.81x + 2100)$ $= 1.59x - 0.81x - 2100$ = ____ $x - 2100$

The expression ____ $x - 2100$ dollars represents the Hamburger Heaven's profit when x hamburgers are sold in one week.

Check your work on page 56. ▶

Trial Run

Change each word expression to a variable expression. Then simplify if possible.

_______ 1. If Julio drives for 6 hours at a constant rate of r miles per hour, write an algebraic expression for his total distance.

_______ 2. Tony bought 10 pounds of chicken parts for his barbecue, some thighs and some legs. If thighs cost \$1.19 per pound and legs cost \$0.99 per pound, write an algebraic expression for the total cost of the chicken parts.

_______ 3. A recent concert was attended by 200 children, 500 adults, and 300 senior citizens. A child's ticket costs half as much as an adult's ticket and a senior citizen's ticket costs \$1 less than an adult's ticket. Write an algebraic expression for the total dollars from concert ticket sales.

Answers are below.

Examples You Completed

Example 3 (*Solution*)

We Write

$$\begin{aligned} \text{Let } h &= \text{number of credit hours} \\ 50h &= \text{total cost} \\ 800 - 50h &= \text{dollars left} \end{aligned}$$

After enrolling, Charlie will have $800 - 50h$ dollars left. What happens if $h > 16$? $800 - 50h$ will be negative. What does that mean to Charlie? He cannot enroll for more than 16 hours.

Example 5 (*Solution*)

We Write

$$\begin{aligned} \text{Total earnings} &= \text{Mark's earnings} \\ &\quad + \text{mother's earnings} \\ &\quad + \text{father's earnings} \\ \text{Let } x &= \text{Mark's earnings} \\ 2x &= \text{mother's earnings} \\ x + 900 &= \text{father's earnings} \\ \text{Total earnings} &= x + 2x + (x + 900) \\ & x + 2x + x + 900 \\ &= 4x + 900 \end{aligned}$$

The expression $4x + 900$ represents the total monthly earnings for Mark's family.

Example 9 (*Solution*)

We Think	***We Write***
Profit is found by *subtracting* costs from revenue.	Profit = revenue − costs
Revenue is found by *multiplying* selling price times the number of units sold.	Revenue = $1.59x$
Costs are found by *adding* variable costs and fixed costs.	Variable costs = $0.81x$ Fixed costs = 2100 Costs = $0.81x + 2100$
Profit = revenue − costs.	Profit = $1.59x - (0.81x + 2100)$ = $1.59x - 0.81x - 2100$ = $0.78x - 2100$

The expression $0.78x - 2100$ dollars represents Hamburger Heaven's profit when x hamburgers are sold in one week.

Answers to Trial Run

page 56 1. $6r$ miles 2. $0.2x + 9.9$ dollars (or $11.90 - 0.2x$ dollars) 3. $900x - 300$

Name ________________________________ Date ____________

EXERCISE SET 1.5

Change the word expressions to variable expressions and simplify.

________ 1. Lydia is depositing \$27.50 a month from her paycheck in a savings account. Write an algebraic expression for her total savings after x months.

________ 2. The width of a rectangle is 5 feet less than the length. Write an algebraic expression for the perimeter of the rectangle.

________ 3. A car rental agency charges a fee of \$18 a day plus 30 cents per mile driven. If Lee rents a car for one day and drives n miles, write an algebraic expression for the amount he owes at the end of one day.

________ 4. Every morning Abdul jogs and walks for 3 hours. If he jogs at 5 miles per hour and walks at 3 miles per hour, write an algebraic expression for his total distance.

________ 5. If Otho wishes to build a pen for his dog that is twice as long as it is wide, write an algebraic expression for the number of feet of fencing needed.

________ 6. Tillie is buying framing for a rectangular picture that is three times as long as it is wide. Write an algebraic expression for the number of inches of framing she should buy.

________ 7. In a physical fitness class Carlos did 7 more than twice as many push-ups as his father. If his father did x push-ups, write an algebraic expression for the number of push-ups Carlos did.

________ 8. At Weight Watchers Gretchen discovered that she had lost 5 pounds more than one-half as many pounds as Jane. If Jane lost p pounds, write an algebraic expression for the number of pounds Gretchen lost.

________ 9. Hector earns a base salary of \$940 per month plus a \$10 bonus for each suit he sells that costs \$100 or more. Write an algebraic expression for Hector's salary in December if he sold n suits costing \$100 or more that month.

________ 10. Tony is saving money for a vacation during spring break. If he deposits \$50 and then saves \$12 a week for x weeks, write an algebraic expression for the amount that he will have at the end of x weeks.

________ 11. When Mardell went to see her sister she made part of the trip by bus and the other part by taxi. The time she traveled by bus was twice the time she rode in a taxi. If she averaged 40 miles per hour on the bus and 30 miles per hour in the taxi, write an algebraic expression for the total distance.

________ 12. When Gary went to college, he drove for 6 hours, part on the interstate and part on two-lane roads. If his rate was 55 miles per hour on the interstate and 45 miles per hour on two-lane roads, write an expression for the total distance Gary drove.

________ 13. Sidney spends twice as much on food as he does on clothing and six times as much for rent as for clothing. If he spends x dollars per month on clothing, write an algebraic expression for the total of these three expenditures.

_________ 14. When baking a cake, Pat notices that the batter requires twice as much butter as the icing and the filling requires half as much butter as the icing. If the recipe uses n tablespoons of butter for the icing, write an algebraic expression for the total number of tablespoons of butter required.

_________ 15. A person burns up about three times as many calories while walking as while sitting. Suppose that Imogene spends 12 hours sitting and 4 hours walking during the day. Write an algebraic expression for the total number of calories burned that day.

_________ 16. For her flower bed, Callie needs four times as many boxes of petunias as boxes of marigolds. If petunias cost \$1 a box and marigolds cost \$0.75 a box, write an algebraic expression for the total cost of the plants.

_________ 17. Marcus is making 20 pounds of party mix, part peanuts and part cashews. If cashews cost \$2.19 a pound and peanuts cost \$1.39 a pound, write an algebraic expression for the total cost of the party mix.

_________ 18. Hagar is buying 30 pounds of apples for her Halloween party, some winesaps and some golden delicious. If the winesaps cost 79 cents a pound and the golden delicious apples cost 99 cents a pound, write an algebraic expression for the total cost of the apples.

_________ 19. The cashier at French's market has a stack of bills, some five-dollar bills and some ten-dollar bills. There are 7 fewer five-dollar bills than ten-dollar bills. Write an algebraic expression for the total value of the stack of bills.

_________ 20. John sold a box of books at a garage sale. The number of paperbacks was 3 less than twice the number of hardback books. If he sold the paperbacks for \$1 each and the hardbacks sold for \$3 each, write an algebraic expression for the amount he received from the sale of his books.

_________ 21. The Basket Place knows that its fixed costs are \$2500 per month and its variable costs are \$3.85 for each basket produced. Write an algebraic expression for the company's costs when x baskets are produced in a month.

_________ 22. Movac manufactures small motors at a cost of \$53.50 each. The company's fixed costs are \$8500 per month. If the motors sell for \$123.95 each, write an expression for the company's profit when x motors are produced and sold in one month.

_________ 23. The Brunson Battery Company can manufacture a battery for \$18.65. The company's fixed costs each week are \$2475. If the batteries sell for \$59.95 each, write an expression for the company's profit when x batteries are produced and sold in one week.

_________ 24. The Pizza Palace makes pizzas for \$2.30 each and sells them for \$9.75 each. The fixed costs each week are \$2800. Write an expression for the Pizza Palace's profit if x pizzas are made and sold in one week.

_________ 25. Mr. Waggoner has a small farm operation. The fixed costs for a year are \$10,250. If he can produce a bushel of corn for \$2.35 and sell it for \$4.10, write an expression for Mr. Waggoner's profit if x bushels are produced and sold that year.

☆ Stretching the Topics

________ **1.** If $n - 2$ represents an integer, write an expression for the sum of the next higher consecutive integer and the next lower consecutive integer.

________ **2.** Doug earns $1\frac{1}{2}$ times as much per hour for overtime as he does regular salary. Write an expression for Doug's earnings if he works a 40-hour week plus 12 hours of overtime.

________ **3.** Drennan has \$12,000 invested, part at $7\frac{1}{2}$ percent interest and the remainder at $8\frac{3}{4}$ percent interest. Write an expression for the total amount of money he will have at the end of one year.

Check your answers in the back of your book.

If you can complete **Checkup 1.5**, you are ready to do the **Review Exercises** for Chapter 1.

Name ______________________ Date __________

✓ CHECKUP 1.5

_______ 1. Carlos earns \$420 a week and \$12 per hour for each hour of overtime. If he worked x hours of overtime last week, write an algebraic expression for his earnings for that week.

_______ 2. The width of a room is 5 feet less than the length. Write an algebraic expression for the number of feet of ceiling border the paperhanger needs to purchase.

_______ 3. The caterer tells the mother of the bride that he will cater the reception for a fee of \$50, plus \$3 per guest. Write an algebraic expression for the total cost of the reception if n guests are expected to attend.

_______ 4. On a recent trip to see her grandmother, Lola drove 4 hours, part through towns and part on the highway. If her rate in town averaged 30 miles per hour and on the highway averaged 55 miles per hour, write an algebraic expression for the total distance she traveled.

_______ 5. When Sue, Ellen, and Sarah went shopping, Sue spent twice as much as Sarah, and Ellen spent \$50 more than Sarah. Write an algebraic expression for the total amount the women spent while shopping.

Check your answers in the back of your book.

If You Missed Problem:	You Should Review Example:
1	1
2	2
3	3
4	4
5	5

Summary

In this chapter we have discussed the basic ideas that form the foundation for all future work in algebra. After reviewing the rules for operations with real numbers, we introduced algebraic expressions and discovered how those same rules can be applied to variables that represent real numbers.

The most important real number properties and definitions are summarized here for your reference, with examples for each. We assume that all variables represent real numbers.

Property	Statement	Examples
Closure	$a + b \in R$ and unique $a \cdot b \in R$ and unique	$x + 3 \in R$ $5x \in R$
Commutative	$a + b = b + a$ $a \cdot b = b \cdot a$	$x + 2 = 2 + x$ $2(x + y) = (x + y)2$
Associative	$a + (b + c) = (a + b) + c$ $a(bc) = (ab)c$	$3 + (5 + x) = (3 + 5) + x$ $2(7x) = (2 \cdot 7)x$
Distributive	$a(b + c) = ab + ac$	$5(x - y) = 5x - 5y$
Identities	$a + 0 = a$ $a \cdot 1 = a$	$2x + 0 = 2x$ $(xy)(1) = xy$
Inverses	$a + (-a) = 0$ $a \cdot \frac{1}{a} = 1 \quad (a \neq 0)$	$6y + (-6y) = 0$ $5 \cdot \frac{1}{5} = 1$

Some special properties of **zero** were also noted.

Property	Statement	Examples
Addition	$a + 0 = a$	$-5x + 0 = -5x$
Subtraction	$a - 0 = a$	$y - 0 = y$
	$0 - a = -a$	$0 - 2x = -2x$
Multiplication	$a \cdot 0 = 0$	$(15x)(0) = 0$
Division	$\frac{0}{a} = 0 \quad (a \neq 0)$	$\frac{0}{23} = 0$
	$\frac{a}{0}$ is undefined	$\frac{x}{0}$ is undefined
	$\frac{0}{0}$ is not uniquely defined	$\frac{3 - 3}{x + (-x)}$ is not uniquely defined

Finally, we made several definitions that allowed us to write one expression in terms of another.

Operation	Definition	Examples
Subtraction	$a - b = a + (-b)$	$2x - 3x = 2x + (-3x)$
Division	$\frac{a}{b} = a \cdot \frac{1}{b}$	$\frac{5}{3} = 5 \cdot \frac{1}{3}$
Exponentiation	$a^n = \underbrace{a \cdot a \cdot \ldots \cdot a}_{n \text{ factors}}$	$2^3 = 2 \cdot 2 \cdot 2$ $x^4 = x \cdot x \cdot x \cdot x$
Absolute value	$\lvert x \rvert = \begin{cases} x & \text{if } x \geq 0 \\ -x & \text{if } x < 0 \end{cases}$	$\lvert 17 \rvert = 17$ $\lvert -5 \rvert = 5$

After much practice using these properties to simplify algebraic expressions, we learned to translate word expressions into variable expressions.

❑ Speaking the Language of Algebra

Complete each sentence with the appropriate word or phrase.

1. In this chapter we learned to perform operations with ________ (numbers whose values are fixed) and ________ (symbols that represent numbers).

2. All numbers that we dealt with in this chapter were ________ numbers.

3. In an addition (or subtraction) problem, the quantities being added (or subtracted) are called ________ . The answer in an addition problem is called a ________ and the answer in a subtraction problem is called a ________ .

4. To add (or subtract) algebraic expressions, we must ________ ________ ________ .

5. In a multiplication problem, the quantities being multiplied are called ________ and the answer is called a ________ .

6. In a division problem, the quantity being divided is called the ________ . The quantity doing the dividing is called the ________ , and the answer is called the ________ .

7. When we divide zero by any nonzero number, the answer is ________ . When we divide any nonzero number by zero, the answer is ________ .

8. The absolute value of a real number is never ________ .

△ Writing About Mathematics

Write your response to each question in complete sentences.

1. Explain how you would use the number line to compare two real numbers.

2. Explain how to find the value of a fraction when its numerator and/or denominator is 0.

3. Explain how you would evaluate a given algebraic expression for a given value of the variable.

4. Describe the order in which you would perform the indicated operations in simplifying the following expressions.
   (a) $(3 + 2)[5x - 2(x - 1)]$
   (b) $3 + 2[5x - 2(x - 1)]$

5. Suppose the expression $5x - 1500$ represents the profit (in dollars) when a company produces and sells x radios. Explain what is happening if $x < 300$.

Name ______________________________ Date ______________

REVIEW EXERCISES for Chapter 1

From the set $\{-4, -2.3, -\sqrt{2}, -\frac{5}{7}, 0, \sqrt{3}, 2.\overline{3}, \frac{1}{2}, 5\}$ *list the members that belong to each of the following sets.*

________ 1. Irrational numbers

________ 2. Integers

Compare the following numbers using $<$, $>$, *or* $=$.

________ 3. $\frac{15}{3}$ and $\sqrt{25}$

________ 4. $-\frac{9}{4}$ and 13

________ 5. -19 and $-\frac{3}{4}$

________ 6. 5 and $\sqrt{7}$

Use an inequality to state whether each number is positive or negative.

________ 7. $-\sqrt{6}$

________ 8. 8.75

Perform the indicated operations.

________ 9. $(-36) + (-19)$

________ 10. $7 + (-21)$

________ 11. $3.9\,(-1.1)$

________ 12. $(-17) - (-13)$

________ 13. $9 - (-8)$

________ 14. $(-5)(-9)$

________ 15. $(-7)(0.8)$

________ 16. $-91 \div \dfrac{7}{2}$

________ 17. $\dfrac{-55}{-11}$

________ 18. $\dfrac{9 - (-9)}{3 + (-3)}$

________ 19. $-12 + 18 + 9 - 6 - 5$

________ 20. $19 - 3 - 21 + 6 - 1$

________ 21. $4(-2)(7)(-5)(3)$

________ 22. $-6(7)(10)(8)(-1)$

________ 23. $-6(15 - 23)$

________ 24. $5 - 7[9 - 2(-1)]$

________ 25. $(8 - 12)[9 - 3(11)]$

________ 26. $\dfrac{4(-3) - \frac{1}{4}(16)}{-8(9 - 5)}$

________ 27. $\dfrac{-11(4) - 5[6 - 2(9)]}{-2[13 - 3(4)]}$

________ 28. $\dfrac{\frac{1}{5}(35) - [13 + 2(-3)]}{\dfrac{44}{-11} - 5(-3)}$

Evaluate each expression.

________ 29. $(2x - 3)(x - 5)$ when $x = 2$

_______ 30. $2x^2 - 7x + 5$ when $x = -1$

_______ 31. $A = p(1 + rt)$ when $p = \$2000$, $r = 0.09$, and $t = 4$.

_______ 32. $S = a + (n - 1)d$ when $a = -5$, $n = 10$, and $d = 3$.

Perform the indicated operations and simplify.

_______ 33. $-6x^2 + 4x - 9 + 4x^2 - 3x + 2$

_______ 34. $17x^2 - 3xy - 4y^2 - 14x^2 + 3xy + y^2$

_______ 35. $9(-3a) - 5(2a)$

_______ 36. $-3(7y) + \frac{32y^2}{8} - 5(3)$

_______ 37. $-7(n - 5) + 9(n + 2)$

_______ 38. $(11 - 8)[9x - 4(2x + 3)]$

_______ 39. $8^2 - 3[x - 5(1 - 3x)]$

_______ 40. $-\frac{1}{5}(10x + 15y) + 3[x - (x + 2y)]$

_______ 41. $-4[3(2x - y) + 5(3x + y)]$

_______ 42. $9[7x - (x - 3y)] - [3(2x - 5y) - 11y]$

Write each expression without absolute value bars.

_______ 43. $|-21|$

_______ 44. $|a|$

_______ 45. $|\sqrt{10} - 6|$

_______ 46. $|9 - 13|$

_______ 47. $|9| - |13|$

_______ 48. $|9| + |-13|$

Change the word expressions to variable expressions and simplify.

_______ 49. If the width of a rectangular tablecloth is 7 feet less than twice the length, write an algebraic expression for the total number of feet of lace needed to make a border.

_______ 50. The Bookrack is having a spring sale on all its paperback books, selling some for \$0.95 and the rest for \$1.75 each. Carlos bought 5 more of the \$1.75 books than he did of the cheaper ones. Write an algebraic expression for the total amount he spent at the sale.

_______ 51. In the last election for city attorney, the Democratic candidate received twice as many votes as the Independent candidate and the Republican candidate received 15 fewer than three times as many as the Independent. Write an algebraic expression for the total number of votes cast in the election.

________ 52. Charlene is using her savings to attend an exercise salon twice a week. If she must pay \$6 each time she attends and she has saved \$500, write an algebraic expression for the amount Charlene will have left after attending for x weeks.

Check your answers in the back of your book.

If You Missed Exercises:	You Should Review Examples:	
1, 2	SECTION 1.1	1–3
3–6		5, 6
7, 8		7, 8
9, 10	SECTION 1.2	5, 6
11–13		7–10
14, 15		11, 12
16, 17		13–16
18		21
19, 20	SECTION 1.3	3, 4
21, 22		5, 6
23–28		7, 8, 15, 16
29, 30	SECTION 1.4	4, 5
31, 32		7
33, 34		8, 9
35–37		10–15
38–42		16, 17
43–46		18–21
47, 48		22
49–52	SECTION 1.5	1–6

If you have completed the **Review Exercises** and corrected your errors, you are ready to take the **Practice Test** for Chapter 1.

Name ______________________ Date ____________

PRACTICE TEST for Chapter 1

	Problem	Section	Examples
________	1. Compare $\sqrt{10}$ and 3 using $<$, $>$, or $=$.	1.1	5, 6
________	2. Use an inequality to state that $-\sqrt{5}$ is a negative number.	1.1	7, 8

Perform the indicated operations.

	Problem	Section	Examples
________	3. $-35 + (-12)$	1.2	2
________	4. $14 + (-17)$	1.2	5
________	5. $-8 - (-23)$	1.2	9
________	6. $-11(-4)$	1.2	12
________	7. $\dfrac{-28}{7}$	1.2	15
________	8. $4(-2)(11)(0)$	1.3	5, 6
________	9. $10 - 6[3 - (-4)]$	1.3	7
________	10. $(7 - 11)[5 - 3(4)]$	1.3	7
________	11. $\dfrac{\dfrac{-18}{6} - 4[-5 - 2(-7)]}{-[8(-2) + 3]}$	1.3	8–10
________	12. Evaluate $A = 2lw + 2wh + 2lh$ when $l = 5$, $w = 3$, and $h = 7$.	1.4	7

Perform the indicated operations and simplify.

	Problem	Section	Examples
________	13. $-2x^2 - 5x - 7 - x^2 + 3x - 4$	1.4	8, 9
________	14. $6a^2 - 2ab + 3b^2 - 4b^2 + ab - 6a^2$	1.4	8, 9
________	15. $3(-m) - 6(-2m)$	1.4	10, 11
________	16. $3(-3x) + \dfrac{16x}{-8} - 5(4)$	1.4	13
________	17. $-5(2 - a) + 7(a - 2)$	1.4	16
________	18. $1 - 4[2y - (6y - 5)]$	1.4	16
________	19. $(1 - 4)[3x - 4(1 - x)]$	1.4	15, 17
________	20. $-\frac{1}{5}(10x - 5y) + 2[3x - (y - x)]$	1.4	15, 17
________	21. $-\lvert 12 - 19 \rvert$	1.4	20

		Section	Example
________	22. $\lvert -8 \rvert - \lvert -14 \rvert$	1.4	22
________	23. $\lvert \sqrt{5} - 3 \rvert$	1.4	20
________	24. On a recent trip, Martina drove for 3 hours, part of the way on interstate highway and part on two-lane roads. If her rate on the interstate was 55 mph and her rate on the two-lane roads was 45 mph, write an algebraic expression for the total distance she traveled.	1.5	4
________	25. Harriet's rectangular quilt is 2 feet longer than it is wide. Write an algebraic expression for the perimeter of the quilt.	1.5	2

CHAPTER

2

Solving First-Degree Equations and Inequalities

Mary left Hometown at 6 P.M., driving due east on Route 22 at 55 miles per hour. At the same time, Todd left Anytown driving at 60 miles per hour due west on Route 22. If Hometown is 368 miles due west of Anytown, at what time will Mary and Todd meet?

Now that we have learned to simplify algebraic expressions, it is only natural to move on to solving equations and inequalities. In Chapter 1 we learned that numbers can be compared using symbols of equality ($=$) or inequality ($<$, $>$, $\neq$). In this chapter we learn how to find those values of the variable that make statements of equality or inequality true. In this chapter we learn how to

1. Solve first-degree equations by addition, subtraction, multiplication, and division.
2. Solve first-degree inequalities by addition, subtraction, multiplication, and division.
3. Graph solutions of equations and inequalities on the real number line.
4. Solve equations and inequalities that contain absolute-value expressions.
5. Switch from word statements to first-degree equations and inequalities.

2.1 Solving First-Degree Equations

An **equation** is a mathematical statement which says that one quantity is equal to another quantity. You can probably solve some equations "on sight" just by substituting a likely number for the variable to see if it "works."

If $x + 3 = 7$ then $x = 4$ because $4 + 3 = 7$

If $x - 9 = -2$ then $x = 7$ because $7 - 9 = -2$

If $2x = 16$ then $x = 8$ because $2 \cdot 8 = 16$

If $\frac{x}{3} = 4$ then $x = 12$ because $\frac{12}{3} = 4$

Remember, to *solve an equation*, we must find the value of the variable that makes the statement true. Such a value is called a **solution** for the equation.

A number is a **solution** for an equation if substituting that number for the variable makes the equation a true statement.

We solved the first-degree equations above on sight, using trial and error, but you should remember that there are many equations for which trial and error would be a very inefficient way to arrive at a solution. In a *first-degree* equation, the variable appears to the *first power* only. For example,

$x - 6 = -7$ is a first-degree equation

$3y + 2 = 17$ is a first-degree equation

$\frac{4y - 3}{5} = 5$ is a first-degree equation

$x^2 + x = 3$ is *not* a first-degree equation

$y^3 = 27$ is *not* a first-degree equation

First-Degree Equation. An equation that can be written in the form

$$ax + b = c$$

where a, b, and c are real numbers (and $a \neq 0$), is called a first-degree equation in x.

In solving first-degree equations, our main goal is to *isolate the variable* on one side of the equation. To accomplish this, we must recall some very important properties of equality.

We may add (or subtract) the same quantity to (or from) both sides of an equation.

In other words, for real numbers a, b, and c, we have

Addition Property of Equality

If $a = b$

then $a + c = b + c$

and $a - c = b - c$

Let's see how this property helps us isolate the variable in an equation.

$$x - 6 = -7 \qquad \text{We wish to isolate } x.$$

$$x - 6 + 6 = -7 + 6 \qquad \text{Add 6 to both sides of the equation.}$$

$$x + 0 = -1 \qquad \text{Find the sums.}$$

$$x = -1 \qquad \text{Simplify the left side.}$$

We can **check** our solution by substituting -1 for x in the *original equation*.

$$x - 6 = -7$$

$$-1 - 6 \stackrel{?}{=} -7$$

$$-7 = -7$$

Our solution checks.

Example 1. Solve $-3 = 5 + y$ and check.

Solution. We must isolate y.

$$-3 = 5 + y$$

$$-5 - 3 = -5 + 5 + y \qquad \text{Subtract 5.}$$

$$-8 = y \qquad \text{Simplify.}$$

CHECK

$$-3 = 5 + y$$

$$-3 \stackrel{?}{=} 5 + (-8) \qquad \text{Substitute } -8 \text{ for } y.$$

$$-3 = -3 \qquad \text{Simplify.}$$

Now you try Example 2.

Example 2. Solve $x - 12 = 0$ and check.

Solution. We must isolate ______ .

$$x - 12 = 0$$

CHECK

Check your work on page 81. ▶

To solve equations in which the numerical coefficient of the variable is different from 1, we need another property of equality.

We may multiply or divide both sides of an equation by the same *nonzero* quantity.

In other words, for real numbers a, b, and c ($c \neq 0$), we have

Multiplication Property of Equality

If $a = b$

then $a \cdot c = b \cdot c$

and $\frac{a}{c} = \frac{b}{c}$

Let's use this property to solve an equation.

		CHECK
$\frac{x}{3} = 11$	We must isolate x.	$\frac{x}{3} = 11$
$\frac{3}{1} \cdot \frac{x}{3} = 3 \cdot 11$	Multiply both sides of the equation by 3.	$\frac{33}{3} \stackrel{?}{=} 11$
$\frac{\overset{1}{\cancel{3}}}{1} \cdot \frac{x}{\underset{1}{\cancel{3}}} = 33$	Simplify.	$11 = 11$
$x = 33$		

Let's try another equation.

		CHECK
$-5y = 13$	We must isolate y.	$-5y = 13$
$\frac{-5y}{-5} = \frac{13}{-5}$	Divide both sides of the equation by -5.	$-5\left(-\frac{13}{5}\right) \stackrel{?}{=} 13$
$y = -\frac{13}{5}$	Simplify.	$13 = 13$

Now you try Example 3.

Example 3. Solve $6 = -12x$.

Solution

$$6 = -12x$$

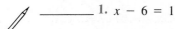

Check your work on page 81. ▶

Example 4. Solve $-\frac{2}{5}x = 10$.

Solution

$$-\frac{2}{5}x = 10$$

$$-\frac{5}{2}\left(-\frac{2}{5}x\right) = -\frac{5}{2}\left(\frac{10}{1}\right)$$

$$x = -25$$

Note from Example 4 that if the coefficient of the variable is a fraction, we may multiply both sides of the equation by the *reciprocal* of the coefficient in order to make the coefficient 1.

Trial Run

Solve each equation and check your solution.

_______ **1.** $x - 6 = 1$

_______ **2.** $a - 5 = 0$

_______ 3. $x + 6 = 6$

_______ 4. $-2 + y = -3$

_______ 5. $-3 = x + 7$

_______ 6. $6x = 42$

_______ 7. $-12x = 4$

_______ 8. $5 = \frac{x}{7}$

_______ 9. $-\frac{3x}{4} = 5$

_______ 10. $0 = \frac{x}{-3}$

Answers are on page 82.

Many first-degree equations require that more than one operation be performed on both sides before the variable can be isolated. Most of the time it is best to use the *addition* property of equality *before* the *multiplication* property of equality in such cases.

		CHECK
$3y + 2 = 17$	We must isolate y.	$3y + 2 = 17$
$3y + 2 - 2 = 17 - 2$	Subtract 2 from both sides.	$3(5) + 2 \stackrel{?}{=} 17$
$3y = 15$	Simplify each side.	$15 + 2 \stackrel{?}{=} 17$
$\frac{3y}{3} = \frac{15}{3}$	Divide both sides by 3.	$17 = 17$
$y = 5$	Simplify the quotients.	

Example 5. Solve $7 - x = 39$.

Solution

$7 - x = 39$

$-7 + 7 - x = -7 + 39$ Subtract 7.

$-x = 32$ Simplify.

$\frac{-x}{-1} = \frac{32}{-1}$ Divide by -1.

$x = -32$ Simplify.

Now try Example 6.

Example 6. Solve $\frac{x}{2} - 8 = 0$.

Solution

$\frac{x}{2} - 8 = 0$

Check your work on page 81. ▶

Trial Run

Solve each equation and check your solution.

_______ 1. $5x + 3 = 13$

_______ 2. $9 - 2x = 6$

_______ 3. $\frac{x}{3} + 6 = 0$

_______ 4. $5 - \frac{a}{2} = -1$

_______ 5. $8 = -3m + 7$

_______ 6. $-3y + 9 = 27$

Answers are on page 82.

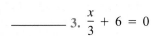

When an equation contains variable terms on *both sides*, we must first perform the additions and/or subtractions needed to get all the variable terms on the *same side* of the equation. For instance, consider the equation $5x = 3x + 10$. To get all the x terms on the left side, we must get rid of the $3x$ on the right. How? By *subtracting* $3x$ from both sides of the equation.

		CHECK
$5x = 3x + 10$		
$5x - 3x = 3x + 10 - 3x$	Subtract $3x$.	$5x = 3x + 10$
$2x = 10$	Simplify.	$5(5) \stackrel{?}{=} 3(5) + 10$
$\frac{2x}{2} = \frac{10}{2}$	Divide by 2.	$25 \stackrel{?}{=} 15 + 10$
		$25 = 25$
$x = 5$	Simplify.	

Example 7. Solve $13 + 7x = 10x - 2$ and check.

Solution

$$13 + 7x = 10x - 2$$
$$13 + 7x - 7x = 10x - 2 - 7x$$
$$13 = 3x - 2$$
$$13 + 2 = 3x - 2 + 2$$
$$15 = 3x$$
$$\frac{15}{3} = \frac{3x}{3}$$
$$5 = x$$

CHECK

$$13 + 7x = 10x - 2$$
$$13 + 7(5) \stackrel{?}{=} 10(5) - 2$$
$$13 + 35 \stackrel{?}{=} 50 - 2$$
$$48 = 48$$

Now try Example 8.

Example 8. Solve $10 - 3x = 16 - x$ and check.

Solution

$$10 - 3x = 16 - x$$
$$10 - 3x + x = 16 - x + x$$
$$10 - 2x = 16$$

CHECK

$$10 - 3x = 16 - x$$
$$10 - 3(\_\_\_\_) \stackrel{?}{=} 16 - (\_\_\_\_)$$

Check your work on page 81. ▶

These examples illustrate that it does *not* matter on which side of the equation you decide to put the variable. However, once you make your decision, you must do what is necessary to get the constant terms to the *other* side.

First-degree equations can also contain *parentheses* or other grouping symbols. First we simplify such an equation by removing the grouping symbols. Then we combine like terms and solve by the usual methods.

Example 9. Solve and check

$$3(y - 7) + 7y = 5 - 2(5y + 3)$$

Solution

$$3(y - 7) + 7y = 5 - 2(5y + 3)$$
$$3y - 21 + 7y = 5 - 10y - 6$$
$$10y - 21 = -10y - 1$$
$$10y + 10y - 21 = 10y - 10y - 1$$
$$20y - 21 = -1$$
$$20y - 21 + 21 = -1 + 21$$
$$20y = 20$$
$$\frac{20y}{20} = \frac{20}{20}$$
$$y = 1$$

CHECK

$$3(y - 7) + 7y = 5 - 2(5y + 3)$$
$$3(1 - 7) + 7(1) \stackrel{?}{=} 5 - 2[5(1) + 3]$$
$$3(-6) + 7 \stackrel{?}{=} 5 - 2[5 + 3]$$
$$-18 + 7 \stackrel{?}{=} 5 - 2 \cdot 8$$
$$-11 \stackrel{?}{=} 5 - 16$$
$$-11 = -11$$

Now you complete Example 10.

Example 10. Solve and check

$$-3(x - 1) = 2(-3 - x)$$

Solution

$$-3(x - 1) = 2(-3 - x)$$
$$-3x + 3 = -6 - 2x$$

CHECK

$$-3(x - 1) = 2(-3 - x)$$
$$-3(\_\_\_\_ - 1) \stackrel{?}{=} 2(-3 - \_\_\_\_)$$

Check your work on page 82. ▶

Example 11. Solve $5x - 2(x + 1) = 3(x + 2)$.

Solution

$5x - 2(x + 1) = 3(x + 2)$	
$5x - 2x - 2 = 3x + 6$	Remove parentheses.
$3x - 2 = 3x + 6$	Combine like terms.
$3x - 3x - 2 = 3x - 3x + 6$	Subtract $3x$.
$-2 = 6$	A false statement.

This statement is *false*, so there is *no solution* for this equation.

Example 12. Solve $5x - 2(x - 3) = 3(x + 2)$.

Solution

$$5x - 2(x - 3) = 3(x + 2)$$
$$5x - 2x + 6 = 3x + 6$$
$$3x + 6 = 3x + 6$$
$$6 = 6$$

This statement is *always true*, so *any real number* will satisfy this equation.

These examples illustrate two important facts about equations.

If, in solving an equation, we arrive at a statement that is *never* true, the equation has *no solution*.

If, in solving an equation, we arrive at a statement that is *always* true, the equation is called an **identity**, and *any real number* is a solution.

Whenever fractions appear in an equation, we can eliminate them by multiplying both sides of the equation by the **lowest common denominator** (LCD) for all the fractions in the equation. Then we use our usual steps to solve for the variable.

Example 13. Solve $\dfrac{x+3}{7} = 11$.

Solution. The LCD is 7.

$$\frac{7}{1}\left(\frac{x+3}{7}\right) = 7 \cdot 11 \qquad \text{Multiply by 7.}$$

$$x + 3 = 77 \qquad \text{Simplify.}$$

$$x = 74 \qquad \text{Subtract 3.}$$

Now you complete Example 14.

Example 14. Solve $\dfrac{x}{7} + 3 = 11$.

Solution. The LCD is _____ .

$$\frac{7}{1}\left(\frac{x}{7}\right) + 7 \cdot 3 = 7 \cdot 11$$

Check your work on page 82. ▶

Example 15. Solve $\dfrac{5x-1}{4} = \dfrac{7}{2}$.

Solution. The LCD is 4.

$$\frac{4}{1}\left(\frac{5x-1}{4}\right) = \frac{4}{1}\left(\frac{7}{2}\right) \qquad \text{Multiply by 4.}$$

$$5x - 1 = 14 \qquad \text{Simplify.}$$

$$5x = 15 \qquad \text{Add 1.}$$

$$x = 3 \qquad \text{Divide by 5.}$$

Example 16. Solve $\dfrac{2x}{3} + \dfrac{4}{3} = 4x$.

Solution. The LCD is 3.

$$\frac{3}{1}\left(\frac{2x}{3}\right) + \frac{3}{1}\left(\frac{4}{3}\right) = 3(4x)$$

$$2x + 4 = 12x$$

$$4 = 10x$$

$$\frac{4}{10} = x$$

$$\frac{2}{5} = x$$

Let's summarize the steps we have developed for solving first-degree equations.

Solving First-Degree Equations

1. Remove fractions by multiplying both sides by the LCD.
2. Remove parentheses and combine like terms on each side of the equation.
3. Get all variable terms on one side of the equation by addition or subtraction.
4. Get all constant terms on the other side of the equation by addition or subtraction.
5. Solve for the variable by multiplication or division.
6. Check your solution in the original equation.

Trial Run

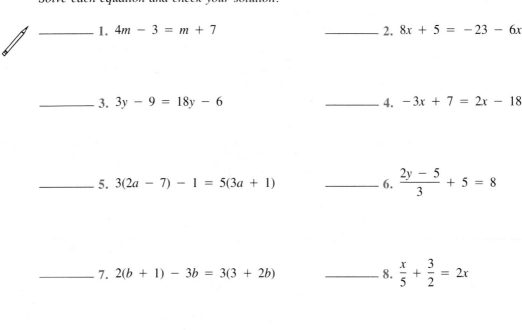

Solve each equation and check your solution.

_______ 1. $4m - 3 = m + 7$

_______ 2. $8x + 5 = -23 - 6x$

_______ 3. $3y - 9 = 18y - 6$

_______ 4. $-3x + 7 = 2x - 18$

_______ 5. $3(2a - 7) - 1 = 5(3a + 1)$

_______ 6. $\dfrac{2y - 5}{3} + 5 = 8$

_______ 7. $2(b + 1) - 3b = 3(3 + 2b)$

_______ 8. $\dfrac{x}{5} + \dfrac{3}{2} = 2x$

_______ 9. $2x - (5 - 3x) = 2[x - (1 - x)]$

Answers are on page 82.

Solving Equations for a Specified Variable

Sometimes algebraic equations or formulas contain more than one variable. If we wish to isolate a particular variable in such an equation, we may do so by performing the usual operations on both sides of the equation.

Example 17. Solve $I = Prt$ for P.

Solution. This is the formula for simple interest. We are asked to isolate P.

$$I = Prt$$

$$\frac{I}{rt} = \frac{Prt}{rt} \qquad \text{Divide by } rt.$$

$$\frac{I}{rt} = P \qquad \text{Simplify.}$$

Example 18. Solve $P = 2l + 2w$ for l.

Solution. This is the formula for the perimeter of a rectangle. We are asked to isolate l.

$$P = 2l + 2w$$

$$P - 2w = 2l + 2w - 2w$$

$$P - 2w = 2l$$

$$\frac{P - 2w}{2} = l$$

This procedure is called **solving an equation for a specified variable**. If the equation contains fractions or symbols of grouping, we continue to follow our usual steps for isolating the variable that has been specified.

Example 19. Solve $C = \frac{5}{9}(F - 32)$ for F.

Solution. This is the formula for changing Fahrenheit temperatures to Celsius temperatures. We must isolate F.

$C = \frac{5}{9}(F - 32)$	LCD: 9
$9 \cdot C = \frac{9}{1} \cdot \frac{5}{9}(F - 32)$	Multiply by LCD.
$9C = 5(F - 32)$	Simplify products.
$9C = 5F - 160$	Remove parentheses.
$9C + 160 = 5F$	Add 160 to both sides.
$\frac{9C + 160}{5} = F$	Divide both sides by 5.

You complete Example 20.

Example 20. Solve $A = \frac{h(a + b)}{2}$ for a.

Solution. This is the formula for the area of a trapezoid. We must isolate ____ .

$$A = \frac{h(a + b)}{2}$$

$$2 \cdot A = \frac{2}{1}\left[\frac{h(a + b)}{2}\right]$$

$$2A = h(a + b)$$

$$2A = ha + hb$$

Check your work on page 82. ▶

Trial Run

Solve each equation for the specified variable.

_______ 1. $C = 2\pi r$ for r

_______ 2. $A = \frac{bh}{2}$ for h

_______ 3. $A = P(1 + r)$ for r

_______ 4. $y = mx + b$ for x

Answers are on page 82.

▶ Examples You Completed

Example 2. Solve $x - 12 = 0$ and check.

Solution. We must isolate x.

$$x - 12 = 0$$
$$x - 12 + 12 = 0 + 12$$
$$x = 12$$

CHECK

$$x - 12 = 0$$
$$12 - 12 \stackrel{?}{=} 0$$
$$0 = 0$$

Example 3. Solve $6 = -12x$.

Solution

$$6 = -12x$$
$$\frac{6}{-12} = \frac{-12x}{-12}$$
$$-\frac{1}{2} = x$$

Example 6. Solve $\frac{x}{2} - 8 = 0$.

Solution

$$\frac{x}{2} - 8 = 0$$
$$\frac{x}{2} - 8 + 8 = 0 + 8$$
$$\frac{x}{2} = 8$$
$$\frac{2}{1}\left(\frac{x}{2}\right) = 2(8)$$
$$x = 16$$

Example 8. Solve $10 - 3x = 16 - x$ and check.

Solution

$$10 - 3x = 16 - x$$
$$10 - 3x + x = 16 - x + x$$
$$10 - 2x = 16$$
$$10 - 2x - 10 = 16 - 10$$
$$-2x = 6$$
$$\frac{-2x}{-2} = \frac{6}{-2}$$
$$x = -3$$

CHECK

$$10 - 3x = 16 - x$$
$$10 - 3(-3) \stackrel{?}{=} 16 - (-3)$$
$$10 + 9 \stackrel{?}{=} 16 + 3$$
$$19 = 19$$

Example 10. Solve and check

$$-3(x - 1) = 2(-3 - x)$$

Solution

$$-3(x - 1) = 2(-3 - x)$$
$$-3x + 3 = -6 - 2x$$
$$-3x + 3 + 2x = -6 - 2x + 2x$$
$$-x + 3 = -6$$
$$-x + 3 - 3 = -6 - 3$$
$$-x = -9$$
$$\frac{-1x}{-1} = \frac{-9}{-1}$$
$$x = 9$$

CHECK

$$-3(x - 1) = 2(-3 - x)$$
$$-3(9 - 1) \stackrel{?}{=} 2(-3 - 9)$$
$$-3(8) \stackrel{?}{=} 2(-12)$$
$$-24 = -24$$

Example 14. Solve $\frac{x}{7} + 3 = 11$

Solution. The LCD is 7.

$$\frac{7}{1}\left(\frac{x}{7}\right) + 7 \cdot 3 = 7 \cdot 11$$
$$x + 21 = 77$$
$$x = 56$$

Example 20. Solve $A = \frac{h(a + b)}{2}$ for a.

Solution. This is the formula for the area of a trapezoid. We must isolate a.

$$A = \frac{h(a + b)}{2}$$
$$2 \cdot A = \frac{2}{1}\left[\frac{h(a + b)}{2}\right]$$
$$2A = h(a + b)$$
$$2A = ha + hb$$
$$2A - hb = ha$$
$$\frac{2A - hb}{h} = a$$

Answers to Trial Runs

page 74 1. $x = 7$ 2. $a = 5$ 3. $x = 0$ 4. $y = -1$ 5. $x = -10$ 6. $x = 7$
7. $x = -\frac{1}{3}$ 8. $x = 35$ 9. $x = -\frac{20}{3}$ 10. $x = 0$

page 75 1. $x = 2$ 2. $x = \frac{3}{2}$ 3. $x = -18$ 4. $a = 12$ 5. $m = -\frac{1}{3}$ 6. $y = -6$

page 79 1. $m = \frac{10}{3}$ 2. $x = -2$ 3. $y = -\frac{1}{5}$ 4. $x = 5$ 5. $a = -3$ 6. $y = 7$
7. $b = -1$ 8. $x = \frac{5}{6}$ 9. $x = 3$

page 80 1. $r = \frac{C}{2\pi}$ 2. $h = \frac{2A}{b}$ 3. $r = \frac{A - P}{P}$ 4. $x = \frac{y - b}{m}$

Name ________________________ Date ____________

EXERCISE SET 2.1

Solve each equation and check the solution.

________ 1. $x - 11 = -6$

________ 2. $x - 9 = -8$

________ 3. $4 = y + 6$

________ 4. $5 = 9 + y$

________ 5. $-3 + m = -9$

________ 6. $-5 + m = -15$

________ 7. $-15y = -10$

________ 8. $-18y = -9$

________ 9. $7a = 0$

________ 10. $9a = 0$

________ 11. $-5 = 9x$

________ 12. $-3 = 7x$

________ 13. $\frac{x}{2} = 6$

________ 14. $\frac{x}{4} = 11$

________ 15. $\frac{-x}{2} = 13$

________ 16. $\frac{-x}{3} = 5$

________ 17. $3x + 5 = 26$

________ 18. $4x + 6 = 26$

________ 19. $7 - 4x = 9$

________ 20. $6 - 15x = 9$

________ 21. $21 = 12y - 15$

________ 22. $17 = 6y + 7$

________ 23. $\frac{x}{6} + 3 = 5$

________ 24. $\frac{x}{7} + 4 = 9$

________ 25. $6 - \frac{x}{4} = 7$

________ 26. $9 - \frac{x}{2} = 12$

________ 27. $-9 = \frac{a}{5} - 9$

________ 28. $-7 = \frac{a}{8} - 7$

________ 29. $3x + 2 = x + 10$

________ 30. $5x + 3 = 3x + 15$

________ 31. $-4y + 2 = y - 13$

________ 32. $-5y + 3 = 2y - 11$

________ 33. $3x - 8 = 5x$

________ 34. $7x - 15 = 12x$

________ 35. $-13 + 4m = -2m - 3$

________ 36. $-18 + 3m = -3m - 8$

________ 37. $7 - 2a = -3 + 8a$

________ 38. $9 - 3a = -6 + 2a$

________ 39. $10 = 2(x - 5) - 5(x + 2)$

________ 40. $-16 = 3(x - 6) - (2x + 5)$

________ 41. $-2[3(x - 2) - 5] = 22$

________ 42. $4[2(x - 2) - (x + 4)] = 0$

________ 43. $4(2x - 3) + 2 = 2(4x - 5)$

________ 44. $3(6x - 5) + 6 = 9(2x - 1)$

______ 45. $\frac{11y - 3}{4} + 6 = 3y$

______ 46. $\frac{7y - 5}{2} + 7 = 4y$

______ 47. $2(x + 1) + 3(x - 2) = 5x - 37$

______ 48. $4(x + 1) + (x - 5) = 5x - 19$

______ 49. $15 = \frac{2x - 1}{3} - \frac{1}{2}$

______ 50. $19 = \frac{4x - 6}{2} - \frac{2}{3}$

______ 51. $\frac{3x - 7}{5} - \frac{7}{3} = 2x$

______ 52. $\frac{3x - 9}{4} - \frac{9}{5} = 3x$

______ 53. $2[x + 3(x - 2)] = 5(x - 4) - 1$

______ 54. $3[2x + (x - 5)] = 2(x - 7) + 6$

______ 55. $5(x - 1) - (x + 2) = 3[5x - (x - 7)]$

______ 56. $3(x - 2) - (5x + 4) = 4[3x - (x + 7)]$

______ 57. $3[2x - 3(2x - 4)] = 5[2(x - 4) + 2]$

______ 58. $2[3x - (x + 4)] = 3[2(5 - x) - 6]$

Solve for the specified variable.

______ 59. $A = \frac{1}{2}bh$ for h.

______ 60. $V = \frac{1}{3}bh$ for b.

______ 61. $d = rt$ for t.

______ 62. $V = lwh$ for h.

______ 63. $A = P + Prt$ for t.

______ 64. $A = \pi R^2 + 2\pi rh$ for h.

______ 65. $3x - 2y = 1$ for x.

______ 66. $2x - 5y = 3$ for x.

______ 67. $ax + by = 5$ for x.

______ 68. $cx + dy = 9$ for y.

______ 69. $3x - 8a = 5(x - 4a)$ for x.

______ 70. $2y - 2b = -5(b - y)$ for y.

______ 71. $l = a + (n - 1)d$ for n.

______ 72. $A = p(1 + rt)$ for r.

______ 73. $d = \frac{V}{2r}$ for r.

______ 74. $W = \frac{wl}{L}$ for L.

☆ Stretching the Topics

______ 1. Solve $3ax + 7b = ax + 10b$ for x.

______ 2. Solve $n = \frac{l - a}{d} + 1$ for a.

______ 3. Solve $\frac{2}{3}(x - 4) = 1 + \frac{x + 5}{3}$ for x.

Check your answers in the back of your book.

If you can complete **Checkup 2.1,** you are ready to go on to Section 2.2.

Name ______________________ Date ______________

✓ CHECKUP 2.1

Solve each equation and check the solution.

______ 1. $x - 35 = -8$

______ 2. $11x = -36$

______ 3. $\frac{-a}{4} = 6$

______ 4. $17 - 6x = -1$

______ 5. $-5x + 1 = 7x - 3$

______ 6. $2(5y - 1) - 5 = 3(7 + y)$

______ 7. $8x - 3(2x + 4) = 2(x - 5)$

______ 8. $\frac{9x - 8}{4} - 2 = 5x$

______ 9. $\frac{4x - 3}{6} = \frac{9}{2}$

______ 10. $A = P(1 + rt)$ for r

Check your answers in the back of your book.

If You Missed Problems:	You Should Review Examples:
1	1, 2
2	3
3, 4	4, 5
5	7, 8
6	9, 10
7	11
8, 9	14, 15
10	19, 20

2.2 Graphing First-Degree Inequalities

A thorough understanding of first-degree equations will help us learn to solve first-degree inequalities. Before we attempt such solutions, however, let's take a look at some inequalities in which the variable has already been isolated.

In each case there will not be just one answer; in fact, there will be an infinite number of solutions for each inequality. Listing all the solutions is an impossible task. The *number line* allows us to picture our answers without listing each one. Using a number line to show solutions is called **graphing** the solutions.

Example 1. Graph the solutions for $x > 2$.

Solution. To satisfy $x > 2$, x must be a real number *greater than* 2. All numbers to the *right* of 2 on the number line are solutions. The number 2 itself is *not* a solution, but it acts as a *boundary*.

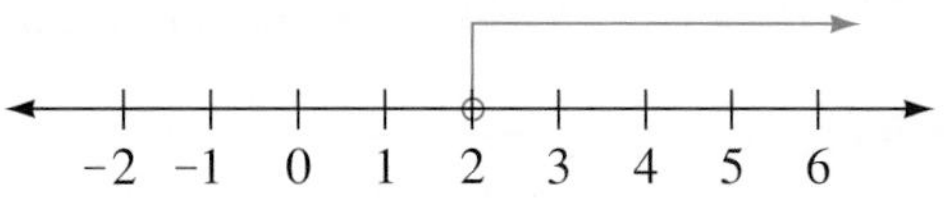

The **open dot** at 2 indicates that the number corresponding to that point is *not* a solution but is a boundary. The arrow at the end of the graph indicates that it extends forever to the right.

Example 2. Graph the solutions for $x \leq 0$.

Solution. To satisfy $x \leq 0$, x must be a real number *less than* 0 *or equal to* 0. All numbers to the *left* of 0 and 0 itself are solutions.

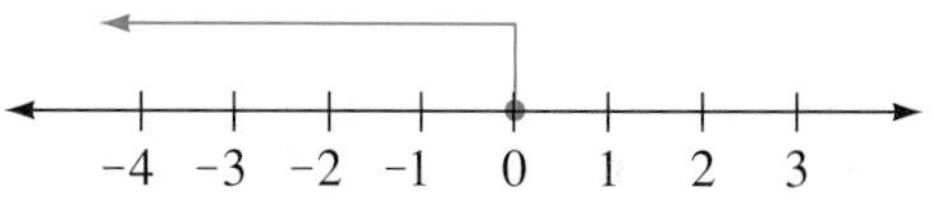

The **closed dot** at 0 indicates that the number corresponding to that point *is* a solution as well as a boundary. The arrow indicates that the graph extends forever to the left.

Mathematicians have invented another method for denoting the solutions for an inequality by describing the **interval** containing the numbers that satisfy that inequality. The numbers at each end of the interval are enclosed in brackets (if the boundary point *is* a solution) or parentheses (if the boundary point is *not* a solution). If an interval extends forever to the *right*, then the right endpoint is symbolized by ∞ (**positive infinity**). If an interval extends forever to the *left*, then the left endpoint is symbolized by $-\infty$ (**negative infinity**). Since ∞ and $-\infty$ are not finite numbers, their end of an interval is always closed by a parenthesis.

For instance, we represent

$$x > 2 \quad \text{by the interval} \quad (2, \infty)$$

$$x \leq 0 \quad \text{by the interval} \quad (-\infty, 0]$$

This way of representing solutions for inequalities is called **interval notation**.

Example 3. Graph the solutions for $x \geq -3$. Then write the solutions using interval notation.

Solution. There will be a closed dot at -3 and a line extending to the right.

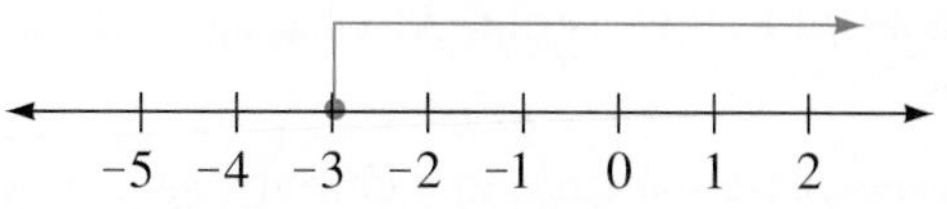

Interval notation: $[-3, \infty)$

You try Example 4.

Example 4. Graph the solutions for $x < \frac{1}{2}$. Then write the solutions using interval notation.

Solution. There will be a(n) ________ dot at _____ and a line extending to the ________.

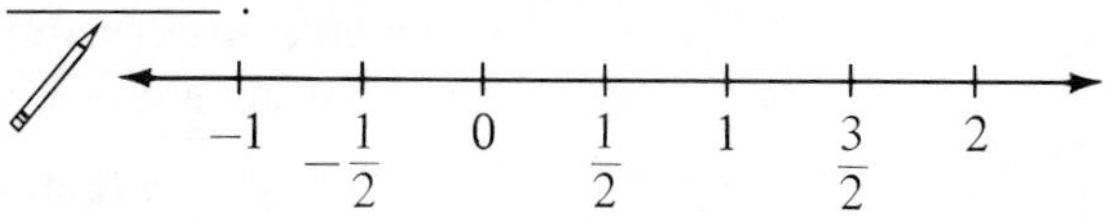

Interval notation: ________

Check your work on page 93. ▶

We make note of the pattern we have observed. Suppose that a represents any real number.

Inequality	Words	Graph	Interval Notation
$x < a$	x is less than a	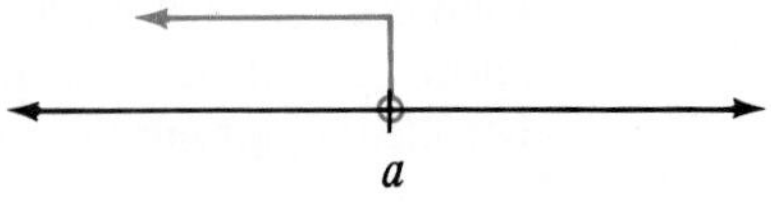	$(-\infty, a)$
$x > a$	x is greater than a	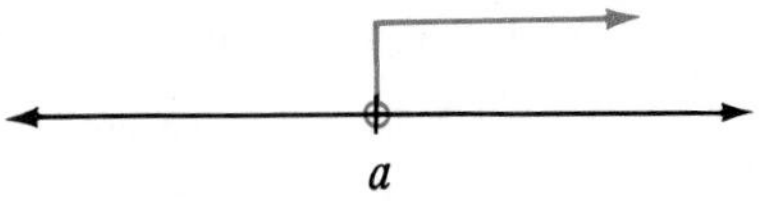	(a, ∞)
$x \leq a$	x is less than or equal to a	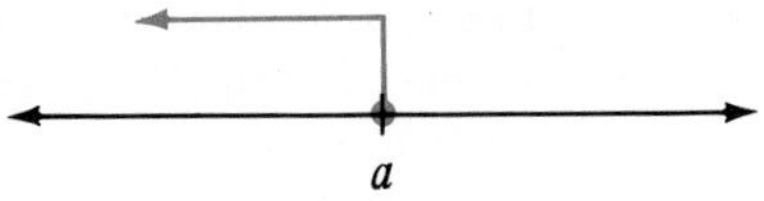	$(-\infty, a]$
$x \geq a$	x is greater than or equal to a	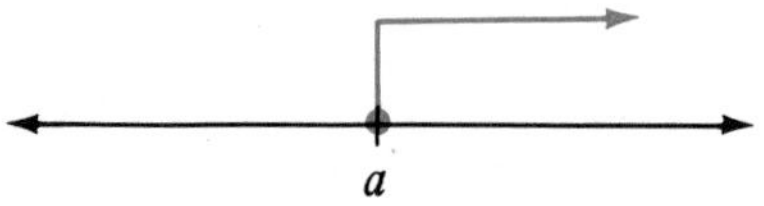	$[a, \infty)$

Trial Run

Graph the solutions on a number line. Then write the solutions using interval notation.

1. $x > 3$
2. $a \leq -2$
3. $y \geq 0$
4. $m < \frac{5}{2}$
5. $a \geq -\frac{1}{3}$

Answers are on page 93.

Finding Intersections of Graphs

Sometimes it is necessary to satisfy more than one inequality at the same time. In each of these examples, there are two inequalities that must be satisfied.

Example 5. A citizen of this state must be at least 16 years old to obtain a driver's license. To vote, a citizen must be at least 18 years old. At what ages will a citizen be able to vote *and* obtain a driver's license?

Solution. Both statements make reference to a citizen's age, so we can let a represent a citizen's age.

First statement	*Second statement*
$a \geq 16$	$a \geq 18$

The word *and* means that *both* qualifications must be met. When will *both* statements be true? For what ages will *both* inequalities be satisfied? Let's graph both inequalities on the same number line.

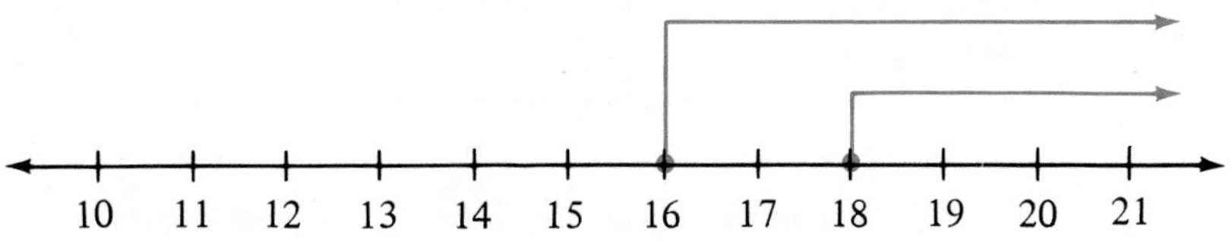

To decide where both inequalities are satisfied, we must find where *both* lines appear in our graph. It should be clear that both lines appear at 18 and to the right of 18. In other words, at all ages greater than or equal to 18, a citizen will be able to drive *and* vote.

We can write our solution as the single inequality $a \geq 18$ or as the interval $[18, \infty)$.

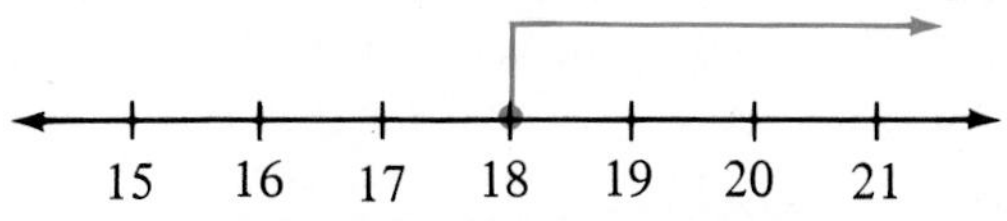

Example 6. To be used in tournament play, a tennis ball must bounce more than 53 inches but fewer than 58 inches (when dropped 100 inches onto a concrete surface). Write a single inequality describing the acceptable bounce heights.

Solution. We can let b represent the bounce of the tennis ball.

The bounce must be more than 53 inches	The bounce must be less than 58 inches
$b > 53$	$b < 58$

Again we can graph both inequalities on the same number line to see where *both* conditions are met.

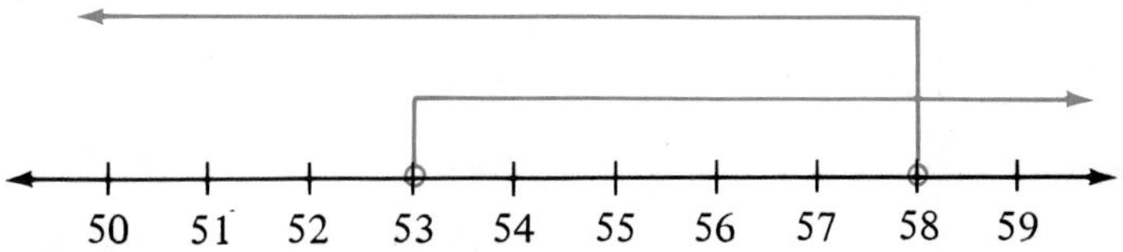

Both lines appear between 53 and 58 but *not* at the points 53 and 58 themselves. When a variable is located *between* two numbers, we write that statement as an inequality and graph it in the following way:

$53 < b < 58$

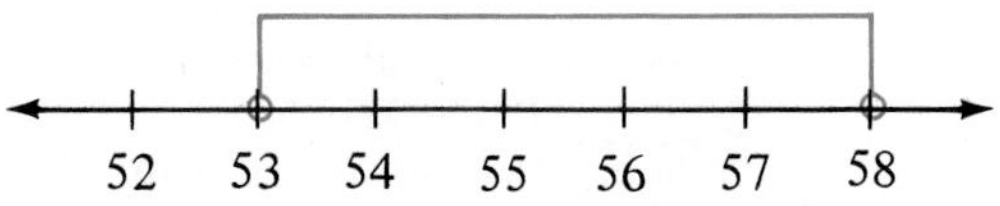

This inequality says:

b is greater than 53 *and* less than 58

or

53 is less than b, and b is less than 58

> The inequality $a < x < c$ means that x is *between* a and c.
> This inequality is written using interval notation as (a, c).

For instance, we may write the inequality $53 < b < 58$ as the interval $(53, 58)$.

In Examples 5 and 6 we were asked to satisfy one inequality *and* another inequality at the same time. In each case we looked for all points *common* to the set of real numbers satisfying the first inequality *and* the set of real numbers satisfying the second inequality. This set of common points is called the **intersection** of the two sets being discussed. The symbol for intersection is $\cap$.

Intersection of Sets A and B ($A \cap B$). The intersection of sets A and B is the set of elements that set A and set B have in common.

Remember that in seeking an intersection, we are demanding that *both* conditions be met at the same time. The best way to find the intersection of two inequalities is to graph both on the same number line and look for numbers that are included in *both* graphs.

Example 7. Solve $x < -3$ and $x \le 2$.

Solution. Graph both inequalities on the same number line.

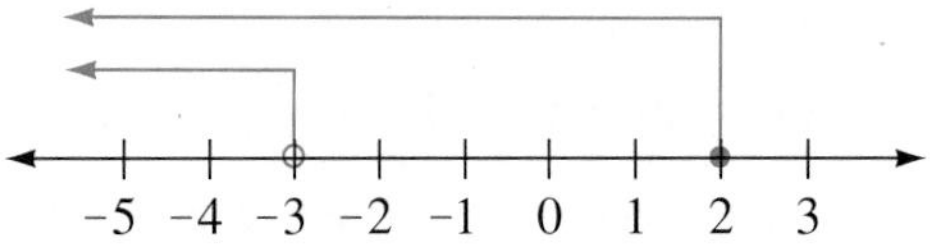

Both lines appear to the left of (and not including) the point -3. Therefore, we describe the solutions by

$$x < -3$$

or by the interval $(-\infty, -3)$.

Example 8. Solve $x \ge -1$ and $x < 4$.

Solution. Graph both inequalities on the same number line.

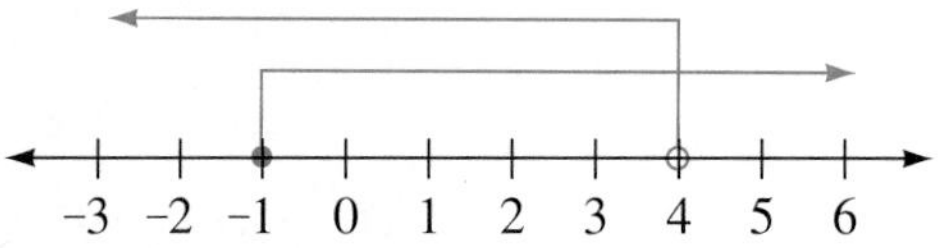

Both lines appear between -1 and 4 (including -1 but not including 4). Therefore, we describe the solutions by

$$-1 \le x < 4$$

or by the interval $[-1, 4)$.

Example 9. Solve $x > 2$ and $x \le 0$.

Solution. Graph both inequalities on the same number line.

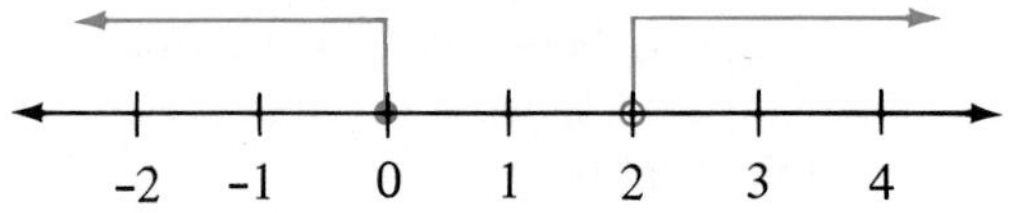

We see that *both* lines appear *nowhere* at the same time on the graph. There are *no* numbers that will satisfy both inequalities at the same time. Therefore, our intersection contains no numbers at all; it is an **empty set**. Mathematicians sometimes use the symbol $\varnothing$ to stand for the empty set. Here we say that the intersection of $x > 2$ and $x \le 0$ is $\varnothing$.

Trial Run

Find the solution for each pair of inequalities by graphing on a number line.

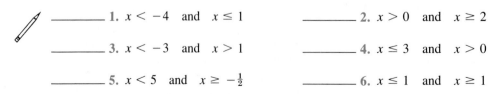

______ 1. $x < -4$ and $x \le 1$

______ 2. $x > 0$ and $x \ge 2$

______ 3. $x < -3$ and $x > 1$

______ 4. $x \le 3$ and $x > 0$

______ 5. $x < 5$ and $x \ge -\frac{1}{2}$

______ 6. $x \le 1$ and $x \ge 1$

Answers are on page 93.

Finding Unions of Graphs

Let's consider some more examples in which two inequalities are discussed.

Example 10. A local movie theater offers tickets at half-price to children (age 12 or below) and to senior citizens (age 62 or above). Describe the set of people eligible for half-price tickets using inequality notation.

Solution. We can let a represent the age of a moviegoer.

Children	*Senior citizens*
$a \leq 12$	$a \geq 62$

To qualify for a half-price ticket, must a moviegoer meet *both* age conditions? Surely not. No one could qualify as a child and as a senior citizen at the same time. We are *not* seeking an *intersection* here.

Instead, we agree that a moviegoer will qualify for a half-price ticket by satisfying the first inequality ($a \leq 12$) *or* the second inequality ($a \geq 62$). Let's graph our inequalities and find the numbers that satisfy one *or* the other of them.

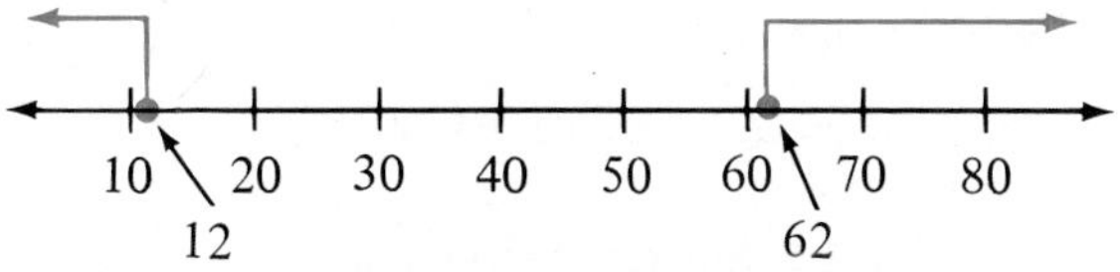

In a situation such as this we are *not* concerned with intervals in which *both* lines appear at the same time. *This is not an intersection problem.* Instead, we are concerned with intervals in which *either* of the lines appears. The solution here is

$$a \leq 12 \quad \text{or} \quad a \geq 62$$

Example 11. A student is eligible for a federal educational grant if the family's income is below \$13,000. A student is eligible for university financial aid if the family's income is below \$10,000. For what incomes is a student eligible for federal aid *or* university aid?

Solution. We can let I represent family income.

Federal grant	*University aid*
$I < 13{,}000$	$I < 10{,}000$

We graph our two inequalities.

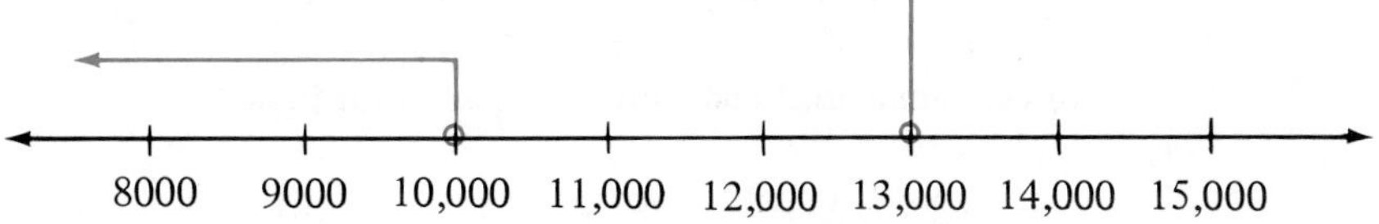

Since we are looking for incomes that will make a student eligible for one kind of aid *or* the other, we look for the numbers appearing in either graph. Our solution is

$$I < \$13{,}000$$

In Examples 10 and 11 we found a set called the **union** of our original sets. The symbol for union is ∪.

Union of Sets A and B ($A \cup B$). The union of sets A and B is the set containing all elements in set A or in set B (or possibly in both).

To find the union of two inequalities we graph them on the same number line and look for numbers appearing in either graph (or possibly in both).

You try to find the union described in Example 12.

Example 12. Solve $x \geq 7$ or $x > 3$.

Solution. We graph both inequalities on the same number line.

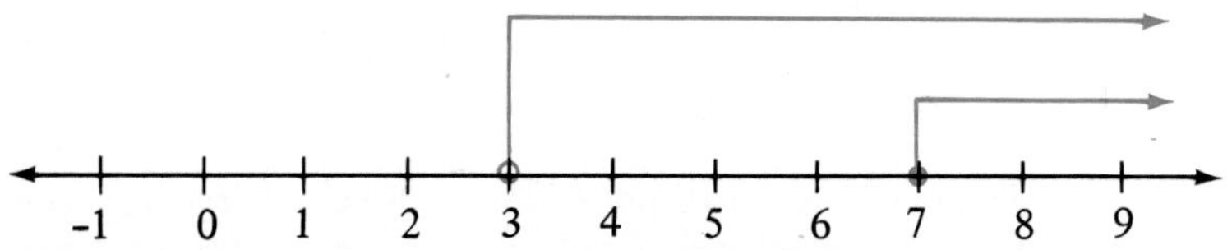

A line appears everywhere to the right of (but not including) the point ____ . Since we are looking for numbers on the graph that satisfy *either* inequality, our solutions are described by the inequality

$$x > \_\_\_\_$$

or by the interval ____ .

Check your work on page 93. ▶

Example 13. Solve $x > 1$ or $x < 1$.

Solution. We graph both inequalities on the number line.

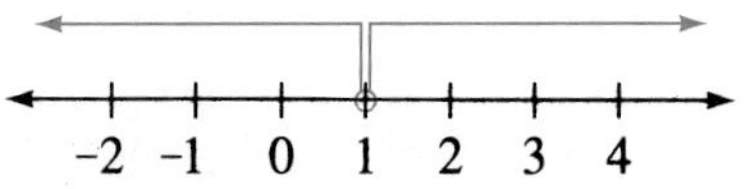

Lines appear everywhere except at the point 1. Therefore, we describe the solutions by the original inequality

$$x > 1 \quad \text{or} \quad x < 1$$

or by using intervals $(-\infty, 1) \cup (1, \infty)$. The solution may also be written as

$$x \in R \quad \text{and} \quad x \neq 1$$

Example 14. Solve $x > -2$ or $x \leq 3$.

Solution. We graph both inequalities on the number line.

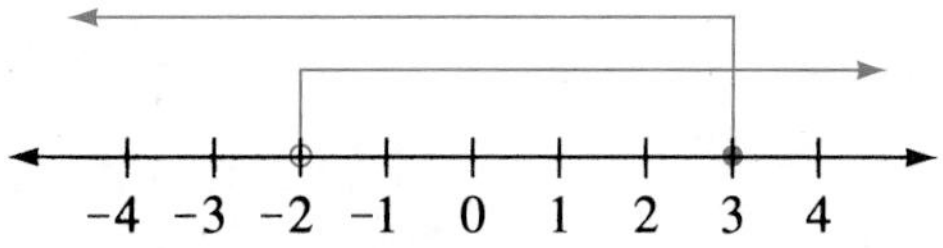

All real numbers are included in one line or the other. Therefore, solutions that satisfy

$$x > -2 \quad \text{or} \quad x \leq 3$$

may be described by

$$x \in R$$

or by the interval $(-\infty, \infty)$.

Let's summarize our work with pairs of inequalities.

To satisfy one inequality **and** another, a number must satisfy *both* inequalities. On the number line we look for numbers appearing in *both* graphs (the *intersection*).

To satisfy one inequality **or** another, a number may satisfy one of the inequalities *or* the other (or possibly both). On the number line we look for numbers appearing in either one graph *or* the other (the *union*).

Trial Run

Find the solution for each pair of inequalities by graphing on a number line.

______ 1. $x \geq 3$ or $x > 5$

______ 2. $x < 0$ or $x \leq -2$

______ 3. $x > 2$ or $x < 2$

______ 4. $x \geq 3$ or $x < -1$

______ 5. $x > -4$ or $x < 2$

______ 6. $x \geq 1$ or $x < 1$

Answers are below.

Examples You Completed

Example 4. Graph the solutions for $x < \frac{1}{2}$. Then write the solutions using interval notation.

Solution. There will be an open dot at $\frac{1}{2}$ and a line extending to the left.

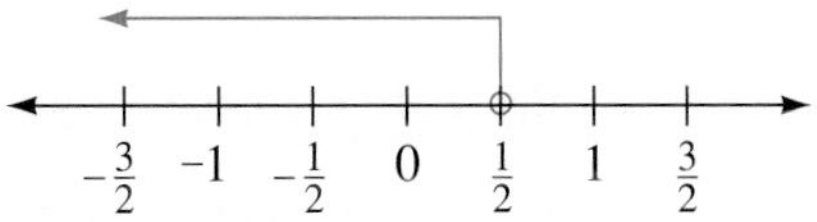

Interval notation: $(-\infty, \frac{1}{2})$

Example 12. Solve $x \geq 7$ or $x > 3$.

Solution. We graph both inequalities on the number line.

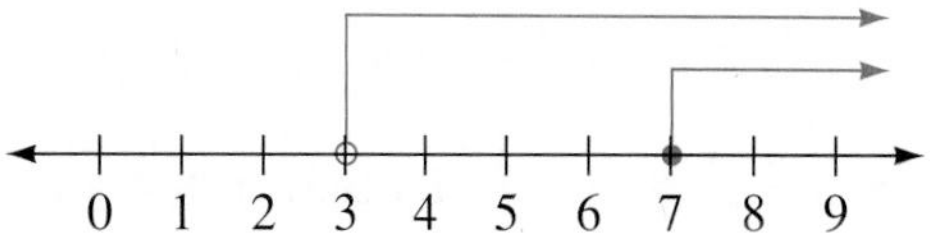

A line appears everywhere to the right of (but not including) the point 3. Therefore, our solutions are described by the inequality

$$x > 3$$

or by the interval $(3, \infty)$.

Answers to Trial Runs

page 88

1.

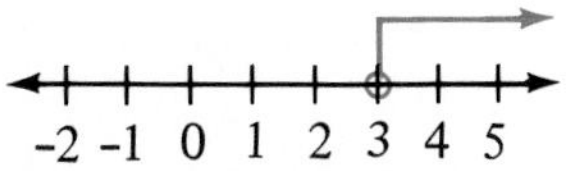

$(3, \infty)$

2.

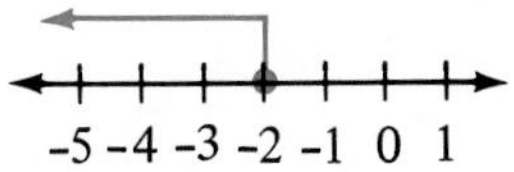

$(-\infty, -2]$

3.

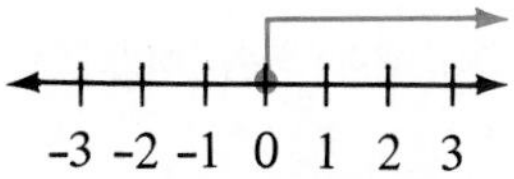

$[0, \infty)$

4.

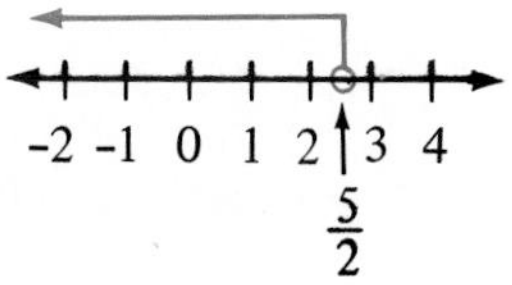

$(-\infty, \frac{5}{2})$

5.

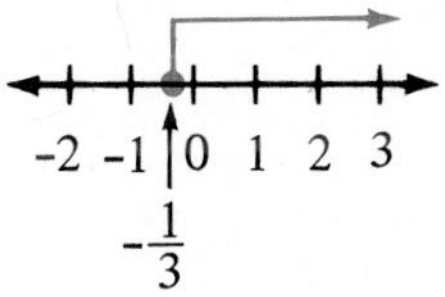

$[-\frac{1}{3}, \infty)$

page 90 1. $x < -4$; $(-\infty, -4)$ 2. $x \geq 2$; $[2, \infty)$ 3. $\varnothing$ 4. $0 < x \leq 3$; $(0, 3]$
5. $-\frac{1}{2} \leq x < 5$; $[-\frac{1}{2}, 5)$ 6. $x = 1$

page 93 1. $x \geq 3$; $[3, \infty)$ 2. $x < 0$; $(-\infty, 0)$ 3. $x \in R, x \neq 2$; $(-\infty, 2) \cup (2, \infty)$
4. $x < -1$ or $x \geq 3$; $(-\infty, -1) \cup [3, \infty)$ 5. $x \in R$; $(-\infty, \infty)$ 6. $x \in R$; $(-\infty, \infty)$

Name ______________________ Date ____________

EXERCISE SET 2.2

Graph each inequality on a number line. Then state solution using interval notation.

1. $x > -5$
2. $x > 1$
3. $a \leq 3$
4. $a \leq -2$
5. $y \geq 0$
6. $y \leq 0$
7. $m > \frac{7}{2}$
8. $m > \frac{5}{3}$
9. $x \leq -\frac{3}{2}$
10. $x \leq -\frac{5}{2}$

Find the solution for each pair of inequalities by graphing on a number line. Then state solutions using interval notation.

______ 11. $x < -3$ and $x \leq 4$

______ 12. $x < -5$ and $x < 1$

______ 13. $x > -1$ and $x \geq 3$

______ 14. $x > -2$ and $x \geq 0$

______ 15. $x < -1$ and $x > 2$

______ 16. $x < 0$ and $x > 2$

______ 17. $x \leq 5$ and $x > 2$

______ 18. $x \leq 1$ and $x > -2$

______ 19. $x \geq -1$ and $x \leq 4$

______ 20. $x \geq -3$ and $x \leq 1$

______ 21. $x < 0$ and $x \geq 0$

______ 22. $x < 2$ and $x \geq 2$

______ 23. $x \geq -3$ and $x \leq -3$

______ 24. $x \geq 4$ and $x \leq 4$

______ 25. $x \geq 2$ or $x > 5$

______ 26. $x \geq -1$ or $x > 1$

______ 27. $x < 4$ or $x \leq -1$

______ 28. $x \leq 3$ or $x < 7$

______ 29. $x > 3$ or $x < 3$

______ 30. $x > -4$ or $x < -4$

______ 31. $x < 2$ or $x \geq -1$

______ 32. $x < 5$ or $x \geq 1$

______ 33. $x < -\frac{1}{2}$ or $x \geq 0$

______ 34. $x \leq 1$ or $x > \frac{3}{2}$

______ 35. $x > -2$ or $x < 3$

______ 36. $x > -4$ or $x < 0$

______ 37. $x < -3$ and $x \leq 0$

______ 38. $x < 4$ and $x \leq \frac{1}{2}$

______ 39. $x > -4$ and $x < \frac{7}{2}$

______ 40. $x > -\frac{5}{2}$ and $x < 1$

______ 41. $x < -1$ or $x > 0$

______ 42. $x < -3$ or $x > 2$

______ 43. $x < 5$ or $x \leq 1$

______ 44. $x \leq 10$ or $x < 5$

______ 45. $x > 0$ and $x \geq \frac{3}{5}$

______ 46. $x > -1$ and $x \geq 0$

______ 47. $x < 5$ or $x > -1$

______ 48. $x < 12$ or $x \geq 2$

______ 49. $x \geq \frac{1}{2}$ and $x < \frac{1}{2}$

______ 50. $x > 9$ and $x \leq 9$

☆ Stretching the Topics

Find the solution by graphing on a number line.

________ 1. $x \geq 3.5$ and $x \leq \frac{7}{2}$

________ 2. $x > -0.75$ or $x \geq -\frac{3}{4}$

________ 3. $x \leq \frac{22}{7}$ and $x \geq \pi$

Check your answers in the back of your book.

If you can complete **Checkup 2.2**, you are ready to go on to Section 2.3.

Name ________________________________ Date ______________

✓ CHECKUP 2.2

Graph each inequality. Then state solutions using interval notation.

________ 1. $x \geq 3$

________ 2. $a \leq -0.5$

________ 3. $y > \frac{7}{2}$

Find the solution for each pair of inequalities by graphing. Then state solutions using interval notation.

________ 4. $x > 0$ and $x \leq 2$

________ 5. $x < 3$ or $x \leq -1$

________ 6. $x < -2$ and $x > \frac{1}{2}$

________ 7. $x > -2$ or $x < 4$

________ 8. $x < 5$ and $x \geq 0$

________ 9. $x < -3$ or $x > 2$

________ 10. $x < \frac{1}{2}$ or $x > \frac{1}{2}$

Check your answers in the back of your book.

If You Missed Problems:	You Should Review Examples:
1–3	1–4
4	5, 7
5	11, 12
6	9
7	14
8	8
9	10
10	13

2.3 Solving First-Degree Inequalities

Having learned to solve first-degree equations in Section 2.1, you will be relieved to know that solving first-degree inequalities requires similar techniques.

Solving Inequalities by Addition or Multiplication

In solving equations, we learned that it was permissible to add (or subtract) the same quantity to (or from) both sides of an equation. Is this also permissible with inequalities? Let's see what happens when we add and subtract some numbers to both sides of the inequality $3 < 7$.

Add 4	Subtract 10	Add -6	Subtract -5
$3 + 4 \overset{?}{<} 7 + 4$	$3 - 10 \overset{?}{<} 7 - 10$	$3 + (-6) \overset{?}{<} 7 + (-6)$	$3 - (-5) \overset{?}{<} 7 - (-5)$
$7 \overset{?}{<} 11$	$-7 \overset{?}{<} -3$	$-3 \overset{?}{<} 1$	$8 \overset{?}{<} 12$
True	True	True	True
$7 < 11$	$-7 < -3$	$-3 < 1$	$8 < 12$

We may add (or subtract) the same quantity to (or from) both sides of an inequality. The inequality is preserved in the same direction.

The statement "The inequality is preserved in the same direction" means that if we start with a less-than inequality ($<$ or $\leq$), we end up with a less-than inequality. If we start with a greater-than inequality ($>$ or $\geq$), we end up with a greater-than inequality.

We can state this fact using a, b, and c to represent real numbers.

Addition Property of Inequality

If $a < b$	If $a > b$
then $a + c < b + c$	then $a + c > b + c$
and $a - c < b - c$	and $a - c > b - c$

This property allows us to use addition or subtraction to isolate the variable in solving some inequalities.

Example 1. Solve $x - 9 \leq 1$ and graph the solutions.

Solution

$$x - 9 \leq 1$$

$$x - 9 + 9 \leq 1 + 9 \qquad \text{Add 9 to both sides.}$$

$$x \leq 10 \qquad \text{Simplify.}$$

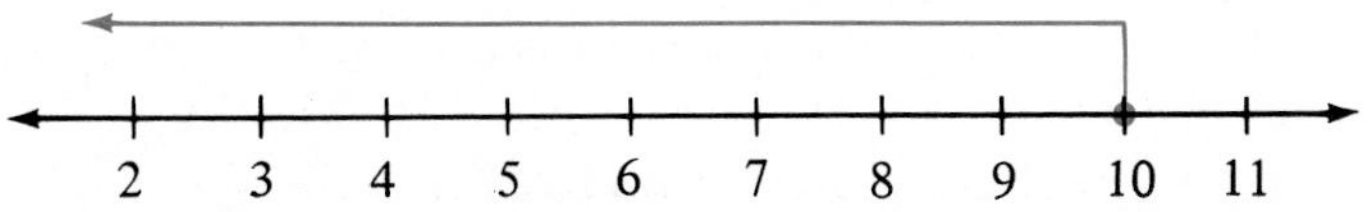

You complete Example 2.

Example 2. Solve $y + 3 > -2$ and graph the solutions.

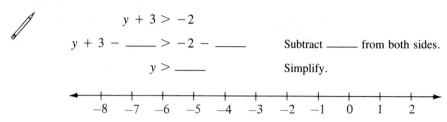

Solution

$$y + 3 > -2$$

$y + 3 - \_\_\_\_ > -2 - \_\_\_\_$ Subtract ____ from both sides.

$y > \_\_\_\_$ Simplify.

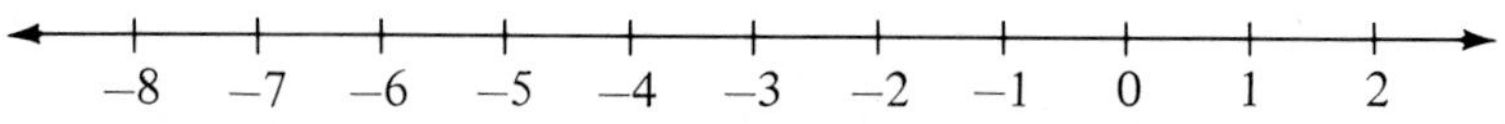

Check your work on page 106. ▶

In solving first-degree equations we were also allowed to multiply or divide both sides by the same nonzero quantity. Let's see if a similar rule can be developed for inequalities by noticing what happens when we multiply and divide both sides of the inequality $8 < 12$.

Multiply by 3	Divide by 2	Multiply by −5	Divide by −4
$3(8) \overset{?}{<} 3(12)$	$\frac{8}{2} \overset{?}{<} \frac{12}{2}$	$-5(8) \overset{?}{<} -5(12)$	$\frac{8}{-4} \overset{?}{<} \frac{12}{-4}$
$24 \overset{?}{<} 36$	$4 \overset{?}{<} 6$	$-40 \overset{?}{<} -60$	$-2 \overset{?}{<} -3$
True	True	False	False
$24 < 36$	$4 < 6$	$-40 > -60$	$-2 > -3$

We may multiply (or divide) both sides of an inequality by the same *positive* quantity, and the direction of the inequality will be *preserved*.

We may multiply (or divide) both sides of an inequality by the same *negative* quantity, but the direction of the inequality will be *reversed*.

Using a, b, and c to represent real numbers, we can state

Multiplication Property of Inequality

If $a < b$ and c is *positive* ($c > 0$), then $a \cdot c < b \cdot c$ and $\frac{a}{c} < \frac{b}{c}$.

If $a > b$ and c is *positive* ($c > 0$), then $a \cdot c > b \cdot c$ and $\frac{a}{c} > \frac{b}{c}$.

If $a < b$ and c is *negative* ($c < 0$), then $a \cdot c > b \cdot c$ and $\frac{a}{c} > \frac{b}{c}$.

If $a > b$ and c is *negative* ($c < 0$), then $a \cdot c < b \cdot c$ and $\frac{a}{c} < \frac{b}{c}$.

These properties allow us to solve first-degree inequalities in which the variable appears with a coefficient different from 1. See if you can complete Example 3.

Example 3. Solve and graph $\frac{x}{3} \geq -2$.

Solution

$$\frac{x}{3} \geq -2$$

$$3\left(\frac{x}{3}\right) \geq 3(-2)$$

$$x \geq \_\_\_\_$$

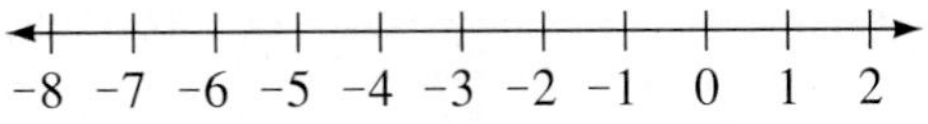

Check your work on page 106. ▶

Example 4. Solve and graph $-2x < -3$.

Solution

$$-2x < -3$$

$$\frac{-2x}{-2} > \frac{-3}{-2}$$

$$x > \frac{3}{2}$$

Notice that the direction of the inequality was *reversed* when we divided by -2.

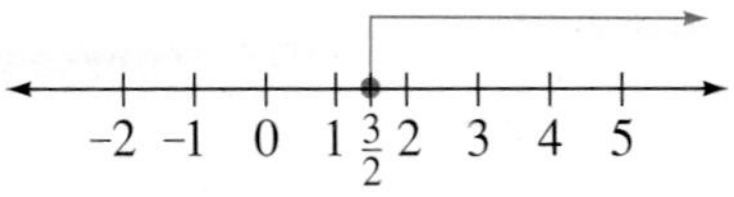

Trial Run

Solve and graph each inequality.

________ 1. $-3 + y < 0$

________ 2. $x + 7 \geq 2$

________ 3. $-2x \leq 6$

________ 4. $3x > 15$

________ 5. $\frac{x}{-5} < 2$

Answers are on page 107.

Now it is time to solve some first-degree inequalities requiring addition and/or subtraction, as well as multiplication and/or division. Such problems should be solved with the same procedure as that used for equations. But remember to be careful whenever you multiply or divide both sides of an inequality by a *negative* number.

Example 5. Solve $-4x + 10 \le 34$.

Solution

$-4x + 10 \le 34$

$-4x \le 24$ Subtract 10.

$\dfrac{-4x}{-4} \ge \dfrac{24}{-4}$ Divide by -4.

$x \ge -6$ Simplify.

Interval notation: $[-6, \infty)$

Now you try Example 6.

Example 6. Solve $\dfrac{x}{2} - 1 > 1$.

Solution

$\dfrac{x}{2} - 1 > 1$

$\dfrac{x}{2} > 2$ Add ______ .

$2\left(\dfrac{x}{2}\right) > 2(2)$ Multiply by ______ .

$x >$ ______ Simplify.

Interval notation: ______

Check your work on page 106. ▶

If a variable appears on both sides of an inequality, we proceed as with equations.

Example 7. Solve and graph $5x + 1 \ge 3x + 7$.

Solution

$5x + 1 \ge 3x + 7$

$2x + 1 \ge 7$ Subtract $3x$ from both sides to isolate variable term.

$2x \ge 6$ Subtract 1 from both sides.

$\dfrac{2x}{2} \ge \dfrac{6}{2}$ Divide both sides by 2.

$x \ge 3$ Simplify.

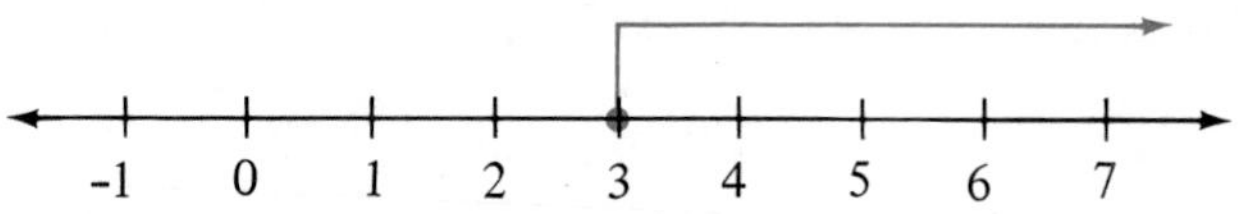

$[3, \infty)$

Now you complete Example 8.

Example 8. Solve and graph

$$\frac{3-2x}{4} \le -1$$

Solution

$\frac{3-2x}{4} \le -1$	LCD: 4.
$4\left(\frac{3-2x}{4}\right) \le 4(-1)$	Multiply by LCD.
$3 - 2x \le$ ______	Simplify.
$-2x \le$ ______	Subtract 3.
$\frac{-2x}{-2} \ge \frac{\quad}{-2}$	Divide by -2.
$x \ge$ ______	Simplify.

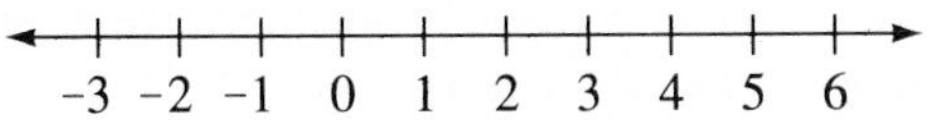

Check your work on page 106. ▶

Example 9. Solve and graph

$$5(x+1) - x \le 1 - 2x$$

Solution

$5(x+1) - x \le 1 - 2x$	
$5x + 5 - x \le 1 - 2x$	Remove parentheses.
$4x + 5 \le 1 - 2x$	Combine like terms.
$6x + 5 \le 1$	Add $2x$.
$6x \le -4$	Subtract 5.
$\frac{6x}{6} \le \frac{-4}{6}$	Divide by 6.
$x \le -\frac{2}{3}$	Simplify.

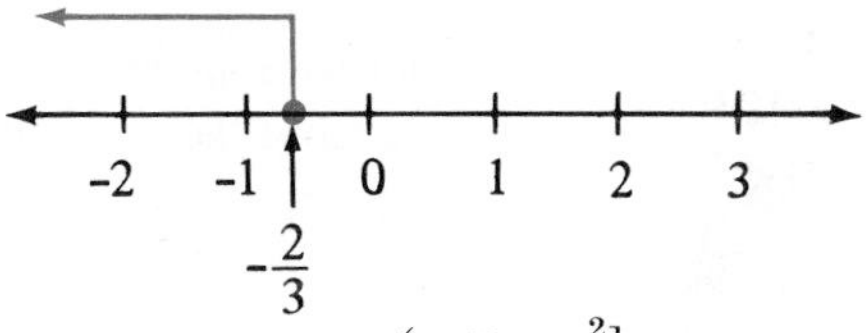

$(-\infty, -\frac{2}{3}]$

Trial Run

Solve. Then write solutions using interval notation.

______ 1. $5a - 3 \ge 12$

______ 2. $\frac{y}{7} - 1 < -2$

______ 3. $\frac{2x}{3} + 4 \le 7$

______ 4. $2a - 4 < 5a - 1$

______ 5. $5x - (x - 4) \le 2x + 5$

______ 6. $2(z - 2) - 3 \le 5(z + 5) - 23$

Answers are on page 107.

Solving Pairs of Inequalities

In Section 2.2 we learned how to find solutions satisfying one inequality *and/or* another. Now we can extend those techniques to solving pairs of inequalities whose solutions are not immediately obvious.

Example 10. Solve $3x - 1 < 2$ *and* $2(5 - x) \le 16$.

Solution. We must find the solutions for each inequality and graph them on the same number line.

$$3x - 1 < 2 \qquad \text{and} \qquad 2(5 - x) \le 16$$

$$3x - 1 + 1 < 2 + 1 \qquad\qquad 10 - 2x \le 16$$

$$3x < 3 \qquad\qquad -2x \le 6$$

$$\frac{3x}{3} < \frac{3}{3} \qquad\qquad \frac{-2x}{-2} \ge \frac{6}{-2}$$

$$x < 1 \qquad \text{and} \qquad x \ge -3$$

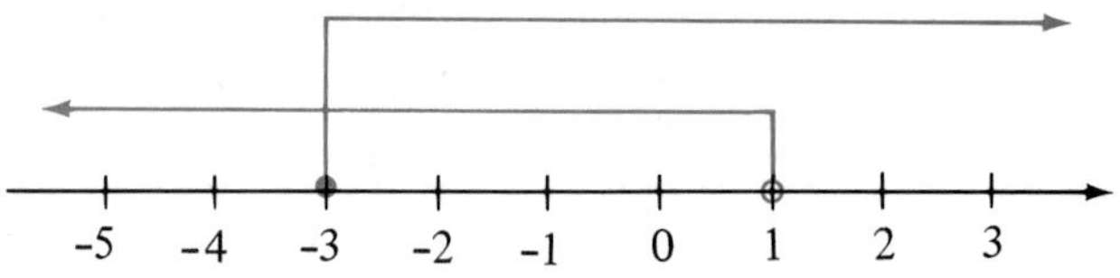

Since we are seeking the *intersection* of our sets of solutions, we must find where both lines appear at the same time. We write the solutions as

$$-3 \le x < 1 \qquad \text{or} \qquad [-3, 1)$$

Example 11. Solve $7 - x > 3$ *and* $4(x + 1) - x < 4$.

Solution

$$7 - x > 3 \qquad \text{and} \qquad 4(x + 1) - x < 4$$

$$-x > -4 \qquad\qquad 4x + 4 - x < 4$$

$$\frac{-x}{-1} < \frac{-4}{-1} \qquad\qquad 3x + 4 < 4$$

$$3x < 0$$

$$x < 4 \qquad \text{and} \qquad x < 0$$

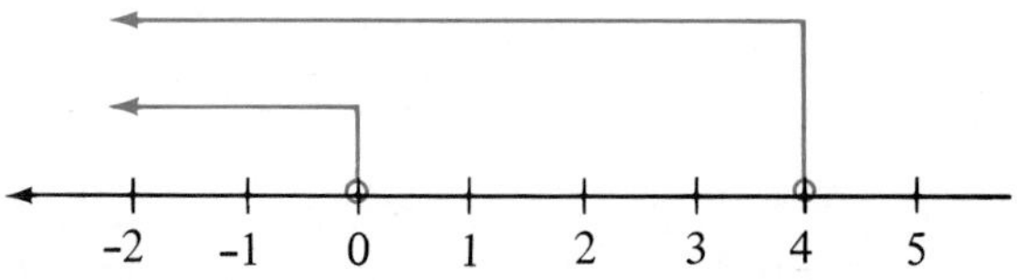

We look for solutions that satisfy $x < 4$ *and* $x < 0$. The intersection is $x < 0$ or $(-\infty, 0)$.

Finding the *union* of two inequalities involves a similar procedure. After solving and graphing the original inequalities, however, we look for those solutions that appear in one graph *or* the other.

Example 12. Solve $3(2x - 4) + 1 > 7$ *or* $7x - 5 \geq 9$.

Solution

$$\begin{array}{rcr}
3(2x - 4) + 1 > 7 & \text{or} & 7x - 5 \geq 9 \\
6x - 12 + 1 > 7 & & 7x \geq 14 \\
6x - 11 > 7 & & \dfrac{7x}{7} \geq \dfrac{14}{7} \\
6x > 18 & & \\
\dfrac{6x}{6} > \dfrac{18}{6} & & \\
x > 3 & \text{or} & x \geq 2
\end{array}$$

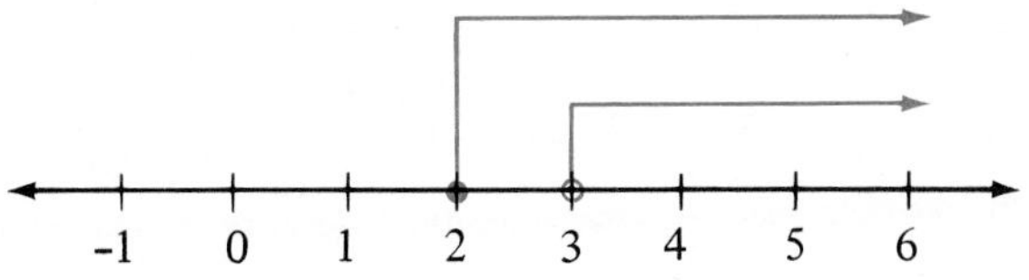

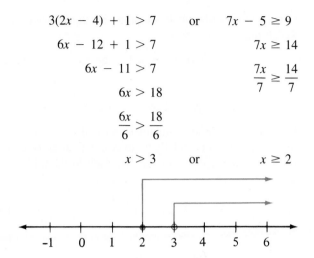

We know that $x \geq 2$ or $x > 3$ must be satisfied. Any number to the right of (and including) 2 will satisfy one inequality or the other. Our union is described by $x \geq 2$ or by $[2, \infty)$.

Example 13. Solve $4x - 7 > 9$ *or* $1 - 2(x - 3) \geq 3$.

Solution

$$\begin{array}{rcr}
4x - 7 > 9 & \text{or} & 1 - 2(x - 3) \geq 3 \\
4x > 16 & & 1 - 2x + 6 \geq 3 \\
\dfrac{4x}{4} > \dfrac{16}{4} & & -2x + 7 \geq 3 \\
& & -2x \geq -4 \\
& & \dfrac{-2x}{-2} \leq \dfrac{-4}{-2} \\
x > 4 & \text{or} & x \leq 2
\end{array}$$

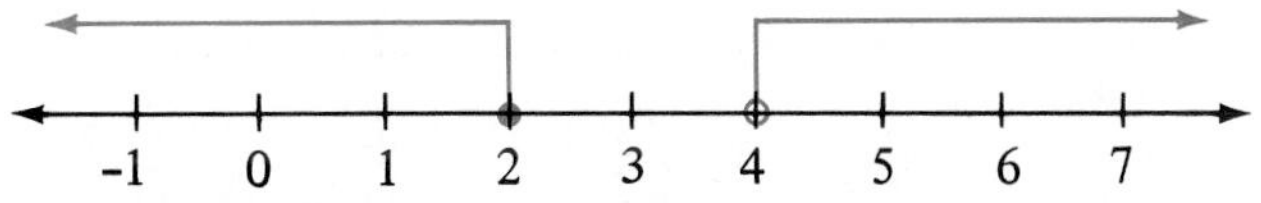

Our union is all solutions satisfying one inequality *or* the other. Here the union is written $x \leq 2$ or $x > 4$. It may be written using intervals as $(-\infty, 2] \cup (4, \infty)$.

As the examples illustrate, sometimes an intersection or union of inequalities will give us an answer that is a single inequality, sometimes it will be a three-part inequality, and sometimes it will be two separate inequalities. There is no sure way to predict which kind of solution will occur. We must solve and graph the original inequalities and then look for the intersection or union.

Finding the values of the variable that satisfy a **three-part inequality** such as

$$-3 < 2x + 1 < 7$$

is not terribly difficult if we stop and think about what three-part inequalities really mean. Recall that

$$a < x < b \quad \text{means} \quad a < x \quad \textit{and} \quad x < b$$

So

$$-3 < 2x + 1 < 7 \quad \text{means} \quad -3 < 2x + 1 \quad \textit{and} \quad 2x + 1 < 7$$

We must solve both inequalities and find the *intersection*. Since both inequalities require the same steps to isolate x, let's try to deal with all three parts of the inequality at once. Remember, we would like to end up with x by itself in the middle.

$$-3 < 2x + 1 < 7$$

$$-3 - 1 < 2x + 1 - 1 < 7 - 1 \qquad \text{Subtract 1 from all three parts.}$$

$$-4 < 2x < 6 \qquad \text{Simplify.}$$

$$\frac{-4}{2} < \frac{2x}{2} < \frac{6}{2} \qquad \text{Divide all three parts by 2.}$$

$$-2 < x < 3 \qquad \text{Simplify.}$$

The graph of our solutions is

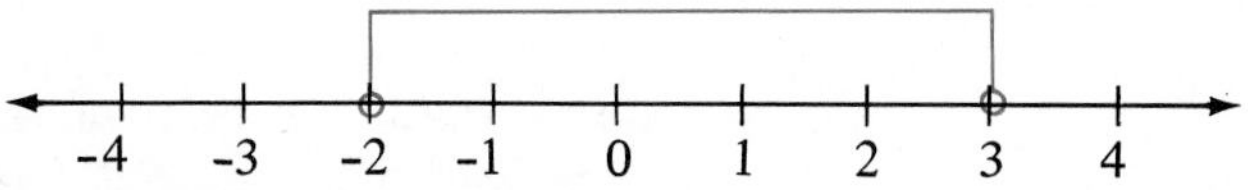

and, using interval notation, we may write $(-2, 3)$.

Now you try completing Example 14.

Example 14. Solve and graph

$$-14 \le 5(x - 3) + 1 \le 11$$

Solution

$$-14 \le 5(x - 3) + 1 \le 11$$

$$-14 \le 5x - 15 + 1 \le 11$$

$$-14 \le 5x - 14 \le 11$$

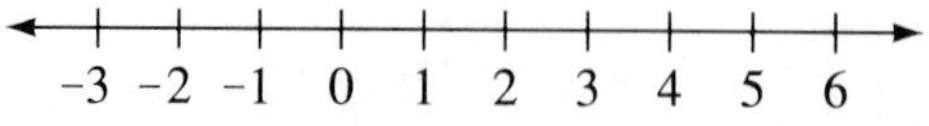

Check your work on page 106. ▶

Example 15. Solve $4 < \dfrac{1 - 3x}{4} \le 7$.

Solution. The LCD is 4.

$$4 \cdot 4 < \frac{4}{1}\left[\frac{1 - 3x}{4}\right] \le 4 \cdot 7$$

$$16 < 1 - 3x \le 28$$

$$16 - 1 < 1 - 3x - 1 \le 28 - 1$$

$$15 < -3x \le 27$$

$$\frac{15}{-3} > \frac{-3x}{-3} \ge \frac{27}{-3}$$

$$-5 > x \ge -9$$

Notice that we reversed directions when we divided by -3. This inequality is correct, but let's write it in better order, starting with the -9.

$$-9 \le x < -5$$

or

$$[-9, -5)$$

From our work with inequalities you should notice that a solution set whose graph is *one continuous interval* on the number line can always be written using a *single inequality* or as a single interval.

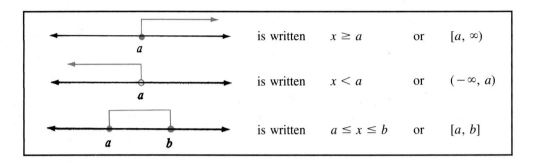

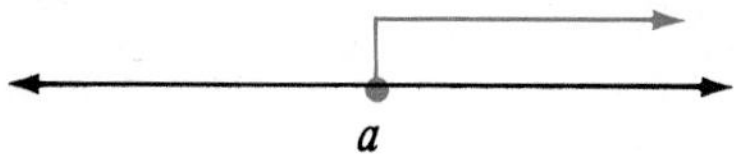	is written	$x \geq a$	or	$[a, \infty)$
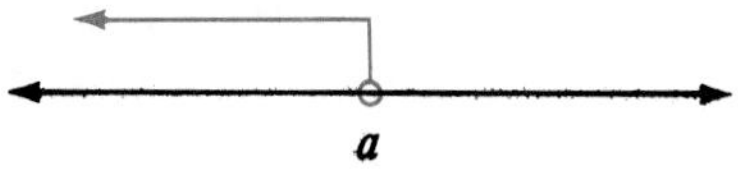	is written	$x < a$	or	$(-\infty, a)$
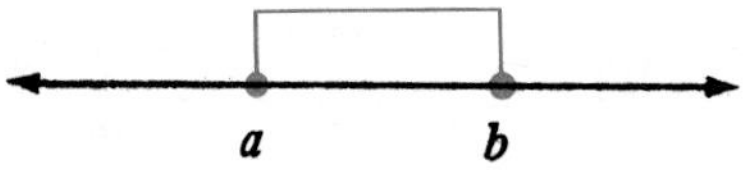	is written	$a \leq x \leq b$	or	$[a, b]$

But solutions whose graph is a *split interval* on the number line must be written using *two separate inequalities* or as the union of two separate intervals.

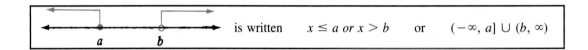

	is written	$x \leq a$ *or* $x > b$	or	$(-\infty, a] \cup (b, \infty)$

This last pair of inequalities should *never* be written $a \geq x > b$ because this statement implies that $x > b$ *and* $x \leq a$ at the same time, which is clearly impossible.

Trial Run

Solve each pair of inequalities by graphing. Then write solutions using interval notation.

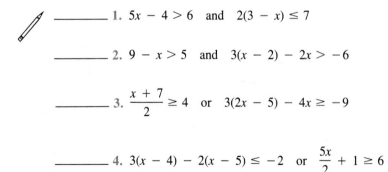

________ 1. $5x - 4 > 6$ and $2(3 - x) \leq 7$

________ 2. $9 - x > 5$ and $3(x - 2) - 2x > -6$

________ 3. $\dfrac{x + 7}{2} \geq 4$ or $3(2x - 5) - 4x \geq -9$

________ 4. $3(x - 4) - 2(x - 5) \leq -2$ or $\dfrac{5x}{2} + 1 \geq 6$

Solve each inequality and graph the solutions. Then write solutions using interval notation.

________ 5. $5 < 1 - 2x \leq 9$

________ 6. $0 \leq 2(x - 3) < 10$

Answers are on page 107.

▶ Examples You Completed

Example 2. Solve $y + 3 > -2$ and graph the solutions.

Solution

$$y + 3 > -2$$

$$y + 3 - 3 > -2 - 3 \qquad \text{Subtract 3 from both sides.}$$

$$y > -5 \qquad \text{Simplify.}$$

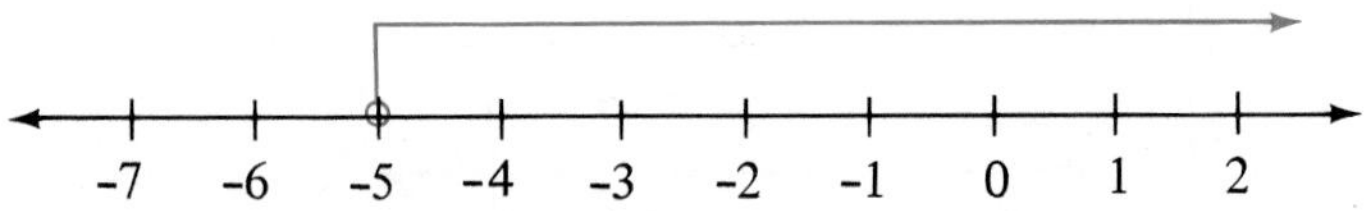

Example 3. Solve and graph $\frac{x}{3} \geq -2$.

Solution

$$\frac{x}{3} \geq -2$$

$$3\left(\frac{x}{3}\right) \geq 3(-2)$$

$$x \geq -6$$

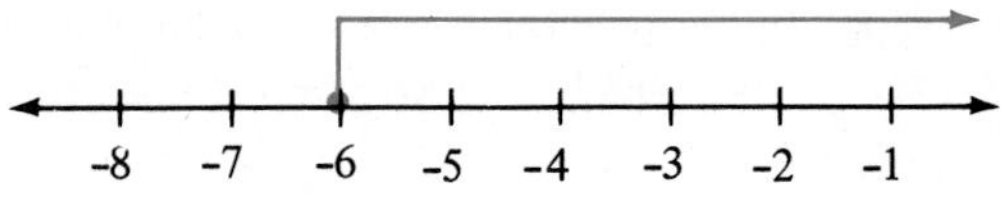

$[-6, \infty)$

Example 6. Solve $\frac{x}{2} - 1 > 1$.

Solution

$$\frac{x}{2} - 1 > 1$$

$$\frac{x}{2} - 1 + 1 > 1 + 1$$

$$\frac{x}{2} > 2$$

$$2\left(\frac{x}{2}\right) > 2(2)$$

$$x > 4$$

Interval notation: $(4, \infty)$

Example 8. Solve and graph $\frac{3 - 2x}{4} \leq -1$.

Solution

$$\frac{3 - 2x}{4} \leq -1$$

$$4\left(\frac{3 - 2x}{4}\right) \leq 4(-1)$$

$$3 - 2x \leq -4$$

$$3 - 2x - 3 \leq -4 - 3$$

$$-2x \leq -7$$

$$\frac{-2x}{-2} \geq \frac{-7}{-2}$$

$$x \geq \frac{7}{2}$$

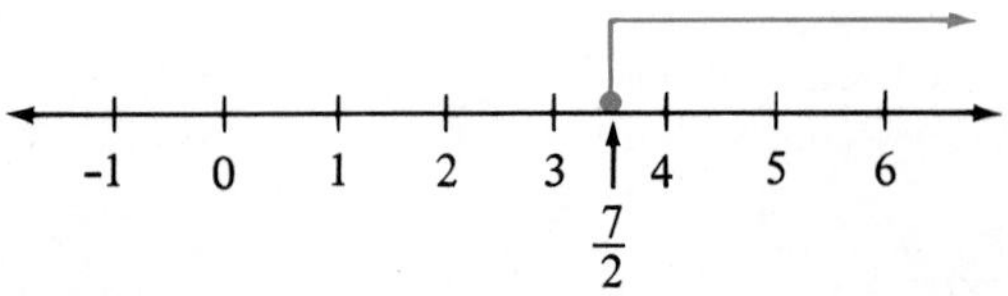

$[\frac{7}{2}, \infty)$

Example 14. Solve and graph

$$-14 \leq 5(x - 3) + 1 \leq 11$$

Solution

$$-14 \leq 5(x - 3) + 1 \leq 11$$

$$-14 \leq 5x - 15 + 1 \leq 11$$

$$-14 \leq 5x - 14 \leq 11$$

$$-14 + 14 \leq 5x - 14 + 14 \leq 11 + 14$$

$$0 \leq 5x \leq 25$$

$$\frac{0}{5} \leq \frac{5x}{5} \leq \frac{25}{5}$$

$$0 \leq x \leq 5$$

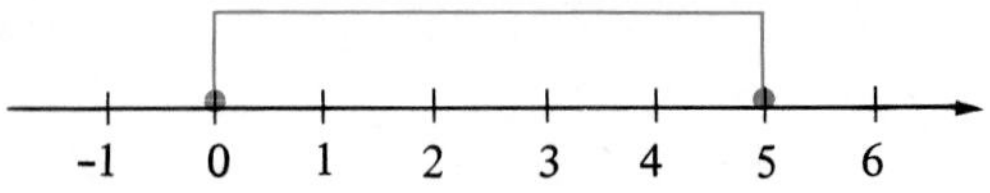

$[0, 5]$

Answers to Trial Runs

page 99 1. $y < 3$

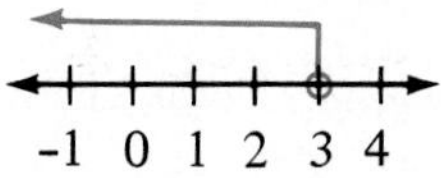

2. $x \geq -5$

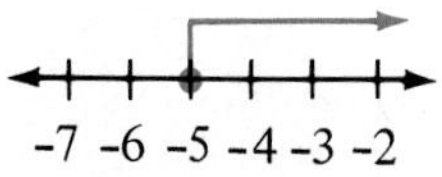

3. $x \geq -3$

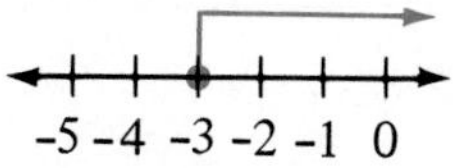

4. $x > 5$

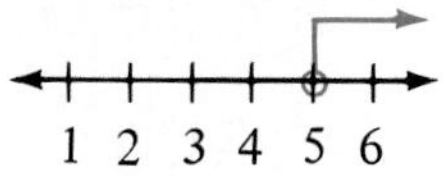

5. $x > -10$

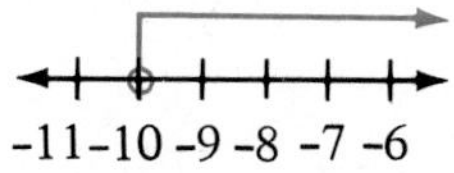

page 101 1. $a \geq 3$; $[3, \infty)$ 2. $y < -7$; $(-\infty, -7)$ 3. $x \leq \frac{9}{2}$; $(-\infty, \frac{9}{2}]$ 4. $a > -1$; $(-1, \infty)$
5. $x \leq \frac{1}{2}$; $(-\infty, \frac{1}{2}]$ 6. $z \geq -3$; $[-3, \infty)$

page 105 1. $x > 2$; $(2, \infty)$ 2. $0 < x < 4$; $(0, 4)$ 3. $x \geq 1$; $[1, \infty)$
4. $x \leq 0$ or $x \geq 2$; $(-\infty, 0] \cup [2, \infty)$

5. $-4 \leq x < -2$; $[-4, -2)$ 6. $3 \leq x < 8$; $[3, 8)$

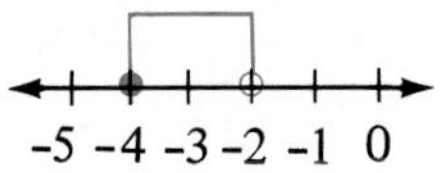

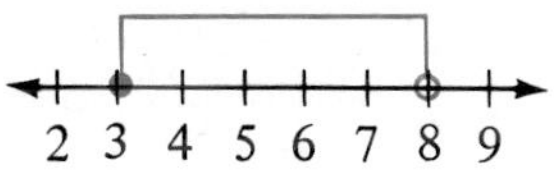

Name ____________________ Date __________

EXERCISE SET 2.3

Solve the inequalities and graph. Then write solutions using interval notation.

_______ 1. $x - 5 \geq -3$

_______ 2. $x - 7 \geq -2$

_______ 3. $3 - y > 2$

_______ 4. $7 - y > 5$

_______ 5. $10a \leq -5$

_______ 6. $9a \leq 3$

_______ 7. $\frac{x}{2} > -1$

_______ 8. $\frac{x}{3} > -2$

_______ 9. $3a + 9 \geq 15$

_______ 10. $7a - 6 \geq -13$

_______ 11. $-\frac{2x}{3} + 5 < 4$

_______ 12. $-\frac{5x}{2} - 7 < 8$

_______ 13. $8(m - 3) \leq 8$

_______ 14. $3(m - 6) \leq -3$

_______ 15. $\frac{2x - 9}{3} > 5$

_______ 16. $\frac{5x + 1}{4} > 9$

_______ 17. $7x - (x + 5) \leq 3x + 2$

_______ 18. $9x - (2x + 3) \leq 5x + 8$

_______ 19. $3(y - 4) - 10 \leq 9(y - 1) + 11$

_______ 20. $5(y - 3) - 7 \leq 8(y + 1) - 3$

_______ 21. $2x - 5 > 1$ and $3(x - 1) < 6$

_______ 22. $5x - 7 > 3$ and $4(x - 2) < 0$

_______ 23. $\frac{x - 7}{5} \leq -2$ and $4(1 - x) < -6$

_______ 24. $\frac{x - 6}{4} \leq -2$ and $5(2 - x) < -15$

_______ 25. $4x - 7 < 3 - x$ and $7x - 1 \geq x - 1$

_______ 26. $3x + 5 \geq x - 1$ and $4x - 3 < 2x + 1$

_______ 27. $\frac{x - 8}{5} < 2 - x$ or $\frac{x}{4} + 3 < 7$

_______ 28. $\frac{2x - 6}{4} < 3 - x$ or $\frac{2x}{9} + 5 < 7$

_______ 29. $x + 4 \leq 2x - 2$ or $3x + 1 < x + 11$

_______ 30. $2x - 1 \geq 2 - x$ or $3x + 4 \leq 2x + 1$

_______ 31. $3(2x - 1) \geq x + 7$ or $2(4 - x) > x - 1$

_______ 32. $5(3x - 4) \le 9x + 4$ or $4(5x - 1) > 12x + 6$

_______ 33. $2 \le m - 5 \le 7$

_______ 34. $-3 < m + 1 \le 5$

_______ 35. $5 \le -2a - 7 < 9$

_______ 36. $1 < -3a - 1 \le 5$

_______ 37. $-5 \le \dfrac{2y - 1}{3} < 3$

_______ 38. $-2 \le \dfrac{3y - 2}{4} < 1$

_______ 39. $0 \le 3(-x + 5) - 9 < 6$

_______ 40. $0 \le 4(-x - 3) - 8 \le 4$

☆ Stretching the Topics

Solve and graph. Then write solutions using interval notation.

_______ 1. $3b - [7 - (2 - b)] \ge b + 8$

_______ 2. $5 + 7x \ge -(3 - 2x) - 3x$ and $-7 + 5x - (x + 3) \ge x + 2$

_______ 3. $-4 < 3x + 5 < 14$ and $-16 \le 4(x - 2) < 13$

Check your answers in the back of your book.

If you can complete **Checkup 2.3**, you are ready to go on to Section 2.4.

Name ______________________ Date ____________

✓ CHECKUP 2.3

Solve the inequalities and graph. Then write solutions using interval notation.

_______ 1. $3x - 5 \geq 4$

_______ 2. $-\frac{a}{4} + 2 < 1$

_______ 3. $3(y - 2) > 4(y + 1)$

_______ 4. $6x - (x + 3) \leq 2x - 9$

_______ 5. $\frac{x - 8}{3} < -2$ and $4(1 - x) < 2$

_______ 6. $3x - 5 \leq 5x + 3$ and $2x - 1 > x + 2$

_______ 7. $\frac{2x - 3}{4} < 1 + x$ or $2(3x - 4) < 7x - 6$

_______ 8. $3(2 - x) \geq 0$ or $5(x - 1) > 2x + 4$

_______ 9. $-3 \leq 2a + 5 \leq 1$

_______ 10. $0 \leq \frac{-y + 2}{6} < 1$

Check your answers in the back of your book.

If You Missed Problems:	You Should Review Examples:
1, 2	5, 6
3, 4	7–9
5, 6	10, 11
7, 8	12, 13
9, 10	14, 15

2.4 Solving Absolute-Value Equations and Inequalities

Solving absolute-value equations and inequalities requires little more than a thorough understanding of the number line and an ability to solve first-degree equations and inequalities.

Absolute-Value Equations

Let's look at some simple absolute-value equations and identify the solutions.

Equation	Solutions	Reasons
$\|x\| = 3$	$x = 3$ or $x = -3$	$\|3\| = 3$ $\|-3\| = 3$
$\|x\| = 0$	$x = 0$	$\|0\| = 0$
$\|x\| = -2$	No solution	Absolute values are *never* negative

We see that an equation of the type $|x| = a$ will have solutions only when a is *not negative*.

If a is *positive* and	If a is *zero* and	If a is *negative* and
$\|x\| = a$	$\|x\| = a$	$\|x\| = a$
the solutions are	the solution is	there are *no* solutions.
$x = a$ or $x = -a$	$x = 0$	

You try completing Example 1.

Example 1. Solve $|x| = \frac{1}{3}$ and graph the solutions.

Solution

$$|x| = \frac{1}{3}$$

$x =$ _____ or $x =$ _____

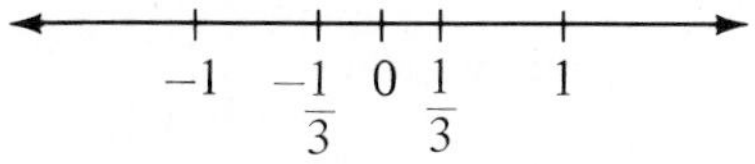

Check your work on page 120. ▶

It should be clear to you that if the absolute value of some *quantity* equals some positive number a, then either that quantity equals a *or* that quantity equals $-a$. Let's use this idea to solve an equation in which the quantity within the absolute value bars contains more terms than x by itself.

Example 2. Solve $|x + 1| = 3$. Check the solutions and graph.

Solution. Since the absolute value of a quantity is 3, we know that the quantity must equal 3 or -3.

$$x + 1 = 3 \quad \text{or} \quad x + 1 = -3$$

$$x = 2 \quad \text{or} \quad x = -4$$

CHECK: $x = 2$	CHECK: $x = -4$
$\|x + 1\| = 3$	$\|x + 1\| = 3$
$\|2 + 1\| \stackrel{?}{=} 3$	$\|-4 + 1\| \stackrel{?}{=} 3$
$\|3\| \stackrel{?}{=} 3$	$\|-3\| \stackrel{?}{=} 3$
$3 = 3$	$3 = 3$

The graph of our solutions is

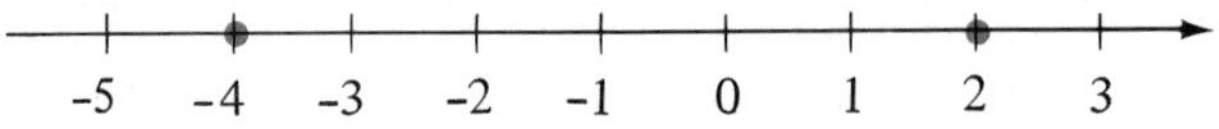

Notice that the two solutions are *not* opposites of each other; usually, they will *not* be.

Now you complete Examples 3 and 4.

Example 3. Solve $|2x - 3| = 7$.

Solution. $|2x - 3| = 7$ means

$2x - 3 = 7$ or $2x - 3 = -7$

$2x =$ ____ $\quad 2x =$ ____

$x =$ ____ or $x =$ ____

Check your work on page 120. ▶

You complete Example 4.

Example 4. Solve $|6 - 9x| = 0$.

Solution

$|6 - 9x| = 0$ means

$6 - 9x = 0$

Check your work on page 120. ▶

Example 5. Solve $\dfrac{|7 - 2x|}{4} = 2$.

Solution

$$\frac{|7 - 2x|}{4} = 2 \qquad \text{First we must isolate the absolute-value quantity.}$$

$$\frac{4}{1} \cdot \frac{|7 - 2x|}{4} = 4(2) \qquad \text{Multiply both sides by 4 (the LCD).}$$

$$|7 - 2x| = 8$$

$$7 - 2x = 8 \quad \text{or} \quad 7 - 2x = -8$$

$$-2x = 1 \qquad\qquad -2x = -15$$

$$\frac{-2x}{-2} = \frac{1}{-2} \qquad\qquad \frac{-2x}{-2} = \frac{-15}{-2}$$

$$x = -\tfrac{1}{2} \quad \text{or} \quad x = \tfrac{15}{2}$$

Trial Run

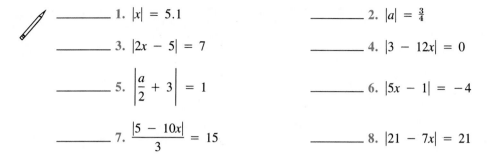

Solve each equation.

_______ 1. $|x| = 5.1$

_______ 2. $|a| = \frac{3}{4}$

_______ 3. $|2x - 5| = 7$

_______ 4. $|3 - 12x| = 0$

_______ 5. $\left|\frac{a}{2} + 3\right| = 1$

_______ 6. $|5x - 1| = -4$

_______ 7. $\frac{|5 - 10x|}{3} = 15$

_______ 8. $|21 - 7x| = 21$

Answers are on page 121.

Absolute-Value Inequalities (Less-Than Type)

Suppose that we wish to determine the location on the number line of all the numbers that satisfy the inequality $|x| < 3$. We know the location of all the numbers for which $|x| = 3$.

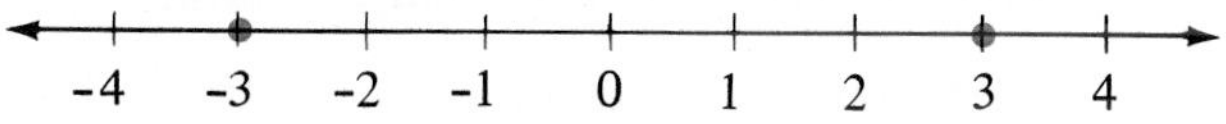

But what about all the other numbers on this line? Where are all those numbers whose absolute values are *less than* 3? Where are the solutions for $|x| < 3$?

Numbers to the Left of -3	Numbers Between -3 and 3	Numbers to the Right of 3
$\|-4\| \overset{?}{<} 3$ No	$\|-2\| \overset{?}{<} 3$ Yes	$\|3.1\| \overset{?}{<} 3$ No
$\|-3.5\| \overset{?}{<} 3$ No	$\|0\| \overset{?}{<} 3$ Yes	$\|5\| \overset{?}{<} 3$ No
	$\|2.95\| \overset{?}{<} 3$ Yes	

It seems that only those numbers *between* -3 and 3 will satisfy our inequality. Therefore, we write the solution for $|x| < 3$ as

$$-3 < x < 3$$

The graph of our solutions is an interval on the number line.

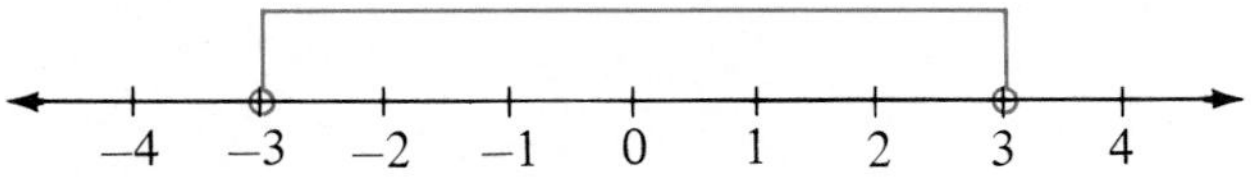

$(-3, 3)$

In general, we may say

If a is *positive*, then $|x| < a$ means that x lies *between* $-a$ and a on the number line.

Try Example 6.

Example 6. Solve $|x| \leq 5$ and graph the solutions.

Solution. $|x| \leq 5$ means ____ $\leq x \leq$ ____.

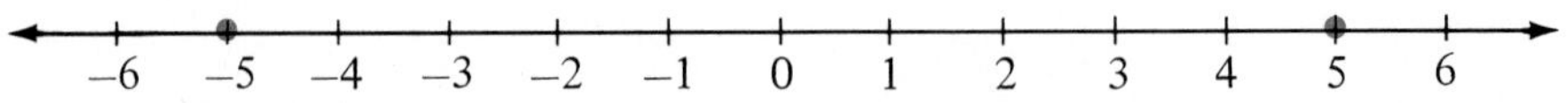

[__, __]

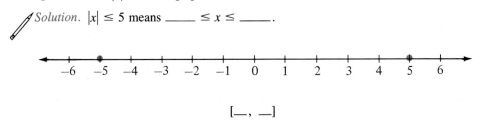

Check your work on page 120. ▶

Example 7. Solve $|x| < -1$.

Solution. $|x| < -1$ has *no solution* because -1 is a negative number and the absolute value of a quantity must be positive or zero.

Let's summarize our observations about less-than type absolute-value inequalities.

If a is positive, then $|x| < a$ means $-a < x < a$	If a is positive, then $|x| \leq a$ means $-a \leq x \leq a$	If a is negative, then $|x| < a$ has no solution.

Example 8. Solve $|x + 1| < 6$ and graph the solutions.

Solution. We realize that the *quantity* within the absolute-value bars must lie between -6 and 6.

$-6 < x + 1 < 6$ — We must isolate x in the middle.

$-6 - 1 < x + 1 - 1 < 6 - 1$ — Subract 1 from all parts.

$-7 < x < 5$ — Simplify.

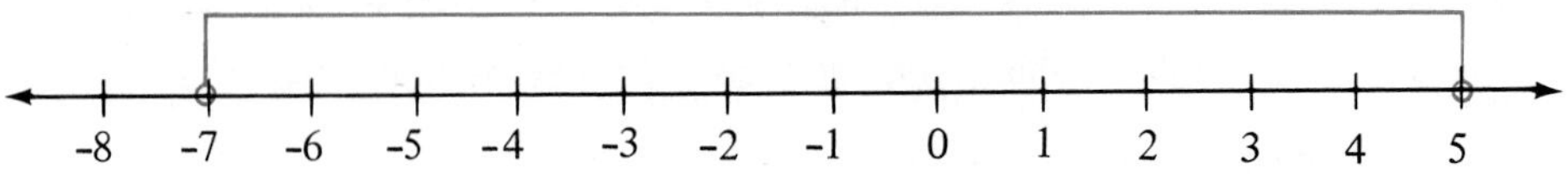

$(-7, 5)$

Now you try Example 9.

Example 9. Solve $|2x - 3| + 5 \leq 12$ and graph the solutions.

Solution. First we isolate the absolute-value expression.

$$|2x - 3| + 5 \leq 12$$

$$|2x - 3| + 5 - 5 \leq 12 - 5$$

$$|2x - 3| \leq 7$$

The quantity within absolute-value bars must lie between ____ and ____ .

$$-7 \leq 2x - 3 \leq 7$$

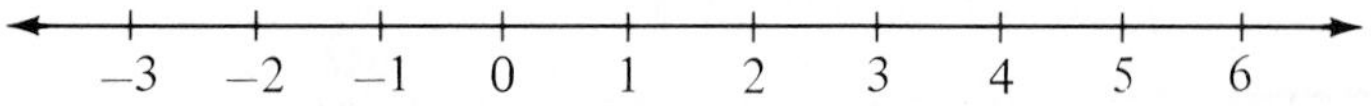

[__, __]

Check your work on page 120. ▶

Example 10. Solve $|5 - x| \leq 3$ and graph the solutions.

Solution. $|5 - x| \leq 3$ means

$$-3 \leq 5 - x \leq 3$$

$$-3 - 5 \leq 5 - x - 5 \leq 3 - 5$$

$$-8 \leq -x \leq -2$$

$$\frac{-8}{-1} \geq \frac{-x}{-1} \geq \frac{-2}{-1}$$

Notice that we reversed the direction of our inequalities. Why?

$$8 \geq x \geq 2$$

Rewriting the inequality in better order (from left to right on the number line), we have

$$2 \leq x \leq 8$$

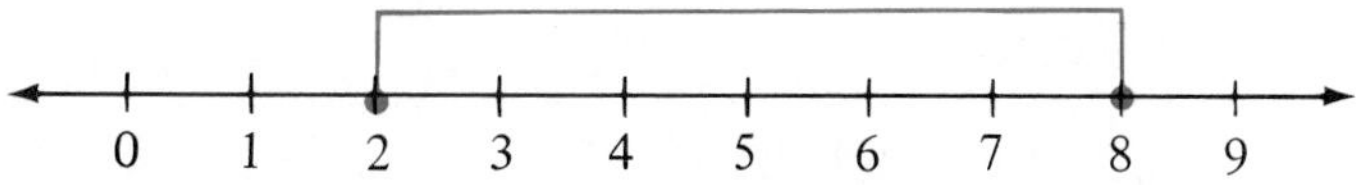

[2, 8]

Trial Run

Solve. Then write solutions using interval notation.

______ 1. $|a| \leq 3.5$

______ 2. $|y| < \frac{4}{5}$

______ 3. $|x| < 0$

______ 4. $|3x| < 9$

______ 5. $|4x + 1| \leq 7$

______ 6. $|7 - 2x| < 11$

_______ 7. $|5x - 3| \le -2$ _______ 8. $\left|\frac{2x - 1}{3}\right| \le 5$

Answers are on page 121.

Absolute-Value Inequalities (Greater-Than Type)

Now let's find all the numbers on the number line that satisfy the inequality $|x| > 3$.

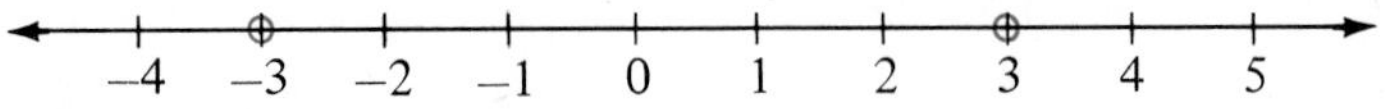

Numbers to the Left of -3	Numbers Between -3 and 3	Numbers to the Right of 3
$\lvert -5\rvert \stackrel{?}{>} 3$ Yes	$\lvert -2\rvert \stackrel{?}{>} 3$ No	$\lvert 3.5\rvert \stackrel{?}{>} 3$ Yes
$\lvert -4\rvert \stackrel{?}{>} 3$ Yes	$\lvert 0\rvert \stackrel{?}{>} 3$ No	$\lvert 4\rvert \stackrel{?}{>} 3$ Yes
$\lvert -3.1\rvert \stackrel{?}{>} 3$ Yes	$\lvert 2.5\rvert \stackrel{?}{>} 3$ No	$\lvert 10\rvert \stackrel{?}{>} 3$ Yes

As we expected, numbers to the *left* of -3 or to the *right* of 3 will satisfy the inequality $|x| > 3$. We write those solutions as

$$x < -3 \quad \textit{or} \quad x > 3$$

and the graph is two separate intervals on the number line.

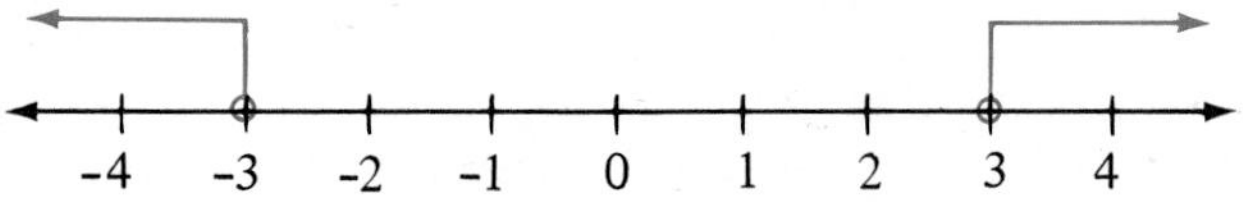

$$(-\infty, -3) \cup (3, \infty)$$

In general, we say

If a is *positive*, then $|x| > a$ means that x lies to the *left* of $-a$ *or* to the *right* of a on the number line.

Try finding the solution for Example 11.

Example 11. Solve $|x| > 2$ and graph the solutions.

Solution. $|x| > 2$ means

$$x < \_\_\_\_ \quad \text{or} \quad x > \_\_\_\_$$

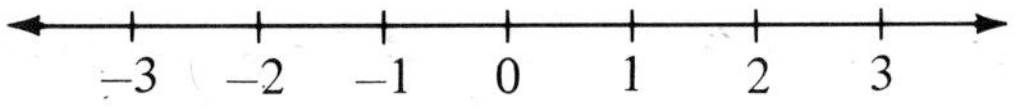

$(-\infty, \_\_) \cup (\_\_, \infty)$

Check your work on page 120. ▶

Example 12. Solve $|x| \geq \frac{1}{2}$ and graph the solutions.

Solution. $|x| \geq \frac{1}{2}$ means

$$x \leq -\tfrac{1}{2} \quad \text{or} \quad x \geq \tfrac{1}{2}$$

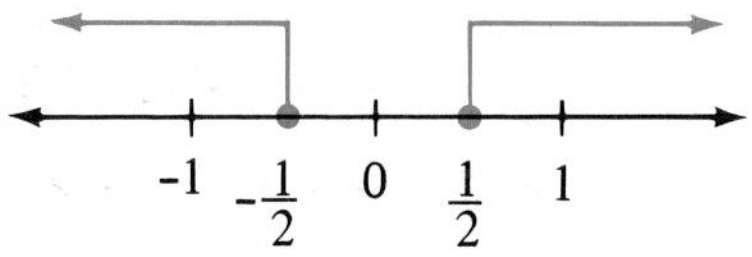

$(-\infty, -\frac{1}{2}] \cup [\frac{1}{2}, \infty)$

If a is positive, then	If a is positive, then
$|x| > a$	$|x| \geq a$
means	means
$x < -a$ or $x > a$	$x \leq -a$ or $x \geq a$

Example 13. Solve $|x - 2| > 7$ and graph the solutions.

Solution. We realize that the *quantity* within the absolute-value bars must lie to the left of -7 *or* to the right of 7.

$$x - 2 < -7 \quad \text{or} \quad x - 2 > 7$$

$$x - 2 + 2 < -7 + 2 \qquad x - 2 + 2 > 7 + 2$$

$$x < -5 \quad \text{or} \quad x > 9$$

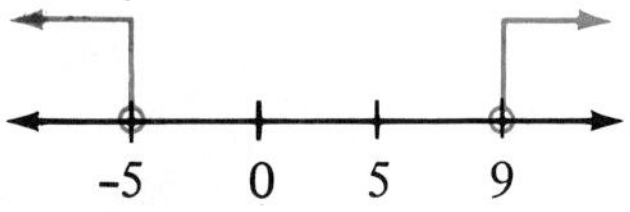

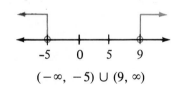

$(-\infty, -5) \cup (9, \infty)$

Now you try Example 14.

Example 14. Solve $|7x - 4| \geq 3$ and graph the solutions.

Solution. The quantity within the absolute-value bars must lie to the left of (and including) ____ *or* to the right of (and including) ____ .

$$7x - 4 \leq \_\_\_\_ \quad \text{or} \quad 7x - 4 \geq \_\_\_\_$$

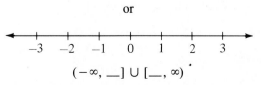

or

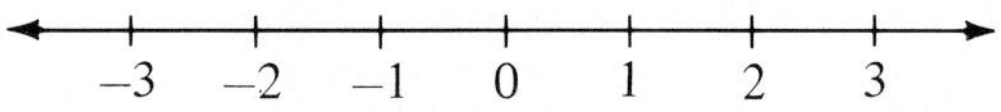

$(-\infty, \_\_] \cup [\_\_, \infty)$

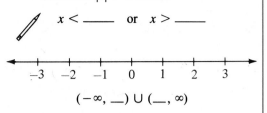

Check your work on page 121. ▶

Example 15. Solve $\left|2 - \frac{x}{3}\right| \geq 4$ and graph the solutions.

Solution. $\left|2 - \frac{x}{3}\right| \geq 4$ means

$2 - \frac{x}{3} \leq -4$	or	$2 - \frac{x}{3} \geq 4$
$3 \cdot 2 - \frac{3}{1} \cdot \frac{x}{3} \leq 3(-4)$	Multiply both sides by LCD: 3.	$3 \cdot 2 - \frac{3}{1} \cdot \frac{x}{3} \geq 3 \cdot 4$
$6 - x \leq -12$	Simplify.	$6 - x \geq 12$
$-x \leq -18$	Subtract 6.	$-x \geq 6$
$\frac{-x}{-1} \geq \frac{-18}{-1}$	Divide by -1	$\frac{-x}{-1} \leq \frac{6}{-1}$
$x \geq 18$	Simplify.	$x \leq -6$

Notice that we reversed the directions of the inequalities when we divided both sides by -1.

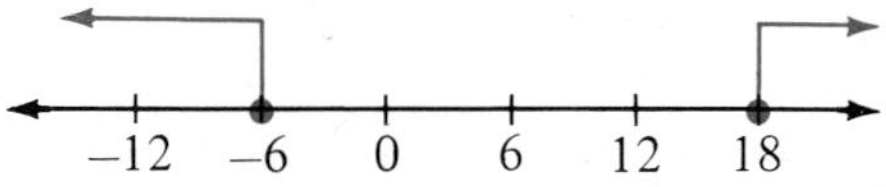

$(-\infty, -6] \cup [18, \infty)$

Now you try Example 16.

Example 16. Solve $|5 - 3x| > 1$ and graph the solutions.

Solution. $|5 - 3x| > 1$ means

$5 - 3x < -1$ or $5 - 3x > 1$

or

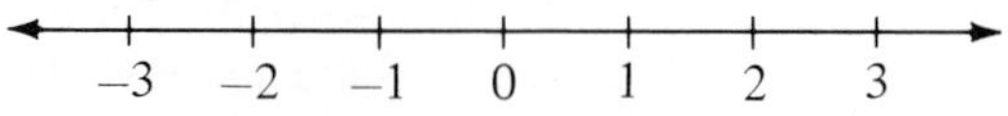

$(-\infty, \_\_) \cup (\_\_, \infty)$

Check your work on page 121. ▶

Example 17. Solve $|x - 2| > -3$.

Solution. Since absolute values are always *positive*, the statement $|x - 2| > -3$ is true for *all* values of the variable. Every real number is a solution and we may state the solutions in interval notation as $(-\infty, \infty)$.

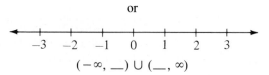

Trial Run

Solve. Write solutions using interval notation.

_______ 1. $|y| > \frac{2}{3}$

_______ 2. $|m| > -11$

_______ 3. $|x| \geq 7$

_______ 4. $|7x| > 14$

_______ 5. $|2x - 3| \geq 5$

_______ 6. $|9 - x| \geq 4$

_______ 7. $\left|\frac{5x}{3}\right| > 10$

_______ 8. $|2 - 3x| > 2$

Answers are on page 121.

A summary of our work with absolute-value equations and inequalities is included here for your reference. We assume that a represents a positive number.

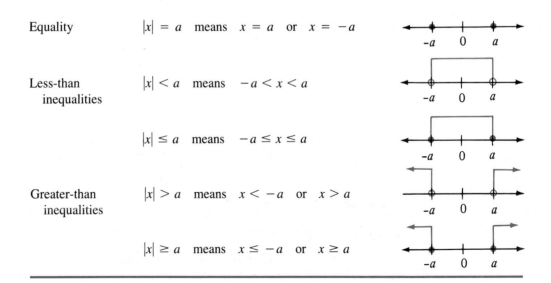

Equality	$\|x\| = a$ means $x = a$ or $x = -a$	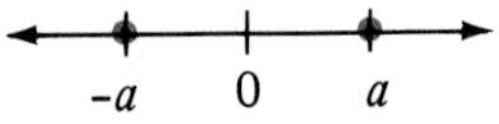
Less-than inequalities	$\|x\| < a$ means $-a < x < a$	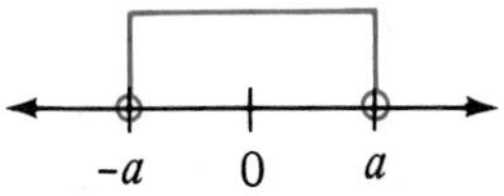
	$\|x\| \leq a$ means $-a \leq x \leq a$	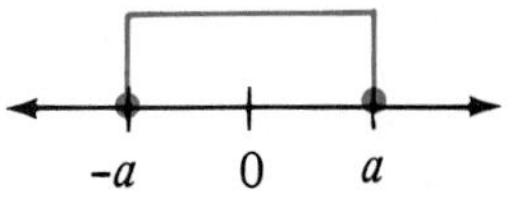
Greater-than inequalities	$\|x\| > a$ means $x < -a$ or $x > a$	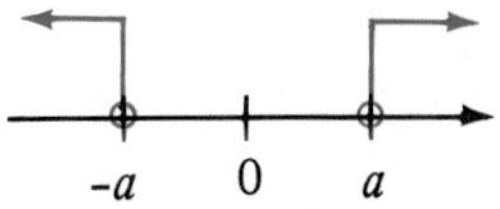
	$\|x\| \geq a$ means $x \leq -a$ or $x \geq a$	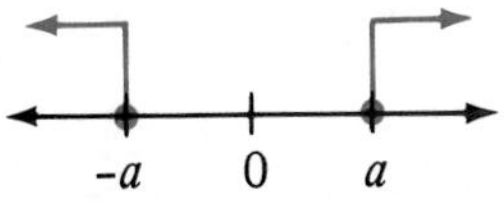

The secret to success with absolute-value equations and inequalities is to visualize the number line and immediately rewrite the original statement as an equivalent statement without absolute-value bars. The solutions can then be found using the usual methods for solving first-degree equations and inequalities.

▶ Examples You Completed

Example 1. Solve $|x| = \frac{1}{3}$. Graph the solutions.

Solution

$$|x| = \tfrac{1}{3}$$

$$x = \tfrac{1}{3} \quad \text{or} \quad x = -\tfrac{1}{3}$$

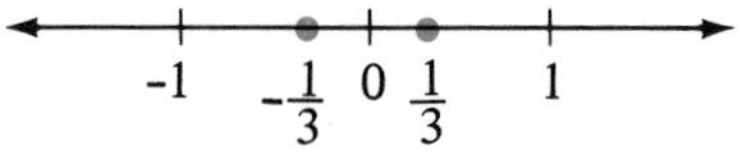

Example 3. Solve $|2x - 3| = 7$.

Solution. $|2x - 3| = 7$ means

$$2x - 3 = 7 \quad \text{or} \quad 2x - 3 = -7$$

$$2x = 10 \qquad 2x = -4$$

$$x = 5 \quad \text{or} \quad x = -2$$

Example 4. Solve $|6 - 9x| = 0$.

Solution. $|6 - 9x| = 0$ means

$$6 - 9x = 0$$

$$6 - 9x - 6 = 0 - 6$$

$$-9x = -6$$

$$\frac{-9x}{-9} = \frac{-6}{-9}$$

$$x = \frac{2}{3}$$

Example 6. Solve $|x| \le 5$ and graph the solutions.

Solution. $|x| \le 5$ means

$$-5 \le x \le 5$$

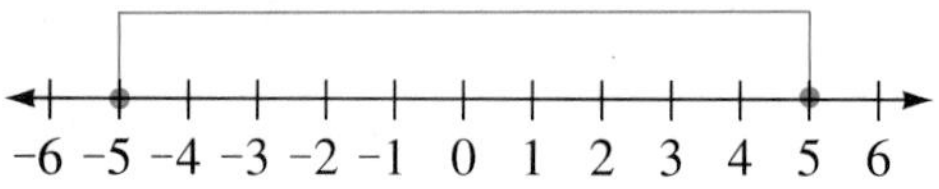

$[-5, 5]$

Example 9. Solve $|2x - 3| + 5 \le 12$ and graph the solutions.

Solution

$$|2x - 3| + 5 \le 12$$

$$|2x - 3| + 5 - 5 \le 12 - 5$$

$$|2x - 3| \le 7$$

The quantity within the absolute-value bars must lie between (and including) -7 and 7.

$$-7 \le 2x - 3 \le 7$$

$$-4 \le 2x \le 10$$

$$\frac{-4}{2} \le \frac{2x}{2} \le \frac{10}{2}$$

$$-2 \le x \le 5$$

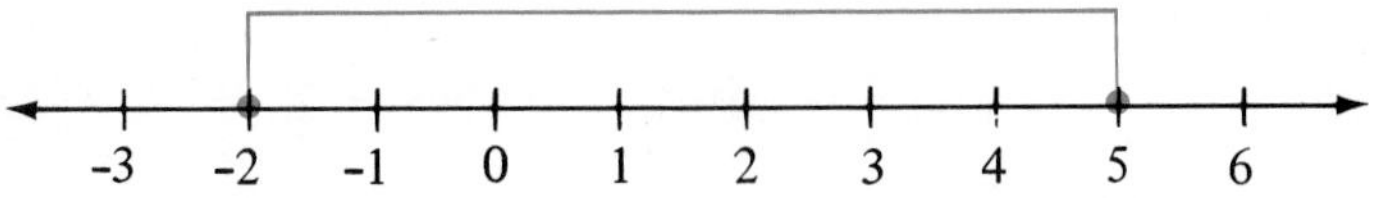

$[-2, 5]$

Example 11. Solve $|x| > 2$ and graph the solutions.

Solution. $|x| > 2$ means

$$x < -2 \quad \text{or} \quad x > 2$$

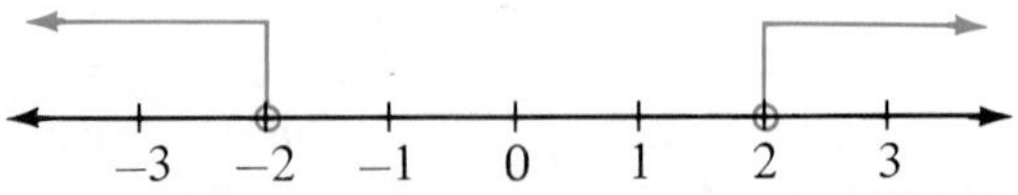

$(-\infty, -2) \cup (2, \infty)$

Example 14. Solve $|7x - 4| \geq 3$ and graph the solutions.

Solution. The quantity within the absolute-value bars must lie to the left of (and including) -3 *or* to the right of (and including) 3.

$$7x - 4 \leq -3 \quad \text{or} \quad 7x - 4 \geq 3$$
$$7x \leq 1 \qquad\qquad 7x \geq 7$$
$$x \leq \frac{1}{7} \quad \text{or} \quad x \geq 1$$

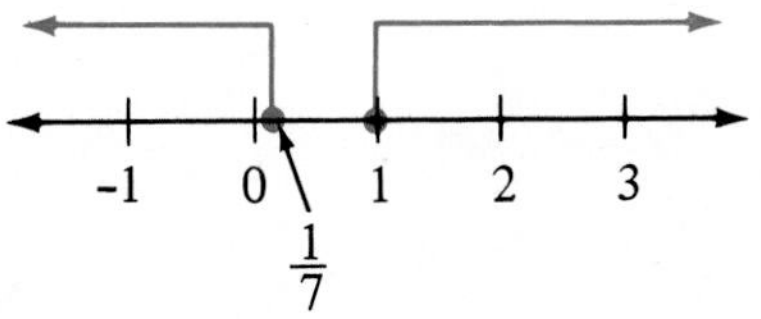

$(-\infty, \frac{1}{7}] \cup [1, \infty)$

Example 16. Solve $|5 - 3x| > 1$ and graph the solutions.

Solution. $|5 - 3x| > 1$ means

$$5 - 3x < -1 \quad \text{or} \quad 5 - 3x > 1$$
$$-3x < -6 \qquad\qquad -3x > -4$$
$$\frac{-3x}{-3} > \frac{-6}{-3} \qquad\qquad \frac{-3x}{-3} < \frac{-4}{-3}$$
$$x > 2 \quad \text{or} \quad x < \frac{4}{3}$$

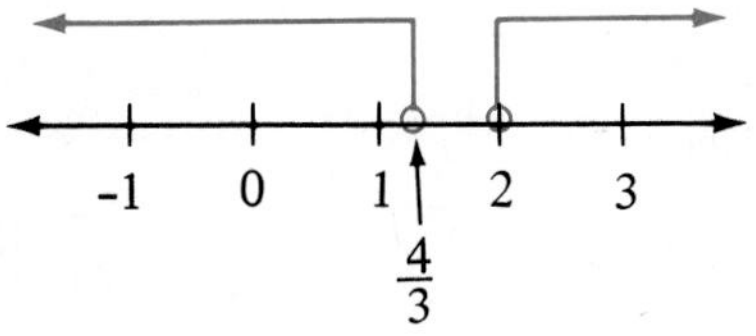

$(-\infty, \frac{4}{3}) \cup (2, \infty)$

Answers to Trial Runs

page 113 1. $x = 5.1$ or $x = -5.1$ 2. $a = \frac{3}{4}$ or $a = -\frac{3}{4}$ 3. $x = 6$ or $x = -1$ 4. $x = \frac{1}{4}$ 5. $a = -4$ or $a = -8$ 6. No solution 7. $x = -4$ or $x = 5$ 8. $x = 0$ or $x = 6$

page 115 1. $-3.5 \leq a \leq 3.5$; $[-3.5, 3.5]$ 2. $-\frac{4}{5} < y < \frac{4}{5}$; $(-\frac{4}{5}, \frac{4}{5})$ 3. No solution 4. $-3 < x < 3$; $(-3, 3)$ 5. $-2 \leq x \leq \frac{3}{2}$; $[-2, \frac{3}{2}]$ 6. $-2 < x < 9$; $(-2, 9)$ 7. No solution 8. $-7 \leq x \leq 8$; $[-7, 8]$

page 119 1. $y < -\frac{2}{3}$ or $y > \frac{2}{3}$; $(-\infty, -\frac{2}{3}) \cup (\frac{2}{3}, \infty)$ 2. All real numbers; $(-\infty, \infty)$ 3. $x \leq -7$ or $x \geq 7$; $(-\infty, -7] \cup [7, \infty)$ 4. $x < -2$ or $x > 2$; $(-\infty, -2) \cup (2, \infty)$ 5. $x \leq -1$ or $x \geq 4$; $(-\infty, -1] \cup [4, \infty)$ 6. $x \leq 5$ or $x \geq 13$; $(-\infty, 5] \cup [13, \infty)$ 7. $x < -6$ or $x > 6$; $(-\infty, -6) \cup (6, \infty)$ 8. $x < 0$ or $x > \frac{4}{3}$; $(-\infty, 0) \cup (\frac{4}{3}, \infty)$

Name ________________________ Date ____________

EXERCISE SET 2.4

Solve. Write solutions for inequalities using interval notation.

_____ 1. $|x| = 9$

_____ 2. $|x| = 13$

_____ 3. $\left|\frac{a}{3}\right| = 7$

_____ 4. $\left|\frac{a}{5}\right| = 3$

_____ 5. $|3x| = 2.1$

_____ 6. $|5x| = 4.5$

_____ 7. $|y - 5| = 0$

_____ 8. $|y + 3| = 0$

_____ 9. $|2x - 9| = -7$

_____ 10. $|3x - 5| = -10$

_____ 11. $|7 - 6x| = 1$

_____ 12. $|12 - 3x| = 9$

_____ 13. $|3x - 1| = -2$

_____ 14. $|3x + 5| = -6$

_____ 15. $\left|\frac{a}{6} + 5\right| = 1$

_____ 16. $\left|\frac{a}{7} + 3\right| = 4$

_____ 17. $|13 - 3x| = 5$

_____ 18. $|9 - 3x| = 6$

_____ 19. $\frac{|7 - x|}{2} = 10$

_____ 20. $\frac{|8 - x|}{3} = 2$

_____ 21. $|y| < \frac{5}{3}$

_____ 22. $|y| < \frac{7}{2}$

_____ 23. $\left|\frac{a}{7}\right| < 5$

_____ 24. $\left|\frac{a}{3}\right| < 9$

_____ 25. $|-3x| < 6$

_____ 26. $|-2x| < 12$

_____ 27. $|3x - 2| \le 4$

_____ 28. $|5x - 1| \le 9$

_____ 29. $\left|5 - \frac{x}{3}\right| \le 1$

_____ 30. $\left|3 - \frac{x}{4}\right| \le 2$

_____ 31. $|2m - 7| \le -8$

_____ 32. $|4m - 9| \le -10$

_____ 33. $\left|\frac{3x - 1}{4}\right| < 7$

_____ 34. $\left|\frac{4x - 8}{2}\right| < 6$

_____ 35. $|a| > 2.5$

_____ 36. $|a| > 3.2$

_____ 37. $|y - 3| > -7$

_____ 38. $|y - 2| > -9$

_____ 39. $|-5x| \ge 10$

_____ 40. $|-9x| \ge 18$

_____ 41. $\left|\frac{3}{5}x\right| > 6$

_____ 42. $\left|\frac{5}{2}x\right| > 10$

_______ **43.** $|5x - 1| - 3 \geq 1$

_______ **44.** $|7x - 1| - 5 \geq 8$

_______ **45.** $\left|1 - \frac{m}{3}\right| > 6$

_______ **46.** $\left|2 - \frac{m}{7}\right| > 1$

_______ **47.** $|3x - 7| > -9$

_______ **48.** $|5x - 3| > -12$

_______ **49.** $|7x - 3| + 4 \geq 14$

_______ **50.** $|8x - 5| + 7 \geq 10$

☆ Stretching the Topics

Solve.

_______ **1.** $|6 - 2[3 - (5 - 4m)]| = 2$

_______ **2.** $4|2x - 2| + 6 < 10$

_______ **3.** $|2x - 8| < 8$ and $|x| + 3 > 4$

Check your answers in the back of your book.

If you can complete **Checkup 2.4**, you are ready to go on to Section 2.5.

Name ______________________ Date ____________

✓ CHECKUP 2.4

Solve. Write solutions for inequalities using interval notation.

_______ 1. $|3x| = 15.6$

_______ 2. $|8 - x| = 9$

_______ 3. $|5y - 2| = -8$

_______ 4. $|-4x| < 12$

_______ 5. $\left|\frac{x}{2} - 7\right| \le 5$

_______ 6. $|3a - 2| \le 7$

_______ 7. $|3x| > 6$

_______ 8. $|9 - m| \ge 3$

_______ 9. $|2x - 1| - 5 > 6$

_______ 10. $\left|\frac{3}{2}x - 9\right| > 12$

Check your answers in the back of your book.

If You Missed Problems:	You Should Review Examples:
1–3	1–3
4–6	8–10
7–10	13–16

2.5 Switching from Words to Equations and Inequalities

In Chapter 1 we practiced switching word expressions to expressions containing variables and constants. Now that we have learned to solve first-degree equations and inequalities, we can put those skills to use in solving problems stated in words.

Once again we shall be orderly in our approach and keep the following steps in mind.

1. Read the problem carefully and sort out the given information.
2. Identify the variable.
3. Switch each word expression in the problem to a variable expression.
4. Use an equation or inequality to relate the variable expressions.
5. Solve the equation or inequality.
6. Write a sentence describing the answer to the problem.

Let's try our thinking, then writing approach to solving some word problems.

Example 1. At an hourly wage of \$3.50, for how many hours must Beth work so that she will earn \$140?

Solution

We Think	***We Write***
The unknown quantity is the number of hours that Beth works.	Let h = number of hours
Earnings are found by *multiplying* rate per hour times number of hours worked.	$3.50h$ = earnings
Beth's earnings must total \$140.	$3.50h = 140$

$$3.50h = 140$$

$$\frac{3.50h}{3.50} = \frac{140}{3.50}$$

$$h = 40 \text{ hours}$$

Beth must work for 40 hours to earn \$140.

Example 2. Ira has scored 80, 62, 75, and 69 on his first four mathematics tests. What is the lowest score that he can earn on his fifth test to have a test average of at least 70?

Solution

We Think	***We Write***
The unknown quantity is Ira's grade on the fifth test.	Let t = Ira's grade on fifth test
To find the average of five tests, we must *add* all the scores and *divide* by 5.	$\frac{80 + 62 + 75 + 69 + t}{5} =$ average of five tests
Ira wants an average score of *at least* 70.	$\frac{80 + 62 + 75 + 69 + t}{5} \geq 70$

$$\frac{80 + 62 + 75 + 69 + t}{5} \geq 70$$

$$\frac{286 + t}{5} \geq 70$$

$$286 + t \geq 350$$

$$t \geq 64$$

If Ira's fifth score is 64 or higher, he will have an average of at least 70 on the five tests.

Now you try completing Example 3.

Example 3. James wishes to enclose a rectangular chicken yard with 150 feet of fencing. He would like the yard to be twice as long as it is wide. What must be the yard's dimensions?

Solution

We Think	***We Write***
The number of feet of fencing is the *perimeter* of the yard.	Perimeter = 2 lengths + 2 widths
We need expressions for length and width. We know less about the width.	Let w = width of yard (feet)
We know that the length is twice the width.	Let ____ = length of yard (feet)

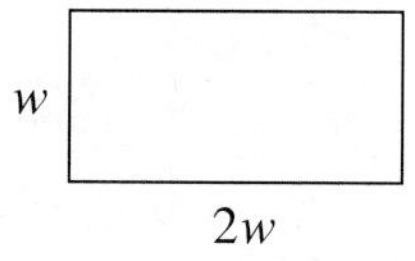

We know that the perimeter is the *sum* of 2 lengths and 2 widths.	$2(\underline{\quad\quad}) + 2(\underline{\quad\quad}) =$ perimeter
The perimeter is 150 feet.	$2(\underline{\quad\quad}) + 2(\underline{\quad\quad}) = 150$

$$2w + 2(2w) = 150$$

$$2w + 4w = 150$$

$$\underline{\quad\quad}\, w = 150$$

$$w = \underline{\quad\quad}$$

The width of the yard is ____ feet and the length is ____ feet.

Check your work on page 130. ▶

Example 4. In her purse Anita has 16 coins (nickels, dimes, and quarters), worth \$2.05. She has three times as many dimes as nickels. How many of each coin does she have?

Solution

We Think	***We Write***
First we concentrate on the *number* of coins.	Let n = number of nickels
We know that there are *three times* as many dimes as nickels.	$3n$ = number of dimes
There are 16 coins in all, so the number of quarters is 16 *minus* the number of dimes and nickels.	$16 - (n + 3n)$ = number of quarters $16 - 4n$ = number of quarters

Now we concentrate on the *value* of the coins. Each nickel is worth 5 cents, each dime is worth 10 cents, each quarter is worth 25 cents.

$5 \cdot n =$ value of nickels
$10(3n) =$ value of dimes
$25(16 - 4n) =$ value of quarters

We know that the total value of the coins is the *sum* of all the coins' values.

$5n + 10(3n) + 25(16 - 4n) =$ total value

We know that the total value is \$2.05, or 205 cents.

$5n + 10(3n) + 25(16 - 4n) = 205$

$$5n + 10(3n) + 25(16 - 4n) = 205$$
$$5n + 30n + 400 - 100n = 205$$
$$400 - 65n = 205$$
$$-65n = -195$$
$$\frac{-65n}{-65} = \frac{-195}{-65}$$
$$n = 3$$

Number of nickels: $n = 3$

Number of dimes: $3n = 3 \cdot 3 = 9$

Number of quarters: $16 - 4n = 16 - 4 \cdot 3 = 16 - 12 = 4$

Anita's purse contains 3 nickels, 9 dimes, and 4 quarters.

Now let's try working the problem stated at the beginning of this chapter.

Example 5. Mary left Hometown at 6 P.M. driving due east on Route 22 at 55 miles per hour. At the same time, Todd left Anytown driving at 60 miles per hour due west on Route 22. If Hometown is 368 miles due west of Anytown, at what time will Mary and Todd meet?

Solution. Let's use an illustration to help understand the situation in this problem.

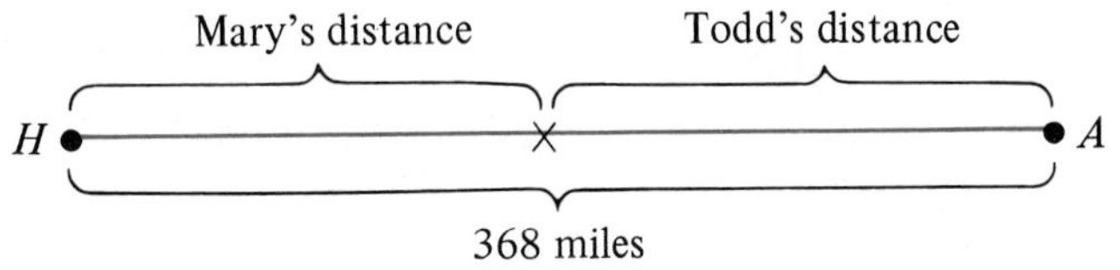

We Think	***We Write***
The total distance will be the *sum* of Mary's distance and Todd's distance.	Total distance = Mary's distance + Todd's distance
We know that the distance is found by *multiplying* rate per hour times hours driven.	Distance = rate $\times$ time
Mary and Todd both drive for the same number of hours.	Let $t =$ hours driven
Mary's rate is 55 mph and Todd's rate is 60 mph.	$55 \cdot t =$ Mary's distance (miles) $60 \cdot t =$ Todd's distance (miles)
The two distances must *sum* to the total distance, 368 miles.	$55t + 60t = 368$

$$55t + 60t = 368$$

$$115t = 368$$

$$\frac{115t}{115} = \frac{368}{115}$$

$$t = 3\frac{23}{115}$$

$$= 3\tfrac{1}{5} \text{ hours}$$

Mary and Todd must drive for $3\frac{1}{5}$ hours (or 3 hours and 12 minutes) before they meet. If they started at 6 P.M., they will meet at 9:12 P.M.

See if you can complete Example 6.

Example 6. A car rental agency charges a daily fee of \$18, plus 30 cents per mile. If Jake rents a car for 1 day, what distance can he drive and keep his total rental charge under \$45?

Solution

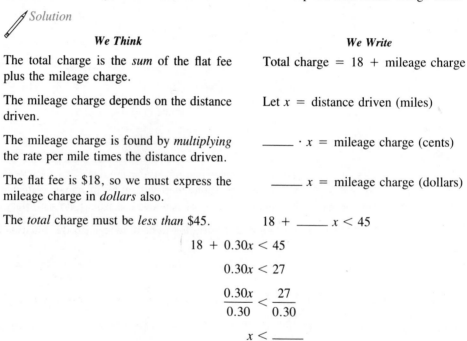

We Think	*We Write*
The total charge is the *sum* of the flat fee plus the mileage charge.	Total charge = 18 + mileage charge
The mileage charge depends on the distance driven.	Let x = distance driven (miles)
The mileage charge is found by *multiplying* the rate per mile times the distance driven.	_____ $\cdot x$ = mileage charge (cents)
The flat fee is \$18, so we must express the mileage charge in *dollars* also.	_____ x = mileage charge (dollars)
The *total* charge must be *less than* \$45.	18 + _____ $x < 45$

$$18 + 0.30x < 45$$

$$0.30x < 27$$

$$\frac{0.30x}{0.30} < \frac{27}{0.30}$$

$x <$ _____

Jake must drive fewer than _____ miles.

Check your work on page 130. ▶

Example 7. The Supercal Calculator Company manufactures calculators at a cost of \$8.45 each and sells them for \$23.95 each. The company's fixed costs are \$7300 per month. How many calculators must the company produce and sell to have a profit of at least \$2000 this month?

Solution

We Think	*We Write*
Profit is found by *subtracting* costs from revenue.	Profit = revenue − costs
Revenue is found by *multiplying* selling price times number of units sold.	Revenue $= 23.95x$
Costs are found by *adding* variable costs and fixed costs.	Variable costs $= 8.45x$ Fixed costs $= 7300$ Costs $= 8.45x + 7300$
Profit = revenue − costs.	Profit $= 23.95x - (8.45x + 7300)$ $= 23.95x - 8.45x - 7300$ $= 15.5x - 7300$
We want profit to be *at least* \$2000.	$15.5x - 7300 \geq 2000$

$$15.5x - 7300 \geq 2000$$

$$15.5x \geq 9300$$

$$\frac{15.5x}{15.5} \geq \frac{9300}{15.5}$$

$$x \geq 600$$

At least 600 calculators must be sold to have a profit of at least \$2000 this month.

Applying the techniques of algebra to solving word problems can be one of the most rewarding parts of your study in mathematics. You will become successful in that skill if you practice a great deal and learn to be organized in your approach.

Trial Run

Use an equation or inequality to solve each problem.

________ 1. Carl rode his bike to the city at a rate of 10 miles per hour and returned immediately on a bus traveling the same route at a rate of 50 miles per hour. If the entire trip took 6 hours, how far is Carl's house from the city?

________ 2. The refrigerator repairman charges a service call fee of \$29, plus \$7.50 per half-hour spent on the job. If James's bill for labor was \$66.50, how long did the repairman work on his refrigerator?

Answers are on page 130.

▶ Examples You Completed

Example 3 (*Solution*)

We Write

Perimeter = 2 lengths + 2 widths

Let w = width of yard

Let $2w$ = length of yard

$2(w) + 2(2w) = \text{perimeter}$

$2(w) + 2(2w) = 150$

$$2w + 2(2w) = 150$$

$$2w + 4w = 150$$

$$6w = 150$$

$$w = 25$$

The width of the yard is 25 feet and the length is 50 feet.

Example 6 (*Solution*)

We Write

Total charge = 18 + mileage charge

Let x = distance driven (miles)

$30 \cdot x$ = mileage charge (cents)

$0.30x$ = mileage charge (dollars)

$$18 + 0.30x < 45$$

$$0.30x < 27$$

$$\frac{0.30x}{0.30} < \frac{27}{0.30}$$

$$x < 90$$

Jake must drive fewer than 90 miles.

Answers to Trial Run

page 129 **1.** Carl's house is 50 miles from the city. **2.** The repairman worked for $2\frac{1}{2}$ hours.

Name ______________________________ Date ______________

EXERCISE SET 2.5

Use an equation or inequality to solve each problem.

________ 1. Steve is saving to buy a car. He has decided he will start looking as soon as he has saved \$2000. If he has \$1100 in savings now and he can save \$150 a month, how many months will it be before Steve can begin looking at cars?

________ 2. Mrs. Bealmear, a retiree, will lose some retirement benefits if she earns over \$8000 per year working as a baby-sitter. If she has already earned \$6500, for how many more hours may she work at \$2.50 per hour without losing any retirement benefits?

________ 3. The Band Boosters Club is selling cookies to buy new uniforms. Chocolate chip cookies sell for \$3 a box and sugar cookies for \$2 a box. They decide that they can sell 650 boxes. How many boxes of each should they sell so that their total sales will be \$1700?

________ 4. Sara is making a silk flower arrangement of roses and carnations. She needs twice as many roses as carnations. Carnations cost 75 cents each and roses are \$1 each. How many of each should she buy if she can spend no more than \$18 for the arrangement?

________ 5. If Don loses 16 pounds, 9 pounds, and 11 pounds during the first three months of his diet, how much will he have to lose during the fourth month so that he will have an average weight loss of 12 pounds per month?

________ 6. If Evelyn earns \$37, \$58, \$61, and \$42 during the first 4 days that she sells vegetables from her garden, how much will she have to sell during the fifth day so that her sales average is \$50 per day?

________ 7. For the toll road, Frank needs 5 more dimes than quarters and the total amount he will have to pay is \$3.65. How many of each will he need?

________ 8. At the end of the day, Alice's tips consist of three times as many nickels as quarters. If the total of the tips is \$8.40, how many of each does she have?

________ 9. Dr. Tucker invested part of \$12,000 at 7 percent and the rest at 9 percent. His investments earn \$940 in a year. How much does he have invested at each rate?

________ 10. Two amounts of money totaling \$25,000 are invested in stocks. One amount earns 10 percent interest and the other earns 13 percent. Find how much is invested at each amount if the total yearly interest is \$2920.

________ 11. Mr. Hsu has \$5000 invested at 6 percent interest. How much additional money must he invest at 9 percent so that his annual income from interest will be the same as if the total amount were invested at 8 percent?

________ 12. The Girls Club raised \$15,000 in contributions. The total amount was invested and the interest was used for scholarships totaling \$1350. If part of the money was invested at 8 percent and the other part was invested at 10 percent, how much was invested at each rate?

_______ 13. Mrs. Young is making meat loaf for a banquet. She bought 20 pounds of meat: some sausage and some ground beef. Ground beef is \$1.75 a pound and sausage is \$1.30 a pound. If the total cost was \$31.40, how many pounds of each did she buy?

_______ 14. Carroll is making ice cream for the math club picnic. He needed twice as much milk as cream. If milk is 90 cents a quart and cream is \$2.10 a quart, how many quarts of each did he buy if the total cost was \$11.70?

_______ 15. Two cars are 315 miles apart and are moving toward each other. They will meet in 3 hours. If their speeds differ by 5 miles per hour, what is the speed of each car?

_______ 16. Richard jogs and walks a total of 10 miles every day. He jogs at 5 miles per hour and walks at 3 miles per hour. If he covers the total distance in 3 hours, how much of the time does he spend walking?

_______ 17. On a recent vacation, Toni drove to Kentucky Lake. The trip to the lake took her 1 hour less than the trip home. If she averaged 55 miles per hour going and only 45 miles per hour returning, how far does Toni live from the lake?

_______ 18. Sam rode the bus to see his father but returned by car with a neighbor. On the bus he averaged 10 miles per hour less than when riding with his neighbor. If it took Sam 3 hours to get to his father's house but only 2 hours to return, how far does he live from his father?

_______ 19. The Book Rack is celebrating National Book Week by having a sale. Charles bought 30 books, some for \$2.50 and some for \$4. If his total bill was \$102, how many of each did he buy?

_______ 20. Kris is buying bags of candy for Christmas, some for 75 cents and some for \$1.25. He bought three times as many bags of 75-cent candy as he did bags costing \$1.25. If he spent \$17.50 for candy, how many bags of each did he buy?

_______ 21. Della is buying a living-room suite on the installment plan. She paid \$150 as a down payment and is paying \$55 a month until it is paid for. If the living-room suite cost \$810, for how many months will she be making payments?

_______ 22. Juan earns a weekly salary of \$380 a week and \$12 per hour for each hour of overtime. If Juan earned \$470 last week, how many hours overtime did he work?

_______ 23. A cashier at First Federal has a stack of bills—some ten-dollar bills, some five-dollar bills, and some ones. There are 8 fewer five-dollar bills than ten-dollar bills and twice as many one-dollar bills as five-dollar bills. If the total value of the stack is \$811, how many bills are there of each denomination?

_______ 24. Miles has a total of \$3.65 in change. He has 2 more nickels than dimes and 3 more quarters than dimes. How many coins of each kind does he have?

_______ 25. Benson sells vacuum cleaners on a commission. He earns \$75 for each vacuum cleaner he sells. How many vacuum cleaners should he sell so that he will have a monthly salary of at least \$1400?

_______ 26. If Thor works for an hourly wage of $4.50, for how many hours must he work so that his gross pay will be at least $162?

_______ 27. The perimeter of a triangle can be no more than 35 meters. One side is 7 meters and the other two sides are of equal length. Find the possible lengths for the two equal sides.

_______ 28. The length of one side of a triangle is 9 inches more than another. The sum of the length of these two sides must be greater than the third side, which is 43 inches. Find the possible lengths for the two sides.

_______ 29. The length of a rectangular playground is to be twice its width. The dimensions must be such that the playground can be enclosed with no more than 360 feet of available fencing. What dimensions can the playground have?

_______ 30. Janyce has only 134 inches of framing. What will be the dimensions of the largest picture frame she can make if she makes the length 5 inches more than the width?

_______ 31. The Miller Theater has only 250 seats. Assuming that the theater is filled and that the theater group charges admissions of $5 for adults and $3 for children, find the possible number of children's tickets they can sell and still cover their production expenses of $1050.

_______ 32. The Women's Political Caucus is having a barbecue for its 150 members. It will cost them $2.50 to serve a chicken plate and $3.75 to serve a ham plate. If they have only $415, use an inequality to state the possible number of ham plates that could be served.

_______ 33. A manufacturer of lighting fixtures produces a certain lamp at a cost of $25.50. The fixed costs for producing lamps each month are $2400. How many lamps can be produced each month if the cost of production is to be kept below $3420?

_______ 34. Candy Land sells 5-pound boxes of candy for $13.25. How many boxes must be sold each week if the weekly revenue from the sale of the 5-pound boxes of candy must be greater than $331.25?

_______ 35. A publishing company finds that the cost of publishing a certain book is $12.50 per copy. The fixed costs for publishing the book are $35,000. How many books must the company publish and sell in order to have a profit of at least $15,000 if each book sells for $25?

☆ Stretching the Topics

______ 1. Mr. Hickman bought a car, making a down payment of \$1000 and financing the remainder. The finance charge is 8.8 percent per year on the amount financed. If he pays the finance company \$15,548 in 48 equal monthly payments, what was the original price of the car?

______ 2. The length of a rectangle is 5 feet less than 3 times the width. The sides of an equilateral triangle are the same as the length of the rectangle. If the sum of the two perimeters is 94 feet, find the dimensions of the rectangle.

______ 3. The width of a rectangle is 4 centimeters more than one-third of the length. If the perimeter must be no more than 64 centimeters, find the possible values for the width.

Check your answers in the back of your book.

If you can complete **Checkup 2.5**, you are ready to do the **Review Exercises** for Chapter 2.

Name ______________________ Date ____________

✓ CHECKUP 2.5

Use an equation or inequality to solve each problem.

________ **1.** Irene has 16 feet of antique lace that she wishes to use as a border for a scarf. She would like the scarf to be 2 feet longer than it is wide. What must be the dimensions of the scarf?

________ **2.** Two cars are 415 miles apart and are on Highway 65 traveling toward each other. One car is traveling at 55 miles per hour and the other is traveling at 65 miles per hour. How long will each car travel before they meet?

________ **3.** The Little Tot's Baby-sitting Service charges a total of $5 for the first two hours, plus $2.75 an hour for each additional hour. For how many hours may Mrs. Hickman shop and keep her total baby-sitting charge under $25?

________ **4.** In the first four basketball games of the season, Lillie scored 23, 19, 26, and 22 points. How many points must she score during the fifth game to have a scoring average of at least 25 points per game?

Check your answers in the back of your book.

If You Missed Problem:	You Should Review Example:
1	3
2	5
3	6
4	2

Summary

In this chapter we learned to find values of the variable that satisfy first-degree equations or inequalities. A **first-degree equation** is an equation that can be written in the form

$$ax + b = c$$

and we learned to follow these steps to find its solution.

1. Remove fractions by multiplying both sides by the LCD.
2. Remove parentheses and combine like terms.
3. Get variable terms on one side of the equation by addition or subtraction.
4. Get constant terms on the other side by addition or subtraction.
5. Solve for the variable by multiplication and/or division.

First-degree inequalities are inequalities of the form

$$ax + b < c \qquad ax + b > c$$

$$ax + b \leq c \qquad ax + b \geq c$$

To solve first-degree inequalities, we use the same steps as those used for solving equations, *except* when we multiply or divide both sides by a *negative* number, in which case we must *reverse* the direction of the inequality. We learned to graph the solutions of inequalities on the number line and to write the solutions using **interval notation**.

$x < a$

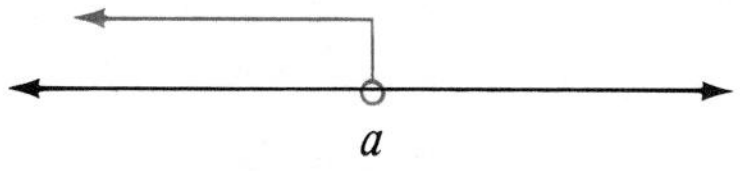

$(-\infty, a)$

$x \leq a$

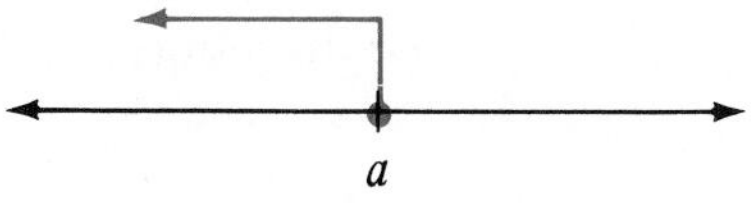

$(-\infty, a]$

$x > a$

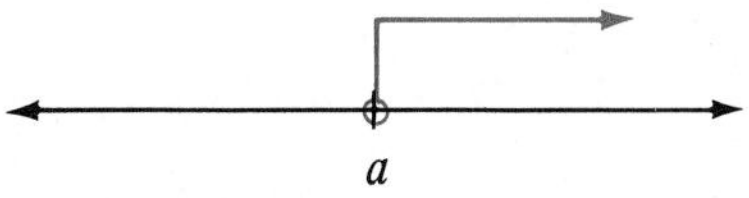

(a, ∞)

$x \geq a$

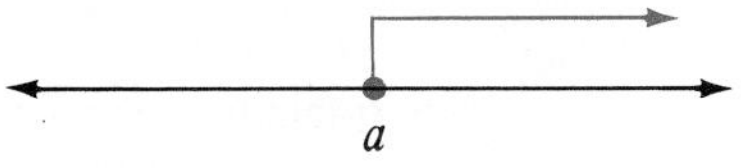

$[a, \infty)$

$a < x < b$

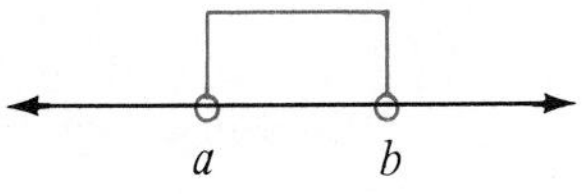

(a, b)

$a \leq x \leq b$

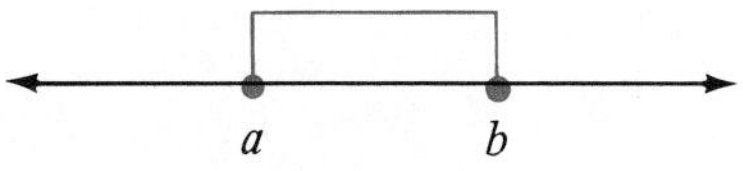

$[a, b]$

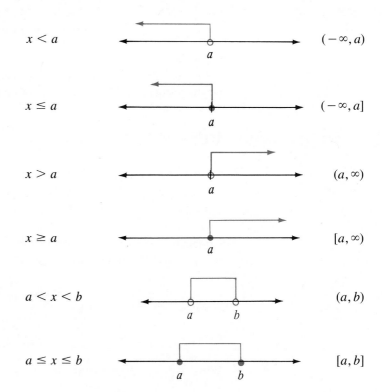

Finally, we discussed **absolute-value equations and inequalities** and learned to solve them by rewriting them as equivalent statements.

$|x| = a$ means $x = a$ or $x = -a$

$|x| < a$ means $-a < x < a$

$|x| > a$ means $x < -a$ or $x > a$

Speaking the Language of Algebra

Complete each sentence with the appropriate word or phrase.

1. In a first-degree equation, the variable appears only with an exponent of ________ .

2. In solving a first-degree equation, our aim is to ________ ________ ________ .

3. If an equation or inequality contains parentheses, we first ________ ________ and ________ ________ ________ .

4. When graphing the solutions for the inequality $x < a$, there will be a(n) ________ dot at a, and a line extending to the ________ .

5. To say that x is between a and b (but not including a or b) we may write the inequality ________ .

6. When seeking the ________ of the solutions for two inequalities, we look for solutions to one inequality *and* the other inequality. When seeking the ________ of the solutions of two inequalities, we look for the solutions to one inequality *or* the other inequality.

7. If we multiply or divide both sides of an inequality by a *positive* number, the direction of the inequality is ________ . If we multiply or divide both sides of an inequality by a *negative* number, the direction of the inequality is ________ .

8. For the equation $|x| = a$, the solutions are ________ or ________ .
   For the inequality $|x| \leq a$, the solutions are ________ .
   For the inequality $|x| \geq a$, the solutions are ________ or ________ .

Writing About Mathematics

Write your response to each question in complete sentences.

1. Show and explain the steps you would use to solve and check the equation $2(x - 3) + 1 = 5x - 2$.

2. Explain how you decide whether to use an open dot or a closed dot in graphing the solutions for an inequality on the number line.

3. Explain the similarities and differences in the properties used to solve first-degree equations and first-degree inequalities.

4. Explain why the absolute-value inequality $|x| < a$ has no solution when a is negative.

5. Write two word problems that would lead you to write the first-degree equation $\dfrac{x + 25 + 12}{5} = 17$.

Name ____________________ Date __________

REVIEW EXERCISES for Chapter 2

Solve each equation and check the solutions.

______ 1. $3x - 5 = 16$

______ 2. $7 - 4x = 31$

______ 3. $\frac{x}{6} + 4 = -3$

______ 4. $9 + \frac{x}{5} = 0$

______ 5. $-4x + 2 = 5x + 29$

______ 6. $2(x - 4) = -3(2x + 4)$

______ 7. $\frac{3x - 7}{3} = 6x - 5$

______ 8. $4 + \frac{5x - 1}{4} = 2x$

______ 9. $\frac{2x}{7} - 9 = 5x - \frac{6}{7}$

______ 10. $5(2y - 1) + 3 = 4(2 - y)$

______ 11. $4[x - 2(x - 3)] = 3(2x + 8)$

______ 12. $F = \frac{9}{5}C + 32$ for C

Graph the solutions for each inequality on a number line.

13. $m \geq \frac{8}{3}$

14. $x < -1$

Solve each pair of inequalities by graphing. Write solutions using interval notation.

______ 15. $x \geq -5$ and $x < 2$

______ 16. $x < -3$ and $x \geq -2$

______ 17. $x > -1$ or $x > 4$

______ 18. $x \geq -5$ or $x < 0$

Solve the inequalities and graph the solutions. Then write solutions using interval notation.

______ 19. $-\frac{3}{5}x + 2 \geq 8$

______ 20. $4(x - 5) < -16$

______ 21. $5x - (x + 2) < 5x + 3$

______ 22. $3(2y - 1) - 5 \geq 4(y - 3) + 5$

______ 23. $5x - 6 < 4 - x$ and $3x - 1 \geq 5x + 3$

______ 24. $\frac{x - 1}{3} < 5 - x$ and $\frac{2x}{3} + 5 > 7$

_______ 25. $3(2x - 1) \le 7x + 2$ or $2(3x - 1) > 7x + 5$

_______ 26. $\frac{2x - 6}{3} < 3 - x$ or $\frac{5x}{3} + 6 < 1$

_______ 27. $3 \le x - 4 \le 9$

_______ 28. $-2 \le \frac{4 - x}{3} < 0$

Solve each equation or inequality.

_______ 29. $|2x - 3| = 7$

_______ 30. $|9 - x| = 4$

_______ 31. $\left|\frac{2}{3}x + 4\right| = -9$

_______ 32. $|3x - 2| \le 13$

_______ 33. $\left|\frac{x}{5} - 3\right| < -1$

_______ 34. $|2x - 3| > 11$

_______ 35. $|3x - 7| > -2$

_______ 36. $\left|9 - \frac{3x}{2}\right| \ge 6$

Use an equation or inequality to solve each problem.

_______ 37. Mickey sells family-size sailboats. He earns \$360 for each one he sells. How many sailboats should he sell each month so that he will have a monthly salary of at least \$1600?

_______ 38. R. J. has two amounts of money invested in stocks. The first amount is \$5000 more than the second amount. The first amount earns 12 percent interest and the other earns 14 percent interest. Find the amount invested at each rate if the total interest from the two sums last year was \$3720.

_______ 39. Two cars leave from the same point on the interstate traveling in opposite directions. One car is traveling 10 miles per hour faster than the other. If in 4 hours they will be 400 miles apart, at what rate of speed is each traveling?

_______ 40. Mr. McDonald has 120 feet of fencing to build a loading pen for cattle. The pen must be three times as long as it is wide. Find the largest dimensions the loading pen can have.

Check your answers in the back of your book.

If You Missed Exercises:	You Should Review Examples:	
1, 2	SECTION 2.1	5
3, 4		6
5		7, 8
6		10
7, 8		13–16
9		11, 12
10, 11		9, 10
12		19
13, 14	SECTION 2.2	3, 4
15, 16		7–9
17, 18		12–14
19–22	SECTION 2.3	5–9
23, 24		10, 11
25, 26		12, 13
27, 28		14, 15
29–31	SECTION 2.4	2–5
32, 33		8–10
34–36		13–17
37–40	SECTION 2.5	1–6

If you have completed the **Review Exercises** and corrected your errors, you are ready to take the **Practice Test** for Chapter 2.

Name ______________________________ Date ______________

PRACTICE TEST for Chapter 2

Solve each equation or inequality.

	Problem	Section	Examples
________	1. $10 - 7x = -4$	2.1	5
________	2. $5y - 2 = 8 - 5y$	2.1	7, 8
________	3. $\frac{3x}{2} - 1 = 5$	2.1	14
________	4. $2(5a - 1) - 6 = -5(2 - a)$	2.1	9
________	5. $\frac{7x - 3}{4} + 1 = 4x$	2.1	16
________	6. $A = \frac{h(a + b)}{2}$ for b	2.1	20
________	7. $-4x + 12 \leq 4$	2.3	5
________	8. $6(y - 3) + 5 > 4(1 - y) + 3$	2.3	9
________	9. $0 \leq \frac{8x + 1}{5} < 5$	2.3	15
________	10. $\lvert 4x - 7\rvert = 2$	2.4	3
________	11. $\left\lvert \frac{5 - 2x}{5} \right\rvert = 3$	2.4	5
________	12. $\lvert 1 - 2x\rvert < 11$	2.4	10
________	13. $\lvert 3x + 2\rvert \geq 5$	2.4	14

Solve each pair of inequalities by graphing. Then write solutions using interval notation.

	Problem	Section	Examples
________	14. $x \leq -2$ and $x > -6$	2.2	8
________	15. $x < 7$ or $x \leq 0$	2.2	12
________	16. $2x - 5 < 3$ and $\frac{5 - x}{2} < 4$	2.3	10
________	17. $5(x + 2) < 2x + 7$ or $4 - (x - 3) \leq 3x - 5$	2.3	13

Use an equation or inequality to solve each problem.	Section	Example
________ **18.** When Janice rides her bike to work each morning, her rate is 2 miles per hour faster than her rate when she rides home in the evening. If her morning ride takes 1 hour and her evening ride takes $1\frac{1}{4}$ hours, find Janice's rate in the evening.	2.5	5
________ **19.** The Zeta Zeta fraternity needs \$2500 for repairs on its fraternity house. If the members have \$1630 in the treasury and an alumnus donates a stereo to be raffled, what is the least number of raffle tickets they must sell at \$2 each in order to finance the repairs?	2.5	6
________ **20.** George has purchased a painting which is 13 inches long and 5 inches wide. What width of framing can he use for the painting if he wishes the framed length to be no more than twice the framed width?	2.5	3

CHAPTER

3

Working with Exponents and Polynomials

Marco's rectangular garden is 10 feet longer than it is wide. If he decreases both the length and width of the garden by 3 feet, what variable expression represents the amount of garden area that he will lose?

So far we have dealt with algebraic expressions, equations, and inequalities involving variables raised only to the first power. Now we shall discover how to work with expressions containing variables raised to any integer powers. In this chapter we learn how to

1. Use the laws of exponents.
2. Identify and simplify polynomials.
3. Operate with polynomials.
4. Switch from word expressions to polynomials.

3.1 Working with Whole-Number Exponents

Using the Definition of a Power

In Chapter 1 we learned that a raised natural number, called an *exponent,* tells us how many times to use its *base* as a factor in a product.

Exponent Form		*Factor Form*
3^2	=	$3 \cdot 3$
$(-1)^3$	=	$(-1)(-1)(-1)$
x^4	=	$x \cdot x \cdot x \cdot x$

In general, we agreed to make this definition.

> **Definition of a Power.** If a is a real number and n is a natural number, then the nth power of a is
>
> $$a^n = \underbrace{a \cdot a \cdot a \cdot \ldots \cdot a}_{n \text{ factors}}$$

Here a is the **base,** n is the **exponent,** and a^n is called the ***n*th power of *a*.** Notice that an exponent applies only to the base to which it is attached unless parentheses indicate otherwise. In other words,

$$2x^3 = 2 \cdot x \cdot x \cdot x \quad \text{but} \quad (2x)^3 = (2x)(2x)(2x) = 8x^3$$

$$-a^2 = -1 \cdot a \cdot a \quad \text{but} \quad (-a)^2 = (-a)(-a) = a^2$$

Try Example 1.

Example 1. Evaluate $x^3 - 5x$ when $x = -2$.

Solution. We substitute -2 for x.

$x^3 - 5x = (\_\_\_\_)^3 - 5(\_\_\_\_)$

Check your work on page 154. ▶

Example 2. Evaluate $-x^4 + 2x^2 - 2x$ when $x = 3$.

Solution. We substitute 3 for x.

$$\begin{aligned} -x^4 + 2x^2 - 2x &= -(3)^4 + 2(3)^2 - 2(3) \\ &= -81 + 2(9) - 6 \\ &= -81 + 18 - 6 \\ &= -69 \end{aligned}$$

When a base is written without an exponent, recall that the exponent is "understood" to be 1. You might wonder whether *zero* can be used as an exponent. The answer is yes. We soon shall see how the following definition for zero used as an exponent fits in neatly with the rules for working with exponents.

> **Definition of Zero Exponent**
>
> $a^0 = 1 \qquad (a \neq 0)$

Any nonzero base raised to the zero power is equal to 1. Assuming that variable bases are not zero, you try Example 3.

Example 3. Simplify $x^0 + 3^0$.

Solution

$$x^0 + 3^0 =$$
$$=$$

Check your work on page 154. ▶

Example 4. Simplify $(x + 3)^0$.

Solution. Parentheses indicate that $x + 3$ is the base and 0 is the exponent.

$$(x + 3)^0 = 1$$

Using the Laws of Exponents

In algebra it is often necessary to multiply several powers of the same base. We need a technique for finding a product such as $x^3 \cdot x^5$. Let's write each power as a product of factors. Notice that x appears as a factor eight times on the right.

$$\begin{aligned} x^3 \cdot x^5 &= \overbrace{x \cdot x \cdot x}^{x^3} \cdot \overbrace{x \cdot x \cdot x \cdot x \cdot x}^{x^5} \\ &= x^8 \end{aligned}$$

The new exponent 8 is just the *sum* of the exponents in the original product and we generalize this rule as the **first law of exponents**.

To *multiply* powers of the same base, we keep the same base and *add* the exponents.

Notice that we apply this law only when the bases are the same.

First Law of Exponents

$$a^m \cdot a^n = a^{m+n}$$

Example 5. Simplify $(xy^2)(xy^3)$.

Solution. We use the commutative and associative properties to rewrite this product with powers of like bases next to each other.

$$\begin{aligned} (xy^2)(xy^3) &= x \cdot x \cdot y^2 \cdot y^3 \\ &= x^1 \cdot x^1 \cdot y^2 \cdot y^3 \\ &= x^{1+1} \cdot y^{2+3} \\ &= x^2y^5 \end{aligned}$$

You complete Example 6.

Example 6. Simplify $3x^2(4xy)$.

Solution. First we rearrange the factors.

$$3x^2(4xy) = 3 \cdot 4 \cdot x^2 \cdot x \cdot y$$

Check your work on page 154. ▶

Example 7. Simplify

$$(-5a^2b)(2ab^3)(\tfrac{1}{2}a^4b)$$

Solution

$$(-5a^2b)(2ab^3)\left(\frac{1}{2}a^4b\right)$$

$$= -5 \cdot 2 \cdot \frac{1}{2} \cdot a^2 \cdot a \cdot a^4 \cdot b \cdot b^3 \cdot b$$

$$= -5a^7b^5$$

Now you try Example 8.

Example 8. Simplify $a^7 \cdot a^0$.

Solution

$$a^7 \cdot a^0$$

$$=$$

Check your work on page 154. ▶

Example 8 should help you see why the definition $a^0 = 1$ makes sense. Here we multiplied a^7 times a^0 and our answer was a^7. The only real number that leaves every number unchanged in multiplication is 1, the *multiplicative identity*.

Trial Run

______ 1. Evaluate $a^4 + 6$ when $a = -3$.

______ 2. Evaluate $y^5 - 3y^2 + y$ when $y = -1$.

Simplify.

______ 3. $5a^0$

______ 4. $(5a)^0$

______ 5. $2x^2(5xy)$

______ 6. $(4ab)(a^4b^5)$

______ 7. $(-2x^2y)(3xy^3)(\frac{1}{3}x^4y)$

______ 8. $7x^3y^2z^0(-2xy)(-3yz)$

Answers are on page 155.

Suppose that we use the first law of exponents to simplify the following expressions.

$(x^2)^3$	$(y^5)^2$
The base is x^2. The exponent is 3.	The base is y^5. The exponent is 2.
$(x^2)^3 = x^2 \cdot x^2 \cdot x^2$	$(y^5)^2 = y^5 \cdot y^5$
$= x^{2+2+2}$	$= y^{5+5}$
$= x^6$	$= y^{10}$

This procedure is called **raising a power to a power.** Notice how the exponent in each answer is related to the exponents in the original problem.

To *raise a power* of a base *to a power,* we keep the same base and *multiply* the exponents.

The **second law of exponents** describes this procedure.

Second Law of Exponents

$$(a^m)^n = a^{m \cdot n}$$

Example 9. Simplify $(x^7)^4$.

Solution

$$(x^7)^4 = x^{7 \cdot 4}$$
$$= x^{28}$$

You try Example 10.

Example 10. Simplify $(a^6)^5$.

Solution

$$(a^6)^5 =$$

Check your work on page 154. ▶

Suppose that we wish to find a **power of a product,** as in the following expressions.

$(2y)^4$	$(x^2y)^3$
The base is $2y$. The exponent is 4.	The base is x^2y. The exponent is 3.
$(2y)^4$	$(x^2y)^3$
$= (2y)(2y)(2y)(2y)$	$= (x^2y)(x^2y)(x^2y)$
$= 2 \cdot 2 \cdot 2 \cdot 2 \cdot y \cdot y \cdot y \cdot y$	$= x^2 \cdot x^2 \cdot x^2 \cdot y \cdot y \cdot y$
$= 2^4 \cdot y^4$	$= (x^2)^3y^3$
$= 16y^4$	$= x^6y^3$

Can you see a pattern in these examples?

To raise a *product* to a power, we raise *each factor* to that power.

This is the **third law of exponents** and it can be stated in general as

Third Law of Exponents

$$(a \cdot b)^n = a^n \cdot b^n$$

Example 11. Simplify $(-3x)^4$.

Solution

$$(-3x)^4 = (-3)^4x^4$$
$$= 81x^4$$

Now you try Example 12.

Example 12. Simplify $(4x^2y^3)^3$.

Solution

$$(4x^2y^3)^3 = 4^3(x^2)^3(y^3)^3$$
$$=$$

Check your work on page 154. ▶

Student errors in raising products to powers often occur in raising *constant* factors to powers too quickly. You may avoid mistakes if you first write the constant with the appropriate exponent as we have done in our examples.

Example 13. Simplify $(-a^4bc^2)^5$.

Solution

$$(-a^4bc^2)^5 = (-1 \cdot a^4bc^2)^5$$
$$= (-1)^5(a^4)^5(b^1)^5(c^2)^5$$
$$= -1 \cdot a^{20} \cdot b^5 \cdot c^{10}$$
$$= -a^{20}b^5c^{10}$$

Now complete Example 14.

Example 14. Simplify $(-a^4bc^2)^6$.

Solution

$$(-a^4bc^2)^6 = (-1 \cdot a^4bc^2)^6$$
$$= (-1)^6(a^4)^6(b^1)^6(c^2)^6$$
$$=$$

Check your work on page 154. ▶

We can use the laws of exponents to simplify more complicated algebraic expressions provided that we review the rules for the order of operations. Remember that we deal with symbols of inclusion (parentheses) first; then we deal with powers, followed by multiplications and/or divisions from left to right, and finally we perform addition and/or subtraction from left to right.

Example 15. Simplify $x^2(xy)^3$.

Solution

$$x^2(xy)^3$$
$$= x^2 \cdot x^3y^3$$
$$= x^5y^3$$

Example 16. Simplify $3a(a^2b)^2 + a^2b(a^3b)$.

Solution

$$3a(a^2b)^2 + a^2b(a^3b)$$
$$= 3a(a^2)^2b^2 + a^2b(a^3b)$$
$$= 3a \cdot a^4 \cdot b^2 + a^2 \cdot a^3 \cdot b \cdot b$$
$$= 3a^5b^2 + a^5b^2$$
$$= 4a^5b^2$$

Notice that we combined like terms in the final step.

Trial Run

Simplify each expression.

_______ 1. $(x^5)^2$

_______ 2. $(3a)^4$

_______ 3. $(-5x^3y^4)^2$

_______ 4. $(-2a^2b^3c)^3$

_______ 5. $y^2(xy)^5$

_______ 6. $(2xy)^2 + 3x(xy^2)$

Answers are on page 155.

Since we have learned to raise a product to a power by raising each factor to that power, it seems logical to consider a method for **raising a fraction to a power.**

$\left(\frac{x}{y}\right)^2$	$\left(\frac{-2}{a}\right)^3$
The base is $\frac{x}{y}$.	The base is $\frac{-2}{a}$.
The exponent is 2.	The exponent is 3.
$\left(\frac{x}{y}\right)^2 = \frac{x}{y} \cdot \frac{x}{y}$	$\left(\frac{-2}{a}\right)^3 = \frac{-2}{a} \cdot \frac{-2}{a} \cdot \frac{-2}{a}$
$= \frac{x \cdot x}{y \cdot y}$	$= \frac{(-2)(-2)(-2)}{a \cdot a \cdot a}$
$= \frac{x^2}{y^2}$	$= \frac{(-2)^3}{a^3}$
	$= \frac{-8}{a^3}$ or $-\frac{8}{a^3}$

It seems that we could have arrived at these answers more easily just by raising the numerator and denominator to the original power.

> To raise a *fraction* to a power, we raise the *numerator* to the power and raise the *denominator* to the power.

Called the **fourth law of exponents,** this rule can be stated in general as follows:

Fourth Law of Exponents

$$\left(\frac{a}{b}\right)^n = \frac{a^n}{b^n} \qquad (b \neq 0)$$

Notice that b cannot be zero because it appears as the denominator and division by zero is undefined.

Example 17. Simplify $\left(\frac{2x^2}{5y}\right)^3$.

Solution

$$\left(\frac{2x^2}{5y}\right)^3 = \frac{(2x^2)^3}{(5y)^3} = \frac{2^3(x^2)^3}{5^3y^3} = \frac{8x^6}{125y^3}$$

You complete Example 18.

Example 18. Simplify $\left(\frac{-xy^3}{2z}\right)^4$.

Solution

$$\left(\frac{-xy^3}{2z}\right)^4 = \frac{(-1xy^3)^4}{(2z)^4} = \frac{(-1)^4x^4(y^3)^4}{2^4z^4} =$$

Check your work on page 155. ▶

Trial Run

Simplify each each expression.

_____ 1. $\left(\frac{-1}{x}\right)^5$

_____ 2. $\left(\frac{3x^2}{2y}\right)^2$

_____ 3. $\left(\frac{-ab^2}{3c}\right)^4$

_____ 4. $\left(\frac{-3x^0y}{5z^2}\right)^3$

_____ 5. $\left(\frac{7a^3b^5}{9c^2}\right)^0$

_____ 6. $\left(\frac{x^2y}{-z^3}\right)^7$

Answers are on page 155.

Now we must learn to simplify quotients in which the numerator and denominator contain powers of the same base. Consider the following quotients.

$$\frac{x^5}{x^2} = \frac{x \cdot x \cdot x \cdot x \cdot x}{x \cdot x} = \frac{x}{x} \cdot \frac{x}{x} \cdot x \cdot x \cdot x = 1 \cdot 1 \cdot x \cdot x \cdot x = x^3$$

$$\frac{a^3y^4}{ay^3} = \frac{a^3}{a} \cdot \frac{y^4}{y^3} = \frac{a \cdot a \cdot a}{a} \cdot \frac{y \cdot y \cdot y \cdot y}{y \cdot y \cdot y} = \frac{a}{a} \cdot a \cdot a \cdot \frac{y}{y} \cdot \frac{y}{y} \cdot \frac{y}{y} \cdot y = 1 \cdot a \cdot a \cdot 1 \cdot 1 \cdot 1 \cdot y = a^2y$$

Notice how the exponent on each variable in the answers is related to the exponents in the original expressions.

To *divide* powers of the same base, we keep the same base and *subtract* the exponent in the denominator from the exponent in the numerator.

This rule for quotients is the **fifth law of exponents.**

Fifth Law of Exponents

$$\frac{a^m}{a^n} = a^{m-n} \qquad (a \neq 0)$$

When variables in a quotient appear with numerical coefficients, we divide the coefficients just as we have always done and then turn our attention to simplifying the variable parts by the fifth law of exponents.

Example 19. Simplify $\frac{-6x^2y}{3xy}$.

Solution

$$\begin{aligned}\frac{-6x^2y}{3xy} &= \frac{-6}{3} \cdot \frac{x^2}{x} \cdot \frac{y}{y} \\ &= -2 \cdot x^{2-1} \cdot y^{1-1} \\ &= -2 \cdot x^1 \cdot y^0 \\ &= -2 \cdot x \cdot 1 \\ &= -2x\end{aligned}$$

Now try Example 20.

Example 20. Simplify $\frac{3a^4b^2c}{2ab}$.

Solution

$$\frac{3a^4b^2c}{2ab} = \frac{3}{2} \cdot \frac{a^4}{a} \cdot \frac{b^2}{b} \cdot \frac{c}{1}$$

Check your work on page 155. ▶

Trial Run

Simplify each expression.

_______ 1. $\frac{x^7}{x^3}$

_______ 2. $\frac{-8a^5}{2a^3}$

_______ 3. $\frac{x^5y^3}{x^3y}$

_______ 4. $\frac{-9x^3y^5z^2}{9xy^5z^0}$

_______ 5. $\frac{5a^3b^7}{15ab^2}$

_______ 6. $\frac{7a^4b^2c}{5a^3b}$

Answers are on page 155.

The ability to work with exponents will be important throughout your study of algebra. We summarize the fundamental definitions and laws of exponents here for your reference. Assume that all variables represent real numbers and that n is a whole number.

Definitions

$$a^n = \underbrace{a \cdot a \cdot a \cdot \ldots \cdot a}_{n \text{ factors}}$$

$$a^0 = 1 \qquad (a \neq 0)$$

Laws of Exponents

(1) $a^m \cdot a^n = a^{m+n}$

(2) $(a^m)^n = a^{m \cdot n}$

(3) $(ab)^n = a^n b^n$

(4) $\left(\dfrac{a}{b}\right)^n = \dfrac{a^n}{b^n} \qquad b \neq 0$

(5) $\dfrac{a^m}{a^n} = a^{m-n} \qquad a \neq 0$

▶ Examples You Completed

Example 1. Evaluate $x^3 - 5x$ when $x = -2$.

Solution. We substitute -2 for x.

$$\begin{aligned} x^3 - 5x &= (-2)^3 - 5(-2) \\ &= -8 + 10 \\ &= 2 \end{aligned}$$

Example 3. Simplify $x^0 + 3^0$.

Solution

$$\begin{aligned} x^0 + 3^0 &= 1 + 1 \\ &= 2 \end{aligned}$$

Example 6. Simplify $3x^2(4xy)$.

Solution. First we rearrange the factors.

$$\begin{aligned} 3x^2(4xy) &= 3 \cdot 4 \cdot x^2 \cdot x \cdot y \\ &= 3 \cdot 4 \cdot x^2 \cdot x^1 \cdot y \\ &= 12x^{2+1} \cdot y \\ &= 12x^3y \end{aligned}$$

Example 8. Simplify $a^7 \cdot a^0$.

Solution

$$\begin{aligned} a^7 \cdot a^0 &= a^{7+0} \\ &= a^7 \end{aligned}$$

Example 10. Simplify $(a^6)^5$.

Solution

$$\begin{aligned} (a^6)^5 &= a^{6 \cdot 5} \\ &= a^{30} \end{aligned}$$

Example 12. Simplify $(4x^2y^3)^3$.

Solution

$$\begin{aligned} (4x^2y^3)^3 &= 4^3(x^2)^3(y^3)^3 \\ &= 4^3x^{2 \cdot 3}y^{3 \cdot 3} \\ &= 64x^6y^9 \end{aligned}$$

Example 14. Simplify $(-a^4bc^2)^6$.

Solution

$$\begin{aligned} (-a^4bc^2)^6 &= (-1 \cdot a^4bc^2)^6 \\ &= (-1)^6(a^4)^6(b^1)^6(c^2)^6 \\ &= 1 \cdot a^{24}b^6c^{12} \\ &= a^{24}b^6c^{12} \end{aligned}$$

Example 18. Simplify $\left(\frac{-xy^3}{2z}\right)^4$.

Solution

$$\begin{aligned}\left(\frac{-xy^3}{2z}\right)^4 &= \frac{(-1 \cdot xy^3)^4}{(2z)^4}\\ &= \frac{(-1)^4x^4(y^3)^4}{2^4z^4}\\ &= \frac{1 \cdot x^4y^{12}}{16z^4}\\ &= \frac{x^4y^{12}}{16z^4}\end{aligned}$$

Example 20. Simplify $\frac{3a^4b^2c}{2ab}$.

Solution

$$\begin{aligned}\frac{3a^4b^2c}{2ab} &= \frac{3}{2} \cdot \frac{a^4}{a} \cdot \frac{b^2}{b} \cdot c\\ &= \frac{3}{2} \cdot a^{4-1} \cdot b^{2-1} \cdot c\\ &= \frac{3}{2} \cdot a^3 \cdot b^1 \cdot c\\ &= \frac{3a^3bc}{2}\end{aligned}$$

Answers to Trial Runs

page 148 1. 87 2. -5 3. 5 4. 1 5. $10x^3y$ 6. $4a^5b^6$ 7. $-2x^7y^5$ 8. $42x^4y^4z$

page 150 1. x^{10} 2. $81a^4$ 3. $25x^6y^8$ 4. $-8a^6b^9c^3$ 5. x^5y^7 6. $7x^2y^2$

page 152 1. $-\frac{1}{x^5}$ 2. $\frac{9x^4}{4y^2}$ 3. $\frac{a^4b^8}{81c^4}$ 4. $-\frac{27y^3}{125z^6}$ 5. 1 6. $-\frac{x^{14}y^7}{z^{21}}$

page 153 1. x^4 2. $-4a^2$ 3. x^2y^2 4. $-x^2z^2$ 5. $\frac{a^2b^5}{3}$ 6. $\frac{7abc}{5}$

Name ______________________ Date ____________

EXERCISE SET 3.1

______ 1. Evaluate $x^5 - 2x^3 - x$ when $x = 3$.

______ 2. Evaluate $x^3 - 4x^2 - 3x$ when $x = 5$.

______ 3. Evaluate $-a^3 + 3a$ when $a = -2$.

______ 4. Evaluate $-a^5 + 5a$ when $a = -1$.

______ 5. Simplify $(x + 5)^0$.

______ 6. Simplify $(x - 7)^0$.

______ 7. Simplify $a^0 + 3^0$.

______ 8. Simplify $a^0 - 9^0$.

Simplify each expression using the laws of exponents.

______ 9. $(-5x^7)(4x)$

______ 10. $(-7x^5)(9x)$

______ 11. $(4x^2y^3)(-3x^5y^2)$

______ 12. $(8x^3y^4)(-2x^2y^5)$

______ 13. $3x(-x^2y)(-2x^4y^3)$

______ 14. $-5x(-x^3y)(4x^4y^3)$

______ 15. $(7xy^2)(-2xy)(-3x^2y)$

______ 16. $(-2xy^2)(8x^2y^3)(-3xy)$

______ 17. $(5a^2b^3c)(7abc^4)$

______ 18. $(9a^2b^3c^5)(3abc^2)$

______ 19. $(7x^0yz^5)(-6x^2y^3)(-\frac{1}{3}y^4z^2)$

______ 20. $(8x^2y^0z)(-5x^5y^3)(-\frac{1}{4}y^5z^2)$

______ 21. $(-8y^2)^2$

______ 22. $(-5y^3)^2$

______ 23. $(-2x^2y^3)^5$

______ 24. $(-3x^4y^3)^3$

______ 25. $5x(2x^3)^2$

______ 26. $6x(3x^4)^2$

______ 27. $9x(7x^5y^2)^0$

______ 28. $11x(5x^4y^3)^0$

______ 29. $(5x^4)^2(-2y^5)$

______ 30. $(2x^2)^3(-y)^4$

______ 31. $(-6a^2b)^2(2ab^3)^3$

______ 32. $(-2a^2b)^4(3a^4b^2)^3$

______ 33. $3a(-2a^4b)^2(a^5b^2)^3$

______ 34. $5a(-5a^7b)^2(a^2b^3)^4$

______ 35. $(7x^2yz^5)(-4x^5y^2z)^3$

______ 36. $(5xy^2z^3)(-3x^4yz^2)^3$

______ 37. $xy^2(3xy^4) + 2y^3(4x^2y^3)$

______ 38. $x^2y^3(5x^2y^4) - 3y^5(x^4y^2)$

______ 39. $-a^2(ab^3)^4 + 3a^2b^5(a^{10}b)$

______ 40. $-a^5(ab^2)^3 + 7a^4b^4(ab^2)^2$

______ 41. $\left(-\dfrac{1}{3y}\right)^2$

______ 42. $\left(-\dfrac{1}{4y}\right)^3$

______ 43. $\left(\dfrac{5a^2}{3b}\right)^3$

______ 44. $\left(\dfrac{7a^3}{2b^2}\right)^2$

______ 45. $\left(\dfrac{-ab^5}{2c^2}\right)^4$

______ 46. $\left(\dfrac{-a^2b}{3c^3}\right)^4$

_______ 47. $\left(\frac{-2x^0y^3}{3z^5}\right)^4$

_______ 48. $\left(\frac{-5x^0y^5}{7z^6}\right)^2$

_______ 49. $\left(\frac{11x^3y^5}{2z}\right)^0$

_______ 50. $\left(\frac{9x^5y^3}{5z^2}\right)^0$

_______ 51. $\frac{x^9}{x^5}$

_______ 52. $\frac{x^{11}}{x^7}$

_______ 53. $\frac{-15a^7}{5a^3}$

_______ 54. $\frac{-16a^8}{8a^5}$

_______ 55. $\frac{x^9y^4}{x^3y^2}$

_______ 56. $\frac{x^{11}y^5}{x^2y^2}$

_______ 57. $\frac{-3x^4y^5z^3}{3xy^2z^2}$

_______ 58. $\frac{-7x^5y^7z^9}{7x^4y^2z^3}$

_______ 59. $\frac{3a^5b^2}{21ab^2}$

_______ 60. $\frac{4a^6b^3}{28a^5b^3}$

_______ 61. $\frac{9a^6b^5c}{6a^0b^3}$

_______ 62. $\frac{12a^7b^4c^2}{8a^3b^0}$

_______ 63. $\left(\frac{2a^2b^3c}{ab^2}\right)^3$

_______ 64. $\left(\frac{5a^5b^4c}{a^3b^3}\right)^2$

_______ 65. $\left(\frac{7ab^3}{b^2}\right)^2\left(\frac{a^2b}{a}\right)^3$

_______ 66. $\left(\frac{9a^4b^2}{ab}\right)^2\left(\frac{a^4b^5}{a^3b^2}\right)^3$

☆ Stretching the Topics

_______ 1. Simplify $(-4ab^2c)^3(-ab^3c)^2(-2a^2b^5c)^0$ and evaluate when $a = -1$, $b = 2$, and $c = 1$.

_______ 2. Simplify $(x^{n-4}y)^3(x^6y^{n+4})^2$.

_______ 3. Find the value of a that will make this equation true.

$$(y^{3a+5})(y^a)^4 = y^{8a+12}$$

Check your answers in the back of your book.

If you can complete **Checkup 3.1,** you are ready to go on to Section 3.2.

Name ______________________________ Date ____________

✓ CHECKUP 3.1

______ 1. Evaluate $y^3 - 2y + 1$ when $y = -4$.

______ 2. Simplify $12a^0 + 1$.

Simplify each expression using the laws of exponents.

______ 3. $3x(-7x^2y)(-2x^5y^2)$

______ 4. $(9a^3b^4c^2)(-5a^2b^2c)$

______ 5. $(-11y^3)^2$

______ 6. $(5x^7)^2(-3x^4)$

______ 7. $\left(\dfrac{-9a^2}{3b^3}\right)^3$

______ 8. $\left(\dfrac{4a^3b^0}{3c^2}\right)^2$

______ 9. $\dfrac{27a^7b^3}{8a^2b^3}$

______ 10. $\dfrac{15a^7b^5c}{-5a^3b}$

Check your answers in the back of your book.

If You Missed Problems:	You Should Review Examples:
1	1
2	5
3, 4	5–8
5	11
6	15
7, 8	17, 18
9, 10	19, 20

3.2 Working with Negative Exponents

Defining Negative Exponents

Let's use two approaches to simplify the expression $\frac{x^2}{x^6}$.

First Approach

$$\frac{x^2}{x^6} = \frac{x \cdot x}{x \cdot x \cdot x \cdot x \cdot x \cdot x}$$

$$= \frac{x}{x} \cdot \frac{x}{x} \cdot \frac{1}{x \cdot x \cdot x \cdot x}$$

$$= 1 \cdot 1 \cdot \frac{1}{x^4}$$

$$= \frac{1}{x^4}$$

Second Approach

Using the fifth law of exponents, we find

$$\frac{x^2}{x^6} = x^{2-6}$$

$$= x^{-4}$$

Since we know the first approach is correct and since we want the fifth law of exponents to work for every quotient, we must agree that our two answers mean the same thing.

Negative Exponent Form	Positive Exponent Form
x^{-4}	$\frac{1}{x^4}$
a^{-2}	$\frac{1}{a^2}$
2^{-1}	$\frac{1}{2}$

These examples lead us to the following definition for a negative exponent.

Definition of Negative Exponent

$$a^{-n} = \frac{1}{a^n} \qquad (a \neq 0)$$

Use this definition to complete Example 1.

Example 1. Evaluate 2^{-3}.

Solution

$$2^{-3} = \frac{1}{2^3}$$

$$=$$

Check your work on page 166. ▶

Example 2. Evaluate $(-2)^{-5}$.

Solution

$$(-2)^{-5} = \frac{1}{(-2)^5}$$

$$= \frac{1}{-32}$$

$$= -\frac{1}{32}$$

Now try Example 3.

Example 3. Evaluate $(-2)^{-4}$.

Solution

$$(-2)^{-4} = \frac{1}{(-2)^4}$$

$$=$$

Check your work on page 166. ▶

Example 4. Evaluate -2^{-4}.

Solution. Remember that an exponent applies only to the base to which it is attached, unless parentheses indicate otherwise.

$$-2^{-4} = -1 \cdot 2^{-4}$$

$$= -1 \cdot \frac{1}{2^4}$$

$$= -1 \cdot \frac{1}{16}$$

$$= -\frac{1}{16}$$

Notice from these examples that *a negative exponent does not affect the sign of the answer*. The negative exponent merely says, "Rewrite the power as 1 divided by the same base raised to the opposite power." The *sign* of your answer will depend on the sign of the *base* and whether the final exponent is *odd* or *even*.

Example 5. Rewrite $x^{-3}y^{-2}$ without negative exponents.

Solution

$$x^{-3}y^{-2}$$

$$= \frac{1}{x^3} \cdot \frac{1}{y^2}$$

$$= \frac{1}{x^3y^2}$$

Example 6. Simplify $\dfrac{x^2y}{x^3y^5}$ and write the answer with positive exponents.

Solution

$$\frac{x^2y}{x^3y^5} = \frac{x^2}{x^3} \cdot \frac{y^1}{y^5}$$

$$= x^{2-3} \cdot y^{1-5}$$

$$= x^{-1}y^{-4}$$

$$= \frac{1}{x} \cdot \frac{1}{y^4}$$

$$= \frac{1}{xy^4}$$

Trial Run

Evaluate each expression.

_______ 1. 3^{-2} _______ 2. 15^{-1}

_______ 3. $(-5)^{-3}$ _______ 4. $(-4)^{-2}$

_______ 5. -2^{-6} _______ 6. $-(-3)^{-3}$

Simplify each expression and write answers with positive exponents.

_______ 7. $a^{-2}b^{-3}$ _______ 8. $x^{-4}yz^{-1}$

_______ 9. $\frac{y^4}{y^{10}}$ _______ 10. $\frac{x^3y^2}{x^4y^5}$

Answers are on page 167.

Using the Laws of Exponents

In simplifying expressions containing negative exponents, you will be relieved to know that

> Negative exponents obey all five laws of exponents.

When you simplify expressions with negative exponents, it is wise to *use the laws of exponents first*. If you are asked to express your answer using positive exponents only, *wait* until you have an answer and then make any necessary changes using the definition of a negative exponent.

Let's practice simplifying some expressions, writing answers using positive exponents.

You try Example 7.

Example 7. Simplify $x^3 \cdot x^{-7} \cdot x^{-1}$.

Solution. Since the bases are the same, we find the product by _______ the exponents.

$$x^3 \cdot x^{-7} \cdot x^{-1}$$
$$= x^{3+(-7)+(-1)}$$

Check your work on page 166. ▶

Example 8. Simplify $(x^4y^{-2})(x^{-5}y^6)$.

Solution

$$(x^4y^{-2})(x^{-5}y^6) = x^4 \cdot x^{-5} \cdot y^{-2} \cdot y^6$$
$$= x^{4+(-5)} \cdot y^{-2+6}$$
$$= x^{-1} \cdot y^4$$
$$= \frac{1}{x} \cdot y^4$$
$$= \frac{y^4}{x}$$

Now try Example 9.

Example 9. Simplify $(x^{-2})^4$.

Solution. To raise a power to a power, we ________ exponents.

$$(x^{-2})^4$$
$$= x^{-2\cdot 4}$$

Check your work on page 166. ▶

Example 10. Simplify $(5x^{-3}y^{-1})^{-2}$.

Solution. To raise a product to a power, we raise each factor to that power.

$$\begin{aligned}(5x^{-3}y^{-1})^{-2} &= 5^{-2}(x^{-3})^{-2}(y^{-1})^{-2}\\ &= 5^{-2}x^6y^2\\ &= \frac{1}{5^2}x^6y^2\\ &= \frac{x^6y^2}{25}\end{aligned}$$

Example 11. Simplify $\left(\frac{x^2y^{-4}}{z^{-3}}\right)^{-1}$.

Solution. To raise a quotient to a power, we raise the numerator and denominator to that power.

$$\begin{aligned}\left(\frac{x^2y^{-4}}{z^{-3}}\right)^{-1} &= \frac{(x^2)^{-1}\cdot(y^{-4})^{-1}}{(z^{-3})^{-1}}\\ &= \frac{x^{-2}y^4}{z^3}\\ &= \frac{1}{x^2}\cdot\frac{y^4}{z^3}\\ &= \frac{y^4}{x^2z^3}\end{aligned}$$

Now you try Example 12.

Example 12. Simplify $\frac{x^{-8}y^4}{x^{-3}y^{-2}}$.

Solution. To divide powers of the same base, we ________ exponents.

$$\begin{aligned}\frac{x^{-8}y^4}{x^{-3}y^{-2}} &= \frac{x^{-8}}{x^{-3}}\cdot\frac{y^4}{y^{-2}}\\ &= x^{-8-(-3)}\cdot y^{4-(-2)}\end{aligned}$$

Check your work on page 166. ▶

Notice that in each example we first simplified the expression using the laws of exponents. We waited until we had our answer before we rewrote that answer using positive exponents.

Trial Run

Simplify and write answers with positive exponents.

________ 1. $x^4 \cdot x^{-5} \cdot x^{-2}$

________ 2. $(-3x^5y^{-3})(2x^{-8}y^5)$

________ 3. $(x^{-3})^2$

________ 4. $(x^{-2}y^{-1})^{-4}$

________ 5. $\left(\frac{4x^3y^{-5}}{z^{-2}}\right)^{-1}$

________ 6. $\frac{x^{-7}y^5}{x^{-2}y^{-3}}$

Answers are on page 167.

Until now we have not encountered an answer with a negative exponent in the denominator of a fraction. How can we express such answers using positive exponents? Let's see what we can do with the fraction $\frac{1}{x^{-3}}$.

$\frac{1}{x^{-3}} = 1 \div x^{-3}$	Meaning of fraction bar.
$= 1 \div \frac{1}{x^3}$	Use definition of negative exponent.
$= \frac{1}{1} \cdot \frac{x^3}{1}$	Use rule for dividing fractions.
$= \frac{x^3}{1}$	Use rule for multiplying fractions.
$\frac{1}{x^{-3}} = x^3$	Simplify.

Similarly we could show

Negative Exponent Form	Positive Exponent Form
$\frac{1}{a^{-2}}$	a^2
$\frac{1}{2^{-4}}$	2^4
$\frac{1}{y^{-1}}$	y

These examples should help you see that, in general

$$\frac{1}{a^{-n}} = a^n \qquad (a \neq 0)$$

This fact allows us to change answers with negative exponents appearing on factors in the denominator to answers in which all factors have positive exponents. Try rewriting the fraction in Example 13.

Example 13. Rewrite $\frac{x^5}{y^{-6}}$ with positive exponents.

Solution

$$\frac{x^5}{y^{-6}} = \frac{x^5}{1} \cdot \frac{1}{y^{-6}}$$

$$=$$

Check your work on page 166. ▶

Example 14. Rewrite $\frac{y^4}{x^{-2}z^{-1}}$ with positive exponents.

Solution

$$\frac{y^4}{x^{-2}z^{-1}} = \frac{y^4}{1} \cdot \frac{1}{x^{-2}} \cdot \frac{1}{z^{-1}}$$

$$= y^4x^2z^1$$

$$= x^2y^4z$$

From our work with negative exponents on factors in the numerators and denominators of fractions, we make the following observations:

Observation	Examples
A factor in the numerator with a negative exponent should be moved to the denominator, and the exponent will be positive.	$\dfrac{3x^{-2}y^{-1}}{5} = \dfrac{3}{5x^2y}$
A factor in the denominator with a negative exponent should be moved to the numerator, and the exponent will be positive.	$\dfrac{10}{x^{-1}} = 10x$ $\dfrac{x^{-2}}{3^{-1}y^{-2}} = \dfrac{3y^2}{x^2}$
A factor with a positive exponent stays where it is.	$\dfrac{2^{-4}a^{-8}b}{c^5} = \dfrac{b}{2^4a^8c^5}$ $= \dfrac{b}{16a^8c^5}$

These observations make it much easier to rewrite a simplified fraction using positive exponents. Just be sure that you first use the laws of exponents to simplify the expression.

You try to rewrite the fraction in Example 15.

Example 15. Rewrite $\dfrac{x^{-6}y^3}{a^{-1}z^{-2}}$ using positive exponents.

Solution. Since each base appears just once, we use our observations to rewrite the fraction.

$$\frac{x^{-6}y^3}{a^{-1}z^{-2}} =$$

Check your work on page 167. ▶
Then do Example 16.

Example 16. Simplify $\left(\dfrac{a^7b^3}{c}\right)^{-3}$ and write the answer with positive exponents.

Solution. We first simplify the expression.

$$\left(\frac{a^7b^3}{c}\right)^{-3} = \frac{(a^7)^{-3}(b^3)^{-3}}{c^{-3}}$$

$$=$$

Check your work on page 167. ▶

Example 17. Simplify $\left(\dfrac{-3x^{-2}}{y^{-3}z}\right)^{-2}$.

Solution

$$\left(\frac{-3x^{-2}}{y^{-3}z}\right)^{-2} = \frac{(-3)^{-2}(x^{-2})^{-2}}{(y^{-3})^{-2}z^{-2}}$$
$$= \frac{(-3)^{-2}x^{4}}{y^{6}z^{-2}}$$
$$= \frac{x^{4}z^{2}}{(-3)^{2}y^{6}}$$
$$= \frac{x^{4}z^{2}}{9y^{6}}$$

Example 18. Simplify $\left(\dfrac{-2x^{6}y^{8}}{x^{-2}y^{10}}\right)^{-3}$.

Solution. Since there are *like bases* within the parentheses, we simplify that quantity *first*.

$$\left(\frac{-2x^{6}y^{8}}{x^{-2}y^{10}}\right)^{-3} = [-2x^{6-(-2)}y^{8-10}]^{-3}$$
$$= [-2x^{8}y^{-2}]^{-3}$$
$$= (-2)^{-3}(x^{8})^{-3}(y^{-2})^{-3}$$
$$= (-2)^{-3}x^{-24}y^{6}$$
$$= \frac{y^{6}}{(-2)^{3}x^{24}}$$
$$= \frac{y^{6}}{-8x^{24}}$$
$$= -\frac{y^{6}}{8x^{24}}$$

In Examples 17 and 18, notice that the powers of the *numerical factors* were not computed until they had been rewritten with *positive* exponents. Trying to compute those powers before rewriting them almost always leads to errors.

$$(-3)^{-2} \neq 9 \qquad (-2)^{-3} \neq 8$$
$$(-3)^{-2} \neq -9 \qquad (-2)^{-3} \neq -8$$
$$(-3)^{-2} = \frac{1}{(-3)^{2}} = \frac{1}{9} \qquad (-2)^{-3} = \frac{1}{(-2)^{3}} = \frac{1}{-8} = -\frac{1}{8}$$

Example 19. Simplify $\dfrac{(x^{-2}y)^{-3}}{(2x^{-1}y^{-4})^{-1}}$.

Solution

$$\frac{(x^{-2}y)^{-3}}{(2x^{-1}y^{-4})^{-1}} = \frac{x^{6}y^{-3}}{2^{-1}x^{1}y^{4}}$$
$$= \frac{1}{2^{-1}} \cdot x^{6-1} \cdot y^{-3-4}$$
$$= \frac{x^{5}y^{-7}}{2^{-1}}$$
$$= \frac{2x^{5}}{y^{7}}$$

You try Example 20.

Example 20. Simplify $\dfrac{(-a^{-3}b^{-1})^{-2}}{(a^{-4}b^{2})^{-3}}$.

Solution

$$\frac{(-a^{-3}b^{-1})^{-2}}{(a^{-4}b^{2})^{-3}} = \frac{(-1 \cdot a^{-3}b^{-1})^{-2}}{(a^{-4}b^{2})^{-3}}$$
$$= \frac{(-1)^{-2}a^{6}b^{2}}{a^{12}b^{-6}}$$
$$=$$

Check your work on page 167. ▶

Trial Run

Simplify and write answers with positive exponents.

_______ 1. $\dfrac{x^7}{y^{-9}}$

_______ 2. $\dfrac{x^{-3}z^{-2}}{y^4}$

_______ 3. $\dfrac{a^{-2}b^{-3}}{2^{-1}c^{-1}}$

_______ 4. $\left(\dfrac{a^2b^5}{c}\right)^{-2}$

_______ 5. $\left(\dfrac{-3x^4y^{-3}}{x^{-2}y^{-5}}\right)^{-2}$

_______ 6. $\dfrac{(2x^2z^{-1})^{-5}}{(-x^{-1}z^4)^{-3}}$

Answers are on page 167.

Examples You Completed

Example 1. Evaluate 2^{-3}.

Solution

$$2^{-3} = \frac{1}{2^3} = \frac{1}{8}$$

Example 3. Evaluate $(-2)^{-4}$.

Solution

$$(-2)^{-4} = \frac{1}{(-2)^4} = \frac{1}{16}$$

Example 7. Simplify $x^3 \cdot x^{-7} \cdot x^{-1}$.

Solution. Since the bases are the same, we find the product by *adding* the exponents.

$$x^3 \cdot x^{-7} \cdot x^{-1} = x^{3+(-7)+(-1)} = x^{-5} = \frac{1}{x^5}$$

Example 9. Simplify $(x^{-2})^4$.

Solution. To raise a power to a power, we *multiply* exponents.

$$(x^{-2})^4 = x^{-2\cdot 4} = x^{-8} = \frac{1}{x^8}$$

Example 12. Simplify $\dfrac{x^{-8}y^4}{x^{-3}y^{-2}}$.

Solution. To divide powers of the same base, we *subtract* exponents.

$$\frac{x^{-8}y^4}{x^{-3}y^{-2}} = \frac{x^{-8}}{x^{-3}} \cdot \frac{y^4}{y^{-2}} = x^{-8-(-3)} \cdot y^{4-(-2)} = x^{-5} \cdot y^6 = \frac{1}{x^5} \cdot y^6 = \frac{y^6}{x^5}$$

Example 13. Rewrite $\dfrac{x^5}{y^{-6}}$ with positive exponents.

Solution

$$\frac{x^5}{y^{-6}} = \frac{x^5}{1} \cdot \frac{1}{y^{-6}} = x^5y^6$$

Example 15. Rewrite $\dfrac{x^{-6}y^3}{a^{-1}z^{-2}}$ with positive exponents.

Solution

$$\frac{x^{-6}y^3}{a^{-1}z^{-2}} = \frac{ay^3z^2}{x^6}$$

Example 16. Simplify $\left(\frac{a^7b^3}{c}\right)^{-3}$ and write the answer with positive exponents.

Solution. We first simplify the expression.

$$\left(\frac{a^7b^3}{c}\right)^{-3} = \frac{(a^7)^{-3}(b^3)^{-3}}{c^{-3}}$$
$$= \frac{a^{-21}b^{-9}}{c^{-3}}$$
$$= \frac{c^3}{a^{21}b^9}$$

Example 20. Simplify $\frac{(-a^{-3}b^{-1})^{-2}}{(a^{-4}b^2)^{-3}}$.

Solution

$$\frac{(-a^{-3}b^{-1})^{-2}}{(a^{-4}b^2)^{-3}} = \frac{(-1\cdot a^{-3}b^{-1})^{-2}}{(a^{-4}b^2)^{-3}}$$
$$= \frac{(-1)^{-2}a^6b^2}{a^{12}b^{-6}}$$
$$= (-1)^{-2}a^{6-12}b^{2-(-6)}$$
$$= (-1)^{-2}a^{-6}b^8$$
$$= \frac{b^8}{(-1)^2a^6}$$
$$= \frac{b^8}{a^6}$$

Answers to Trial Runs

page 161 1. $\frac{1}{9}$ 2. $\frac{1}{15}$ 3. $-\frac{1}{125}$ 4. $\frac{1}{16}$ 5. $-\frac{1}{64}$ 6. $\frac{1}{27}$ 7. $\frac{1}{a^2b^3}$ 8. $\frac{y}{x^4z}$ 9. $\frac{1}{y^6}$ 10. $\frac{1}{xy^3}$

page 162 1. $\frac{1}{x^3}$ 2. $-\frac{6y^2}{x^3}$ 3. $\frac{1}{x^6}$ 4. x^8y^4 5. $\frac{y^5}{4x^3z^2}$ 6. $\frac{y^8}{x^5}$

page 166 1. x^7y^9 2. $\frac{1}{x^3y^4z^2}$ 3. $\frac{2c}{a^2b^3}$ 4. $\frac{c^2}{a^4b^{10}}$ 5. $\frac{1}{9x^{12}y^4}$ 6. $-\frac{z^{17}}{32x^{13}}$

Name ________________________________ Date ______________

EXERCISE SET 3.2

Evaluate each expression.

_______ 1. 4^{-2}

_______ 2. 7^{-2}

_______ 3. $(-3)^{-5}$

_______ 4. $(-4)^{-3}$

_______ 5. $(-9)^{-2}$

_______ 6. $(-3)^{-4}$

_______ 7. -5^{-2}

_______ 8. -2^{-4}

_______ 9. $-(-6)^{-1} + 6^{-2}$

_______ 10. $-(-2)^{-3} + 2^{-4}$

_______ 11. $\left(\frac{3}{4}\right)^{-2}$

_______ 12. $\left(\frac{2}{3}\right)^{-4}$

Simplify and write each answer with positive exponents.

_______ 13. $x^3 \cdot x^{-5} \cdot x^{-4}$

_______ 14. $x^{-3} \cdot x^4 \cdot x^{-8}$

_______ 15. $(a^5b^{-8})(a^{-3}b^2)$

_______ 16. $(a^{-7}b^3)(a^{10}b^{-5})$

_______ 17. $(x^3y^{-5}z^{-9})(x^{-4}y^5z^3)$

_______ 18. $(x^5y^{-5}z^{-7})(x^{-6}y^2z^7)$

_______ 19. $(a^{-4}b^3c^{-2})(a^5b^{-7}c^5)$

_______ 20. $(a^{-9}b^2c^{-4})(a^{10}b^{-8}c^7)$

_______ 21. $(x^{-3})^2$

_______ 22. $(x^{-4})^2$

_______ 23. $(2x)^{-5}$

_______ 24. $(3x)^{-3}$

_______ 25. $9a^{-2}$

_______ 26. $7a^{-3}$

_______ 27. $(x^{-5}y^{-1})^{-3}$

_______ 28. $(x^{-7}y^{-2})^{-2}$

_______ 29. $(-3x^{-1}y^2)^{-3}$

_______ 30. $(-2x^{-2}y^3)^{-5}$

_______ 31. $(7x^{-5}y^2)^2$

_______ 32. $(8x^{-7}y^3)^2$

_______ 33. $\frac{a^9}{b^{-7}}$

_______ 34. $\frac{a^{10}}{b^{-8}}$

_______ 35. $\frac{y^9}{x^{-1}z^{-3}}$

_______ 36. $\frac{y^7}{x^{-4}z^{-1}}$

_______ 37. $\left(\frac{5}{a^4}\right)^{-3}$

_______ 38. $\left(\frac{6}{a^6}\right)^{-2}$

_______ 39. $\left(\frac{x^6y^{-3}}{z^{-3}}\right)^{-1}$

_______ 40. $\left(\frac{x^7y^{-4}}{z^{-5}}\right)^{-1}$

_______ 41. $\frac{x^{-9}y^3}{x^{-4}y^{-3}}$

_______ 42. $\frac{x^{-11}y^5}{x^{-5}y^{-5}}$

_______ 43. $\left(\frac{a^3b^4}{c^2}\right)^{-3}$

_______ 44. $\left(\frac{a^3b^5}{c^4}\right)^{-2}$

_______ 45. $\left(\frac{x^9y^5}{x^{-1}y^{-7}}\right)^{-1}$

_______ 46. $\left(\frac{x^{10}y^7}{x^{-3}y^{-4}}\right)^{-1}$

_______ 47. $\left(\frac{-3x^5y^{-2}}{x^{-3}y^{-7}}\right)^{-2}$

_______ 48. $\left(\frac{-5x^7y^{-4}}{x^{-4}y^{-9}}\right)^{-2}$

_______ 49. $\left(\frac{-4a^{-5}b^{-4}}{ac^{-3}}\right)^{3}$

_______ 50. $\left(\frac{-2a^{-8}b^{-5}}{a^2c^{-4}}\right)^{5}$

_______ 51. $\left(\frac{2a^{-2}b}{b^{-3}}\right)^{2}\left(\frac{3a^4b^{-1}}{b^3}\right)^{-3}$

_______ 52. $\left(\frac{5a^{-1}b^{-2}}{a^{-4}}\right)^{3}\left(\frac{9a^{-3}b}{b^{-2}}\right)^{-2}$

_______ 53. $\frac{(a^{-1}b^2)^{-2}}{(2a^{-2}b^{-3})^{-1}}$

_______ 54. $\frac{(a^2b^{-2})^{-3}}{(3a^{-4}b^3)^{-2}}$

_______ 55. $\frac{(-x^{-3}y^{-2})^{-2}}{(x^{-2}y^2)^{-3}}$

_______ 56. $\frac{(-2x^{-1}y^{-3})^{-2}}{(x^{-4}y^3)^{-1}}$

☆ Stretching the Topics

_______ 1. Write x^{2a-3b} with positive exponents if $a < 0$ and $b > 0$.

_______ 2. Simplify $\left(\frac{-ab^{-2}c^3}{2d^{-3}ef^2}\right)^{-2}\left(\frac{2a^{-1}b^3c^{-1}}{-d^2e^{-2}f^2}\right)^{3}$ and write the answer with positive exponents.

_______ 3. Simplify $\frac{x^{6a+4}y^{3b-2}}{x^{2a-2}y^{b+3}}$ and write the answer without a denominator.

Check your answers in the back of your book.

If you can complete **Checkup 3.2,** you are ready to go on to Section 3.3.

Name ______________________ Date ____________

✓ CHECKUP 3.2

Evaluate each expression.

______ 1. 8^{-2}

______ 2. $(-4)^{-3}$

Simplify and write each answer with positive exponents.

______ 3. $(x^5y^{-7}z^{-3})(x^{-4}y^7z^8)$

______ 4. $(x^{-3})^3$

______ 5. $(6x)^{-2}$

______ 6. $(x^{-3}y^7z^3)^{-2}$

______ 7. $\dfrac{x^{-13}y^7}{x^{-8}y^{-7}}$

______ 8. $\left(\dfrac{x^5y^{-4}}{z^{-3}}\right)^{-1}$

______ 9. $\left(\dfrac{3x^5y^{-2}}{x^{-4}y^3}\right)^{-2}$

______ 10. $\dfrac{(3x^{-8}y)^{-1}}{(x^{-6}y^{-2})^{-2}}$

Check your answers in the back of your book.

If You Missed Problems:	You Should Review Examples:
1	1
2	2
3	8
4	9
5, 6	10
7	12
8, 9	18
10	19, 20

3.3 Operating with Polynomials

Now that we have gained skill working with all integers used as exponents, we turn our attention to algebraic expressions that involve *sums* of constant terms and/or variable terms whose variable parts appear with *whole-number exponents*. Such expressions are called **polynomials.**

Definition of a Polynomial. A polynomial in x is a sum of one or more terms of the form

$$ax^n$$

where a is a constant and n is a whole number.

Identifying Polynomials

A polynomial is named according to the *number of terms* it contains.

A **monomial** is a polynomial containing *one* term.
A **binomial** is a polynomial containing *two* terms.
A **trinomial** is a polynomial containing *three* terms.

Polynomials containing more than three terms do not have special names; we label them by stating the number of terms they contain. We also describe a polynomial by naming the variable it contains and by identifying the highest power of that variable that occurs in the polynomial. The highest power in a polynomial of one variable determines the **degree** of the polynomial.

Example	Number of Terms	Degree	Description
x	One	1	First-degree monomial in x
$-3x^2$	One	2	Second-degree monomial in x
$\frac{1}{3}y^5$	One	5	Fifth-degree monomial in y
3	One	0	Constant monomial
$x + 3$	Two	1	First-degree binomial in x
$-3x^2 + 2x$	Two	2	Second-degree binomial in x
$2z - 5z^4$	Two	4	Fourth-degree binomial in z
$x^2 + 5x + 7$	Three	2	Second-degree trinomial in x
$y^3 - 4y^2 + y$	Three	3	Third-degree trinomial in y
$x^3 + 5x^2 - 7x + 2$	Four	3	Third-degree four-term polynomial in x

If a polynomial contains two or more variables, it becomes awkward (and not very useful) to identify its degree. For this reason, we usually label such a polynomial according to its variables and number of terms only.

a^2bc is a monomial in a, b, and c

$xy^2 + 3xy$ is a binomial in x and y

$x^2 - y^2 + z^2$ is a trinomial in x, y, and z

We have dealt with many algebraic expressions that are *not* polynomials because they cannot be written as *sums* of terms whose variable parts appear with *whole-number* exponents. Expressions containing negative exponents or variable denominators are *not* polynomials.

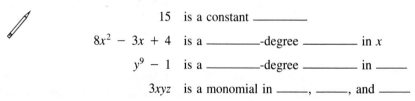

$x^{-3} + 2$ is *not* a polynomial because the exponent is negative.

$\frac{x}{y^2z}$ is *not* a polynomial because variables appear in the denominator.

$\frac{2}{x + 1}$ is *not* a polynomial because a variable appears in the denominator.

Let's practice naming some polynomials. You try Example 1.

Example 1. Name each polynomial.

15 is a constant ________

$8x^2 - 3x + 4$ is a ________-degree ________ in x

$y^9 - 1$ is a ________-degree ________ in ____

$3xyz$ is a monomial in ____, ____, and ____

Check your work on page 176. ▶

As you may have noticed, we usually write polynomial terms so that the exponents on the variable decrease from left to right. This is called writing a polynomial in **descending powers of the variable.**

Example 2. Rewrite $2x + 5x^2 + 1 - x^3$ in descending powers of the variable.

Solution

$$2x + 5x^2 + 1 - x^3$$
$$= -x^3 + 5x^2 + 2x + 1$$

Now try Example 3.

Example 3. Rewrite $y - 3 + y^5$ in descending powers of the variable.

Solution

$$y - 3 + y^5$$
$$=$$

Check your work on page 176. ▶

Adding and Subtracting Polynomials

We learned to add and subtract polynomials in Chapter 1 when we discovered how to **combine like terms.** Recall that like terms are terms that contain exactly the same variable part (same variable raised to same power). We combine like terms by combining the numerical coefficients and keeping the same variable. In adding polynomials, you may use the commutative and associative properties to rearrange the terms if this is helpful.

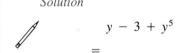

Example 4. Find

$$(6x^2 + 3x - 1) + (-4x^2 - 5x + 3)$$

Solution

$$\begin{aligned} &(6x^2 + 3x - 1) + (-4x^2 - 5x + 3) \\ &= 6x^2 + 3x - 1 - 4x^2 - 5x + 3 \\ &= 6x^2 - 4x^2 + 3x - 5x - 1 + 3 \\ &= 2x^2 - 2x + 2 \end{aligned}$$

Now you try Example 5.

Example 5. Find

$$(a^3 + 2a) + (5a^2 - 2a + 1)$$

Solution

$$\begin{aligned} &(a^3 + 2a) + (5a^2 - 2a + 1) \\ &= a^3 + 2a + 5a^2 - 2a + 1 \end{aligned}$$

Check your work on page 176. ▶

Subtraction is performed in a similar manner. From Chapter 1 recall that the subtraction sign between polynomials means that we must find the *opposite* of each term in the second polynomial, and *add*.

Example 6. Find

$$(3x^2 - 4x + 7) - (2x^2 - x + 3)$$

Solution

$$\begin{aligned} &(3x^2 - 4x + 7) - (2x^2 - x + 3) \\ &= 3x^2 - 4x + 7 - 2x^2 + x - 3 \\ &= 3x^2 - 2x^2 - 4x + x + 7 - 3 \\ &= x^2 - 3x + 4 \end{aligned}$$

You complete Example 7.

Example 7. Find

$$(x^3 - 2x^2 + 1) - (2x^2 + 3x + 1)$$

Solution

$$\begin{aligned} &(x^3 - 2x^2 + 1) - (2x^2 + 3x + 1) \\ &= x^3 - 2x^2 + 1 - 2x^2 - 3x - 1 \end{aligned}$$

Check your work on page 176. ▶

In Chapter 1 we also learned how to multiply a constant times a polynomial using the **distributive property.**

Example 8. Find

$$3(a^2 - 5a) - 5(2a^2 + 4a - 3)$$

Solution

$$\begin{aligned} &3(a^2 - 5a) - 5(2a^2 + 4a - 3) \\ &= 3a^2 - 15a - 10a^2 - 20a + 15 \\ &= 3a^2 - 10a^2 - 15a - 20a + 15 \\ &= -7a^2 - 35a + 15 \end{aligned}$$

You try Example 9.

Example 9. Find

$$-2(5 - x^2) + 4(x^2 - 2x + 3)$$

Solution

$$\begin{aligned} &-2(5 - x^2) + 4(x^2 - 2x + 3) \\ &= \end{aligned}$$

Check your work on page 176. ▶

Example 10. Find $x^2 - 3[x - 2(x^2 - x + 6)]$.

Solution

$x^2 - 3[x - 2(x^2 - x + 6)]$	
$= x^2 - 3[x - 2x^2 + 2x - 12]$	Remove parentheses (distributive property).
$= x^2 - 3[-2x^2 + 3x - 12]$	Combine like terms within brackets.
$= x^2 + 6x^2 - 9x + 36$	Remove brackets (distributive property).
$= 7x^2 - 9x + 36$	Combine like terms.

From these examples we see that in simplifying a polynomial expression containing symbols of grouping, we must remember to

> **1.** Deal with symbols of grouping, working from the inside out.
> **2.** Perform multiplications (and/or divisions) from left to right.
> **3.** Perform additions (and/or subtractions) from left to right.

Trial Run

Perform the indicated operations and simplify.

_______ 1. $(x^3 - 9) + (x^2 - 6x + 9)$

_______ 2. $(5x^2 - 3x + 1) - (3x^2 + x - 2)$

_______ 3. $5(1 - y^2) + 2(y^2 - 3y)$

_______ 4. $2(a^2 - 7a) - 3(-a^2 - a + 1)$

_______ 5. $9x^2 - 2x - 3(2x^2 - 5) + 6$

_______ 6. $7a - 3[2(a - 5b) + 2b]$

Answers are on page 176.

Multiplying Monomials Times Polynomials

In Section 3.1 we learned to multiply monomials using the first law of exponents. Now we turn our attention to products of monomials and other polynomials. We would like to learn how to find a product such as

$$x(x^2 - 2x + 3)$$

Once again the **distributive property** provides us with the method for performing such multiplications. According to the distributive property for multiplication over addition, we must multiply each term inside the parentheses by the factor outside the parentheses. Sometimes students like to use arrows to organize their work in finding such a product.

$$\begin{aligned} x(x^2 - 2x + 3) &= x(x^2 - 2x + 3) \\ &= x \cdot x^2 + x(-2x) + x \cdot 3 \\ &= x^3 - 2x^2 + 3x \end{aligned}$$

As you can see, once you have used the distributive property, the final steps of simplification merely require that you know how to multiply monomials.

Now you try Example 11.

Example 11. Find $-6a(2a^3 - 4a)$.

Solution

$$-6a(2a^3 - 4a)$$
$$= -6a(2a^3) - 6a(-4a)$$
$$=$$

Check your work on page 176. ▶

Example 12. Find $(x^2 + xy - y^2)5xy$.

Solution

$$(x^2 + xy - y^2)5xy$$
$$= 5xy \cdot x^2 + 5xy \cdot xy + 5xy(-y^2)$$
$$= 5x^3y + 5x^2y^2 - 5xy^3$$

Now try Example 13.

Example 13. Find

$$2a^3b(4a^2b - 7ab - 2ab^2)$$

Solution

$$2a^3b(4a^2b - 7ab - 2ab^2)$$
$$= 2a^3b \cdot 4a^2b + 2a^3b(-7ab) + 2a^3b(-2ab^2)$$
$$=$$

Check your work on page 176. ▶

Example 14. Find $6x - 2x(x - 4)$.

Solution

$$6x - 2x(x - 4)$$
$$= 6x - 2x^2 + 8x$$
$$= -2x^2 + 14x$$

Example 15. Find $x^2 - 5[x(x - 1) + 2x(x^2 - x)]$.

Solution

$x^2 - 5[x(x - 1) + 2x(x^2 - x)]$	
$= x^2 - 5[x^2 - x + 2x^3 - 2x^2]$	Remove parentheses (distributive property).
$= x^2 - 5[2x^3 - x^2 - x]$	Combine like terms within brackets.
$= x^2 - 10x^3 + 5x^2 + 5x$	Remove brackets (distributive property).
$= -10x^3 + 6x^2 + 5x$	Combine like terms.

Trial Run

Perform the indicated operations and simplify.

_______ 1. $a(a^2 - 3a - 10)$

_______ 2. $-3x(2x^2 - 5x)$

_______ 3. $(x^2 - 7xy + 12y^2)7x$

_______ 4. $-5a^2b(7a^2 - 5ab + 3b^2)$

_______ 5. $7x - 2x[x^2 - x(4x - 2)]$

_______ 6. $a^2 - 9[a(a - 5) + 3a(a^2 - 1)]$

Answers are on page 176.

Examples You Completed

Example 1. Name each polynomial.

15	is a constant monomial
$8x^2 - 3x + 4$	is a second-degree trinomial in x
$y^9 - 1$	is a ninth-degree binomial in y
$3xyz$	is a monomial in x, y, and z

Example 3. Rewrite $y - 3 + y^5$ in descending powers of the variable.

Solution

$$y - 3 + y^5$$
$$= y^5 + y - 3$$

Example 5. Find

$$(a^3 + 2a) + (5a^2 - 2a + 1)$$

Solution

$$(a^3 + 2a) + (5a^2 - 2a + 1)$$
$$= a^3 + 2a + 5a^2 - 2a + 1$$
$$= a^3 + 5a^2 + 2a - 2a + 1$$
$$= a^3 + 5a^2 + 1$$

Example 7. Find

$$(x^3 - 2x^2 + 1) - (2x^2 + 3x + 1)$$

Solution

$$(x^3 - 2x^2 + 1) - (2x^2 + 3x + 1)$$
$$= x^3 - 2x^2 + 1 - 2x^2 - 3x - 1$$
$$= x^3 - 2x^2 - 2x^2 - 3x + 1 - 1$$
$$= x^3 - 4x^2 - 3x$$

Example 9. Find

$$-2(5 - x^2) + 4(x^2 - 2x + 3)$$

Solution

$$-2(5 - x^2) + 4(x^2 - 2x + 3)$$
$$= -10 + 2x^2 + 4x^2 - 8x + 12$$
$$= 2x^2 + 4x^2 - 8x - 10 + 12$$
$$= 6x^2 - 8x + 2$$

Example 11. Find $-6a(2a^3 - 4a)$.

Solution

$$-6a(2a^3 - 4a)$$
$$= -6a(2a^3) - 6a(-4a)$$
$$= -12a^4 + 24a^2$$

Example 13. Find $2a^3b(4a^2b - 7ab - 2ab^2)$.

Solution

$$2a^3b(4a^2b - 7ab - 2ab^2)$$
$$= 2a^3b \cdot 4a^2b + 2a^3b(-7ab) + 2a^3b(-2ab^2)$$
$$= 8a^5b^2 - 14a^4b^2 - 4a^4b^3$$

Answers to Trial Runs

page 174 **1.** $x^3 + x^2 - 6x$ **2.** $2x^2 - 4x + 3$ **3.** $-3y^2 - 6y + 5$ **4.** $5a^2 - 11a - 3$ **5.** $3x^2 - 2x + 21$ **6.** $a + 24b$

page 175 **1.** $a^3 - 3a^2 - 10a$ **2.** $-6x^3 + 15x^2$ **3.** $7x^3 - 49x^2y + 84xy^2$ **4.** $-35a^4b + 25a^3b^2 - 15a^2b^3$ **5.** $6x^3 - 4x^2 + 7x$ **6.** $-27a^3 - 8a^2 + 72a$

Name ______________________ Date ____________

EXERCISE SET 3.3

Name each polynomial.

______ 1. $5x^5 + 10x^4$

______ 2. $3x^7 - 9x^5$

______ 3. $x^3 - 1$

______ 4. $y^3 - 8$

______ 5. -15

______ 6. -8

______ 7. $4y^2 - 4y + 1$

______ 8. $9y^2 - 6y - 1$

______ 9. $x^4 - 3x^3 - x^2 + 3x$

______ 10. $5x^5 + 10x^4 - x + 2$

Rewrite each polynomial in descending powers of the variable.

______ 11. $3x - 4x^2 + 7 + x^3$

______ 12. $5x^2 - 3 + 2x + x^3$

______ 13. $8x^2 + 12 + x^4$

______ 14. $-9x^3 + 20 + x^6$

______ 15. $1 - a + a^2 - a^3$

______ 16. $2 - 2a + 3a^2 - 3a^3$

______ 17. $5 + 3x$

______ 18. $8 - 3y$

Perform the indicated operations and simplify.

______ 19. $(x^3 + 3x^2) + (x^2 - 3x + 2)$

______ 20. $(x^2 - 3x) + (4x + 2)$

______ 21. $(3x^2 + 2x - 1) - (4x^2 - 2x + 5)$

______ 22. $(x^2 - 3x - 28) - (2x^2 + 3x - 1)$

______ 23. $(4a^2 - 3ab + 2b^2) - (a^2 + 5ab - 14b^2)$

______ 24. $(5a^2 - 2ab + b^2) - (a^2 - 7ab - 9b^2)$

______ 25. $4(a^2 - 3) - 5(a^2 - 2a - 1)$

______ 26. $5(a^2 - 9) - 3(a^2 + 7a - 11)$

______ 27. $7(-a^2 + 2a) + 3(2a^2 - 7a + 2)$

______ 28. $9(-a^2 + 2a) + 5(3a^2 - 4a + 2)$

______ 29. $4(a + b) - 5(2a - b) + 6(a - 2b)$

______ 30. $5(a - b) - 4(2a - 3b) + 3(a + 5b)$

______ 31. $2(x^2 + 2xy - y^2) - (3x^2 + 4xy - y^2)$

______ 32. $3(x^2 - 10xy + 5y^2) - (5x^2 - xy + y^2)$

______ 33. $-3(a^2 + a) - 5(3a + 7)$

______ 34. $-5(a^2 - 7) - 6(2a + 3)$

______ 35. $5(y^2 - 2y - 1) - 2(3y^2 - 5y - 4)$

______ 36. $3(x^2 - 2x + 1) - 5(x^2 + x - 1)$

______ 37. $3x^2 - 2x + 2(3x^2 - 1) + 9$

______ 38. $5x^2 - 7x + 5(4x^2 - 2) + 3$

______ 39. $5x^2 - 2(4x^2 - x) - 4(2x + 1)$

______ 40. $7x^2 - (3x^2 - 2x) + 3(5x - 1)$

_______ 41. $4a - 3[2a - 3(a + 2b) - 1] + 5b$

_______ 42. $7a - 2[3a - 2(a + 5b) + 1] + 3b$

_______ 43. $9x^2 - 3[2(x^2 - 3x) + 5x] - 4$

_______ 44. $10x^2 - 5[3(x^2 - 4x) + 2x] - 6$

_______ 45. $2m(m^2 - 5m + 6)$

_______ 46. $9m(m^2 - m - 6)$

_______ 47. $-5xy(3x^2y^2 - xy - 4)$

_______ 48. $-4xy(5x^2y^2 - xy - 6)$

_______ 49. $(7a^2 - b^2)2a^2b^2$

_______ 50. $(a^2 - 2b^2)6a^2b^2$

_______ 51. $4mn(8m^3 - 2m^2n + mn^2 - 3n^3)$

_______ 52. $5mn(6m^3 - m^3n + 2mn^2 - 7n^3)$

_______ 53. $5xy(3x - 4y) - 2xy(4x - 5y)$

_______ 54. $8xy(2x - y) - xy(x - 2y)$

_______ 55. $9x - 3x[x^2 - 2x(x - 1)]$

_______ 56. $8x - 5x[x^2 - 7x(2x - 1)]$

_______ 57. $a^2 - 7[a(a - 3) + 2a(a^2 - 1)]$

_______ 58. $2a^2 - 3[a(2a - 1) + a(a^2 - 11)]$

_______ 59. $-2x^2y^3(3x + 5xy^4) + x^3y(y^2 - 3y^6)$

_______ 60. $-x^2y^3(8x - 3xy^4) + 2x^3y(4y^2 - 5y^6)$

☆ Stretching the Topics

_______ 1. From $-2x^3 - x^2 + 5x - 4$ subtract the sum of $5x^3 - 4x^2 - 5x + 2$ and $-7x^3 + 3x^2 + x - 7$.

_______ 2. Simplify $3ab(a^2 - 5ab + 4b^2) - 2ab(6a^2 - 7ab - 8b^2)$ and evaluate when $a = 1$ and $b = -1$.

_______ 3. Simplify $x^{-2}y^{-1}(3x^2y^3 - x^3y^2 - 2x^4y^{-1}) - xy^{-1}(x^{-1}y^3 - 3y^2 + 4xy)$.

Check your answers in the back of your book.

If you can complete **Checkup 3.3,** you are ready to go on to Section 3.4.

Name ______________________ Date ____________

✓ CHECKUP 3.3

Name each polynomial.

________ 1. $x^4 - 16$

________ 2. $5y^2 - 2y - 7$

Rewrite each polynomial in descending powers of the variable.

________ $x^2 - x^4 + 2 - 2x^3$

________ 4. $8a - 9 - 7a^2$

Perform the indicated operations and simplify.

________ 5. $4(2a^2 - ab - 3b^2) + 3(-a^2 + 2ab - b^2)$

________ 6. $9x^2 - 3(x^2 - 4x) - 5(-x + 2) + 3$

________ 7. $11a - 4[2a - 3(a - 7b) - 4b] + 3b$

________ 8. $5x(x^2 - 3x + 5)$

________ 9. $7mn(3m + 5n) - 2mn(4m - 5n)$

________ 10. $5a^2 - 3[2a(a - 1) - 5a(a^2 - 2)]$

Check your answers in the back of your book.

If You Missed Problems:	You Should Review Examples:
1, 2	1
3, 4	2, 3
5, 6	8, 9
7	10
8, 9	11–14
10	15

3.4 Multiplying More Polynomials

Multiplying Binomials

Suppose that we wish to find the product of two binomials such as $(x + 5)(2x + 3)$. According to the distributive property, the second binomial $(2x + 3)$ can be multiplied times each term in the first binomial $(x + 5)$. Then we combine any like terms.

$$\begin{aligned}(x + 5)(2x + 3) &= x(2x + 3) + 5(2x + 3)\\ &= 2x^2 + 3x + 10x + 15\\ &= 2x^2 + 13x + 15\end{aligned}$$

> To multiply binomials, we multiply the first term in the first binomial times each term in the second binomial; then we multiply the second term in the first binomial times each term in the second binomial. Finally, we combine like terms.

You multiply the binomials in Example 1.

Example 1. Find $(7x - 1)(x - 2)$.

Solution

$$\begin{aligned}&(7x - 1)(x - 2)\\ &= 7x(x - 2) - 1(x - 2)\\ &=\end{aligned}$$

Check your work on page 188. ▶

Example 2. Find $(5 - 2x)(8 - x)$.

Solution

$$\begin{aligned}&(5 - 2x)(8 - x)\\ &= 5(8 - x) - 2x(8 - x)\\ &= 40 - 5x - 16x + 2x^2\\ &= 2x^2 - 21x + 40\end{aligned}$$

Example 3. Find $(3x - 4y)^2$.

Solution

$$\begin{aligned}&(3x - 4y)^2\\ &= (3x - 4y)(3x - 4y)\\ &= 3x(3x - 4y) - 4y(3x - 4y)\\ &= 9x^2 - 12xy - 12xy + 16y^2\\ &= 9x^2 - 24xy + 16y^2\end{aligned}$$

Now you complete Example 4.

Example 4. Find $(a + 6)(a - 6)$.

Solution

$$\begin{aligned}&(a + 6)(a - 6)\\ &= a(\underline{\qquad\qquad}) + 6(\underline{\qquad\qquad})\\ &=\end{aligned}$$

Check your work on page 188. ▶

Notice that there is no first-degree term (no a-term) in the final answer in Example 4.

Some students prefer to use a slightly shorter method to find products of binomials. As long as we realize that the distributive property still forms the basis for this shorter method, we are safe in using it. This shortcut is sometimes called the **FOIL method** because its steps require that we

1. Multiply the **F**irst terms in the binomials.
2. Multiply the **O**uter terms in the binomials.
3. Multiply the **I**nner terms in the binomials.
4. Multiply the **L**ast terms in the binomials.
5. Combine the like terms resulting from those multiplications.

Let's find a product using the FOIL method.

F: $x \cdot 2x$, L: $5 \cdot 3$, I: $5 \cdot 2x$, O: $x \cdot 3$

$$(x + 5)(2x + 3) = \underset{F}{2x^2} + \underset{O}{3x} + \underset{I}{10x} + \underset{L}{15}$$

$$= 2x^2 + 13x + 15$$

Notice that this is the same product that we found when we did our multiplication the longer way. We are doing exactly the same thing, skipping one step of writing.

You try using the FOIL method in Example 5.

Example 5. Find $(7x - 3)(2x + 4)$.

Solution

$$(7x - 3)(2x + 4)$$

(F, O, I, L marked)

F O I L

= ______ + ______ − ______ − ______

= ________

Check your work on page 188. ▶

Example 6. Find $(8x - 3y)(8x + 3y)$.

Solution

$$(8x - 3y)(8x + 3y)$$

(F, O, I, L marked)

$$= \underset{F}{64x^2} + \underset{O}{24xy} - \underset{I}{24xy} - \underset{L}{9y^2}$$

$$= 64x^2 - 9y^2$$

Now try Example 7.

Example 7. Find $(5x - 2)^2$.

Solution

$$(5x - 2)^2 = (5x - 2)(5x - 2)$$

(F, O, I, L marked)

F O I L

= ______ − ______ − ______ + ______

= ________

Check your work on page 188. ▶

Trial Run

Find each product.

_______ 1. $(2a - 5)(a + 7)$

_______ 2. $(7x - 2y)^2$

_______ 3. $(7a - 3)(7a + 3)$

_______ 4. $(9 - y)(3 - 2y)$

_______ 5. $(5x + 2y)(5x - 2y)$

_______ 6. $(3a - 5b)(a + 2b)$

Answers are on page 189.

When we simplify expressions involving several polynomial operations, we must remember to use the rules governing the order of operations.

You try to complete Example 8.

Example 8. Simplify

$$x(2x + 1) + (x + 3)(4x + 2)$$

Solution

$$x(2x + 1) + (x + 3)(4x + 2)$$
$$= 2x^2 + x + 4x^2 + 2x + 12x + 6$$
$$=$$

Check your work on page 188. ▶

Example 9. Simplify

$$(x - 3)(2x - 1) - (6x + 1)(3x - 2)$$

Solution

$$(x - 3)(2x - 1) - (6x + 1)(3x - 2)$$
$$= 2x^2 - x - 6x + 3$$
$$- (18x^2 - 12x + 3x - 2)$$
$$= 2x^2 - 7x + 3 - (18x^2 - 9x - 2)$$
$$= 2x^2 - 7x + 3 - 18x^2 + 9x + 2$$
$$= -16x^2 + 2x + 5$$

In Example 9, notice that the entire second product was being subtracted from the first product. Therefore, we used parentheses to keep that second product together before taking its opposite and adding.

Example 10. Simplify $(x + 1)^2 - 3(x - 5)^2$.

Solution

$$(x + 1)^2 - 3(x - 5)^2$$
$$= (x + 1)(x + 1) - 3(x - 5)(x - 5)$$
$$= x^2 + x + x + 1 - 3(x^2 - 5x - 5x + 25)$$
$$= x^2 + 2x + 1 - 3(x^2 - 10x + 25)$$
$$= x^2 + 2x + 1 - 3x^2 + 30x - 75$$
$$= -2x^2 + 32x - 74$$

Skipping steps in these problems is not a good idea because it can lead to errors, especially in signs.

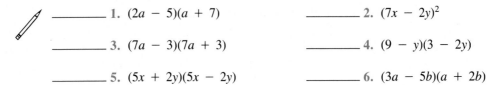

Trial Run

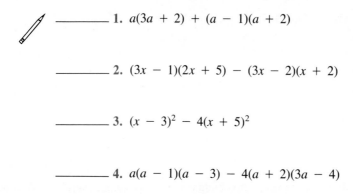

Perform the indicated operations and simplify.

______ 1. $a(3a + 2) + (a - 1)(a + 2)$

______ 2. $(3x - 1)(2x + 5) - (3x - 2)(x + 2)$

______ 3. $(x - 3)^2 - 4(x + 5)^2$

______ 4. $a(a - 1)(a - 3) - 4(a + 2)(3a - 4)$

Answers are on page 189.

Multiplying Polynomials

Once again the distributive property will be our tool for finding a product such as

$$(x + 3)(x^2 + 8x - 1)$$

We must multiply each term in the first polynomial times each term in the second polynomial and then combine like terms.

Example 11. Find $(x + 3)(x^2 + 8x - 1)$.

Solution

$$\begin{aligned} & (x + 3)(x^2 + 8x - 1) \\ &= x(x^2 + 8x - 1) + 3(x^2 + 8x - 1) \\ &= x^3 + 8x^2 - x + 3x^2 + 24x - 3 \\ &= x^3 + 11x^2 + 23x - 3 \end{aligned}$$

Now complete Example 12.

Example 12. Find $(x + y)(x^2 - xy + y^2)$.

Solution

$$\begin{aligned} & (x + y)(x^2 - xy + y^2) \\ &= x(x - xy + y^2) + y(x^2 - xy + y^2) \\ &= \end{aligned}$$

Check your work on page 188. ▶

Example 13. Find $(x - 2)^3$.

Solution

$(x - 2)^3 = (x - 2)(x - 2)(x - 2)$	Meaning of exponent.
$= (x - 2)[(x - 2)(x - 2)]$	Use associative property.
$= (x - 2)(x^2 - 2x - 2x + 4)$	Multiply within brackets.
$= (x - 2)(x^2 - 4x + 4)$	Combine like terms.
$= x(x^2 - 4x + 4) - 2(x^2 - 4x + 4)$	Use distributive property.
$= x^3 - 4x^2 + 4x - 2x^2 + 8x - 8$	Use distributive property.
$= x^3 - 6x^2 + 12x - 8$	Combine like terms.

Now you complete Example 14.

Example 14. Find $(2x^2 + x - 3)(x^2 - 5x + 1)$.

Solution

$$
\begin{aligned}
&(2x^2 + x - 3)(x^2 - 5x + 1) \\
&= 2x^2(x^2 - 5x + 1) + x(x^2 - 5x + 1) - 3(x^2 - 5x + 1) \\
&=
\end{aligned}
$$

Check your work on page 189. ▶

Trial Run

Perform the indicated operation and simplify.

_______ 1. $(x - 1)(x^2 - 3x - 40)$

_______ 2. $(a - 3)^3$

_______ 3. $(2x^2 - x + 1)(3x^2 - 2x + 2)$

_______ 4. $(2x + y)(x^2 - 5xy + 4y^2)$

Answers are on page 189.

Recognizing Special Products

There are certain products of polynomials that crop up over and over again in algebra. Because they occur so frequently, it is worth our while to recognize these special products and learn to compute them quickly.

Let's find a few products and see if we can recognize a pattern.

$(x + 3)(x - 3)$	$(5 - ay)(5 + ay)$	$(6x - 7y)(6x + 7y)$
$= x^2 - 3x + 3x - 9$	$= 25 + 5ay - 5ay - a^2y^2$	$= 36x^2 + 42xy - 42xy - 49y^2$
$= x^2 - 9$	$= 25 - a^2y^2$	$= 36x^2 - 49y^2$

In each of these products, the factors are binomials that are *the sum and the difference of the same two terms*. Each answer is the *difference of the squares* of those same two terms.

same first term → $(x + 3)(x - 3)$ ← same second term; opposite signs

$$(x + 3)(x - 3) = x^2 - 3^2 = x^2 - 9$$

(square of first term: x^2; square of second term: 3^2)

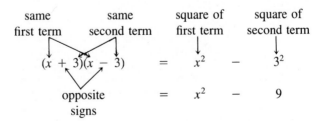

In fact, this result will always occur.

> If we multiply the sum of two terms by the difference of the same two terms, the product will be the difference of the squares of those two terms.

We call this special product the **difference of two squares** and we note that for any algebraic expressions A and B

Product of Sum and Difference of Two Terms

$$(A + B)(A - B) = A^2 - B^2$$

You use this formula to complete Example 15.

Example 15. Find $(3x + \frac{2}{5})(3x - \frac{2}{5})$.

Solution

$$\left(3x + \frac{2}{5}\right)\left(3x - \frac{2}{5}\right)$$

$$= (3x)^2 - (\_\_\_\_)^2$$

$$= 9x^2 - \_\_\_\_$$

Check your work on page 189. ▶
Then complete Example 16.

Example 16. Find $(8 - xy)(8 + xy)$.

Solution

$$(8 - xy)(8 + xy)$$

$$= (\_\_\_\_)^2 - (\_\_\_\_)^2$$

$$= \_\_\_\_ - x^2y^2$$

Check your work on page 189. ▶

We know how to find the *square* of a binomial by multiplying that binomial times itself. Can you observe a pattern in squaring the following binomials?

$(x + 5)^2$	$(7 - x)^2$	$(a + b)^2$
$= (x + 5)(x + 5)$	$= (7 - x)(7 - x)$	$= (a + b)(a + b)$
$= x^2 + 5x + 5x + 25$	$= 49 - 7x - 7x + x^2$	$= a^2 + ab + ab + b^2$
$= x^2 + 10x + 25$	$= 49 - 14x + x^2$	$= a^2 + 2ab + b^2$

Look closely at the first of these examples.

first term ↘ second term ↙		square of first term ↓		twice product of first and second terms ↓		square of second term ↓
$(x + 5)^2$	$=$	x^2	$+$	$2(5 \cdot x)$	$+$	5^2
	$=$	x^2	$+$	$10x$	$+$	25

When we square a binomial, the product is a trinomial in which

1. The first term is the square of the first term in the binomial.
2. The third term is the square of the second term in the binomial.
3. The middle term is twice the product of the first and second terms in the binomial.

Using A and B to represent any algebraic expressions, we state the following rule for finding the **square of a binomial.**

Square of a Binomial

$(A + B)^2 = A^2 + 2AB + B^2$

$(A - B)^2 = A^2 - 2AB + B^2$

Such products are called **perfect square trinomials.** To be honest, it is not crucial that students learn the rule for squaring a binomial, since other methods for multiplying binomials are almost as quick as using the rule. However, it is helpful when squaring complicated binomials to have such a rule available.

You practice this rule for squaring a binomial in Example 17.

Example 17. Find $(3x + 8)^2$.

Solution. Here $A = 3x$ and $B = 8$.

$$(3x + 8)^2$$
$$= (3x)^2 + 2(3x)(8) + 8^2$$
$$=$$

Check your work on page 189. ▶

Example 18. Find $(6 - 0.5a)^2$.

Solution. Here $A = 6$ and $B = 0.5a$.

$$(6 - 0.5a)^2$$
$$= 6^2 - 2(6)(0.5a) + (0.5a)^2$$
$$= 36 - 2(3a) + 0.25a^2$$
$$= 36 - 6a + 0.25a^2$$

Now complete Example 19.

Example 19. Find $(x^2y - z)^2$.

Solution. Here $A = x^2y$ and $B = z$.

$$(x^2y - z)^2$$
$$= (x^2y)^2 - 2(x^2y)(z) + z^2$$
$$=$$

Check your work on page 189. ▶

Example 20. Find $[(x + 1) + c]^2$.

Solution. Here $A = (x + 1)$ and $B = c$.

$$[(x + 1) + c]^2$$
$$= (x + 1)^2 + 2(x + 1)(c) + c^2$$
$$= x^2 + 2x + 1 + 2c(x + 1) + c^2$$
$$= x^2 + 2x + 1 + 2cx + 2c + c^2$$

Example 21. Find $[(x + 3) + 5y][(x + 3) - 5y]$.

Solution. We must multiply the sum and difference of the same two quantities. Here $A = (x + 3)$ and $B = 5y$. Our answer will be the *difference of two squares,* $A^2 - B^2$.

$$[(x + 3) + 5y][(x + 3) - 5y] = (x + 3)^2 - (5y)^2$$
$$= x^2 + 6x + 9 - 25y^2$$

Students sometimes forget the rule for squaring binomials and try shortcuts that simply will **not** work. *Squaring a binomial yields a trinomial; we do not square a binomial simply by squaring each term.*

$$(x + 3)^2 \neq x^2 + 9 \qquad (a - 5)^2 \neq a^2 - 25$$

$$(x + 3)^2 = x^2 + 6x + 9 \qquad (a - 5)^2 = a^2 - 10a + 25$$

Trial Run

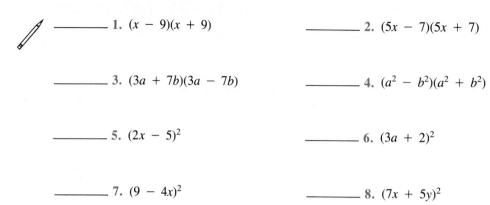

Find each special product.

______ 1. $(x - 9)(x + 9)$

______ 2. $(5x - 7)(5x + 7)$

______ 3. $(3a + 7b)(3a - 7b)$

______ 4. $(a^2 - b^2)(a^2 + b^2)$

______ 5. $(2x - 5)^2$

______ 6. $(3a + 2)^2$

______ 7. $(9 - 4x)^2$

______ 8. $(7x + 5y)^2$

Answers are on page 189.

Examples You Completed

Example 1. Multiply $(7x - 1)(x - 2)$.

Solution

$$
\begin{aligned}
&(7x - 1)(x - 2) \\
&= 7x(x - 2) - 1(x - 2) \\
&= 7x^2 - 14x - x + 2 \\
&= 7x^2 - 15x + 2
\end{aligned}
$$

Example 4. Find $(a + 6)(a - 6)$.

Solution

$$
\begin{aligned}
&(a + 6)(a - 6) \\
&= a(a - 6) + 6(a - 6) \\
&= a^2 - 6a + 6a - 36 \\
&= a^2 - 36
\end{aligned}
$$

Example 5. Find $(7x - 3)(2x + 4)$.

Solution

F L

$(7x - 3)(2x + 4)$

I

O

$$
\begin{aligned}
&\quad\ \text{F} \qquad \text{O} \qquad \text{I} \qquad \text{L} \\
&= 14x^2 + 28x - 6x - 12 \\
&= 14x^2 + 22x - 12
\end{aligned}
$$

Example 7. Find $(5x - 2)^2$.

Solution

F L

$(5x - 2)^2 = (5x - 2)(5x - 2)$

I

O

$$
\begin{aligned}
&\quad\ \text{F} \qquad \text{O} \qquad \text{I} \qquad \text{L} \\
&= 25x^2 - 10x - 10x + 4 \\
&= 25x^2 - 20x + 4
\end{aligned}
$$

Example 8. Simplify

$$x(2x + 1) + (x + 3)(4x + 2)$$

Solution

$$
\begin{aligned}
&x(2x + 1) + (x + 3)(4x + 2) \\
&= 2x^2 + x + 4x^2 + 2x + 12x + 6 \\
&= 6x^2 + 15x + 6
\end{aligned}
$$

Example 12. Find $(x + y)(x^2 - xy + y^2)$.

Solution

$$
\begin{aligned}
&(x + y)(x^2 - xy + y^2) \\
&= x(x^2 - xy + y^2) + y(x^2 - xy + y^2) \\
&= x^3 - x^2y + xy^2 + x^2y - xy^2 + y^3 \\
&= x^3 + y^3
\end{aligned}
$$

Example 14. Find $(2x^2 + x - 3)(x^2 - 5x + 1)$.

Solution

$$\begin{aligned} &(2x^2 + x - 3)(x^2 - 5x + 1) \\ &= 2x^2(x^2 - 5x + 1) + x(x^2 - 5x + 1) - 3(x^2 - 5x + 1) \\ &= 2x^4 - 10x^3 + 2x^2 + x^3 - 5x^2 + x - 3x^2 + 15x - 3 \\ &= 2x^4 - 9x^3 - 6x^2 + 16x - 3 \end{aligned}$$

Example 15. Find $(3x + \frac{2}{5})(3x - \frac{2}{5})$.

Solution

$$\begin{aligned} &\left(3x + \frac{2}{5}\right)\left(3x - \frac{2}{5}\right) \\ &= (3x)^2 - \left(\frac{2}{5}\right)^2 \\ &= 9x^2 - \frac{4}{25} \end{aligned}$$

Example 16. Find $(8 - xy)(8 + xy)$.

Solution

$$\begin{aligned} &(8 - xy)(8 + xy) \\ &= (8)^2 - (xy)^2 \\ &= 64 - x^2y^2 \end{aligned}$$

Example 17. Find $(3x + 8)^2$.

Solution

$$\begin{aligned} &(3x + 8)^2 \\ &= (3x)^2 + 2(3x)(8) + 8^2 \\ &= 9x^2 + 2(24x) + 64 \\ &= 9x^2 + 48x + 64 \end{aligned}$$

Example 19. Find $(x^2y - z)^2$.

Solution

$$\begin{aligned} &(x^2y - z)^2 \\ &= (x^2y)^2 - 2(x^2y)(z) + z^2 \\ &= x^4y^2 - 2x^2yz + z^2 \end{aligned}$$

Answers to Trial Runs

page 183 **1.** $2a^2 + 9a - 35$ **2.** $49x^2 - 28xy + 4y^2$ **3.** $49a^2 - 9$ **4.** $27 - 21y + 2y^2$
5. $25x^2 - 4y^2$ **6.** $3a^2 + ab - 10b^2$

page 184 **1.** $4a^2 + 3a - 2$ **2.** $3x^2 + 9x - 1$ **3.** $-3x^2 - 46x - 91$ **4.** $a^3 - 16a^2 - 5a + 32$

page 185 **1.** $x^3 - 4x^2 - 37x + 40$ **2.** $a^3 - 9a^2 + 27a - 27$ **3.** $6x^4 - 7x^3 + 9x^2 - 4x + 2$
4. $2x^3 - 9x^2y + 3xy^2 + 4y^3$

page 188 **1.** $x^2 - 81$ **2.** $25x^2 - 49$ **3.** $9a^2 - 49b^2$ **4.** $a^4 - b^4$ **5.** $4x^2 - 20x + 25$
6. $9a^2 + 12a + 4$ **7.** $81 - 72x + 16x^2$ **8.** $49x^2 + 70xy + 25y^2$

Name ______________________ Date ____________

EXERCISE SET 3.4

Perform the indicated operations and simplify.

_______ 1. $(4y - 1)(3y - 5)$

_______ 2. $(5y - 2)(3y - 1)$

_______ 3. $(6x - 5)(2x + 3)$

_______ 4. $(5x - 3)(3x + 4)$

_______ 5. $(9 - 2y)(2 - y)$

_______ 6. $(8 - 3y)(3 - y)$

_______ 7. $(4x + 3y)(7x + 2y)$

_______ 8. $(9x + y)(5x + 7y)$

_______ 9. $(a + 3b)^2$

_______ 10. $(a + 5b)^2$

_______ 11. $(2x - 3y)(2x + 3y)$

_______ 12. $(5x - 4y)(5x + 4y)$

_______ 13. $(xy - 10)(xy + 5)$

_______ 14. $(xy - 7)(2xy + 1)$

_______ 15. $(3xy - 2z)^2$

_______ 16. $(4xy - 3z)^2$

_______ 17. $(0.6a - 5b)(0.6a + 5b)$

_______ 18. $(1.1a - 3b)(1.1a + 3b)$

_______ 19. $(x^2 - 3y^2)(3x^2 + 8y^2)$

_______ 20. $(2x^2 - y^2)(4x^2 + 5y^2)$

_______ 21. $(a^2 + \frac{1}{2}b^2)^2$

_______ 22. $(\frac{1}{3}a^2 - b^2)^2$

_______ 23. $(x - 3)(x^2 - 2x + 1)$

_______ 24. $(x - 5)(x^2 - 6x + 9)$

_______ 25. $-y^2(8 - y)(7 + y)$

_______ 26. $-y^2(5 - y)(4 + y)$

_______ 27. $x(2x - 1) + (x - 3)(x + 5)$

_______ 28. $2x(x - 3) + (x - 2)(x + 4)$

_______ 29. $(3a + 2)(2a + 5) - (a - 3)(2a + 1)$

_______ 30. $(7a - 1)(a + 2) - (a - 4)(a + 2)$

_______ 31. $(x + 1)^2 - 5(x - 3)^2$

_______ 32. $(x + 2)^2 - 7(x - 1)^2$

_______ 33. $2(x - 1)(2x - 3) + 3(5x - 2)^2$

_______ 34. $4(x - 7)(3x + 1) + 2(3x - 4)^2$

_______ 35. $a(a - 4)(a + 4) - 11a(a - 2)$

_______ 36. $a(a - 1)(a - 8) - 3a(a - 9)$

_______ 37. $(a - b)(a + 2b)(a - 5b)$

_______ 38. $(a - 2b)(a + b)(a - 3b)$

_______ 39. $(x + 2)(x - 3)^2$

_______ 40. $(x - 4)(x - 2)^2$

_______ 41. $(a + 2)^3$

_______ 42. $(a + 5)^3$

_______ 43. $(x^2 - 1)(x^3 - 2x^2 + x - 5)$

_______ 44. $(x^2 - 2)(x^3 - 7x^2 + 3x - 1)$

_______ 45. $(3x^2 - x - 5)(x^2 + 2x - 1)$

_______ 46. $(2x^2 - x - 3)(x^2 + 3x - 7)$

_______ 47. $(a^2 - 5b^2)^2$

_______ 48. $(a^2 + 7b^2)^2$

_______ 49. $(x^2 - 2y^2)(x^2 + 2y^2)$

_______ 50. $(3x^2 - y^2)(3x^2 + y^2)$

_______ 51. $[(x - y) - 5]^2$

_______ 52. $[(x - 3y) - 9]^2$

_______ 53. $[(a + 4) - 3b][(a + 4) + 3b]$

_______ 54. $[(a + 2) - 5b][(a + 2) + 5b]$

_______ 55. $(2 - x)^4$

_______ 56. $(3 - x)^4$

☆ Stretching the Topics

Perform the operations and simplify.

_______ 1. $(2a - b - 3c)^2$

_______ 2. $(4x^{3a} - 7y^b)(2x^{3a} - y^b)$

_______ 3. $(x^{-1} - 2y^{-1})(x^{-2} + 2x^{-1}y^{-1} + 4y^{-2})$

Check your answers in the back of your book.

If you can complete **Checkup 3.4,** you are ready to go on to Section 3.5.

Name ________________________ Date ____________

✓ CHECKUP 3.4

Perform the indicated operations and simplify.

________ **1.** $(2x - 3)(x + 7)$

________ **2.** $(a - 10b)^2$

________ **3.** $(3x - 5y)(3x + 5y)$

________ **4.** $(x - 3)(x^2 - 5x - 6)$

________ **5.** $2x(7x - 3y) - (x - 2y)^2$

________ **6.** $4(a - 1)(a - 9) - (a - 9)(2a - 3)$

________ **7.** $(x - y)(x + y)(x - 2y)$

________ **8.** $(a - 5)^3$

________ **9.** $(x^2 - x + 1)(x^2 - 3x + 2)$

________ **10.** $[(a + b) - 9]^2$

Check your answers in the back of your book.

If You Missed Problems:	You Should Review Examples:
1	1
2	3, 17, 18
3	4, 15, 16
4	11
5	10
6	8, 9
7, 8	13
9	14
10	20

3.5 Dividing Polynomials

Dividing Polynomials by Monomials

To divide a polynomial by a monomial, we return to the definition of division stated in Chapter 1. Recall that division was defined as multiplication by the reciprocal of the divisor. Consider the division problem

$$\frac{6x - 12y}{3}$$

and see how the definition of division and the distributive property are used in Approach 1. Then look at Approach 2.

Approach 1

$$\frac{6x - 12y}{3}$$

$$= \frac{1}{3}(6x - 12y)$$

$$= \frac{1}{3}(6x) - \frac{1}{3}(12y)$$

$$= 2x - 4y$$

Approach 2

$$\frac{6x - 12y}{3}$$

$$= \frac{6x}{3} - \frac{12y}{3}$$

$$= 2x - 4y$$

Since our quotient is the same using either approach and Approach 2 is simpler, we agree that

> To divide a polynomial by a monomial, we may divide each term of the polynomial dividend by the monomial divisor.

This technique can be used to divide any polynomial by a monomial. It changes such divisions into a sum of monomial fractions that can be simplified using the laws of exponents.

Complete Example 1.

Example 1. Find

$$(6x^3y - 3x^2y^2 + 9xy^2) \div 3xy$$

Solution

$$\frac{6x^3y - 3x^2y^2 + 9xy^2}{3xy}$$

$$= \frac{6x^3y}{3xy} - \frac{3x^2y^2}{3xy} + \frac{9xy^2}{3xy}$$

$$=$$

Check your work on page 200. ▶

Example 2. Find $\dfrac{-2ab^2 + 2a^2b - a^2}{-2a^2b}$.

Solution

$$\frac{-2ab^2 + 2a^2b - a^2}{-2a^2b}$$

$$= \frac{-2ab^2}{-2a^2b} + \frac{2a^2b}{-2a^2b} - \frac{a^2}{-2a^2b}$$

$$= \frac{b}{a} - 1 + \frac{1}{2b}$$

Trial Run

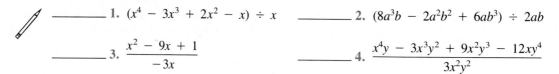

Find each quotient.

_______ 1. $(x^4 - 3x^3 + 2x^2 - x) \div x$

_______ 2. $(8a^3b - 2a^2b^2 + 6ab^3) \div 2ab$

_______ 3. $\dfrac{x^2 - 9x + 1}{-3x}$

_______ 4. $\dfrac{x^4y - 3x^3y^2 + 9x^2y^3 - 12xy^4}{3x^2y^2}$

Answers are on page 200.

Dividing Polynomials Using Long Division

When dividing a polynomial by another polynomial that is not a monomial, we use **long-division** techniques which are similar to long-division techniques used for dividing whole numbers. Let's briefly review long division with whole numbers to remind ourselves exactly what is done in each step.

$$\begin{array}{r} 26 \\ 21\overline{)547} \\ \underline{42} \\ 127 \\ \underline{126} \\ 1 \end{array} \qquad\qquad \begin{array}{r} 27 \\ 13\overline{)351} \\ \underline{26} \\ 91 \\ \underline{91} \\ 0 \end{array}$$

$$547 \div 21 = 26 \text{ remainder } 1 \qquad\qquad 351 \div 13 = 27$$
$$= 26\tfrac{1}{21}$$

To check a long-division problem in arithmetic, recall that we must multiply the divisor times the quotient, then add the remainder to see whether the dividend is obtained. Here

$$\begin{aligned} 21(26) + 1 &\stackrel{?}{=} 547 \\ 546 + 1 &\stackrel{?}{=} 547 \\ 547 &= 547 \end{aligned} \qquad\qquad \begin{aligned} 13(27) + 0 &\stackrel{?}{=} 351 \\ 351 + 0 &\stackrel{?}{=} 351 \\ 351 &= 351 \end{aligned}$$

In general, it will be true that

dividend = (divisor × quotient) + remainder

What is the significance of a *zero* remainder in arithmetic? When no remainder occurs, we know that the divisor divides the dividend exactly. Therefore, the divisor and quotient are *factors* of the dividend. Here, for example,

21 is *not* a factor of 547

13 *is* a factor of 351
27 *is* a factor of 351

Let's keep these ideas in mind as we approach long division of polynomials, starting with the quotient $(x^2 + 5x - 14) \div (x - 2)$.

$$x - 2\overline{)x^2 + 5x - 14}$$

Set up quotient as long-division problem.

$$\begin{array}{r} x \\ x - 2\overline{)x^2 + 5x - 14} \end{array}$$

Divide first term (x) of divisor into first term (x^2) of dividend to get first term (x) of quotient.

$$\begin{array}{r} x \\ x - 2\overline{)x^2 + 5x - 14} \\ \underline{x^2 - 2x} \end{array}$$

Multiply quotient term times *entire* divisor and write product beneath like terms in dividend.

$$\begin{array}{r} x \\ x - 2\overline{)x^2 + 5x - 14} \\ \underline{{}^{\ominus}x^2 \overset{\oplus}{-} 2x } \\ 7x - 14 \end{array}$$

Subtract carefully (change the sign and add) and bring down the next term of the dividend (-14).

$$\begin{array}{r} x + 7 \\ x - 2\overline{)x^2 + 5x - 14} \\ \underline{x^2 - 2x } \\ 7x - 14 \\ \underline{{}^{\ominus}7x \overset{\oplus}{-} 14} \\ 0 \end{array}$$

Divide first term (x) of divisor into first term ($7x$) of new dividend. Then multiply and subtract as before.

We say that

$$\frac{x^2 + 5x - 14}{x - 2} = x + 7$$

so

$$x^2 + 5x - 14 = (x - 2)(x + 7)$$

and since the remainder is zero, we conclude that

$$x - 2 \text{ is a factor of } x^2 + 5x - 14$$

$$x + 7 \text{ is a factor of } x^2 + 5x - 14$$

In performing long division of polynomials, your work will be simplified if you always remember to write the terms of the dividend and divisor in descending powers of the same variable and to use a coefficient of 0 with any missing terms in the dividend. Keep these ideas in mind as you look at Example 3.

Example 3. Divide $5x^2 + x^4 + 10x - 8$ by $x^2 + x - 1$.

Solution. We arrange the terms in descending powers of x and note the missing x^3-term in the dividend.

$$\begin{array}{r} x^2 - \; x + 7 \\ x^2 + x - 1\overline{)x^4 + 0x^3 + 5x^2 + 10x - 8} \\ \underline{{}^{\ominus}x^4 \overset{\ominus}{+} x^3 \overset{\oplus}{-} x^2 } \\ - x^3 + 6x^2 + 10x \\ \underline{\overset{\oplus}{-} x^3 \overset{\oplus}{-} x^2 \overset{\ominus}{+} x } \\ 7x^2 + 9x - 8 \\ \underline{{}^{\ominus}7x^2 \overset{\ominus}{+} 7x \overset{\oplus}{-} 7} \\ 2x - 1 \end{array}$$

Since the degree of the last dividend $(2x - 1)$ is less than the degree of the divisor $(x^2 + x - 1)$, we stop the long-division process, writing $2x - 1$ as the remainder.

$$\frac{5x^2 + x^4 + 10x - 8}{x^2 + x - 1} = x^2 - x + 7, \quad \text{remainder } 2x - 1$$

$$= x^2 - x + 7 + \frac{2x - 1}{x^2 + x - 1}$$

We have noted four important points to keep in mind when performing long division of polynomials.

1. Write terms of the dividend and divisor in *descending* powers of the same variable.
2. Use a *zero* coefficient for any missing terms in the dividend.
3. Always carefully *subtract* the product of the divisor and the new quotient term after each division.
4. When the difference is an expression with degree less than the degree of the divisor, that difference is the *remainder*.

Example 4. Find $(8a^3 - 10a^2b + 5ab^2 + b^3) \div (2a - b)$.

Solution

$$\begin{array}{r l}
 & 4a^2 - 3ab + b^2 \\
2a - b \,\big)\!\! & \overline{8a^3 - 10a^2b + 5ab^2 + b^3} \\
 & \underline{\overset{\ominus}{8a^3} \overset{\oplus}{-} 4a^2b} \\
 & -6a^2b + 5ab^2 \\
 & \underline{\overset{\oplus}{-} 6a^2b \overset{\ominus}{+} 3ab^2} \\
 & 2ab^2 + b^3 \\
 & \underline{\overset{\ominus}{2ab^2} \overset{\oplus}{-} b^3} \\
 & 2b^3
\end{array}$$

$$\frac{8a^3 - 10a^2b + 5ab^2 + b^3}{2a - b} = 4a^2 - 3ab + b^2 + \frac{2b^3}{2a - b}$$

Now you carefully complete Example 5.

Example 5. Find $(4x^4 - 2x^3 + 5x - 1) \div (x + 1)$.

Solution

$$\begin{array}{r l}
 & 4x^3 - 6x^2 \\
x + 1 \,\big)\!\! & \overline{4x^4 - 2x^3 + 0x^2 + 5x - 1} \\
 & \underline{\overset{\ominus}{4x^4} \overset{\ominus}{+} 4x^3} \\
 & -6x^3 + 0x^2 \\
 & \underline{\overset{\oplus}{-} 6x^3 \overset{\oplus}{-} 6x^2}
\end{array}$$

$$(4x^4 - 2x^3 + 5x - 1) \div (x + 1) = \underline{\qquad\qquad}$$

Is $x + 1$ a factor of $4x^4 - 2x^3 + 5x - 1$? ________

Check your work on page 200. ▶

Trial Run

Find each quotient by long division.

________ **1.** $(x^3 - 5x^2 + 2x + 10) \div (x - 2)$

________ **2.** $(2x^3 - 11x^2 + 10x + 3) \div (2x - 3)$

________ **3.** $(3x^4 - 2x^2 + 9) \div (x + 3)$

Answers are on page 200.

Dividing Polynomials Using Synthetic Division

Long division can be a tedious process, but there is a shortcut that can be used under certain circumstances. When a polynomial in x is to be divided by a divisor of the form $x - c$, where c is a constant, we may use the technique of **synthetic division** to find the quotient.

Recall this long-division problem worked earlier.

$$\begin{array}{rl} & \underline{x + 7} \quad \leftarrow \text{quotient} \\ \text{divisor} \rightarrow x - 2 & \big)\, x^2 + 5x - 14 \leftarrow \text{dividend} \\ & \underline{\ominus x^2 \oplus 2x} \\ & 7x - 14 \\ & \underline{\ominus 7x \oplus 14} \\ & 0 \leftarrow \text{remainder} \end{array}$$

This was not a difficult problem, but we had to be very careful with our subtractions. In synthetic division, the subtractions are replaced by additions. Moreover, the polynomial dividend is replaced by its numerical coefficients only, and the binomial divisor is replaced by the constant c (called the **synthetic divisor**).

The steps for finding the quotient $(x^2 + 5x - 14) \div (x - 2)$ by synthetic division are outlined below. Note here that the synthetic divisor is 2.

$$\begin{array}{r|rrr} 2 & 1 & 5 & -14 \\ & & & \\ \hline \end{array}$$

Write the synthetic divisor and the coefficients of the dividend.

$$\begin{array}{r|rrr} 2 & 1 & 5 & -14 \\ & \downarrow & & \\ \hline & 1 & & \end{array}$$

Bring down the first term of the synthetic dividend.

$$\begin{array}{r|rrr} 2 & 1 & 5 & -14 \\ & \downarrow & 2 & \\ \hline & 1 & 7 & \end{array}$$

Multiply the synthetic divisor (2) times the first term in the quotient (1) and write that product (2) under the second term in the dividend (5). Then *add*.

$$\begin{array}{r|rrr} 2 & 1 & 5 & -14 \\ & \downarrow & 2 & 14 \\ \hline & 1 & 7 & 0 \end{array}$$

Multiply the synthetic divisor (2) times the next term in the quotient (7) and write that product (14) under the next term in the dividend (-14). *Add*.

synthetic divisor → 2; coefficients of dividend → 1, 5, −14

$$\begin{array}{r|rrr} 2 & 1 & 5 & -14 \\ & \downarrow & 2 & 14 \\ \hline & 1 & 7 & \boxed{0} \end{array}$$

1 7 ← coefficients of quotient; 0 ← remainder

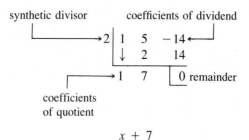

The last term in the quotient (0) is the *remainder*. The other terms are the coefficients of the quotient polynomial.

$x + 7$

The *degree* of the quotient polynomial is always *one less* than the degree of the dividend.

Using synthetic division, we again have found that

$$(x^2 + 5x - 14) \div (x - 2) = x + 7$$

Example 6. Use synthetic division to find $(x^3 + 3x^2 + 3x + 1) \div (x + 1)$.

Solution. The divisor is $x + 1$, which can be written $x - (-1)$, so the synthetic divisor is -1. The quotient will be of *second* degree.

$$\begin{array}{r|rrrr} -1 & 1 & 3 & 3 & 1 \\ & \downarrow & -1 & -2 & -1 \\ \hline & 1 & 2 & 1 & \boxed{0} \\ & 1x^2 + & 2x + & 1 & \end{array}$$

So $(x^3 + 3x^2 + 3x + 1) \div (x + 1) = x^2 + 2x + 1$.

Example 7. Use synthetic division to find $(x^4 + 5x^2 + 100x - 30) \div (x + 4)$.

Solution. The divisor is $x + 4$, which can be written $x - (-4)$, so the synthetic divisor is -4. The quotient will be of *third* degree. Notice the missing x^3-term in the dividend for which we must use a *zero* coefficient.

$$\begin{array}{r|rrrrr} -4 & 1 & 0 & 5 & 100 & -30 \\ & \downarrow & -4 & 16 & -84 & -64 \\ \hline & 1 & -4 & 21 & 16 & \boxed{-94} \\ & x^3 & -4x^2 + & 21x + & 16 & \text{R } -94 \end{array}$$

$$(x^4 + 5x^2 + 100x - 30) \div (x + 4) = x^3 - 4x^2 + 21x + 16 - \frac{94}{x + 4}$$

For emphasis we repeat that synthetic division can be used only when a polynomial is to be divided by a divisor of the form $x - c$. It can be used only for a divisor that is a first-degree binomial with a leading coefficient of 1.

Now try Example 8.

Example 8. Use synthetic division to find $(x^3 + 2x^2 - 3x - 1) \div (x - 3)$.

Solution. The synthetic divisor is ________ . The quotient will be of ________ degree.

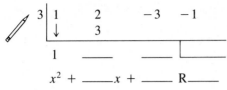

$$\begin{array}{r|rrrr} 3 & 1 & 2 & -3 & -1 \\ & \downarrow & 3 & & \\ \hline & 1 & \_\_\_ & \_\_\_ & \boxed{\_\_\_} \end{array}$$

$x^2 +$ ____ $x +$ ____ R____

So

$$(x^3 + 2x^2 - 3x - 1) \div (x - 3)$$

$$= x^2 + \_\_\_\_ x + \_\_\_\_ + \frac{\_\_\_\_}{x - 3}$$

Check your work on page 200. ▶

Example 9. Use synthetic division to find $(y^5 + 1) \div (y + 1)$.

Solution. The synthetic divisor is -1. The quotient will be of fourth degree.

$$\begin{array}{r|rrrrrr} -1 & 1 & 0 & 0 & 0 & 0 & 1 \\ & \downarrow & -1 & 1 & -1 & 1 & -1 \\ \hline & 1 & -1 & 1 & -1 & 1 & \boxed{0} \end{array}$$

$$y^4 - y^3 + y^2 - y + 1$$

So

$$(y^5 + 1) \div (y + 1)$$

$$= y^4 - y^3 + y^2 - y + 1$$

Trial Run

Tell whether synthetic division can be used to find these quotients. If not, tell why not.

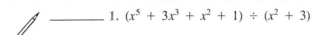

________ **1.** $(x^5 + 3x^3 + x^2 + 1) \div (x^2 + 3)$

________ **2.** $(3x^3 + 5x^2 - 2x + 3) \div (3x + 2)$

Use synthetic division to find each quotient.

________ **3.** $(x^2 - 3x - 4) \div (x - 2)$

________ **4.** $(2x^4 - x^3 - 20x^2 - 9) \div (x + 3)$

________ **5.** $(a^4 - 1) \div (a - 1)$

Answers are on page 200.

▶ Examples You Completed

Example 1. Find

$$(6x^3y - 3x^2y^2 + 9xy^2) \div 3xy$$

Solution

$$\frac{6x^3y - 3x^2y^2 + 9xy^2}{3xy}$$

$$= \frac{6x^3y}{3xy} - \frac{3x^2y^2}{3xy} + \frac{9xy^2}{3xy}$$

$$= 2x^2 - xy + 3y$$

Example 8. Use synthetic division to find $(x^3 + 2x^2 - 3x - 1) \div (x - 3)$.

Solution. The synthetic divisor is 3. The quotient will be of second degree.

3	1	2	−3	−1
	↓	3	15	36
	1	5	12	35

$x^2 + 5x + 12$ R 35

So

$$(x^3 + 2x^2 - 3x - 1) \div (x - 3)$$

$$= x^2 + 5x + 12 + \frac{35}{x - 3}$$

Example 5. Find

$$(4x^4 - 2x^3 + 5x - 1) \div (x + 1)$$

Solution

$$\begin{array}{r} 4x^3 - 6x^2 + 6x - 1 \\ x + 1 \overline{)\, 4x^4 - 2x^3 + 0x^2 + 5x - 1} \\ \underline{4x^4 + 4x^3} \qquad\qquad\qquad\quad \\ -6x^3 + 0x^2 \qquad\qquad \\ \underline{-6x^3 - 6x^2} \qquad\qquad \\ 6x^2 + 5x \qquad \\ \underline{6x^2 + 6x} \qquad \\ -x - 1 \\ \underline{-x - 1} \\ 0 \end{array}$$

$$(4x^4 - 2x^3 + 5x - 1) \div (x + 1) = 4x^3 - 6x^2 + 6x - 1$$

Is $x + 1$ a factor of $4x^4 - 2x^3 + 5x - 1$? Yes.

Answers to Trial Runs

page 194 **1.** $x^3 - 3x^2 + 2x - 1$ **2.** $4a^2 - ab + 3b^2$ **3.** $-\frac{x}{3} + 3 - \frac{1}{3x}$
4. $\frac{x^2}{3y} - x + 3y + \frac{4y^2}{x}$

page 197 **1.** $x^2 - 3x - 4 + \frac{2}{x - 2}$ **2.** $x^2 - 4x - 1$ **3.** $3x^3 - 9x^2 + 25x - 75 + \frac{234}{x + 3}$

page 199 **1.** No; divisor is of second degree. **2.** No; coefficient of x is not 1.
3. $x - 1 - \frac{6}{x - 2}$ **4.** $2x^3 - 7x^2 + x - 3$ **5.** $a^3 + a^2 + a + 1$

Name ______________________ Date ____________

EXERCISE SET 3.5

Find each quotient.

_______ 1. $(x^4 - 7x^3 + 3x^2 + 8x) \div x$

_______ 2. $(3x^4 - 2x^3 - x^2 + 2x) \div x$

_______ 3. $(9a^3b - 6a^2b^2 + 15ab^3) \div 3ab$

_______ 4. $(16a^4b^2 - 8a^3b^3 - 24a^2b^4) \div 4ab$

_______ 5. $(3x^2 - 6x + 5) \div 3x$

_______ 6. $(5x^2 - 10x + 1) \div 5x$

_______ 7. $(2xy - 8xyz + 10x^2y^2) \div (-2xy)$

_______ 8. $(5xy - 15xyz - 20x^2y^2) \div (-5xy)$

_______ 9. $(x^6 - 3x^5 + 2x^4 - x^3 + 3x^2) \div x^3$

_______ 10. $(4x^6 - 8x^5 + 2x^4 - 10x^3 - 5x^2) \div 2x^2$

_______ 11. $(x^2 - 12x + 32) \div (x - 4)$

_______ 12. $(x^2 - 9x - 10) \div (x + 1)$

_______ 13. $(8x^2 + 10x + 3) \div (2x + 1)$

_______ 14. $(12x^2 - 4x - 1) \div (2x - 1)$

_______ 15. $(2x^2 + x - 12) \div (2x + 5)$

_______ 16. $(12x^2 + 17x - 6) \div (4x + 7)$

_______ 17. $(x^3 - 3x^2 + 5x + 7) \div (x + 1)$

_______ 18. $(2x^3 + 4x^2 + x - 24) \div (x - 2)$

_______ 19. $(2a^3 + 5a^2 - 11a + 4) \div (2a - 1)$

_______ 20. $(3a^3 + a^2 - a - 15) \div (3a - 5)$

_______ 21. $(x^3 - 3x^2y + 5xy^2 - 6y^3) \div (x - 2y)$

_______ 22. $(x^3 - x^2y + 2xy^2 + 4y^3) \div (x + y)$

_______ 23. $(x^4 - 17x^2 + 20x - 6) \div (x^2 + 4x - 3)$

_______ 24. $(x^4 - 4x^2 + 12x - 9) \div (x^2 - 2x + 3)$

_______ 25. $(x^3 - 8) \div (x - 2)$

_______ 26. $(x^3 + 27) \div (x + 3)$

_______ 27. $(2x^4 - 2x^2 + 1) \div (2x - 4)$

_______ 28. $(2x^3 - x^2 - 7x + 2) \div (2x + 1)$

_______ 29. $(8x^4 - 6x^3 + 11x^2 - 9) \div (4x^2 - x - 3)$

_______ 30. $(10x^4 - 11x^3 + 12x^2 - 4x - 2) \div (2x^2 - x - 1)$

Use synthetic division to find each quotient.

_______ 31. $(x^4 - 2x^3 - x^2 - 1) \div (x + 1)$

_______ 32. $(x^4 + x^3 - 3x^2 + 7x - 6) \div (x + 2)$

_______ 33. $(3x^4 - 16x^3 + 6x^2 - 5x) \div (x - 5)$

_______ 34. $(2x^4 - 6x^3 + 4x - 12) \div (x - 3)$

_______ 35. $(x^4 + 16) \div (x + 2)$

_______ 36. $(x^5 + 32) \div (x + 2)$

_______ 37. $(x^5 + 3x^4 - 10x^3 + x^2 + 3x - 12) \div (x + 5)$

_______ 38. $(x^5 + 2x^4 + 8x^3 + x^2 + x - 15) \div (x + 4)$

_______ 39. $(x^6 - 2x^5 + 2x^4 - 2x^3 + 2x^2 - 2) \div (x - 1)$

_______ 40. $(x^6 - 4x^5 + 5x^4 - 4x^3 + 5x^2 - 4x + 5) \div (x - 2)$

☆ Stretching the Topics

Find each quotient.

_______ 1. $(8x^3 - 6x^2y + 6x^2y^2 - 5xy^2 + xy^3) \div (2x + y)$

_______ 2. $(x^{3a} + 3x^{2a} - x^a - 3) \div (x^a + 1)$

_______ 3. Use synthetic division to determine if $n - 2$ is a factor of the polynomial $5n^5 + 10n^4 - 10n^3 - 34n + 33$. Evaluate the polynomial when $n = 2$.

Check your answers in the back of your book.

If you can complete **Checkup 3.5,** you are ready to go on to Section 3.6.

Name ______________________________ Date ______________

✓ CHECKUP 3.5

Find each quotient.

________ 1. $(14x^3y - 7x^2y^2 + 21xy^3) \div 7xy$

________ 2. $(x^4 - 2x^3 + 4x^2 - 8x) \div (-2x^2)$

________ 3. $(6x^2 - 5x - 21) \div (3x - 7)$

________ 4. $(x^3 - x^2 - 17x - 17) \div (x + 3)$

________ 5. $(2x^4 - 11x^2 - x^3 - 11x - 2) \div (2x^2 - 5x - 3)$

Use synthetic division to find each quotient.

________ 6. $(a^5 - 32) \div (a - 2)$

________ 7. $(x^3 + x^2 - 10x - 10) \div (x + 3)$

________ 8. $(2x^4 + 3x^3 - x^2 - 4) \div (x + 2)$

Check your answers in the back of your book.

If You Missed Problems:	You Should Review Examples:
1	1
2	1, 2
3, 4	5
5	3
6	9
7, 8	6–8

3.6 Switching from Word Expressions to Polynomials

Now that we have learned to operate with polynomials, let's see if we can use our knowledge to translate some word expressions into polynomials.

Example 1. The length of a rectangle is 3 feet longer than twice the width. Write an expression for the area of the rectangle.

Solution

We Think	***We Write***
The area of a rectangle is the *product* of its length and width.	Area = (length)(width)
We know less about the width.	Let w = width (feet)
The length is 3 feet *more* than *twice* the width.	$2w$ = twice the width (feet) $2w + 3$ = length (feet)
We can illustrate this rectangle and label its dimensions.	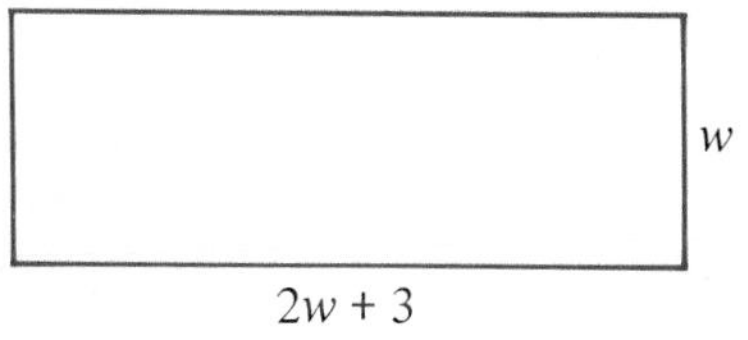
Area is length *times* width.	Area $= (2w + 3)w$ $= 2w^2 + 3w$

The expression $2w^2 + 3w$ square feet represents the area of this rectangle.

Now you try Example 2.

Example 2. If the length of each side of a square is increased by 5 meters, write an algebraic expression for the area of the new square.

Solution

We Think	***We Write***
In a square, the length and width are the *same*.	Let s = length of side of old square (meters)
Each side of the new square is 5 meters *more* than each side of the old square.	$s +$ ____ = length of side of new square (meters)
The area of a square is found by *squaring* the length of its side.	$(s +$ ____$)^2$ = area of new square (square meters)
	$(s +$ ____$)^2$ = ________

The expression ________ square meters represents the area of the new square.

Check your work on page 209. ▶

Now let's tackle the problem stated at the beginning of this chapter.

Example 3. Marco's rectangular garden is 10 feet longer than it is wide. If he decreases both the length and width of the garden by 3 feet, what variable expression represents the amount of garden area that he will lose?

Solution

We Think	***We Write***
The area lost will be the *difference* between old area and new area.	Area lost = old area − new area
The *area* of a rectangle is length times width.	Area = (length)(width)
For the old garden, we know less about the width. The old length is 10 feet *more* than the old width.	Let x = old width (feet) $x + 10$ = old length (feet)

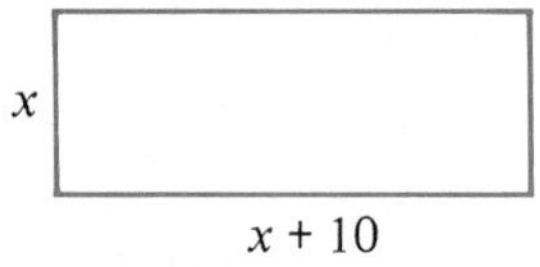

Now we know the *old area*.	$x(x + 10)$ = old area (square feet)
For the new garden, each dimension is *decreased* by 3 feet.	$x - 3$ = new width (feet) $(x + 10) - 3$ = new length (feet) $x + 7$ = new length (feet)

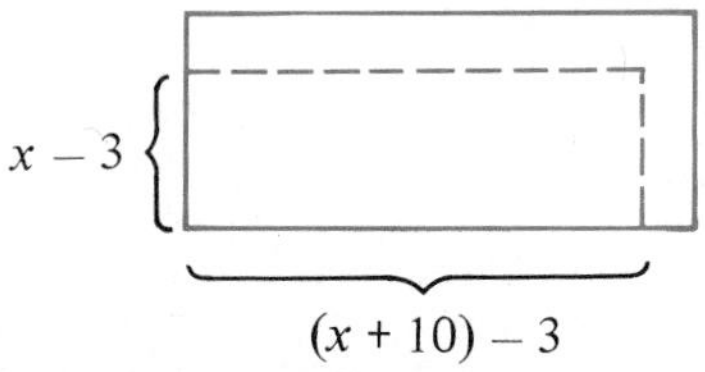

Now we know the *new area*.	$(x - 3)(x + 7)$ = new area (square feet)
The area lost is found by *subtracting* new area from old area.	$x(x + 10) - (x - 3)(x + 7)$ $= (x^2 + 10x) - (x^2 + 4x - 21)$ $= x^2 + 10x - x^2 - 4x + 21$ $= 6x + 21$

The expression $6x + 21$ square feet represents the garden area that Marco will lose.

Now you complete Example 4.

Example 4. The width of a shoe box is 3 inches more than its height, and its length is 10 inches more than its height. Write an expression for the volume of the box.

Solution

We Think	***We Write***
We recall that the *volume* of a rectangular solid (box) is the *product* of the length, width, and height.	Volume = (length)(width)(height)
We know the least about the height.	Let h = height (inches)
The width is 3 inches *more* than the height.	h + ____ = width (inches)
The length is 10 inches *more* than the height.	h + ____ = length (inches)

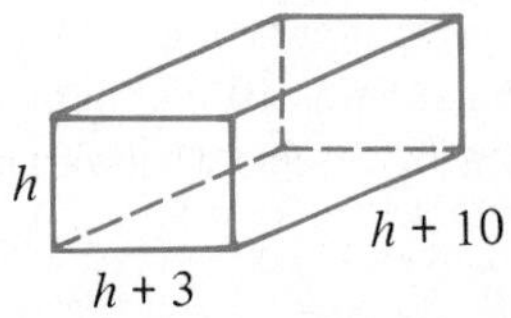

The volume is the *product* of the dimensions.	$h(h + \underline{\qquad})(h + \underline{\qquad})$ cubic inches $= \underline{\qquad\qquad\qquad}$

The expression for the volume of the shoe box is ________ cubic inches.

Check your work on page 210. ▶

A **right triangle** is a triangle in which two of the sides (called the **legs**) meet at a **right angle** (a 90° angle). The two legs are said to be *perpendicular* to each other. The third side, which is opposite the right angle, is called the **hypotenuse** of the right triangle. The parts of a right triangle are often labeled in the following way.

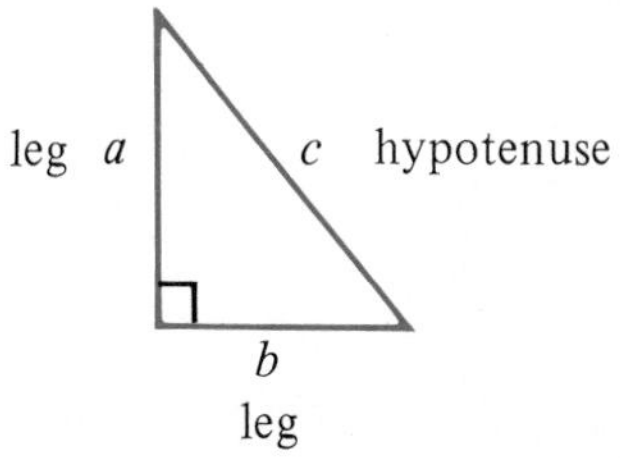

The area of a right triangle is found by multiplying the lengths of the legs and dividing that product by 2.

Area of Right Triangle

$$A = \frac{a \cdot b}{2}$$

See if you can complete Example 5.

Example 5. One leg of a right triangle is 3 inches shorter than the other leg. Write an algebraic expression for the area of the right triangle.

Solution

We Think	***We Write***
The area of a right triangle depends on the lengths of its legs, a and b.	$\text{Area} = \dfrac{a \cdot b}{2}$
We know less about the longer leg.	Let a = length of longer leg (inches)
The other leg is 3 inches *shorter*.	$a - \underline{\qquad}$ = length of shorter leg (inches)
The area is found by multiplying the lengths of the legs and dividing by 2.	$\text{Area} = \dfrac{a(a - \qquad)}{2}$ square inches $= \underline{\qquad\qquad\qquad}$

The expression ________ square inches represents the area of the right triangle.

Check your work on page 210. ▶

One of the most important mathematical discoveries was made by followers of the Greek mathematician **Pythagoras.** They observed that if we square the length of each leg of a right triangle and add those squares together, that sum will equal the square of the length of the hypotenuse. Using the usual letters to represent the parts of a right triangle, we state this important result.

Pythagorean Theorem

$$a^2 + b^2 = c^2$$

Example 6. If the legs of a right triangle measure 3 inches and 4 inches, find the length of the hypotenuse.

Solution. We must use the Pythagorean theorem with $a = 3$ inches and $b = 4$ inches.

$$a^2 + b^2 = c^2$$
$$3^2 + 4^2 = c^2$$
$$9 + 16 = c^2$$
$$25 = c^2$$
$$5 = c \quad \text{or} \quad -5 = c$$

But a dimension cannot be negative, so we conclude that the length of the hypotenuse is 5 inches.

Example 7. One leg of a right triangle is 1 centimeter (cm) more than three times the other leg. Write an algebraic expression for the square of the hypotenuse.

Solution

We Think	***We Write***
We must use the Pythagorean theorem.	$a^2 + b^2 = c^2$
We know less about the shorter leg.	Let x = length of shorter leg (cm)
The longer leg is 1 cm *more* than 3 *times* the other.	$3x + 1$ = length of longer leg (cm)
We must *square* the length of each leg and *sum* those squares.	$x^2 + (3x + 1)^2$ $= x^2 + (9x^2 + 6x + 1)$ $= 10x^2 + 6x + 1$

The square of the hypotenuse is the expression $10x^2 + 6x + 1$.

Consecutive integers are integers that immediately follow each other in the list of integers. For example,

17, 18, 19	are three consecutive integers
-6, -5	are two consecutive integers

You should see that each integer is just *1 more than* the integer immediately preceding it.

Example 8. Write an algebraic expression to represent the sum of any three consecutive integers.

Solution

We Think	***We Write***
We know nothing about the first integer.	Let x = first integer
The next consecutive integer is 1 *more than* the first integer.	$x + 1$ = second integer
The third consecutive integer is 1 *more than* the second integer.	$(x + 1) + 1$ = third integer $x + 2$ = third integer

The *sum* of the three integers is found by adding.	$x + (x + 1) + (x + 2)$ $= 3x + 3$

The expression $3x + 3$ represents the sum of any three consecutive integers.

Now you try to complete Example 9.

Example 9. Write an expression for the product of any three consecutive integers.

Solution

We Think	***We Write***
We know how to represent each of three consecutive integers (from Example 8).	Let x = first integer x + ____ = second integer x + ____ = third integer
To find the *product* of the three integers, we must ________ .	$x(x$ + ____ $)(x$ + ____ $)$ =

The expression ________ represents the product of any three consecutive integers.

Check your work on page 210. ▶

Trial Run

________ 1. If one leg of a right triangle is 8 inches shorter than twice the other leg, write an expression for the square of the hypotenuse. Simplify your answer.

________ 2. If two opposite sides of a square are increased by 3 meters and the other two sides are doubled, write an expression for the area of the resulting rectangle. Simplify your answer.

________ 3. Write an expression for four times the product of three consecutive integers. Simplify your answer.

Answers are on page 210.

▶ Examples You Completed

Example 2 (*Solution*)

We Write

Let s = length of side of old square.
$s + 5$ = length of side of new square.

$(s + 5)^2$ = area of new square

$(s + 5)^2 = s^2 + 10s + 25$

The expression $s^2 + 10s + 25$ square meters represents the area of the new square.

Example 4 *(Solution)*

We Write

Volume = (length)(width)(height)

Let h = height

$h + 3$ = width

$h + 10$ = length

$$h(h + 3)(h + 10)$$
$$= h(h^2 + 13h + 30)$$
$$= h^3 + 13h^2 + 30h$$

The expression for the volume of the shoe box is $h^3 + 13h^2 + 30h$ cubic inches.

Example 5 *(Solution)*

We Write

$$\text{Area} = \frac{a \cdot b}{2}$$

Let a = length of longer leg

$a - 3$ = length of shorter leg

$$\text{Area} = \frac{a(a-3)}{2}$$
$$= \frac{a^2 - 3a}{2}$$

The expression $\frac{a^2 - 3a}{2}$ square inches represents the area of the right triangle.

Example 9. Write an expression for the product of any three consecutive integers.

Solution

We Think	*We Write*
We know how to represent each of three consecutive integers (from Example 8).	Let x = first integer $x + 1$ = second integer $x + 2$ = third integer
To find the *product* of the three integers, we must multiply.	$x(x + 1)(x + 2)$ $= x(x^2 + 3x + 2)$ $= x^3 + 3x^2 + 2x$

The expression $x^3 + 3x^2 + 2x$ represents the product of any three consecutive integers.

Answers to Trial Run

page 209 1. $5x^2 - 32x + 64$ 2. $2x^2 + 6x$ 3. $4x^3 + 12x^2 + 8x$

Name ______________________________ Date ______________

EXERCISE SET 3.6

Write the indicated polynomial expressions.

________ 1. One integer is 12 more than twice another. Write an expression for their product.

________ 2. One integer is 4 less than three times another. Write an expression for their product.

________ 3. One integer is 7 less than another. Write an expression for the sum of the square of the larger integer and five times the product of the two integers.

________ 4. One integer is 2 more than another. Write an expression for the difference of the square of the larger integer and three times the product of the two integers.

________ 5. Write an expression for nine times the sum of the squares of two consecutive integers.

________ 6. Write an expression for four times the product of two consecutive integers.

________ 7. Write an expression for seven times the product of three consecutive integers.

________ 8. Write an expression for the sum of the squares of three consecutive integers.

________ 9. If the width of a rectangle is 8 feet less than twice the length, write an expression for the area of the rectangle.

________ 10. If the length of a rectangle is 6 feet more than three times the width, write an expression for the area of the rectangle.

________ 11. The width of a rectangle is 8 inches less than the length. Write an expression for the perimeter of the rectangle.

________ 12. The length of a rectangle is 12 inches more than the width. Write an expression for the perimeter of the rectangle.

________ 13. If each side of a square is increased by 10 centimeters, write an expression for the difference between the new area and the old area.

________ 14. If each side of a square is decreased by 9 centimeters, write an expression for the difference between the old area and the new area.

________ 15. One leg of a right triangle is 5 feet more than twice the other leg. Write an expression for the square of the hypotenuse.

________ 16. One leg of a right triangle is 1 yard more than four times the other leg. Write an expression for the square of the hypotenuse.

________ 17. Write an expression for the area of a right triangle if one leg is 7 inches less than the other.

_______ 18. Write an expression for the area of a right triangle if one leg is 8 meters more than the other.

_______ 19. If one number is 3 less than twice another, write an expression for the product of the two numbers decreased by three times their sum.

_______ 20. If one number is 4 more than three times another, write an expression for five times the sum of their squares.

_______ 21. The Caseys plan to buy a rectangular lot that is 50 feet longer than it is wide. Write an expression for the area of the lot.

_______ 22. The Brinkleys plan to fence a dog pen that is to be 3 feet longer than twice its width. Write an expression for the area of the pen.

_______ 23. A flower bed is in the shape of a right triangle with one leg 9 feet shorter than the other leg. Write an expression for the area of the flower bed.

_______ 24. Donita is building a corner table for her kitchen. The top will be a right triangle with equal legs. Write an expression for the area of the tabletop.

_______ 25. A homeowner is planning to enlarge the recreation room of her home. The room now measures 4 meters by 5 meters. If she enlarges the room by adding the same amount to each side, write an expression for the area of the room after it has been enlarged.

_______ 26. Mr Abell's garden plot is in the shape of a square. Next year, he plans to enlarge his garden by increasing two opposite sides by 6 feet and the other two sides by 4 feet. Write an expression for the area of the garden plot after it has been enlarged.

_______ 27. The Hickman's rectangular patio is 8 feet longer than it is wide. If they decide to increase each dimension by 5 feet, write an expression for the amount of patio area they will gain.

_______ 28. Irene plans to enlarge her kitchen. The room now measures 9 feet by 12 feet. If she enlarges the room by adding the same number of feet to each dimension, write an expression for the amount of area she will gain.

_______ 29. The floor of a shed is a square. If the height of the shed is 4 feet more than the width of the floor, write an expression for the volume of the shed.

_______ 30. Mrs. Worth has a square cake pan that is 2 inches deep. She has a larger rectangular cake pan 2 inches deep whose width is 1 inch more than one side of the square pan and whose length is 3 inches more than one side of the square pan. Write an expression for the difference in the volumes of the two pans.

☆ Stretching the Topics

_______ 1. A rectangular cattle pen next to a barn is to be enclosed using 300 feet of fence on three sides and the barn wall on the fourth side. Write an expression for the area that will be enclosed.

________ 2. A rectangular piece of cardboard is three times as long as it is wide. A 6-centimeter square is cut from each corner and the sides are turned upward to form a box. Write an expression for the volume of the box.

________ 3. Write an expression for the area of a right triangle if the hypotenuse is 25 inches and the perimeter is 60 inches.

Check your answers in the back of your book.

If you can complete **Checkup 3.6,** you are ready to do the **Review Exercises** for Chapter 3.

Name ______________________ Date ____________

✓ CHECKUP 3.6

Write the indicated polynomial expressions.

________ **1.** The length of a rectangle is 3 feet shorter than four times the width. Write an expression for the area of the rectangle.

________ **2.** Write an expression for the area of a right triangle if one leg is 5 feet more than twice the other leg.

________ **3.** One leg of a right triangle is 6 inches less than the other. Write an expression for the square of the hypotenuse.

________ **4.** Write an expression for twice the product of two consecutive integers.

________ **5.** If the side of a square is increased by 15 centimeters, write an expression for the difference between the old area and the new area.

Check your answers in the back of your book.

If You Missed Problem:	You Should Review Examples:
1	1
2	5
3	7
4	9
5	2, 3

Summary

In this chapter we discussed definitions and rules for working with *integer* **exponents.**

	Symbols	Examples
Definition	$a^n = \underbrace{a \cdot a \cdot \ldots \cdot a}_{n \text{ factors}}$	$(-3)^4 = (-3)(-3)(-3)(-3) = 81$ $x^3 = x \cdot x \cdot x$
Definition	$a^0 = 1 \quad (a \neq 0)$	$x^0 = 1$ $(3x + y)^0 = 1$
Definition	$a^{-n} = \dfrac{1}{a^n} \quad (a \neq 0)$	$(-3)^{-2} = \dfrac{1}{(-3)^2} = \dfrac{1}{9}$ $x^{-1}y^{-3} = \dfrac{1}{xy^3}$
First law of exponents	$a^m \cdot a^n = a^{m+n}$	$x^4 \cdot x^3 = x^7$ $y^5 \cdot y^{-6} = y^{-1} = \dfrac{1}{y}$
Second law of exponents	$(a^m)^n = a^{m \cdot n}$	$(x^4)^3 = x^{12}$ $(y^5)^{-6} = y^{-30} = \dfrac{1}{y^{30}}$
Third law of exponents	$(ab)^n = a^n b^n$	$(3x)^2 = 3^2x^2 = 9x^2$ $(-2x^{-2}y)^{-3} = (-2)^{-3}x^6y^{-3}$ $= \dfrac{x^6}{(-2)^3y^3} = -\dfrac{x^6}{8y^3}$
Fourth law of exponents	$\left(\dfrac{a}{b}\right)^n = \dfrac{a^n}{b^n} \quad (b \neq 0)$	$\left(\dfrac{5}{a}\right)^3 = \dfrac{5^3}{a^3} = \dfrac{125}{a^3}$ $\left(\dfrac{xy^{-1}}{2z^{-2}}\right)^2 = \dfrac{x^2y^{-2}}{2^2z^{-4}} = \dfrac{x^2z^4}{4y^2}$
Fifth law of exponents	$\dfrac{a^m}{a^n} = a^{m-n} \quad (a \neq 0)$	$\dfrac{x^7}{x^2} = x^5$ $\dfrac{x^3y^{-1}}{x^{-2}y} = x^{3-(-2)}y^{-1-1} = x^5y^{-2} = \dfrac{x^5}{y^2}$

We learned to recognize **polynomials** as sums of constant terms and/or variable terms whose variable parts appear with natural number exponents. To add and subtract polynomials, we used the techniques for combining like terms. To multiply polynomials, we used the distributive property and the laws of exponents. In multiplication we learned to recognize certain special products.

Factors	Special Product	Example
$(A + B)(A - B)$	$= A^2 - B^2$	$(x + 3y)(x - 3y) = x^2 - 9y^2$
$(A + B)^2$	$= A^2 + 2AB + B^2$	$(3x + 5z)^2 = 9x^2 + 30xz + 25z^2$
$(A - B)^2$	$= A^2 - 2AB + B^2$	$(2x - yz)^2 = 4x^2 - 4xyz + y^2z^2$

To divide polynomials, we learned to use **long division**. When the divisor is of the form $x - c$, we discovered that the method of **synthetic division** can be used to find a quotient.

Finally, we learned to use polynomials to switch from words to variable expressions, and we noted the very important **Pythagorean theorem** for right triangles.

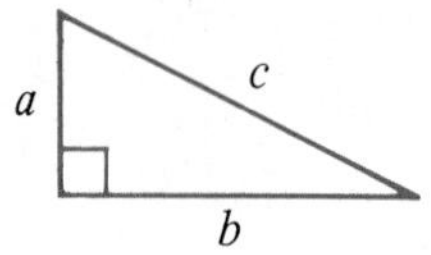

$$a^2 + b^2 = c^2$$

Speaking the Language of Algebra

Complete each sentence with the appropriate word or phrase.

1. In the expression a^n, a is called the ________ and n is called the ________ . The expression a^n tells us to use ________ as a factor in a product ________ times.

2. To multiply powers of the same base, we keep the base and ________ the exponents.

3. To raise a power of a base to another power, we keep the base and ________ the exponents.

4. To raise a product to a power, we raise each ________ to that power.

5. To raise a fraction to a power, we raise the ________ to the power and raise the ________ to the power.

6. To divide powers of the same base, we ________ the exponent in the ________ from the exponent in the ________ .

7. The expression $3x + 2$ is called a ________-degree ________ .

8. The expression $x^2 + 7x - 1$ is called a ________-degree ________ .

9. If the sum of two terms is multiplied by the difference of the same two terms, the product is the ________ ________ ________ ________ of the two terms.

10. According to the Pythagorean theorem, if the square of one leg of a right triangle is added to the square of the other leg, the sum equals the ________ ________ ________ ________ .

Writing About Mathematics

Write your response to each question in complete sentences.

1. Explain the difference between the expressions.
   (a) -3^2 and $(-3)^2$ **(b)** -2^{-4} and $(-2)^{-4}$

2. Explain the short cuts you could use to rewrite the expression $\dfrac{3^{-2}x^3}{y^{-1}z}$ without negative exponents.

3. Explain the relationship between the distributive property and the FOIL method in finding the product of two binomials. (Use a specific example if you wish.)

4. Explain what is *wrong* with each of the following statements.
   (a) $(3x^2y)^4 = 3x^8y^4$ **(b)** $x^{-2} + 1$ is a binomial.
   (c) $(x + 4)^2 = x^2 + 16$ **(d)** $\dfrac{x^2 - 5}{x} = x - 5$

Name ________________________ Date ____________

REVIEW EXERCISES for Chapter 3

Use the laws of exponents to simplify each expression.

________ 1. $2x(-5xy^2)(5x^4y^3)$

________ 2. $(-7y^5)^2$

________ 3. $(-2x^3)^5(-3x)^2$

________ 4. $\left(\frac{-15x^4}{5y^2}\right)^3$

________ 5. $\left(\frac{2a^3b^0}{3a^2c}\right)^3$

________ 6. $\frac{28a^5b^4}{6a^2b^4}$

________ 7. Evaluate $(4)^{-2}$.

________ 8. Evaluate $(-2)^{-3}$.

Simplify each expression and write the answer with positive exponents.

________ 9. $(x^3y^{-2}z^{-1})(x^{-5}y^2z^4)$

________ 10. $5x^{-3}$

________ 11. $(x^{-3}y^5z^{-2})^{-2}$

________ 12. $\frac{x^{-12}y^6}{x^{-9}y^{-6}}$

________ 13. $\left(\frac{2x^3y^{-1}}{x^{-4}y^3}\right)^{-3}$

________ 14. $\left(\frac{3x^{-7}y}{x^{-5}y^{-3}}\right)^2$

Name each polynomial.

________ 15. $6xy$

________ 16. $3x^3 - 2x + 1$

Rewrite each polynomial in descending powers of the variable.

________ 17. $y^3 - y^2 + 2y^4 + 6$

________ 18. $3x - 5 - 2x^2$

Perform the indicated operations and simplify.

________ 19. $7x^2 - 2(x^2 - 3x) + 6(-x + 5) - 7$

________ 20. $7a - 3[2a - 4(a - 2b) - 6b] + 5b$

________ 21. $3x(x^2 - 5x + 6)$

________ 22. $3mn(7m - 2n) - mn(4m - 7n)$

________ 23. $6a^2 - 2[4a(a - 2) - 3a(a^2 - 1)]$

________ 24. $-2xy^3(3x - 4x^2y^2) + x^2y(4y^2 - xy^4)$

________ 25. $(3x - 1)(x + 5)$

________ 26. $(a - 7b)^2$

________ 27. $(4x - 3y)(4x + 3y)$

_______ 28. $(x - 2)(x^2 - 4x - 3)$

_______ 29. $4x(2x - y) - (x - 3y)^2$

_______ 30. $3(a - 1)(a - 7) - (a + 2)(2a - 3)$

_______ 31. $(a - 2b)(a + b)(2a - b)$

_______ 32. $(a - 2)^3$

_______ 33. $(x^2 - x - 1)(x^2 + 2x - 3)$

_______ 34. $[(a + 2b) + 3]^2$

_______ 35. $(5a^3b - 15a^2b^2 + 10ab^3) \div 5ab$

_______ 36. $(3x^2 + 17x - 9) \div (x + 6)$

_______ 37. $(x^3 + 64) \div (x + 4)$

_______ 38. $(x^3 + 5x^2y - 4xy^2 - 20y^3) \div (x + 2y)$

Use synthetic division to find each quotient.

_______ 39. $(x^3 + 27) \div (x + 3)$

_______ 40. $(2x^3 - x^2 + x - 2) \div (x - 2)$

_______ 41. $(3x^4 - x^3 - 2x^2 + 2) \div (x + 1)$

Write the indicated polynomial expressions.

_______ 42. If one leg of a right triangle is five times the other leg, write an algebraic expression to represent the area of the triangle.

_______ 43. Write an expression for five times the product of two consecutive integers.

_______ 44. If two opposite sides of a square are increased by 5 centimeters and the other two sides are decreased by 3 centimeters, write an expression for the area of the resulting rectangle.

_______ 45. If one number is 2 more than three times the other, write an expression for the sum of the two numbers decreased by twice their product.

Check your answers in the back of your book.

If You Missed Exercises:	You Should Review Examples:	
1	SECTION 3.1	5–7
2, 3		11–16
4, 5		17, 18
6		19, 20
7, 8	SECTION 3.2	1–4
9		7, 8
10		5
11		9, 10
12		12
13, 14		16–18
15, 16	SECTION 3.3	1
17, 18		2, 3
19, 20		8–10
21–24		11–15
25–27	SECTION 3.4	1–7
28		11, 12
29, 30		8, 9
31–33		13, 14
34		21
35	SECTION 3.5	1, 2
36–38		3–5
39–41		6–9
42	SECTION 3.6	5
43		8, 9
44		2
45		8, 9

If you have completed the **Review Exercises** and corrected your errors, you are ready to take the **Practice Test** for Chapter 3.

Name ______________________ Date ____________

PRACTICE TEST for Chapter 3

	Problem	Section	Examples
________	1. Evaluate $-x^2 + 4x$ when $x = -8$.	3.1	1, 2
________	2. Evaluate x^{-5} when $x = -2$.	3.2	2

Perform the indicated operations and simplify. Write answers with positive exponents.

	Problem	Section	Examples
________	3. $(-4a^7b)(-2a^5b^2)$	3.1	5
________	4. $-6x^2y^2(x^4y)(\frac{1}{3}xy)$	3.1	7
________	5. $(-5x^3y)^3$	3.1	12
________	6. $(-a^4b^2c^3)^3$	3.1	14
________	7. $\left(\dfrac{-9x^3}{y^2z^3}\right)^2$	3.1	17
________	8. $\dfrac{-18a^4b^5c^5}{9a^3b^2c^5}$	3.1	19
________	9. $\left(\dfrac{x^5yz^2}{2x^2yz}\right)^3$	3.1	17, 19
________	10. $-4x(x^{-3}y^2)(-2x^4y^{-3})$	3.2	8
________	11. $(a^{-1}b^3c^{-2})^{-3}$	3.2	10
________	12. $\left(\dfrac{-x^{-3}y^{-1}}{z}\right)^{-3}$	3.2	17
________	13. $\dfrac{-2x^{-1}y^{-3}}{x^5y^{-7}}$	3.2	12
________	14. $x^2 - 7(x^2 - 9x) - 3(2 - 4x) - 6x$	3.3	8
________	15. $5a - 6[2(1 - 3a + a^2) - 4(a^2 - 1)]$	3.3	10
________	16. $(4x - 7)(3x + 5)$	3.4	1
________	17. $(3c - 7d)^2$	3.4	3
________	18. $(9y - 2z)(9y + 2z)$	3.4	4
________	19. $(x - 5)(3x^2 - 2x - 1)$	3.4	11
________	20. $(a - 5)^3$	3.4	13

		Section	Examples
________	21. $(3x + 1)(x - 2) - (x + 4)(x - 4)$	3.4	9
________	22. $(-20x^3y^3 - 5x^2y^2 + 10xy) \div (5xy)$	3.5	1
________	23. $(7x^2 + 3x - 2) \div (x - 3)$	3.5	6
________	24. $(y^3 + 27) \div (y + 3)$	3.5	7

Write the indicated polynomial expression.

________	25. If one leg of a right triangle is 8 inches shorter than the other leg, write an expression for the square of the hypotenuse.	3.6	7
________	26. Write an expression for 3 times the sum of two consecutive integers.	3.6	8, 9
________	27. If the length of a rectangle is 2 feet longer than three times the width, write an expression for the area of the rectangle.	3.6	1

Name ________________________________ **Date** ______________

SHARPENING YOUR SKILLS after Chapters 1–3

		SECTION
________	1. Compare $-\frac{11}{2}$ and 4.5 using <, >, or =.	1.1
________	2. Use an inequality to state whether 8.2 is positive or negative.	1.1
________	3. Evaluate $4a^2 - 7a - 20$ when $a = -1$.	1.4
________	4. Evaluate $\|8 - 15\|$.	1.4
________	5. Name the property illustrated by $(-3) + 8 = 8 + (-3)$.	1.3

Perform the indicated operations and simplify.

________	6. $\dfrac{0}{9 - 4}$	1.2
________	7. $7 - 11[5 - 3(2)]$	1.3
________	8. $-\frac{1}{7}(7x - 21y) + 3[x - (x - 2y)]$	1.4
________	9. $-9x^2 + 4x - 3 + 7x^2 - 4x + 1$	1.4

Solve for the variable.

________	10. $-3x + 1 = 7x - 4$	2.1
________	11. $3(3y - 1) - 9 = 5(4 + y)$	2.1
________	12. $5(x + 1) > 6(x + 2)$	2.3
________	13. $0 \leq \dfrac{-y + 5}{2} < 3$	2.3
________	14. $\|9 - x\| = 13$	2.4
________	15. $\|5a - 3\| \leq 7$	2.4

Solve each pair of inequalities by graphing. Write answers in interval notation.

________	16. $x < -3$ and $x > \frac{3}{2}$	2.2
________	17. $\dfrac{x - 2}{5} \geq -1$ and $5(2 - x) > 20$	2.3
________	18. $2(3 - x) \geq 0$ or $6(x - 2) > 3x + 6$	2.3
________	19. Clara has $500 in a savings account. She finds that her present salary is not adequate to meet her expenses, so she has to use $15 per month from her savings. Write an algebraic expression for the amount left in savings after n months.	1.5

SECTION

________ **20.** Simon walks and jogs every evening for 2 hours. He jogs at 3 miles per hour and walks at 2 miles per hour. If the total distance covered this evening was $5\frac{1}{2}$ miles, how much time did Simon spend walking? 2.5

C H A P T E R

4

Factoring

A company has found that its *revenue* from the sale of x units of a product is $2x^2 - x + 10$ dollars, while its *costs* for producing x units are $x^2 - 3x + 109$ dollars. Write an expression for the company's *profit* when x units are produced and sold. Then write that expression in factored form and find the profit when the company produces and sells 6 units; when it sells 19 units.

Having learned to multiply algebraic expressions, we must now learn how to go backward. We would like to look at an algebraic expression and try to decide what factors might have been multiplied together to yield that expression. We call this process **factoring** an algebraic expression.

Factoring is the process by which we rewrite a sum of terms as a *product*.

Factoring is one of the most useful and important skills to be mastered in the study of algebra. In this chapter we learn how to

1. Factor out a common factor.
2. Factor special products.
3. Factor trinomials.

4.1 Common Factors

Common Monomial Factors

The first type of polynomial multiplication we studied was multiplication by a monomial, and we learned to use the *distributive property* to find products such as these:

$$3(x + 2y) = 3x + 6y \qquad x(5 - x) = 5x - x^2 \qquad -7a(a^2 + 2a - 3) = -7a^3 - 14a^2 + 21a$$

Now we would like to rewrite a polynomial such as $5a + 5b$ as a *product* of a monomial and a polynomial. We see that both terms contain the numerical factor 5. We say that 5 is a **common factor** for both terms of the polynomial and we write

$$5a + 5b = 5(a + b)$$

The polynomial on the left has now been written as a product of the factors on the right. In other words, we have *factored* our polynomial, and we can check our factors by multiplying to see if their product is actually the original polynomial.

Consider another polynomial to be rewritten in factored form: $6x^2 + 9x$. We agree to look for *all* factors common to all terms of the polynomial. Here we see the common numerical factor 3 *and* the common variable factor x. In other words, the **greatest common monomial factor** is $3x$, and the factored form for our polynomial is

$$6x^2 + 9x = 3x(2x + 3)$$

We repeat here that we should always try to "factor out" the greatest common factor. Keep these ideas in mind:

1. First look for the largest numerical factor common to *all* terms of the polynomial.
2. If *all* terms in the polynomial contain the same variable (or variables), we must factor out the *lowest* power of that variable appearing in the polynomial.

Since factors can always be checked by multiplication, there is no reason for you to factor incorrectly. If the product of your factors is not the original polynomial, you know you have made an error.

You try factoring some polynomials.

Example 1. Factor $2x^9 - 10x^7$.

Solution. The common monomial factor is ____ .

$$2x^9 - 10x^7 = 2x^7(\_\_\_\_ - \_\_\_\_)$$

Check your work on page 231. ▶

Example 2. Factor $14x^3y + 7xy^2 - 21xy$.

Solution. The common monomial factor is ____ .

$$14x^3y + 7xy^2 - 21xy = 7xy(\_\_\_\_ + \_\_\_\_ - \_\_\_\_)$$

Check your work on page 231. ▶

Example 3. Factor $-3x^2 - 6x + 3$.

Solution. Each term contains the numerical factor 3 or -3, but which should be factored out? In general, we agree that if the first numerical coefficient in the polynomial (sometimes called the **leading coefficient**) is negative, then we'll factor out a negative common factor.

$$-3x^2 - 6x + 3 = -3(x^2 + 2x - 1)$$

Notice that we were very careful about the signs in the second factor.

Now try completing Examples 4 and 5.

Example 4. Factor $4x^2 - 16y^3$.

Solution. The common monomial factor is ______.

$$4x^2 - 16y^3$$
$$=$$

Check your work on page 231. ▶

Example 5. Factor $-x^7y^3 + 4x^5y^4 - x^3y^3$.

Solution. The common monomial factor is ______ .

$$-x^7y^3 + 4x^5y^4 - x^3y^3$$
$$= -x^3y^3(\underline{\qquad\qquad\qquad})$$

Check your work on page 231. ▶

Often the terms in a polynomial will contain no common factors. In the expression $3x^2 + 5y$, there is *no* common factor; we cannot rewrite this polynomial as a product.

When a polynomial cannot be written as a product of factors, we say that the polynomial is **prime** (or **irreducible**).

Although we shall discuss other types of factoring, the common monomial factor is *always* the first thing we look for in factoring polynomials.

Trial Run

Rewrite each expression as a product of factors.

______ 1. $3x^6 - 4x^8$

______ 2. $12a^2 - 18a + 36$

______ 3. $9x^2z - 45y^2z$

______ 4. $10x^3y + 5xy^2 - 15xy$

______ 5. $-7x^2 + 7x$

______ 6. $-x^8y^5 + 3x^6y^3 - 5x^4y^2$

Answers are on page 232.

Other Common Factors

Factoring an expression such as

$$a(x + y) + 2(x + y)$$

requires the same skill that you have just mastered. Notice that this expression has two terms and that the factor $(x + y)$ is *common* to both terms. Therefore, we use the distributive property to factor the expression as

$$a(x + y) + 2(x + y)$$
$$= (x + y)(a + 2)$$

Now try Example 6.

Example 6. Factor $c(x - 3y) - 5(x - 3y)$.

Solution. The common binomial factor is $(x - 3y)$.

$$c(x - 3y) - 5(x - 3y)$$
$$= (x - 3y)(\underline{\qquad} - \underline{\qquad})$$

Check your work on page 232. ▶

Example 7. Factor $(b + c)^2 + (b + c)$.

Solution. The common binomial factor is $(b + c)$.

$$(b + c)^2 + (b + c)$$
$$= (b + c)[(b + c) + 1]$$
$$= (b + c)(b + c + 1)$$

Example 8. Factor $4(x - y)^3 - (x - y)^4$.

Solution. The common binomial factor is $(x - y)^3$.

$$4(x - y)^3 - (x - y)^4$$
$$= (x - y)^3[4 - (x - y)]$$
$$= (x - y)^3(4 - x + y)$$

You complete Example 9.

Example 9. Factor

$$(x - 1)^3 - 4(x - 1)^2 + 2(x - 1)$$

Solution. The common binomial factor is $\underline{\qquad}$.

$$(x - 1)^3 - 4(x - 1)^2 + 2(x - 1)$$
$$= (x - 1)[(\underline{\qquad})^2 - 4(\underline{\qquad}) + \underline{\qquad}]$$

Check your work on page 232. ▶

As you can see, removing common factors requires only that you are familiar with the distributive property and the laws of exponents.

Trial Run

Rewrite each expression as a product of factors.

$\underline{\qquad}$ 1. $x(x + 5) - 3(x + 5)$

$\underline{\qquad}$ 2. $a(x - y) + b(x - y)$

$\underline{\qquad}$ 3. $(a + b)^2 + 9(a + b)$

$\underline{\qquad}$ 4. $(x + y)(m + n) - z(m + n)$

Answers are on page 232.

Factoring by Grouping

Suppose that we use the idea of removing common factors to factor expressions containing *four terms*. In the expression

$$ax + 3x + ay + 3y$$

there is no factor common to all four terms. However, we notice that the first pair of terms contains the common factor x, and the second pair of terms contains the common factor y.

$ax + 3x + ay + 3y = (ax + 3x) + (ay + 3y)$ — We can group the expression into two pairs of terms.

$= x(a + 3) + y(a + 3)$ — Remove the common factor from each pair of terms. Now the expression is a sum of *two* terms.

$= x(a + 3) + y(a + 3)$ — Notice the common binomial factor $(a + 3)$ in both terms.

$= (a + 3)(x + y)$ — Remove the common binomial factor $(a + 3)$.

This procedure is called **factoring by grouping** because it requires that we group the terms two by two so that each twosome contains a common factor.

You try completing Example 10.

Example 10. Factor $5x - 5 + xy - y$.

Solution

$$5x - 5 + xy - y$$
$$= (5x - 5) + (xy - y)$$
$$= 5(x - 1) + y(x - 1)$$
$$= (x - 1)(\_\_\_\_ + \_\_\_\_)$$

Check your work on page 232. ▶

Example 11. Factor $ax + y + ay + x$.

Solution. We must rearrange the terms.

$$ax + y + ay + x$$
$$= ax + ay + x + y$$
$$= (ax + ay) + (x + y)$$
$$= a(x + y) + 1(x + y)$$
$$= (x + y)(a + 1)$$

Example 12. Factor $ab - 3b - 2a + 6$.

Solution

$$ab - 3b - 2a + 6$$
$$= (ab - 3b) + (-2a + 6)$$
$$= b(a - 3) - 2(a - 3)$$
$$= (a - 3)(b - 2)$$

Notice that we factored -2 from the second pair of terms so that $(a - 3)$ would be a common factor.

Now you complete Example 13.

Example 13. Factor $x^2 + ax - x - a$.

Solution

$$x^2 + ax - x - a$$
$$= (x^2 + ax) + (-x - a)$$
$$= x(\_\_\_\_\_\_) - 1(\_\_\_\_\_\_)$$
$$=$$

Check your work on page 232. ▶

Trial Run

Rewrite each expression as a product of factors.

_______ **1.** $7x + 7y + ax + ay$

_______ **2.** $3ax - 15ay + 2x - 10y$

_______ **3.** $8a^2 + 4a - 2ab - b$

_______ **4.** $12x - 4x^3 - 15 + 5x^2$

Answers are on page 232.

▶ Examples You Completed

Example 1. Factor $2x^9 - 10x^7$.

Solution. The common monomial factor is $2x^7$.

$$2x^9 - 10x^7 = 2x^7(x^2 - 5)$$

Example 2. Factor $14x^3y + 7xy^2 - 21xy$.

Solution. The common monomial factor is $7xy$.

$$14x^3y + 7xy^2 - 21xy = 7xy(2x^2 + y - 3)$$

Example 4. Factor $4x^2 - 16y^3$.

Solution. The common monomial factor is 4.

$$4x^2 - 16y^3 = 4(x^2 - 4y^3)$$

Example 5. Factor $-x^7y^3 + 4x^5y^4 - x^3y^3$.

Solution. The common monomial factor is $-x^3y^3$.

$$-x^7y^3 + 4x^5y^4 - x^3y^3$$
$$= -x^3y^3(x^4 - 4x^2y + 1)$$

Example 6. Factor $c(x - 3y) - 5(x - 3y)$.

Solution. The common binomial factor is $(x - 3y)$.

$$c(x - 3y) - 5(x - 3y)$$
$$= (x - 3y)(c - 5)$$

Example 9. Factor

$$(x - 1)^3 - 4(x - 1)^2 + 2(x - 1)$$

Solution. The common binomial factor is $x - 1$.

$$(x - 1)^3 - 4(x - 1)^2 + 2(x - 1)$$
$$= (x - 1)[(x - 1)^2 - 4(x - 1) + 2]$$

Example 10. Factor $5x - 5 + xy - y$.

Solution

$$5x - 5 + xy - y$$
$$= (5x - 5) + (xy - y)$$
$$= 5(x - 1) + y(x - 1)$$
$$= (x - 1)(5 + y)$$

Example 13. Factor $x^2 + ax - x - a$.

Solution

$$x^2 + ax - x - a$$
$$= (x^2 + ax) + (-x - a)$$
$$= x(x + a) - 1(x + a)$$
$$= (x + a)(x - 1)$$

Answers to Trial Runs

page 229 1. $x^6(3 - 4x^2)$ 2. $6(2a^2 - 3a + 6)$ 3. $9z(x^2 - 5y^2)$ 4. $5xy(2x^2 + y - 3)$ 5. $-7x(x - 1)$ 6. $-x^4y^2(x^4y^3 - 3x^2y + 5)$

page 230 1. $(x + 5)(x - 3)$ 2. $(x - y)(a + b)$ 3. $(a + b)(a + b + 9)$ 4. $(m + n)(x + y - z)$

page 231 1. $(x + y)(7 + a)$ 2. $(x - 5y)(3a + 2)$ 3. $(2a + 1)(4a - b)$ 4. $(3 - x^2)(4x - 5)$

Name ________________ Date ________

EXERCISE SET 4.1

Rewrite each expression as a product of factors.

_____ 1. $50x^2 - 10xy + 20y^2$

_____ 2. $36x^2 - 24xy + 48y^2$

_____ 3. $2b^3 - 6b^2 - 4b$

_____ 4. $5b^3 - 10b^2 + 25b$

_____ 5. $-y^3 + y^2 - 2y$

_____ 6. $-y^3 - y^2 + 4y$

_____ 7. $3x^3 - 6x^2y + 12xy^2$

_____ 8. $5x^3 + 15x^2y + 55xy^2$

_____ 9. $-a^4b + 8a^3b^2 - 5a^2b^3$

_____ 10. $-a^5b - 4a^2b^2 + 5a^3b^3$

_____ 11. $5x^3y^2 - 80x^2y^3$

_____ 12. $7x^4y^2 - 56x^3y^3$

_____ 13. $9x^3y - 12x^2y^2 + 24xy^3$

_____ 14. $27x^3y + 45x^2y^2 - 63xy^3$

_____ 15. $72a^4 - 40a^3 + 8a^2$

_____ 16. $72a^5 - 54a^4 - 6a^3$

_____ 17. $12x^4y^2 + 8x^3y^3 + 4x^2y^4$

_____ 18. $18x^4y^2 - 12x^3y^3 - 3x^2y^4$

_____ 19. $9x^4yz + 18x^3y^2z - 27x^2y^3z - 45xy^4z$

_____ 20. $35xy^4z - 28x^2y^3z - 14x^3y^2z + 7x^4yz$

_____ 21. $y(y - 1) - 3(y - 1)$

_____ 22. $y(y + 5) - 4(y + 5)$

_____ 23. $2x(x + 3) - (x + 3)$

_____ 24. $3x(x + 8) + (x + 8)$

_____ 25. $2a(3x - y) - b(3x - y)$

_____ 26. $5a(2x + y) + 3b(2x + y)$

_____ 27. $(x - y)^2 - 3(x - y)$

_____ 28. $(2x + y)^2 + 5(2x + y)$

_____ 29. $(a + b)(x - y) - 2c(x - y)$

_____ 30. $(3a - b)(x + y) - 4c(x + y)$

_____ 31. $(3x - y)(2a + b) - (x - y)(2a + b)$

_____ 32. $(4x + y)(3a + 2b) - (x + y)(3a + 2b)$

_____ 33. $5(x - 2)^3 - 4(x - 2)^2 + (x - 2)$

_____ 34. $9(x - 3)^3 - 2(x - 3)^2 - (x - 3)$

_____ 35. $4(x + 2y)^3 - 6(x + 2y)^2 - 2(x + 2y)$

_____ 36. $9(x - y)^3 - 12(x - y)^2 - 3(x - y)$

_____ 37. $6a + ac + 6b + bc$

_____ 38. $9x + 9y + xz + yz$

_____ 39. $x^3 - 7x^2 + 6x - 42$

_____ 40. $x^3 - 11x^2 + 3x - 33$

_____ 41. $14ax + 21ay - 10x - 15y$

_____ 42. $36ax + 45ay - 28x - 35y$

_____ 43. $70a^2 - 40ab + 21a - 12b$

_____ 44. $30a^2 - 75ab + 4a - 10b$

_____ 45. $6a^2x + 3a^2y + 6b^2x + 3b^2y$

_____ 46. $5a^2x + 10a^2y + 5b^2x + 10b^2y$

_____ 47. $x^3 + x^2 + x + 1$

_____ 48. $x^3 - 3x^2 + x - 3$

_____ 49. $9ab + 7c - 9ac - 7b$

_____ 50. $8xy + 3z - 8xz - 3y$

_____ 51. $2ax^2 - 3b - 3bx^2 + 2a$

_____ 52. $ax + 5by - bx - 5ay$

☆ Stretching the Topics

_______ 1. Factor $x^{5a} + 4x^{4a}$.

_______ 2. Factor $x^{3a+2} + 3x^{2a+2} - 5x^2$.

_______ 3. From the expression $x^{-3} + 5x^{-2} + x^{-1}$, factor out x^{-3}.

Check your answers in the back of your book.

If you can factor the problems in **Checkup 4.1,** you are ready to go on to Section 4.2.

Name ______________________ Date ____________

✓ CHECKUP 4.1

Rewrite each expression as a product of factors.

________ **1.** $35x^2 - 14x + 7$

________ **2.** $7b^3 - 14b^2 + 28b$

________ **3.** $5x^2y - 25y^3$

________ **4.** $-a^6b^2 + 3a^5b^4 + 6a^4b^6$

________ **5.** $9x^5y^3z - 27x^4y^2z - 18x^3yz$

________ **6.** $2x(x - 9) - 3(x - 9)$

________ **7.** $(a + b)^2 - 12(a + b)$

________ **8.** $6(x - y)^2 - (x - y)^3$

________ **9.** $2ax + 2bx - 3ay - 3by$

________ **10.** $x^3 - 7x^2 + x - 7$

Check your answers in the back of your book.

If You Missed Problems:	You Should Review Examples:
1–3	1–4
4	5
5	2
6	6
7	7
8	8
9, 10	10–13

4.2 Factoring Special Products

In the section on multiplying polynomials, we learned that when certain kinds of factors are multiplied, they yield certain **special products.** In factoring, it is often possible to spot such a special product and know immediately what its factors are.

Factoring the Difference of Two Squares

From multiplication we know that

$(x + 3)(x - 3)$	$(y - x)(y + x)$	$(2a + \frac{5}{7}b)(2a - \frac{5}{7}b)$
$= x^2 - 9$	$= y^2 - x^2$	$= 4a^2 - \frac{25}{49}b^2$

In other words, we recall that the product of the sum and difference of the same two terms is the **difference of the squares** of those terms. Going backwards with this rule, you should agree that a polynomial that is a difference of the squares of two terms can be factored into a product of the sum and difference of those two terms.

Therefore, we see that

$x^2 - 0.16$	$a^2 - 9y^2$	$x^2y^2 - z^2$
$= (x + 0.4)(x - 0.4)$	$= (a + 3y)(a - 3y)$	$= (xy + z)(xy - z)$

and in general we may say

Difference of Two Squares

$$A^2 - B^2 = (A + B)(A - B)$$

You try Example 1.

Example 1. Factor $4a^2 - 49y^2$.

Solution

$4a^2 - 49y^2$

$= (\_\_\_\_\_\_\_\_)(\_\_\_\_\_\_\_\_)$

Check your work on page 241. ▶

Example 2. Factor $5x^2 - 20$.

Solution. First we notice a common monomial factor.

$5x^2 - 20$

$= 5(x^2 - 4)$ Factor out 5.

$= 5(x + 2)(x - 2)$ Factor difference of squares.

Now try Example 3.

Example 3. Factor $9 - 81y^2$.

Solution

$9 - 81y^2$

$= 9(\_\_\_\_\_\_\_\_)$ Factor out 9.

$= 9(\_\_\_\_\_\_\_\_)(\_\_\_\_\_\_\_\_)$ Factor difference of squares.

Check your work on page 241. ▶

Example 4. Factor $-x^3 + x$.

Solution

$-x^3 + x$

$= -x(x^2 - 1)$ Factor out $-x$.

$= -x(x + 1)(x - 1)$ Factor difference of squares.

Let's tackle a problem that looks more complicated.

Example 5. Factor $(x + 1)^2 - 25$.

Solution. Is this polynomial a difference of two squares? Does it fit into the form $A^2 - B^2$?

$$\begin{array}{ccc} (x+1)^2 & - & 25 \\ \uparrow & & \uparrow \\ A^2 & & -B^2 \end{array}$$

We see that
$A = x + 1$
$B = 5$

So our factoring should look like this:

$$\begin{array}{ccccccccc} A^2 & - B^2 & = & (A & + & B) & (A & - & B) \\ \downarrow & \downarrow & & \downarrow & & \downarrow & \downarrow & & \downarrow \\ (x+1)^2 & - 25 & = & [(x+1) & + & 5][(x+1) & & - & 5] \end{array}$$

$$= [x + 1 + 5][x + 1 - 5]$$

$$(x + 1)^2 - 25 = (x + 6)(x - 4)$$

You try Example 6, being sure to identify A and B.

Example 6. Factor $4y^2 - (x - 3)^2$.

Solution. Is $4y^2 - (x - 3)^2$ a difference of two squares? ________

Here A = ______
B = (________)

So

$$4y^2 - (x - 3)^2$$
$$= [2y + (x - 3)][2y - (x - 3)]$$
$$= (\_\_\_\_\_\_\_\_)(\_\_\_\_\_\_\_\_)$$

Check your work on page 242. ▶

Example 7. Factor $x^4 - 16$.

Solution. Is $x^4 - 16$ a difference of two squares? Yes.

Here $A = x^2$
$B = 4$

So

$$x^4 - 16$$
$$= (x^2 + 4)(x^2 - 4)$$
$$= (x^2 + 4)(x + 2)(x - 2)$$

In Example 7, notice that we were able to factor $x^2 - 4$ as a difference of two squares, but we were *not* able to factor $x^2 + 4$. As long as we are working with real numbers and common factors have been removed, we note that

> The *sum* of two squares cannot be factored.
> The sum of two squares is a prime polynomial.

Example 8. Factor $x^3 - 3x^2 - x + 3$.

Solution. This polynomial contains four terms, so we start to factor it by *grouping*.

$x^3 - 3x^2 - x + 3$	
$= (x^3 - 3x^2) + (-x + 3)$	Group terms two by two.
$= x^2(x - 3) - 1(x - 3)$	Remove factor from each twosome.
$= (x - 3)(x^2 - 1)$	Remove common binomial factor.
$= (x - 3)(x + 1)(x - 1)$	Factor the difference of the two squares.

Trial Run

Rewrite each expression as as product of factors.

_______ 1. $x^2 - 36$

_______ 2. $25a^2 - 9b^2$

_______ 3. $2y^2 - 98$

_______ 4. $400 - 4x^2$

_______ 5. $x^4 - 81y^4$

_______ 6. $(a + b)^2 - 16c^2$

Answers are on page 242.

Factoring the Difference or Sum of Two Cubes

Just as there is a special rule for factoring the difference of two squares, there is a special rule for factoring the **difference of two cubes.** A cube is a quantity that is the *third* power of some base.

x^3 is the cube of x

a^3 is the cube of a

8 is the cube of 2, because $8 = 2^3$

1 is the cube of 1, because $1 = 1^3$

In this section we seek a method for factoring the difference of any two cubes

$$A^3 - B^3$$

Recalling that $A - B$ was one of the factors of the difference of two *squares*, $A^2 - B^2$, perhaps we should see whether $A - B$ might be a factor for the difference of two *cubes*, $A^3 - B^3$. If $A - B$ is a factor of $A^3 - B^3$, we should be able to use *long division* to find the other factor. Let's try dividing $A^3 - B^3$ by $A - B$, using long division.

$$\begin{array}{r|l}
 & A^2 + AB + B^2 \\
A - B & A^3 - 0A^2B + 0AB^2 - B^3 \\
 & \ominus A^3 \oplus\!\!\!\!- A^2B \\ \hline
 & \qquad A^2B + 0AB^2 \\
 & \qquad \ominus A^2B \oplus\!\!\!\!- AB^2 \\ \hline
 & \qquad\qquad AB^2 - B^3 \\
 & \qquad\qquad \ominus AB^2 \oplus\!\!\!\!- B^3 \\ \hline
 & \qquad\qquad\qquad 0
\end{array}$$

Since the remainder is 0, we know that $A - B$ divides exactly into $A^3 - B^3$. Therefore, $A - B$ is indeed a factor of $A^3 - B^3$ and the other factor is the quotient, $A^2 + AB + B^2$. We have discovered the factors for the difference of two cubes.

Difference of Two Cubes

$$A^3 - B^3 = (A - B)(A^2 + AB + B^2)$$

To factor a polynomial that is the difference of two cubes, we first identify A and B, and then substitute into the formula.

Example 9. Factor $y^3 - 1$.

Solution. Here we see that

$$A^3 = y^3 \qquad \text{so } A = y$$
$$B^3 = 1 \qquad \text{so } B = 1$$

We substitute in the formula.

$$A^3 - B^3 = (A - B)(A^2 + AB + B^2)$$
$$y^3 - 1 = (y - 1)(y^2 + y \cdot 1 + 1^2)$$
$$y^3 - 1 = (y - 1)(y^2 + y + 1)$$

Now you try Example 10.

Example 10. Factor $2x^3 - 54y^3$.

Solution. We see a common monomial factor of 2.

$$2x^3 - 54y^3 = 2(x^3 - 27y^3)$$

Now we see a difference of two cubes in which $A =$ ____ and $B =$ ____ .

$$2x^3 - 54y^3$$
$$= 2(x^3 - 27y^3)$$
$$= 2(x - 3y)[x^2 + x \cdot 3y + (3y)^2]$$
$$= 2(x - 3y)(\underline{\qquad\qquad})$$

Check your work on page 242. ▶

The formula for factoring the **sum of two cubes** is similar to the formula for factoring the difference of two cubes. We present it without proof.

Sum of Two Cubes

$$A^3 + B^3 = (A + B)(A^2 - AB + B^2)$$

Example 11. Factor $x^3 + 64$.

Solution. This is a sum of two cubes and we see that

$$A^3 = x^3 \qquad \text{so } A = x$$
$$B^3 = 64 \qquad \text{so } B = 4$$

We substitute in the formula.

$$A^3 + B^3 = (A + B)(A^2 - AB + B^2)$$
$$x^3 + 64 = (x + 4)(x^2 - x \cdot 4 + 4^2)$$
$$= (x + 4)(x^2 - 4x + 16)$$

Now you try completing Example 12.

Example 12. Factor $27a^3 + 8c^3$.

Solution. This is a sum of two cubes and we see that

$$A^3 = 27a^3 \qquad \text{so } A = 3a$$
$$B^3 = 8c^3 \qquad \text{so } B = 2c$$

We substitute in the formula.

$$A^3 + B^3 = (A + B)(A^2 - AB + B^2)$$
$$27a^3 + 8c^3 = (3a + 2c)[\underline{\qquad\qquad}]$$
$$=$$

Check your work on page 242. ▶

The formulas for factoring the sum or difference of two cubes should be memorized. There is no quick way to arrive at the correct factors through trial and error. If you write each formula when you use it in the exercises, you will soon find that you have memorized both of them.

Example 13. Factor $125 + (x + 1)^3$.

Solution. Is this a sum of two cubes? Yes, and we see that

$$A^3 = 125 \qquad \text{so } A = 5$$
$$B^3 = (x + 1)^3 \qquad \text{so } B = (x + 1)$$

We substitute in the formula.

$$A^3 + B^3 = (A + B)(A^2 - AB + B^2)$$

$$\begin{aligned} 125 + (x + 1)^3 &= [5 + (x + 1)][5^2 - 5(x + 1) + (x + 1)^2] \\ &= [5 + x + 1][25 - 5x - 5 + x^2 + 2x + 1] \\ &= (x + 6)(x^2 - 3x + 21) \end{aligned}$$

We simplified the right-hand side because we could see that it would contain like terms to be combined.

Example 14. Factor $x^6 - a^6$.

Solution. Is this a difference of two *cubes*? Yes, but it is also a difference of two *squares*. Let's treat it as a difference of two squares.

$$x^6 - a^6 = (x^3 + a^3)(x^3 - a^3)$$

Notice on the right-hand side that one factor is a sum of two cubes and the other is a difference of two cubes. We continue.

$$\begin{aligned} x^6 - a^6 &= (x^3 + a^3)(x^3 - a^3) \\ &= (x + a)(x^2 - ax + a^2)(x - a)(x^2 + ax + a^2) \end{aligned}$$

From Example 14 we observe that whenever an expression is a difference of two squares *and* a difference of two cubes, we should begin by factoring it as a difference of two squares.

Trial Run

Rewrite each expression as a product of factors.

_______ 1. $x^3 - 8$

_______ 2. $125 + y^3$

_______ 3. $64a^3 - 27b^3$

_______ 4. $3x^3 - 81y^3$

_______ 5. $5c^3 + 5$

_______ 6. $a^6 - 64$

Answers are on page 242.

Examples You Completed

Example 1. Factor $4a^2 - 49y^2$.

Solution

$$\begin{aligned} &4a^2 - 49y^2 \\ &= (2a + 7y)(2a - 7y) \end{aligned}$$

Example 3. Factor $9 - 81y^2$.

Solution

$$\begin{aligned} &9 - 81y^2 \\ &= 9(1 - 9y^2) \\ &= 9(1 + 3y)(1 - 3y) \end{aligned}$$

Example 6. Factor $4y^2 - (x - 3)^2$.

Solution. Is $4y^2 - (x - 3)^2$ a difference of two squares? Yes.

Here $A = 2y$
$B = x - 3$

So

$$\begin{aligned} &4y^2 - (x - 3)^2 \\ &= [2y + (x - 3)][2y - (x - 3)] \\ &= (2y + x - 3)(2y - x + 3) \end{aligned}$$

Example 10. Factor $2x^3 - 54y^3$.

Solution. We see a common monomial factor of 2.

$$2x^3 - 54y^3 = 2(x^3 - 27y^3)$$

Now we see a difference of two cubes in which

$$A = x \quad \text{and} \quad B = 3y$$

$$\begin{aligned} 2x^3 - 54y^3 &= 2(x^3 - 27y^3) \\ &= 2(x - 3y)[x^2 + x \cdot 3y + (3y)^2] \\ &= 2(x - 3y)(x^2 + 3xy + 9y^2) \end{aligned}$$

Example 12. Factor $27a^3 + 8c^3$.

Solution. This is a sum of two cubes and we see that

$$A^3 = 27a^3 \quad \text{so} \quad A = 3a$$

$$B^3 = 8c^3 \quad \text{so} \quad B = 2c$$

We substitute in the formula.

$$\begin{aligned} A^3 + B^3 &= (A + B)(A^2 - AB + B^2) \\ 27a^3 + 8c^3 &= (3a + 2c)[(3a)^2 - 3a \cdot 2c + (2c)^2] \\ &= (3a + 2c)(9a^2 - 6ac + 4c^2) \end{aligned}$$

Answers to Trial Runs

page 239 **1.** $(x + 6)(x - 6)$ **2.** $(5a + 3b)(5a - 3b)$ **3.** $2(y + 7)(y - 7)$ **4.** $4(10 + x)(10 - x)$
5. $(x^2 + 9y^2)(x + 3y)(x - 3y)$ **6.** $(a + b + 4c)(a + b - 4c)$

page 241 **1.** $(x - 2)(x^2 + 2x + 4)$ **2.** $(5 + y)(25 - 5y + y^2)$ **3.** $(4a - 3b)(16a^2 + 12ab + 9b^2)$
4. $3(x - 3y)(x^2 + 3xy + 9y^2)$ **5.** $5(c + 1)(c^2 - c + 1)$
6. $(a + 2)(a^2 - 2a + 4)(a - 2)(a^2 + 2a + 4)$

Name ______________________________ Date ______________

EXERCISE SET 4.2

Rewrite each expression as a product of factors.

_______ 1. $9y^2 - 1$

_______ 2. $144y^2 - 1$

_______ 3. $\frac{1}{4}a^2 - \frac{25}{36}$

_______ 4. $\frac{1}{9}b^2 - \frac{4}{25}$

_______ 5. $m^2n^2 - 225$

_______ 6. $m^2n^2 - 16$

_______ 7. $25x^2y^2 - 16z^2$

_______ 8. $36x^2y^2 - 49z^2$

_______ 9. $3x^2 - 0.75$

_______ 10. $2x^2 - 0.32$

_______ 11. $6m^3 - 150m$

_______ 12. $7m^3 - 112m$

_______ 13. $2x^2y^2 - 72y^4$

_______ 14. $3x^2y^2 - 48y^4$

_______ 15. $25x^4 - 121$

_______ 16. $9x^4 - 64$

_______ 17. $m^4 - 16$

_______ 18. $m^4 - 625$

_______ 19. $3m^4 - 243$

_______ 20. $5m^4 - 80$

_______ 21. $(x + y)^2 - 49$

_______ 22. $(x + y)^2 - 81$

_______ 23. $(3a + 2b)^2 - 4c^2$

_______ 24. $(a + 2b)^2 - 9c^2$

_______ 25. $2x^3y - 8xy^3$

_______ 26. $3x^3y - 27xy^3$

_______ 27. $256 - (x + 2y)^2$

_______ 28. $289 - (x + 3y)^2$

_______ 29. $(x - 2)^2 - (y + 3)^2$

_______ 30. $(x + 6)^2 - (y - 1)^2$

_______ 31. $x^3 - 9x^2 - 9x + 81$

_______ 32. $x^3 - 4x^2 - 4x + 16$

_______ 33. $x^3 - 2x^2 - 25x + 50$

_______ 34. $x^3 - 3x^2 - 4x + 12$

_______ 35. $x^3 - y^3$

_______ 36. $m^3 - n^3$

_______ 37. $a^3 + 8$

_______ 38. $a^3 + 27$

_______ 39. $64 - m^3$

_______ 40. $125 - m^3$

_______ 41. $x^3y^3 + 1$

_______ 42. $x^3y^3 + 64$

_______ 43. $125x^3 - 27y^3$

_______ 44. $8x^3 - 343y^3$

_______ 45. $2x^4 + 16x$

_______ 46. $3x^4 + 81x$

_______ 47. $a^3b^3 - 64c^3$

_______ 48. $216a^3b^3 - c^3$

_______ 49. $4x^5y^2 + 32x^2y^5$

_______ 50. $5x^4y + 135xy^4$

_______ 51. $7x^4y^2 - 56xy^5$

_______ 52. $9x^6y^2 - 72x^3y^5$

_______ 53. $(x - y)^3 + 125$

_______ 54. $(x - 3y)^3 + 216$

_______ 55. $8 - (x - 3)^3$

_______ 56. $27 - (x + 4)^3$

_______ 57. $3x^6 - 3y^6$

_______ 58. $5x^6 - 5y^6$

_______ 59. $x^4 - x^3 + x - 1$

_______ 60. $x^4 + 2x^3 - 8x - 16$

☆ Stretching the Topics

Rewrite each expression as a product of factors.

_______ 1. $4x^{2a} - 36y^{2a}$

_______ 2. $x^8 - 256$

_______ 3. $x^{4a} - x^a$

Check your answers in the back of your book.

If you can do **Checkup 4.2,** then you are ready to go on to Section 4.3.

Name ______________________________ Date ______________

✓ CHECKUP 4.2

Rewrite each expression as a product of factors.

________ 1. $25a^2 - 4$

________ 2. $100x^2y^2 - 121z^2$

________ 3. $3x^2 - 432$

________ 4. $20m^3 - 45mn^2$

________ 5. $x^4 - 256$

________ 6. $x^3 + 64y^3$

________ 7. $3a^3 - 24b^3$

________ 8. $(x - 2)^2 - 9y^2$

________ 9. $(x - y)^3 + 216$

________ 10. $a^6 - b^6$

Check your answers in the back of your book.

If You Missed Problems:	You Should Review Examples:
1	1
2–4	2, 3
5	7
6	11
7	10
8	6
9	13
10	14

4.3 Factoring Trinomials

When we multiplied two first-degree binomials (that were not the sum and difference of the same two terms), recall that our result was usually a second-degree trinomial. Keeping that in mind, we seek a method for factoring trinomials.

Factoring Trinomials (Leading Coefficient of 1)

The FOIL method helped us organize the multiplication of binomials. For example,

$$\overset{\text{F}\quad\text{L}}{(x + 3)(x - 5)} = x^2 - 5x + 3x - 15$$

$$\text{I} \qquad \text{O}$$

$$= \underset{\text{F}}{x^2} \underset{\text{O + I}}{- 2x} \underset{\text{L}}{- 15}$$

Notice that the first term (the F term) in the trinomial comes from multiplying the first terms of the binomials. The last term (the L term) in the trinomial comes from multiplying the second terms of the binomials. The middle term in the trinomial is the sum of the O and I terms.

With these facts in mind let us see if we can discover what factors were multiplied to yield the trinomial

$$x^2 + 7x + 12$$

We are looking for two binomials. $\qquad (x \quad\quad)(x \quad\quad)$

Since the last term of the trinomial is 12, there are several possibilities for the second terms of the binomials.

$$\begin{array}{ll} (x + 1)(x + 12) & (x - 1)(x - 12) \\ (x + 2)(x + 6) & (x - 2)(x - 6) \\ (x + 3)(x + 4) & (x - 3)(x - 4) \end{array}$$

Since the middle term of the trinomial is *positive* $(+7x)$, we may reject three of the possibilities. We can find out which of the remaining pairs is correct by multiplying.

$$\begin{aligned} (x + 1)(x + 12) &= x^3 + 13x + 12 \\ (x + 2)(x + 6) &= x^2 + 8x + 12 \\ (x + 3)(x + 4) &= x^2 + 7x + 12 \end{aligned}$$

Only the last pair of factors gives the correct middle term, so

$$x^2 + 7x + 12 = (x + 3)(x + 4)$$

Example 1. Factor $x^2 - 8x + 12$.

Solution

$$x^2 - 8x + 12 = (x \quad\quad)(x \quad\quad)$$

Since the last term is again 12, the possible factors are the same as in the preceding example. In this case, however, we reject the *first* three possibilities since their middle terms will be *positive* and we need a *negative* middle term.

$$\begin{aligned} (x - 1)(x - 12) &= x^2 - 13x + 12 \\ (x - 2)(x - 6) &= x^2 - 8x + 12 \\ (x - 3)(x - 4) &= x^2 - 7x + 12 \end{aligned}$$

We conclude that the correct factorization is

$$x^2 - 8x + 12 = (x - 2)(x - 6)$$

As you can see from these examples, we can best factor trinomials by first concentrating on the first and last terms of the trinomial. Then we can check the possible binomials by multiplying to see when the correct middle term appears. After some practice, you will develop a good sense about likely factors and it will not be necessary to try every possibility.

You try factoring the trinomial in Example 2.

Example 2. Factor $x^2 - 11x + 10$.

Solution. Since the last term (10) is positive, but the middle term $(-11x)$ is negative, we need only consider the possibilities

$$(x - 5)(x - 2)$$

$$(x - 10)(x - 1)$$

After multiplying, we see that

$x^2 - 11x + 10 =$ (______)(______)

Check your work on page 255. ▶

Then complete Example 3.

Example 3. Factor $x^2 + 5xy + 6y^2$.

Solution. Since the last term $(6y^2)$ is ______ , and the middle term $(5xy)$ is ______ , we consider the possibilities

$(x + 6y)(x +$ ______)

$(x + 3y)(x +$ ______)

After multiplying, we see that

$x^2 + 5xy + 6y^2 =$ (______)(______)

Check your work on page 255. ▶

Now consider factoring the trinomial

$$x^2 + 3x - 4$$

We need two binomials.

$$(x \quad\quad)(x \quad\quad)$$

Since the last term of the trinomial is -4, there are several choices for the second terms of the binomials.

$$(x + 2)(x - 2)$$
$$(x + 1)(x - 4)$$
$$(x + 4)(x - 1)$$

We find the correct pair by multiplying.

$$(x + 2)(x - 2) = x^2 - 4$$
$$(x + 1)(x - 4) = x^2 - 3x - 4$$
$$(x + 4)(x - 1) = x^2 + 3x - 4$$

The correct factorization is

$$x^2 + 3x - 4 = (x + 4)(x - 1)$$

Example 4. Factor $y^2 - 3y - 10$.

Solution. Since the last term of the trinomial is negative (-10), we start out with

$$y^2 - 3y - 10 = (y + \quad\quad)(y - \quad\quad)$$

Considering the factors of 10, the possible binomials are

$$(y + 10)(y - 1)$$

$$(y + 1)(y - 10)$$

$$(y + 5)(y - 2)$$

$$(y + 2)(y - 5)$$

After multiplying, we see that

$$y^2 - 3y - 10 = (y + 2)(y - 5)$$

Now you try Example 5.

Example 5. Factor $a^2 + 3a - 18$.

Solution. Since the last term of the trinomial is negative (-18), we start out with

$$a^2 + 3a - 18 = (a + \quad\quad)(a - \quad\quad)$$

Considering the factors of 18, we find that the possible binomial factors are

$(a + 18)(a -$ ______)

$(a + 1)(a -$ ______)

$(a + 9)(a -$ ______)

$(a + 2)(a -$ ______)

$(a + 6)(a -$ ______)

$(a + 3)(a -$ ______)

After multiplying, we see that

$a^2 + 3a - 18 =$ (______)(______)

Check your work on page 255. ▶

After a bit of practice you will learn to spot the likely factors without listing every possibility. Look at Example 6.

Example 6. Factor $x^2 - x - 30$.

Solution. Since the last term is negative (-30), we start out with

$$x^2 - x - 30 = (x + \quad)(x - \quad)$$

There are several ways to factor 30, but our middle term of $-1x$ tells us that the factors are just 1 unit apart. The only logical choices are

$$(x + 6)(x - 5)$$
$$(x + 5)(x - 6)$$

After multiplying, we see that

$$x^2 - x - 30 = (x + 5)(x - 6)$$

You try completing Example 7.

Example 7. Factor $20 + 8x - x^2$.

Solution. Since the last term is negative $(-x^2)$, we start out with

$$20 + 8x - x^2 = (\quad + x)(\quad - x)$$

There are several ways to factor 20, but 20 and 1 are too far apart, and 5 and 4 are too close together. The logical choices are

$$(10 + x)(2 - x)$$
$$(2 + x)(10 - x)$$

After multiplying, we see that

$$20 + 8x - x^2 = (\_\_\_\_\_\_)(\_\_\_\_\_\_)$$

Check your factorization on page 255. ▶

From our work so far, we make some observations about **factoring second-degree trinomials.**

1. If the first and last terms of the trinomial are *positive* and the middle term is *positive*, both terms of both binomial factors will be positive.

   Example: $x^2 + 5x + 6 = (x + 3)(x + 2)$

2. If the first and last terms of the trinomial are *positive*, but the middle term is *negative*, the first terms of both binomial factors will be *positive*, but the second terms of both binomial factors will be *negative*.

   Example: $x^2 - 5x + 6 = (x - 3)(x - 2)$

3. If the first term of the trinomial is *positive*, but the last term is *negative*, the first terms of both binomials will be *positive*, but one of the second terms will be *positive* and the other second term will be *negative*.

   Example: $x^2 - 5x - 6 = (x - 6)(x + 1)$

   Example: $x^2 + 5x - 6 = (x + 6)(x - 1)$

If you are thoroughly familiar with the multiplication of binomials, factoring trinomials will be like undoing a puzzle using wise trial and error. Rules are not really necessary since you can (and should) always check your factors by multiplying them.

Trial Run

Rewrite each expression as a product of factors.

______ 1. $x^2 - 6x + 5$

______ 2. $a^2 - 7a - 18$

________ 3. $15 + 2x - x^2$

________ 4. $x^2 + 14x + 49$

________ 5. $x^2 + 11x - 12$

________ 6. $24 - 10x + x^2$

Answers are on page 256.

Factoring Trinomials (Leading Coefficient Not 1)

Let's look at more trinomials in which the coefficient of the second-degree term is different from 1. Now there will be more choices to consider, but the basic principles are exactly the same and our choices can still be checked by multiplication.

Example 8. Factor $5x^2 + 7x + 2$.

Solution

There is no common factor, so we know that the first terms in the binomials must multiply to give $5x^2$. The signs are all positive.

$$5x^2 + 7x + 2$$
$$(5x + \quad)(x + \quad)$$

Since the last term is 2, we know there are only two possible pairs of factors.

$$(5x + 2)(x + 1)$$
$$(5x + 1)(x + 2)$$

After checking by multiplication we decide that

$$5x^2 + 7x + 2 = (5x + 2)(x + 1)$$

Example 9. Factor $6x^2 - 29xy - 5y^2$.

Solution

Since the last term is negative, we know what signs the binomial factors must contain.

$$6x^2 - 29xy - 5y^2$$
$$(\_\_\_ + \_\_\_)(\_\_\_ - \_\_\_)$$

The last term is $-5y^2$, so we know what the second terms of the binomials can be.

$$(\_\_\_ + 5y)(\_\_\_ - y)$$
$$(\_\_\_ - 5y)(\_\_\_ + y)$$

Since the first term is $6x^2$, there are several possible pairs of factors.

$(3x + 5y)(2x - y)$	$(3x - 5y)(2x + y)$
$(2x + 5y)(3x - y)$	$(2x - 5y)(3x + y)$
$(6x + 5y)(x - y)$	$(6x - 5y)(x + y)$
$(x + 5y)(6x - y)$	$(x - 5y)(6x + y)$

After checking by multiplication we decide that

$$6x^2 - 29xy - 5y^2 = (x - 5y)(6x + y)$$

We found earlier that it will rarely be necessary for you to list *all* possible pairs of factors. You should *test each pair as you proceed rather than write them all down and then test*. In your search, you should never erase a pair that you have tried, even though it is not correct. When there are several possibilities, you will never remember the pairs you have already tested or see a pattern in your work if you erase your factorizations. Usually, you will hit upon the correct pair of factors before exhausting the entire list of possibilities.

You try factoring the trinomial in Example 10.

Example 10. Factor $6x^2 + 29x + 9$.

Solution. We know that all signs must be positive.

$$6x^2 + 29x + 9$$
$$= (\_\_\_\_ + \_\_\_\_)(\_\_\_\_ + \_\_\_\_)$$

Check your work on page 255. ▶

Example 11. Factor $15x^2 - 9x - 42$.

Solution. We spot the *common monomial factor* 3.

$$15x^2 - 9x - 42$$
$$= 3(5x^2 - 3x - 14)$$
$$= 3(5x + 7)(x - 2)$$

Now try to factor the trinomial in Example 12.

Example 12. Factor $18xy^3 - xy^2 - 4xy$.

Solution. We spot the common factor xy.

$$18xy^3 - xy^2 - 4xy$$
$$= xy(\_\_\_\_\_\_\_\_)$$
$$= xy(\_\_\_\_\_\_)(\_\_\_\_\_\_)$$

Check your work on page 255. ▶

Example 13. Factor $-8x^3 + 22x^2 - 15x$.

Solution. We spot the common factor $-x$.

$$-8x^3 + 22x^2 - 15x$$
$$= -x(8x^2 - 22x + 15)$$
$$= -x(2x - 3)(4x - 5)$$

Trial Run

Rewrite each expression as a product of factors.

______ 1. $3y^2 + 36y + 105$

______ 2. $30x^2 - x - 42$

______ 3. $6x^2 + 5x - 56$

______ 4. $21x^2 + 20x + 4$

______ 5. $-2m^3 + 3m^2n - mn^2$

______ 6. $12x^3y - 26x^2y^2 - 10xy^3$

Answers are on page 256.

Factoring Perfect Square Trinomials

In Chapter 3 we learned that the square of a binomial is a perfect square trinomial. If we recognize a perfect square trinomial, we may factor it by reversing the formulas learned earlier.

Perfect Square Trinomials

$$A^2 + 2AB + B^2 = (A + B)^2$$
$$A^2 - 2AB + B^2 = (A - B)^2$$

In a perfect square trinomial, the first term must be the square of some quantity A, the last term must be the square of some quantity B, and the middle term must be twice the product of A times B. As always, factors should be checked by multiplying.

Example 14. Factor $x^2 + 10x + 25$.

Solution. Is $x^2 + 10x + 25$ a perfect square trinomial? Yes.

$$A = x \quad \text{and} \quad B = 5$$

$$2AB = 2(x)(5) = 10x$$

Therefore,

$$x^2 + 10x + 25 = (x + 5)^2$$

You try Example 15.

Example 15. Factor $9a^2 - 60ab + 100b^2$.

Solution. Is $9a^2 - 60ab + 100b^2$ a perfect square trinomial? Yes.

$$A = 3a \quad \text{and} \quad B = 10b$$

$$2AB = 2(\underline{\qquad})(\underline{\qquad}) = \underline{\qquad}$$

Therefore,

$$9a^2 - 60ab + 100b^2 = (\underline{\qquad} - \underline{\qquad})^2$$

Check your work on page 255. ▶

Trial Run

Rewrite each expression as a product of factors.

______ 1. $x^2 + 12x + 36$	______ 2. $x^2y^2 - 4xy + 4$
______ 3. $x^3 + 14x^2 + 49x$	______ 4. $-12x^2 - 12x - 3$
______ 5. $x^2 - 0.6xy + 0.09y^2$	______ 6. $a^2 - \frac{2}{5}a + \frac{1}{25}$

Answers are on page 256.

Factoring More Trinomials

So far we have discovered the general method for factoring a second-degree trinomial as a product of first-degree binomials. Now consider the polynomial

$$(x + y)^2 + 3(x + y) + 2$$

Notice that the *square* of the variable part $(x + y)$ of the middle term is the variable part of the first term. Letting A represent the quantity $(x + y)$, we can write the original trinomials using the variable A and then factor it.

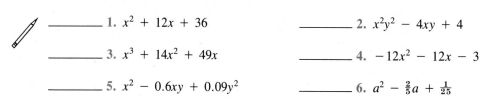

$$\begin{array}{ccccccccccc} A^2 & + & 3A & + & 2 & = & (A & + 2) & (A & + 1) \\ \downarrow & & \downarrow & & \downarrow & & \downarrow & \downarrow & \downarrow & \downarrow \\ (x + y)^2 & + & 3(x + y) & + & 2 & = & [(x + y) & + 2] & [(x + y) & + 1] \end{array}$$

$$= (x + y + 2)(x + y + 1)$$

Polynomials in which the square of the middle term's variable part is the same as the first term's variable part are called **quadratic in form** because they behave like second-degree (quadratic) polynomials.

Example 16. Factor $2(y - h)^2 - 5(y - h) - 12$.

Solution. Notice that the square of the middle term's variable part, $(y - h)$, is the same as the first term's variable part, $(y - h)^2$.

$$\begin{array}{ccccccccccc} 2A^2 & - & 5A & - & 12 & = & (2A & + 3) & (A & - 4) \\ \downarrow & & \downarrow & & \downarrow & & \downarrow & \downarrow & \downarrow & \downarrow \\ 2(y - h)^2 & - & 5(y - h) & - & 12 & = & [2(y - h) & + 3] & [(y - h) & - 4] \end{array}$$

$$= (2y - 2h + 3)(y - h - 4)$$

Try to complete Example 17.

Example 17. Factor $x^{10} + 7x^5 + 6$.

Solution. We notice that the square of the variable part (x^5) of the second term is the variable part (x^{10}) of the first term.

$$x^{10} + 7x^5 + 6$$
$$= (x^5 + \_\_\_\_)(x^5 + \_\_\_\_)$$

Check your work on page 255. ▶

Example 18. Factor $x^4 - 5x^2 + 4$.

Solution

$$x^4 - 5x^2 + 4$$
$$= (x^2 - 4)(x^2 - 1)$$
$$= (x + 2)(x - 2)(x + 1)(x - 1)$$

Notice that we continued factoring when we saw the difference of two squares.

Trial Run

Rewrite each expression as a product of factors.

________ 1. $(x - y)^2 - (x - y) - 56$

________ 2. $2(a - 3)^2 + 5(a - 3) - 12$

________ 3. $36(x - 2y)^2 + 17(x - 2y) + 2$

________ 4. $x^6 + 4x^3 + 4$

________ 5. $x^4 - 13x^2 + 36$

________ 6. $4x^4 - 21x^2 - 25$

Answers are on page 256.

Using Factored Forms of Polynomials

When it is necessary to *evaluate* a formula or a polynomial expression, often it is easier to make the required substitution in the *factored form* of the polynomial rather than in the original expression.

For example, consider this physics problem.

Example 19. If an object is thrown upward from the ground with a velocity of 1600 feet per second, its distance (s) above the ground after t seconds is given by the formula

$$s = 1600t - 16t^2$$

Find the object's distance above the ground when $t = 50$ seconds.

Solution. We must evaluate the expression on the right side to find s when $t = 50$. Let's use two approaches.

Original Form	***Factored Form***
$s = 1600t - 16t^2$	$s = 1600t - 16t^2$
	$= 16t(100 - t)$
When $t = 50$:	When $t = 50$:
$s = 1600(50) - 16(50)^2$	$s = 16(50)[100 - 50]$
$= 80,000 - 16(2500)$	$= 800(50)$
$= 80,000 - 40,000$	$= 40,000$ ft
$= 40,000$ ft	

Do you agree that the computation was simpler when the original expression was rewritten in factored form before the substitution was made?

You try Example 20.

Example 20. When an amount of money (P) is invested in an account at 8.5 percent simple annual interest for t years, the total amount (A) in the account is given by the formula

$$A = P + 0.085Pt$$

Write the expression on the right in factored form and evaluate it when $P = \$6700$ and $t = 10$ years.

Solution

$$A = P + 0.085Pt$$

$$A = P(1 + 0.085t)$$

When $P = 6700$ and $t = 10$,

$$\begin{aligned} A &= 6700[1 + 0.085(10)] \\ &= 6700[1 + \underline{\qquad}] \\ &= 6700[\underline{\qquad}] \\ &= \underline{\qquad} \end{aligned}$$

There will be \$________ in the account.

Check your work on page 255. ▶

Now let's try the problem stated at the beginning of this chapter.

Example 21. A company has found that its *revenue* from the sale of x units of a product is $2x^2 - x + 10$ dollars, while its *costs* for producing x units are $x^2 - 3x + 109$ dollars. Write an expression for the company's *profit* when x units are produced and sold. Then write that expression in factored form, and find the profit when the company produces and sells 6 units; when it sells 19 units.

Solution. Remember from earlier chapters that

$$\begin{aligned} \text{Profit} &= \text{revenue} - \text{costs} \\ \text{Profit} &= (2x^2 - x + 10) - (x^2 - 3x + 109) \\ &= 2x^2 - x + 10 - x^2 + 3x - 109 \\ &= x^2 + 2x - 99 \end{aligned}$$

In factored form,

$$\text{Profit} = (x + 11)(x - 9)$$

When 6 units are sold, $x = 6$ and

$$\begin{aligned} \text{Profit} &= (x + 11)(x - 9) \\ &= (6 + 11)(6 - 9) \\ &= 17(-3) \\ &= -51 \text{ dollars} \end{aligned}$$

The company will *lose* \$51 when it produces and sells 6 units.

When 19 units are sold, $x = 19$ and

$$\begin{aligned} \text{Profit} &= (x + 11)(x - 9) \\ &= (19 + 11)(19 - 9) \\ &= 30(10) \\ &= 300 \text{ dollars} \end{aligned}$$

The company will make \$300 profit when it produces and sells 19 units.

Trial Run

_______ 1. If a stone is thrown upward from a height of 35 feet with a velocity of 46 feet per second, its height (s) above the ground after t seconds is given by the formula: $s = 35 + 46t - 16t^2$. Rewrite the right-hand expression in factored form and find the stone's height above the ground after 2.5 seconds.

_______ 2. The perimeter (P) of a rectangle with length l and width w is given by the formula $P = 2l + 2w$. Rewrite the right-hand expression in factored form and find the perimeter of a rectangle that is 2.37 centimeters long and 1.23 centimeters wide.

Answers are on page 256.

Examples You Completed

Example 2 *(Solution)*

$$x^2 - 11x + 10 = (x - 10)(x - 1)$$

Example 3. Factor $x^2 + 5xy + 6y^2$.

Solution. Since the last term ($6y^2$) is positive, and the middle term ($5xy$) is positive, we consider the possibilities

$$(x + 6y)(x + y)$$

$$(x + 3y)(x + 2y)$$

After multiplying, we see that

$$x^2 + 5xy + 6y^2 = (x + 3y)(x + 2y)$$

Example 7 *(Solution)*

$$20 + 8x - x^2 = (10 - x)(2 + x)$$

Example 10 *(Solution)*

$$6x^2 + 29x + 9$$

$$= (3x + 1)(2x + 9)$$

Example 5. Factor $a^2 + 3a - 18$.

Solution. Since the last term of the trinomial is negative (-18), we start out with

$$a^2 + 3a - 18 = (a + \quad)(a - \quad)$$

Considering the factors of 18, we find that the possible binomial factors are

$$(a + 18)(a - 1)$$

$$(a + 1)(a - 18)$$

$$(a + 9)(a - 2)$$

$$(a + 2)(a - 9)$$

$$(a + 6)(a - 3)$$

$$(a + 3)(a - 6)$$

After multiplying, we see that

$$a^2 + 3a - 18 = (a + 6)(a - 3)$$

Example 12 *(Solution)*

$$18xy^3 - xy^2 - 4xy$$

$$= xy(18y^2 - y - 4)$$

$$= xy(9y + 4)(2y - 1)$$

Example 15. Factor $9a^2 - 60ab + 100b^2$.

Solution. Is $9a^2 - 60ab + 100b^2$ a perfect square trinomial? Yes.

$$A = 3a \quad \text{and} \quad B = 10b$$

$$2AB = 2(3a)(10b) = 60ab$$

Therefore,

$$9a^2 - 60ab + 100b^2 = (3a - 10b)^2$$

Example 17 *(Solution)*

$$x^{10} + 7x^5 + 6 = (x^5 + 6)(x^5 + 1)$$

Example 20 *(Solution)*

$$A = P + 0.085Pt$$

$$= P(1 + 0.085t)$$

When $P = 6700$ and $t = 10$,

$$A = 6700[1 + 0.085(10)]$$

$$= 6700[1 + 0.85]$$

$$= 6700[1.85]$$

$$= 12{,}395$$

There will be $12,395 in the account.

Answers to Trial Runs

page 249 1. $(x - 5)(x - 1)$ 2. $(a - 9)(a + 2)$ 3. $(5 - x)(3 + x)$ 4. $(x + 7)(x + 7)$
5. $(x + 12)(x - 1)$ 6. $(6 - x)(4 - x)$

page 251 1. $3(y + 5)(y + 7)$ 2. $(6x + 7)(5x - 6)$ 3. $(3x - 8)(2x + 7)$ 4. $(7x + 2)(3x + 2)$
5. $-m(2m - n)(m - n)$ 6. $2xy(3x + y)(2x - 5y)$

page 252 1. $(x + 6)^2$ 2. $(xy - 2)^2$ 3. $x(x + 7)^2$ 4. $-3(2x + 1)^2$ 5. $(x - 0.3y)^2$
6. $(a - \frac{1}{5})^2$

page 253 1. $(x - y - 8)(x - y + 7)$ 2. $(2a - 9)(a + 1)$ 3. $(9x - 18y + 2)(4x - 8y + 1)$
4. $(x^3 + 2)^2$ 5. $(x + 3)(x - 3)(x + 2)(x - 2)$ 6. $(2x - 5)(2x + 5)(x^2 + 1)$

page 255 1. $s = 50$ ft 2. $P = 7.2$ cm

Name ______________________ Date ____________

EXERCISE SET 4.3

Rewrite each expression as a product of factors. When the expression cannot be factored using integers, write "prime."

_______ 1. $x^2 - 24x + 144$

_______ 2. $x^2 - 18x + 81$

_______ 3. $y^2 - y - 30$

_______ 4. $y^2 - 11y - 26$

_______ 5. $b^2 + 8b - 33$

_______ 6. $b^2 + 4b - 21$

_______ 7. $9a^2 - 24ab + 16b^2$

_______ 8. $4a^2 + 20ab + 25b^2$

_______ 9. $x^2 - 10xy + 25y^2$

_______ 10. $x^2 - 16xy + 64y^2$

_______ 11. $39 - 16m + m^2$

_______ 12. $72 - 17m + m^2$

_______ 13. $x^2y^2 + 2xy - 15$

_______ 14. $x^2y^2 + 5xy - 24$

_______ 15. $3x^2 - 3x - 18$

_______ 16. $5x^2 - 20x + 15$

_______ 17. $5a^3 - 40a^2 + 80a$

_______ 18. $4a^3 + 56a^2 + 196a$

_______ 19. $20y + 6y^2 - 2y^3$

_______ 20. $27y - 24y^2 - 3y^3$

_______ 21. $-10ab^4 + 60ab^2 - 90a$

_______ 22. $-3ab^4 - 12ab^2 - 12a$

_______ 23. $x^3y - 18x^2y^2 + 80xy^3$

_______ 24. $x^3y - 15x^2y^2 + 26xy^3$

_______ 25. $a^3bc - 2a^2b^2c - 63ab^3c$

_______ 26. $a^3bc - a^2b^2c - 56ab^3c$

_______ 27. $(a + 1)^2 - 12(a + 1) + 36$

_______ 28. $(a + 1)^2 - 10(a + 1) + 25$

_______ 29. $(x + y)^2 - 7(x + y) + 12$

_______ 30. $(x + y)^2 - 7(x + y) + 10$

_______ 31. $(x + 5)^2 + 4(x + 5) - 32$

_______ 32. $(x + 5)^2 - 7(x + 5) - 18$

_______ 33. $2y^2 + 7y + 3$

_______ 34. $3y^2 + 16y + 5$

_______ 35. $10x^2 - 19x + 6$

_______ 36. $6x^2 - 25x + 25$

_______ 37. $6a^2 - 5ab - 56b^2$

_______ 38. $16a^2 + 6ab - 27b^2$

_______ 39. $30c^2 - 145cd + 45d^2$

_______ 40. $20c^2 - 68cd + 24d^2$

_______ 41. $-9x^4 + 33x^3 - 28x^2$

_______ 42. $-4a^4 + 32a^3 - 63a^2$

_______ 43. $12m^2 + 13mn - 55n^2$

_______ 44. $12m^2 - 7mn - 10n^2$

_______ 45. $24x^3y + 26x^2y - 70xy$

_______ 46. $105x^3y + 57x^2y - 72xy$

_______ 47. $5(a - b)^2 + 7(a - b) + 2$

_______ 48. $7(a - b)^2 + 22(a - b) + 3$

_______ 49. $25(x + y)^2 - 20(x + y) + 4$

_______ 50. $16(x + 2y)^2 - 24(x + 2y) + 9$

_______ 51. $8 - 38(x - 2y) + 39(x - 2y)^2$

_______ 52. $35 - 89(2x + y) + 30(2x + y)^2$

_______ 53. $x^4 - x^2 - 72$

_______ 54. $x^4 - 6x^2 - 27$

_______ 55. $3x^4 - 27x^2 + 60$

_______ 56. $2x^4 - 6x^2 - 8$

_______ 57. $3x^6 - 30x^4y^2 + 75x^2y^4$

_______ 58. $4x^4y^2 - 88x^2y^4 + 484y^6$

_______ 59. $2x^{10} + 16x^5 - 66$

_______ 60. $3x^{10} - 9x^5 - 84$

_______ 61. The area of a trapezoid is given by $A = \frac{1}{2}ah + \frac{1}{2}bh$, where h represents the height of the trapezoid and a and b are the lengths of the two bases. Write the expression on the right in factored form and find the area when h is 7 inches, a is 5 inches, and b is 3 inches.

_______ 62. The difference in the areas of two circles is given by the formula $d = \pi R^2 - \pi r^2$. Write the expression on the right in factored form and find the difference in areas when R is 7.5 meters and r is 2.5 meters. (Leave π in your answer.)

_______ 63. If an object is thrown upward from a height of 30 feet with a velocity of 38 feet per second, its height (s) above the ground after t seconds is given by $s = 30 + 38t - 16t^2$. Write the expression on the right in factored form and find the object's height above the ground after 2.2 seconds.

_______ 64. A company has found that its revenue from the sale of x units of a product is $R = 3x^2 + 13x - 10$ dollars. Write the expression on the right in factored form and find the revenue when 20 units are produced.

_______ 65. The Close Knit Sweater Company has found that its costs for producing x sweaters is $C = 2x^2 + x - 6$ dollars. Write the expression on the right in factored form and find the cost of producing 50 sweaters.

_______ 66. A certain deodorant manufacturer has found that its revenue from the sale of x units of its product is $3x^2 + 3x + 18$ dollars, while its costs for producing x units are $2x^2 - x + 50$ dollars. Write an expression for the company's profit when x units are produced and sold. Then write the expression in factored form, and find the profit when the company produces and sells 92 units.

☆ Stretching the Topics

Rewrite each expression in completely factored form.

_______ 1. $4x^{2a} + 44x^a + 121$

_______ 2. $6(x + y)^{2m} + 19(x + y)^m + 15$

_______ 3. $x^{2a+6} + 12x^{a+3}y^{a-2} + 36y^{2a-4}$

Check your answers in the back of your book.

If you can do the problems in **Checkup 4.3,** you are ready to go on to Section 4.4.

Name ______________________ Date __________

✓ CHECKUP 4.3

Rewrite each expression as a product of factors.

______ 1. $x^2 - 9xy + 18y^2$

______ 2. $2y^2 + 6y - 108$

______ 3. $10x^2 - 3xy - 27y^2$

______ 4. $49a^2 - 42ab + 9b^2$

______ 5. $-6y^3 + 6y^2 + 12y$

______ 6. $x^4 - 2x^2 - 63$

______ 7. $4x^4 - 37x^2 + 9$

______ 8. $(x + y)^2 + 8(x + y) - 20$

______ 9. $2(x + y)^2 - 13(x + y) + 21$

______ 10. If an object is thrown upward from the ground with a velocity of 800 feet per second, its distance (s) above the ground after t seconds is given by $s = 800t - 16t^2$. Rewrite the right-hand expression in factored form and find the object's distance above the ground when $t = 25$ seconds.

Check your answers in the back of your book.

If You Missed Problems:	You Should Review Examples:
1	3
2, 3	9–11
4	15
5	12, 13
6, 7	18
8, 9	16
10	19

4.4 Using All Types of Factoring

Factoring is not a random process. To be successful at rewriting expressions in factored form, you should be systematic in your approach and learn to ask yourself the right questions.

1. Do all the terms contain a common factor?

If so, use the distributive property to rewrite the expression

$$AB + AC = A(B + C)$$

2. Does the expression consist of just *two* terms?

Is it a *difference* of two *squares*? If so, identify A and B and rewrite the expression

$$A^2 - B^2 = (A + B)(A - B)$$

Is it a *sum* or *difference* of two *cubes*? If so, identify A and B and rewrite the expression

$$A^3 + B^3 = (A + B)(A^2 - AB + B^2)$$

$$A^3 - B^3 = (A - B)(A^2 + AB + B^2)$$

3. Does the expression contain exactly *three* terms?

Is it a *perfect square* trinomial that is the square of a binomial? If so, identify A and B and rewrite the expression

$$A^2 + 2AB + B^2 = (A + B)^2$$

$$A^2 - 2AB + B^2 = (A - B)^2$$

Is it a trinomial that is the product of two different binomials? If so, find those binomials by *trial and error*.

4. Does the expression contain exactly *four* terms?

If so, try to *group* the terms two by two, factor each pair of terms, and remove the common binomial factor.

In many cases you will be able to use more than one factoring technique. After each step you should ask yourself these same questions. In other cases you will discover that the expression cannot be factored; it is prime (irreducible).

Example 1. Factor completely $5x^2y + 15x^2 - 20y - 60$.

Solution

$5x^2y + 15x^2 - 20y - 60$	
$= 5(x^2y + 3x^2 - 4y - 12)$	Remove the common factor 5.
$= 5(\underbrace{x^2y + 3x^2} - \underbrace{4y - 12})$	Mentally group the four terms.
$= 5[x^2(y + 3) - 4(y + 3)]$	Factor each pair of terms.
$= 5(y + 3)(x^2 - 4)$	Remove the common binomial factor.
$= 5(y + 3)(x + 2)(x - 2)$	Factor the difference of two squares.

Example 2. Factor completely $-a^3 + 12a^2 - 36a$.

Solution

$-a^3 + 12a^2 - 36a$	
$= -a(a^2 - 12a + 36)$	Remove the common factor $-a$.
$= -a(a - 6)^2$	Factor the perfect square trinomial.

Name ____________________ Date __________

EXERCISE SET 4.4

Rewrite each expression as a product of factors.

_____ 1. $4(x - y)^3 - 3(x - y)^2 - (x - y)$

_____ 2. $5(a + b)^3 + 13(a + b)^2 - 6(a + b)$

_____ 3. $2m^3 - 50m$

_____ 4. $3m^3 - 48m$

_____ 5. $6a^3 - 30a^2 - 144a$

_____ 6. $5a^3 - 30a^2 - 135a$

_____ 7. $a^3bc - 3a^2b^2c - 54ab^3c$

_____ 8. $x^3yz + 2x^2y^2z - 63xy^3z$

_____ 9. $3x^6 - 25x^3 + 8$

_____ 10. $2x^6 - 55x^3 + 27$

_____ 11. $7(a - b)^2 + 4(a - b) - 3$

_____ 12. $6(a + b)^2 + (a + b) - 12$

_____ 13. $x^3 - 5x^2 + x - 5$

_____ 14. $2x^3 - 3x^2 + 8x - 12$

_____ 15. $3x^2 - 3x - 270$

_____ 16. $5x^2 - 5x - 210$

_____ 17. $-2x^4 + 128x$

_____ 18. $-3x^4 + 192x$

_____ 19. $24x^4 - 40x^3y^3 + 56x^2y^4$

_____ 20. $32x^4 - 80x^3y^3 + 56x^2y^4$

_____ 21. $81x^2 - 36xy + 4y^2$

_____ 22. $169x^2 - 130xy + 25y^2$

_____ 23. $11(x + y)^2 - 35(x + y) + 6$

_____ 24. $7(x + y)^2 + 26(x + y) - 45$

_____ 25. $4x^6 - 24x^4y^2 + 20x^2y^4$

_____ 26. $3x^6 - 24x^4y^2 - 27x^2y^4$

_____ 27. $x^2 - 30xy + 225y^2$

_____ 28. $x^2 - 40xy + 144y^2$

_____ 29. $3x^2 - 6x - 270$

_____ 30. $4x^2 - 12x - 224$

_____ 31. $x^3y^3 + 216$

_____ 32. $x^3y^3 + 125$

_____ 33. $3b^2 + 5b - 82$

_____ 34. $4b^2 - 3b + 7$

_____ 35. $4a^2x - 8a^2y - 4b^2x + 8b^2y$

_____ 36. $18a^2x - 9a^2y - 18b^2x + 9b^2y$

_____ 37. $9x^4 - 7x^2 - 16$

_____ 38. $4x^4 - 21x^2 - 25$

_____ 39. $24x^3 - 375$

_____ 40. $54x^3 - 128$

_____ 41. $36a^3b + 120a^2b^2 + 100ab^3$

_____ 42. $96a^3b - 144a^2b^2 + 54ab^3$

_____ 43. $0.25x^2y^2 - 1.21z^2$

_____ 44. $0.64x^2y^2 - 1.69z^2$

_____ 45. $4a^6b^4 + 28a^3b^2 + 49$

_____ 46. $9a^6b^4 - 48a^3b^2 + 64$

_____ 47. $-18x^2 + 32$

_____ 48. $-150x^2 + 24$

_____ 49. $44x^3yz^4 - 100x^3y^2z^3 - 96x^3y^3z^2$

_____ 50. $35x^4yz^3 - 20x^4y^2z^2 - 55x^4y^3z$

☆ Stretching the Topics

Rewrite each expression as a product of factors.

________ 1. $2x^{3a-4} + 6x^{2a-2} + 18x^{a}$

________ 2. $a^{3x+1}b - 729ab^{3x+1}$

________ 3. $27(x - y)^3 - 64(x + y)^3$

Check your answers in the back of your book.

If you can do the problems in **Checkup 4.4,** you are ready to go on to the **Review Exercises** for Chapter 4.

Name ______________________ Date __________

✓ CHECKUP 4.4

Rewrite each expression as a product of factors.

_______ 1. $25x^2 - 90xy + 81y^2$

_______ 2. $3x^2y^2 - 192y^4$

_______ 3. $4x^5y^2 - 12x^2y^2$

_______ 4. $15ax - 10ay - 9x + 6y$

_______ 5. $9x^2 + 48x + 64$

_______ 6. $x^4 - 35x^2 - 36$

_______ 7. $7(a - b)^2 + 4(a - b) - 3$

_______ 8. $x^6 - 64y^6$

_______ 9. $24x^2 - 46xy + 21y^2$

_______ 10. $x^5 - x^3 - x^2 + 1$

Check your answers in the back of your book.

If You Missed Problem:	You Should Review Examples:	
1	SECTION 4.3	15
2	4.2	2
3	4.1	1–3
4	4.1	11
5	4.3	14
6	4.3	18
7	4.3	16
8	4.2	14
9	4.3	9
10	4.2	8, 9

Summary

For reasons that will become very clear in the following chapters, we have spent a great deal of time sharpening our factoring skills in this chapter. We learned that **factoring** is the process by which we rewrite a *sum* (or difference) of terms as a *product*.

In attempting to factor polynomials we learned to check mentally the list of different types of factoring techniques available to us. We agreed to always look first for a common factor and then for a special product before resorting to trial and error.

Type	Polynomial	Factors	Example
Common factor	$AB + AC$	$A(B + C)$	$3x^2 - 6x = 3x(x - 2)$
Difference of two squares	$A^2 - B^2$	$(A + B)(A - B)$	$9x^2 - 25y^2$ $= (3x + 5y)(3x - 5y)$
Difference of two cubes	$A^3 - B^3$	$(A - B)(A^2 + AB + B^2)$	$8a^3 - 1$ $= (2a - 1)(4a^2 + 2a + 1)$
Sum of two cubes	$A^3 + B^3$	$(A + B)(A^2 - AB + B^2)$	$x^3 + 27c^3$ $= (x + 3c)(x^2 - 3cx + 9c^2)$
General trinomial	$Ax^2 + Bx + C$	Trial and error	$6x^2 + 13x - 5$ $= (2x + 5)(3x - 1)$
Perfect square trinomial	$A^2 + 2AB + B^2$ $A^2 - 2AB + B^2$	$(A + B)(A + B)$ $(A - B)(A - B)$	$x^2 + 14x + 49$ $= (x + 7)(x + 7)$
Four-term polynomial	$AB + AC + DB + DC$	$A(B + C) + D(B + C)$ $= (B + C)(A + D)$	$2x^3 - x^2 + 10x - 5$ $= x^2(2x - 1) + 5(2x - 1)$ $= (2x - 1)(x^2 + 5)$

❑ Speaking the Language of Algebra

Complete each sentence with the appropriate word or phrase.

1. Factoring is the process by which we rewrite a sum (or difference) of terms as a ______ .

2. When factoring a polynomial, we always look first for a ______ ______ .

3. To factor a polynomial containing four terms, we begin by ______ the terms.

4. To factor a second-degree trinomial that is not a special product, we use ______ and ______ .

5. If a polynomial cannot be rewritten as a product of factors, we say that it is ______ .

Writing About Mathematics

Write your response to each question in complete sentences.

1. Explain the steps you would use to factor the expression

$$ax + a + bx + b$$

2. Tell how you recognize an expression that can be factored as
   (a) The product of the sum and difference of the same two terms.
   (b) The square of a binomial.

3. Describe the questions you ask yourself when you begin to factor a polynomial.

4. Tell what is wrong with each of the following factorizations.
   (a) $-2x - 2a = -2(x - a)$
   (b) $x^2 - 16 = (x - 4)^2$
   (c) $x^3 + 27 = (x + 3)^3$
   (d) $x^2 + 3x - 10 = (x - 5)(x + 2)$

Name ______________________________ Date ______________

REVIEW EXERCISES for Chapter 4

Rewrite each expression as a product of factors.

________ 1. $5a^3 - 15a^2 + 10a$

________ 2. $-x^5y^2 + 5x^4y^3 - 2x^3y^4$

________ 3. $3x^4y^3z - 6x^3y^2z + 3x^2yz$

________ 4. $3x(x - 2) + 4(x - 2)$

________ 5. $(a - 2b)^2 - 3(a - 2b)$

________ 6. $(2a - 1)(b + 2) - (a + 1)(b + 2)$

________ 7. $5a + 5b + ac + bc$

________ 8. $3ax + 3ay - 2bx - 2by$

________ 9. $x^3 - 9x^2 + 2x - 18$

________ 10. $6xy + 10z - 6xz - 10y$

________ 11. $9a^2 - 16$

________ 12. $4a^2b^2 - 9c^2$

________ 13. $3x^2 - 48$

________ 14. $24m^3 - 150mn^2$

________ 15. $x^4 - 81$

________ 16. $x^3 + 27y^3$

________ 17. $27a^3 - 8b^3$

________ 18. $3x^3y^3z - 24z$

________ 19. $(x + y)^3 + 27$

________ 20. $a^6 + 63a^3 - 64$

________ 21. $x^2 - 12xy + 35y^2$

________ 22. $3a^2 - 3a - 168$

________ 23. $-2y^3 - 4y^2 + 30y$

________ 24. $x^4 - 15x^2 - 16$

________ 25. $(x + y)^2 + 2(x + y) - 63$

________ 26. $18x^2 + 23x - 6$

________ 27. $9x^2 - 30xy + 25y^2$

________ 28. $36x^4 + 12x^3 - 63x^2$

________ 29. $3(x - 1)^2 - 22(x - 1) + 24$

________ 30. $4a^4 - 101a^2 + 25$

Check your answers in the back of your book.

If You Missed Exercises:	You Should Review Examples:	
1–3	SECTION 4.1	1–5
4–6		6–9
7–10		10–13
11, 12	SECTION 4.2	1
13, 14		2–4
15		7
16		11, 12
17, 18		9
19		13
20		14
21	SECTION 4.3	1–5
22, 23		8, 9
24		18
25		16
26–28		9–13
29		16
30		15

If you have completed the **Review Exercises** and corrected your errors, you are ready to take the **Practice Test** for Chapter 4.

Name ______________________ Date ____________

PRACTICE TEST for Chapter 4

Rewrite each expression as a product of factors.

		Section	Examples
______	1. $8w^4 - 4w^2 + 12w$	4.1	2
______	2. $-5a^2b^3 + 15a^2b - 25ab^2$	4.1	5
______	3. $7(3x - 4) - 6x(3x - 4)$	4.1	6
______	4. $5(3y - 2z) + (3y - 2z)^2$	4.1	7
______	5. $2a + 2b + ac + bc$	4.1	10
______	6. $5a - ax - 5 + x$	4.1	12, 13
______	7. $121x^2 - 4y^2$	4.2	1
______	8. $4x^2 - 324$	4.2	2
______	9. $a^4 - 1$	4.2	7
______	10. $(2a - 5b)^2 - 100$	4.2	6
______	11. $2x^3 + x^2 - 50x - 25$	4.2	8
______	12. $125x^3 + y^3$	4.2	11
______	13. $5a^3 - 40$	4.2	10
______	14. $30x^2 - 13x - 3$	4.3	8
______	15. $7x^2 - 7xy - 42y^2$	4.3	9, 11
______	16. $25a^2 - 30ab + 9b^2$	4.3	9
______	17. $2x^5 - 5x^4 - 12x^3$	4.3	12, 13
______	18. $40x^2 + 26x - 9$	4.3	10
______	19. $6(x + y)^2 - 5(x + y) - 4$	4.3	16
______	20. $x^4 - 37x^2 + 36$	4.3	18

Name ______________________ Date ______________

SHARPENING YOUR SKILLS after Chapters 1–4

		Section
________	1. List the members of the set $\{-9, -3.2, -\sqrt{2}, \frac{1}{3}, 0, \sqrt{3}, \frac{5}{2}, 4, 5.\overline{13}\}$ that belong to the set of rational numbers.	1.1
________	2. Rewrite $4x^2 - 3x^4 - 5x + 3x^3$ in descending powers of the variable.	3.3

Evaluate each expression.

________	3. $(x - 7)(3x - 2)$ when $x = 2$	1.4
________	4. $-5x^2 + 2xy + 9y^2$ when $x = 3$ and $y = -2$	1.4
________	5. $\lvert 2 - \pi \rvert$	1.5
________	6. $-x^4 - 4$ when $x = -2$	3.1
________	7. a^{-2} when $a = -4$	3.2
________	8. $(-3a)^0$ when $a = 2$	3.1

Perform the indicated operations and simplify. Write answers with positive exponents.

________	9. $\dfrac{-7(5) - 4[9 - 3(5)]}{-11[15 - 4(4)]}$	1.3
________	10. $-7(3n - 1) + 6(n + 2)$	1.4
________	11. $12a - 2[3a - 4(a - 11b) - 9b] + 20b$	1.4
________	12. $(4x - 3y)(2x + 7y)$	3.4
________	13. $(x + 2)(x^2 + 4x + 4)$	3.4
________	14. $(a - 2)^3$	3.4
________	15. $(x^3 + 125) \div (x + 5)$	3.5
________	16. $4x(-3x^5y^2)(-\frac{2}{3}x^3y^3)$	3.1
________	17. $\left(\dfrac{-3a^2b^7c^3}{ab^5}\right)^2$	3.1
________	18. $(x^9y^{-3}z^{-4})(x^{-7}y^8z^4)$	3.2
________	19. $(2x^{-4}y^6z^2)^{-3}$	3.2
________	20. $\dfrac{7x^6y^{-3}}{x^{-5}y^2}$	3.2

Solve each equation or inequality. SECTION

________	21. $\frac{7x-1}{5} - 4 = 2x$	2.1
________	22. $7x - (x + 5) \leq 3x - 1$	2.3
________	23. $\lvert 13 - m \rvert \geq 2$	2.4
________	24. Solve and graph $\frac{3x-1}{7} \leq 1 + x$ or $4(2x - 1) < 9x + 2$.	2.3
________	25. If each side of a square is increased by 6 feet, write an expression for the difference between the new area and the old area.	3.6

C H A P T E R

5

Solving Quadratic Equations and Inequalities

The three support beams of a picnic shelter form a right triangle. The longest beam measures 25 feet and the shortest is 5 feet less than the other. Find the length of the piece of lumber that would be needed to cut all three beams.

In Chapter 2 we learned to solve first-degree equations and inequalities. Now that we are thoroughly familiar with the algebra of polynomials and the techniques of factoring, we are ready to solve equations and inequalities of higher degree. We shall concentrate on equations of the second degree, also called **quadratic equations**, and in this chapter we learn how to

1. Use the zero product rule.
2. Solve quadratic equations by factoring.
3. Solve quadratic equations of the form $ax^2 = c$.
4. Solve quadratic equations by completing the square.
5. Solve quadratic equations by using the quadratic formula.
6. Solve quadratic inequalities.
7. Switch from word statements to quadratic equations.

5.1 Solving Equations by Factoring

We have solved many first-degree equations that were of the form $ax + b = c$. Now we would like to find techniques for solving equations containing the *square* of a variable, that is, equations of the form $ax^2 + bx + c = 0$.

Using the Zero Product Rule

Suppose you are told that the product of two numbers equals zero. In other words, suppose you know that $A \cdot B = 0$. What conclusion can be drawn from such a statement?

If the product of two factors is zero, then at least one of the factors must be zero.

This statement is called the **zero product rule** and it can be expressed in symbols as

Zero Product Rule. If $A \cdot B = 0$, then $A = 0$ or $B = 0$ (or both).

The zero product rule will be our key to solving equations of degree greater than 1. It can be used whenever a product of two factors or three factors or any number of factors is zero.

Example 1. Solve $x(x - 3) = 0$.

Solution. From the zero product rule we know that

$$x = 0 \quad \text{or} \quad x - 3 = 0$$

and we must isolate x in each of these first-degree equations.

$$x = 0 \quad \text{or} \quad x - 3 = 0$$

$$x = 0 \quad \text{or} \quad x = 3$$

Let's check each solution in the original equation.

CHECK: $x = 0$	CHECK: $x = 3$
$x(x - 3) = 0$	$x(x - 3) = 0$
$0(0 - 3) \stackrel{?}{=} 0$	$3(3 - 3) \stackrel{?}{=} 0$
$0(-3) \stackrel{?}{=} 0$	$3(0) \stackrel{?}{=} 0$
$0 = 0$	$0 = 0$

You complete Example 2.

Example 2. Solve $(2x + 1)(3x - 7) = 0$.

Solution. From the zero product rule we know that

$$2x + 1 = 0 \quad \text{or} \quad 3x - 7 = 0$$

We must isolate x in each of these first-degree equations.

$2x + 1 = 0$ or $3x - 7 = 0$

$2x =$ ____ $\quad 3x =$ ____

$x =$ ____ or $x =$ ____

Check your work on page 282. ▶

Example 3. Solve $8(2 + x)(2 + x) = 0$.

Solution. There are three factors to consider here.

$$8(2 + x)(2 + x) = 0$$

$8 \neq 0 \quad 2 + x = 0 \quad$ or $\quad 2 + x = 0$

$x = -2 \qquad x = -2$

Since 8 can never equal zero, that factor does not contribute a solution. We do not write the same solution twice, so our solution is $x = -2$.

Trial Run

Solve each equation.

_______ 1. $(x + 3)(x - 7) = 0$

_______ 2. $(5 - a)(5 - a) = 0$

_______ 3. $(5x - 2)(3x + 1) = 0$

_______ 4. $(8 - 3x)(2 + x) = 0$

_______ 5. $3(x - 4)(x + 4) = 0$

_______ 6. $10x(x - 5)(3x - 2) = 0$

Answers are on page 283.

Solving Quadratic Equations by Factoring

Each of the equations in the preceding section was presented in factored form. To discover the degree of such equations, we can multiply the factors. For example,

$$(2x + 1)(3x - 7) = 0 \quad \text{becomes} \quad 6x^2 - 11x - 7 = 0$$

After multiplying the factors we have a **second-degree equation.** All second-degree equations can be written in standard form as $ax^2 + bx + c = 0$, where a, b, and c are real numbers, but $a \neq 0$. Such second-degree equations are also called **quadratic equations.**

> **Quadratic Equation.** Any equation that can be written in the form
>
> $$ax^2 + bx + c = 0 \qquad (a \neq 0)$$
>
> is called a quadratic equation.

Now that we have learned to recognize quadratic equations, we must develop a method for solving such equations. From our work with the zero product rule, you may have realized that to solve a quadratic equation, we should try to *rewrite the quadratic expression in factored form.*

Example 4. Solve $3x^2 + 5x - 2 = 0$.

Solution. We rewrite the left side in factored form.

$$3x^2 + 5x - 2 = 0$$

$$(3x - 1)(x + 2) = 0$$

Now we use the zero product rule.

$$3x - 1 = 0 \quad \text{or} \quad x + 2 = 0$$

$$3x = 1$$

$$x = \frac{1}{3} \quad \text{or} \quad x = -2$$

Let's check our solutions in the original equation.

CHECK: $x = \frac{1}{3}$

$$x^2 + 5x - 2 = 0$$

$$3\left(\frac{1}{3}\right)^2 + 5 \cdot \frac{1}{3} - 2 \stackrel{?}{=} 0$$

$$3\left(\frac{1}{9}\right) + \frac{5}{3} - 2 \stackrel{?}{=} 0$$

$$\frac{3}{9} + \frac{15}{9} - \frac{18}{9} \stackrel{?}{=} 0$$

$$0 = 0$$

CHECK: $x = -2$

$$3x^2 + 5x - 2 = 0$$

$$3(-2)^2 + 5(-2) - 2 \stackrel{?}{=} 0$$

$$3(4) - 10 - 2 \stackrel{?}{=} 0$$

$$12 - 10 - 2 \stackrel{?}{=} 0$$

$$0 = 0$$

You try solving the equation in Example 5.

Example 5. Solve $0 = 5y^2 - 20$.

Solution

$$0 = 5y^2 - 20$$
$$0 = 5(y^2 - 4)$$
$$0 = 5(\_\_\_\_\_)(\_\_\_\_\_)$$

The solutions are $y =$ ____ or $y =$ ____ .

Check your work on page 282. ▶
Now try Example 6.

Example 6. Solve $7x - 3x^2 = 0$.

Solution

$$7x - 3x^2 = 0$$
$$x(7 - 3x) = 0$$
$$x = \_\_\_\_ \quad \text{or} \quad 7 - 3x = 0$$

The solutions are $x =$ ____ or $x =$ ____ .

Check your work on page 282. ▶

Let's summarize our method for solving a quadratic equation in standard form.

Solving Quadratic Equations in Standard Form by Factoring

1. Factor the quadratic expression into a product of first-degree factors.
2. Using the zero product rule, set each of the factors equal to zero, and solve.

Trial Run

Solve each equation.

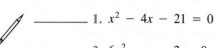

_____ 1. $x^2 - 4x - 21 = 0$

_____ 2. $x^2 + 10x + 25 = 0$

_____ 3. $6x^2 - x - 2 = 0$

_____ 4. $0 = 6y^2 - 24$

Answers are on page 283.

Solving More Quadratic Equations

To use the zero product rule, we must express a quadratic equation as a product of factors equal to zero. What happens when an equation contains a second-degree term on one side but the other side is not zero? We must investigate equations such as $x^2 - 3x = 4$. This is a second-degree equation because it contains a variable raised to the second power. However, it is *not* written in the standard quadratic form $ax^2 + bx + c = 0$. Before we can solve such an equation by factoring, we must put it into standard form, with all nonzero terms on one side and zero by itself on the other side. We can accomplish this by adding and/or subtracting the same terms to or from both sides of the original equation.

Example 7. Solve $x^2 - 3x = 4$ and check.

Solution

$$x^2 - 3x = 4$$
$$x^2 - 3x - 4 = 0 \qquad \text{Rewrite equation in standard form.}$$

$(x - 4)(x + 1) = 0$	Factor the quadratic expression.
$x - 4 = 0$ or $x + 1 = 0$	Use the zero product rule.
$x = 4$ or $x = -1$	Solve each equation.

Let's check both solutions in the original equation.

CHECK: $x = 4$	CHECK: $x = -1$
$x^2 - 3x = 4$	$x^2 - 3x = 4$
$4^2 - 3(4) \stackrel{?}{=} 4$	$(-1)^2 - 3(-1) \stackrel{?}{=} 4$
$16 - 12 \stackrel{?}{=} 4$	$1 + 3 \stackrel{?}{=} 4$
$4 = 4$	$4 = 4$

Now you try Example 8. Remember to put the equation into standard form before solving it.

Example 8. Solve $x^2 = 10 - 9x$.

Solution

$$x^2 = 10 - 9x$$

$$x^2 + 9x = 10$$

$$x^2 + 9x - 10 = 0$$

The solutions are $x =$ _____ or $x =$ _____.

Check your work on page 282. ▶

Example 9. Solve $5y = 3y^2$.

Solution

$$5y = 3y^2$$

$$0 = 3y^2 - 5y$$

$$0 = y(3y - 5)$$

$$y = 0 \quad \text{or} \quad 3y - 5 = 0$$

$$3y = 5$$

$$y = \frac{5}{3}$$

The solutions are $y = 0$ or $y = \frac{5}{3}$.

Trial Run

Solve each equation.

_______ 1. $y^2 - 6y = 7$

_______ 2. $6y^2 = 15y$

_______ 3. $2x^2 = 19x - 24$

_______ 4. $-32 + 16x = 2x^2$

Answers are on page 283.

As was the case with first-degree equations, second-degree equations sometimes contain parentheses or other symbols of grouping that must be removed before we can get down to the business of finding solutions. Suppose, for instance, that we must solve

$$x(x + 3) - 7 = 11$$

It may not be clear to you whether or not this is a quadratic equation at all. At this point, however, that is not important. Whenever we see an equation containing a sum of terms and/or symbols of inclusion, we automatically proceed to remove the parentheses and combine like terms. Then we can decide what kind of equation we are solving and continue with the correct method.

Example 10. Solve $x(x + 3) - 7 = 11$ and check.

Solution

$x(x + 3) - 7 = 11$	
$x^2 + 3x - 7 = 11$	Remove parentheses.
$x^2 + 3x - 18 = 0$	Rewrite equation in standard form.
$(x + 6)(x - 3) = 0$	Factor quadratic expression.
$x + 6 = 0$ or $x - 3 = 0$	Use zero product rule.
$x = -6$ or $x = 3$	Solve for x.

CHECK: $x = -6$	CHECK: $x = 3$
$x(x + 3) - 7 = 11$	$x(x + 3) - 7 = 11$
$-6(-6 + 3) - 7 \stackrel{?}{=} 11$	$3(3 + 3) - 7 \stackrel{?}{=} 11$
$-6(-3) - 7 \stackrel{?}{=} 11$	$3(6) - 7 \stackrel{?}{=} 11$
$18 - 7 \stackrel{?}{=} 11$	$18 - 7 \stackrel{?}{=} 11$
$11 = 11$	$11 = 11$

Example 11. Solve $(x - 6)(x + 1) = x(x - 8) + 5$.

Solution

$(x - 6)(x + 1) = x(x - 8) + 5$	
$x^2 - 5x - 6 = x^2 - 8x + 5$	Remove parentheses.
$-5x - 6 = -8x + 5$	Move x^2-terms to same side.

Notice that the x^2-terms have dropped out. This is a *first-degree equation* that we solve by getting the variable on one side and constant terms on the other.

$3x - 6 = 5$	Add $8x$ to both sides.
$3x = 11$	Add 6 to both sides.
$x = \dfrac{11}{3}$	Divide both sides by 3.

You try Example 12.

Example 12. Solve $(3x - 1)(x + 7) = -3(x^2 + x + 9)$.

Solution

$(3x - 1)(x + 7) = -3(x^2 + x + 9)$	
$3x^2 + 20x - 7 = -3x^2 - 3x - 27$	Remove parentheses.
$6x^2 + 20x - 7 = -3x - 27$	Add $3x^2$ to both sides.
$6x^2 + 23x - 7 = -27$	Add $3x$ to both sides.
$6x^2 + 23x + 20 = 0$	Add 27 to both sides.
(________)(________) = 0	Factor the quadratic expression.

$x =$ ____ or $x =$ ____

Check your work on page 282. ▶

In general, we have discovered that any quadratic equation in which the quadratic expression can be factored should be solved using the following steps.

Solving Quadratic Equations by Factoring

1. Deal with symbols of grouping and combine like terms.
2. Get all nonzero terms on one side of the equation and zero on the other side.
3. Factor the quadratic expression.
4. Set each factor equal to zero and solve.
5. Check your work by substituting each solution into the original equation.

Trial Run

Solve each equation.

________ 1. $9x^2 - 11x + 9 = 8x^2 - 9x + 12$

________ 2. $5x(x - 2) - 6 = 4x^2 - 2(5x + 1)$

________ 3. $3(x^2 - 1) - 9 = x(x + 10)$

________ 4. $(3x - 2)(x - 4) = -3(x^2 + x - 6)$

Answers are on page 283.

Solving Other Equations by Factoring

Any equation that can be written with a product of factors on one side and zero on the other side can be solved using the zero product rule. You try using the zero product rule in Example 13.

Example 13. Solve

$$(x - 2)(x + 3)(2x - 7) = 0$$

Solution

$$(x - 2)(x + 3)(2x - 7) = 0$$

$x - 2 = 0$ or $x + 3 = 0$ or $2x - 7 = 0$

$x =$ _____ $x =$ _____ $2x =$ _____

$x =$ _____

The three solutions are $x =$ _____ or $x =$ _____ or $x =$ _____ .

Check your answers on page 282. ▶

Example 14. Solve

$$x^3 + 8x^2 = 20x$$

Solution

$$x^3 + 8x^2 = 20x$$

$$x^3 + 8x^2 - 20x = 0$$

$$x(x^2 + 8x - 20) = 0$$

$$x(x + 10)(x - 2) = 0$$

$x = 0$ or $x + 10 = 0$ or $x - 2 = 0$

$x = -10$ $x = 2$

The three solutions are $x = 0$ or $x = -10$ or $x = 2$.

Example 15. Solve $x^4 - 29x^2 + 100 = 0$.

Solution. Since the square of the variable part of the second term (x^2) is identical to the variable part of the first term (x^4), we can factor this polynomial just as we factor a quadratic trinomial.

$$x^4 - 29x^2 + 100 = 0$$

$$(x^2 - 25)(x^2 - 4) = 0$$

$$(x + 5)(x - 5)(x + 2)(x - 2) = 0$$

$x + 5 = 0$ or $x - 5 = 0$ or $x + 2 = 0$ or $x - 2 = 0$

$x = -5$ or $x = 5$ or $x = -2$ or $x = 2$

Any one of these four solutions satisfies the original equation.

Now you try Example 16.

Example 16. Solve $x^3 + 3x^2 - 4x - 12 = 0$.

Solution. Since this polynomial contains four terms, we can try factoring by grouping.

$$x^3 + 3x^2 - 4x - 12 = 0$$

$$x^2(x + 3) - 4(x + 3) = 0$$

$$(x + 3)(x^2 - 4) = 0$$

$(x + 3)($________$)($________$) = 0$

$x =$ _____ or $x =$ _____ or $x =$ _____

Check your work on page 283. ▶

Trial Run

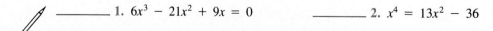

Solve each equation.

________ 1. $6x^3 - 21x^2 + 9x = 0$

________ 2. $x^4 = 13x^2 - 36$

_______ 3. $a^3 - 7a^2 - a + 7 = 0$ _______ 4. $5x^5 - 10x^3 + 5x = 0$

Answers are on page 283.

Finding Equations with Given Solutions

In solving equations by factoring, we found the solutions by setting each first-degree factor equal to zero. Now we would like to reverse our procedure to find an equation for which we know the solutions.

Let's find a quadratic equation with the solutions $x = 2$ and $x = 3$. Throughout this process, think of what the *previous* step would have been in solving the quadratic equation. Work backwards!

$$x = 2 \quad \text{or} \quad x = 3$$
$$x - 2 = 0 \quad \text{or} \quad x - 3 = 0$$
$$(x - 2)(x - 3) = 0$$
$$x^2 - 5x + 6 = 0$$

We are making use of an important fact.

If r is a *solution* for a polynomial equation, then $x - r$ must be a *factor* of the polynomial.

Example 17. Write a third-degree equation with solutions $x = 0$, $x = -1$, and $x = 2$.

Solution. We must work backwards.

$x = 0$ so x is a factor

$x = -1$ so $x + 1$ is a factor

$x = 2$ so $x - 2$ is a factor

$$x(x + 1)(x - 2) = 0$$
$$x(x^2 - x - 2) = 0$$
$$x^3 - x^2 - 2x = 0$$

Example 18. Write a fourth-degree equation with solutions $x = 1$, $x = -1$, $x = 5$, and $x = -5$.

Solution. We must work backwards.

$x = 1$ so $x - 1$ is a factor

$x = -1$ so $x + 1$ is a factor

$x = 5$ so $x - 5$ is a factor

$x = -5$ so $x + 5$ is a factor

$$(x - 1)(x + 1)(x - 5)(x + 5) = 0$$
$$(x^2 - 1)(x^2 - 25) = 0$$
$$x^4 - 26x^2 + 25 = 0$$

We must point out that in each of these examples, we did not find the *only* equation with the given solutions. Any *constant multiple* of the equations we found will also have the same solutions. For example, the equation

$$x^2 - 5x + 6 = 0 \quad \text{has solutions} \quad x = 2 \quad \text{or} \quad x = 3$$

and the equation

$$10x^2 - 50x + 60 = 0$$
$$10(x^2 - 5x + 6) = 0 \quad \text{has solutions} \quad x = 2 \quad \text{or} \quad x = 3$$

Trial Run

Find an equation satisfying the given conditions.

________ **1.** A quadratic equation with solutions $x = 0$ and $x = 5$.

________ **2.** A third-degree equation with solutions $x = 2$, $x = -2$, and $x = 1$.

Answers are on page 283.

Examples You Completed

Example 2. Solve $(2x + 1)(3x - 7) = 0$.

Solution. From the zero product rule we know that

$$2x + 1 = 0 \quad \text{or} \quad 3x - 7 = 0$$

We must isolate x in each of these first-degree equations.

$$\begin{aligned} 2x + 1 &= 0 & \text{or} \quad 3x - 7 &= 0 \\ 2x &= -1 & 3x &= 7 \\ x &= -\frac{1}{2} & \text{or} \quad x &= \frac{7}{3} \end{aligned}$$

Example 5. Solve $0 = 5y^2 - 20$.

Solution

$$\begin{aligned} 0 &= 5y^2 - 20 \\ 0 &= 5(y^2 - 4) \\ 0 &= 5(y + 2)(y - 2) \end{aligned}$$

$$5 \neq 0 \quad 0 = y + 2 \quad \text{or} \quad 0 = y - 2$$

$$-2 = y \quad \text{or} \quad 2 = y$$

The solutions are $y = -2$ or $y = 2$.

Example 6. Solve $7x - 3x^2 = 0$.

Solution

$$\begin{aligned} 7x - 3x^2 &= 0 \\ x(7 - 3x) &= 0 \end{aligned}$$

$$x = 0 \quad \text{or} \quad 7 - 3x = 0$$

$$-3x = -7$$

$$x = 0 \quad \text{or} \quad x = \frac{7}{3}$$

Example 8. Solve $x^2 = 10 - 9x$.

Solution

$$\begin{aligned} x^2 &= 10 - 9x \\ x^2 + 9x &= 10 \\ x^2 + 9x - 10 &= 0 \\ (x + 10)(x - 1) &= 0 \end{aligned}$$

$$x + 10 = 0 \quad \text{or} \quad x - 1 = 0$$

$$x = -10 \quad \text{or} \quad x = 1$$

The solutions are $x = -10$ or $x = 1$.

Example 12 (*Solution*)

$$\begin{aligned} 6x^2 + 23x + 20 &= 0 \\ (3x + 4)(2x + 5) &= 0 \end{aligned}$$

$$\begin{aligned} 3x + 4 &= 0 & \text{or} \quad 2x + 5 &= 0 \\ 3x &= -4 & 2x &= -5 \\ x &= -\frac{4}{3} & \text{or} \quad x &= -\frac{5}{2} \end{aligned}$$

Example 13. Solve

$$(x - 2)(x + 3)(2x - 7) = 0$$

Solution

$$(x - 2)(x + 3)(2x - 7) = 0$$

$$x - 2 = 0 \quad \text{or} \quad x + 3 = 0 \quad \text{or} \quad 2x - 7 = 0$$

$$x = 2 \qquad x = -3 \qquad 2x = 7$$

$$x = \frac{7}{2}$$

The three solutions are $x = 2$ or $x = -3$ or $x = \frac{7}{2}$.

Example 16. Solve $x^3 + 3x^2 - 4x - 12 = 0$.

Solution

$$x^3 + 3x^2 - 4x - 12 = 0$$
$$x^2(x + 3) - 4(x + 3) = 0$$
$$(x + 3)(x^2 - 4) = 0$$
$$(x + 3)(x + 2)(x - 2) = 0$$
$$x + 3 = 0 \quad \text{or } x + 2 = 0 \quad \text{or } x - 2 = 0$$
$$x = -3 \text{ or } \quad x = -2 \text{ or } \quad x = 2$$

Answers to Trial Runs

page 275 **1.** $x = -3, x = 7$ **2.** $a = 5$ **3.** $x = \frac{2}{5}, x = -\frac{1}{3}$ **4.** $x = \frac{8}{3}, x = -2$
5. $x = 4, x = -4$ **6.** $x = 0, x = 5, x = \frac{2}{3}$

page 276 **1.** $x = 7, x = -3$ **2.** $x = -5$ **3.** $x = -\frac{1}{2}, x = \frac{2}{3}$ **4.** $y = 2, y = -2$

page 277 **1.** $y = 7, y = -1$ **2.** $y = 0, y = \frac{5}{2}$ **3.** $x = 8, x = \frac{3}{2}$ **4.** $x = 4$

page 279 **1.** $x = 3, x = -1$ **2.** $x = -2, x = 2$ **3.** $x = 6, x = -1$ **4.** $x = \frac{5}{2}, x = -\frac{2}{3}$

page 280 **1.** $x = 0, x = \frac{1}{2}, x = 3$ **2.** $x = 3, x = -3, x = 2, x = -2$ **3.** $a = 7, a = 1, a = -1$
4. $x = 0, x = -1, x = 1$

page 282 **1.** $x^2 - 5x = 0$ **2.** $x^3 - x^2 - 4x + 4 = 0$

Name ______________________ Date ____________

EXERCISE SET 5.1

Solve each equation.

______ 1. $(x - 9)(x + 11) = 0$

______ 2. $(x - 7)(x + 12) = 0$

______ 3. $(3 - 5x)(3 - 5x) = 0$

______ 4. $(4 - 9x)(4 - 9x) = 0$

______ 5. $(8x - 1)(3x + 5) = 0$

______ 6. $(10x - 3)(3x + 8) = 0$

______ 7. $2(9 - 2x)(7 + x) = 0$

______ 8. $3(1 - 5x)(6 - x) = 0$

______ 9. $(3x - 1)(3x + 1) = 0$

______ 10. $(5x - 2)(5x + 2) = 0$

______ 11. $8x(x - 2)(3x - 7) = 0$

______ 12. $-5x(x - 7)(2x - 9) = 0$

______ 13. $x^2 - 11x + 30 = 0$

______ 14. $x^2 + 20x + 99 = 0$

______ 15. $x^2 + 8x - 20 = 0$

______ 16. $x^2 - x - 12 = 0$

______ 17. $25 + 10x + x^2 = 0$

______ 18. $64 - 16x + x^2 = 0$

______ 19. $m^2 + 7m - 44 = 0$

______ 20. $m^2 + 3m - 54 = 0$

______ 21. $5y^2 + 7y + 2 = 0$

______ 22. $7y^2 + 22y + 3 = 0$

______ 23. $6a^2 - 34a + 20 = 0$

______ 24. $33a^2 - 105a + 18 = 0$

______ 25. $a^2 = 16a - 64$

______ 26. $a^2 = 15a - 36$

______ 27. $5x^2 = 45$

______ 28. $3x^2 = 48$

______ 29. $2y^2 = 26y$

______ 30. $4y^2 = 60y$

______ 31. $8m^2 = 15 - 14m$

______ 32. $8m^2 = 35 - 18m$

______ 33. $2y^2 + 21 = 13y$

______ 34. $4y^2 + 15 = 23y$

______ 35. $18m + 16 = 9m^2$

______ 36. $22 - 25y = 12y^2$

______ 37. $7x^2 + 3x - 10 = 5x^2 + 7x + 20$

______ 38. $8x^2 - 10x - 16 = 5x^2 - 13x + 20$

______ 39. $3x^2 - x = 2(x^2 - 5x)$

______ 40. $9x^2 - 2x = 4(2x^2 - 3x)$

______ 41. $(x - 5)(x - 6) = x(x + 3) + 7$

______ 42. $(x - 11)(x + 9) = x(x - 8) - 3$

______ 43. $x(2x - 3) - 4(3x - 5) = 29 - 8x$

______ 44. $9x(x - 2) + 6(2x - 1) = 9x + 8$

______ 45. $3x(x - 2) + 2(2x^2 - 5) = 2(4x - 5)$

______ 46. $13x(x - 8) + 3(x^2 + 8) = 8(3 - 4x)$

______ 47. $(3x - 7)(x + 1) = -3(x^2 + 10x + 9)$

______ 48. $(4x - 3)(2x - 1) = -8(x^2 - x) + 1$

______ 49. $(x - 4)(x + 2)(2x - 3) = 0$

______ 50. $(2x - 1)(3x - 2)(x + 5) = 0$

________ 51. $25x^3 - 50x^2 + 16x = 0$

________ 52. $4x^3 - 20x^2 + 9x = 0$

________ 53. $x^4 - 17x^2 + 16 = 0$

________ 54. $x^4 - 25x^2 + 144 = 0$

________ 55. $3x^4 = 30x^2 - 27$

________ 56. $5x^4 = 145x^2 - 500$

________ 57. $3x^5 - 24x^3 + 48x = 0$

________ 58. $2x^5 - 36x^3 + 162x = 0$

________ 59. $2x^5 = 128x^3$

________ 60. $3x^5 = 147x^3$

________ 61. $x^3 + 3x^2 - 16x - 48 = 0$

________ 62. $x^3 + 5x^2 - 4x - 20 = 0$

________ 63. $8x^3 - 4x^2 - 18x + 9 = 0$

________ 64. $9x^3 - 18x^2 - 4x + 8 = 0$

Find an equation satisfying the given conditions.

________ 65. A quadratic equation with solutions $x = -2$ and $x = -9$.

________ 66. A quadratic equation with solutions $x = -7$ and $x = 2$.

________ 67. A third-degree equation with solutions $x = 0$, $x = 3$, and $x = -3$.

________ 68. A third-degree equation with solutions $x = 2$, $x = 1$, and $x = -1$.

________ 69. A quadratic equation with solutions $x = \frac{2}{3}$ and $x = 5$.

________ 70. A quadratic equation with solutions $x = \frac{1}{2}$ and $x = \frac{2}{5}$.

________ 71. A fourth-degree equation with solutions $x = 0$, $x = -2$, $x = 3$, and $x = \frac{1}{2}$.

________ 72. A fourth-degree equation with solutions $x = 1$, $x = -1$, $x = 3$, and $x = -2$.

☆ Stretching the Topics

Solve each equation.

________ 1. $(x + 3)(2x + 3) = 77$

________ 2. $(x + 3)^2 + (5x + 2)^2 = (7x - 1)^2$

________ 3. $3(x - 2)(3x - 1) - 2(2x - 3)(2x + 3) + 5x - 3 = -5(2x + 3) + 27$

Check your answers in the back of your book.

If you can solve the equations in **Checkup 5.1,** you are ready to go on to Section 5.2.

Name ______________________________ Date ______________

✓ CHECKUP 5.1

Solve each equation.

________ **1.** $(10 - 3x)(1 + 5x) = 0$

________ **2.** $x^2 - 2x - 63 = 0$

________ **3.** $6y^2 = 11y + 7$

________ **4.** $12a^2 - 20 = 22a$

________ **5.** $(2x - 1)(x + 1) = -x(x - 11) - 4$

________ **6.** $x(x + 3) + 3(x^2 - 9) = -4(x + 3)$

________ **7.** $26x^3 - 43x^2 + 6x = 0$

________ **8.** $x^4 - 40x^2 + 144 = 0$

________ **9.** $2x^5 - 130x^4 + 128x^3 = 0$

________ **10.** Write a quadratic equation with solutions $x = -1$ and $x = 4$.

Check your answers in the back of your book.

If You Missed Problems:	You Should Review Exercises:
1	2, 3
2	4
3, 4	7, 8
5, 6	10–12
7	14
8	15
9	16
10	17

5.2 Solving Quadratic Equations of the Form $ax^2 = c$

Although the method of factoring is usually the best way to solve a quadratic equation, there are many quadratic expressions that cannot be factored to yield rational number solutions. In order to develop a method for solving *any* quadratic equation, we take a slight detour here and look at some special kinds of quadratic equations.

Using Square-Root Notation

We learned to solve an equation such as $x^2 - 4 = 0$ by factoring the quadratic expression as a difference of two squares. However, there is another approach that we might have used on this problem; it is a special kind of quadratic equation in which *the first-degree term is missing*. Let's try using two approaches to solve $x^2 - 4 = 0$.

First Approach

$$x^2 - 4 = 0$$

$$(x + 2)(x - 2) = 0$$

$$x + 2 = 0 \quad \text{or} \quad x - 2 = 0$$

$$x = -2 \quad \text{or} \quad x = 2$$

Second Approach

$$x^2 - 4 = 0$$

$$x^2 = 4$$

$$x = 2 \quad \text{or} \quad x = -2$$

because

$$(2)^2 = 4 \quad \text{and} \quad (-2)^2 = 4$$

Let's practice the second approach on some more equations.

Example 1. Solve $4x^2 = 25$.

Solution

$$4x^2 = 25$$

$$x^2 = \frac{25}{4}$$

$$x = \frac{5}{2} \quad \text{or} \quad x = -\frac{5}{2}$$

because

$$\left(\frac{5}{2}\right)^2 = \frac{25}{4} \quad \text{and} \quad \left(-\frac{5}{2}\right)^2 = \frac{25}{4}$$

You try Example 2.

Example 2. Solve $x^2 - 1 = 0$.

Solution

$$x^2 - 1 = 0$$

$$x^2 = 1$$

$x =$ _____ or $x =$ _____

because

(_____)$^2 = 1$ and (_____)$^2 = 1$

Check your work on page 294. ▶

In each of these examples we solved quadratic equations of the form $ax^2 = c$. After dividing both sides of the equation by the coefficient of the squared term we were left with $x^2 = \frac{c}{a}$. Then we asked the question: What numbers, when squared, would equal $\frac{c}{a}$? In answering that question, we were actually finding the **square roots** of $\frac{c}{a}$.

Every positive number has two square roots—one positive and one negative. The positive square root of a number is called its **principal square root.**

Principal Square Root. The positive square root of a number n is called its principal square root. It is denoted by $\sqrt{n}$.

The symbol $\sqrt{}$ is called a **radical**. Let's practice this notation on some numbers.

Number	Principal Square Root	Square Roots
9	$\sqrt{9} = 3$	$3, -3$
36	$\sqrt{36} = 6$	$6, -6$
$\frac{1}{16}$	$\sqrt{\frac{1}{16}} = \frac{1}{4}$	$\frac{1}{4}, -\frac{1}{4}$
3	$\sqrt{3} = \sqrt{3}$	$\sqrt{3}, -\sqrt{3}$

Notice that the square root of 3 is *not a rational number*. We cannot write $\sqrt{3}$ as a quotient of integers. We can *approximate* $\sqrt{3}$ using a calculator or the table of square roots inside the front cover of this book, but using either of those methods, we can obtain only an estimate ($\sqrt{3} \doteq 1.732051$).* We conclude that $\sqrt{3}$ is an irrational number.

Let's practice using our principal-square-root notation to write solutions for some equations of the form $ax^2 = c$.

If $ax^2 = c$, the solutions can be found by

$$ax^2 = c$$

$$x^2 = \frac{c}{a}$$

$$x = \sqrt{\frac{c}{a}} \quad \text{or} \quad x = -\sqrt{\frac{c}{a}}$$

where $\frac{c}{a}$ is nonnegative.

Example 3. Solve $x^2 = 100$ and state the nature of the solutions.

Solution

$$x^2 = 100$$

$$x = \sqrt{100} \quad \text{or} \quad x = -\sqrt{100}$$

$$x = 10 \quad \text{or} \quad x = -10$$

We note that both solutions are rational.

Example 4. Solve $9x^2 - 1 = 0$ and state the nature of the solutions.

Solution

$$9x^2 - 1 = 0$$

$$9x^2 = 1$$

$$x^2 = \frac{1}{9}$$

$$x = \frac{1}{3} \quad \text{or} \quad x = -\frac{1}{3}$$

Here both solutions are rational.

Now you try Example 5.

*The symbol $\doteq$ means "is approximately equal to."

Example 5. Solve $x^2 - 49 = 0$ and state the nature of the solutions.

Solution

$$x^2 - 49 = 0$$

$$x^2 = 49$$

$$x = \sqrt{\underline{\qquad}} \quad \text{or} \quad x = -\sqrt{\underline{\qquad}}$$

$$x = \underline{\qquad} \quad \text{or} \quad x = \underline{\qquad}$$

Both solutions are __________.

Check your work on page 294. ▶

Example 6. Solve $5x^2 = 65$ and state the nature of the solutions.

Solution

$$5x^2 = 65$$

$$\frac{5x^2}{5} = \frac{65}{5}$$

$$x^2 = 13$$

$$x = \sqrt{13} \quad \text{or} \quad x = -\sqrt{13}$$

Both solutions are irrational.

Suppose that we wish to solve the equation

$$x^2 + 4 = 0$$

$$x^2 = -4$$

Can you think of any numbers whose square equals -4? You know that the square of a real number can never be *negative*. Our conclusion is

$x^2 + 4 = 0$ has *no real number* solutions

Although we can write that $x = \sqrt{-4}$ or $x = -\sqrt{-4}$, these solutions are *not real numbers*. We shall discuss numbers that are not real numbers in Chapter 7.

Example 7. Solve $-5x^2 = 35$.

Solution

$$-5x^2 = 35$$

$$\frac{-5x^2}{-5} = \frac{35}{-5}$$

$$x^2 = -7$$

$$x = \sqrt{-7} \quad \text{or} \quad x = -\sqrt{-7}$$

There are *no* real number solutions.

Example 8. Solve $2x^2 + 3 = 5x^2 - 3$.

Solution

$$2x^2 + 3 = 5x^2 - 3$$

$$2x^2 + 3 - 5x^2 = -3$$

$$-3x^2 + 3 = -3$$

$$-3x^2 = -3 - 3$$

$$-3x^2 = -6$$

$$x^2 = 2$$

$$x = \sqrt{2} \quad \text{or} \quad x = -\sqrt{2}$$

Both solutions are real, but irrational, numbers.

Let's summarize our work with the principal square root as it applies to solving quadratic equations of the form $x^2 = n$.

If $x^2 = n$, and n is *positive* and a *perfect square*, the solutions are $x = \sqrt{n}$ or $x = -\sqrt{n}$ and both solutions are *rational* numbers.

Example: $x^2 = 81$

$x = \sqrt{81}$ or $x = -\sqrt{81}$

$x = 9$ or $x = -9$

If $x^2 = n$, and n is *positive* but *not a perfect square*, the solutions are $x = \sqrt{n}$ or $x = -\sqrt{n}$ and both solutions are *irrational* numbers.

Example: $x^2 = 5$

$x = \sqrt{5}$ or $x = -\sqrt{5}$

If $x^2 = n$, and n is *negative*, there are *no real* solutions.

Example: $x^2 + 3 = 0$

$x^2 = -3$

No real solutions.

If $x^2 = n$, and n is 0, the only solution is $x = 0$.

Example: $x^2 + 2 = 2$

$x^2 = 0$

$x = 0$

Trial Run

Solve each equation and state the nature of the solutions.

_______ 1. $x^2 - 2 = 0$

_______ 2. $16x^2 = 9$

_______ 3. $7x^2 - 28 = 0$

_______ 4. $5a^2 = 30$

_______ 5. $-4x^2 = 36$

_______ 6. $3x^2 - 7 = 6x^2 + 5$

Answers are on page 295.

Simplifying Square Roots

The set of irrational numbers contains many numbers besides square roots of numbers that are not perfect squares. The algebra of irrational numbers will be discussed thoroughly in Chapter 7. In this chapter we are concerned only with irrational numbers of the form $\sqrt{n}$, and there are a few rules that we might find handy for simplifying square roots of *nonnegative* numbers.

Let's use two approaches to simplify $\sqrt{36}$.

First Approach	***Second Approach***
$\sqrt{36} = 6$	$\sqrt{36} = \sqrt{4 \cdot 9}$
because $6^2 = 36$	$= \sqrt{4} \cdot \sqrt{9}$
	$= 2 \cdot 3$
	$= 6$

Notice that we obtained the same value for $\sqrt{36}$ using both approaches.

We have discovered that for positive factors,

> The square root of a product of factors is equal to the product of the square roots of the factors.

Using A and B to represent nonnegative numbers, we have

> **Product Rule for Square Roots**
>
> $\sqrt{A \cdot B} = \sqrt{A} \cdot \sqrt{B} \qquad \text{where } A \geq 0,\ B \geq 0$

This rule for simplifying square roots is even more useful when the original quantity is not a perfect square. For instance, to simplify $\sqrt{12}$ we must try to write 12 as a product of factors so that at least one factor is a perfect square.

$$\begin{aligned} \sqrt{12} &= \sqrt{4 \cdot 3} \\ &= \sqrt{4} \cdot \sqrt{3} \\ &= 2\sqrt{3} \end{aligned}$$

Notice that we factored 12 as $4 \cdot 3$ rather than $2 \cdot 6$ because 4 is a perfect square; neither 2 nor 6 is a perfect square. It is customary to write the rational factor before the irrational factor in our final answers. The answer $2\sqrt{3}$ is read "2 times the square root of 3." The number $2\sqrt{3}$ is *irrational* because $\sqrt{3}$ is irrational. Now you try Example 9.

Example 9. Simplify $\sqrt{700}$.

Solution

$$\begin{aligned} \sqrt{700} &= \sqrt{100 \cdot \underline{\hspace{2em}}} \\ &= \sqrt{100} \cdot \sqrt{\underline{\hspace{2em}}} \\ &= \end{aligned}$$

Check your work on page 294. ▶

Example 10. Simplify $\sqrt{252}$.

Solution. Sometimes you will not see the largest square factor immediately. Just keep going.

$$\begin{aligned} \sqrt{252} &= \sqrt{4 \cdot 63} \\ &= \sqrt{4 \cdot 9 \cdot 7} \\ &= \sqrt{4} \cdot \sqrt{9} \cdot \sqrt{7} \\ &= 2 \cdot 3\sqrt{7} \\ &= 6\sqrt{7} \end{aligned}$$

Square roots of quotients (fractions) can be simplified in a similar way.

> The square root of a quotient of numbers is equal to the quotient of the square roots of those numbers.

Again using A and B to represent numbers, we have

> **Quotient Rule for Square Roots**
>
> $\sqrt{\dfrac{A}{B}} = \dfrac{\sqrt{A}}{\sqrt{B}} \qquad \text{where } A \geq 0,\ B > 0$

Example 11. Simplify $\sqrt{\frac{9}{16}}$.

Solution

$$\sqrt{\frac{9}{16}} = \frac{\sqrt{9}}{\sqrt{16}}$$
$$= \frac{3}{4}$$

You try Example 12.

Example 12. Simplify $\sqrt{\frac{81}{25}}$.

Solution

$$\sqrt{\frac{81}{25}} =$$

Check your work on page 294. ▶

Example 13. Simplify $\sqrt{\frac{68}{9}}$.

Solution

$$\sqrt{\frac{68}{9}} = \frac{\sqrt{68}}{\sqrt{9}}$$
$$= \frac{\sqrt{4 \cdot 17}}{3}$$
$$= \frac{\sqrt{4}\sqrt{17}}{3}$$
$$= \frac{2\sqrt{17}}{3}$$

Now you try Example 14.

Example 14. Simplify $\sqrt{\frac{48}{49}}$.

Solution

$$\sqrt{\frac{48}{49}} = \frac{\sqrt{48}}{\sqrt{49}}$$
$$=$$

Check your work on page 295. ▶

Trial Run

Simplify each square root.

_______ 1. $\sqrt{48}$

_______ 2. $\sqrt{150}$

_______ 3. $\sqrt{72}$

_______ 4. $\sqrt{76}$

_______ 5. $\sqrt{\frac{25}{16}}$

_______ 6. $\sqrt{\frac{5}{49}}$

_______ 7. $\sqrt{\frac{12}{25}}$

_______ 8. $\sqrt{\frac{52}{121}}$

Answers are on page 295.

Our ability to work with square roots now allows us to solve *any* quadratic equation of the form

$$ax^2 = c \qquad \left(\frac{c}{a} \geq 0\right)$$

and to write the solutions in simplified form.

Example 15. Solve $121x^2 = 81$.

Solution

$$121x^2 = 81$$

$$x^2 = \frac{81}{121}$$

$$x = \sqrt{\frac{81}{121}} \quad \text{or} \quad x = -\sqrt{\frac{81}{121}}$$

$$= \frac{\sqrt{81}}{\sqrt{121}} \qquad\quad = -\frac{\sqrt{81}}{\sqrt{121}}$$

$$x = \frac{9}{11} \quad \text{or} \quad x = -\frac{9}{11}$$

Both solutions are rational.

You try Example 16.

Example 16. Solve $2y^2 - 54 = 0$.

Solution

$$2y^2 - 54 = 0$$

$$2y^2 = 54$$

$$y^2 = 27$$

$$y = \sqrt{27} \quad \text{or} \quad y = -\sqrt{27}$$

$$= \sqrt{\_\_\_\_ \cdot 3} \qquad = -\sqrt{\_\_\_\_ \cdot 3}$$

$$y = \_\_\_\_\sqrt{3} \quad \text{or} \quad y = \_\_\_\_\sqrt{3}$$

Both solutions are ________ .

Check your work on page 295. ▶

In each of these examples we were forced to simplify the same square root twice. To avoid such duplication, mathematicians sometimes write both solutions at the same time using the symbol ± before the square root. The ± symbol is read "plus or minus." It is just a shorthand way of denoting the two different solutions.

Example 17. Solve $18x^2 - 28 = 0$.

Solution

$$18x^2 - 28 = 0$$

$$18x^2 = 28$$

$$x^2 = \frac{28}{18}$$

$$x^2 = \frac{14}{9}$$

$$x = \pm\sqrt{\frac{14}{9}}$$

$$= \pm\frac{\sqrt{14}}{\sqrt{9}}$$

$$= \pm\frac{\sqrt{14}}{3}$$

$$x = \frac{\sqrt{14}}{3} \quad \text{or} \quad x = -\frac{\sqrt{14}}{3}$$

Both solutions are irrational.

You try this notation in Example 18.

Example 18. Solve $100y^2 = 243$.

Solution

$$100y^2 = 243$$

$$y^2 = \frac{243}{100}$$

$$y = \pm\sqrt{\frac{243}{100}}$$

$$= \pm\frac{\sqrt{243}}{\sqrt{100}}$$

$$=$$

$$y = \_\_\_\_ \quad \text{or} \quad y = \_\_\_\_$$

Both solutions are ________ .

Check your work on page 295. ▶

Example 19. Solve the formula $V = \frac{1}{3}\pi r^2 h$ for r.

Solution

$$V = \frac{1}{3}\pi r^2 h$$

$$3V = \pi r^2 h$$

$$\frac{3V}{\pi h} = r^2$$

$$\pm\sqrt{\frac{3V}{\pi h}} = r$$

Now you try Example 20.

Example 20. Solve the Pythagorean theorem, $a^2 + b^2 = c^2$, for a.

Solution

$$a^2 + b^2 = c^2$$

$$a^2 = c^2 - \underline{\qquad}$$

$$a = \pm\sqrt{c^2 - \underline{\qquad}}$$

Check your work on page 295. ▶

Trial Run

Solve each equation and state the nature of the solutions.

_______ 1. $x^2 - 28 = 0$

_______ 2. $2x^2 = 96$

_______ 3. $49y^2 = 50$

_______ 4. $25x^2 - 7 = 0$

_______ 5. Solve $A = \pi r^2$ for r.

Answers are on page 295.

▶ Examples You Completed

Example 2. Solve $x^2 - 1 = 0$.

Solution

$$x^2 - 1 = 0$$

$$x^2 = 1$$

$$x = 1 \quad \text{or} \quad x = -1$$

because

$$(1)^2 = 1 \quad \text{and} \quad (-1)^2 = 1$$

Example 5. Solve $x^2 - 49 = 0$ and state the nature of the solutions.

Solution

$$x^2 - 49 = 0$$

$$x^2 = 49$$

$$x = \sqrt{49} \quad \text{or} \quad x = -\sqrt{49}$$

$$x = 7 \quad \text{or} \quad x = -7$$

Both solutions are rational.

Example 9. Simplify $\sqrt{700}$.

Solution

$$\sqrt{700} = \sqrt{100 \cdot 7}$$

$$= \sqrt{100} \cdot \sqrt{7}$$

$$= 10\sqrt{7}$$

Example 12. Simplify $\sqrt{\frac{81}{25}}$.

Solution

$$\sqrt{\frac{81}{25}} = \frac{\sqrt{81}}{\sqrt{25}}$$

$$= \frac{9}{5}$$

Example 14. Simplify $\sqrt{\frac{48}{49}}$.

Solution

$$\sqrt{\frac{48}{49}} = \frac{\sqrt{48}}{\sqrt{49}} = \frac{\sqrt{16 \cdot 3}}{7} = \frac{\sqrt{16} \cdot \sqrt{3}}{7} = \frac{4\sqrt{3}}{7}$$

Example 16. Solve $2y^2 - 54 = 0$.

Solution

$$2y^2 - 54 = 0$$
$$2y^2 = 54$$
$$y^2 = 27$$
$$y = \sqrt{27} \quad \text{or} \quad y = -\sqrt{27}$$
$$= \sqrt{9 \cdot 3} \qquad = -\sqrt{9 \cdot 3}$$
$$y = 3\sqrt{3} \quad \text{or} \quad y = -3\sqrt{3}$$

Both solutions are irrational.

Example 18. Solve $100y^2 = 243$.

Solution

$$100y^2 = 243$$
$$y^2 = \frac{243}{100}$$
$$y = \pm\sqrt{\frac{243}{100}} = \pm\frac{\sqrt{243}}{\sqrt{100}} = \pm\frac{\sqrt{81 \cdot 3}}{10} = \pm\frac{9\sqrt{3}}{10}$$
$$y = \frac{9\sqrt{3}}{10} \quad \text{or} \quad y = -\frac{9\sqrt{3}}{10}$$

Both solutions are irrational.

Example 20. Solve the Pythagorean theorem, $a^2 + b^2 = c^2$, for a.

Solution

$$a^2 + b^2 = c^2$$
$$a^2 = c^2 - b^2$$
$$a = \pm\sqrt{c^2 - b^2}$$

Answers to Trial Runs

page 290 **1.** $x = \sqrt{2}$, $x = -\sqrt{2}$; irrational **2.** $x = \frac{3}{4}$, $x = -\frac{3}{4}$; rational **3.** $x = 2$, $x = -2$; rational **4.** $a = \sqrt{6}$, $a = -\sqrt{6}$; irrational **5.** No real solutions **6.** No real solutions

page 292 **1.** $4\sqrt{3}$ **2.** $5\sqrt{6}$ **3.** $6\sqrt{2}$ **4.** $2\sqrt{19}$ **5.** $\frac{5}{4}$ **6.** $\frac{\sqrt{5}}{7}$ **7.** $\frac{2\sqrt{3}}{5}$ **8.** $\frac{2\sqrt{13}}{11}$

page 294 **1.** $x = 2\sqrt{7}$, $x = -2\sqrt{7}$; irrational **2.** $x = 4\sqrt{3}$, $x = -4\sqrt{3}$; irrational **3.** $y = \frac{5\sqrt{2}}{7}$, $y = -\frac{5\sqrt{2}}{7}$; irrational **4.** $x = \frac{\sqrt{7}}{5}$, $x = -\frac{\sqrt{7}}{5}$; irrational **5.** $r = \pm\sqrt{\frac{A}{\pi}}$

Name ______________________ Date ____________

EXERCISE SET 5.2

Simplify each expression.

______ 1. $\sqrt{54}$

______ 2. $\sqrt{75}$

______ 3. $\sqrt{242}$

______ 4. $\sqrt{288}$

______ 5. $\sqrt{180}$

______ 6. $\sqrt{300}$

______ 7. $\sqrt{507}$

______ 8. $\sqrt{578}$

______ 9. $\sqrt{225}$

______ 10. $\sqrt{256}$

______ 11. $\sqrt{\frac{9}{16}}$

______ 12. $\sqrt{\frac{25}{49}}$

______ 13. $\sqrt{\frac{121}{4}}$

______ 14. $\sqrt{\frac{169}{81}}$

______ 15. $\sqrt{\frac{7}{25}}$

______ 16. $\sqrt{\frac{3}{121}}$

______ 17. $\sqrt{\frac{8}{9}}$

______ 18. $\sqrt{\frac{27}{16}}$

______ 19. $\sqrt{\frac{98}{289}}$

______ 20. $\sqrt{\frac{72}{225}}$

Solve each equation and state the nature of the solutions.

______ 21. $x^2 - 8 = 0$

______ 22. $x^2 - 12 = 0$

______ 23. $x^2 + 9 = 0$

______ 24. $x^2 + 25 = 0$

______ 25. $2x^2 = 98$

______ 26. $3x^2 = 75$

______ 27. $27y^2 - 3 = 0$

______ 28. $16y^2 - 4 = 0$

______ 29. $225x^2 = 49$

______ 30. $81x^2 = 16$

______ 31. $32a^2 - 2 = 0$

______ 32. $18x^2 - 2 = 0$

______ 33. $6 - 5x^2 = 11$

______ 34. $9 - 7x^2 = 16$

______ 35. $3x^2 + 22 = 5x^2 + 2$

______ 36. $8x^2 + 21 = 10x^2 + 9$

______ 37. $2(x^2 - 5) = 11(x^2 - 10)$

______ 38. $3(x^2 - 2) = 7(x^2 - 8)$

______ 39. $4x(x - 6) = 12(5 - 2x)$

______ 40. $3x(x - 5) = 15(4 - x)$

Solve each equation for the specified variable.

_________ 41. $x = 3a^2 - y$ for a

_________ 42. $x = 5a^2 + 2y$ for a

_________ 43. $A = 6s^2$ for s

_________ 44. $A = \frac{1}{2}h^2$ for h

_________ 45. $S = \frac{1}{2}gt^2$ for t

_________ 46. $S = 4\pi r^2$ for r

_________ 47. $V = \pi r^2 h$ for r

_________ 48. $E = mc^2$ for c

☆ Stretching the Topics

Solve each equation.

_________ 1. $4x^4 - 17x^2 + 15 = 0$

_________ 2. $x^4 - 16 = 0$

_________ 3. Solve $x^4 - 10a^2x^2 + 9a^4 = 0$ for x in terms of a.

Check your answers in the back of your book.

If you can complete **Checkup 5.2,** you are ready to go on to Section 5.3.

Name ______________________ Date ______________

✓ CHECKUP 5.2

Simplify each square root.

______ 1. $\sqrt{12}$ ______ 2. $\sqrt{175}$ ______ 3. $\sqrt{\frac{25}{81}}$

______ 4. $\sqrt{\frac{125}{144}}$ ______ 5. $\sqrt{\frac{32}{25}}$

Solve each equation and state the nature of the solutions.

______ 6. $x^2 - 24 = 0$ ______ 7. $x^2 + 36 = 0$

______ 8. $25x^2 = 81$ ______ 9. $18x^2 - 40 = 0$

______ 10. Solve $a^2 + b^2 = c^2$ for b.

Check your answers in the back of your book.

If You Missed Problems:	You Should Review Examples:
1, 2	9, 10
3, 4	11, 12
5	13, 14
6	16
7	7
8, 9	15–18
10	19, 20

5.3 Solving Quadratic Equations by Other Methods

Solving Quadratic Equations by Completing the Square

In Section 5.2 we became thoroughly familiar with the method for solving quadratic equations of the form

$$ax^2 = c$$

by isolating x^2 and using the square root notation. A similar method can be used to solve an equation such as

$$(x + 1)^2 = 4$$

By treating $(x + 1)$ as the variable quantity, we solve the equation by using square root notation.

$$(x + 1)^2 = 4$$
$$x + 1 = \pm\sqrt{4}$$
$$x + 1 = \pm 2$$
$$x + 1 = 2 \quad \text{or} \quad x + 1 = -2$$
$$x = 1 \qquad\qquad x = -3$$

Notice that we solved for x in the last step. You should check to see that both solutions satisfy the original equation.

Example 1. Solve $(x + 6)^2 - 7 = 0$.

Solution

$(x + 6)^2 - 7 = 0$

$(x + 6)^2 = 7$ — Isolate the squared quantity.

$x + 6 = \pm\sqrt{7}$ — Solve for $x + 6$.

$x = -6 \pm\sqrt{7}$ — Isolate x.

$x = -6 + \sqrt{7}$ or $x = -6 - \sqrt{7}$

Here both solutions are irrational.

You try Example 2.

Example 2. Solve $(x - 2)^2 = 6$.

Solution

$(x - 2)^2 = 6$

$x =$ ________ or $x =$ ________

Both solutions are ________ .

Check your work on page 309. ▶

Notice what these equations have in common. Each of them was expressed as "the square of a first-degree binomial equals a constant." If we can rewrite a quadratic equation in this form, we can use this method to solve it.

Remember that the *square of a binomial is a perfect square trinomial* in which
- The first term is the *square* of the first binomial term.
- The last term is the *square* of the second binomial term.
- The middle term is *twice the product* of the first and second binomial terms.

Example 3. Solve $x^2 + 2x + 1 = 5$.

Solution. You should see that the quantity on the left is a perfect square trinomial.

$$x^2 + 2x + 1 = 5$$
$$(x + 1)^2 = 5$$
$$x + 1 = \pm\sqrt{5}$$
$$x = -1 \pm \sqrt{5}$$
$$x = -1 + \sqrt{5} \quad \text{or} \quad x = -1 - \sqrt{5}$$

Both solutions are irrational.

You try Example 4.

Example 4. Solve $x^2 - 10x + 25 = 3$.

Solution

$$x^2 - 10x + 25 = 3$$
$$(x - 5)^2 = 3$$

$x =$ ________ or $x =$ ________

Both solutions are ________ .

Check your work on page 309. ▶

Let's try this method on the equation

$$x^2 + 6x = 1$$

Wait a minute. The left-hand side here is not a trinomial at all, much less a *perfect square trinomial*. Let's take a closer look at the left-hand side of this equation and see if we can make it a perfect square trinomial.

$x^2 + 6x$	We certainly need a constant term.
$x^2 + 6x +$ ____	We want this trinomial to be the square of a binomial.
$(x +$ ____$)^2$	We know the middle term of the trinomial ($6x$) must be *twice* the product of the first and second binomial terms.
$(x + 3)^2$	Since the first term is x, the second term must be 3, because twice $3x$ is $6x$.
$x^2 + 6x + 9$	We square this binomial to decide on the constant term.

Now we return to the original equation

$$x^2 + 6x = 1$$

We know we must add 9 to the left-hand side to make it a perfect square trinomial, but if we add 9 on the left, we must also add 9 on the right.

$x^2 + 6x + 9 = 1 + 9$	Add 9 to each side.
$x^2 + 6x + 9 = 10$	Simplify the right side.
$(x + 3)^2 = 10$	Rewrite the left side as the square of a binomial.
$x + 3 = \pm\sqrt{10}$	Solve for $x + 3$.
$x = -3 \pm \sqrt{10}$	Isolate x.

$$x = -3 + \sqrt{10} \quad \text{or} \quad x = -3 - \sqrt{10}$$

This technique is called **completing the square** and it can be diagrammed for the last problem in the following way:

$$x^2 + 6x = 1$$

$$x^2 + 6x + 9 = 1 + 9$$

half square

$$(x + 3)^2 = 10$$

Solving a Quadratic Equation by Completing the Square

1. Keep the variable terms on one side and combine the constant terms on the other side.
2. Be sure that the coefficient of the squared term is 1.
3. Find half of the coefficient of x, square it, and add that number to *both* sides of the equation.
4. Rewrite the perfect square trinomial as the square of a binomial.
5. Solve the resulting equation $(x + r)^2 = c$ by solving for $x + r$ and then isolating x.

Example 5. Solve $x^2 - 14x + 3 = 8$ by completing the square.

Solution

$$x^2 - 14x + 3 = 8$$

$$x^2 - 14x = 5$$

$$x^2 - 14x + 49 = 5 + 49$$

half square

$$(x - 7)^2 = 54$$

$$x - 7 = \pm\sqrt{54}$$

$$x - 7 = \pm\sqrt{9 \cdot 6}$$ Simplify the square root.

$$x - 7 = \pm 3\sqrt{6}$$

$$x = 7 \pm 3\sqrt{6}$$

$$x = 7 + 3\sqrt{6} \quad \text{or} \quad x = 7 - 3\sqrt{6}$$

Now you carefully complete Example 6.

Example 6. Solve $x^2 + 4x = -2$ by completing the square.

Solution

$$x^2 + 4x = -2$$

$$x^2 + 4x + \underline{\qquad} = -2 + \underline{\qquad}$$

half square

$$(x + \underline{\qquad})^2 = 2$$

$x = \underline{\qquad\qquad}$ or $x = \underline{\qquad\qquad}$

Check your work on page 309. ▶

Example 7. Solve $4y^2 - 20y = 2$ by completing the square.

Solution. Notice that the coefficient of the squared term is not 1. We must divide both sides of the equation by 4 to make the leading coefficient 1.

$$4y^2 - 20y = 2$$

$$y^2 - 5y = \frac{2}{4}$$ Divide both sides by 4.

$$y^2 - 5y + \frac{25}{4} = \frac{2}{4} + \frac{25}{4}$$ Add the necessary constant term to complete the square.

half ↘ ↗ square

$$\left(y - \frac{5}{2}\right)^2 = \frac{27}{4}$$

$$y - \frac{5}{2} = \pm\sqrt{\frac{27}{4}}$$ Use square root notation.

$$y - \frac{5}{2} = \pm\frac{3\sqrt{3}}{2}$$ Simplify the radical expression.

$$y = \frac{5}{2} \pm \frac{3\sqrt{3}}{2}$$ Isolate the variable.

$$y = \frac{5 \pm 3\sqrt{3}}{2}$$ Combine fractions on the right.

$$y = \frac{5 + 3\sqrt{3}}{2} \quad \text{or} \quad y = \frac{5 - 3\sqrt{3}}{2}$$ Write the solutions.

Trial Run

Solve by completing the square.

_______ **1.** $x^2 - 8x = 1$

_______ **2.** $x^2 + 20x + 50 = 0$

_______ **3.** $z^2 - 24z - 1 = 5$

_______ **4.** $2x^2 + 6x + 5 = 6$

Answers are on page 310.

Using the Quadratic Formula

The method of completing the square can be used to find a formula that allows us to solve any quadratic equation. Consider the standard quadratic equation

$$ax^2 + bx + c = 0 \qquad (a \neq 0)$$

and use the method of completing the square to solve for x in terms of a, b, and c.

$$ax^2 + bx = -c$$ Move constant to the right side.

$$x^2 + \frac{b}{a}x = \frac{-c}{a}$$ Divide both sides by a so that the coefficient of x^2 is 1.

$$x^2 + \frac{b}{a}x + \frac{b^2}{4a^2} = \frac{-c}{a} + \frac{b^2}{4a^2}$$

Take half the coefficient of x, square it, and add the square to both sides.

half square

$$\left(x + \frac{1}{2}\cdot\frac{b}{a}\right)^2 = \frac{-c}{a} + \frac{b^2}{4a^2}$$

$$\left(x + \frac{b}{2a}\right)^2 = \frac{-4ac}{4a^2} + \frac{b^2}{4a^2}$$

Add the fractions on the right using $4a^2$ as the common denominator.

$$\left(x + \frac{b}{2a}\right)^2 = \frac{b^2 - 4ac}{4a^2}$$

$$x + \frac{b}{2a} = \pm\sqrt{\frac{b^2 - 4ac}{4a^2}}$$

Use square root notation.

$$x + \frac{b}{2a} = \pm\frac{\sqrt{b^2 - 4ac}}{\sqrt{4a^2}}$$

Simplify the square-root expression on the right.

$$x + \frac{b}{2a} = \pm\frac{\sqrt{b^2 - 4ac}}{2a}$$

$$x = \frac{-b}{2a} \pm \frac{\sqrt{b^2 - 4ac}}{2a}$$

Subtract $\frac{b}{2a}$ from both sides to isolate x on the left.

$$x = \frac{-b \pm \sqrt{b^2 - 4ac}}{2a}$$

Combine the fractions on the right.

Our solutions for $ax^2 + bx + c = 0$ are

$$x = \frac{-b + \sqrt{b^2 - 4ac}}{2a} \quad \text{or} \quad x = \frac{-b - \sqrt{b^2 - 4ac}}{2a}$$

We have derived a formula that allows us to find the solutions of a quadratic equation using only the coefficients of the terms in the original equation. This important and useful formula is called the **quadratic formula.**

Quadratic Formula. If $ax^2 + bx + c = 0$, then

$$x = \frac{-b \pm \sqrt{b^2 - 4ac}}{2a}$$

are the solutions.

You must memorize the quadratic formula unless you intend to complete the square whenever you solve a quadratic equation that does not factor easily. As with every new formula, you will learn it best if you write it down every time you use it.

Let's practice solving a few quadratic equations using the quadratic formula. *Be sure to identify* ***a, b****, and* ***c*** *after writing the equation in the form* $\boldsymbol{ax^2 + bx + c = 0}$.

Example 8. Solve $x^2 + 5x - 6 = 0$ using the quadratic formula.

Solution. Here $a = 1$, $b = 5$, and $c = -6$.

$$x = \frac{-b \pm \sqrt{b^2 - 4ac}}{2a}$$

$$x = \frac{-5 \pm \sqrt{5^2 - 4(1)(-6)}}{2(1)}$$

$$= \frac{-5 \pm \sqrt{25 + 24}}{2}$$

$$= \frac{-5 \pm \sqrt{49}}{2}$$

$$x = \frac{-5 \pm 7}{2}$$

$$x = \frac{-5 + 7}{2} \quad \text{or} \quad x = \frac{-5 - 7}{2}$$

$$= \frac{2}{2} \qquad\qquad = \frac{-12}{2}$$

$$x = 1 \quad \text{or} \quad x = -6$$

Example 9. Solve $x^2 - 2x = 1$ using the quadratic formula.

Solution. We rewrite the equation in standard form.

$$x^2 - 2x - 1 = 0$$

Here $a = 1$, $b = -2$, and $c = -1$.

$$x = \frac{-b \pm \sqrt{b^2 - 4ac}}{2a}$$

$$x = \frac{-(-2) \pm \sqrt{(-2)^2 - 4(1)(-1)}}{2(1)}$$ Substitute for a, b, c.

$$= \frac{2 \pm \sqrt{4 + 4}}{2}$$ Remove parentheses.

$$= \frac{2 \pm \sqrt{8}}{2}$$ Find the sum under the radical.

$$= \frac{2 \pm \sqrt{4 \cdot 2}}{2}$$ Simplify the square root.

$$= \frac{2 \pm 2\sqrt{2}}{2}$$

$$= \frac{2(1 \pm \sqrt{2})}{2}$$ Factor the numerator.

$$x = 1 \pm \sqrt{2}$$ Reduce the fraction.

$$x = 1 + \sqrt{2} \quad \text{or} \quad x = 1 - \sqrt{2}$$

In Example 8 you should notice that the original trinomial could have been factored. The solutions found by factoring are the same as those found using the quadratic formula, but the method of factoring would be much quicker. *Whenever possible, you should solve a second-degree equation by factoring.* You should use the quadratic formula only when you cannot see any way in which to factor the quadratic expression.

Now try Example 10.

Example 10. Solve $3x^2 + 5x + 1 = 0$ using the quadratic formula.

Solution. The polynomial does not factor. Here $a = 3$, $b = 5$, and $c = 1$.

$$x = \frac{-b \pm \sqrt{b^2 - 4ac}}{2a}$$

$$x = \frac{-5 \pm \sqrt{5^2 - 4(3)(1)}}{2(3)}$$

$$= \frac{-5 \pm \sqrt{\underline{\qquad} - \underline{\qquad}}}{6}$$

$$= \frac{-5 \pm \sqrt{\underline{\qquad}}}{6}$$

$$x = \frac{-5 + \sqrt{\underline{\qquad}}}{6} \quad \text{or}$$

$$x = \frac{-5 - \sqrt{\underline{\qquad}}}{6}$$

Both solutions are _________ .

Check your work on page 309. ▶

Example 11. Solve $2y^2 - y + 3 = 0$ using the quadratic formula.

Solution. The polynomial does not factor. Here $a = 2$, $b = -1$, and $c = 3$.

$$y = \frac{-b \pm \sqrt{b^2 - 4ac}}{2a}$$

$$y = \frac{-(-1) \pm \sqrt{(-1)^2 - 4(2)(3)}}{2(2)}$$

$$= \frac{1 \pm \sqrt{1 - 24}}{4}$$

$$= \frac{1 \pm \sqrt{-23}}{4}$$

Since $\sqrt{-23}$ is not a real number, this quadratic equation has *no* real solutions.

Example 12. Solve the formula $d = vt - 16t^2$ for t, using the quadratic formula.

Solution. We put the equation in standard form and identify a, b, and c.

$$d = vt - 16t^2$$

$$16t^2 - vt + d = 0$$

Add $16t^2$, subtract vt to put equation in standard form.

$$t = \frac{-b \pm \sqrt{b^2 - 4ac}}{2a}$$

Here $a = 16$, $b = -v$, and $c = d$.

$$t = \frac{-(-v) \pm \sqrt{(-v)^2 - 4(16)(d)}}{2(16)}$$

Substitute into the quadratic formula.

$$t = \frac{v \pm \sqrt{v^2 - 64d}}{32}$$

Simplify.

$$t = \frac{v + \sqrt{v^2 - 64d}}{32} \quad \text{or} \quad t = \frac{v - \sqrt{v^2 - 64d}}{32}$$

Trial Run

Solve each equation using the quadratic formula.

_______ 1. $x^2 - 4x + 4 = 0$

_______ 2. $x^2 - 4x = 21$

_______ 3. $2y^2 = 5y - 1$

_______ 4. $x^2 - 3x = 1$

_______ 5. $5a^2 + 2a = 4$

_______ 6. $2x^2 + x + 1 = 0$

Answers are on page 310.

Working with the Discriminant

In using the quadratic formula, we discovered that the solutions of a quadratic equation might be rational numbers or irrational numbers or not real numbers. The kind of solutions obtained depends entirely on the quantity under the radical in the quadratic formula

$$x = \frac{-b \pm \sqrt{b^2 - 4ac}}{2a}$$

That quantity $b^2 - 4ac$ is called the **discriminant** for a quadratic equation and it can be used to decide what kind of solutions the equation will have.

If $b^2 - 4ac$ is *zero,* then the square root is zero and the only solution to the equation is $x = \dfrac{-b}{2a}$.

If $b^2 - 4ac$ is *positive and a perfect square,* then the square root is a rational number, and the solutions to the equation are both rational numbers.

If $b^2 - 4ac$ is *positive but not a perfect square,* then the square root is an irrational number, and the solutions to the equation are both irrational numbers.

If $b^2 - 4ac$ is *negative,* then the square root is not a real number, and the equation has no real solutions.

We can summarize these facts about the discriminant in the following table.

Discriminant $b^2 - 4ac$	Kind of Solutions for $ax^2 + bx + c = 0$	Example
$b^2 - 4ac = 0$	One rational solution	$x^2 + 6x + 9 = 0$ $b^2 - 4ac = 6^2 - 4(1)(9)$ $= 36 - 36$ $= 0$ (*Solution*: $x = -3$)
$b^2 - 4ac > 0$ and perfect square	Two different rational solutions	$x^2 + 3x - 4 = 0$ $b^2 - 4ac = 3^2 - 4(1)(-4)$ $= 9 + 16$ $= 25$ (*Solutions*: $x = -4$, $x = 1$)
$b^2 - 4ac > 0$ and *not* perfect square	Two different irrational solutions	$x^2 + 4x - 3 = 0$ $b^2 - 4ac = 4^2 - 4(1)(-3)$ $= 16 + 12$ $= 28$ (*Solutions*: $x = -2 + \sqrt{7}$, $x = -2 - \sqrt{7}$)
$b^2 - 4ac < 0$	No real solutions	$x^2 - 2x + 4 = 0$ $b^2 - 4ac = (-2)^2 - 4(1)(4)$ $= 4 - 16$ $= -12$ (*Solutions*: $x = 1 + \sqrt{-3}$, $x = 1 - \sqrt{-3}$; not real numbers)

You practice using the discriminant to describe the solutions for the equations in Examples 13 and 14.

Example 13. Describe the solutions for $2x^2 - 5x - 3 = 0$.

Solution

$$2x^2 - 5x - 3 = 0$$
$$b^2 - 4ac = (-5)^2 - 4(2)(-3)$$

There are ________ solutions.

Check your work on page 309. ▶

Example 14. Describe the solutions for $x^2 + x + 1 = 0$.

Solution

$$x^2 + x + 1 = 0$$
$$b^2 - 4ac = 1^2 - 4(1)(1)$$

There are ________ solutions.

Check your work on page 309. ▶

Trial Run

Describe the solutions for each equation.

________ **1.** $x^2 - 3x - 10 = 0$ ________ **2.** $x^2 + 5x + 2 = 0$

________ **3.** $x^2 - 10x + 25 = 0$ ________ **4.** $x^2 + 2x + 5 = 0$

Answers are on page 310.

Choosing the Best Method

Having learned four ways to solve quadratic equations, you may wonder how to choose the most direct method to use. The most efficient method depends on the particular quadratic equation you wish to solve. In any case, first set the quadratic expression equal to zero and look at the expression $ax^2 + bx + c$.

If the Quadratic Expression:	Then the Equation Can Be Solved by:	Example
Factors readily using integers	Factoring and using the zero product rule	$3x^2 + 2x - 1 = 0$ $(3x - 1)(x + 1) = 0$ $3x - 1 = 0$ or $x + 1 = 0$ $x = \frac{1}{3}$ or $x = -1$
Contains no first-degree term ($b = 0$)	Isolating the second-degree term, then using square root notation	$2x^2 - 6 = 0$ $2x^2 = 6$ $x^2 = 3$ $x = \pm\sqrt{3}$ $x = \sqrt{3}$ or $x = -\sqrt{3}$
Has a first-degree term with an *even* coefficient	Completing the square	$x^2 - 4x + 2 = 0$ $x^2 - 4x + 4 = -2 + 4$ $(x - 2)^2 = 2$ $x - 2 = \pm\sqrt{2}$ $x = 2 \pm \sqrt{2}$ $x = 2 + \sqrt{2}$ or $x = 2 - \sqrt{2}$
Does not meet any special conditions	Using the quadratic formula $x = \dfrac{-b \pm \sqrt{b^2 - 4ac}}{2a}$	$x^2 + 3x - 5 = 0$ $x = \dfrac{-3 \pm \sqrt{3^2 - 4(1)(-5)}}{2(1)}$ $= \dfrac{-3 \pm \sqrt{9 + 20}}{2}$ $= \dfrac{-3 \pm \sqrt{29}}{2}$ $x = \dfrac{-3 + \sqrt{29}}{2}$ or $x = \dfrac{-3 - \sqrt{29}}{2}$

In reality, few students choose to solve a quadratic equation by completing the square unless instructed to do so. If a quadratic expression does not factor or the first-degree term is not missing, most students turn to the quadratic formula.

▶ Examples You Completed

Example 2. Solve $(x - 2)^2 = 6$.

Solution

$$(x - 2)^2 = 6$$
$$x - 2 = \pm\sqrt{6}$$
$$x = 2 \pm \sqrt{6}$$
$$x = 2 + \sqrt{6} \quad \text{or} \quad x = 2 - \sqrt{6}$$

Both solutions are irrational.

Example 4. Solve $x^2 - 10x + 25 = 3$.

Solution

$$x^2 - 10x + 25 = 3$$
$$(x - 5)^2 = 3$$
$$x - 5 = \pm\sqrt{3}$$
$$x = 5 \pm \sqrt{3}$$
$$x = 5 + \sqrt{3} \quad \text{or} \quad x = 5 - \sqrt{3}$$

Both solutions are irrational.

Example 6. Solve $x^2 + 4x = -2$ by completing the square.

Solution

$$x^2 + 4x = -2$$
$$x^2 + 4x + 4 = -2 + 4$$

half ↘ ↗ square

$$(x + 2)^2 = 2$$
$$x + 2 = \pm\sqrt{2}$$
$$x = -2 \pm \sqrt{2}$$
$$x = -2 + \sqrt{2} \quad \text{or} \quad x = -2 - \sqrt{2}$$

Example 10. Solve $3x^2 + 5x + 1 = 0$ using the quadratic formula.

Solution. The polynomial does not factor. Here $a = 3$, $b = 5$, and $c = 1$.

$$x = \frac{-b \pm \sqrt{b^2 - 4ac}}{2a}$$
$$x = \frac{-5 \pm \sqrt{5^2 - 4(3)(1)}}{2(3)}$$
$$= \frac{-5 \pm \sqrt{25 - 12}}{6}$$
$$= \frac{-5 \pm \sqrt{13}}{6}$$
$$x = \frac{-5 + \sqrt{13}}{6} \quad \text{or} \quad x = \frac{-5 - \sqrt{13}}{6}$$

Both solutions are irrational.

Example 13. Describe the solutions for $2x^2 - 5x - 3 = 0$.

Solution

$$2x^2 - 5x - 3 = 0$$
$$b^2 - 4ac = (-5)^2 - 4(2)(-3)$$
$$= 25 + 24$$
$$= 49$$

There are two different rational solutions.

Example 14. Describe the solutions for $x^2 + x + 1 = 0$.

Solution

$$x^2 + x + 1 = 0$$
$$b^2 - 4ac = 1^2 - 4(1)(1)$$
$$= 1 - 4$$
$$= -3$$

There are no real solutions.

Answers to Trial Runs

page 302 1. $x = 4 + \sqrt{17}, x = 4 - \sqrt{17}$ 2. $x = -10 + 5\sqrt{2}, x = -10 - 5\sqrt{2}$
3. $z = 12 + 5\sqrt{6}, z = 12 - 5\sqrt{6}$ 4. $x = \frac{-3 + \sqrt{11}}{2}, x = \frac{-3 - \sqrt{11}}{2}$

page 306 1. $x = 2$ 2. $x = 7, x = -3$ 3. $y = \frac{5 + \sqrt{17}}{4}, y = \frac{5 - \sqrt{17}}{4}$
4. $x = \frac{3 + \sqrt{13}}{2}, x = \frac{3 - \sqrt{13}}{2}$ 5. $a = \frac{-1 + \sqrt{21}}{5}, a = \frac{-1 - \sqrt{21}}{5}$
6. $x = \frac{-1 \pm \sqrt{-7}}{4}$; no real solutions

page 308 1. Two different rational solutions 2. Two different irrational solutions
3. One rational solution 4. No real solutions

Name ______________________ Date __________

EXERCISE SET 5.3

Solve each equation by completing the square.

______ 1. $x^2 + 2x - 15 = 0$

______ 2. $x^2 - 2x - 8 = 0$

______ 3. $x^2 + 4x = 0$

______ 4. $x^2 - 10x = 0$

______ 5. $3x^2 + x - 3 = 0$

______ 6. $3x^2 + 7x + 3 = 0$

______ 7. $4x^2 + 9x + 4 = 0$

______ 8. $5x^2 - 11x + 5 = 0$

______ 9. $x^2 + 4x + 7 = 0$

______ 10. $x^2 - 2x + 5 = 0$

Solve each equation by using the quadratic formula.

______ 11. $x^2 - 4x + 3 = 0$

______ 12. $x^2 - 5x + 6 = 0$

______ 13. $x^2 + 8x = 0$

______ 14. $x^2 + 7x = 0$

______ 15. $x^2 + x = 11$

______ 16. $x^2 + x = 3$

______ 17. $2x^2 + x - 2 = 0$

______ 18. $2x^2 - 3x - 1 = 0$

______ 19. $25x^2 = 10x + 2$

______ 20. $9x^2 = 6x + 1$

Solve each equation by using the best method.

______ 21. $x^2 - 9x + 20 = 0$

______ 22. $x^2 - 7x + 12 = 0$

______ 23. $2x^2 - 5x - 3 = 0$

______ 24. $2x^2 - x - 6 = 0$

______ 25. $2x^2 + x + 4 = 0$

______ 26. $2x^2 - x + 1 = 0$

______ 27. $x^2 + 3x - 17 = 0$

______ 28. $x^2 + 2x - 7 = 0$

______ 29. $6x^2 + 8x + 1 = 0$

______ 30. $9x^2 + 12x - 1 = 0$

______ 31. $x^2 + 6x + 9 = 0$

______ 32. $x^2 + 10x + 25 = 0$

______ 33. $x^2 - 2 = 0$

______ 34. $x^2 - 5 = 0$

______ 35. $x^2 - 4 = 2x$

______ 36. $x^2 - 7 = 2x$

______ 37. $3x^2 + 7x + 1 = 0$

______ 38. $3x^2 + 5x + 1 = 0$

______ 39. $x^2 - 2x + 5 = 0$

______ 40. $x^2 - 4x + 5 = 0$

______ 41. $x^2 + 2 = 4x$

______ 42. $x^2 + 6 = 6x$

______ 43. $x(4x - 3) = 5 - 2x$

______ 44. $3x(x - 3) = 4 - 7x$

______ 45. $-4(x^2 + 4) = -12x - 9$

______ 46. $-2(x^2 + 9) = -12x - 10$

________ 47. $M = ax^2 + cx$ for x.

________ 48. $P = kx^2 - nx$ for x.

________ 49. $A = \pi r^2 + 2\pi rh$ for r.

________ 50. $A = 2s^2 + 4sh$ for s.

Describe the solutions for each equation after evaluating the discriminant.

________ 51. $x^2 + 64 = 16x$

________ 52. $x^2 + 25 = 10x$

________ 53. $2x^2 - 9x - 5 = 0$

________ 54. $3x^2 + 13x + 4 = 0$

________ 55. $x^2 - 4x - 6 = 0$

________ 56. $x^2 - 7x - 1 = 0$

________ 57. $2x^2 - 3x - 5 = 0$

________ 58. $4x^2 - x + 2 = 0$

________ 59. $6x^2 - 7x = 3$

________ 60. $8x^2 + 10x = 3$

☆ Stretching the Topics

________ 1. Solve $x^2 + 2\sqrt{3}x - 24 = 0$ by completing the square.

________ 2. Solve $x^2 + 6\sqrt{5}x + 45 = 0$ by using the quadratic formula.

________ 3. Use the discriminant to describe the solutions for the equation $\sqrt{6}x^2 - 4x - 2\sqrt{6} = 0$.

Check your answers in the back of your book.

If you can solve the equations in **Checkup 5.3,** you are ready to go on to Section 5.4.

Name ______________________ Date ______________

✓ CHECKUP 5.3

Solve by completing the square.

______ 1. $y^2 - 2y - 9 = 2$

Solve using the quadratic formula.

______ 2. $2a^2 - 2a = -1$

Solve each equation by the best method.

______ 3. $x^2 - 2x - 4 = 0$

______ 4. $x^2 + 3x = 1$

______ 5. $5a^2 - 8a + 2 = 0$

______ 6. $2x^2 + 2x - 3 = 0$

______ 7. Solve $A = 2\pi r^2 + 2\pi rh$ for r.

______ 8. Describe the solutions for $5x^2 + 14x = 3$.

Check your answers in the back of your book.

If You Missed Problems:	You Should Review Examples:
1	5
2	7
3–5	8–10
6	11
7	12
8	13, 14

5.4 Solving Quadratic Inequalities

Now that we have a thorough understanding of the methods for solving quadratic equations, we turn our attention to **quadratic inequalities.** We learned how to solve first-degree inequalities in Chapter 2, but second-degree inequalities require a slightly different approach.

Using the Case Method

We would like to solve inequalities such as

$$x^2 + 3x - 10 > 0$$

You recall that quadratic (second-degree) equations required the use of factoring whenever possible. Perhaps we should consider factoring as a good starting place in dealing with quadratic inequalities also. Let's rewrite the quadratic expression in our inequality in factored form.

$$x^2 + 3x - 10 > 0$$

$$(x + 5)(x - 2) > 0$$

This inequality now says that we want the product of the quantities $(x + 5)$ and $(x - 2)$ to be greater than zero. In other words, we want that product of factors to be *positive*. When will the product of two factors be positive? If both factors are positive *or* if both factors are negative.

We agree, then, that there are two different cases to be considered here.

Case 1: Suppose that both factors are positive.
Case 2: Suppose that both factors are negative.

Each of these cases must be treated separately, using our knowledge of solutions of pairs of inequalities.

Case 1: Suppose both factors are *positive*.

$$x + 5 > 0 \quad \text{and} \quad x - 2 > 0$$

$$x > -5 \quad \text{and} \quad x > 2$$

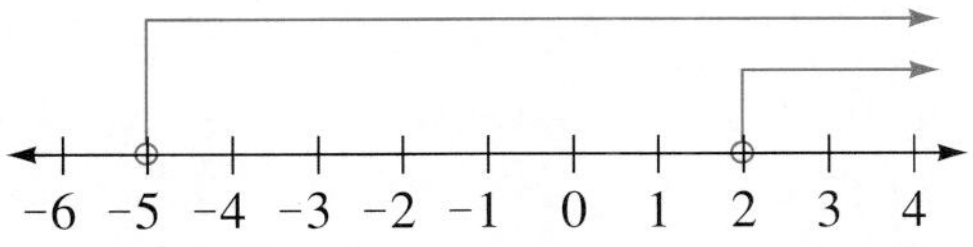

Both conditions must be met at the same time. Our solutions from Case 1 are

$$x > 2$$
$$(2, \infty)$$

Case 2: Suppose both factors are *negative*.

$$x + 5 < 0 \quad \text{and} \quad x - 2 < 0$$

$$x < -5 \quad \text{and} \quad x < 2$$

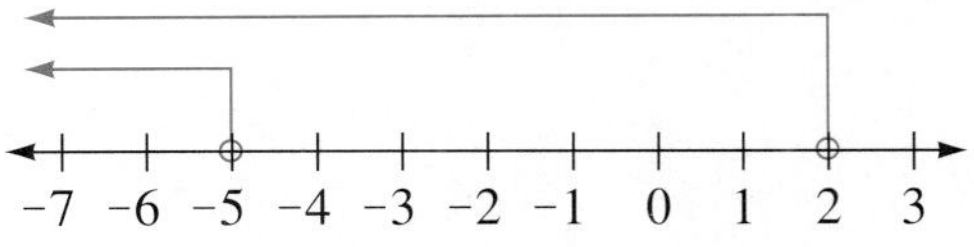

Both conditions must be met at the same time. Our solutions from Case 2 are

$$x < -5$$
$$(-\infty, -5)$$

Since the inequality $x^2 + 3x - 10 > 0$ is satisfied by either Case 1 solutions *or* Case 2 solutions, our final solutions are $x > 2$ or $x < -5$.

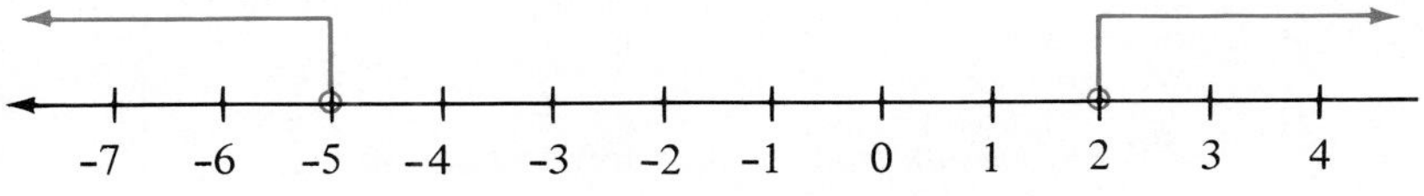

$$(-\infty, -5) \cup (2, \infty)$$

Example 1. Solve $3y^2 - 8y \leq 35$.

Solution. We first get the quadratic expression on one side and write it in factored form.

$$3y^2 - 8y - 35 \leq 0$$

$$(3y + 7)(y - 5) \leq 0$$

Here we have a product that is *less than* or equal to zero. In other words, the product of $(3y + 7)$ and $(y - 5)$ is a *negative* number or zero. When will the product of two factors be negative? When one factor is positive and the other factor is negative. Since we do not know which is positive and which is negative, we must consider two cases.

Case 1: Suppose that the first factor is positive (or zero) and the second factor is negative (or zero).

$$3y + 7 \geq 0 \quad \text{and} \quad y - 5 \leq 0$$

$$3y \geq -7$$

$$y \geq -\frac{7}{3} \quad \text{and} \quad y \leq 5$$

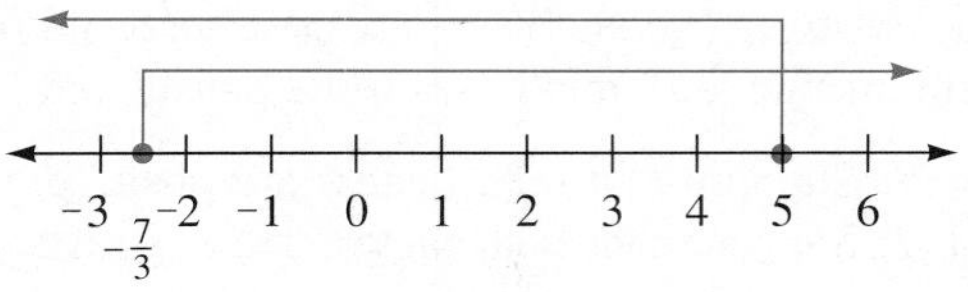

Both conditions must be met at the same time, so our solutions from Case 1 are

$$-\tfrac{7}{3} \leq y \leq 5$$

Case 2: Suppose that the first factor is negative (or zero) and the second factor is positive (or zero).

$$3y + 7 \leq 0 \quad \text{and} \quad y - 5 \geq 0$$

$$3y \leq -7$$

$$y \leq -\frac{7}{3} \quad \text{and} \quad y \geq 5$$

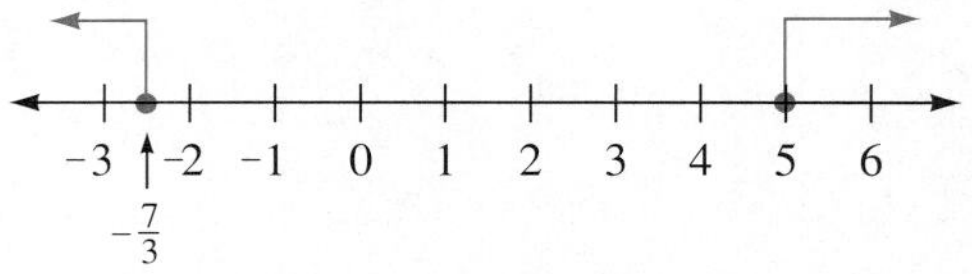

Both conditions must be met at the same time, so Case 2 gives *no solutions*.

Our original inequality $3y^2 - 8y \leq 35$ will be satisfied by Case 1 solutions *or* Case 2 solutions, so our final solutions are

$$-\tfrac{7}{3} \leq y \leq 5$$

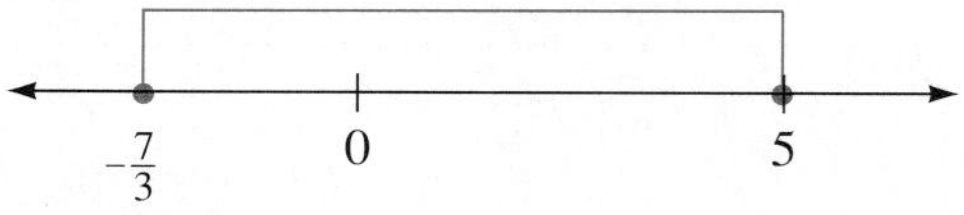

$$[-\tfrac{7}{3}, 5]$$

In other words, if we choose any y-value from the interval between $-\frac{7}{3}$ and 5 (including the endpoints), our original inequality will be satisfied.

Example 2. Solve $1 - x^2 > 0$.

Solution

$$1 - x^2 > 0$$

$$(1 + x)(1 - x) > 0$$

We want our product of factors to be *positive*.

Case 1: Suppose both factors are *positive*.

$$1 + x > 0 \quad \text{and} \quad 1 - x > 0$$
$$-x > -1$$
$$x > -1 \quad \text{and} \quad x < 1$$

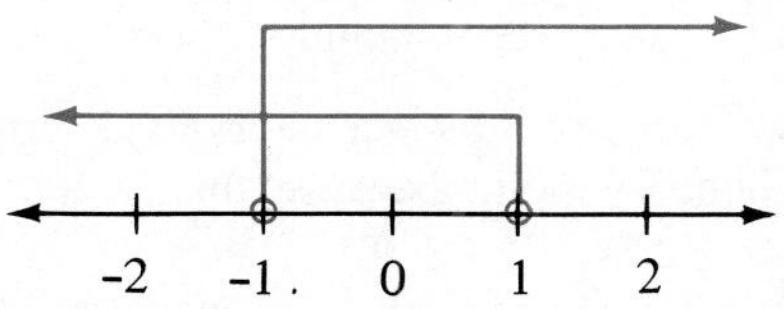

Solutions from Case 1 are

$$-1 < x < 1$$

Case 2: Suppose both factors are *negative*.

$$1 + x < 0 \quad \text{and} \quad 1 - x < 0$$
$$-x < -1$$
$$x < -1 \quad \text{and} \quad x > 1$$

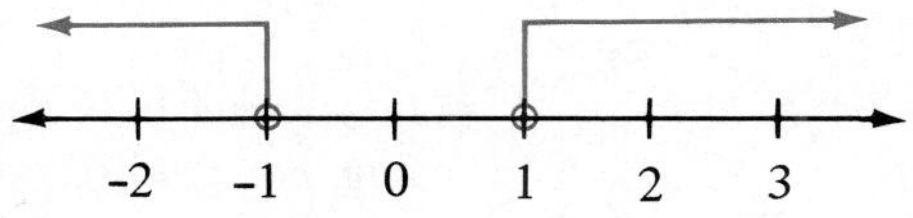

There are *no solutions* from Case 2.

Our final solutions for the inequality $1 - x^2 > 0$ are $-1 < x < 1$

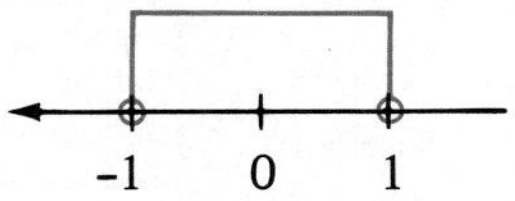

$$(-1, 1)$$

Trial Run

Solve each inequality by considering the cases.

_______ 1. $x^2 + x - 12 > 0$

_______ 2. $x^2 - 2x \leq 24$

_______ 3. $2x^2 - x \geq 0$

_______ 4. $4 - x^2 > 0$

Answers are on page 321.

Using Boundary Points on the Number Line

As you may have noticed, in each of the last examples there were certain key points (or **boundary points**) on the number line where we concentrated our attention.

Inequality	Factored Form	Boundary Points
$x^2 + 3x - 10 > 0$	$(x + 5)(x - 2) > 0$	$x = -5, x = 2$
$3y^2 - 8y \leq 35$	$(3y + 7)(y - 5) \leq 0$	$y = -\frac{7}{3}, y = 5$
$1 - x^2 > 0$	$(1 + x)(1 - x) > 0$	$x = -1, x = 1$

The boundary points for an inequality separate the number line into distinct intervals. Each of our inequalities was satisfied by all the numbers in one or more of these intervals. Perhaps we could use a number-line approach to solve quadratic inequalities without using cases at all.

For the inequality

$$x^2 + 3x - 10 > 0$$
$$(x + 5)(x - 2) > 0$$

the boundary points on the number line are

$$x = -5 \quad \text{and} \quad x = 2$$

We separate the number line into three distinct intervals determined by the boundary points, using *open* dots at each boundary point because this is a $>$ inequality (rather than a $\geq$ inequality).

Remember, to satisfy our original inequality, the *product* of the factors must be *positive*. We must check the *signs* of each of the factors $(x + 5)$ and $(x - 2)$ in each of the intervals to determine in which intervals their product is positive.

To decide on the sign of a factor in a particular interval, we simply choose some number in the interval, substitute it into each factor, and draw a conclusion about that factor's *sign* in that interval. We organize our work in the following diagram.

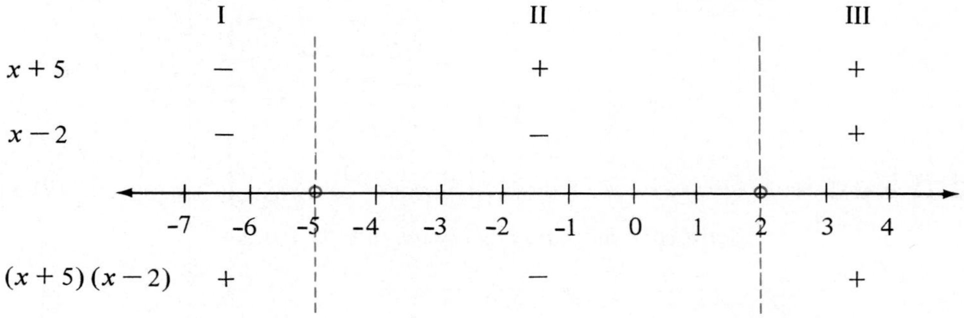

Notice in interval I that both factors are negative, so their product is *positive*. In interval II, the factor $(x + 5)$ is positive, but the factor $(x - 2)$ is negative, so their product is *negative*. In interval III both factors are positive, so their product is *positive*.

What is the solution? Since we wanted the product to be *positive,* the original inequality is satisfied in interval I or in interval III. We can describe those intervals as follows

Interval I: $x < -5$

Interval III: $x > 2$

Our solutions are

$$x < -5 \quad \textit{or} \quad x > 2$$
$$(-\infty, -5) \cup (2, \infty)$$

which are the same solutions obtained earlier using the case method.

Solving Quadratic Inequalities (Boundary-Point Method)

1. Rewrite the inequality with 0 on one side and note the sign that you wish the expression to have.
2. Factor the quadratic expression.
3. Find the boundary points by finding the values of x that make any factor 0.
4. Separate the number line into intervals determined by the boundary points. If the inequality is $<$ or $>$, the boundary points are *open*; if the inequality is $\leq$ or $\geq$, the boundary points are *solid*.

5. Choose a number in each interval and substitute it into each factor to determine the *sign* of each factor in each interval.
6. Use the signs of the factors to determine the sign of the quadratic expression in each interval.
7. Decide in which intervals the expression has the required sign (from step 1).
8. Write your solution using inequalities or interval notation.

Example 3. Solve $x^2 + 10x + 21 \le 0$ using the boundary-point method.

Solution

$$x^2 + 10x + 21 \le 0$$

$$(x + 7)(x + 3) \le 0$$

The boundary points are

$$x = -7 \quad \text{and} \quad x = -3$$

and we use *solid* dots because this is a $\le$ inequality (rather than a $<$ inequality).

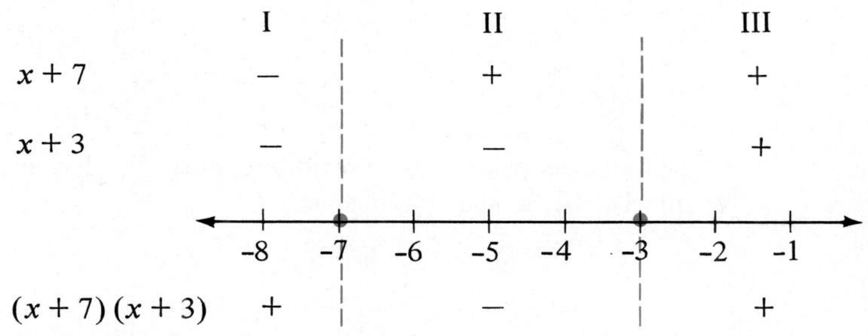

Since we want the product to be *negative* or zero (≤ 0), the solutions lie in interval II.

$$-7 \le x \le -3$$
$$[-7, -3]$$

Example 4. Solve $2x^2 + 3x \geq -1$.

Solution

$$2x^2 + 3x \geq -1$$
$$2x^2 + 3x + 1 \geq 0$$
$$(2x + 1)(x + 1) \geq 0$$

Boundary points are

$$x = -\tfrac{1}{2} \quad \text{and} \quad x = -1$$

The dots should be solid, and we want the product to be positive or zero (≥ 0).

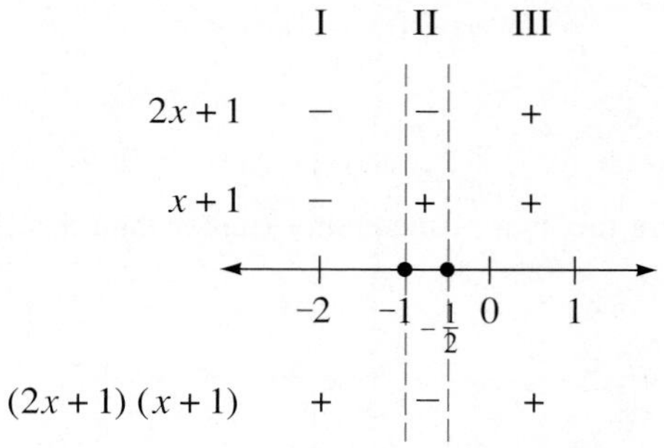

The product is positive or zero in interval I or interval III, so the solutions are

$$x \leq -1 \quad \text{or} \quad x \geq -\tfrac{1}{2}$$
$$(-\infty, -1] \cup [-\tfrac{1}{2}, \infty)$$

You try completing Example 5.

Example 5. Solve $x^2 < 2x - 1$.

Solution

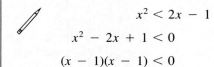

$$x^2 < 2x - 1$$
$$x^2 - 2x + 1 < 0$$
$$(x - 1)(x - 1) < 0$$

The only boundary point is

$$x = \_\_\_\_$$

The dot should be ________ and we want the product to be ________ (< 0).

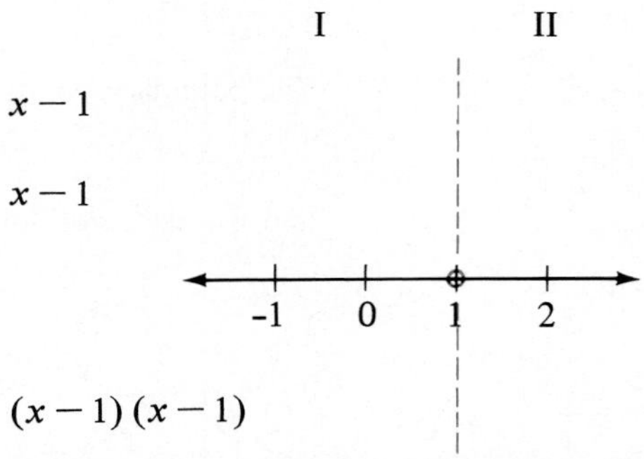

The product is *never* negative, so there is ________ for this inequality.

Check your work on page 321. ▶

It should not surprise you that there was no solution for the inequality in Example 5.

$$x^2 - 2x + 1 < 0$$
$$(x - 1)(x - 1) < 0$$
$$(x - 1)^2 < 0$$

We know that the square of a real quantity can *never* be *negative*.

On the other hand, if we had been asked to solve

$$x^2 - 2x + 1 \geq 0$$
$$(x - 1)^2 \geq 0$$

the solutions would have been *all* real numbers, since $(x - 1)^2$ is *always* positive or zero.

Trial Run

Solve each inequality by the boundary-point method.

________ 1. $x^2 - 5x < 0$

________ 2. $3x^2 + 11x - 4 \geq 0$

_______ 3. $6 - x - x^2 < 0$ _______ 4. $x^2 \le 25$

Answers are below.

▶ Example You Completed

Example 5. Solve $x^2 < 2x - 1$.

Solution

$$x^2 - 2x + 1 < 0$$

$$(x - 1)(x - 1) < 0$$

The only boundary point is

$$x = 1$$

The dot should be open and we want the product to be negative (< 0).

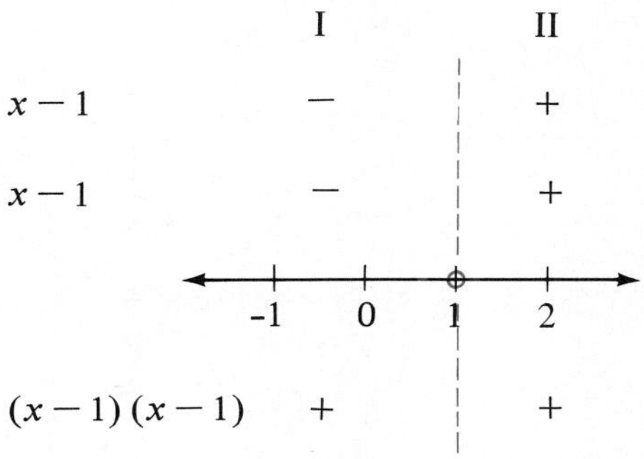

The product is *never* negative, so there is no solution for this inequality.

Answers to Trial Runs

page 317 **1.** $x < -4$ or $x > 3$; $(-\infty, -4) \cup (3, \infty)$ **2.** $-4 \le x \le 6$; $[-4, 6]$ **3.** $x \le 0$ or $x \ge \frac{1}{2}$; $(-\infty, 0] \cup [\frac{1}{2}, \infty)$ **4.** $-2 < x < 2$; $(-2, 2)$

page 320 **1.** $0 < x < 5$; $(0, 5)$ **2.** $x \le -4$ or $x \ge \frac{1}{3}$; $(-\infty, -4] \cup [\frac{1}{3}, \infty)$ **3.** $x < -3$ or $x > 2$; $(-\infty, -3) \cup (2, \infty)$ **4.** $-5 \le x \le 5$; $[-5, 5]$

Name ______________________ Date ____________

EXERCISE SET 5.4

Solve each inequality.

______ 1. $x^2 - 49 < 0$

______ 2. $x^2 - 16 < 0$

______ 3. $4x^2 > 25$

______ 4. $9x^2 > 4$

______ 5. $x^2 + 3x \geq 0$

______ 6. $x^2 + 9x \geq 0$

______ 7. $2x^2 - x \leq 0$

______ 8. $3x^2 - 2x \leq 0$

______ 9. $x^2 + 3x + 2 < 0$

______ 10. $x^2 + 7x + 10 < 0$

______ 11. $x^2 + 3x \leq 4$

______ 12. $x^2 - 2x \leq 8$

______ 13. $x^2 + 4x > 5$

______ 14. $x^2 - x > 42$

______ 15. $x^2 + 21 \leq 10x$

______ 16. $x^2 + 20 \leq 9x$

______ 17. $25 - x^2 > 0$

______ 18. $36 - x^2 > 0$

______ 19. $3x^2 + 17x > 6$

______ 20. $6x^2 - x > 12$

______ 21. $6x^2 - 13x \leq 8$

______ 22. $3x^2 + 17x \leq 6$

______ 23. $4x^2 - 20x + 25 \geq 0$

______ 24. $9x^2 - 12x + 4 \geq 0$

______ 25. $3x(x + 5) \geq 5(x + 1) - 4x$

______ 26. $2x(x + 5) \geq 3(x + 6) - 2x$

______ 27. $(3x - 1)(2x + 5) < -x(6 + x) + 1$

______ 28. $(4x - 1)(2x - 3) < 3(4 - x) + 1$

☆ Stretching the Topics

Solve each inequality.

______ 1. $x^2 - 2 \geq 0$

______ 2. $x^3 < 4x$

______ 3. $y^3 - y^2 > 9y - 9$

Check your answers in the back of your book.

If you can solve the inequalities in **Checkup 5.4,** you are ready to go on to Section 5.5.

Name ______________________ Date ____________

✓ CHECKUP 5.4

Solve each inequality.

________ 1. $x^2 + 8x + 15 \geq 0$

________ 2. $x^2 - 4x < 0$

________ 3. $x^2 + 9x > 36$

________ 4. $2x^2 + 9 < 19x$

________ 5. $4x^2 \geq 25x + 21$

________ 6. $18 < 9x - x^2$

Check your answers in the back of your book.

If You Missed Problems:	You Should Review Example:
1, 2	3
3, 4	4
5, 6	5

5.5 Switching from Word Statements to Quadratic Equations

Now that we have learned to solve quadratic equations, it is time to put that skill to use in solving word problems.

Example 1. Mr. Green's rectangular garden is 10 feet longer than it is wide. If the area of the garden is 264 square feet, what are the dimensions of the garden?

Solution

We Think	***We Write***
The area of a rectangle is found by *multiplying* length times width.	Area = (length)(width)
We know less about the width.	Let w = width
The length is 10 feet *more than* the width.	$w + 10$ = length
We can illustrate the garden with dimensions labeled.	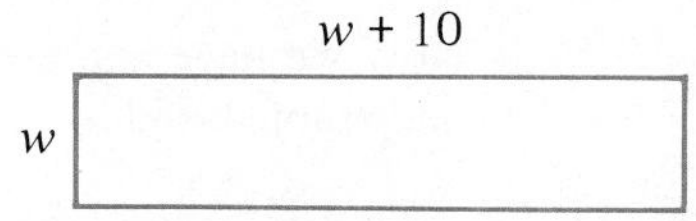
Area is the product of length times width.	Area = $(w + 10)w$
Here the area is 264 square feet.	$(w + 10)w = 264$ $w^2 + 10w = 264$

Now we have a quadratic equation to solve.

$$w^2 + 10w = 264$$

$$w^2 + 10w - 264 = 0$$

$$(w + 22)(w - 12) = 0$$

$$w + 22 = 0 \quad \text{or} \quad w - 12 = 0$$

$$w = -22 \qquad\qquad w = 12$$

Since width cannot be *negative*, we conclude that

$$\text{Width:} \qquad w = 12 \text{ feet}$$

$$\text{Length: } w + 10 = 12 + 10$$

$$= 22 \text{ feet}$$

The garden is 12 feet wide and 22 feet long.

Example 2. What must be the width of a frame around a square painting measuring 11 inches on each side if the framed painting is to cover exactly 289 square inches on a wall?

Solution

We Think	***We Write***
The area covered by the framed painting will be the *product* of its length and width.	Area = (framed length)(framed width)
We do not know the width of the frame.	Let x = width of frame

We Think

We can use an illustration to find an expression for the length and width of the framed painting.

We Write

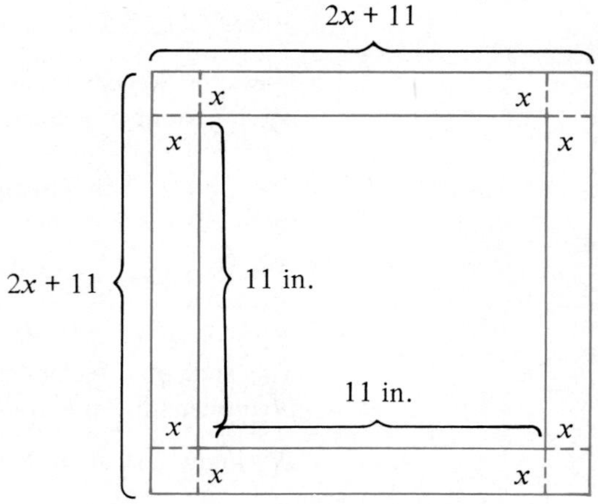

Now we have expressions for framed length and framed width.

Framed length $= 2x + 11$
Framed width $= 2x + 11$

Area is the *product* of framed length and width.

Area $= (2x + 11)(2x + 11)$

We know that the area covered by the framed painting is 289 square inches.

$$(2x + 11)(2x + 11) = 289$$
$$4x^2 + 44x + 121 = 289$$

Now we have a quadratic equation to solve.

$$4x^2 + 44x + 121 = 289$$
$$4x^2 + 44x - 168 = 0$$
$$4(x^2 + 11x - 42) = 0$$
$$4(x + 14)(x - 3) = 0$$
$$x + 14 = 0 \quad \text{or} \quad x - 3 = 0$$
$$x = -14 \qquad\qquad x = 3$$

Only a positive solution makes sense here. The frame must be 3 inches wide.

Now you help solve Example 3, which is the problem stated at the beginning of this chapter.

Example 3. The three support beams of a picnic shelter form a right triangle. The longest beam measures 25 feet and the shortest is 5 feet less than the other. Find the length of the piece of lumber that would be needed to cut all three beams.

Solution

We Think

A right triangle has two legs and a hypotenuse. The hypotenuse is always the longest side of a right triangle.

We Write

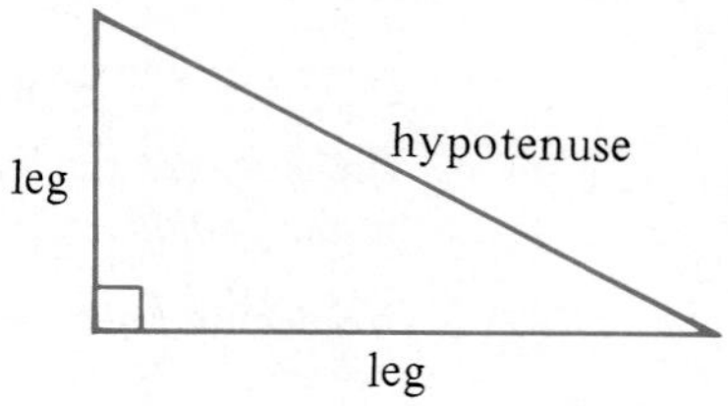

We Think	***We Write***
The length of the piece of lumber must equal the *sum* of the lengths of the two legs and the hypotenuse.	Length of lumber = first leg + second leg + hypotenuse
The shorter leg is 5 feet *less* than the other leg.	Let x = longer leg length x − ____ = shorter leg length 25 = hypotenuse length
According to the Pythagorean theorem, the square of the hypotenuse of a right triangle is equal to the *sum* of the squares of its legs.	$(25)^2 = x^2 + (x - \_\_\_\_)^2$ $625 = x^2 + x^2 - \_\_\_\_x + \_\_\_\_$ $625 = \_\_\_\_x^2 - \_\_\_\_x + \_\_\_\_$

Now we have a quadratic equation to solve.

$$2x^2 - 10x + 25 = 625$$

$$2x^2 - 10x - 600 = 0$$

$$2(x^2 - 5x - 300) = 0$$

$$2(x - \_\_\_\_)(x + \_\_\_\_) = 0$$

$$x = \_\_\_\_ \quad \text{or} \quad x = -\_\_\_\_$$

Only a positive length makes sense here.

Length of longer leg: x = ____ ft

Length of shorter leg: $x - 5$ = ____ − 5

= ____ ft

Length of hypotenuse: 25 ft

We have not yet answered the question. How long must the piece of lumber be?

Length of lumber = first leg + second leg + hypotenuse

= ____ + ____ + 25

= ____ ft

The piece of lumber for all three beams must be ____ feet long.

Check your work on page 328. ▶

Example 4. A company has found that when x units of a product are manufactured and sold, its revenue is $x^2 + 100x$ dollars and its costs are $240x + 500$ dollars. How many units must be produced and sold to make a profit of \$1000?

Solution

We Think	***We Write***
We know that profit is found by *subtracting* costs from revenue.	Profit = revenue − costs
We have expressions for revenue and costs.	Profit = $(x^2 + 100x) - (240x + 500)$
We want profit to *equal* \$1000.	$(x^2 + 100x) - (240x + 500) = 1000$

We Think ***We Write***

We have a quadratic equation to solve.

$$x^2 + 100x - (240x + 500) = 1000$$

$$x^2 + 100x - 240x - 500 = 1000$$

$$x^2 - 140x - 500 = 1000$$

$$x^2 - 140x - 1500 = 0$$

$$(x - 150)(x + 10) = 0$$

$$x - 150 = 0 \quad \text{or} \quad x + 10 = 0$$

$$x = 150 \qquad\qquad x = -10$$

Only a positive number of units makes sense here. The company must manufacture and sell 150 units of its product to make a profit of \$1000.

Trial Run

Use an equation to solve each problem.

_______ 1. A gardener has decided to decrease the area of his square garden to 2700 square yards by decreasing the length by 6 yards and the width by 2 yards. What were the dimensions of the square garden?

_______ 2. The total revenue from producing and selling q units of a product is given by the expression $q^2 + 50q$. How many units must be produced and sold to have \$5000 in revenue?

Answers are on page 329.

Example You Completed

Example 3 *(Solution)*

We Write

Let x = longer leg length

$x - 5$ = shorter leg length

25 = hypotenuse length

$$(25)^2 = x^2 + (x - 5)^2$$

$$625 = x^2 + x^2 - 10x + 25$$

$$625 = 2x^2 - 10x + 25$$

Now we have a quadratic equation to solve.

$$2x^2 - 10x + 25 = 625$$

$$2x^2 - 10x - 600 = 0$$

$$2(x^2 - 5x - 300) = 0$$

$$2(x - 20)(x + 15) = 0$$

$$x - 20 = 0 \quad \text{or} \quad x + 15 = 0$$

$$x = 20 \qquad\qquad x = -15$$

We Write

Only a positive length makes sense here.

$$\text{Length of longer leg:} \quad x = 20 \text{ ft}$$

$$\text{Length of shorter leg:} \quad x - 5 = 20 - 5$$

$$= 15 \text{ ft}$$

Length of hypotenuse: 25 ft

We have not yet answered the question. How long must the piece of lumber be?

$$\text{Length of lumber} = \text{first leg} + \text{second leg} + \text{hypotenuse}$$

$$= 20 + 15 + 25$$

$$= 60 \text{ ft}$$

The piece of lumber for all three beams must be 60 feet long.

Answers to Trial Run

page 328 **1.** 56 yards by 56 yards **2.** 50

Name ______________________ Date __________

EXERCISE SET 5.5

Use a quadratic equation to solve each problem.

_______ 1. The length of a rectangle is 5 feet less than twice its width. If the area is 88 square feet, find the dimensions of the rectangle.

_______ 2. The width of a rectangle is 18 feet less than three times the length and the area is 48 square feet. Find the dimensions of the rectangle.

_______ 3. Mr. Mrogenski plans to buy a rectangular lot that is 50 feet longer than it is wide. If the area of the lot is 30,000 square feet, find its dimensions.

_______ 4. Mr. Collins is building a cattle loading pen that is to be 15 feet longer than it is wide. He has enough fence to enclose an area of 1750 square feet. Find the dimensions of the pen.

_______ 5. Ella has a vegetable garden that is a square. She plans to increase the area of the garden by increasing the length by 10 feet and decreasing the width by 2 feet. If the area of the enlarged garden is 1900 square feet, find its dimensions. By how many square feet did Ella increase the size of her garden?

_______ 6. Eudena has a square family room that she plans to enlarge by increasing the length by 6 feet and increasing the width by 2 feet. The area of the enlarged family room will be 252 square feet. Find the dimensions of the new room. By how many square feet did Eudena increase the size of her room?

_______ 7. One square has a side 5 centimeters shorter than the side of a second square. The area of the larger is four times the area of the smaller. Find the dimensions of each square.

_______ 8. One square has a side 8 centimeters longer than the side of a second square. The area of the larger is nine times the area of the smaller. Find the dimensions of each square.

_______ 9. Find the dimensions of a room that is 6 feet longer than it is wide, if its floor can be covered by 216 square feet of carpeting.

_______ 10. The length of a rectangular lot is 6 feet less than twice the width. If it takes 1008 square feet of sod to cover the lot, find its dimensions.

_______ 11. Jim wishes to build an open tin storage box. The box is to have a square bottom, sides 4 inches high, and a volume of 576 cubic inches. Find the dimensions of the box. What size piece of tin should Jim buy if he wishes to build the box from one piece by cutting squares with 4-inch sides from each corner?

_______ 12. Olivia has a square piece of cardboard that she wishes to make into an open box by cutting a 3-inch square out of each corner and folding up the sides. If the box will have a volume of 675 cubic inches, find the dimensions of the piece of cardboard.

________ 13. Mr. Olsen is plowing a field that is 60 rods on each side. He wishes to plow the field in two different time periods. How wide a strip must he plow around the field the first time so that the second time he will have only 1600 square rods to plow?

________ 14. What must be the width of a sidewalk around a swimming pool that is 54 feet by 34 feet if the total area allowed for the pool and walk is 2400 square feet?

________ 15. The product of two consecutive integers is 24 more than nine times the larger. Find the two numbers.

________ 16. The sum of the squares of two consecutive integers is 73 more than the product of the two. Find the two numbers.

________ 17. One leg of a right triangle is 7 centimeters less than the other leg and the hypotenuse is 17 centimeters. Find the lengths of the two legs.

________ 18. One leg of a right triangle is 3 centimeters more than the other leg and the hypotenuse is 15 centimeters. Find the lengths of the two legs.

________ 19. The altitude of a triangle is 4 feet less than twice the base and the area is 63 square feet. Find the altitude of the triangle.

________ 20. The base of a triangle is 6 centimeters less than four times the altitude and the area is 104 square centimeters. Find the base of the triangle.

________ 21. A television antenna was bent over by a storm in such a way that the part bending over was 13 feet. The part left standing was 7 feet more than the distance from the foot of the antenna to the point where the broken part touched the ground. What was the height of the antenna before the storm?

________ 22. There are two pictures on Margaret's wall; one is a square and one is a rectangle. The length of the rectangular picture is 3 inches more than one side of the square, and the width of the rectangular picture is 1 inch less than one side of the square. If the two pictures occupy 109 square inches of area, find the dimensions of the rectangular picture.

________ 23. In a craft shop the total cost in dollars for producing Christmas wreaths is given by the expression $x^2 - 8x + 25$, where x represents the number of wreaths. If the shop can spend \$45 on this project, how many wreaths can be produced?

________ 24. The Basketball Boosters Club is planning a trip for fans to the tournament in San Diego. The profit to be made by the club in dollars is given by the expression $x^2 + 5x - 150$, where x represents the number of people. How many people must make the trip if the club wants to make a profit of \$600?

________ 25. The Bronco Boot Company has determined that the revenue received from selling x pairs of boots is given in dollars by the expression $x^2 - 29x + 390$. If the company wishes the revenue for one week to be \$600, how many pairs of boots must be sold?

________ **26.** A manufacturer of toy chests has found that when x chests are manufactured and sold, its revenue is $x^2 + 220x$ dollars and its costs are $300x + 2500$ dollars. How many units must be produced and sold in order to make a profit of $4000?

________ **27.** The distance an object falls from rest is given by $s = 16t^2$, where s is the distance and t is the time in seconds. How long does it take an object dropped from a height of 144 feet to reach the ground?

________ **28.** The relationship of the distance s to the velocity v and the time t in seconds when an object is thrown downward is given by $s = vt + 16t^2$. How long does it take an object thrown downward from a height of 156 feet with a velocity of 4 feet per second to hit the ground?

☆ Stretching the Topics

Use an equation to solve each problem.

________ **1.** The length of a rectangular room is 12 feet less than twice its width. The width of a carpet on the floor is 4 feet less than the width of the room. The length of the carpet is 6 feet more than its width. If 152 square feet of the floor are not covered by carpet, find the dimensions of the room.

________ **2.** Find three consecutive integers such that the square of the second is 51 less than the sum of the squares of the other two.

Check your answers in the back of your book.

If you can complete **Checkup 5.5,** you are ready to do the **Review Exercises** for Chapter 5.

Name ____________________ Date __________

✓ CHECKUP 5.5

Use an equation to solve each problem.

________ 1. The length of Mr. Schroder's rectangular garden is 8 feet less than twice the width. If the area of the garden is 192 square feet, what are the dimensions of the garden?

________ 2. What must be the width of a border around a rectangular quilt that is 7 feet wide and 8 feet long if the total area the quilt must cover is 72 square feet?

________ 3. The hypotenuse of a right triangle is 2 centimeters more than the longer leg. The shorter leg is 7 centimeters less than the longer leg. Find the length of the hypotenuse.

Check your answers in the back of your book.

If You Missed Problem:	You Should Review Example:
1	1
2	2
3	3

Summary

In this chapter we discussed solutions for **quadratic** (second-degree) **equations** and **inequalities.** We learned several methods for solving quadratic equations of the form $ax^2 + bx + c = 0$ and agreed to use the most appropriate method for a particular equation.

Method	When a Good Choice	Example
Factoring	When quadratic expression can be factored using integers	$2x^2 - 5x - 3 = 0$ $(2x + 1)(x - 3) = 0$ $2x + 1 = 0$ or $x - 3 = 0$ $x = -\frac{1}{2}$ or $x = 3$
Taking square roots	When first-degree term is missing ($b = 0$)	$9x^2 - 2 = 0$ $9x^2 = 2$ $x^2 = \frac{2}{9}$ $x = \sqrt{\frac{2}{9}}$ or $x = -\sqrt{\frac{2}{9}}$ $x = \frac{\sqrt{2}}{3}$ or $x = -\frac{\sqrt{2}}{3}$
Completing the square	When coefficient of middle term is even, but quadratic expression does not factor	$x^2 - 2x - 5 = 0$ $x^2 - 2x = 5$ $x^2 - 2x + 1 = 5 + 1$ $(x - 1)^2 = 6$ $x - 1 = \pm\sqrt{6}$ $x = 1 \pm \sqrt{6}$ $x = 1 + \sqrt{6}$ or $x = 1 - \sqrt{6}$
Quadratic formula	When other methods cannot be used	$x^2 + 3x + 1 = 0$ $x = \frac{-3 \pm \sqrt{3^2 - 4(1)(1)}}{2(1)}$ $= \frac{-3 \pm \sqrt{5}}{2}$ $x = \frac{-3 + \sqrt{5}}{2}$ or $x = \frac{-3 - \sqrt{5}}{2}$

After solving equations by factoring, we learned to write equations with given solutions using the fact that if r is a solution for a polynomial equation, then $x - r$ is a factor of the polynomial.

Our work with the **quadratic formula**

$$x = \frac{-b \pm \sqrt{b^2 - 4ac}}{2a}$$

led us to a discussion of square roots and we learned two rules for simplifying square roots.

Product Rule for Square Roots

$$\sqrt{A \cdot B} = \sqrt{A} \cdot \sqrt{B} \qquad \begin{matrix} A \geq 0 \\ B \geq 0 \end{matrix}$$

Example: $\sqrt{150} = \sqrt{25 \cdot 6}$

$= 5\sqrt{6}$

Quotient Rule for Square Roots

$$\sqrt{\frac{A}{B}} = \frac{\sqrt{A}}{\sqrt{B}} \qquad \begin{matrix} A \geq 0 \\ B > 0 \end{matrix}$$

Example: $\sqrt{\frac{3}{16}} = \frac{\sqrt{3}}{\sqrt{16}}$

$= \frac{\sqrt{3}}{4}$

The quadratic formula also led us to a discussion of the **discriminant,** $b^2 - 4ac$, from which we can determine what kind of solutions a quadratic equation has.

Discriminant	Kind of Solutions
$b^2 - 4ac = 0$	One rational solution
$b^2 - 4ac > 0$ and perfect square	Two different rational solutions
$b^2 - 4ac > 0$, *not* perfect square	Two different irrational solutions
$b^2 - 4ac < 0$	No real number solutions

In our work with **quadratic inequalities** we learned to find solutions by using cases and by using boundary points that separate the real number line into distinct intervals. Determining the sign of each factor of the quadratic expression in each interval allowed us to determine the intervals on which the inequality was satisfied.

Finally, we learned to switch from word statements to quadratic equations.

❑ Speaking the Language of Algebra

Complete each sentence with the appropriate word or phrase.

1. An equation of the form $ax^2 + bx + c = 0$ is called a second-degree or ________ equation.

2. According to the zero product rule, if a product of factors is zero, then ________ .

3. If $x = r$ is a solution for a polynomial equation, we know that ________ is a ________ of the polynomial.

4. If $ax^2 + bx + c = 0$, the solutions for the equation are $x =$ ________ .

5. The expression $b^2 - 4ac$ is called the ________ . If $b^2 - 4ac < 0$, the equation $ax^2 + bx + c = 0$ has ________ ________ solutions.

6. If $(x - r)(x - s) > 0$, we know that $x - r$ and $x - s$ are both ________ or both ________ .

△ Writing About Mathematics

Write your response to each question in complete sentences.

1. Explain, in your own words, the zero product rule.

2. Explain how you would recognize a quadratic equation and the process you would use to solve such an equation when the quadratic expression can be factored.

3. Describe the different kinds of solutions you might find when solving a quadratic equation of the form $x^2 = n$.

4. Describe the process you would use to solve the equation $x^2 + 4x + 1 = 0$ by the method of completing the square.

5. Explain the steps you would use to solve the quadratic inequality $x^2 + 7x - 18 > 0$ by the boundary-point method.

Name ______________________ Date ____________

REVIEW EXERCISES for Chapter 5

Solve each equation by factoring.

_______ 1. $(3x - 5)(x + 7) = 0$

_______ 2. $x^2 - 5x - 24 = 0$

_______ 3. $2x^2 + 9x - 35 = 0$

_______ 4. $x^2 + 12x + 36 = 0$

_______ 5. $0 = 7x^2 - 63$

_______ 6. $32x - 8x^2 = 0$

_______ 7. $3a^2 = 14a - 8$

_______ 8. $10a^2 + 6 = 32a$

_______ 9. $3x^2 = 48$

_______ 10. $7y = 5y^2$

_______ 11. $(2x - 1)(x - 5) = -x(x - 14) - 23$

_______ 12. $x(x - 5) + 4(x^2 - 4) = 2(13x - 11)$

_______ 13. $12x^3 + 31x^2 - 15x = 0$

_______ 14. $x^4 - 29x^2 + 100 = 0$

_______ 15. $3x^5 + 3x^4 - 168x^3 = 0$

_______ 16. Write a quadratic equation with solutions $x = 8$ and $x = -3$.

Simplify each radical expression.

_______ 17. $\sqrt{75}$

_______ 18. $\sqrt{162}$

_______ 19. $\sqrt{\frac{5}{9}}$

_______ 20. $\sqrt{\frac{18}{49}}$

Solve each equation and state the nature of the solutions.

_______ 21. $x^2 - 32 = 0$

_______ 22. $x^2 + 49 = 0$

_______ 23. $9x^2 = 25$

_______ 24. $16x^2 - 1 = 0$

Solve each equation by completing the square.

_______ 25. $x^2 - 2x - 48 = 0$

_______ 26. $x^2 - 3x - 6 = 0$

_______ 27. $2x^2 - 7x + 2 = 0$

_______ 28. $x^2 + 4x - 1 = 0$

Solve each equation by using the quadratic formula.

_______ 29. $9x^2 - x = 1$

_______ 30. $3y^2 - 2y - 5 = 0$

_______ 31. $x^2 + 2x = 6$

_______ 32. $4x^2 - 3x = 2$

_______ 33. Solve $2d = n^2 - 3n$ for n.

_______ 34. Solve $S = gt + 16t^2$ for t.

Use the discriminant to describe the solutions for each equation.

________ 35. $2x^2 + x - 2 = 0$

________ 36. $3x^2 + 2x - 1 = 0$

________ 37. $x^2 + 81 = 0$

________ 38. $x^2 = 18x - 81$

Solve each inequality.

________ 39. $x^2 < 144$

________ 40. $x^2 - 3x \geq 0$

________ 41. $2x^2 - 9x \leq 5$

________ 42. $x^3 - 4x > 0$

Use an equation to solve each problem.

________ 43. Mr. Crawford is building a rectangular table that is 24 inches longer than it is wide. If the area of the table is 5040 square inches, find its dimensions.

________ 44. One square has a side 6 centimeters longer than the side of a second square. The area of the larger is four times the area of the smaller. Find the dimensions of each square.

________ 45. Find the width of a frame around a picture that is 12 inches by 14 inches if the total area allowed for the framed picture is 360 square inches.

________ 46. The altitude of a triangle is 5 feet less than the base and the area is 42 square feet. Find the altitude of the triangle.

Check your answers in the back of your book.

If You Missed Exercises:	You Should Review Examples:	
1	SECTION 5.1	1–3
2–6		4–6
7–10		7–9
11, 12		10–12
13		14
14		15
15		16
16		17
17, 18	SECTION 5.2	9, 10
19, 20		11–14
21–24		15–18
25–28	SECTION 5.3	1–7
29–32		8–11
33, 34		12
35–38		13, 14
39–42	SECTION 5.4	3–5
43–46	SECTION 5.5	1–4

If you have completed the **Review Exercises** and corrected your errors, you are ready to take the **Practice Test** for Chapter 5.

Name ____________________ Date ____________

PRACTICE TEST for Chapter 5

		Section	Examples
	Solve each equation.		
______	1. $x^2 + 5x - 36 = 0$	5.1	4
______	2. $9x^2 = 30x - 25$	5.1	8
______	3. $7x^2 = 21x$	5.1	9
______	4. $5x^3 - 7x^2 - 6x = 0$	5.1	14
______	5. $(x + 6)(x - 1) = 8$	5.1	11, 12
______	6. $x^4 - 20x^2 + 64 = 0$	5.1	15
______	7. $2x^3 - x^2 - 2x + 1 = 0$	5.1	16
______	8. Write a quadratic equation with solutions $x = 4$ and $x = -4$.	5.1	17
	Solve each equation.		
______	9. $81a^2 = 4$	5.2	15
______	10. $9x^2 - 20 = 0$	5.2	16
______	11. $x^2 + 7x + 3 = 0$	5.3	9, 10
______	12. $3x^2 - 2x - 2 = 0$	5.3	9, 10
______	13. $y^2 + 9y + 25 = 0$	5.3	11
______	14. Solve $K = 3ay^2 - y$ for y.	5.3	12
______	15. Use the discriminant to describe the solutions for $x^2 + 16 = 8x$.	5.3	13, 14
	Solve each inequality.		
______	16. $2x^2 - 7x \leq 0$	5.4	4
______	17. $5x^2 > 6x + 8$	5.4	5
	Use an equation to solve each problem.		
______	18. The length of the shorter leg of a right triangle is 7 inches less than the longer leg. The hypotenuse is 1 inch longer than the longer leg. Find the lengths of the sides of the triangle.	5.5	3
______	19. The width of a rectangle is 5 centimeters less than the length. If the area of the rectangle is 36 square centimeters, find the dimensions of the rectangle.	5.5	1

		Section	Example
________	20. A company has found that when x units of a product are manufactured and sold, its revenue is $x^2 + 70x$ dollars and its costs are $120x + 100$ dollars. How many units must be produced and sold to make a profit of $500?	5.5	4

Name ______________________ Date ____________

SHARPENING YOUR SKILLS after Chapters 1–5

		Section
______	1. Use an inequality to state that $x - 2$ is a negative number.	1.1
______	2. Compare $-\frac{16}{3}$ and $-\sqrt{36}$ using $<$, $>$, or $=$.	1.1

Perform the indicated operation and simplify.

______	3. $\dfrac{-12(\frac{1}{4}) + 8[5 - (-7)]}{-7[36 - 9(4)]}$	1.3
______	4. $\lvert 8 - 12\rvert - \lvert 7 - 5\rvert$	1.4
______	5. -4^{-2}	3.2
______	6. $(x + 3)^0$	3.1
______	7. $-\frac{1}{4}(8x - 16y) + 3[x - (x + 2y)]$	1.4
______	8. $(9x^2yz^5)(-7x^4y^3z)$	3.1
______	9. $\left(\dfrac{-3x^5y^0}{5z^3}\right)^2$	3.1
______	10. $\left(\dfrac{x^{12}y^{-2}}{x^{-5}y^7}\right)^{-1}$	3.2
______	11. $-5x^5y^2(3xy^4 + 2y^3) + 3x^3y^4(x^2y - 2x^3y^2)$	3.3
______	12. $(a - 2b)(a + 5b)(a - 7b)$	3.4
______	13. $[(x + y) - 6]^2$	3.4
______	14. $(10a^3b - 20a^2b^2 + 15ab^3) \div 5ab$	3.5
______	15. $(x^3 - 9x + 10) \div (x - 2)$	3.5

Rewrite each expression in completely factored form.

______	16. $a^2 + 12ab + 27b^2$	4.3
______	17. $5x^4 - 80$	4.3
______	18. $27x^3 - 64$	4.2
______	19. $x^3 - 5x^2 + 9x - 45$	4.2
______	20. $-18x^4 + 9x^3 + 27x^2$	4.3

		Section
________	21. $x^2 - 18x + 81$	4.3
________	22. Evaluate $A = P + Prt$ when $P = \$5000$, $r = 9.5\%$, and $t = 4$ years.	4.3

Solve each equation or inequality.

________	23. $4[3x + (x - 7)] = 3(x - 8) + 9$	2.1
________	24. $x < 3$ or $x \leq 0$	2.2
________	25. $3x - 4 > 2$ and $2(x - 3) < -10$	2.3
________	26. $-1 \leq 2(-x + 4) - 3 < 7$	2.3
________	27. $\|3x - 1\| < 11$	2.4
________	28. Solve $ax + by = ab$ for x.	2.1
________	29. Salesman Henri earns a base salary of \$700 per month plus a \$30 commission for each water bed he sells. How many water beds must he sell to earn at least \$1000 this month?	2.5
________	30. If two opposite sides of a square are increased by 3 inches and the other two sides are decreased by 2 inches, write an expression for the area of the resulting rectangle. Simplify your answer.	3.6

C H A P T E R

6

Working with Rational Expressions

Mr. Romano uses two hoses to fill his swimming pool. By itself, the larger hose can fill the pool in 7 hours; the smaller hose requires 10 hours by itself. While using both hoses to fill his pool last weekend, Mr. Romano discovered after 2 hours that the pool drain had been left open. If the pool normally drains in 12 hours, how long after he closed the drain was the pool filled?

From Chapter 1 we recall that a **rational number** is any real number that can be expressed as a fraction with an integer numerator and a *nonzero* integer denominator. In this chapter we discuss the algebra of **rational expressions,** which are fractions with variables and/or constants in the numerator and *nonzero* variable and/or constant expressions in the denominator. For the most part, we shall concentrate on **rational algebraic expressions,** which are fractions with a polynomial in the numerator and a *nonzero* polynomial in the denominator. In this chapter we learn how to

1. State restrictions on the domain of the variable in a rational expression.
2. Reduce rational expressions to lowest terms.
3. Multiply and divide rational expressions.
4. Add and subtract rational expressions.
5. Simplify complex fractions.
6. Solve fractional equations and inequalities.
7. Switch from words to fractional equations.

6.1 Simplifying Rational Algebraic Expressions

Before we learn to simplify rational algebraic expressions, it is important that we discuss thoroughly the fact that the denominators of such expressions can *never* be zero.

Restricting the Domain of a Variable

When dealing with an algebraic fraction such as $\frac{5}{x}$, we know that its numerical value depends on the value that we choose to substitute for the variable. We may choose *almost* any value for x in evaluating this fraction, but there is one number that we may *not* substitute for x. That number is zero, because

$$\text{if } x = 0 \quad \text{then} \quad \frac{5}{0} \text{ is undefined}$$

The set of all possible replacements for a variable in an expression is called the **domain** of the variable. Not allowing the replacement of a particular value for the variable is called **restricting the domain** of the variable.

> For a rational algebraic expression, we must restrict the domain of the variable to exclude any values that will make the denominator equal zero.

To determine the necessary restrictions on the domain of the variable in a rational expression, we consider the factors in the denominator and decide what values would make those factors zero. Then we exclude those values from the domain.

Rational Expression	Denominator	Restrictions
$\frac{11}{3x}$	$3x$	$x \neq 0$
$\frac{2a}{a-3}$	$a - 3$	$a \neq 3$
$\frac{x+2}{x(x+6)}$	$x(x + 6)$	$x \neq 0, x \neq -6$
$\frac{3x+5}{x^2+3x+2}$	$x^2 + 3x + 2$ $= (x + 2)(x + 1)$	$x \neq -2, x \neq -1$

Notice that we do not concern ourselves with the nature of the numerator of a fraction. The numerator of a fraction can certainly be zero, as long as the denominator is *not* zero.

> If the numerator of a fraction is zero but the denominator is not zero, then the value of the fraction is zero.
>
> If the denominator of a fraction is zero, then the fraction is undefined.

Example 1. Restrict the domain for $\dfrac{7x + 1}{2x^3 - 7x^2 - 15x}$.

Solution. We factor the denominator.

$$\frac{7x + 1}{2x^3 - 7x^2 - 15x} = \frac{7x + 1}{x(2x^2 - 7x - 15)}$$

$$= \frac{7x + 1}{x(2x + 3)(x - 5)}$$

The denominator will be zero whenever any one of its factors is zero.

$$x = 0 \qquad 2x + 3 = 0 \qquad x - 5 = 0$$

$$2x = -3$$

$$x = 0 \qquad x = -\frac{3}{2} \qquad x = 5$$

Restrictions: $x \neq 0$, $x \neq -\frac{3}{2}$, $x \neq 5$.

You complete Example 2.

Example 2. Restrict the domain for $\dfrac{y + 3}{y^2 - 4}$.

Solution. We factor the denominator.

$$\frac{y + 3}{y^2 - 4} =$$

Restrictions:

Check your work on page 352. ▶

Reducing Rational Algebraic Expressions

We say that two fractions are *equal* if they represent the same number. We can decide if two fractions are equal using the following definition.

Equality of Fractions

$$\frac{A}{B} = \frac{C}{D} \quad \text{provided that} \quad A \cdot D = B \cdot C$$

where $B \neq 0$, $D \neq 0$.

For example,

$$\frac{2}{3} = \frac{8}{12} \quad \text{because} \quad 2 \cdot 12 = 3 \cdot 8$$

$$\frac{15}{20} = \frac{3}{4} \quad \text{because} \quad 15 \cdot 4 = 20 \cdot 3$$

We know from arithmetic that a numerical fraction can be **reduced** by dividing its numerator and denominator by any factors common to both.

Fraction	Factored Form	Reduced Form
$\frac{3}{12}$	$\frac{3 \cdot 1}{3 \cdot 2 \cdot 2} = \frac{\overset{1}{\cancel{3}} \cdot 1}{\underset{1}{\cancel{3}} \cdot 2 \cdot 2}$	$\frac{1}{4}$
$\frac{10}{15}$	$\frac{2 \cdot 5}{3 \cdot 5} = \frac{2 \cdot \overset{1}{\cancel{5}}}{3 \cdot \underset{1}{\cancel{5}}}$	$\frac{2}{3}$
$\frac{42}{28}$	$\frac{2 \cdot 3 \cdot 7}{2 \cdot 2 \cdot 7} = \frac{\overset{1}{\cancel{2}} \cdot 3 \cdot \overset{1}{\cancel{7}}}{\underset{1}{\cancel{2}} \cdot 2 \cdot \underset{1}{\cancel{7}}}$	$\frac{3}{2}$

To reduce a fraction, divide the numerator and denominator by any nonzero *factors* common to both. A fraction has been *reduced to lowest terms* when the numerator and denominator contain no common factor (except the number 1).

Letting A, B, and C represent algebraic expressions, we may say

Reducing Fractions

$$\frac{A \cdot C}{B \cdot C} = \frac{A}{B} \qquad B \neq 0,\ C \neq 0$$

To reduce a fraction with a monomial numerator and denominator, we may use either the laws of exponents or the method of common factors to reduce the fraction to lowest terms. Let's use both approaches to reduce

$$\frac{-3x^4y^3z}{5x^2y^5z}$$

after noting that the restrictions are $x \neq 0$, $y \neq 0$, $z \neq 0$.

First Approach

Using the laws of exponents, we have

$$\frac{-3x^4y^3z}{5x^2y^5z} = \frac{-3x^{4-2}y^{3-5}z^{1-1}}{5}$$

$$= \frac{-3x^2y^{-2}z^0}{5}$$

$$= -\frac{3x^2}{5y^2}$$

Second Approach

Using common factors, we have

$$\frac{-3x^4y^3z}{5x^2y^5z} = \frac{-3 \cdot x^2 \cdot x^2 \cdot y^3 \cdot z}{5 \cdot x^2 \cdot y^3 \cdot y^2 \cdot z}$$

$$= \frac{-3 \cdot \overset{1}{\cancel{x^2}} \cdot x^2 \cdot \overset{1}{\cancel{y^3}} \cdot \overset{1}{\cancel{z}}}{5 \cdot \underset{1}{\cancel{x^2}} \cdot \underset{1}{\cancel{y^3}} \cdot y^2 \cdot \underset{1}{\cancel{z}}}$$

$$= -\frac{3x^2}{5y^2}$$

Our answer is the same using either approach.

You complete Example 3.

Example 3. Use the laws of exponents to reduce $\dfrac{2x^3y^7z}{x^2yz^2}$.

Solution. We note that the restrictions are $x \neq 0$, $y \neq 0$, $z \neq 0$.

$$\frac{2x^3y^7z}{x^2yz^2} = 2x^{3-2}y^{7-1}z^{1-2}$$

Check your work on page 352. ▶

Then complete Example 4.

Example 4. Use common factors to reduce $\dfrac{2x^3y^7z}{x^2yz^2}$.

Solution. We note that the restrictions are $x \neq 0$, $y \neq 0$, $z \neq 0$.

$$\frac{2x^3y^7z}{x^2yz^2} = \frac{2 \cdot x^2 \cdot x \cdot y \cdot y^6 \cdot z}{x^2 \cdot y \cdot z \cdot z}$$

$$= \underline{\qquad\qquad}$$

Check your work on page 353. ▶

In dealing with a fraction whose numerator and/or denominator are *not* monomials, we must rewrite the numerator and denominator in completely factored form before looking for common factors. Moreover, *we must state restrictions on the domain of the variable before dividing numerator and denominator by any common factors.*

For example, suppose that we reduce this fraction.

$$\frac{3x + 15}{x^2 + 7x + 10} = \frac{3(x + 5)}{(x + 5)(x + 2)} \qquad x \neq -5, x \neq -2$$

$$= \frac{3\overset{1}{\cancel{(x + 5)}}}{\underset{1}{\cancel{(x + 5)}}(x + 2)}$$

$$= \frac{3}{x + 2} \qquad x \neq -5, x \neq -2$$

The reduced version of our fraction is equivalent to the original fraction, *provided* that we avoid those values of the variable that have been excluded from the domain. Either version of the fraction will yield the same numerical value when any number (except -5 or -2) is substituted for the variable x.

Steps for Reducing Fractions

1. Factor the numerator and denominator completely.
2. State necessary restrictions on the domain of the variable.
3. Divide the numerator and denominator by any factors common to both.

Stating restrictions should become an automatic procedure whenever we are working with rational algebraic expressions.

Example 5. Reduce $\dfrac{x^2 - 3x}{x^2 - 9}$.

Solution

$$\frac{x^2 - 3x}{x^2 - 9}$$

$$= \frac{x(x - 3)}{(x + 3)(x - 3)} \qquad x \neq -3, x \neq 3$$

$$= \frac{x\overset{1}{\cancel{(x - 3)}}}{(x + 3)\underset{1}{\cancel{(x - 3)}}}$$

$$= \frac{x}{x + 3} \qquad x \neq -3, x \neq 3$$

Now you complete Example 6.

Example 6. Reduce $\dfrac{x + 4}{x^2 + 8x + 16}$.

Solution

$$\frac{x + 4}{x^2 + 8x + 16}$$

$$= \qquad\qquad x \neq \underline{\qquad}$$

$$= \qquad\qquad x \neq \underline{\qquad}$$

Check your work on page 353. ▶

In Example 5 you should not be tempted to "cross out" the x's in the numerator and denominator of $\dfrac{x}{x + 3}$. In that denominator, x is a *term*, not a factor. *We can reduce a fraction only by dividing out common factors.*

Example 7. Reduce $\dfrac{x^3 - 8}{3x - 6}$.

Solution

$$\frac{x^3 - 8}{3x - 6}$$ The numerator is a difference of two cubes.

$$= \frac{(x - 2)(x^2 + 2x + 4)}{3(x - 2)} \qquad x \neq 2$$

$$= \frac{\overset{1}{\cancel{(x - 2)}}(x^2 + 2x + 4)}{3\underset{1}{\cancel{(x - 2)}}}$$

$$= \frac{x^2 + 2x + 4}{3} \qquad x \neq 2$$

You complete Example 8.

Example 8. Reduce $\dfrac{x^2 + 11x + 30}{2x^3 + 11x^2 - 6x}$.

Solution

$$\frac{x^2 + 11x + 30}{2x^3 + 11x^2 - 6x}$$

$$= \frac{(x + 5)(x + 6)}{x(2x^2 + 11x - 6)}$$

$$= \frac{(x + 5)(x + 6)}{x(\underline{\qquad\qquad})(\underline{\qquad\qquad})} \qquad x \neq 0,\ x \neq \underline{\qquad},\ x \neq \underline{\qquad}$$

Check your work on page 353. ▶

From Example 8, note that we agree to *leave a reduced fraction with the numerator and denominator in factored form.*

Example 9. Reduce $\dfrac{x^2 + 2xy - x - 2y}{x^2 + 5xy + 6y^2}$.

Solution

$$\frac{x^2 + 2xy - x - 2y}{x^2 + 5xy + 6y^2}$$

$$= \frac{x(x + 2y) - 1(x + 2y)}{(x + 2y)(x + 3y)} \qquad x \neq -2y,\ x \neq -3y$$

$$= \frac{\overset{1}{\cancel{(x + 2y)}}(x - 1)}{\underset{1}{\cancel{(x + 2y)}}(x + 3y)}$$

$$= \frac{x - 1}{x + 3y} \qquad x \neq -2y,\ x \neq -3y$$

Example 10. Reduce

$$\frac{5(x - 3)^2 - 10x(x - 3)}{(x - 3)^4}$$

Solution

$$\frac{5(x - 3)^2 - 10x(x - 3)}{(x - 3)^4}$$

$$= \frac{5(x - 3)[(x - 3) - 2x]}{(x - 3)^4} \qquad x \neq 3$$

$$= \frac{5(x - 3)(x - 3 - 2x)}{(x - 3)^4}$$

$$= \frac{5(x - 3)(-x - 3)}{(x - 3)^4}$$

$$= \frac{5(-x - 3)}{(x - 3)^3}$$

$$= \frac{-5(x + 3)}{(x - 3)^3} \qquad x \neq 3$$

At first glance, fractions such as those that follow may appear to be in lowest terms. Suppose we take a closer look at them.

$\dfrac{x - 2}{2 - x}$		$\dfrac{4 - y}{y - 4}$
$= \dfrac{x - 2}{-x + 2}$	Write numerators and denominators with the variable first.	$= \dfrac{-y + 4}{y - 4}$
$= \dfrac{x - 2}{-1(x - 2)} \quad x \neq 2$	Factor -1 from the binomial in which the leading coefficient is negative. State restrictions.	$= \dfrac{-1(y - 4)}{y - 4} \quad y \neq 4$
$= \dfrac{\overset{1}{\cancel{x - 2}}}{-1\underset{1}{\cancel{(x - 2)}}}$	Divide numerator and denominator by the common factor.	$= \dfrac{-1\overset{1}{\cancel{(y - 4)}}}{\underset{1}{\cancel{y - 4}}}$
$= \dfrac{1}{-1}$		$= \dfrac{-1}{1}$
$= -1$	Simplify your answer.	$= -1$

Indeed, it is always the case that

$$\frac{A - B}{B - A} = -1 \qquad A \neq B$$

and this is often a handy fact to use in reducing fractions.

Example 11. Reduce $\dfrac{25 - x^2}{x^2 - 2x - 15}$.

Solution

$$\frac{25 - x^2}{x^2 - 2x - 15}$$

$$= \frac{(5 + x)(5 - x)}{(x - 5)(x + 3)} \qquad x \neq 5, x \neq -3$$

$$= \frac{(5 + x)\overset{-1}{\cancel{(5 - x)}}}{\underset{1}{\cancel{(x - 5)}}(x + 3)}$$

$$= \frac{-(5 + x)}{x + 3} \qquad x \neq 5, x \neq -3$$

Now you try Example 12.

Example 12. Reduce $\dfrac{5 - 10x}{2x^2 + 7x - 4}$.

Solution

$$\frac{5 - 10x}{2x^2 + 7x - 4}$$

$=$ $\qquad$ $x \neq$ _____

$x \neq$ _____

Check your work on page 353. ▶

Trial Run

State restrictions and reduce the fractions.

______ 1. $\dfrac{14a^2bc^4}{-7a^3bc}$

______ 2. $\dfrac{18xy - 14x}{2x}$

______ 3. $\dfrac{81x^4}{9x^2 + 9x}$

______ 4. $\dfrac{x^2 - 49}{4x + 28}$

______ 5. $\dfrac{y - 5}{y^2 - 10y + 25}$

______ 6. $\dfrac{x^2 - 5x - 36}{2x^2 - 19x + 9}$

______ 7. $\dfrac{x^2 - 4xy - 3x + 12y}{x^2 - 3xy - 4y^2}$

______ 8. $\dfrac{16 - a^2}{3a^2 - 7a - 20}$

Answers are on page 353.

▶ Examples You Completed

Example 2. Restrict the domain for $\dfrac{y + 3}{y^2 - 4}$.

Solution. We factor the denominator.

$$\frac{y + 3}{y^2 - 4} = \frac{y + 3}{(y + 2)(y - 2)}$$

The denominator will be zero whenever any one of its factors is zero.

$$y + 2 = 0 \qquad y - 2 = 0$$

$$y = -2 \qquad y = 2$$

Restrictions: $y \neq -2$, $y \neq 2$.

Example 3. Use the laws of exponents to reduce $\dfrac{2x^3y^7z}{x^2yz^2}$.

Solution. We note that the restrictions are $x \neq 0$, $y \neq 0$, $z \neq 0$.

$$\frac{2x^3y^7z}{x^2yz^2} = 2x^{3-2}y^{7-1}z^{1-2}$$

$$= 2x^1y^6z^{-1}$$

$$= \frac{2xy^6}{z}$$

Example 4. Use common factors to reduce $\frac{2x^3y^7z}{x^2yz^2}$.

Solution. We note that the restrictions are $x \neq 0$, $y \neq 0$, $z \neq 0$.

$$\frac{2x^3y^7z}{x^2yz^2} = \frac{2 \cdot \cancel{x^2} \cdot x \cdot \cancel{y} \cdot y^6 \cdot \cancel{z}}{\cancel{x^2}\cancel{y} \cdot \cancel{z} \cdot z}$$

$$= \frac{2xy^6}{z}$$

Example 6. Reduce $\frac{x + 4}{x^2 + 8x + 16}$.

Solution

$$\frac{x + 4}{x^2 + 8x + 16}$$

$$= \frac{x + 4}{(x + 4)(x + 4)} \qquad x \neq -4$$

$$= \frac{1}{x + 4} \qquad x \neq -4$$

Example 8. Reduce $\frac{x^2 + 11x + 30}{2x^3 + 11x^2 - 6x}$.

Solution

$$\frac{x^2 + 11x + 30}{2x^3 + 11x^2 - 6x}$$

$$= \frac{(x + 5)(x + 6)}{x(2x^2 + 11x - 6)}$$

$$= \frac{(x + 5)\cancel{(x + 6)}}{x(2x - 1)\cancel{(x + 6)}} \qquad x \neq 0,\ x \neq \tfrac{1}{2},\ x \neq -6$$

$$= \frac{x + 5}{x(2x - 1)} \qquad x \neq 0,\ x \neq \tfrac{1}{2},\ x \neq -6$$

Example 12. Reduce $\frac{5 - 10x}{2x^2 + 7x - 4}$.

Solution

$$\frac{5 - 10x}{2x^2 + 7x - 4}$$

$$= \frac{5\overset{-1}{\cancel{(1 - 2x)}}}{\underset{1}{\cancel{(2x - 1)}}(x + 4)} \qquad x \neq \tfrac{1}{2},\ x \neq -4$$

$$= \frac{-5}{x + 4} \qquad x \neq \tfrac{1}{2},\ x \neq -4$$

Answers to Trial Run

page 352

1. $-\frac{2c^3}{a}$; $a \neq 0$, $b \neq 0$, $c \neq 0$
2. $9y - 7$; $x \neq 0$
3. $\frac{9x^3}{x + 1}$; $x \neq 0$, $x \neq -1$
4. $\frac{x - 7}{4}$; $x \neq -7$
5. $\frac{1}{y - 5}$; $y \neq 5$
6. $\frac{x + 4}{2x - 1}$; $x \neq 9$, $x \neq \frac{1}{2}$
7. $\frac{x - 3}{x + y}$; $x \neq 4y$, $x \neq -y$
8. $\frac{-(4 + a)}{3a + 5}$; $a \neq 4$, $a \neq -\frac{5}{3}$

Name ______________________ Date ____________

EXERCISE SET 6.1

State restrictions on the domain of the variables and reduce the fractions.

_______ 1. $\dfrac{-10a^3b^2}{5ab^5}$

_______ 2. $\dfrac{-9a^5b^3}{3a^2b^7}$

_______ 3. $\dfrac{8a^2b^3c^2}{24a^4b^5c}$

_______ 4. $\dfrac{6a^3b^4c^3}{18a^8b^3c^4}$

_______ 5. $\dfrac{x^3y^2z^7}{x^5y^4z^7}$

_______ 6. $\dfrac{x^5y^3z^2}{x^7y^3z^4}$

_______ 7. $\dfrac{12xy - 6x}{3x}$

_______ 8. $\dfrac{15xy - 10x}{5x}$

_______ 9. $\dfrac{5y}{25y^2 - 20y}$

_______ 10. $\dfrac{3y}{9y^2 - 15y}$

_______ 11. $\dfrac{-12a^2b - 15abc}{18ab}$

_______ 12. $\dfrac{-10a^2b - 18abc}{22ab}$

_______ 13. $\dfrac{9x + 9y}{18x + 18y}$

_______ 14. $\dfrac{5x + 5y}{15x + 15y}$

_______ 15. $\dfrac{3x - 3y}{6x + 6y}$

_______ 16. $\dfrac{4x - 4y}{12x + 12y}$

_______ 17. $\dfrac{x^2 - 81}{3x + 27}$

_______ 18. $\dfrac{x^2 - 64}{4x + 32}$

_______ 19. $\dfrac{x^2 + 9x + 20}{x^2 - 16}$

_______ 20. $\dfrac{x^2 + 7x + 12}{x^2 - 9}$

_______ 21. $\dfrac{x^2 + 6x + 8}{x^2 + 7x + 10}$

_______ 22. $\dfrac{x^2 + 8x + 15}{x^2 + 5x + 6}$

_______ 23. $\dfrac{25y^2 - 225}{5y^2 - 50y + 105}$

_______ 24. $\dfrac{16y^2 - 144}{4y^2 - 20y + 24}$

_______ 25. $\dfrac{2b - 1}{4b^2 - 4b + 1}$

_______ 26. $\dfrac{3b + 1}{9b^2 + 6b + 1}$

_______ 27. $\dfrac{3x^2 + 2x}{6x^2 - 17x - 14}$

_______ 28. $\dfrac{5x^2 - 4x}{5x^2 - 14x + 8}$

_______ 29. $\dfrac{a^2 + 12a + 35}{2a^3 + 9a^2 - 5a}$

_______ 30. $\dfrac{a^2 - 6a - 27}{2a^3 + 3a^2 - 9a}$

_______ 31. $\dfrac{x^3 + 27}{4x^2 + 7x - 15}$

_______ 32. $\dfrac{x^3 + 8}{8x^2 + 13x - 6}$

Reduce each fraction. Do not state restrictions.

_______ 33. $\dfrac{3a^2 + 15a + 75}{a^3 - 125}$

_______ 34. $\dfrac{4a^2 + 16a + 64}{a^3 - 64}$

_______ 35. $\dfrac{16a^4 - 1}{20a^2 + 5}$

_______ 36. $\dfrac{81a^4 - 16}{27a^2 + 12}$

_______ 37. $\dfrac{x^3 - x^2 + x - 1}{x^4 - 1}$

_______ 38. $\dfrac{x^3 - 2x^2 + 4x - 8}{x^4 - 16}$

_______ 39. $\dfrac{3x^2 + 14xy - 5y^2}{3x^2 + 2xy - y^2}$

_______ 40. $\dfrac{6x^2 + xy - y^2}{2x^2 - xy - y^2}$

_______ 41. $\dfrac{2a - b}{b^2 - 4ab + 4a^2}$

_______ 42. $\dfrac{3a - b}{b^2 - 6ab + 9a^2}$

_______ 43. $\dfrac{x^2 - 25}{20 - 19x + 3x^2}$

_______ 44. $\dfrac{x^2 - 49}{14 - 23x + 3x^2}$

_______ 45. $\dfrac{3x^2 + 2xy - y^2}{4y^2 - 17xy + 15x^2}$

_______ 46. $\dfrac{2x^2 - 5xy - 7y^2}{14y^2 - 39xy + 10x^2}$

_______ 47. $\dfrac{a^2 - 5a - ab + 5b}{3a^2 - 75}$

_______ 48. $\dfrac{a^2 - 6a - ab + 6b}{2a^2 - 72}$

_______ 49. $\dfrac{4x^3 - 9x + 16x^2 - 36}{2x^2 + 11x + 12}$

_______ 50. $\dfrac{25x^3 - 16x + 50x^2 - 32}{5x^2 + 6x - 8}$

_______ 51. $\dfrac{4(x - 2)^2 + 8x(x - 2)}{(x - 2)^4}$

_______ 52. $\dfrac{3(x - 5)^2 + 6x(x - 5)}{(x - 5)^4}$

_______ 53. $\dfrac{a^4 - 29a^2 + 100}{a^2 - 3a - 10}$

_______ 54. $\dfrac{a^4 - 20a^2 + 64}{a^2 - 2a - 8}$

☆ Stretching the Topics

Reduce each fraction.

_______ 1. $\dfrac{6x^3 - 24x^2 + 24x}{12x^3 - 4x^4 - x^5}$

_______ 2. $\dfrac{x^{4n} - x^{2n} - 12}{x^{3n} - 2x^{2n} + 3x^n - 6}$

_______ 3. State the restrictions on the variable for $\dfrac{3x + 4y}{x^3 - 4x^2y - 5xy^2}$.

Check your answers in the back of your book.

If you can reduce the fractions in **Checkup 6.1,** you are ready to go on to Section 6.2.

Name ______________________ Date ____________

✓ CHECKUP 6.1

State restrictions and reduce each fraction.

_______ 1. $\dfrac{-34x^5y^{10}z^3}{17x^4y^{10}z^5}$

_______ 2. $\dfrac{9y}{18y^2 - 27y}$

_______ 3. $\dfrac{9x + 9y}{18x - 18y}$

_______ 4. $\dfrac{-10a^2b - 15ab}{20ab}$

_______ 5. $\dfrac{x^2 - 3x - 54}{x^2 - 81}$

_______ 6. $\dfrac{3x^2 - 18x}{x^2 - 36}$

Reduce each fraction. Do not state restrictions.

_______ 7. $\dfrac{36x^2 - 23xy - 3y^2}{4x^2 + xy - 3y^2}$

_______ 8. $\dfrac{3a - 2b}{4b^2 - 12ab + 9a^2}$

_______ 9. $\dfrac{a^4 - 625}{a^3 - 5a^2 + 25a - 125}$

_______ 10. $\dfrac{8(x - y)^2 + 24x(x - y)}{4x^2 - 5xy + y^2}$

Check your answers in the back of your book.

If You Missed Problems:	You Should Review Examples:
1	3, 4
2–6	5–8
7	9
8	11, 12
9	9
10	10

6.2 Multiplying and Dividing Rational Expressions

Multiplying Rational Expressions

A look at some products of numerical fractions might give us a hint about how to multiply rational expressions.

$$\frac{1}{3} \cdot \frac{2}{5} = \frac{1 \cdot 2}{3 \cdot 5} = \frac{2}{15}$$

$$\frac{5}{6} \cdot \frac{3}{10} = \frac{5}{2 \cdot 3} \cdot \frac{3}{2 \cdot 5} = \frac{\overset{1}{\cancel{5}}}{2 \cdot \underset{1}{\cancel{3}}} \cdot \frac{\overset{1}{\cancel{3}}}{2 \cdot \underset{1}{\cancel{5}}} = \frac{1 \cdot 1}{2 \cdot 2} = \frac{1}{4}$$

Notice that we reduced in our second example even before we found the product of the numerators and denominators. In general, we agree that

> To find the product of fractions, we multiply the numerators and multiply the denominators, reducing if possible.

Or, using A, B, C, and D to represent algebraic expressions, we say

Product of Rational Expressions

$$\frac{A}{B} \cdot \frac{C}{D} = \frac{A \cdot C}{B \cdot D} \qquad B \neq 0, D \neq 0$$

Whenever we multiply fractions, we should first factor the numerators and denominators completely, so that we can easily spot common factors.

Example 1. Find

$$\frac{3x + 6}{x^2 - 9} \cdot \frac{5x - 15}{x^2 + 5x + 6}$$

Solution

$$\frac{3x + 6}{x^2 - 9} \cdot \frac{5x - 15}{x^2 + 5x + 6}$$

$$= \frac{3\overset{1}{\cancel{(x + 2)}}}{\underset{1}{\cancel{(x - 3)}}(x + 3)} \cdot \frac{5\overset{1}{\cancel{(x - 3)}}}{(x + 3)\underset{1}{\cancel{(x + 2)}}} \qquad x \neq 3, x \neq -3, x \neq -2$$

$$= \frac{3 \cdot 5}{(x + 3)(x + 3)}$$

$$= \frac{15}{(x + 3)^2}$$

Example 2. Find

$$\frac{x^2 + 4x + 3}{x^2 + 2x + 1} \cdot \frac{x^2 - 7x - 8}{x^2 - 5x - 24}$$

Solution

$$\frac{x^2 + 4x + 3}{x^2 + 2x + 1} \cdot \frac{x^2 - 7x - 8}{x^2 - 5x - 24}$$

$$= \frac{\overset{1}{\cancel{(x + 1)}}\overset{1}{\cancel{(x + 3)}}}{\underset{1}{\cancel{(x + 1)}}\underset{1}{\cancel{(x + 1)}}} \cdot \frac{\overset{1}{\cancel{(x - 8)}}\overset{1}{\cancel{(x + 1)}}}{\underset{1}{\cancel{(x + 3)}}\underset{1}{\cancel{(x - 8)}}} \qquad x \neq -1, x \neq -3, x \neq 8$$

$$= 1$$

Notice that we do *not* multiply the final variable factors in the numerator and denominator. They are best left in factored form.

Now you complete Example 3.

Example 3. Find $\dfrac{x^2 - 9x}{6x^4} \cdot \dfrac{x^2}{x^2 - 9} \cdot \dfrac{4x + 12}{9 + 17x - 2x^2}$.

Solution

$$\frac{x^2 - 9x}{6x^4} \cdot \frac{x^2}{x^2 - 9} \cdot \frac{4x + 12}{9 + 17x - 2x^2}$$

$$= \frac{x(x - 9)}{6x^4} \cdot \frac{x^2}{(x + 3)(x - 3)} \cdot \frac{4(x + 3)}{(9 - x)(1 + 2x)}$$

$x \neq$ _____, $x \neq$ _____,
$x \neq$ _____, $x \neq$ _____,
$x \neq$ _____

$= $ ________

Check your work on page 362. ▶

Dividing Rational Expressions

Once again, we recall from arithmetic the technique for dividing numerical fractions.

$$\frac{5}{6} \div \frac{2}{7} = \frac{5}{6} \cdot \frac{7}{2}$$
$$= \frac{35}{12}$$

Before stating the rule for dividing rational expressions, we must recall the names of the parts of a division problem.

$$\frac{5}{6} \div \frac{2}{7} = \frac{35}{12}$$

dividend (5/6) — **divisor** (2/7) — **quotient** (35/12)

Now the rule for division can be stated as follows:

To divide rational expressions, we multiply the dividend by the reciprocal of the divisor.

Letting A, B, C, and D represent algebraic expressions, we say

Division of Rational Expressions

$$\frac{A}{B} \div \frac{C}{D} = \frac{A}{B} \cdot \frac{D}{C} \qquad B \neq 0,\ D \neq 0,\ C \neq 0$$

Notice that we place restrictions on D because it is the denominator of the original divisor. Restrictions are also necessary on C because it is the denominator of the *reciprocal* of the divisor. Rather than get bogged down stating restrictions in these more complicated expressions, let us agree that values of the variable that make the denominators equal zero will be understood to be excluded. Until further notice, *we will not state restrictions unless specifically asked to do so*.

Example 4. Find $\frac{x}{x+1} \div \frac{x^2}{2x+2}$.

Solution

$$\frac{x}{x+1} \div \frac{x^2}{2x+2}$$

$$= \frac{x}{x+1} \cdot \frac{2x+2}{x^2} \quad \text{Find reciprocal of divisor and multiply.}$$

$$= \frac{\overset{1}{\cancel{x}}}{\underset{1}{\cancel{x+1}}} \cdot \frac{2\overset{1}{\cancel{(x+1)}}}{\underset{x}{\cancel{x^2}}} \quad \text{Factor and reduce.}$$

$$= \frac{2}{x} \quad \text{Simplify.}$$

Example 5. Find $\frac{x^2y}{x^3 + 5x^2} \div 5xy^2$.

Solution

$$\frac{x^2y}{x^3 + 5x^2} \div \frac{5xy^2}{1}$$

$$= \frac{\overset{1}{\cancel{x^2}}\overset{1}{\cancel{y}}}{\underset{1}{\cancel{x^2}}(x+5)} \cdot \frac{1}{5x\underset{y}{\cancel{y^2}}}$$

$$= \frac{1}{5xy(x+5)}$$

Now you try Example 6.

Example 6. Find $\frac{x^2 + 6x + 8}{x^2 - 4} \div \frac{5x + 15}{x^2 - 2x}$.

Solution

$$\frac{x^2 + 6x + 8}{x^2 - 4} \div \frac{5x + 15}{x^2 - 2x} = \frac{x^2 + 6x + 8}{x^2 - 4} \cdot \frac{x^2 - 2x}{5x + 15}$$

$$= \underline{\qquad\qquad}$$

Check your work on page 362. ▶

Example 7. Find $\dfrac{2x^2 - 11x - 21}{x^2 + 10x + 25} \div \dfrac{98 - 2x^2}{20x + 4x^2}$.

Solution

$\dfrac{2x^2 - 11x - 21}{x^2 + 10x + 25} \div \dfrac{98 - 2x^2}{20x + 4x^2}$	
$= \dfrac{2x^2 - 11x - 21}{x^2 + 10x + 25} \cdot \dfrac{20x + 4x^2}{98 - 2x^2}$	Find reciprocal of divisor and multiply.
$= \dfrac{(2x + 3)(x - 7)}{(x + 5)(x + 5)} \cdot \dfrac{4x(5 + x)}{2(49 - x^2)}$	Factor.
$= \dfrac{(2x + 3)\overset{-1}{\cancel{(x - 7)}}}{(x + 5)\underset{1}{\cancel{(x + 5)}}} \cdot \dfrac{\overset{2}{\cancel{4}}x\overset{1}{\cancel{(5 + x)}}}{\underset{1}{\cancel{2}}(7 + x)\underset{1}{\cancel{(7 - x)}}}$	Reduce.
$= \dfrac{-2x(2x + 3)}{(x + 5)(x + 7)}$	Write the products.

Trial Run

Perform the indicated operation.

_______ 1. $\dfrac{2x + 2}{x + 3} \cdot \dfrac{1}{x + 1}$

_______ 2. $\dfrac{x^3 - 8}{6x - 12} \cdot \dfrac{36}{3x^3 + 6x^2 + 12x}$

_______ 3. $\dfrac{x^2 + 6x + 5}{x^2 + 10x + 25} \cdot \dfrac{x^2 + 2x - 15}{x^2 - 6x - 7}$

_______ 4. $\dfrac{16 - x^2}{2x^2 - 11x + 12} \cdot \dfrac{6x^2 - 9x}{9x^2 + 37x + 4}$

_______ 5. $\dfrac{x^3 - 49x}{xy} \div (x + 7)$

_______ 6. $\dfrac{x^2 + x - 6}{x^2 - 9} \div \dfrac{5x - 10}{x^2 + 2x}$

_______ 7. $\dfrac{x + 2}{4x - 8} \div \dfrac{x^2 - 7x - 18}{2x - 4}$

Answers are on page 363.

We use the same methods to multiply or divide with three fractions. When necessary, symbols of grouping tell us which operation to perform first.

Example 8. Find $\dfrac{x^2 - 1}{6x + 12} \cdot \dfrac{x^2 + 2x}{x^3 + 2x^2 + x} \div \dfrac{x^2 + 3x - 4}{x^2 + 5x + 4}$.

Solution

$$\frac{x^2 - 1}{6x + 12} \cdot \frac{x^2 + 2x}{x^3 + 2x^2 + x} \div \frac{x^2 + 3x - 4}{x^2 + 5x + 4}$$

$$= \frac{x^2 - 1}{6x + 12} \cdot \frac{x^2 + 2x}{x^3 + 2x^2 + x} \cdot \frac{x^2 + 5x + 4}{x^2 + 3x - 4}$$

Find reciprocal of divisor and multiply.

$$= \frac{(x + 1)(x - 1)}{6(x + 2)} \cdot \frac{x(x + 2)}{x(x^2 + 2x + 1)} \cdot \frac{(x + 4)(x + 1)}{(x + 4)(x - 1)}$$

Factor.

$$= \frac{\overset{1}{\cancel{(x + 1)}}\overset{1}{\cancel{(x - 1)}}}{6\underset{1}{\cancel{(x + 2)}}} \cdot \frac{\overset{1}{\cancel{x}}\overset{1}{\cancel{(x + 2)}}}{\underset{1}{\cancel{x}}\underset{1}{\cancel{(x + 1)}}\underset{1}{\cancel{(x + 1)}}} \cdot \frac{\overset{1}{\cancel{(x + 4)}}\overset{1}{\cancel{(x + 1)}}}{\underset{1}{\cancel{(x + 4)}}\underset{1}{\cancel{(x - 1)}}}$$

Reduce.

$$= \frac{1}{6}$$

Write the product.

Now you complete Example 9.

Example 9. Find $\dfrac{3x^2}{(2x + 1)^2} \div \left[\dfrac{6x^2 - 3x}{5x} \div \dfrac{4x^2 - 1}{20x^3}\right]$.

Solution

$$\frac{3x^2}{(2x + 1)^2} \div \left[\frac{6x^2 - 3x}{5x} \div \frac{4x^2 - 1}{20x^3}\right]$$

$$= \frac{3x^2}{(2x + 1)^2} \div \left[\frac{6x^2 - 3x}{5x} \cdot \frac{20x^3}{4x^2 - 1}\right]$$

$$= \frac{3x^2}{(2x + 1)^2} \div \left[\frac{3x(\underline{\qquad\quad})}{5x} \cdot \frac{20x^3}{(\underline{\qquad\quad})(\underline{\qquad\quad})}\right]$$

Check your work on page 363. ▶

Trial Run

Perform the indicated operations.

______ 1. $\left[\dfrac{xy^2}{4x^2 + 4x} \div \dfrac{x^2y}{3x^2 - 3}\right] \cdot \dfrac{4x}{3y}$

______ 2. $\dfrac{x^2 - 25}{10x} \cdot \dfrac{50}{3x + 15} \div \dfrac{2x - 10}{3x}$

______ 3. $\left[\dfrac{x^3 - 8}{5x - 15} \div \dfrac{x^2 + x - 6}{15x^2 + 40x - 15}\right] \div \dfrac{2x^3 + 4x^2 + 8x}{4x^2 - 13x + 3}$

Answers are on page 363.

Examples You Completed

Example 3. Find $\dfrac{x^2 - 9x}{6x^4} \cdot \dfrac{x^2}{x^2 - 9} \cdot \dfrac{4x + 12}{9 + 17x - 2x^2}$.

Solution

$$\frac{x^2 - 9x}{6x^4} \cdot \frac{x^2}{x^2 - 9} \cdot \frac{4x + 12}{9 + 17x - 2x^2}$$

$$= \frac{\cancel{x}(\cancel{x - 9})}{\cancel{6x^4}_{\,3x}} \cdot \frac{\cancel{x^2}}{(\cancel{x + 3})(x - 3)} \cdot \frac{\cancel{4}(\cancel{x + 3})}{(\cancel{9 - x})(1 + 2x)} \qquad x \neq 0,\ x \neq -3,\ x \neq 3,\ x \neq 9,\ x \neq -\tfrac{1}{2}$$

$$= \frac{-2}{3x(x - 3)(1 + 2x)}$$

Example 6. Find $\dfrac{x^2 + 6x + 8}{x^2 - 4} \div \dfrac{5x + 15}{x^2 - 2x}$.

Solution

$$\frac{x^2 + 6x + 8}{x^2 - 4} \div \frac{5x + 15}{x^2 - 2x} = \frac{x^2 + 6x + 8}{x^2 - 4} \cdot \frac{x^2 - 2x}{5x + 15}$$

$$= \frac{(x + 4)\cancel{(x + 2)}}{\cancel{(x + 2)}\cancel{(x - 2)}} \cdot \frac{x\cancel{(x - 2)}}{5(x + 3)}$$

$$= \frac{x(x + 4)}{5(x + 3)}$$

Example 9. Find $\dfrac{3x^2}{(2x+1)^2} \div \left[\dfrac{6x^2-3x}{5x} \div \dfrac{4x^2-1}{20x^3}\right]$.

Solution

$$\frac{3x^2}{(2x+1)^2} \div \left[\frac{6x^2-3x}{5x} \div \frac{4x^2-1}{20x^3}\right] = \frac{3x^2}{(2x+1)^2} \div \left[\frac{6x^2-3x}{5x} \cdot \frac{20x^3}{4x^2-1}\right]$$

$$= \frac{3x^2}{(2x+1)^2} \div \left[\frac{3\cancel{x}\cancel{(2x-1)}}{\cancel{5}\cancel{x}} \cdot \frac{\cancel{20}x^3}{(2x+1)\cancel{(2x-1)}}\right]$$

$$= \frac{3x^2}{(2x+1)^2} \div \frac{12x^3}{2x+1}$$

$$= \frac{\cancel{3x^2}}{(2x+1)\cancel{(2x+1)}} \cdot \frac{\cancel{2x+1}}{\cancel{12x^3}}$$

$$= \frac{1}{4x(2x+1)}$$

Answers to Trial Runs

page 360 **1.** $\dfrac{2}{x+3}$ **2.** $\dfrac{2}{x}$ **3.** $\dfrac{x-3}{x-7}$ **4.** $\dfrac{-3x}{9x+1}$ **5.** $\dfrac{x-7}{y}$ **6.** $\dfrac{x(x+2)}{5(x-3)}$ **7.** $\dfrac{1}{2(x-9)}$

page 362 **1.** $\dfrac{x-1}{x}$ **2.** $\dfrac{5}{2}$ **3.** $\dfrac{(3x-1)(4x-1)}{2x}$

Name ______________________ Date ____________

EXERCISE SET 6.2

Perform the indicated operations.

_____ 1. $\frac{2x + 18}{3x^2} \cdot \frac{9x}{x + 9}$

_____ 2. $\frac{5x + 20}{4x^2} \cdot \frac{12x}{x + 4}$

_____ 3. $\frac{a^2 + b^2}{6} \cdot \frac{42}{7a + 7b}$

_____ 4. $\frac{a^2 + b^2}{19} \cdot \frac{38}{2a + 2b}$

_____ 5. $\frac{x^2 - 25}{3x - 21} \cdot \frac{2x - 14}{x^2 + 10x + 25}$

_____ 6. $\frac{x^2 - 36}{2x - 10} \cdot \frac{3x - 15}{x^2 - 12x + 36}$

_____ 7. $\frac{x^2 - 81y^2}{x^2 + 9y^2} \cdot \frac{3x^2 + 27y^2}{x - 9y}$

_____ 8. $\frac{x^2 - 64y^2}{x^2 + 8y^2} \cdot \frac{5x^2 + 40y^2}{x - 8y}$

_____ 9. $\frac{(2x + 5)^2}{64yz^3} \cdot \frac{40y^2z}{4x^2 - 25}$

_____ 10. $\frac{(3x - 2)^2}{81yz^3} \cdot \frac{72y^5z}{9x^2 - 4}$

_____ 11. $\frac{2x^2 + 3x + 1}{2x^2 + 5x + 3} \cdot \frac{2x^2 + 15x + 18}{2x^2 + 13x + 6}$

_____ 12. $\frac{20x^2 - 7x - 3}{4x^2 - 7x - 2} \cdot \frac{3x^2 - 5x - 2}{15x^2 - 4x - 3}$

_____ 13. $\frac{6x^2 + xy - y^2}{2x^2 - 9xy - 5y^2} \cdot \frac{x^2 - 25y^2}{15x - 5y}$

_____ 14. $\frac{8x^2 + 18xy - 5y^2}{16x^2 - 8xy + y^2} \cdot \frac{4x^2 + 3xy - y^2}{12x + 30y}$

_____ 15. $\frac{x^3 - 27}{6x - 18} \cdot \frac{36}{2x^2 + 6x + 18}$

_____ 16. $\frac{x^3 - 64}{7x - 28} \cdot \frac{42}{3x^2 + 12x + 48}$

_____ 17. $\frac{25 - x^2}{3x^2 + 7x - 40} \cdot \frac{3x^2 + 12x}{2x^3 - 2x^2 - 40x}$

_____ 18. $\frac{49 - x^2}{5x^2 + 32x - 21} \cdot \frac{5x^2 - 30x}{2x^3 - 26x^2 + 84x}$

_____ 19. $\frac{6 - 23x - 4x^2}{3x^2 + 10x - 48} \cdot \frac{3x^2 + x - 24}{8x^2 - 6x + 1}$

_____ 20. $\frac{9 - 3x - 2x^2}{2x^2 + 11x - 21} \cdot \frac{2x^2 + 9x - 35}{2x^2 + 11x + 15}$

_____ 21. $\frac{3x^3 - 75x}{9x^4} \cdot \frac{3x - 24}{x^2 - 3x - 40} \cdot \frac{x^2}{5 - 6x + x^2}$

_____ 22. $\frac{2x^3 - 98x}{10x^4} \cdot \frac{x^2}{x^2 - 2x - 63} \cdot \frac{5x - 45}{14 - 9x + x^2}$

_____ 23. $\frac{x^4 - 13x^2 + 36}{x^4 - 14x^2 - 32} \cdot \frac{x^3 - 4x^2 + 2x - 8}{x^3 + 3x^2 - 4x - 12}$

_____ 24. $\frac{x^4 - 26x^2 + 25}{x^4 - 6x^2 - 27} \cdot \frac{x^3 - 3x^2 + 3x - 9}{x^3 + 5x^2 - x - 5}$

_____ 25. $\frac{x^5}{x + 5} \div \frac{x^2}{3x + 15}$

_____ 26. $\frac{x^6}{x - 6} \div \frac{x^3}{2x - 12}$

_____ 27. $\frac{x^2y^5}{x + 9} \div 9x^2y^3$

_____ 28. $\frac{x^3y^4}{x + 5} \div 5x^3y$

_____ 29. $\frac{x^2 + 7x + 10}{63x} \div \frac{7x + 14}{9x}$

_____ 30. $\frac{x^2 - 12x + 32}{60x} \div \frac{5x - 20}{12x}$

_____ 31. $\frac{2x^2 - 19x + 24}{x^2 - 64} \div \frac{3x - 24}{x^2 + 8x}$

_____ 32. $\frac{3x^2 - 22x + 7}{x^2 - 49} \div \frac{5x - 35}{x^2 + 7x}$

_____ 33. $\frac{x + 3}{5x - 20} \div \frac{x^2 - 9x - 36}{3x - 12}$

_____ 34. $\frac{x + 7}{8x - 56} \div \frac{x^2 + 4x - 21}{4x - 28}$

_____ 35. $\frac{x^2 - 8x + 15}{x^2 - 25} \div \frac{x^2 - 9}{x + 5}$

_____ 36. $\frac{x^2 - 7x - 18}{x^2 - 81} \div \frac{x^2 + 4x + 4}{x + 9}$

_____ 37. $\frac{6x^2 - 11x + 5}{x^3 - 1} \div \frac{6x^2 - 23x + 15}{3x^3 + 3x^2 + 3x}$

_____ 38. $\frac{3x^2 - 19x + 28}{x^3 - 64} \div \frac{3x^2 - 22x + 35}{5x^3 + 20x^2 + 80x}$

_____ 39. $\frac{xy - y^2}{x^2 - 2xy + y^2} \div \frac{5xy - 3y^2}{5x^2 - 8xy + 3y^2}$

_____ 40. $\frac{x^2 - 7xy}{x^2 - 14xy + 49y^2} \div \frac{3x^2 + 2xy}{3x^2 - 19xy - 14y^2}$

_____ 41. $\frac{25 - x^2}{x^2 - 11x + 30} \div \frac{2x^2 + 7x - 15}{2x^2 - 15x + 18}$

_____ 42. $\frac{9 - x^2}{x^2 - 11x + 24} \div \frac{4x^2 + 13x + 3}{4x^2 - 31x - 8}$

_____ 43. $\frac{4x^2 + 19x - 5}{6x^2 - 11x - 2} \div \frac{6 - 19x - 20x^2}{5x^2 - 4x - 12}$

_____ 44. $\frac{3x^2 + 26x - 9}{3x^2 - 14x + 8} \div \frac{1 + 5x - 24x^2}{8x^2 - 31x - 4}$

_____ 45. $\frac{x^2 - 25}{10x} \cdot \frac{50}{3x + 15} \div \frac{2x - 10}{3x}$

_____ 46. $\frac{x^2 - 9}{3x} \cdot \frac{45}{4x - 12} \div \frac{5x + 15}{7x}$

_____ 47. $\frac{4x^2 - 36}{x^2 + 2x - 15} \cdot \left[\frac{2x^2 + 11x + 5}{16x^2 + 16x - 96} \div \frac{6x^2 + 7x + 2}{3x^2 - 4x - 4}\right]$

_____ 48. $\frac{9x^2 - 1}{3x^2 - 26x - 9} \cdot \left[\frac{x^2 - 13x + 36}{2x^2 - 21x + 55} \div \frac{3x^2 - 13x + 4}{x^2 - 10x + 25}\right]$

_____ 49. $\frac{8x^3 - 27}{(x + 2)^2} \cdot \frac{4x^2 + 8x}{2x^2 - 7x + 6} \div \frac{8x^4 + 12x^3 + 18x^2}{x^2 - 4}$

_____ 50. $\frac{x^3 + 125}{(x + 3)^2} \cdot \frac{8x^2 + 24x}{x^2 + 2x - 15} \div \frac{6x^4 - 30x^3 + 150x^2}{x^2 - 9}$

☆ Stretching the Topics

Perform the indicated operation.

_____ 1. $\frac{10x^3 - 4x^2 + 15x - 6}{10x^2 - 24x + 8} \cdot \frac{2x^3 - 4x^2 - 3x + 6}{4x^4 - 9}$

_____ 2. $\frac{2x^{2a} - 2x^a - 12}{6x^{2a} + 18x^a} \div \frac{x^{3a} - 3x^{2a} - 4x^a + 12}{3x^{2a} - 5x^a - 2}$

_____ 3. $\frac{2x^{3a} + 6x^{2a}}{x^{2b+c}} \cdot \frac{x^{a+2b} - 3x^{2b}}{x^{2a+c} - 9x^c}$

Check your answers in the back of your book.

If you can complete **Checkup 6.2,** you are ready to go on to Section 6.3.

Name ______________________ Date ____________

✓ CHECKUP 6.2

Perform the indicated operation. Leave the answers in factored form.

______ 1. $\frac{4x + 32}{5x^2} \cdot \frac{15x}{x + 8}$

______ 2. $\frac{x^2 - 49}{8x - 24} \cdot \frac{x^2 - 3x}{x^2 + 3x - 28}$

______ 3. $\frac{(2x - 7)^2}{36y^4z^2} \cdot \frac{30y^3z^2}{4x^2 - 49}$

______ 4. $\frac{7x^2 + 5xy - 2y^2}{x^2 - 8xy - 9y^2} \cdot \frac{x^2 - 81y^2}{14xy - 4y^2}$

______ 5. $\frac{x^2 - 225}{x^7y^3} \div \frac{3x + 45}{x^2y^6}$

______ 6. $\frac{x^7y^5}{x^3 + 4x^2} \div 4x^5y^2$

______ 7. $\frac{5x^2 - 49x + 36}{x^2 - 81} \div \frac{25x^2 - 16}{x + 9}$

______ 8. $\frac{5x^2 - 47x - 30}{x^3 + 8} \div \frac{5x^2 - 32x - 21}{7x^3 - 14x^2 + 28x}$

______ 9. $\frac{2x^2 - 6xy}{x^2 - xy - 6y^2} \div \frac{5x^2y - 10x^3}{2x^2 + 3xy - 2y^2}$

______ 10. $\frac{x^2 - 169}{8x} \cdot \frac{16}{3x - 39} \div \frac{6x + 78}{9x}$

Check your answers in the back of your book.

If You Missed Problems:	You Should Review Examples:
1	1
2, 3	2, 3
4	2
5	4
6	5
7	6
8, 9	7
10	8

6.3 Adding and Subtracting Rational Expressions

Combining Fractions with Like Denominators

From earlier work with addition and subtraction of algebraic expressions, you remember that only *like terms* could be combined into a single term. The same conditions hold for addition and subtraction of rational expressions (fractions). Only *like fractions* can be combined into a single fraction. But what do we mean by like fractions?

> Fractions are called *like fractions* if they have the *same denominator*.

Now we recall the rule for adding or subtracting like fractions.

> To add (or subtract) fractions having the same denominator, we add (or subtract) the numerators and keep the same denominator.
>
> $$\frac{A}{C} + \frac{B}{C} = \frac{A + B}{C} \qquad C \neq 0$$

You try Examples 1 and 2. Remember to try reducing *after* you have found the sum or difference.

Example 1. Find $\frac{7}{2x} - \frac{1}{2x}$.

Solution

$$\frac{7}{2x} - \frac{1}{2x} = \frac{7 - 1}{2x}$$

$$=$$

Check your work on page 375. ▶

Example 2. Find $\frac{x}{x + 3} + \frac{2x}{x + 3}$.

Solution

$$\frac{x}{x + 3} + \frac{2x}{x + 3} = \frac{x + 2x}{x + 3}$$

$$=$$

Check your work on page 375. ▶

Example 3. Find $\frac{8x}{2x + 1} + \frac{4}{2x + 1}$.

Solution

$$\frac{8x}{2x + 1} + \frac{4}{2x + 1}$$

$= \frac{8x + 4}{2x + 1}$ Write the sum.

$= \frac{4\cancel{(2x + 1)}^{1}}{\cancel{2x + 1}_{1}}$ Factor and reduce.

$= 4$

Now complete Example 4.

Example 4. Find $\frac{2x + 3}{x - 1} + \frac{4 - x}{x - 1}$.

Solution

$$\frac{2x + 3}{x - 1} + \frac{4 - x}{x - 1}$$

$$= \frac{2x + 3 + 4 - x}{x - 1}$$

Check your work on page 375. ▶

You will avoid unnecessary errors in signs when *subtracting* fractions if you will remember to use *parentheses* around the numerator being subtracted.

Try Example 5.

Example 5. Find $\frac{4x - 7}{x^2 - 4} - \frac{3x - 9}{x^2 - 4}$.

Solution

$$\frac{4x - 7}{x^2 - 4} - \frac{3x - 9}{x^2 - 4}$$

$$= \frac{4x - 7 - (3x - 9)}{x^2 - 4}$$

$$= \frac{4x - 7 - 3x + 9}{x^2 - 4}$$

Check your work on page 375. ▶

Example 6. Find $\frac{5(x - 3)}{2x + 11} - \frac{2(3x + 1)}{2x + 11}$.

Solution

$$\frac{5(x - 3)}{2x + 11} - \frac{2(3x + 1)}{2x + 11}$$ Denominators are the same.

$$= \frac{5(x - 3) - 2(3x + 1)}{2x + 11}$$ Write the difference.

$$= \frac{5x - 15 - 6x - 2}{2x + 11}$$ Remove parentheses.

$$= \frac{-x - 17}{2x + 11}$$ Combine like terms.

Example 7. Find $\frac{(x + 1)(x - 6)}{2x + 1} - \frac{x(4 - x)}{2x + 1} + \frac{1}{2x + 1}$.

Solution

$$\frac{(x + 1)(x - 6)}{2x + 1} - \frac{x(4 - x)}{2x + 1} + \frac{1}{2x + 1} = \frac{(x + 1)(x - 6) - x(4 - x) + 1}{2x + 1}$$

$$= \frac{x^2 - 5x - 6 - 4x + x^2 + 1}{2x + 1}$$

$$= \frac{2x^2 - 9x - 5}{2x + 1}$$

$$= \frac{\overset{1}{\cancel{(2x + 1)}}(x - 5)}{\underset{1}{\cancel{2x + 1}}}$$

$$= x - 5$$

Trial Run

Perform the indicated operation and reduce if possible.

_______ 1. $\frac{5}{3x} + \frac{7}{3x}$

_______ 2. $\frac{2x}{x + 7} + \frac{5x}{x + 7}$

_______ 3. $\frac{6x}{3x - 2} - \frac{4}{3x - 2}$

_______ 4. $\frac{x}{x^2 - 16} + \frac{4}{x^2 - 16}$

______ 5. $\dfrac{6x + 3}{x^2 - 25} - \dfrac{5x + 8}{x^2 - 25}$

______ 6. $\dfrac{(2x - 3)(x - 1)}{x - 2} + \dfrac{x(x + 3)}{x - 2} - \dfrac{11}{x - 2}$

Answers are on page 376.

Building Fractions

We have learned to reduce fractions by *dividing* the numerator and denominator by any common nonzero factor. It is also possible to build a new fraction from an old fraction by *multiplying* the numerator and denominator by any nonzero factor.

For example,

$$\frac{1}{2} = \frac{1 \cdot 5}{2 \cdot 5} = \frac{5}{10}$$

Here we multiplied the numerator and denominator by 5 to change $\frac{1}{2}$ to the equivalent fraction $\frac{5}{10}$. Notice that in performing that multiplication we were actually multiplying the entire fraction by $\frac{5}{5}$ or 1.

The process of **building fractions** is really the opposite of reducing fractions and uses the same basic property. Letting *A*, *B*, and *C* represent algebraic expressions, we note that

Building Fractions

$$\frac{A}{B} = \frac{A \cdot C}{B \cdot C} \qquad B \neq 0, C \neq 0$$

In the next section you will need some skill at building fractions to new fractions having specified denominators. Let's practice a bit here.

Suppose that we wish to change $\frac{3}{4}$ to a fraction with a denominator of 24.

$\frac{3}{4} = \frac{?}{24}$	We must find a numerator for the fraction on the right.
$\frac{3}{4} = \frac{?}{4(6)}$	The denominator on the right is 6 times the denominator on the left.
$\frac{3}{4} = \frac{3(6)}{4(6)}$	The numerator on the right must *also* be 6 times the numerator on the left.
$\frac{3}{4} = \frac{18}{24}$	We have found our new fraction.

Let's state this procedure in words.

To change a fraction to an equivalent fraction with a new denominator, we must decide by what factor the original denominator was multiplied to arrive at the new denominator. Then we must multiply the original numerator by that same factor to arrive at the new numerator.

$$\frac{4}{3x} = \frac{?}{3x(x + 1)}$$

By what factor was the original denominator multiplied to arrive at the new denominator? The factor must be $(x + 1)$.

$$\frac{4}{3x} = \frac{4(x + 1)}{3x(x + 1)}$$

We must multiply the original numerator by $(x + 1)$ also.

Example 8. Find the new fraction.

$$\frac{5}{x^2} = \frac{?}{2x^5y}$$

Solution

$$\frac{5}{x^2} = \frac{?}{2x^5y}$$

$$\frac{5}{x^2} = \frac{?}{x^2(2x^3y)}$$

$$\frac{5}{x^2} = \frac{5(2x^3y)}{x^2(2x^3y)}$$

$$\frac{5}{x^2} = \frac{10x^3y}{2x^5y}$$

Example 9. Find the new fraction.

$$\frac{x}{x + 3} = \frac{?}{x(x + 1)(x + 3)}$$

Solution

$$\frac{x}{x + 3} = \frac{?}{x(x + 1)(x + 3)}$$

$$= \frac{x \cdot x(x + 1)}{x(x + 1)(x + 3)}$$

$$= \frac{x^2(x + 1)}{x(x + 1)(x + 3)}$$

Example 10. Find the new fraction.

$$\frac{x + 1}{3x} = \frac{?}{6x^2(2x + 5)}$$

Solution

$$\frac{x + 1}{3x} = \frac{?}{6x^2(2x + 5)}$$

$$= \frac{?}{3x \cdot 2x(2x + 5)}$$

$$\frac{x + 1}{3x} = \frac{(x + 1)(2x)(2x + 5)}{3x \cdot 2x(2x + 5)}$$

$$\frac{x + 1}{3x} = \frac{2x(x + 1)(2x + 5)}{6x^2(2x + 5)}$$

You complete Example 11.

Example 11. Find the new fraction.

$$\frac{4}{x + 1} = \frac{?}{3(x + 1)^2}$$

Solution

$$\frac{4}{x + 1} = \frac{?}{3(x + 1)^2}$$

Check your work on page 376. ▶

Trial Run

Find the new fraction.

_______ 1. $\frac{-2}{7} = \frac{?}{91}$

_______ 2. $\frac{5}{xy} = \frac{?}{3x^2y^3}$

_______ 3. $\frac{x}{2x + 1} = \frac{?}{(2x + 1)(x - 3)}$

_______ 4. $\frac{7x}{x - 3} = \frac{?}{2x^2(x + 3)(x - 3)}$

_______ 5. $\dfrac{x + 2}{x + 3} = \dfrac{?}{5(x + 3)^2}$

_______ 6. $\dfrac{x + 3}{2x + 3} = \dfrac{?}{2x^2 - 7x - 15}$

Answers are on page 376.

Combining Fractions with Different Denominators

We have agreed that we can combine fractions by addition and subtraction only when their denominators are exactly the same. To find sums of fractions that do not share the same denominator, we must somehow *make* the denominators match by finding a **common denominator.** A common denominator is a quantity into which each of the original denominators will divide exactly (with zero remainder).

> The **lowest common denominator (LCD)** for a group of fractions is the *smallest* quantity into which each of the original denominators will divide exactly (with 0 remainder).

To work the addition problem $\frac{1}{2} + \frac{4}{3}$, we must convert each of the original fractions to a new fraction having the LCD as its denominator. To do this, we must use the method of building fractions that we have just learned. For $\frac{1}{2}$ and $\frac{4}{3}$, the LCD is $2 \cdot 3$.

$$\begin{aligned}\frac{1}{2} + \frac{4}{3} &= \frac{1(\ \ )}{2 \cdot 3} + \frac{4(\ \ )}{2 \cdot 3} \\ &= \frac{1 \cdot 3}{2 \cdot 3} + \frac{4 \cdot 2}{2 \cdot 3} \\ &= \frac{3}{6} + \frac{8}{6} \\ &= \frac{3 + 8}{6} \\ &= \frac{11}{6}\end{aligned}$$

It should be clear that in finding the LCD for a group of fractions, we must consider the *prime factors* of the original denominators. **The LCD must contain any factor that appears in any of the original denominators.** If there is overlap among factors, we must use any factor the *most* number of times that it appears in any *single* denominator.

Consider the sum

$$\frac{1}{25x} + \frac{4}{5x^2} = \frac{1}{5 \cdot 5 \cdot x} + \frac{4}{5 \cdot x \cdot x}$$

The only denominator factors are 5 and x. But 5 appears *twice* in the first denominator and x appears *twice* in the second denominator. Each of those factors must therefore appear *twice* in the LCD.

LCD: $5 \cdot 5 \cdot x \cdot x$

$$\frac{1}{25x} + \frac{4}{5x^2} = \frac{1}{5 \cdot 5 \cdot x} + \frac{4}{5 \cdot x \cdot x}$$

$$= \frac{1(x)}{5 \cdot 5 \cdot x \cdot x} + \frac{4(5)}{5 \cdot 5 \cdot x \cdot x}$$ Multiply each fraction by its building factor.

$$= \frac{x}{25x^2} + \frac{20}{25x^2}$$ Simplify the numerators and denominators.

$$= \frac{x + 20}{25x^2}$$ Combine the numerators.

The lowest common denominator for a group of fractions is the product of all the different prime factors appearing in the original denominators, with any single factor appearing the *most* number of times that it appears in any *single* denominator.

The first task in adding and subtracting fractions is finding the lowest common denominator. Obviously, factoring plays an important role, and the first step is to factor the original denominators completely.

You try Example 12.

Example 12. Find $\frac{6}{xy} + \frac{5}{2x^2}$.

Solution

$$\frac{6}{xy} + \frac{5}{2x^2}$$ LCD: $2x^2y$

$$= \frac{6(\quad)}{xy \cdot 2x} + \frac{5(\quad)}{2x^2 \cdot y}$$ Multiply each fraction by its building factor.

$$= \frac{\quad}{2x^2y} + \frac{\quad}{2x^2y}$$ Simplify each numerator.

$$= \frac{\quad + \quad}{2x^2y}$$ Combine numerators.

Check your work on page 376. ▶

Then try Example 13.

Example 13. Find $\frac{5}{x + 3} + \frac{x}{3}$.

Solution. The denominators are in factored form.

$$\frac{5}{x + 3} + \frac{x}{3}$$ LCD: $3(x + 3)$

$$= \frac{5(\quad)}{3(x + 3)} + \frac{x(\quad)}{3(x + 3)}$$

Check your work on page 376. ▶

Example 14. Find $\dfrac{2}{x(x+1)^2} + \dfrac{7}{x^3(x+1)}$.

Solution. Note how many times each factor appears in a single denominator. The LCD is $x^3(x+1)^2$.

$$\frac{2}{x(x+1)^2} + \frac{7}{x^3(x+1)}$$

$$= \frac{2x^2}{x^3(x+1)^2} + \frac{7(x+1)}{x^3(x+1)^2} \qquad \text{Multiply each fraction by its building factor.}$$

$$= \frac{2x^2}{x^3(x+1)^2} + \frac{7x+7}{x^3(x+1)^2} \qquad \text{Simplify each numerator.}$$

$$= \frac{2x^2+7x+7}{x^3(x+1)^2} \qquad \text{Combine numerators.}$$

Example 15. Find $x + \dfrac{2}{x}$.

Solution

$$x + \frac{2}{x} = \frac{x}{1} + \frac{2}{x} \qquad \text{LCD: } x$$

$$= \frac{x \cdot x}{x} + \frac{2}{x}$$

$$= \frac{x^2}{x} + \frac{2}{x}$$

$$= \frac{x^2+2}{x}$$

Before doing a few more examples, let's summarize the steps to be used in adding or subtracting fractions with different denominators.

Steps for Adding and Subtracting Fractions

1. Write all the denominators in factored form.
2. Choose the LCD as the product of all the different prime factors of the original denominators, with each factor appearing the most number of times that it appears in any single denominator.
3. Change each of the original fractions to a new fraction having the LCD as its denominator.
4. Simplify each numerator by multiplying its factors.
5. Combine the numerators, keeping the common denominator (in factored form).
6. Reduce if possible.

Example 16. Find $\dfrac{x+1}{x^2+5x+6} - \dfrac{2x}{x^2-9}$.

Solution. First we must factor the denominators.

$$\frac{x+1}{x^2+5x+6} - \frac{2x}{x^2-9}$$

$$= \frac{x+1}{(x+3)(x+2)} - \frac{2x}{(x+3)(x-3)} \qquad \text{LCD: } (x+3)(x+2)(x-3)$$

$$= \frac{(x+1)(x-3)}{(x+3)(x+2)(x-3)} - \frac{2x(x+2)}{(x+3)(x+2)(x-3)} \qquad \text{Multiply each fraction by its building factor.}$$

$$= \frac{x^2-2x-3}{(x+3)(x+2)(x-3)} - \frac{2x^2+4x}{(x+3)(x+2)(x-3)} \qquad \text{Simplify each numerator.}$$

$$= \frac{x^2-2x-3-(2x^2+4x)}{(x+3)(x+2)(x-3)} \qquad \text{Combine numerators.}$$

$$= \frac{x^2 - 2x - 3 - 2x^2 - 4x}{(x + 3)(x + 2)(x - 3)}$$ Remove parentheses.

$$= \frac{-x^2 - 6x - 3}{(x + 3)(x + 2)(x - 3)}$$ Simplify by combining like terms.

Notice that we *never* multiply the factors in the LCD; it is always left in factored form so that we can decide whether the final fraction can be reduced.

You try Example 17.

Example 17. Find $\frac{3}{x^2 - 2x - 3} - \frac{4}{x^2 + x}$.

Solution. First we factor the denominators.

$$\frac{3}{x^2 - 2x - 3} - \frac{4}{x^2 + x}$$

$$= \frac{3}{(x - 3)(x + 1)} - \frac{4}{x(x + 1)}$$ LCD: $x(x - 3)(x + 1)$

$$= \frac{3(\quad)}{x(x - 3)(x + 1)} - \frac{4(\quad)}{x(x - 3)(x + 1)}$$ Multiply each fraction by its building factor.

$$= \frac{3x}{x(x - 3)(x + 1)} - \frac{4x - 12}{x(x - 3)(x + 1)}$$ Simplify each numerator.

$$= \frac{3x - (4x - 12)}{x(x - 3)(x + 1)}$$ Combine numerators.

$$= \frac{-x}{x(x - 3)(x + 1)}$$ Remove parentheses.

$$= \frac{x + 12}{x(x - 3)(x + 1)}$$ Simplify by combining like terms.

Check your work on page 376. ▶

Example 18. Find $\frac{3}{2x - 2} + \frac{5}{1 - x}$.

Solution. Factor the denominators first.

$$\frac{3}{2x - 2} + \frac{5}{1 - x} = \frac{3}{2(x - 1)} + \frac{5}{1 - x}$$

Recalling the relationship between $x - 1$ and $1 - x$, we may rewrite the second denominator

$$\frac{3}{2(x - 1)} + \frac{5}{-1(x - 1)}$$ Rewrite $1 - x$ as $-1(x - 1)$.

$$= \frac{3}{2(x - 1)} + \frac{-5}{x - 1}$$ Let the numerator carry the sign. LCD: $2(x - 1)$.

$$= \frac{3}{2(x - 1)} + \frac{-5(2)}{2(x - 1)}$$ Multiply the second fraction by its building factor.

$$= \frac{3}{2(x - 1)} + \frac{-10}{2(x - 1)}$$ Simplify each numerator.

$$= \frac{3 - 10}{2(x - 1)}$$ Combine the numerators.

$$= \frac{-7}{2(x - 1)}$$ Simplify by combining like terms.

Trial Run

Combine the fractions and reduce if possible.

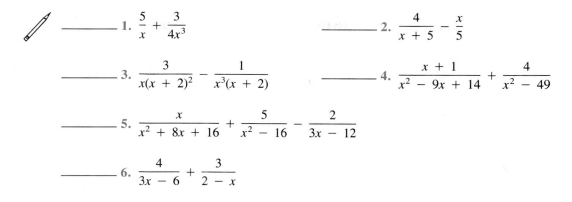

_______ 1. $\frac{5}{x} + \frac{3}{4x^3}$

_______ 2. $\frac{4}{x+5} - \frac{x}{5}$

_______ 3. $\frac{3}{x(x+2)^2} - \frac{1}{x^3(x+2)}$

_______ 4. $\frac{x+1}{x^2-9x+14} + \frac{4}{x^2-49}$

_______ 5. $\frac{x}{x^2+8x+16} + \frac{5}{x^2-16} - \frac{2}{3x-12}$

_______ 6. $\frac{4}{3x-6} + \frac{3}{2-x}$

Answers are on page 376.

Examples You Completed

Example 1. Find $\frac{7}{2x} - \frac{1}{2x}$.

Solution

$$\frac{7}{2x} - \frac{1}{2x} = \frac{7-1}{2x}$$
$$= \frac{6}{2x}$$
$$= \frac{2 \cdot 3}{2x}$$
$$= \frac{3}{x}$$

Example 2. Find $\frac{x}{x+3} + \frac{2x}{x+3}$.

Solution

$$\frac{x}{x+3} + \frac{2x}{x+3}$$
$$= \frac{x+2x}{x+3}$$
$$= \frac{3x}{x+3}$$

Example 4. Find $\frac{2x+3}{x-1} + \frac{4-x}{x-1}$.

Solution

$$\frac{2x+3}{x-1} + \frac{4-x}{x-1}$$
$$= \frac{2x+3+4-x}{x-1}$$
$$= \frac{x+7}{x-1}$$

Example 5. Find $\frac{4x-7}{x^2-4} - \frac{3x-9}{x^2-4}$.

Solution

$$\frac{4x-7}{x^2-4} - \frac{3x-9}{x^2-4} = \frac{4x-7-(3x-9)}{x^2-4}$$
$$= \frac{4x-7-3x+9}{x^2-4}$$
$$= \frac{x+2}{x^2-4}$$
$$= \frac{\overset{1}{\cancel{x+2}}}{\underset{1}{\cancel{(x+2)}}(x-2)}$$
$$= \frac{1}{x-2}$$

Example 11. Find the new fraction.

$$\frac{4}{x+1} = \frac{?}{3(x+1)^2}$$

Solution

$$\frac{4}{x+1} = \frac{?}{3(x+1)^2}$$
$$= \frac{?}{3(x+1)(x+1)}$$
$$= \frac{4 \cdot 3(x+1)}{3(x+1)(x+1)}$$
$$\frac{4}{x+1} = \frac{12(x+1)}{3(x+1)(x+1)}$$

Example 12. Find $\frac{6}{xy} + \frac{5}{2x^2}$.

Solution. The LCD is $2x^2y$.

$$\frac{6}{xy} + \frac{5}{2x^2} = \frac{6(2x)}{xy \cdot 2x} + \frac{5(y)}{2x^2y}$$
$$= \frac{12x}{2x^2y} + \frac{5y}{2x^2y}$$
$$= \frac{12x + 5y}{2x^2y}$$

Example 13. Find $\frac{5}{x+3} + \frac{x}{3}$.

Solution. The LCD is $3(x + 3)$.

$$\frac{5}{x+3} + \frac{x}{3}$$
$$= \frac{5(3)}{3(x+3)} + \frac{x(x+3)}{3(x+3)}$$
$$= \frac{15}{3(x+3)} + \frac{x^2+3x}{3(x+3)}$$
$$= \frac{15 + x^2 + 3x}{3(x+3)}$$
$$= \frac{x^2 + 3x + 15}{3(x+3)}$$

Example 17. Find $\frac{3}{x^2 - 2x - 3} - \frac{4}{x^2 + x}$.

Solution

$$\frac{3}{x^2 - 2x - 3} - \frac{4}{x^2 + x}$$
$$= \frac{3}{(x-3)(x+1)} - \frac{4}{x(x+1)}$$
$$= \frac{3(x)}{x(x-3)(x+1)} - \frac{4(x-3)}{x(x-3)(x+1)}$$
$$= \frac{3x}{x(x-3)(x+1)} - \frac{4x - 12}{x(x-3)(x+1)}$$
$$= \frac{3x - (4x - 12)}{x(x-3)(x+1)}$$
$$= \frac{3x - 4x + 12}{x(x-3)(x+1)}$$
$$= \frac{-x + 12}{x(x-3)(x+1)}$$

Answers to Trial Runs

page 368 1. $\frac{4}{x}$ 2. $\frac{7x}{x+7}$ 3. 2 4. $\frac{1}{x-4}$ 5. $\frac{1}{x+5}$ 6. $3x + 4$

page 370 1. $\frac{-26}{91}$ 2. $\frac{15xy^2}{3x^2y^3}$ 3. $\frac{x(x-3)}{(2x+1)(x-3)}$ 4. $\frac{14x^3(x+3)}{2x^2(x+3)(x-3)}$ 5. $\frac{5(x+2)(x+3)}{5(x+3)^2}$ 6. $\frac{(x+3)(x-5)}{(2x+3)(x-5)}$

page 375 1. $\frac{20x^2+3}{4x^3}$ 2. $\frac{-x^2-5x+20}{5(x+5)}$ 3. $\frac{3x^2-x-2}{x^3(x+2)^2}$ 4. $\frac{x^2+12x-1}{(x-2)(x-7)(x+7)}$ 5. $\frac{x^2-13x+28}{3(x+4)^2(x-4)}$ 6. $\frac{-5}{3(x-2)}$

Name ______________________ Date ____________

EXERCISE SET 6.3

Combine the fractions and reduce if possible.

_____ 1. $\frac{5}{7} + \frac{11}{7} - \frac{2}{7}$

_____ 2. $\frac{8}{3} + \frac{11}{3} - \frac{4}{3}$

_____ 3. $\frac{5}{3x} + \frac{1}{3x}$

_____ 4. $\frac{5}{4x} + \frac{11}{4x}$

_____ 5. $\frac{y}{x} + \frac{3}{x}$

_____ 6. $\frac{a}{b} + \frac{5}{b}$

_____ 7. $\frac{3x}{2x - 1} - \frac{5x}{2x - 1}$

_____ 8. $\frac{4x}{3x + 2} - \frac{7x}{3x + 2}$

_____ 9. $\frac{12x}{4x + 3} + \frac{9}{4x + 3}$

_____ 10. $\frac{15x}{5x - 2} - \frac{6}{5x - 2}$

_____ 11. $\frac{x}{x^2 - 81} + \frac{9}{x^2 - 81}$

_____ 12. $\frac{x}{x^2 - 100} - \frac{10}{x^2 - 100}$

_____ 13. $\frac{5x - 1}{2x - 3} - \frac{x + 5}{2x - 3}$

_____ 14. $\frac{10x - 5}{3x - 4} - \frac{x + 7}{3x - 4}$

_____ 15. $\frac{5x - 9}{x^2 - 16} - \frac{4x - 5}{x^2 - 16}$

_____ 16. $\frac{9x - 7}{x^2 - 9} - \frac{8x - 10}{x^2 - 9}$

_____ 17. $\frac{2a^2}{a^2 + 7a} + \frac{11a}{a^2 + 7a} - \frac{21}{a^2 + 7a}$

_____ 18. $\frac{2a^2}{a^2 + 3a} + \frac{a}{a^2 + 3a} - \frac{15}{a^2 + 3a}$

_____ 19. $\frac{2x^2 - 3x}{2x^2 - 9x + 4} + \frac{4x - 1}{2x^2 - 9x + 4}$

_____ 20. $\frac{x^2 - 11x}{x^2 - 3x - 4} + \frac{x + 24}{x^2 - 3x - 4}$

_____ 21. $\frac{(3x - 5)(x - 1)}{2x - 3} - \frac{x(x + 3)}{2x - 3} + \frac{7}{2x - 3}$

_____ 22. $\frac{(2x + 3)(x + 7)}{x + 5} + \frac{x(3x + 2)}{x + 5} - \frac{51}{x + 5}$

_____ 23. $\frac{3}{x} - \frac{5}{x^2} - \frac{2}{5x^2}$

_____ 24. $\frac{5}{x} + \frac{6}{x^2} - \frac{7}{3x^2}$

_____ 25. $\frac{2}{x} - \frac{3}{y} + \frac{2}{5x} - \frac{1}{3y}$

_____ 26. $\frac{3}{x} - \frac{2}{y} + \frac{1}{4x} - \frac{4}{3y}$

_____ 27. $\frac{3}{x} - \frac{2}{x + 1}$

_____ 28. $\frac{2}{x} + \frac{3}{x - 2}$

_____ 29. $\frac{x - 5}{x^2 - 3x} + \frac{5}{2x^2}$

_____ 30. $\frac{x + 1}{x^2 + 5x} - \frac{4}{3x^2}$

_____ 31. $\frac{7}{x + 2} - \frac{5}{(x + 2)^2}$

_____ 32. $\frac{11}{x - 3} - \frac{4}{(x - 3)^2}$

_____ 33. $\frac{3}{x + 4} - \frac{5}{x - 4}$

_____ 34. $\frac{4}{x + 5} - \frac{3}{x - 5}$

______ 35. $\dfrac{4}{3x + 3} - \dfrac{5}{x + 1}$

______ 36. $\dfrac{7}{4x - 4} - \dfrac{4}{x - 1}$

______ 37. $\dfrac{5}{6 - 2x} + \dfrac{4}{x - 3}$

______ 38. $\dfrac{4}{14 - 7x} + \dfrac{2}{x - 2}$

______ 39. $\dfrac{6}{x^2 - 3x} + \dfrac{2}{3 - x}$

______ 40. $\dfrac{12}{x^2 - 4x} + \dfrac{3}{4 - x}$

______ 41. $\dfrac{x}{x^2 - 25} - \dfrac{1}{x^2 - 5x}$

______ 42. $\dfrac{x}{x^2 - 49} - \dfrac{3}{x^2 - 7x}$

______ 43. $\dfrac{1}{x^3 - 4x^2} - \dfrac{16}{x^5 - 4x^4}$

______ 44. $\dfrac{1}{x^3 - 5x^2} - \dfrac{25}{x^5 - 5x^4}$

______ 45. $\dfrac{x + 6}{x^2 + 2x - 8} + \dfrac{3}{x + 4}$

______ 46. $\dfrac{x - 8}{x^2 - 11x + 18} - \dfrac{4}{x - 9}$

______ 47. $\dfrac{2x - 6}{2x^2 - 7x + 3} - \dfrac{5}{2x^2 - x}$

______ 48. $\dfrac{3x - 21}{3x^2 - 22x + 7} - \dfrac{4}{3x^2 - x}$

______ 49. $\dfrac{5}{x + 3} + \dfrac{1}{x - 2} - \dfrac{3x}{x^2 + x - 6}$

______ 50. $\dfrac{4}{3x + 1} - \dfrac{3}{x - 1} + \dfrac{2x}{3x^2 - 2x - 1}$

______ 51. $\dfrac{4}{x - 2} - \dfrac{3}{x + 5} - \dfrac{28}{x^2 + 3x - 10}$

______ 52. $\dfrac{5}{x - 1} - \dfrac{2}{x + 5} - \dfrac{12}{x^2 + 4x - 5}$

______ 53. $\dfrac{2x + 1}{x^2 + x - 20} + \dfrac{3x + 8}{x^2 + 3x - 10} - \dfrac{3x - 10}{x^2 - 6x + 8}$

______ 54. $\dfrac{x - 4}{2x^2 - 11x + 5} - \dfrac{2x - 1}{x^2 - 8x + 15} + \dfrac{x + 3}{2x^2 - 7x + 3}$

______ 55. $\dfrac{1}{x - 2} - \dfrac{6x}{x^3 - 8}$

______ 56. $\dfrac{1}{x - 3} - \dfrac{9x}{x^3 - 27}$

☆ Stretching the Topics

Perform the indicated operations.

______ 1. $\dfrac{3x^a}{5y^{2b}} - \dfrac{4}{10x^a y^b} + \dfrac{y^b}{2x^{2a}}$

______ 2. $\dfrac{x^a}{x^a + 8} - \dfrac{8 - 7x^a}{x^{2a} + 8x^a}$

______ 3. $\left(x - 1 + \dfrac{1}{x + 1}\right)\left(\dfrac{1}{x} - \dfrac{1}{x^2}\right)$

Check your answers in the back of your book.

If you can complete **Checkup 6.3,** you are ready to go on to Section 6.4.

6.4 Simplifying Complex Fractions

A **complex fraction** is a fraction that contains fractions in its numerator and/or denominator. Success with complex fractions depends on skills already learned in adding, subtracting, multiplying, and dividing fractions. As before, we assume that the variables have been restricted to avoid division by zero.

To simplify a complex fraction we must realize that it is, after all, a *division* problem.

Example 1. Find $\dfrac{\dfrac{x^2}{y^3}}{\dfrac{3x}{y^8}}$.

Solution

$$\frac{\frac{x^2}{y^3}}{\frac{3x}{y^8}} = \frac{x^2}{y^3} \div \frac{3x}{y^8} \qquad \text{Write the division problem.}$$

$$= \frac{x^2}{y^3} \cdot \frac{y^8}{3x} \qquad \text{Find reciprocal of divisor and multiply.}$$

$$= \frac{\overset{x}{\cancel{x^2}}}{\underset{1}{\cancel{y^3}}} \cdot \frac{\overset{y^5}{\cancel{y^8}}}{\underset{1}{3\cancel{x}}} \qquad \text{Reduce.}$$

$$= \frac{xy^5}{3} \qquad \text{Find the product.}$$

Example 2. Find $\dfrac{\dfrac{3x^2 - 12}{x^3}}{\dfrac{5x + 10}{x^2 + x}}$.

Solution

$$\frac{\frac{3x^2 - 12}{x^3}}{\frac{5x + 10}{x^2 + x}} = \frac{3x^2 - 12}{x^3} \div \frac{5x + 10}{x^2 + x} \qquad \text{Write the division problem.}$$

$$= \frac{3(x^2 - 4)}{x^3} \cdot \frac{x^2 + x}{5x + 10} \qquad \text{Find reciprocal of the divisor and multiply.}$$

$$= \frac{3\overset{1}{\cancel{(x+2)}}(x - 2)}{\underset{x^2}{\cancel{x^3}}} \cdot \frac{\overset{1}{\cancel{x}}(x + 1)}{5\underset{1}{\cancel{(x+2)}}} \qquad \text{Factor and reduce.}$$

$$= \frac{3(x - 2)(x + 1)}{5x^2} \qquad \text{Write the product.}$$

> To simplify a complex fraction, we rewrite it as a division problem, and divide.

Trial Run

Simplify the complex fractions.

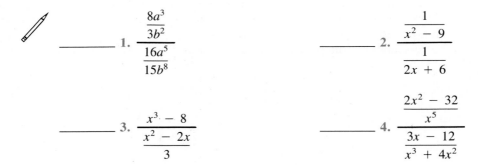

________ 1. $\dfrac{\dfrac{8a^3}{3b^2}}{\dfrac{16a^5}{15b^8}}$

________ 2. $\dfrac{\dfrac{1}{x^2 - 9}}{\dfrac{1}{2x + 6}}$

________ 3. $\dfrac{x^3 - 8}{\dfrac{x^2 - 2x}{3}}$

________ 4. $\dfrac{\dfrac{2x^2 - 32}{x^5}}{\dfrac{3x - 12}{x^3 + 4x^2}}$

Answers are on page 387.

Sometimes a complex fraction contains a numerator and/or denominator that is a sum or difference of fractional expressions. We must develop a method for simplifying such complex fractions.

Example 3. Simplify $\dfrac{x - \dfrac{1}{x}}{x - 1}$.

Solution

$$\frac{x - \dfrac{1}{x}}{x - 1} = \left[x - \frac{1}{x}\right] \div [x - 1]$$
We want each bracketed quantity to be a *single fraction.*

$$= \left[\frac{x^2}{x} - \frac{1}{x}\right] \div \left[\frac{x - 1}{1}\right]$$
Choose the LCD for each bracketed quantity and build the fractions.

$$= \frac{x^2 - 1}{x} \div \frac{x - 1}{1}$$
Combine the fractions within the brackets.

$$= \frac{x^2 - 1}{x} \cdot \frac{1}{x - 1}$$
Find reciprocal of divisor and multiply.

$$= \frac{(x + 1)\overset{1}{\cancel{(x - 1)}}}{x} \cdot \frac{1}{\underset{1}{\cancel{x - 1}}}$$
Factor and reduce.

$$= \frac{x + 1}{x}$$
Write the product.

> **Simplifying Complex Fractions (Method 1)**
>
> **1.** Rewrite the complex fraction as a division problem.
> **2.** Express each part of the division problem as a **single** fraction.
> **3.** Proceed with the method for dividing fractions.

It is important to stress the fact that each part of the division problem *must* be expressed as a single fraction before we proceed with the actual division. In other words, before we can invert the divisor and multiply, that divisor *must* be a single fraction.

Now you complete Example 4.

Example 4. Simplify $\dfrac{2x + \dfrac{5x - 3}{x}}{\dfrac{x}{3} - \dfrac{3}{x}}$.

Solution

$$\frac{2x + \dfrac{5x - 3}{x}}{\dfrac{x}{3} - \dfrac{3}{x}} = \left[\frac{2x}{1} + \frac{5x - 3}{x}\right] \div \left[\frac{x}{3} - \frac{3}{x}\right]$$ Write the division problem.

$$= \left[\frac{2x \cdot x}{x} + \frac{5x - 3}{x}\right] \div \left[\frac{x \cdot x}{3x} - \frac{3 \cdot 3}{3x}\right]$$ Choose the LCD for each bracketed quantity and build the fractions.

$$= \left[\frac{2x^2}{x} + \frac{5x - 3}{x}\right] \div \left[\frac{x^2}{3x} - \frac{9}{3x}\right]$$ Simplify the numerators.

$$= \frac{2x^2 + 5x - 3}{x} \div \frac{x^2 - 9}{3x}$$ Combine the fractions within the brackets.

$$= \frac{2x^2 + 5x - 3}{x} \cdot \frac{3x}{x^2 - 9}$$ Find reciprocal of divisor and multiply.

$=$ Factor and reduce.

$=$ Write the product.

Check your work on page 387. ▶

There is a second approach that some prefer for simplifying complex fractions. It is especially useful if similar denominators appear in the numerator and denominator of the complex fraction. Example 4 is a good candidate for the second approach. In this method, the numerator and denominator are *both* multiplied by the quantity that is the LCD for *all* the fractions in the numerator and denominator. Let's rework Example 4 by the second method.

$$\frac{2x + \dfrac{5x - 3}{x}}{\dfrac{x}{3} - \dfrac{3}{x}}$$ Choose the LCD for *both* numerator and denominator. LCD: $3x$.

$$= \frac{3x\left[2x + \dfrac{5x - 3}{x}\right]}{3x\left[\dfrac{x}{3} - \dfrac{3}{x}\right]}$$ Multiply numerator and denominator by $3x$.

$$= \frac{3x(2x) + \frac{\overset{1}{3\cancel{x}}}{1}\left(\frac{5x - 3}{\underset{1}{\cancel{x}}}\right)}{\frac{\overset{1}{\cancel{3}x}}{1}\left(\frac{x}{\underset{1}{\cancel{3}}}\right) - \frac{\overset{1}{3\cancel{x}}}{1}\left(\frac{3}{\underset{1}{\cancel{x}}}\right)}$$ Use the distributive property.

$$= \frac{6x^2 + 15x - 9}{x^2 - 9}$$ Remove parentheses.

$$= \frac{3(2x^2 + 5x - 3)}{(x + 3)(x - 3)}$$ Factor numerator and denominator.

$$= \frac{3(2x - 1)\overset{1}{\cancel{(x + 3)}}}{\underset{1}{\cancel{(x + 3)}}(x - 3)}$$ Factor numerator completely.

$$= \frac{3(2x - 1)}{x - 3}$$ Reduce the fraction.

Notice that this result is the same as that obtained earlier.

Simplifying Complex Fractions (Method 2)

1. Choose the LCD for *both* the numerator and the denominator.
2. Multiply both numerator and denominator by that LCD.
3. Simplify the numerator and denominator, and reduce if possible.

You may use either method to simplify a complex fraction. Some students always choose the second method, but for fractions containing several different denominators, the first method is actually less complicated. Consider Example 5.

Example 5. Simplify $\dfrac{\dfrac{3}{x} + \dfrac{2}{x - 1}}{x + \dfrac{4x^2 - 5x}{x + 2}}$.

Solution

$$\frac{\dfrac{3}{x} + \dfrac{2}{x - 1}}{x + \dfrac{4x^2 - 5x}{x + 2}} = \left[\frac{3}{x} + \frac{2}{x - 1}\right] \div \left[\frac{x}{1} + \frac{4x^2 - 5x}{x + 2}\right]$$

$$= \left[\frac{3(x - 1)}{x(x - 1)} + \frac{2 \cdot x}{x(x - 1)}\right] \div \left[\frac{x(x + 2)}{x + 2} + \frac{4x^2 - 5x}{x + 2}\right]$$

$$= \left[\frac{3x - 3 + 2x}{x(x - 1)}\right] \div \left[\frac{x^2 + 2x + 4x^2 - 5x}{x + 2}\right]$$

$$= \left[\frac{5x - 3}{x(x - 1)}\right] \div \left[\frac{5x^2 - 3x}{x + 2}\right]$$

$$= \frac{5x - 3}{x(x - 1)} \cdot \frac{x + 2}{5x^2 - 3x}$$

$$= \frac{\overset{1}{\cancel{5x - 3}}}{x(x - 1)} \cdot \frac{x + 2}{x(\underset{1}{\cancel{5x - 3}})}$$

$$= \frac{x + 2}{x^2(x - 1)}$$

Trial Run

Simplify the complex fractions.

_______ 1. $\dfrac{x - \dfrac{4}{x}}{x + 2}$

_______ 2. $\dfrac{3 - \dfrac{1}{x}}{9 - \dfrac{1}{x^2}}$

_______ 3. $\dfrac{\dfrac{1}{x^2} - \dfrac{1}{16}}{\dfrac{1}{x} + \dfrac{1}{4}}$

_______ 4. $\dfrac{\dfrac{-2}{x + 5} + \dfrac{1}{2}}{\dfrac{1}{x} + \dfrac{3}{x + 4}}$

Answers are on page 387.

A complex fraction can appear in disguise when it contains negative exponents. Recall the definition of a negative exponent.

$$a^{-n} = \frac{1}{a^n}$$

Example 6. Simplify $\dfrac{x + 2x^{-1}}{x - 4x^{-3}}$.

Solution

$$\frac{x + 2x^{-1}}{x - 4x^{-3}} = \frac{x + \dfrac{2}{x}}{x - \dfrac{4}{x^3}}$$

Let's multiply numerator and denominator by the LCD for both. LCD: x^3.

$$\frac{x^3\left[x + \dfrac{2}{x}\right]}{x^3\left[x - \dfrac{4}{x^3}\right]}$$

$$= \frac{x^4 + 2x^2}{x^4 - 4}$$

$$= \frac{x^2\overset{1}{\cancel{(x^2 + 2)}}}{(x^2 - 2)\underset{1}{\cancel{(x^2 + 2)}}}$$

$$= \frac{x^2}{x^2 - 2}$$

Now complete Example 7.

Example 7. Simplify $\dfrac{xy^{-1}}{y^{-1} + x}$.

Solution

$$\frac{xy^{-1}}{y^{-1} + x} = \frac{\dfrac{x}{y}}{\dfrac{1}{y} + x}$$

Multiply numerator and denominator by the LCD for both. LCD: y.

$$\frac{y\left[\dfrac{x}{y}\right]}{y\left[\dfrac{1}{y} + x\right]}$$

$$=$$

Check your work on page 387. ▶

Trial Run

Simplify each fraction.

_______ 1. $\dfrac{1 + 3x^{-1}}{1 - 9x^{-2}}$

_______ 2. $\dfrac{1 + 2x^{-1} - 3x^{-2}}{1 + 5x^{-1} + 6x^{-2}}$

_______ 3. $\dfrac{x^{-1} - y^{-1}}{x^{-1} + y^{-1}}$

_______ 4. $\dfrac{x + 3x^{-1}}{x - 9x^{-3}}$

Answers are on page 387.

▶ Examples You Completed

Example 4. Simplify $\dfrac{2x + \dfrac{5x-3}{x}}{\dfrac{x}{3} - \dfrac{3}{x}}$.

Solution

$$\frac{2x + \frac{5x-3}{x}}{\frac{x}{3} - \frac{3}{x}} = \left[\frac{2x}{1} + \frac{5x-3}{x}\right] \div \left[\frac{x}{3} - \frac{3}{x}\right]$$

$$= \left[\frac{2x \cdot x}{x} + \frac{5x-3}{x}\right] \div \left[\frac{x \cdot x}{3x} - \frac{3 \cdot 3}{3x}\right]$$

$$= \left[\frac{2x^2}{x} + \frac{5x-3}{x}\right] \div \left[\frac{x^2}{3x} - \frac{9}{3x}\right]$$

$$= \frac{2x^2 + 5x - 3}{x} \div \frac{x^2 - 9}{3x}$$

$$= \frac{2x^2 + 5x - 3}{x} \cdot \frac{3x}{x^2 - 9}$$

$$= \frac{(2x-1)\overset{1}{\cancel{(x+3)}}}{\underset{1}{\cancel{x}}} \cdot \frac{3\overset{1}{\cancel{x}}}{\underset{1}{\cancel{(x+3)}}(x-3)}$$

$$= \frac{3(2x-1)}{x-3}$$

Example 7. Simplify $\dfrac{xy^{-1}}{y^{-1} + x}$.

Solution

$$\frac{xy^{-1}}{y^{-1} + x} = \frac{\frac{x}{y}}{\frac{1}{y} + x}$$

$$= \frac{y\left[\frac{x}{y}\right]}{y\left[\frac{1}{y} + x\right]}$$

$$= \frac{x}{1 + xy}$$

Answers to Trial Runs

page 382 1. $\dfrac{5b^6}{2a^2}$ 2. $\dfrac{2}{x-3}$ 3. $\dfrac{3(x^2+2x+4)}{x}$ 4. $\dfrac{2(x+4)^2}{3x^3}$

page 385 1. $\dfrac{x-2}{x}$ 2. $\dfrac{x}{3x+1}$ 3. $\dfrac{4-x}{4x}$ 4. $\dfrac{x(x+4)}{8(x+5)}$

page 386 1. $\dfrac{x}{x-3}$ 2. $\dfrac{x-1}{x+2}$ 3. $\dfrac{y-x}{y+x}$ 4. $\dfrac{x^2}{x^2-3}$

Name ______________________________ Date ______________

EXERCISE SET 6.4

Simplify each fraction.

______ 1. $\dfrac{\frac{3}{8}}{\frac{15}{4}}$

______ 2. $\dfrac{\frac{5}{9}}{\frac{10}{3}}$

______ 3. $\dfrac{\frac{x^7}{y^2}}{\frac{3x^5}{y}}$

______ 4. $\dfrac{\frac{x^5}{y^3}}{\frac{4x^2}{y^2}}$

______ 5. $\dfrac{\frac{x + y}{3x^2}}{\frac{x^2 - y^2}{15x^4}}$

______ 6. $\dfrac{\frac{2x - y}{5y^2}}{\frac{4x^2 - y^2}{20y^5}}$

______ 7. $\dfrac{\frac{1}{x^2 - 16}}{\frac{5}{3x - 12}}$

______ 8. $\dfrac{\frac{1}{x^2 - 49}}{\frac{2}{4x + 28}}$

______ 9. $\dfrac{x^2 - 2x}{\frac{x^2 - 4}{x}}$

______ 10. $\dfrac{x^2 - 5x}{\frac{x^2 - 25}{x^3}}$

______ 11. $\dfrac{\frac{x^2 - 3x}{2}}{x^3 - 27}$

______ 12. $\dfrac{\frac{x^2 + 5x}{4}}{x^3 + 125}$

______ 13. $\dfrac{\frac{9x^2 - 16}{7x^2}}{\frac{3x - 4}{14x^5}}$

______ 14. $\dfrac{\frac{4x^2 - 9}{3x^2}}{\frac{2x + 3}{18x^3}}$

______ 15. $\dfrac{3 + \frac{1}{5}}{1 - \frac{3}{7}}$

______ 16. $\dfrac{6 + \frac{2}{3}}{1 - \frac{5}{9}}$

______ 17. $\dfrac{x - \frac{9}{x}}{x + 3}$

______ 18. $\dfrac{x - \frac{25}{x}}{x - 5}$

______ 19. $\dfrac{\frac{1}{x}}{1 + \frac{1}{x}}$

______ 20. $\dfrac{1 - \frac{1}{x}}{\frac{1}{x}}$

______ 21. $\dfrac{4 - \frac{1}{x}}{16 - \frac{1}{x^2}}$

______ 22. $\dfrac{7 + \frac{1}{x}}{49 - \frac{1}{x^2}}$

______ 23. $\dfrac{8 - \frac{1}{a^3}}{2 - \frac{1}{a}}$

______ 24. $\dfrac{64 + \frac{1}{a^3}}{4 + \frac{1}{a}}$

______ 25. $\dfrac{\frac{1}{3} - \frac{1}{4}}{\frac{1}{2} + \frac{1}{8}}$

______ 26. $\dfrac{\frac{4}{5} - \frac{1}{5}}{\frac{7}{10} - \frac{1}{2}}$

______ 27. $\dfrac{\frac{1}{x^2} - \frac{1}{81}}{\frac{1}{x} + \frac{1}{9}}$

_______ 28. $\dfrac{\dfrac{1}{x^2} - \dfrac{1}{100}}{\dfrac{1}{x} - \dfrac{1}{10}}$

_______ 29. $\dfrac{10 - \dfrac{x + 3}{x^2}}{\dfrac{x}{3} - \dfrac{1}{5}}$

_______ 30. $\dfrac{4 - \dfrac{17x + 15}{x^2}}{\dfrac{x}{3} + \dfrac{1}{4}}$

_______ 31. $\dfrac{x + \dfrac{x - 30}{x}}{\dfrac{x}{5} - \dfrac{5}{x}}$

_______ 32. $\dfrac{x + \dfrac{x - 72}{x}}{\dfrac{x}{8} - \dfrac{8}{x}}$

_______ 33. $\dfrac{x + \dfrac{21}{x + 10}}{x + 6 - \dfrac{12}{x + 7}}$

_______ 34. $\dfrac{x + \dfrac{12}{x + 7}}{x + 8 + \dfrac{4}{x + 3}}$

_______ 35. $\dfrac{\dfrac{2x}{x^2 - 4} + \dfrac{3}{x - 2}}{\dfrac{5}{x + 2} + \dfrac{4}{x - 2}}$

_______ 36. $\dfrac{\dfrac{3x}{x^2 - 25} + \dfrac{1}{x - 5}}{\dfrac{4}{x - 5} + \dfrac{3}{x + 5}}$

_______ 37. $\dfrac{\dfrac{a}{a + 4} - \dfrac{3}{a + 2}}{\dfrac{2a + 7}{a^2 + 6a + 8} - \dfrac{1}{a + 2}}$

_______ 38. $\dfrac{\dfrac{a}{a + 3} - \dfrac{2}{a + 1}}{\dfrac{3a + 5}{a^2 + 4a + 3} - \dfrac{1}{a + 3}}$

_______ 39. $\dfrac{\dfrac{a + b}{a - b} - \dfrac{a - b}{a + b}}{\dfrac{1}{a - b} + \dfrac{1}{a + b}}$

_______ 40. $\dfrac{\dfrac{a - b}{a + b} + \dfrac{a + b}{a - b}}{\dfrac{a}{a + b} + \dfrac{b}{a - b}}$

_______ 41. $\dfrac{\dfrac{3}{x - 2} - \dfrac{2}{x - 1}}{\dfrac{5}{x - 1} - \dfrac{4}{x + 2}}$

_______ 42. $\dfrac{\dfrac{4}{x - 3} - \dfrac{1}{x + 4}}{\dfrac{2}{x + 4} - \dfrac{3}{x + 3}}$

_______ 43. $\dfrac{1 + 4x^{-1}}{1 - 16x^{-2}}$

_______ 44. $\dfrac{1 + 5x^{-1}}{1 - 25x^{-2}}$

_______ 45. $\dfrac{x^{-3} + 8}{x^{-1} + 2}$

_______ 46. $\dfrac{x^{-3} + 125}{x^{-1} + 5}$

_______ 47. $\dfrac{2 + 9x^{-1} - 5x^{-2}}{1 + 3x^{-1} - 10x^{-2}}$

_______ 48. $\dfrac{3 - 2x^{-1} - x^{-2}}{1 + 2x^{-1} - 3x^{-2}}$

_______ 49. $\dfrac{x^{-1} - 2y^{-1}}{x^{-1} + 2y^{-1}}$

_______ 50. $\dfrac{3x^{-1} + y^{-1}}{3x^{-1} - y^{-1}}$

_______ 51. $\dfrac{4x^2y^{-3}}{y^{-3} + 2x^2}$

_______ 52. $\dfrac{5xy^{-1}}{y^{-1} + 5x}$

_______ 53. $\dfrac{x + 4x^{-1}}{x - 16x^{-3}}$

_______ 54. $\dfrac{x + 5x^{-1}}{x - 25x^{-3}}$

☆ Stretching the Topics

Simplify the complex fractions.

_______ 1. $$\dfrac{\dfrac{x+1}{x-4}+\dfrac{x-5}{x+5}}{\dfrac{x-1}{x+5}-\dfrac{x-2}{x+3}}$$

_______ 2. $$\dfrac{\dfrac{x+d-1}{x+d}-\dfrac{x-1}{x}}{d}$$

_______ 3. $$\dfrac{\dfrac{12x^{2n}}{x^{2n}-25}-\dfrac{3}{x^n-5}}{\dfrac{4}{x^n+5}+\dfrac{15}{x^{2n}-25}}$$

Check your answers in the back of your book.

If you can simplify the fractions in **Checkup 6.4,** you are ready to do Section 6.5.

Name ______________________ Date __________

✓ CHECKUP 6.4

Simplify each fraction.

______ 1. $\dfrac{\dfrac{x^{10}}{y^4}}{\dfrac{5x^3}{y^7}}$

______ 2. $\dfrac{\dfrac{2x-3}{3x^2}}{\dfrac{4x^2-9}{15x^5}}$

______ 3. $\dfrac{\dfrac{x^2+11x}{x^2-121}}{x^2}$

______ 4. $\dfrac{x-\dfrac{81}{x}}{x+9}$

______ 5. $\dfrac{8-\dfrac{27}{a^3}}{2-\dfrac{3}{a}}$

______ 6. $\dfrac{\dfrac{1}{x^2}-\dfrac{1}{49}}{\dfrac{1}{x}-\dfrac{1}{7}}$

______ 7. $\dfrac{x-\dfrac{35}{x+2}}{\dfrac{x}{7}-\dfrac{7}{x}}$

______ 8. $\dfrac{\dfrac{6}{a+3}-\dfrac{4}{a-4}}{\dfrac{2}{a-4}+\dfrac{5}{a+3}}$

______ 9. $\dfrac{12+11x^{-1}+2x^{-2}}{3x^{-1}-13x^{-2}-10x^{-3}}$

______ 10. $\dfrac{x^{-1}+3y^{-1}}{x^{-2}-9y^{-2}}$

Check your answers in the back of your book.

If You Missed Problems:	You Should Review Examples:
1	1
2, 3	2
4	3
5, 6	4
7, 8	5
9, 10	6, 7

6.5 Solving Fractional Equations and Inequalities

To solve fractional equations and inequalities, we recall previously learned skills.

Solving Fractional Equations

To solve a fractional equation, we try to write it as an equation *without* fractions. To find solutions to equations such as

$$\frac{x}{3} + \frac{2x+1}{5} = \frac{1}{2} \qquad \text{or} \qquad \frac{3}{x+1} - \frac{5}{x} = \frac{19}{x^2+x}$$

we must first find a way to eliminate the denominators.

> To eliminate the denominators in a fractional equation, we may multiply both sides of the equation by the LCD for *all* the fractions in the equation.

Of course, we may multiply both sides of an equation only by a nonzero quantity, so we must *state restrictions* after selecting the LCD.

Example 1. Solve $\frac{2}{x} - \frac{5}{4} = \frac{9}{2x}$.

Solution

$$\frac{2}{x} - \frac{5}{4} = \frac{9}{2x}$$

$$\frac{2}{x} - \frac{5}{2 \cdot 2} = \frac{9}{2 \cdot x} \qquad \text{Factor all denominators.}$$

The LCD is $2 \cdot 2 \cdot x = 4x$, and $x \neq 0$. Choose LCD. State restriction.

$$\frac{4x}{1}\left[\frac{2}{x}\right] - \frac{4x}{1}\left[\frac{5}{4}\right] = \frac{4x}{1}\left[\frac{9}{2x}\right] \qquad \text{Multiply both sides by LCD: } 4x$$

$$\frac{\overset{1}{4\cancel{x}}}{1}\left[\frac{2}{\underset{1}{\cancel{x}}}\right] - \frac{\overset{1}{\cancel{4}x}}{1}\left[\frac{5}{\underset{1}{\cancel{4}}}\right] = \frac{\overset{2}{\cancel{4x}}}{1}\left[\frac{9}{\underset{1}{\cancel{2x}}}\right] \qquad \text{Reduce before finding products.}$$

$$8 - 5x = 18 \qquad \text{Find products. Notice that the fractions have been eliminated.}$$

$$-5x = 10 \qquad \text{Isolate } x\text{-term.}$$

$$x = -2 \qquad \text{Isolate } x.$$

Since the restriction was $x \neq 0$, the solution $x = -2$ is acceptable.

Example 2. Solve $\frac{y}{2} = \frac{x}{x-3}$ for x.

Solution

$$\frac{y}{2} = \frac{x}{x-3} \qquad \text{LCD: } 2(x-3) \quad x \neq 3 \qquad \text{Choose LCD. State restriction.}$$

$$\frac{\overset{1}{\cancel{2}}(x-3)}{1}\left[\frac{y}{\underset{1}{\cancel{2}}}\right] = \frac{2\overset{1}{\cancel{(x-3)}}}{1}\left[\frac{x}{\underset{1}{\cancel{x-3}}}\right] \qquad \text{Multiply both sides by LCD: } 2(x-3).$$

$$(x-3)y = 2x \qquad \text{Write the products.}$$

$$xy - 3y = 2x \qquad \text{Remove parentheses.}$$

$$xy - 2x = 3y \qquad \text{Isolate } x\text{-terms.}$$

$$x(y-2) = 3y \qquad \text{Factor } x \text{ from terms on left.}$$

$$x = \frac{3y}{y-2} \quad y \neq 2,\ x \neq 3 \qquad \text{Divide by } y-2 \text{ to isolate } x.$$

> **Solving Fractional Equations**
>
> 1. Write all denominators in factored form.
> 2. Choose the LCD for all the fractions and state restrictions on the variable.
> 3. Multiply both sides of the equation by the LCD and simplify.
> 4. Observe whether the equation is first-degree or second-degree and solve by the appropriate method.
> 5. Check to be sure that no solution is a restricted value.

You complete Example 3.

Example 3. Solve $\dfrac{3}{x+1} - \dfrac{5}{x} = \dfrac{19}{x^2+x}$.

Solution

$$\frac{3}{x+1} - \frac{5}{x} = \frac{19}{x^2+x}$$

$$\frac{3}{x+1} - \frac{5}{x} = \frac{19}{x(x+1)} \qquad \text{LCD: } x(x+1) \text{ and } x \neq \underline{\quad\quad}, x \neq \underline{\quad\quad}.$$

$$\frac{x\overset{1}{\cancel{(x+1)}}}{1}\left[\frac{3}{\underset{1}{\cancel{x+1}}}\right] - \frac{\overset{1}{\cancel{x}}(x+1)}{1}\left[\frac{5}{\underset{1}{\cancel{x}}}\right] = \frac{\overset{1}{\cancel{x}}\overset{1}{\cancel{(x+1)}}}{1}\left[\frac{19}{\underset{1}{\cancel{x}}\underset{1}{\cancel{(x+1)}}}\right]$$

$$3x - 5(x+1) = 19$$

Since the restrictions were $x \neq$ _____ , $x \neq$ _____ , the solution $x =$ _____ is acceptable.

Check your work on page 400. ▶

Example 4. Solve $\dfrac{x+1}{x-5} = \dfrac{20}{x^2-25} + \dfrac{2}{x+5}$.

Solution

$$\frac{x+1}{x-5} = \frac{20}{x^2-25} + \frac{2}{x+5}$$

$$\frac{x+1}{x-5} = \frac{20}{(x+5)(x-5)} + \frac{2}{x+5} \qquad \text{LCD: } (x+5)(x-5) \text{ and } x \neq 5,\ x \neq -5.$$

$$\frac{(x+5)\overset{1}{\cancel{(x-5)}}}{1}\left[\frac{x+1}{\underset{1}{\cancel{x-5}}}\right] = \frac{\overset{1}{\cancel{(x+5)}}\overset{1}{\cancel{(x-5)}}}{1}\left[\frac{20}{\underset{1}{\cancel{(x+5)}}\underset{1}{\cancel{(x-5)}}}\right] + \frac{\overset{1}{\cancel{(x+5)}}(x-5)}{1}\left[\frac{2}{\underset{1}{\cancel{x+5}}}\right]$$

$$(x+5)(x+1) = 20 + 2(x-5)$$

$$x^2 + 6x + 5 = 20 + 2x - 10$$

$$x^2 + 6x + 5 = 10 + 2x$$

We recognize that this is a second-degree (quadratic) equation that must be written with zero on one side of the equation.

$$x^2 + 4x - 5 = 0$$

$$(x+5)(x-1) = 0$$

$$x + 5 = 0 \quad \text{or} \quad x - 1 = 0$$

$$x = -5 \quad \text{or} \quad x = 1$$

Since the restrictions were $x \neq 5$, $x \neq -5$, one of our solutions is unacceptable. The solution $x = -5$ must be rejected; it will *not* satisfy the original equation. The only solution is

$$x = 1$$

Example 4 should help you realize the importance of stating restrictions on the domain of the variable as soon as you have chosen your LCD. Having done so, you can easily see whether your solutions are acceptable. Solutions that are obtained correctly but do not satisfy the original equation are called **extraneous solutions.**

Trial Run

Solve each equation.

_______ 1. $\dfrac{x}{3} + \dfrac{5}{6} = \dfrac{9x}{2}$

_______ 2. $\dfrac{3x-2}{x^2} - \dfrac{4}{3x} = 0$

_______ 3. $\dfrac{5}{2x} = \dfrac{2}{x+7}$

_______ 4. $\dfrac{x}{x-3} + \dfrac{x-6}{x-3} = 1$

_______ 5. $\dfrac{x-5}{x-3} = \dfrac{18}{x^2-9} + \dfrac{5}{x+3}$

_______ 6. $\dfrac{x+y}{x} = \dfrac{2y}{5}$ for x.

Answers are on page 400.

Solving Fractional Inequalities

Just as the signs of the factors in a *product* determine whether it is positive or negative, you should recall that the signs of the numerator and denominator of a *fraction* determine whether the fraction is positive or negative.

> If its numerator is positive and its denominator is positive, a fraction will be *positive*.
>
> If its numerator is negative and its denominator is negative, a fraction will be *positive*.
>
> If its numerator is positive and its denominator is negative, a fraction will be *negative*.
>
> If its numerator is negative and its denominator is positive, a fraction will be *negative*.

Let's keep these facts in mind as we take a look at some fractional inequalities.

Example 5. Solve $\frac{x}{3} > 0$.

Solution. We wish this fraction to be *positive*. Since the denominator is positive 3, the numerator (x) must also be *positive*. The solution is

$$x > 0$$
$$(0, \infty)$$

Now you complete Example 6.

Example 6. Solve $\frac{5}{3 - x} < 0$.

Solution. We wish this fraction to be ________ . Since the numerator is *positive* 5, the denominator ($3 - x$) must be ________ . The solution is found by

$$3 - x < 0$$
$$-x < \_\_\_\_$$
$$\frac{-1x}{-1} > \frac{\_\_\_\_}{-1}$$
$$x > \_\_\_\_$$

Check your work on page 400. ▶

Example 7. Solve $\frac{-6}{2x + 1} \leq 0$.

Solution. We wish this fraction to be *negative* (or zero). Since the numerator is a *negative* constant, the denominator ($2x + 1$) must be *positive*. Shall we say

$$2x + 1 \geq 0?$$

Can we allow the denominator to be greater than or *equal* to zero? The denominator of a fraction can *never equal zero*. We find our only solutions from

$$2x + 1 > 0$$
$$2x > -1$$
$$x > -\frac{1}{2} \quad \text{or} \quad (-\tfrac{1}{2}, \infty)$$

We conclude that if $x > -\frac{1}{2}$, then $\frac{-6}{2x + 1} < 0$. This fraction can never *equal* zero.

From Example 7 and our earlier work with fractions we must remember

> If the numerator of a fraction is zero and the denominator is *not* zero, the fraction will equal zero.
>
> If the denominator of a fraction is zero, the fraction will be undefined.

Trial Run

Solve each inequality.

_______ 1. $\frac{x}{2} < 0$

_______ 2. $\frac{3}{x + 2} > 0$

_______ 3. $\frac{x - 1}{-5} \geq 0$

_______ 4. $\frac{8}{3 - 4x} \geq 0$

Answers are on page 400.

What happens when a variable appears in more than one factor of the numerator and/or denominator of a fractional inequality? As you may have guessed, fractional inequalities containing several variable factors can be treated using the **boundary-point method** discussed in Chapter 5. We approach them in the same way that we approached inequalities that contained *products* of variable factors.

Example 8. Solve $\frac{7}{x(x - 3)} < 0$.

Solution. We locate boundary points on the number line by deciding what values of the variable will make each factor zero. Here boundary points are $x = 0$ and $x = 3$. Our dots should be *open*, and we want our fraction to be *negative*.

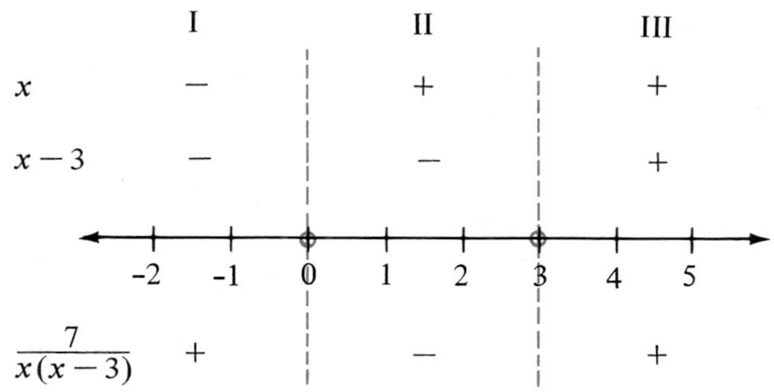

Since the fraction is negative in interval II, we write our solutions as

$$0 < x < 3$$
$$(0, 3)$$

Now you try Example 9.

Example 9. Solve $\dfrac{x}{x+2} > 0$.

Solution. Our boundary points are $x =$ ____ and $x =$ ____ . The dots should be ________ , and we want the fraction to be ________ .

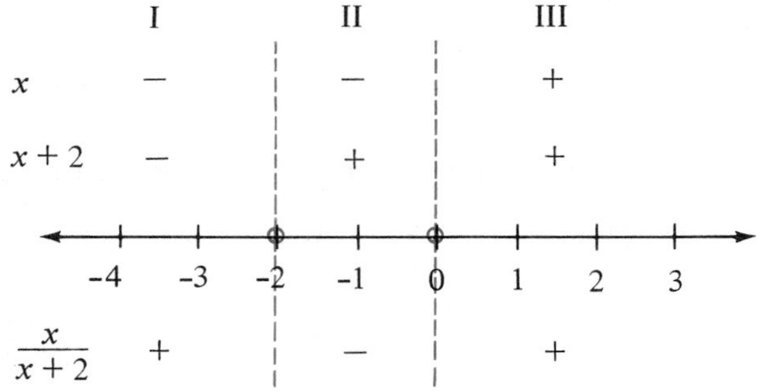

Since the fraction is positive in interval ____ or interval ____ , we write our solutions as

$$x < \text{\_\_\_\_} \quad \text{or} \quad x > \text{\_\_\_\_}$$

Using interval notation, we write the solutions as ____ ∪ ____ .

Check your work on page 400. ▶

The key to solving fractional inequalities lies in knowing that the sign of a fraction is determined by the signs of the factors in its numerator and denominator. To solve an inequality such as

$$\frac{x}{x-2} \geq \frac{2}{3}$$

we must rewrite it in a form that will allow us to make judgments about *signs*. In other words, we must *rewrite the inequality with a single fraction on one side and zero on the other side.*

> We may *not* multiply both sides of any inequality by a *variable* quantity.

Why? Because we do not know the *sign* of such a quantity, so we cannot judge whether or not to reverse the direction of the inequality.

Example 10. Solve $\dfrac{x}{x-2} \geq \dfrac{2}{3}$.

Solution. We must rewrite the inequality with 0 on one side.

$$\frac{x}{x-2} - \frac{2}{3} \geq 0$$ Subtract $\frac{2}{3}$ from both sides. LCD: $3(x - 2)$ and $x \neq 2$.

$$\frac{3 \cdot x}{3(x-2)} - \frac{2(x-2)}{3(x-2)} \geq 0$$ Build each fraction to a fraction containing the LCD.

$$\frac{3x}{3(x-2)} - \frac{2x-4}{3(x-2)} \geq 0$$ Simplify each numerator.

$$\frac{3x - 2x + 4}{3(x - 2)} \geq 0 \qquad \text{Combine the fractions.}$$

$$\frac{x + 4}{3(x - 2)} \geq 0 \qquad \text{Simplify the numerator by combining like terms.}$$

Boundary points are

$x = -4$ (solid dot) — Solid dot because if $x = -4$ the inequality is satisfied.

$x = 2$ (open dot) — Open dot because if $x = 2$ the fraction is undefined.

We want the fraction to be positive (or zero).

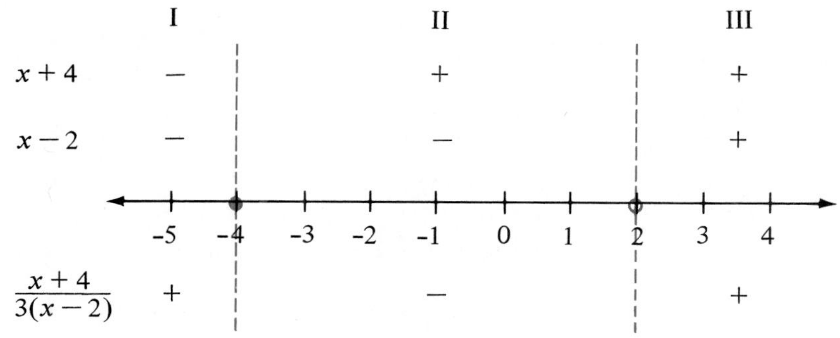

Since the fraction is positive in intervals I and III, our solutions are

$$x \leq -4 \quad \text{or} \quad x > 2$$
$$(-\infty, -4] \quad \cup \quad (2, \infty)$$

Trial Run

Solve each inequality.

_______ 1. $\dfrac{x}{x - 3} > 0$

_______ 2. $\dfrac{5}{x(x - 2)} < 0$

_______ 3. $\dfrac{x - 2}{x} \geq \dfrac{1}{2}$

_______ 4. $\dfrac{x - 2}{x + 1} < 2$

Answers are on page 400.

▶ Examples You Completed

Example 3. Solve $\frac{3}{x+1} - \frac{5}{x} = \frac{19}{x^2 + x}$.

Solution

$$\frac{3}{x+1} - \frac{5}{x} = \frac{19}{x^2 + x}$$

$$\frac{3}{x+1} - \frac{5}{x} = \frac{19}{x(x+1)} \qquad \text{LCD: } x(x+1) \text{ and } x \neq 0,\ x \neq -1.$$

$$\frac{\overset{1}{x(\cancel{x+1})}}{1}\left[\frac{3}{\underset{1}{\cancel{x+1}}}\right] - \frac{\overset{1}{\cancel{x}(x+1)}}{1}\left[\frac{5}{\underset{1}{\cancel{x}}}\right] = \frac{\overset{1\ \ 1}{\cancel{x}(\cancel{x+1})}}{1}\left[\frac{19}{\underset{1\ \ 1}{\cancel{x}(\cancel{x+1})}}\right]$$

$$3x - 5(x + 1) = 19$$

$$3x - 5x - 5 = 19$$

$$-2x - 5 = 19$$

$$-2x = 24$$

$$x = -12$$

Since the restrictions were $x \neq 0$, $x \neq -1$, the solution $x = -12$ is acceptable.

Example 6. Solve $\frac{5}{3 - x} < 0$.

Solution. We wish this fraction to be negative. Since the numerator is *positive* 5, the denominator $(3 - x)$ must be negative. The solution is found by

$$3 - x < 0$$

$$-x < -3$$

$$\frac{-1x}{-1} > \frac{-3}{-1}$$

$$x > 3$$

$$(3, \infty)$$

Example 9. Solve $\frac{x}{x+2} > 0$.

Solution. Our boundary points are $x = 0$ and $x = -2$. The dots should be open, and we want the fraction to be positive.

Since the fraction is positive in interval I or interval III, we write our solutions as

$$x < -2 \quad \text{or} \quad x > 0$$

$$(-\infty, -2) \quad \cup \quad (0, \infty)$$

Answers to Trial Runs

page 395 **1.** $x = \frac{1}{5}$ **2.** $x = \frac{6}{5}$ **3.** $x = -35$ **4.** No solution **5.** $x = 9$, $x = -2$
6. $x = \frac{-5y}{5 - 2y}$

page 397 **1.** $x < 0$; $(-\infty, 0)$ **2.** $x > -2$; $(-2, \infty)$ **3.** $x \leq 1$; $(-\infty, 1]$ **4.** $x < \frac{3}{4}$; $(-\infty, \frac{3}{4})$

page 399 **1.** $x < 0$ or $x > 3$; $(-\infty, 0) \cup (3, \infty)$ **2.** $0 < x < 2$; $(0, 2)$
3. $x < 0$ or $x \geq 4$; $(-\infty, 0) \cup [4, \infty)$ **4.** $x < -4$ or $x > -1$; $(-\infty, -4) \cup (-1, \infty)$

Name ______________________ Date ____________

EXERCISE SET 6.5

Solve each equation.

_______ 1. $\frac{2x}{7} = \frac{x - 2}{6}$

_______ 2. $\frac{3x}{5} = \frac{x - 1}{4}$

_______ 3. $\frac{7}{x} - \frac{3}{2} = \frac{5}{2x}$

_______ 4. $\frac{11}{x} - \frac{4}{3} = \frac{1}{3x}$

_______ 5. $\frac{2x - 1}{x^2} - \frac{5}{3x} = 0$

_______ 6. $\frac{3x + 1}{x^2} - \frac{7}{5x} = 0$

_______ 7. $\frac{11}{4x} = \frac{5}{x + 4}$

_______ 8. $\frac{3}{7x} = \frac{2}{x + 7}$

_______ 9. $\frac{4}{x + 7} = \frac{5}{x^2 - 49}$

_______ 10. $\frac{3}{x - 5} = \frac{2}{x^2 - 25}$

_______ 11. $\frac{4}{2x - 1} + \frac{1}{2x + 1} = \frac{2}{4x^2 - 1}$

_______ 12. $\frac{5}{3x + 1} - \frac{2}{3x - 1} = \frac{1}{9x^2 - 1}$

_______ 13. $\frac{3}{x^2 - x - 12} - \frac{2}{x - 4} = \frac{1}{2x + 6}$

_______ 14. $\frac{1}{x^2 - 6x - 7} - \frac{1}{x - 7} = \frac{2}{3x + 3}$

_______ 15. $\frac{5}{x + 1} = \frac{3x - 2}{x^2 - 15x - 16}$

_______ 16. $\frac{3}{x - 2} = \frac{7x - 4}{x^2 - 8x + 12}$

_______ 17. $\frac{5x + 3}{x - 7} = 2$

_______ 18. $\frac{6x + 5}{x - 9} = 3$

_______ 19. $\frac{a - 2}{a + 5} = \frac{a - 1}{a + 1}$

_______ 20. $\frac{a - 7}{a - 2} = \frac{a + 3}{a + 2}$

_______ 21. $\frac{2x}{2x - 3} - \frac{6}{2x - 3} = 2$

_______ 22. $\frac{3x}{3x - 1} - \frac{4}{3x - 1} = 4$

_______ 23. $\frac{x}{x - 3} - \frac{1}{2x + 3} = \frac{2x^2 + 9}{2x^2 - 3x - 9}$

_______ 24. $\frac{x}{x + 5} - \frac{2}{3x - 1} = \frac{3x^2 + 5}{3x^2 + 14x - 5}$

_______ 25. $\frac{x + 2}{x - 3} = \frac{30}{x^2 - 9} + \frac{1}{x + 3}$

_______ 26. $\frac{x + 4}{x - 2} + \frac{20}{x^2 - 4} = -\frac{5}{x + 2}$

_______ 27. $\frac{x}{x + 1} + \frac{2x}{x - 1} = \frac{10}{x^2 - 1}$

_______ 28. $\frac{3x}{x - 2} + \frac{2x}{x + 2} = \frac{7}{x^2 - 4}$

_______ 29. $\frac{y + 1}{2y - 5} = \frac{2y + 5}{y - 1}$

_______ 30. $\frac{2y - 1}{3y + 4} = \frac{3y - 4}{2y + 1}$

_______ 31. $\frac{x}{2x - 6} - \frac{3}{x^2 - 6x + 9} = \frac{x - 2}{3x - 9}$

_______ 32. $\frac{x}{3x - 6} - \frac{4}{x^2 - 4} = \frac{3}{2x - 4}$

_______ 33. $\frac{3x - y}{4x} = \frac{2 - xy}{4}$ for y.

_______ 34. $\frac{2m - n}{3m} = \frac{1 - mn}{3}$ for n.

_______ 35. $\frac{1}{2x} + \frac{2}{y} = \frac{1}{z}$ for y.

_______ 36. $\frac{2}{a} - \frac{3}{b} = \frac{1}{c}$ for a.

_______ 37. $A = P(1 + rt)$ for t.

_______ 38. $S = \frac{1}{2}g(2t - 1)$ for t.

Solve each inequality.

_______ 39. $\frac{x}{5} < 0$

_______ 40. $\frac{x}{7} \geq 0$

_______ 41. $\frac{5}{x - 3} > 0$

_______ 42. $\frac{9}{x - 7} < 0$

_______ 43. $\frac{x + 3}{-7} \geq 0$

_______ 44. $\frac{x + 6}{-5} \geq 0$

_______ 45. $\frac{-3}{2x - 5} \leq 0$

_______ 46. $\frac{-1}{3x - 7} \leq 0$

_______ 47. $\frac{6 - 2x}{5} \geq 0$

_______ 48. $\frac{9 - 3x}{4} \geq 0$

_______ 49. $\frac{9}{5 - x} \leq 0$

_______ 50. $\frac{10}{7 - x} \leq 0$

_______ 51. $\frac{x}{x - 4} > 0$

_______ 52. $\frac{x}{x - 6} > 0$

_______ 53. $\frac{3}{x(x - 1)} < 0$

_______ 54. $\frac{5}{x(x - 5)} < 0$

_______ 55. $\frac{x - 3}{x} \geq \frac{1}{3}$

_______ 56. $\frac{x - 5}{2x} \geq \frac{1}{4}$

_______ 57. $\frac{x - 3}{x - 1} < 2$

_______ 58. $\frac{x - 5}{x + 3} < 3$

☆ Stretching the Topics

Solve each equation for x.

_______ 1. $\dfrac{2x}{x^2 - 4} - \dfrac{2x}{x^2 + x - 2} + \dfrac{x}{x^2 - 3x + 2} = 0$

_______ 2. $\dfrac{x - 1}{x^2 + 3x + 2} + \dfrac{x + 1}{x^2 + x - 2} = \dfrac{x + 3}{x^2 - 1}$

_______ 3. $\dfrac{10k^2}{x^2 - kx - 6k^2} + \dfrac{4x - 5k}{x - 3k} = \dfrac{4x + 3k}{x + 2k}$

Check your answers in the back of your book.

If you can solve the equations and inequalities in **Checkup 6.5,** you are ready to go on to Section 6.6.

Name ______________________ Date __________

✓ CHECKUP 6.5

Solve each equation or inequality.

_______ 1. $\dfrac{x+5}{2} = \dfrac{x-4}{3}$

_______ 2. $\dfrac{6}{a} - \dfrac{4}{9} = \dfrac{8}{9a}$

_______ 3. $\dfrac{5}{2b+1} = \dfrac{-4}{b-1}$

_______ 4. $\dfrac{3}{a-5} + \dfrac{4a}{a^2-a-20} = 0$

_______ 5. $\dfrac{x-2}{x-4} - \dfrac{16}{x^2-16} = \dfrac{1}{x+4}$

_______ 6. $\dfrac{x}{5} = \dfrac{y}{y-1}$ for y

_______ 7. $\dfrac{6}{x-5} > 0$

_______ 8. $\dfrac{x}{x-10} \leq 0$

_______ 9. $\dfrac{x-6}{x} \geq 3$

_______ 10. $\dfrac{x-1}{x+1} > \dfrac{1}{2}$

Check your answers in the back of your book.

If You Missed Problems:	You Should Review Examples:
1, 2	1
3–5	3, 4
6	2
7	9
8	7
9, 10	10

6.6 Switching from Words to Fractional Equations

There are many instances in which the information given in a word problem requires the use of a fractional equation.

Working with Ratio and Proportion

A **ratio** is a fraction that is used to compare two numbers. Ratios are handy tools to use in solving many everyday mathematical problems.

Suppose that your take-home pay is \$975 per month and you save \$150 from that amount. Then the ratio of your saving to your take-home pay is given by

$$\frac{\text{Saving}}{\text{Take-home pay}} = \frac{\$150}{\$975}$$

$$= \frac{2}{13}$$

Suppose that your take-home pay is increased to \$1170 but you still wish to maintain the *same* ratio of saving to take-home pay. To decide how much you should save (x), you must solve an equation.

$$\frac{\text{Saving}}{\text{Take-home pay}} \qquad \frac{2}{13} = \frac{x}{1170}$$

Such a fractional equation, which states that two ratios are equal, is called a **proportion.** It is most easily solved using the definition of equality of fractions that was stated earlier in this chapter.

$$\frac{A}{B} = \frac{C}{D} \text{ if and only if } A \cdot D = B \cdot C \qquad (B \neq 0, D \neq 0)$$

Example 1. Solve the proportion $\frac{2}{13} = \frac{x}{\$1170}$.

Solution

$$\frac{2}{13} = \frac{x}{1170}$$

$$2(1170) = 13 \cdot x$$

$$2340 = 13x$$

$$\frac{2340}{13} = x$$

$$180 = x$$

You should save \$180 from your take-home pay of \$1170.

Example 2. The admissions policy at Hometown University requires that 7 in-state students be admitted for every 2 out-of-state students. If 1854 freshmen have been admitted in the required ratio, how many are in-state students and how many are from other states?

Solution

We Think	***We Write***
The ratio of in-state to out-of-state students is 7 to 2.	$\frac{\text{In-state}}{\text{Out-of-state}} = \frac{7}{2}$
We do not know how many of either type student there are.	Let x = number of in-state students
We know that the total is 1854.	$1854 - x$ = number of out-of-state students
The ratio is 7 to 2.	$\frac{x}{1854 - x} = \frac{7}{2}$

$$\frac{x}{1854 - x} = \frac{7}{2}$$

$$2x = 7(1854 - x)$$

$$2x = 12{,}978 - 7x$$

$$9x = 12{,}978$$

$$x = \frac{12{,}978}{9}$$

$$x = 1442$$

In-state students: $x = 1442$

Out-of-state students: $1854 - x = 1854 - 1442$

$= 412$

There are 1442 in-state students and 412 out-of-state students in the freshman class.

Working with Distance Problems

In earlier chapters we solved distance problems using the fact that

> Distance = rate × time
>
> $d = r \cdot t$

Now let us consider distance problems that lead to fractional equations.

Example 3. On long trips, Bob drives his car at a constant rate that is 5 miles per hour faster than his brother George's constant rate. If Bob can drive 312 miles in the same length of time that George drives 282 miles, how fast does each one drive?

Solution

We Think	***We Write***
We know that distance equals rate times time.	Distance = (rate)(time)
We know the brothers' distances.	Bob's distance = 312 miles George's distance = 282 miles
We do not know their rates, but we know that Bob drives 5 mph *faster* than George.	Let r = George's rate $r + 5$ = Bob's rate
We know that Bob's time to drive 312 miles is the *same* as George's time to drive 282 miles.	Bob's time = George's time
We do not know their times, but we can isolate *time* in the distance formula.	$\dfrac{\text{distance}}{\text{rate}} = \text{time}$
Now we can write an expression for each brother's time.	$\text{Bob's time} = \dfrac{\text{distance}}{\text{rate}} = \dfrac{312}{r+5}$ $\text{George's time} = \dfrac{\text{distance}}{\text{rate}} = \dfrac{282}{r}$
Since their time to go these distances is the *same,* we may equate the expressions for time. This gives us a proportion to solve.	Bob's time = George's time $\dfrac{312}{r+5} = \dfrac{282}{r}$

$$\frac{312}{r+5} = \frac{282}{r}$$

$$312r = 282(r + 5)$$

$$312r = 282r + 1410$$

$$30r = 1410$$

$$r = \frac{1410}{30}$$

$$r = 47$$

$$\text{George's rate:} \quad r = 47 \text{ mph}$$

$$\text{Bob's rate:} \quad r + 5 = 47 + 5$$

$$= 52 \text{ mph}$$

George's rate is 47 miles per hour and Bob's rate is 52 miles per hour.

Example 4. Helen drove to the airport 120 miles from home to pick up her father. Her father drove home 10 miles per hour faster than Helen had driven. If the return trip took 24 fewer minutes, how fast did Helen and her father each drive?

Solution

We Think	***We Write***
We do not know either rate, but we know Helen's father's rate was 10 mph faster than her rate.	Let r = Helen's rate (mph) $r + 10$ = Father's rate (mph)

We Think	*We Write*
Each driver drove a distance of 120 miles.	Helen's distance = 120 mi Father's distance = 120 mi
Each driver's time is found by $\text{Time} = \dfrac{\text{distance}}{\text{rate}}$	$\text{Helen's time} = \dfrac{120}{r}$ $\text{Father's time} = \dfrac{120}{r+10}$
The drivers' times are *not* the same. We know that her father's time was 24 minutes *less* than Helen's time.	Father's time = Helen's time − 24 min
But rates are expressed in miles per hour, so we must express 24 *minutes* as hours.	$\dfrac{24}{60} = \dfrac{2}{5}$ hour
Now we can write an equation that relates the drivers' times.	$\text{Father's time} = \text{Helen's time} - \dfrac{2}{5}\text{ hour}$ $\dfrac{120}{r+10} = \dfrac{120}{r} - \dfrac{2}{5}$

We can solve this fractional equation by our usual methods.

$$\frac{120}{r+10} = \frac{120}{r} - \frac{2}{5} \qquad \text{LCD: } 5r(r+10)$$

$$5r\overset{1}{\cancel{(r+10)}}\left[\frac{120}{\underset{1}{\cancel{r+10}}}\right] = \frac{5\overset{1}{\cancel{r}}(r+10)}{1}\left[\frac{120}{\underset{1}{\cancel{r}}}\right] - \frac{\overset{1}{\cancel{5}}r(r+10)}{1}\left[\frac{2}{\underset{1}{\cancel{5}}}\right]$$

$$120(5r) = 5 \cdot 120(r+10) - 2r(r+10)$$

$$600r = 600r + 6000 - 2r^2 - 20r$$

$$2r^2 + 20r - 6000 = 0$$

$$2(r^2 + 10r - 3000) = 0$$

$$2(r+60)(r-50) = 0$$

$$r + 60 = 0 \quad \text{or} \quad r - 50 = 0$$

$$r = -60 \qquad\qquad r = 50$$

A negative rate does not make sense here, so we conclude that

Helen's rate: $r = 50$ mph

Father's rate: $r + 10 = 50 + 10 = 60$ mph

Helen's rate was 50 miles per hour and her father's rate was 60 miles per hour.

Solving Work Problems

If two or more people or machines work together to complete a job, we expect that less time will be required than if one person or machine were to complete the job alone. We can always find out how much of a total job has been done by a person or machine by multiplying the rate of work times the amount of time worked.

Amount of work = (rate of work)(time worked)

Notice the similarity to the distance formula.

Example 5. It takes Linda 4 hours to rake the lawn, but it takes her older sister Nancy only 2 hours to rake the same lawn. How long will it take them to rake the lawn together if each sister rakes at her usual rate?

Solution

We Think	*We Write*
We realize that the total work done (1 lawn) will equal the *sum* of the amounts of work (parts of lawn) done by each sister.	Linda's work + Nancy's work = 1 lawn
Each sister's work equals her rate multiplied by her time.	Linda's work = (rate)(time) Nancy's work = (rate)(time)
The time is the same, but we do not know what it is.	Let t = time worked (hours)
If Linda can rake 1 lawn in 4 hours, she can rake $\frac{1}{4}$ of the lawn in 1 hour.	Linda's rate = $\frac{1}{4}$ lawn per hour
If Nancy can rake 1 lawn in 2 hours, she can rake $\frac{1}{2}$ of the lawn in 1 hour.	Nancy's rate = $\frac{1}{2}$ lawn per hour
Each sister's work is the *product* of her rate and her time.	Linda's work = $\frac{1}{4} \cdot t$ Nancy's work = $\frac{1}{2} \cdot t$
The sum of the sisters' work must equal 1 whole job (the whole lawn). This gives us a fractional equation to solve.	$\frac{1}{4}t + \frac{1}{2}t = 1$

$$\frac{t}{4} + \frac{t}{2} = 1 \qquad \text{LCD: } 4$$

$$\frac{4}{1} \cdot \frac{t}{4} + \frac{4}{1} \cdot \frac{t}{2} = 4 \cdot 1$$

$$t + 2t = 4$$

$$3t = 4$$

$$t = \frac{4}{3}$$

If the sisters rake together, they can complete the lawn in $1\frac{1}{3}$ hours (or 1 hour 20 minutes). Notice that the time needed is less than either sister's time alone. This should *always* be true.

Now let's consider the problem stated at the beginning of this chapter.

Example 6. Mr. Romano uses two hoses to fill his swimming pool. By itself, the larger hose can fill the pool in 7 hours; the smaller hose requires 10 hours by itself. While using both hoses to fill his pool last weekend, Mr. Romano discovered after 2 hours that the pool drain had been left open. If the pool normally drains in 12 hours, how long after he closed the drain would the pool be filled?

Solution

We Think	***We Write***
We know each hose's rate.	Large-hose rate $= \frac{1}{7}$ pool per hour Small-hose rate $= \frac{1}{10}$ pool per hour
We know the drain's rate, but it acts *against* the filling of the pool.	Drain rate $= -\frac{1}{12}$ pool per hour
We know the draining time.	Time drain was open = 2 hours
We seek the unknown amount of time *after* the drain was closed.	Let t = time after closing drain.
The hoses had been filling for two hours *before* the drain was closed.	$t + 2$ = large-hose time $t + 2$ = small-hose time
Each hose's work and the drain's work are found by multiplying rate times time.	$\frac{1}{7}(t + 2)$ = large-hose work $\frac{1}{10}(t + 2)$ = small-hose work $-\frac{1}{12}(2)$ = drain work
The total work is the *sum* of all work done and must equal 1 filled pool.	$\frac{1}{7}(t + 2) + \frac{1}{10}(t + 2) - \frac{1}{12}(2) = 1$

$$\frac{t+2}{7} + \frac{t+2}{10} - \frac{2}{12} = 1$$

$$\frac{t+2}{7} + \frac{t+2}{10} - \frac{1}{6} = 1 \qquad \text{LCD: } 210$$

$$\frac{\overset{30}{\cancel{210}}}{1}\left[\frac{t+2}{\underset{1}{\cancel{7}}}\right] + \frac{\overset{21}{\cancel{210}}}{1}\left[\frac{t+2}{\underset{1}{\cancel{10}}}\right] - \frac{\overset{35}{\cancel{210}}}{1}\left[\frac{1}{\underset{1}{\cancel{6}}}\right] = 210(1)$$

$$30(t + 2) + 21(t + 2) - 35 = 210$$

$$30t + 60 + 21t + 42 - 35 = 210$$

$$51t + 67 = 210$$

$$51t = 143$$

$$t = \frac{143}{51}$$

$$t \doteq 2.8$$

The pool would be filled about 2.8 hours after Mr. Romano closed the drain or about 4.8 hours after he began filling it.

Trial Run

Use an equation to solve each problem.

_______ **1.** Gary has decided that he would like the ratio of his saving to his spending to be 3 to 10. If his take-home pay is \$1300 per month, how much should he save and how much should he spend?

_______ **2.** It takes train A one hour longer to travel 400 miles than it takes train B. If train B travels 20 miles per hour faster than train A, find each train's rate.

_______ **3.** Mr. Kretchmar can wax the gym floor in 6 hours and Mr. Freeman can do the same job in 4 hours. If Mr. Kretchmar starts working at 7 A.M. and is joined by Mr. Freeman at 8 A.M., when will the floor be finished?

Answers are below.

Answers to Trial Run

page 411 **1.** Save \$300; spend \$1000 **2.** Train A, 80 mph; train B, 100 mph **3.** 10 A.M.

Name ______________________________ Date ______________

EXERCISE SET 6.6

Use an equation to solve each problem.

________ 1. On a road map, 1 inch represents 15 miles. If the distance between Nashville and Louisville is 185 miles, how many inches are needed to represent this on the map?

________ 2. If an $8000 investment earns $480 in 6 months, how much will $15,000 earn when invested at the same rate for the same period of time?

________ 3. If Kyle drives 660 miles in $1\frac{1}{2}$ days, how many days would it take him to drive 1760 miles at the same rate?

________ 4. The sales tax on a $395 television set is $19.75. If the sales tax on a refrigerator is $39.50, what is the selling price of the refrigerator?

________ 5. Ann and Merideth each owns an interest in the Specialty Shop. The ratio of Ann's income from the investment to Merideth's income is 4 to 5. Together one month they received $1530. How much did each receive that month?

________ 6. In a certain math class the ratio of success to failure was 20 to 3. If 100 people passed the class, what was the total enrollment?

________ 7. At the baseball cap factory, Burt and Johnny together can sew bills on 55 caps in 1 hour. The ratio of the number Burt can sew to the number Johnny can sew is 2 to 3. Find how many caps each can complete in 1 hour.

________ 8. Maria jogs 2 miles per hour faster than her sister Lola. If Maria can jog 9 miles in the same length of time that Lola can jog 6 miles, how fast does each one jog?

________ 9. It takes Gretchen 6 minutes longer to ride her bike 6 miles than it does for George to ride the same distance. Find the rate of speed at which each rides if George rides 2 miles per hour faster than Gretchen.

________ 10. Lucy drives her car 15 miles per hour faster than Mitch rides his motorcycle. If Lucy travels 275 miles in the same amount of time Mitch travels 200 miles, what is the rate of speed of each vehicle?

________ 11. When Michelle and Carolyn leave for work, they each drive at the same rate. If Michelle travels 20 miles in 6 minutes less time than Carol drives 25 miles, how long does it take each of them to get to work?

________ 12. Lyle took his boat 144 miles up the Ohio River and then back again. The total time for the trip was 15 hours. Find the speed each way if the speed down the river was 8 miles per hour faster than the speed up the river.

________ 13. Victoria rode the train part of the way to college and then rode the bus the remainder of the trip. The train traveled 30 miles per hour faster than the bus. The train and the bus each traveled 400 miles, but the time spent on the train was 3 hours less than the time on the bus. Find the speed of each.

______ **14.** Denise can clean the dog pens at the Humane Shelter in 8 hours and Barry can do it in 6 hours. How long will it take them to clean the pens if they work together?

______ **15.** Gus can wallpaper Bernice's house in 5 days. If Bernice helps him, they can do the work in 3 days. How long would it take Bernice to paper the house by herself?

______ **16.** Mel can clean all the ovens at the Dipsy Donut Shop in 5 hours and Gus can do the same job in 3 hours. If Mel starts cleaning and Gus joins him in an hour, how long will it be before the ovens are clean?

______ **17.** Suzie agreed to make draperies for her mother's apartment. She estimated it would take her 15 hours, but with her mother helping her they were finished in 8 hours. How long would it have taken her mother to make the draperies?

______ **18.** It takes 4 hours to fill a cattle watering tank using a pump and a hose. When a pump is used alone, the tank can be filled in 6 hours. If the pump were not working, how long would it take to fill the tank using only the hose?

______ **19.** A bathtub can be filled by the faucet in 5 minutes, but it takes 7 minutes for the drain to empty the tub. If the tub is empty and the faucet and drain are left open, how long would it take before the tub is overflowing?

______ **20.** Mr. Donosky uses two pipes to fill his reservoir. Pipe A takes twice as long as pipe B. A third pipe empties the tank in 2 hours. It takes 4 hours to fill the tank to half its capacity when all three are open. How long would it take pipe A to fill the tank if the other two were closed?

☆ Stretching the Topics

______ **1.** Two pipes are pouring water into a reservoir and one is taking water out. The first pipe can fill it in 20 days, the second pipe can fill it in n days, and the third pipe can empty it in 12 days. What must be the value of n in order that the reservoir be filled in 90 days?

______ **2.** One plane flew 382 miles with the help of a tailwind in $\frac{3}{4}$ of the time that another plane flew 480 miles in the opposite direction. If each plane can fly 500 miles per hour in still air, find the rate of the tailwind to the nearest tenth of a mile per hour.

Check your answers in the back of your book.

If you can complete **Checkup 6.6,** you are ready to do the **Review Exercises** for Chapter 6.

Name ______________________________ Date ______________

✓ CHECKUP 6.6

Use an equation to solve each problem.

________ **1.** In the Hillsboro City School System, 2 out of every 5 teachers belong to a union. How many teachers belong to the union if the school system has 700 teachers?

________ **2.** Working together, Erma and Bob can install the plumbing in their new house in 16 hours. Working alone, it would take Erma 40 hours to do it. How long would it take Bob to do the installation working alone?

________ **3.** It takes Richard the same amount of time to drive 420 miles as it takes Sam to drive 385 miles. If Richard's speed is 5 miles per hour faster than Sam's speed, how fast does Sam drive?

Check your answers in the back of your book.

If You Missed Problem:	You Should Review Example:
1	1
2	5
3	2

Summary

In this chapter we learned to reduce, add, subtract, multiply, and divide rational algebraic expressions.

If We Wish to	We Must	Example
Reduce a fraction	Divide numerator and denominator by any common factors.	$\frac{x^2 - 4}{3x + 6} = \frac{(x + 2)(x - 2)}{3(x + 2)} = \frac{x - 2}{3}$
Multiply fractions	Multiply numerators and multiply denominators, reducing if possible. $\frac{A}{B} \cdot \frac{C}{D} = \frac{AC}{BD}$	$\frac{x + 2}{3x + 3} \cdot \frac{x^2 - 1}{x^2 + 5x + 6}$ $= \frac{\overset{1}{\cancel{x + 2}}}{3(\underset{1}{\cancel{x + 1}})} \cdot \frac{(\overset{1}{\cancel{x + 1}})(x - 1)}{(\underset{1}{\cancel{x + 2}})(x + 3)} = \frac{x - 1}{3(x + 3)}$
Divide fractions	Find the reciprocal of the divisor and multiply. $\frac{A}{B} \div \frac{C}{D} = \frac{A}{B} \cdot \frac{D}{C} = \frac{AD}{BC}$	$\frac{-x^2y}{3} \div \frac{xy^2}{15} = \frac{-x^2y}{3} \cdot \frac{15}{xy^2} = -\frac{5x}{y}$
Add or subtract fractions	Choose the LCD for the fractions. Change all fractions to fractions having the LCD. Then add or subtract numerators, keeping the LCD. Reduce if possible.	$\frac{3}{x^2} + \frac{2}{5x + 10} - \frac{1}{x^2 + 2x}$ $= \frac{3}{x^2} + \frac{2}{5(x + 2)} - \frac{1}{x(x + 2)}$ LCD: $5x^2(x + 2)$ $= \frac{15(x + 2)}{5x^2(x + 2)} + \frac{2x^2}{5x^2(x + 2)} - \frac{5x}{5x^2(x + 2)}$ $= \frac{15x + 30 + 2x^2 - 5x}{5x^2(x + 2)} = \frac{2(x^2 + 5x + 15)}{5x^2(x + 2)}$
Simplify complex fractions	*Method 1*: Rewrite the fraction as a division problem.	$\frac{\frac{1}{x} + \frac{1}{2}}{\frac{2}{x + 2}} = \left[\frac{1}{x} + \frac{1}{2}\right] \div \left[\frac{2}{x + 2}\right]$ $= \left[\frac{2}{2x} + \frac{x}{2x}\right] \div \left[\frac{2}{x + 2}\right]$ $= \frac{2 + x}{2x} \cdot \frac{x + 2}{2}$ $= \frac{(x + 2)^2}{4x}$
	Method 2: Multiply both numerator and denominator by the LCD for both.	$\frac{\frac{1}{x} + \frac{3}{5}}{\frac{2}{x^2}} = \frac{5x^2\left(\frac{1}{x} + \frac{3}{5}\right)}{5x^2\left(\frac{2}{x^2}\right)}$ $= \frac{5x + 3x^2}{10}$

Throughout our work with algebraic fractions, we were careful to state **restrictions** on the domain of the variable in order to avoid division by zero. In solving fractional equations we agreed to multiply both sides of the equation by the LCD for all the fractions *after* stating restrictions. Using this procedure, we were able to reject **extraneous** solutions.

To solve **fractional inequalities** we use the same boundary-point method used in solving quadratic inequalities. Once again we were careful to avoid solutions that made denominators equal zero.

Finally, we used our knowledge of fractional equations to solve problems stated in words. In particular we learned to solve proportions, distance problems, and work problems.

❑ Speaking the Language of Algebra

Complete each sentence with the appropriate word or phrase.

1. A fraction whose numerator and denominator are polynomials is called a ________ ________ ________ .

2. We restrict the domain of the variable in a rational algebraic expression in order to avoid ________ ________ ________ .

3. To reduce fractions we may ________ the numerator and denominator by any ________ ________ .

4. To look for the LCD for a group of fractions, we first ________ each of the original denominators.

5. To solve a fractional equation, we multiply both sides of the equation by the ________ .

6. If a fraction is positive and we know its numerator is negative, we conclude that its denominator is ________ .

7. An equation that states that two ratios are equal is called a ________ .

8. The amount of work done by a person or a machine can be found by multiplying the ________ of work times the ________ ________ .

△ Writing About Mathematics

Write your response to each question in complete sentences.

1. Explain why it is often necessary to restrict the domain of the variable when dealing with rational algebraic expressions. Why should any restrictions be stated *before* reducing a rational expression?

2. Explain why it is easier to multiply or divide fractions with different denominators than it is to add or subtract those same fractions.

3. Explain the process you would use to
   (a) Reduce a fraction to lowest terms.
   (b) Build a fraction to a new fraction having a specified denominator.

4. How do you recognize a complex fraction? Explain how to simplify a complex fraction.

5. Explain the difference in your approach to solving fractional equations and fractional inequalities.

Name ______________________________ Date ______________

REVIEW EXERCISES for Chapter 6

State the restrictions and reduce each fraction.

______ 1. $\dfrac{-9x^5y^3z}{6x^2y^3z^4}$

______ 2. $\dfrac{15x}{15x^2 - 30x}$

______ 3. $\dfrac{3x + 3y}{12x - 12y}$

______ 4. $\dfrac{x^2 + 2x - 35}{x^2 - 49}$

______ 5. $\dfrac{21x^2 - 17xy + 2y^2}{7x^2 + 13xy - 2y^2}$

______ 6. $\dfrac{a^3 - 8b^3}{3a^2 - 11ab + 10b^2}$

______ 7. $\dfrac{4(x - y)^2 + 3y(x - y)}{4x^2 - 5xy + y^2}$

______ 8. $\dfrac{64 - x^2}{2x^2 - 19x + 24}$

Perform the indicated operations and simplify. (Do not state restrictions.)

______ 9. $\dfrac{3x + 15}{21x^2} \cdot \dfrac{14x}{x + 5}$

______ 10. $\dfrac{x^2 - 36}{5x + 20} \cdot \dfrac{x^2 + 4x}{x^2 - 4x - 12}$

______ 11. $\dfrac{(3x - 5y)^2}{2x^2 - 7xy - 4y^2} \cdot \dfrac{3x^2 - 7xy - 20y^2}{9x^2 - 25y^2}$

______ 12. $\dfrac{x^2 - 121y^2}{x^9y^2} \div \dfrac{5x - 55y}{x^5y^3}$

______ 13. $\dfrac{3x^2 - 6xy}{3x^2 + 2xy - y^2} \div \dfrac{6x^2y - 3x^3}{9x^2 + 8xy - y^2}$

______ 14. $\left(\dfrac{x^2 - 144}{7x} \cdot \dfrac{14}{2x - 24}\right) \div \dfrac{3x + 36}{8x}$

______ 15. $\dfrac{5x}{2x - 3} + \dfrac{3x}{2x - 3}$

______ 16. $\dfrac{4x + 1}{x^2 - 9} - \dfrac{3x - 2}{x^2 - 9}$

______ 17. $\dfrac{5}{x + 10} + \dfrac{3}{x}$

______ 18. $\dfrac{2x}{x - 5} + \dfrac{x}{x + 5}$

______ 19. $\dfrac{3x + 6}{x^2 - x - 56} - \dfrac{2}{x - 8}$

______ 20. $\dfrac{1}{x - 4} - \dfrac{3}{x + 2} + \dfrac{2(2x - 11)}{x^2 - 2x - 8}$

______ 21. $\dfrac{\dfrac{x^{12}}{y^2}}{\dfrac{3x^9}{y}}$

______ 22. $\dfrac{\dfrac{3x - 1}{4x^2}}{\dfrac{9x^2 - 1}{12x^3}}$

______ 23. $\dfrac{x - \dfrac{64}{x}}{x - 8}$

______ 24. $\dfrac{\dfrac{1}{x^2} - \dfrac{1}{25}}{\dfrac{1}{x} - \dfrac{1}{5}}$

_______ 25. $\dfrac{1 - 8a^{-3}}{1 - 2a^{-1}}$

_______ 26. $\dfrac{6 - 7x^{-1} - 5x^{-2}}{3 - 5x^{-1}}$

Solve each equation or inequality.

_______ 27. $\dfrac{x + 3}{4} = \dfrac{x - 5}{3}$

_______ 28. $\dfrac{2}{a - 6} - \dfrac{4a}{a^2 + a - 42} = 0$

_______ 29. $\dfrac{x + 5}{x - 3} + \dfrac{2x}{x^2 - 9} = \dfrac{1}{x + 3}$

_______ 30. $\dfrac{3}{x - 9} > 0$

_______ 31. $\dfrac{2x - 5}{x} \geq -3$

_______ 32. $\dfrac{x - 2}{x + 1} < \dfrac{1}{2}$

Use an equation to solve each problem.

_______ 33. On the architect's drawing of Lillian's house, $\frac{1}{4}$ inch represents 2 feet. If the length of her living room on the drawing is $2\frac{1}{2}$ inches, find the actual length of the living room.

_______ 34. Alex can clean the ovens at the Pizza Parlor in 6 hours. If Ron helps him, they can do the work in $2\frac{1}{2}$ hours. How long would it take Ron to clean the ovens by himself?

_______ 35. Becky drives her car 10 miles per hour faster than Mac drives his truck. If Becky drives 385 miles in the same amount of time Mac drives 315 miles, find the rate of speed of each vehicle.

Check your answers in the back of your book.

If You Missed Exercises:	You Should Review Examples:	
1	SECTION 6.1	3, 4
2–6		5–8
7		10
8		11, 12
9–11	SECTION 6.2	1–3
12, 13		4–7
14		8, 9
15, 16	SECTION 6.3	3–5
17, 18		13–15
19, 20		16–18
21, 22	SECTION 6.4	1, 2
23, 24		3–5
25, 26		6, 7
27	SECTION 6.5	1
28, 29		3, 4
30		5–7
31, 32		8–10
33	SECTION 6.6	2
34		5
35		3, 4

If you have completed the **Review Exercises** and corrected your errors, you are ready to take the **Practice Test** for Chapter 6.

Name ______________________ Date __________

PRACTICE TEST for Chapter 6

State the restrictions on the variables and reduce each fraction.

	Section	Examples
______ 1. $\dfrac{27x^4y^5c}{18x^3y^5c^3}$	6.1	3, 4
______ 2. $\dfrac{x^2 + 2x - 15}{x^2 + 3x - 10}$	6.1	8
______ 3. $\dfrac{y^3 - 1}{y^2 - 1}$	6.1	7
______ 4. $\dfrac{m^2 + 2mn - m - 2n}{m^2 - mn - 6n^2}$	6.1	9

Perform the indicated operations and simplify. (Do not state restrictions.)

	Section	Examples
______ 5. $\dfrac{x^2 + 6x - 16}{3x^2 - 9x} \cdot \dfrac{6x - 18}{x^2 - x - 2}$	6.2	1
______ 6. $\dfrac{x^2 - 49}{5x} \cdot \dfrac{50}{2x - 14} \cdot \dfrac{4x^3}{5x + 35}$	6.2	3
______ 7. $\dfrac{3a^2 - 75b^2}{a^4b^5} \div \dfrac{4a - 20b}{ab^6}$	6.2	4, 5
______ 8. $\dfrac{x^2 - 4xy + 4y^2}{x^3 - 2x^2y} \div \dfrac{x^2 + xy - 6y^2}{7x + 21y}$	6.2	6, 7
______ 9. $\dfrac{6x - 5}{x^2 - 49} - \dfrac{5x - 12}{x^2 - 49}$	6.3	5
______ 10. $\dfrac{3}{5x + 2} - \dfrac{1}{5x}$	6.3	13
______ 11. $\dfrac{y}{y + 6} - \dfrac{6}{6 - y}$	6.3	18
______ 12. $\dfrac{x + 1}{x^2 + 7x + 12} + \dfrac{2}{x + 3}$	6.3	16
______ 13. $\dfrac{10}{3x - 4} - \dfrac{x}{x + 6} + \dfrac{x + 2}{3x^2 + 14x - 24}$	6.3	16, 17
______ 14. $\dfrac{\dfrac{1 - 2x}{7x^2}}{\dfrac{1 - 4x^2}{2x}}$	6.4	2

		Section	Examples
________	15. $\dfrac{2x + 7 - \dfrac{15}{x}}{\dfrac{x}{5} - \dfrac{5}{x}}$	6.4	4
________	16. $\dfrac{2 - x^{-1} - 10x^{-2}}{1 + 2x^{-1}}$	6.4	6, 7

Solve each equation or inequality.

________	17. $\dfrac{8x + 1}{15} = \dfrac{2 - x}{6}$	6.5	1
________	18. $\dfrac{2x - 6}{x^2 - 6x - 7} - \dfrac{1}{x - 7} = 0$	6.5	3, 4
________	19. $\dfrac{5x - 1}{x + 4} \leq 0$	6.5	8, 9

Use an equation to solve each problem.

________	20. A large and small hose together can fill a pool in 7 hours. If the large hose alone can fill the pool in 10 hours, how long will it take the small hose alone to fill the pool?	6.6	6
________	21. The sales tax on a new car costing \$15,300 is \$1071. If the tax rate is the same, what will be the sales tax on a new car costing \$11,740?	6.6	1
________	22. Chico drives 5 miles per hour faster than Vagas drives. If Chico can drive 310 miles in the same amount of time that Vagas can drive 285 miles, how fast does each drive?	6.6	3

Name ______________________ Date ____________

SHARPENING YOUR SKILLS after Chapters 1–6

		Section
________	1. Use an inequality to state whether $-\sqrt{13}$ is positive or negative.	1.1
________	2. Evaluate $(2x - 1)(4x - 3)$ when $x = \frac{1}{2}$.	1.4
________	3. Evaluate $-a^3$ when $a = -3$.	3.1
________	4. Evaluate -4^{-2}.	3.1
________	5. Simplify $\sqrt{768}$.	5.2
________	6. Use the discriminant to describe the solutions for $3x^2 + 3x = 1$.	5.3
________	7. Write a quadratic equation with the solutions $x = 3$ and $x = -2$.	5.1

Perform the indicated operations and simplify.

________	8. $9[5x - (3x - y)] - [4(x + y) - 4y]$	1.4
________	9. $a^3b(a^4b^2)^3 + 3ab^7(a^3b^4)^2$	3.1
________	10. $\left(\frac{a^{-3}b}{4b^{-2}}\right)^2\left(\frac{a^2b^{-1}}{3b^5}\right)^{-3}$	3.2
________	11. $a(a^2 - 7) - 2(a - 3)$	3.3
________	12. $-y^2(7 - y)(9 - y)$	3.4
________	13. $(x^5 - 9x^4 + 5x^3 - 11x^2) \div x^3$	3.5
________	14. $(x^3 - 4x^2y - 11xy^2 - 6y^3) \div (x + y)$	3.5

Rewrite each expression in completely factored form.

________	15. $-3x^6y^3 + 21x^4y^2 - 27x^2y$	4.1
________	16. $(a - b)^2 - 5(a - b)$	4.1
________	17. $9x^2y^2 - 121$	4.2
________	18. $10y^2 - 7y - 12$	4.3
________	19. $120 - 9y - 3y^2$	4.3

Solve each equation or inequality. SECTION

________	**20.** $\frac{9y - 2}{5} + 8 = -2y$	2.1
________	**21.** $\lvert 2x - 11 \rvert = -7$	2.4
________	**22.** $\left\lvert 2 - \frac{m}{5} \right\rvert > 3$	2.4
________	**23.** $-3x(x + 8)(9x - 5) = 0$	5.1
________	**24.** $30m^2 = 7m + 2$	5.3
________	**25.** $121x^2 = 64$	5.2
________	**26.** $x(x - 5) = 4 - 3x$	5.3
________	**27.** $2(y - 1) - 9 \leq 8(y + 2) + 13$	2.3
________	**28.** $-2 \leq -a + 4 < 7$	2.3
________	**29.** $5x^2 + 18x \leq 8$	5.4
________	**30.** Solve $M = 2ax^2 + x$ for x.	5.3

CHAPTER

7

Working with Rational Exponents and Radicals

One leg of a right triangle is 1 foot longer than the other. If the length of the shorter leg is increased by 4 feet and the length of the longer leg is multiplied by 6, the length of the hypotenuse of the enlarged triangle is five times the length of the original hypotenuse. Find the dimensions of both triangles.

In Chapter 3 we learned to work with integer exponents, and in Chapter 5 we discussed square roots of numbers. Now it is time to discuss both exponents and roots in more depth and see how the two concepts are related. We must discuss the use of rational numbers as exponents and learn to deal with roots other than square roots. In this chapter we learn to

1. Work with rational exponents.
2. Relate rational exponents to radicals.
3. Simplify radical expressions.
4. Operate with radical expressions.
5. Rationalize denominators.
6. Solve radical equations.
7. Work with complex numbers.

7.1 Introducing Rational Exponents and Radicals

Working with Rational Exponents

From earlier work with exponents and square roots, you should recall the following definitions.

Exponent Definitions	Principal Square Root
$a^n = \underbrace{a \cdot a \cdot \ldots \cdot a}_{n \text{ factors}}$ $a^0 = 1 \quad (a \neq 0)$ $a^{-n} = \dfrac{1}{a^n} \quad (a \neq 0)$	$\sqrt{a} = b$ if $b^2 = a$ where $b \geq 0$

You should also recall the five basic laws of exponents that allowed us to perform operations with integer powers of bases.

Laws of Exponents

1. $a^m \cdot a^n = a^{m+n}$
2. $(a^m)^n = a^{mn}$
3. $(ab)^n = a^n b^n$
4. $\left(\dfrac{a}{b}\right)^n = \dfrac{a^n}{b^n} \quad (b \neq 0)$
5. $\dfrac{a^m}{a^n} = a^{m-n} \quad (a \neq 0)$

You will be relieved to know that these same laws of exponents "work" for any *rational numbers* used as exponents. (Remember that rational numbers are real numbers that are expressible as fractions of integers.)

The five laws of exponents can be used for rational exponents as well as for integer exponents.

For example,

1. $x^{\frac{2}{3}} \cdot x^{\frac{5}{3}} = x^{\frac{2}{3}+\frac{5}{3}} = x^{\frac{7}{3}}$
2. $\left(x^{\frac{3}{4}}\right)^8 = x^{\frac{3}{4}\cdot\frac{8}{1}} = x^6$
3. $(xy)^{\frac{1}{2}} = x^{\frac{1}{2}} \cdot y^{\frac{1}{2}}$
4. $\left(\dfrac{a}{b}\right)^{\frac{2}{3}} = \dfrac{a^{\frac{2}{3}}}{b^{\frac{2}{3}}}$
5. $\dfrac{x^{\frac{3}{4}}}{x^{\frac{1}{4}}} = x^{\frac{3}{4}-\frac{1}{4}} = x^{\frac{2}{4}} = x^{\frac{1}{2}}$

It is also true that negative rational exponents can be interpreted in the same way that we interpreted integer exponents.

$$a^{-\frac{p}{q}} = \frac{1}{a^{\frac{p}{q}}} \qquad (a \neq 0)$$

For example,

$$x^{-\frac{2}{3}} = \frac{1}{x^{\frac{2}{3}}} \qquad \text{and} \qquad \frac{x^{\frac{5}{6}}}{x^{\frac{7}{6}}} = x^{\frac{5}{6}-\frac{7}{6}} = x^{-\frac{2}{6}} = x^{-\frac{1}{3}} = \frac{1}{x^{\frac{1}{3}}}$$

You may be concerned by the fact that we have not yet given any meaning to a rational exponent. We shall arrive at an interpretation for a rational exponent in the third part of this section. For now, however, we need only know that rational exponents obey the five laws of exponents, so that expressions involving rational exponents should be simplified using exactly the same techniques that we used for integer exponents.

You try Example 1.

Example 1. Simplify $x^{\frac{3}{5}} \cdot x^{\frac{4}{5}} \cdot x^{-\frac{2}{5}}$.

Solution. To multiply, we must ______ exponents.

$$x^{\frac{3}{5}} \cdot x^{\frac{4}{5}} \cdot x^{-\frac{2}{5}}$$
$$= x^{\frac{3}{5}+\frac{4}{5}-\frac{2}{5}}$$

Check your work on page 432. ▶

Example 2. Simplify $x^{\frac{1}{3}} \cdot x^{-\frac{1}{6}} \cdot x^{\frac{3}{4}}$.

Solution. We must add exponents, but here we must use an LCD for the fractional exponents.

$$x^{\frac{1}{3}} \cdot x^{-\frac{1}{6}} \cdot x^{\frac{3}{4}}$$
$$= x^{\frac{1}{3}-\frac{1}{6}+\frac{3}{4}}$$
$$= x^{\frac{4}{12}-\frac{2}{12}+\frac{9}{12}}$$
$$= x^{\frac{11}{12}}$$

Now try Example 3.

Example 3. Simplify $\left(x^{\frac{3}{4}}y^{-\frac{1}{8}}\right)^4$.

Solution. To raise a product to a power, we must raise each factor to that power.

$$\left(x^{\frac{3}{4}}y^{-\frac{1}{8}}\right)^4$$
$$= \left(x^{\frac{3}{4}}\right)^4\left(y^{-\frac{1}{8}}\right)^4$$
$$= x^{\frac{3}{4}\cdot\frac{4}{1}}y^{-\frac{1}{8}\cdot\frac{4}{1}}$$

Check your work on page 432. ▶

Example 4. Simplify $\left(\dfrac{x^{\frac{1}{6}}z^{-2}}{y^{\frac{3}{4}}}\right)^{\frac{1}{3}}$.

Solution. To raise a fraction to a power, we raise the numerator and denominator to that power.

$$\left(\frac{x^{\frac{1}{6}}z^{-2}}{y^{\frac{3}{4}}}\right)^{\frac{1}{3}} = \frac{\left(x^{\frac{1}{6}}z^{-2}\right)^{\frac{1}{3}}}{\left(y^{\frac{3}{4}}\right)^{\frac{1}{3}}}$$
$$= \frac{x^{\frac{1}{6}\cdot\frac{1}{3}}z^{-\frac{2}{1}\cdot\frac{1}{3}}}{y^{\frac{3}{4}\cdot\frac{1}{3}}}$$
$$= \frac{x^{\frac{1}{18}}z^{-\frac{2}{3}}}{y^{\frac{1}{4}}}$$
$$= \frac{x^{\frac{1}{18}}}{y^{\frac{1}{4}}z^{\frac{2}{3}}}$$

Example 5. Simplify $\dfrac{x^{\frac{7}{8}}y^{-\frac{1}{12}}}{x^{-\frac{3}{4}}y^{\frac{9}{2}}}$.

Solution. To divide, we must subtract exponents.

$$\begin{aligned}\frac{x^{\frac{7}{8}}y^{-\frac{1}{12}}}{x^{-\frac{3}{4}}y^{\frac{9}{2}}} &= x^{\frac{7}{8}-(-\frac{3}{4})}y^{-\frac{1}{12}-\frac{9}{2}} \\ &= x^{\frac{7}{8}+\frac{6}{8}}y^{-\frac{1}{12}-\frac{54}{12}} \\ &= x^{\frac{13}{8}}y^{-\frac{55}{12}} \\ &= \frac{x^{\frac{13}{8}}}{y^{\frac{55}{12}}}\end{aligned}$$

Remember that we agreed in our earlier work with positive and negative exponents that it is less confusing if we use the laws of exponents *first*. Once we arrive at our simplified expression, we then use the definition of a negative exponent to rewrite the expression using positive exponents.

Trial Run

Perform the indicated operation and simplify.

_______ 1. $x^{\frac{3}{4}} \cdot x^{\frac{9}{4}}$

_______ 2. $a^{\frac{1}{2}} \cdot a^{\frac{3}{4}} \cdot a^{-\frac{1}{4}}$

_______ 3. $x^{\frac{2}{5}} \cdot x^{\frac{3}{10}} \cdot x^{-\frac{1}{2}}$

_______ 4. $\left(x^{\frac{2}{3}}y^{-\frac{1}{2}}\right)^6$

_______ 5. $\left(\dfrac{x^{\frac{5}{2}}y^{-15}}{a^{10}}\right)^{\frac{1}{5}}$

_______ 6. $\dfrac{x^{\frac{5}{8}}y^{\frac{5}{6}}}{x^{\frac{3}{4}}y^{\frac{1}{12}}}$

Answers are on page 432.

Introducing Radical Expressions

Earlier we were reminded of the square-root notation used in Chapter 5. We recalled that if a is a positive number or zero, then

$$\sqrt{a} = b \quad \text{means} \quad b^2 = a$$

We also agreed that $\sqrt{a}$ would always represent the positive square root of a, called the *principal square root of a*. The important restriction that a must be a *positive* number or zero was necessary because

> The square root of a negative number is *not* a real number.

When we are asked to find the *square root* of a, we are being asked to find a number whose *square* is the number a. Suppose that we are asked to find a number whose *cube* (or third power) is the number a, or a number whose *fourth power* is a, and so on. Here we are looking for a **cube root,** or a **fourth root,** and so on.

Root	Symbol	Examples	Reasons
Square root	$\sqrt{a}$	$\sqrt{4} = 2$	$2^2 = 4$
		$\sqrt{-4}$ is not real	
Cube root	$\sqrt[3]{a}$	$\sqrt[3]{8} = 2$	$2^3 = 8$
		$\sqrt[3]{-8} = -2$	$(-2)^3 = -8$
Fourth root	$\sqrt[4]{a}$	$\sqrt[4]{16} = 2$	$2^4 = 16$
		$\sqrt[4]{-16}$ is not real	
Fifth root	$\sqrt[5]{a}$	$\sqrt[5]{32} = 2$	$2^5 = 32$
		$\sqrt[5]{-32} = -2$	$(-2)^5 = -32$

Do you see the pattern? The *odd* root of any real number will be real, but the *even* root of a *negative* number is *not* a real number.

Roots of Real Numbers

$$\sqrt[n]{a} = b \quad \text{if} \quad b^n = a$$

where n is a natural number and a is real.

1. If a is positive, then

$$\sqrt[n]{a} \text{ is real and positive}$$

2. If a is negative, and

if n is odd, then $\sqrt[n]{a}$ is real and negative

if n is even, then $\sqrt[n]{a}$ is *not* a real number

Let's pause a moment and identify the parts of a **radical expression.**

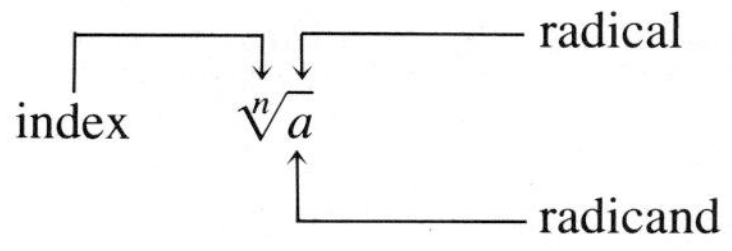

The natural number n is the **index;** it indicates what root we are trying to find. If no index number appears, it is understood to be a 2. The $\sqrt{}$ symbol is called the **radical** and the quantity within the radical is called the **radicand.**

Let's practice using radical notation by finding the values of some radical expressions.

Example 6. Find $\sqrt[3]{1000}$.

Solution

$\sqrt[3]{1000} = 10$ because $10^3 = 1000$.

You try Example 7.

Example 7. Find $\sqrt[4]{16}$.

Solution

$\sqrt[4]{16} =$ _____ because (_____)$^4 = 16$.

Check your work on page 432. ▶

Example 8. Find $\sqrt[3]{-\frac{1}{27}}$.

Solution

$$\sqrt[3]{-\frac{1}{27}} = -\frac{1}{3} \text{ because } \left(-\frac{1}{3}\right)^3 = -\frac{1}{27}.$$

Now try Example 9.

Example 9. Find $\sqrt[5]{\frac{1}{32}}$.

Solution

$$\sqrt[5]{\frac{1}{32}} = \_\_\_\_ \text{ because } (\_\_\_\_)^5 = \frac{1}{32}.$$

Check your work on page 432. ▶

Example 10. Find $\sqrt[6]{-1}$.

Solution. $\sqrt[6]{-1}$ is not real. An even root of a negative number is not a real number.

You try Example 11.

Example 11. Find $\sqrt[7]{-1}$.

Solution

$$\sqrt[7]{-1} = \_\_\_\_ \text{ because } (\_\_\_\_)^7 = -1.$$

Check your work on page 432. ▶

Trial Run

Find the value of each radical expression.

______ 1. $\sqrt[5]{243}$

______ 2. $\sqrt[3]{-125}$

______ 3. $\sqrt{-9}$

______ 4. $\sqrt[5]{-32}$

______ 5. $\sqrt[4]{\frac{1}{16}}$

______ 6. $\sqrt[3]{-\frac{1}{8}}$

Answers are on page 432.

Relating Rational Exponents and Radicals

Suppose that we return to rational exponents and see if we can come up with an interpretation for exponents that are fractions. From the second law of exponents we know that

$$\left(a^{\frac{1}{2}}\right)^2 = a$$

So $a^{\frac{1}{2}}$ is the number whose square is a. But we have already agreed that $\sqrt{a}$ is the number whose square is a. Therefore, we conclude that

$$a^{\frac{1}{2}} = \sqrt{a}$$

$$\left(a^{\frac{1}{3}}\right)^3 = a$$

So $a^{\frac{1}{3}}$ is the number whose cube is a. But we have already agreed that $\sqrt[3]{a}$ is the number whose cube is a. Therefore, we conclude that

$$a^{\frac{1}{3}} = \sqrt[3]{a}$$

In fact, for any natural number n, we define

$$a^{\frac{1}{n}} = \sqrt[n]{a} \quad \text{provided that the root is real}$$

To avoid having to make restrictions upon a when n is even, we often precede a list of exercises with the following statement: *Assume that all radical expressions represent real numbers*. Let's make this assumption as we practice switching back and forth between radical and exponential expressions.

Example 12. Write $\sqrt[3]{4x}$ in exponential form.

Solution

$$\sqrt[3]{4x} = (4x)^{\frac{1}{3}}$$

You try Example 13.

Example 13. Write $3\sqrt[5]{y}$ in exponential form.

Solution

$$3\sqrt[5]{y} =$$

Check your work on page 432. ▶

Example 14. Write $(x + y)^{\frac{1}{3}}$ in radical form.

Solution

$$(x + y)^{\frac{1}{3}} = \sqrt[3]{x + y}$$

Now try Example 15.

Example 15. Write $x^{\frac{1}{3}} + y^{\frac{1}{3}}$ in radical form.

Solution

$$x^{\frac{1}{3}} + y^{\frac{1}{3}} =$$

Check your work on page 432. ▶

Example 16. Write $9^{\frac{1}{2}} + 16^{\frac{1}{2}}$ in radical form and simplify if possible.

Solution

$$9^{\frac{1}{2}} + 16^{\frac{1}{2}} = \sqrt{9} + \sqrt{16}$$
$$= 3 + 4$$
$$= 7$$

You complete Example 17.

Example 17. Write $(9 + 16)^{\frac{1}{2}}$ in radical form and simplify if possible.

Solution

$$(9 + 16)^{\frac{1}{2}} = \sqrt{9 + 16}$$
$$=$$

Check your work on page 432. ▶

From Examples 16 and 17, we note that

$$\sqrt{9} + \sqrt{16} \neq \sqrt{9 + 16}$$
$$3 + 4 \neq \sqrt{25}$$
$$7 \neq 5$$

and we state in general that **the *n*th root of a sum is *not* equal to the sum of the *n*th roots.**

$$\sqrt[n]{a + b} \neq \sqrt[n]{a} + \sqrt[n]{b}$$

Now that we know that

$$a^{\frac{1}{n}} = \sqrt[n]{a}$$

perhaps we can use this definition and the laws of exponents to find meaning for expressions such as $x^{\frac{2}{3}}$ and $a^{\frac{3}{4}}$.

$$x^{\frac{2}{3}} = \left(x^{\frac{1}{3}}\right)^2 = (\sqrt[3]{x})^2 \quad \text{or} \quad x^{\frac{2}{3}} = (x^2)^{\frac{1}{3}} = \sqrt[3]{x^2}$$

$$a^{\frac{3}{4}} = \left(a^{\frac{1}{4}}\right)^3 = (\sqrt[4]{a})^3 \quad \text{or} \quad a^{\frac{3}{4}} = (a^3)^{\frac{1}{4}} = \sqrt[4]{a^3}$$

For each of these examples we arrived at two possible interpretations. Indeed, if we assume that our radicals represent real numbers, then either interpretation is acceptable.

Definition of Rational Exponent

$$a^{\frac{m}{n}} = \sqrt[n]{a^m} = (\sqrt[n]{a})^m$$

where m and n are natural numbers, and $a \geq 0$ whenever n is *even.*

Notice that in switching between fractional exponent form and radical form

$$a^{\frac{m}{n}} = \sqrt[n]{a^m}$$

The denominator of the fractional exponent tells us the index of the radical.
The numerator of the fractional exponent tells us the exponent on the base.

Example 18. Write $8^{\frac{2}{3}}$ in radical form and simplify.

Solution

$$8^{\frac{2}{3}} = \sqrt[3]{8^2} = \sqrt[3]{64} = 4 \quad \text{or} \quad 8^{\frac{2}{3}} = (\sqrt[3]{8})^2 = 2^2 = 4$$

Example 19. Write $32^{\frac{7}{5}}$ in radical form and simplify.

Solution

$$32^{\frac{7}{5}} = (\sqrt[5]{32})^7 = 2^7 = 128 \quad \text{or} \quad 32^{\frac{7}{5}} = \sqrt[5]{32^7} = \sqrt[5]{34{,}359{,}738{,}368} = 128$$

You should agree that in Example 19 it was much easier to find the root first, then the power. Quite often this is the case.

Example 20. Write $(x^2 - 1)^{\frac{4}{7}}$ in radical form.

Solution

$$(x^2 - 1)^{\frac{4}{7}} = \sqrt[7]{(x^2 - 1)^4}$$

or

$$(x^2 - 1)^{\frac{4}{7}} = (\sqrt[7]{x^2 - 1})^4$$

You try Example 21.

Example 21. Write $3x^{\frac{5}{11}}$ in radical form.

Solution

$$3x^{\frac{5}{11}} = 3\_\_\_\_\_$$

or

$$3x^{\frac{5}{11}} = 3(\_\_\_\_\_)$$

Check your work on page 432. ▶

Example 22. Write $(x - 5)^{-\frac{2}{3}}$ in radical form.

Solution

$$(x - 5)^{-\frac{2}{3}} = \frac{1}{(x - 5)^{\frac{2}{3}}} = \frac{1}{\sqrt[3]{(x - 5)^2}}$$

or

$$(x - 5)^{-\frac{2}{3}} = \frac{1}{(\sqrt[3]{x - 5})^2}$$

Example 23. Write $\frac{1}{\sqrt[3]{x^2y}}$ in exponential form.

Solution

$$\frac{1}{\sqrt[3]{x^2y}} = \frac{1}{(x^2y)^{\frac{1}{3}}} = \frac{1}{x^{\frac{2}{3}}y^{\frac{1}{3}}}$$

Now try Example 24.

Example 24. Write $-\sqrt{x}$ in exponential form. Assume that $x \geq 0$.

Solution

$$-\sqrt{x} = -1 \cdot \sqrt{x}$$
$$= -x^{\_\_\_}$$

Check your work on page 432. ▶

Example 25. Write $\sqrt{-x}$ in exponential form. Assume that $x \leq 0$.

Solution

$$\sqrt{-x} = (-x)^{\frac{1}{2}}$$

Trial Run

Assume that all radical expressions represent real numbers.

________ 1. Simplify $27^{\frac{2}{3}}$

________ 2. Write $x^{\frac{3}{8}}$ in radical form.

________ 3. Write $\frac{1}{\sqrt[5]{x^2y^3}}$ in exponential form.

________ 4. Write $-\sqrt[3]{a}$ in exponential form.

________ 5. Write $(x - 3)^{\frac{3}{4}}$ in radical form.

________ 6. Write $(x - y)^{-\frac{2}{5}}$ in radical form.

Answers are on page 432.

▶ Examples You Completed

Example 1. Simplify $x^{\frac{3}{5}} \cdot x^{\frac{4}{5}} \cdot x^{-\frac{2}{5}}$.

Solution. To multiply, we must add exponents.

$$x^{\frac{3}{5}} \cdot x^{\frac{4}{5}} \cdot x^{-\frac{2}{5}}$$
$$= x^{\frac{3}{5}+\frac{4}{5}-\frac{2}{5}}$$
$$= x^{\frac{5}{5}}$$
$$= x$$

Example 3 *(Solution)*

$$\left(x^{\frac{3}{4}}\right)^4\left(y^{-\frac{1}{8}}\right)^4 = x^{\frac{3}{4}\cdot\frac{4}{1}}y^{-\frac{1}{8}\cdot\frac{4}{1}}$$
$$= x^3y^{-\frac{1}{2}}$$
$$= \frac{x^3}{y^{\frac{1}{2}}}$$

Example 7. Find $\sqrt[4]{16}$.

Solution

$\sqrt[4]{16} = 2$ because $(2)^4 = 16$.

Example 9. Find $\sqrt[5]{\frac{1}{32}}$.

Solution

$\sqrt[5]{\frac{1}{32}} = \frac{1}{2}$ because $(\frac{1}{2})^5 = \frac{1}{32}$.

Example 11. Find $\sqrt[7]{-1}$.

Solution

$\sqrt[7]{-1} = -1$ because $(-1)^7 = -1$.

Example 13. Write $3\sqrt[5]{y}$ in exponential form.

Solution

$$3\sqrt[5]{y} = 3y^{\frac{1}{5}}$$

Example 15. Write $x^{\frac{1}{3}} + y^{\frac{1}{3}}$ in radical form.

Solution

$$x^{\frac{1}{3}} + y^{\frac{1}{3}}$$
$$= \sqrt[3]{x} + \sqrt[3]{y}$$

Example 17. Write $(9 + 16)^{\frac{1}{2}}$ in radical form.

Solution

$$(9 + 16)^{\frac{1}{2}} = \sqrt{9 + 16}$$
$$= \sqrt{25}$$
$$= 5$$

Example 21. Write $3x^{\frac{5}{11}}$ in radical form.

Solution

$$3x^{\frac{5}{11}} = 3\sqrt[11]{x^5}$$

or

$$3x^{\frac{5}{11}} = 3(\sqrt[11]{x})^5$$

Example 24. Write $-\sqrt{x}$ in exponential form. Assume that $x \geq 0$.

Solution

$$-\sqrt{x} = -1 \cdot \sqrt{x}$$
$$= -x^{\frac{1}{2}}$$

Answers to Trial Runs

page 426 1. x^3 2. a 3. $x^{\frac{1}{5}}$ 4. $\frac{x^4}{y^3}$ 5. $\frac{x^{\frac{1}{2}}}{y^3a^2}$ 6. $\frac{y^{\frac{3}{4}}}{x^{\frac{1}{8}}}$

page 428 1. 3 2. -5 3. Not a real number 4. -2 5. $\frac{1}{2}$ 6. $-\frac{1}{2}$

page 431 1. 9 2. $\sqrt[8]{x^3}$ or $(\sqrt[8]{x})^3$ 3. $\frac{1}{x^{\frac{2}{5}}y^{\frac{3}{5}}}$ 4. $-a^{\frac{1}{3}}$ 5. $\sqrt[4]{(x-3)^3}$ or $(\sqrt[4]{x-3})^3$

6. $\frac{1}{\sqrt[5]{(x-y)^2}}$ or $\frac{1}{(\sqrt[5]{x-y})^2}$

Name ______________________ Date __________

EXERCISE SET 7.1

Perform the indicated operation and simplify. Write answers with positive exponents.

_____ 1. $x^{\frac{2}{5}} \cdot x^{\frac{8}{5}}$

_____ 2. $x^{\frac{3}{4}} \cdot x^{\frac{5}{4}}$

_____ 3. $a^{\frac{2}{3}} \cdot a^{\frac{7}{3}} \cdot a^{-\frac{4}{3}}$

_____ 4. $a^{\frac{2}{7}} \cdot a^{-\frac{1}{7}} \cdot a^{\frac{4}{7}}$

_____ 5. $\left(x^{\frac{2}{3}}\right)^{6}$

_____ 6. $\left(x^{\frac{3}{4}}\right)^{8}$

_____ 7. $\left(x^{\frac{3}{4}} \cdot y^{-3}\right)^{\frac{4}{3}}$

_____ 8. $\left(x^{\frac{3}{5}} \cdot y^{-6}\right)^{\frac{5}{3}}$

_____ 9. $\left(a^{\frac{3}{4}} \cdot b^{-\frac{1}{3}}\right)^{-2}$

_____ 10. $\left(a^{\frac{5}{6}} \cdot b^{-\frac{2}{5}}\right)^{-3}$

_____ 11. $x^{\frac{2}{3}} \cdot x^{\frac{2}{5}}$

_____ 12. $x^{\frac{1}{3}} \cdot x^{\frac{3}{4}}$

_____ 13. $\left(\dfrac{x^{\frac{3}{5}}y^{-\frac{3}{10}}}{z^{\frac{2}{5}}}\right)^{\frac{5}{6}}$

_____ 14. $\left(\dfrac{x^{\frac{3}{4}}y^{-\frac{1}{8}}}{z^{\frac{5}{6}}}\right)^{\frac{4}{5}}$

_____ 15. $\dfrac{x^{\frac{3}{4}}y^{\frac{2}{3}}}{x^{\frac{1}{4}}y^{\frac{1}{3}}}$

_____ 16. $\dfrac{x^{\frac{5}{6}}y^{\frac{3}{5}}}{x^{\frac{1}{6}}y^{\frac{2}{5}}}$

_____ 17. $\dfrac{a^{\frac{1}{4}}b^{3}}{a^{\frac{5}{8}}b^{\frac{1}{2}}}$

_____ 18. $\dfrac{a^{\frac{2}{3}}b^{4}}{a^{\frac{5}{6}}b^{\frac{2}{5}}}$

_____ 19. $\left(x^{\frac{4}{3}} \cdot y^{-\frac{3}{5}} \cdot z^{\frac{6}{7}}\right)^{\frac{1}{12}}$

_____ 20. $\left(x^{\frac{3}{4}} \cdot y^{-\frac{5}{2}} \cdot z^{\frac{15}{2}}\right)^{\frac{1}{15}}$

Find the value of each radical expression.

_____ 21. $\sqrt[3]{-27}$

_____ 22. $\sqrt[3]{-8}$

_____ 23. $\sqrt[4]{81}$

_____ 24. $\sqrt[4]{256}$

_____ 25. $\sqrt[4]{-16}$

_____ 26. $\sqrt[4]{-1}$

_____ 27. $\sqrt[3]{-\dfrac{64}{125}}$

_____ 28. $\sqrt[3]{-\dfrac{8}{27}}$

_____ 29. $\sqrt[5]{32}$

_____ 30. $\sqrt[5]{243}$

_____ 31. $\sqrt{\dfrac{25}{81}}$

_____ 32. $\sqrt{\dfrac{36}{49}}$

Write each radical expression in exponential form.

_____ 33. $\sqrt[4]{2x}$

_____ 34. $\sqrt{3x}$

_____ 35. $5\sqrt{a}$

_____ 36. $3\sqrt{y}$

_____ 37. $\sqrt{x^2 + 16}$

_____ 38. $\sqrt{x^2 + 25}$

_____ 39. $-\sqrt[5]{y}$

_____ 40. $-\sqrt[3]{x}$

_______ 41. $\sqrt[3]{-a}$

_______ 42. $\sqrt[5]{-b}$

_______ 43. $\sqrt[3]{x} + \sqrt[3]{y}$

_______ 44. $\sqrt{a} + \sqrt{b}$

_______ 45. $\sqrt[5]{x^2y}$

_______ 46. $\sqrt[4]{x^3y}$

_______ 47. $\sqrt[5]{(x - y)^2}$

_______ 48. $\sqrt[3]{(x + y)^2}$

_______ 49. $\dfrac{1}{\sqrt[3]{x^4y^2}}$

_______ 50. $\dfrac{1}{\sqrt[5]{x^4y^3}}$

Write each expression in radical form and simplify as much as possible. Assume that all radical expressions represent real numbers.

_______ 51. $(3y^3)^{\frac{1}{5}}$

_______ 52. $(2x^3)^{\frac{1}{4}}$

_______ 53. $4x^{\frac{1}{2}}$

_______ 54. $16x^{\frac{1}{4}}$

_______ 55. $(a + b)^{\frac{1}{3}}$

_______ 56. $(a - b)^{\frac{1}{2}}$

_______ 57. $x^{\frac{1}{5}} + y^{\frac{1}{5}}$

_______ 58. $x^{\frac{1}{6}} - y^{\frac{1}{6}}$

_______ 59. $(9 + 16)^{\frac{1}{2}}$

_______ 60. $(25 + 144)^{\frac{1}{2}}$

_______ 61. $(4 + 9)^{\frac{1}{2}}$

_______ 62. $(25 + 4)^{\frac{1}{2}}$

_______ 63. $y^{\frac{2}{3}}$

_______ 64. $x^{\frac{5}{8}}$

_______ 65. $(x - 3)^{\frac{2}{3}}$

_______ 66. $(x + 1)^{\frac{3}{4}}$

_______ 67. $(a - b)^{-\frac{3}{5}}$

_______ 68. $(a + b)^{-\frac{2}{3}}$

_______ 69. $8^{\frac{2}{3}}$

_______ 70. $32^{\frac{2}{5}}$

_______ 71. $81^{-\frac{3}{4}}$

_______ 72. $27^{-\frac{2}{3}}$

_______ 73. $8^{\frac{1}{3}} + 27^{\frac{1}{3}}$

_______ 74. $25^{\frac{1}{2}} + 100^{\frac{1}{2}}$

_______ 75. $\left(\dfrac{25}{36}\right)^{-\frac{1}{2}}$

_______ 76. $\left(\dfrac{4}{49}\right)^{-\frac{1}{2}}$

☆ Stretching the Topics

_______ 1. Find the value of $25^{-\frac{3}{2}} \cdot 16^{\frac{3}{4}} \cdot 8^{-\frac{4}{3}}$.

_______ 2. Simplify $\left(\dfrac{64^{-3}}{27n^{-6}}\right)^{\frac{2}{3}}$.

_______ 3. Simplify $\dfrac{27^{\frac{3}{4}}x^{\frac{5}{6}}y^{\frac{1}{6}}}{27^{\frac{1}{12}}x^{-\frac{1}{6}}y^{-\frac{1}{4}}}$.

Check your answers in the back of your book.

If you can complete **Checkup 7.1,** you are ready to go on to Section 7.2.

Name ______________________ Date ____________

✓ CHECKUP 7.1

Perform the indicated operation and simplify. Write answers with positive exponents.

______ 1. $a^{\frac{3}{4}} \cdot a^{\frac{1}{2}} \cdot a^{-\frac{1}{4}}$

______ 2. $\left(x^{\frac{1}{10}} \cdot y^{-\frac{2}{5}}\right)^{-5}$

______ 3. $\dfrac{x^{\frac{3}{4}} \cdot y^{\frac{2}{5}}}{x^{\frac{1}{2}} \cdot y^{\frac{7}{2}}}$

______ 4. $\left(x^{\frac{1}{9}} \cdot y^{-\frac{15}{4}} \cdot z^{\frac{5}{6}}\right)^{\frac{3}{5}}$

Find the value of each radical expression.

______ 5. $\sqrt{-81}$

______ 6. $\sqrt[5]{-\dfrac{1}{32}}$

Write in exponential form.

______ 7. $\sqrt[7]{x^2y^3}$

______ 8. $\sqrt{(x + y)^3}$

Write in radical form and simplify if possible.

______ 9. $(x - y)^{\frac{2}{3}}$

______ 10. $64^{\frac{5}{6}}$

Check your answers in the back of your book.

If You Missed Problems:	You Should Review Examples:
1, 2	1–3
3, 4	3–5
5	10
6	8, 9
7	12, 13
8	14
9, 10	18–20

7.2 Working with Radical Expressions

Simplifying Radical Expressions

In Chapter 5 we learned to simplify radicals that were square roots of perfect squares. Let's look again at $\sqrt{a^2}$.

Suppose that $a > 0$	Suppose that $a < 0$
$\sqrt{4^2} = \sqrt{16} = 4$	$\sqrt{(-4)^2} = \sqrt{16} = 4$
$\sqrt{(\frac{1}{3})^2} = \sqrt{\frac{1}{9}} = \frac{1}{3}$	$\sqrt{(-\frac{1}{3})^2} = \sqrt{\frac{1}{9}} = \frac{1}{3}$
$\sqrt{11^2} = \sqrt{121} = 11$	$\sqrt{(-11)^2} = \sqrt{121} = 11$

Because we know that the principal square root of a number is always nonnegative, we have found that $\sqrt{a^2}$ equals a only when a is *not negative*. (See the first column in the table.) If a *is negative*, then $\sqrt{a^2}$ equals the *opposite* of a. (See the second column.) In other words,

$$\sqrt{a^2} = \begin{cases} a & \text{if } a \text{ is positive or zero} \\ -a & \text{if } a \text{ is negative} \end{cases}$$

Doesn't this definition seem familiar? From Chapter 1 you should recognize it as the definition for the **absolute value of *a*.**

$$\sqrt{a^2} = |a|$$

As a matter of fact, this definition should be used whenever we work with an *even* nth root of an nth power. Of course, if n is *odd,* it is not correct to use absolute value notation.

$$\sqrt[5]{2^5} = \sqrt[5]{32} = 2$$

$$\sqrt[3]{(-4)^3} = \sqrt[3]{-64} = -4$$

We summarize our discussion by noting that

$$\sqrt[n]{a^n} = \begin{cases} a & \text{if } n \text{ is odd} \\ |a| & \text{if } n \text{ is even} \end{cases}$$

Example 1. Simplify $\sqrt[7]{x^7}$.

Solution. Here n is odd.

$$\sqrt[7]{x^7} = x$$

You try Example 2.

Example 2. Simplify $\sqrt[4]{y^4}$.

Solution. Here n is ________ .

$$\sqrt[4]{y^4} = \_\_\_\_$$

Check your work on page 442. ▶

Example 3. Simplify $\sqrt{(x-3)^2}$.

Solution. Here n is even.

$$\sqrt{(x-3)^2} = |x-3|$$

Now complete Example 4.

Example 4. Simplify $\sqrt[5]{(a-1)^5}$.

Solution. Here n is ______ .

$$\sqrt[5]{(a-1)^5} = \underline{\qquad}$$

Check your work on page 442. ▶

In Chapter 5 we also learned to simplify square roots whose radicands were products or quotients. These same properties can be used to simplify radicals with any index, n.

Properties of Radicals

$$\sqrt[n]{a \cdot b} = \sqrt[n]{a} \cdot \sqrt[n]{b}$$

$$\sqrt[n]{\frac{a}{b}} = \frac{\sqrt[n]{a}}{\sqrt[n]{b}} \qquad (b \neq 0)$$

where each radical expression is a real number.

We say that a radical expression is in **simplest form** when no factor appears under the radical with an exponent that is greater than or equal to the index on the radical. To simplify a radical expression with index n, we follow these steps.

Simplifying Radical Expressions

1. Write the radicand as a product of factors, with the most possible factors written as nth powers.
2. Find the nth roots of all the factors that are nth powers.
3. Leave the remaining factors in the radicand.

For simplicity's sake, **we shall assume throughout the following examples that if the index is even, then all variable factors in the radicand represent nonnegative numbers.** Such an assumption allows us to dispense with absolute value notation.

Example 5. Simplify $\sqrt[4]{32x^5y}$.

Solution

$\sqrt[4]{32x^5y}$

$= \sqrt[4]{16 \cdot 2 \cdot x^4 \cdot x \cdot y}$ — Look for perfect fourth powers.

$= \sqrt[4]{16} \cdot \sqrt[4]{x^4} \cdot \sqrt[4]{2xy}$

$= 2x\sqrt[4]{2xy}$ — Remove fourth roots.

Example 6. Simplify $\sqrt[3]{\dfrac{8x^3}{y^6}}$.

Solution

$\sqrt[3]{\dfrac{8x^3}{y^6}} = \dfrac{\sqrt[3]{8x^3}}{\sqrt[3]{y^6}}$ — Separate the radicals.

$= \dfrac{\sqrt[3]{8} \cdot \sqrt[3]{x^3}}{\sqrt[3]{y^3} \cdot \sqrt[3]{y^3}}$ — Look for perfect cubes.

$= \dfrac{2 \cdot x}{y \cdot y}$ — Remove cube roots.

$= \dfrac{2x}{y^2}$ — Simplify.

A quotient of two radical expressions should be written as a *single* radical whenever the numerator and denominator contain common factors. Once the numerator and denominator have no more common factors, they should be considered as the quotient of two *separate* radical expressions.

You try Example 7.

Example 7. Simplify $\sqrt[3]{\dfrac{81x^4}{8y^3}}$.

Solution

$$\sqrt[3]{\frac{81x^4}{8y^3}} = \frac{\sqrt[3]{81x^4}}{\sqrt[3]{8y^3}} \qquad \text{Separate the radicals.}$$

$$= \frac{\sqrt[3]{27 \cdot 3 \cdot x^3 \cdot x}}{\sqrt[3]{8 \cdot y^3}} \qquad \text{Look for perfect cubes.}$$

$$= \frac{\sqrt[3]{27}\,\sqrt[3]{x^3}\,\sqrt[3]{3x}}{\sqrt[3]{8}\,\sqrt[3]{y^3}}$$

$$= \qquad \text{Remove cube roots.}$$

Check your work on page 442. ▶

Example 8. Simplify $\dfrac{\sqrt[5]{-64x^8}}{\sqrt[5]{2x}}$.

Solution

$$\frac{\sqrt[5]{-64x^8}}{\sqrt[5]{2x}} = \sqrt[5]{\frac{-64x^8}{2x}} \qquad \text{Write as one radical.}$$

$$= \sqrt[5]{-32x^7} \qquad \text{Reduce.}$$

$$= \sqrt[5]{-32 \cdot x^5 \cdot x^2}$$

$$= -2x\sqrt[5]{x^2} \qquad \text{Remove fifth roots.}$$

Now try Example 9.

Example 9. Simplify $\sqrt[7]{(x + 3)^8}$.

Solution

$$\sqrt[7]{(x + 3)^8}$$

$$= \sqrt[7]{(x + 3)^7 \cdot (x + 3)}$$

$$=$$

Check your work on page 442. ▶

Example 10. Simplify $\sqrt{x^2 - 10x + 25}$.

Solution

$$\sqrt{x^2 - 10x + 25}$$

$$= \sqrt{(x - 5)^2}$$

$$= x - 5$$

Trial Run

Simplify. Assume that any variable expression under a radical with an even index represents a nonnegative number.

_______ 1. $\sqrt{48x^3y^2}$

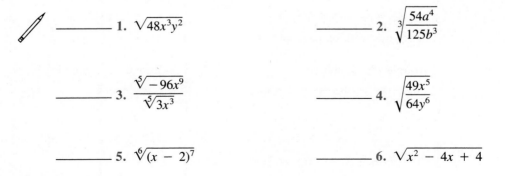

_______ 2. $\sqrt[3]{\dfrac{54a^4}{125b^3}}$

_______ 3. $\dfrac{\sqrt[5]{-96x^9}}{\sqrt[5]{3x^3}}$

_______ 4. $\sqrt{\dfrac{49x^5}{64y^6}}$

_______ 5. $\sqrt[6]{(x - 2)^7}$

_______ 6. $\sqrt{x^2 - 4x + 4}$

Answers are on page 442.

Adding and Subtracting Radical Expressions

Very early in our study of algebra we learned that in addition and subtraction we may combine only *like terms*. The same rule applies to radical expressions; we may combine only **like radicals.**

> Like radicals have the same index *and* the same radicand.

Like Radical Terms	Not Like Radical Terms
$\sqrt{3}$ and $6\sqrt{3}$	$\sqrt{5}$ and $\sqrt{3}$
$4\sqrt[3]{2}$ and $-3\sqrt[3]{2}$	$\sqrt[3]{2}$ and $\sqrt{2}$
$\sqrt{2x}$ and $5\sqrt{2x}$	$3\sqrt{2x}$ and $2\sqrt{3x}$
$x\sqrt{y}$ and $-2x\sqrt{y}$	$x\sqrt{y}$ and $y\sqrt{x}$

We learned to combine like variable terms by combining their numerical coefficients (using the distributive property). Similarly,

> To combine like radicals, we combine their numerical coefficients and keep the same radical.

You try combining the radicals in Example 11.

Example 11. Simplify

$$\sqrt{2} + 5\sqrt{2} - 8\sqrt{2}$$

Solution

$$\sqrt{2} + 5\sqrt{2} - 8\sqrt{2}$$
$$= 1\sqrt{2} + 5\sqrt{2} - 8\sqrt{2}$$
$$= \_\_\_\_\sqrt{2}$$

Check your work on page 442. ▶

Example 12. Simplify

$$\sqrt{2} - 3\sqrt{5} - 9\sqrt{2} + 3\sqrt{5}$$

Solution

$$\sqrt{2} - 3\sqrt{5} - 9\sqrt{2} + 3\sqrt{5}$$
$$= 1\sqrt{2} - 9\sqrt{2} - 3\sqrt{5} + 3\sqrt{5}$$
$$= -8\sqrt{2}$$

Now try Example 13.

Example 13. Simplify

$$6 + x\sqrt[3]{4} + 7 - 5x\sqrt[3]{4}$$

Solution

$$6 + x\sqrt[3]{4} + 7 - 5x\sqrt[3]{4}$$
$$= 6 + 7 + x\sqrt[3]{4} - 5x\sqrt[3]{4}$$
$$= \_\_\_\_ - \_\_\_\_\sqrt[3]{4}$$

Check your work on page 442. ▶

Example 14. Simplify

$$2(\sqrt{5} + 1) - 3(8 - 2\sqrt{5})$$

Solution

$$2(\sqrt{5} + 1) - 3(8 - 2\sqrt{5})$$
$$= 2\sqrt{5} + 2 - 24 + 6\sqrt{5}$$ Use the distributive property.
$$= 2\sqrt{5} + 6\sqrt{5} + 2 - 24$$
$$= 8\sqrt{5} - 22$$

Sometimes we shall be asked to combine radicals that do not appear to be alike. In such cases we must try to simplify each radical expression before deciding whether or not we have like radicals.

Try Example 15.

Example 15. Simplify

$$\sqrt{50} + \sqrt{8}$$

Solution

$$\sqrt{50} + \sqrt{8}$$
$$= \sqrt{25 \cdot 2} + \sqrt{4 \cdot 2}$$
$$= \_\_\_\_\sqrt{2} + \_\_\_\_\sqrt{2}$$
$$= \_\_\_\_\sqrt{2}$$

Check your work on page 442. ▶

Example 16. Simplify

$$-6\sqrt[3]{24x} + 10\sqrt[3]{81x}$$

Solution

$$-6\sqrt[3]{24x} + 10\sqrt[3]{81x}$$
$$= -6\sqrt[3]{8 \cdot 3x} + 10\sqrt[3]{27 \cdot 3x}$$ Look for perfect cubes.
$$= -6 \cdot 2\sqrt[3]{3x} + 10 \cdot 3\sqrt[3]{3x}$$ Remove cube roots.
$$= -12\sqrt[3]{3x} + 30\sqrt[3]{3x}$$ Simplify coefficients.
$$= 18\sqrt[3]{3x}$$ Combine coefficients.

Example 17. Simplify $\sqrt{\frac{8}{9}} + \frac{5\sqrt{2}}{2} - \sqrt{32}$.

Solution

$$\sqrt{\frac{8}{9}} + \frac{5\sqrt{2}}{2} - \sqrt{32} = \frac{\sqrt{8}}{\sqrt{9}} + \frac{5\sqrt{2}}{2} - \frac{\sqrt{32}}{1}$$ Rewrite terms as fractions.
$$= \frac{\sqrt{4 \cdot 2}}{\sqrt{9}} + \frac{5\sqrt{2}}{2} - \frac{\sqrt{16 \cdot 2}}{1}$$ Look for perfect squares.
$$= \frac{2\sqrt{2}}{3} + \frac{5\sqrt{2}}{2} - \frac{4\sqrt{2}}{1}$$ Remove square roots. LCD: $3 \cdot 2 = 6$
$$= \frac{2 \cdot 2\sqrt{2}}{6} + \frac{3 \cdot 5\sqrt{2}}{6} - \frac{6 \cdot 4\sqrt{2}}{6}$$ Change fractions to new fractions having LCD.
$$= \frac{4\sqrt{2} + 15\sqrt{2} - 24\sqrt{2}}{6}$$ Write sum of numerators.
$$= \frac{-5\sqrt{2}}{6}$$ Combine like radicals.

Example 18. Simplify $4\sqrt[5]{-32x^6} + 8x\sqrt[5]{-x}$.

Solution

$$4\sqrt[5]{-32x^6} + 8x\sqrt[5]{-x} = 4\sqrt[5]{-32 \cdot x^5 \cdot x} + 8x\sqrt[5]{-1 \cdot x}$$
$$= 4(-2)x\sqrt[5]{x} + 8x(-1)\sqrt[5]{x}$$
$$= -8x\sqrt[5]{x} - 8x\sqrt[5]{x}$$
$$= -16x\sqrt[5]{x}$$

Trial Run

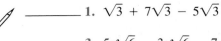

Find the sums or differences.

_______ 1. $\sqrt{3} + 7\sqrt{3} - 5\sqrt{3}$

_______ 2. $\sqrt{2} - 3\sqrt{5} + 4\sqrt{2} + 2\sqrt{5}$

_______ 3. $5x\sqrt{6} - 3x\sqrt{6} - 7x\sqrt{6}$

_______ 4. $3(\sqrt{2} - 4\sqrt[3]{2}) - (\sqrt{2} - 3\sqrt[3]{2})$

_______ 5. $\sqrt{18} + \sqrt{2}$

_______ 6. $5x\sqrt{8} + 3\sqrt{32x^2}$

_______ 7. $4\sqrt[3]{24} - 5\sqrt[3]{81}$

_______ 8. $\sqrt{\dfrac{27}{16}} + \sqrt{\dfrac{48}{25}}$

Answers are below.

Examples You Completed

Example 2. Simplify $\sqrt[4]{y^4}$.

Solution. Here n is even.

$$\sqrt[4]{y^4} = |y|$$

Example 4. Simplify $\sqrt[5]{(a-1)^5}$.

Solution. Here n is odd.

$$\sqrt[5]{(a-1)^5} = a - 1$$

Example 7. Simplify $\sqrt[3]{\dfrac{81x^4}{8y^3}}$.

Solution

$$\begin{aligned}\sqrt[3]{\frac{81x^4}{8y^3}} &= \frac{\sqrt[3]{81x^4}}{\sqrt[3]{8y^3}}\\ &= \frac{\sqrt[3]{27 \cdot 3 \cdot x^3 \cdot x}}{\sqrt[3]{8 \cdot y^3}}\\ &= \frac{\sqrt[3]{27}\,\sqrt[3]{x^3}\,\sqrt[3]{3x}}{\sqrt[3]{8}\,\sqrt[3]{y^3}}\\ &= \frac{3x\sqrt[3]{3x}}{2y}\end{aligned}$$

Example 9. Simplify $\sqrt[7]{(x+3)^8}$.

Solution

$$\begin{aligned}&\sqrt[7]{(x+3)^8}\\ &= \sqrt[7]{(x+3)^7 \cdot (x+3)}\\ &= (x+3)\sqrt[7]{x+3}\end{aligned}$$

Example 11. Simplify $\sqrt{2} + 5\sqrt{2} - 8\sqrt{2}$.

Solution

$$\begin{aligned}&\sqrt{2} + 5\sqrt{2} - 8\sqrt{2}\\ &= 1\sqrt{2} + 5\sqrt{2} - 8\sqrt{2}\\ &= -2\sqrt{2}\end{aligned}$$

Example 13. Simplify

$$6 + x\sqrt[3]{4} + 7 - 5x\sqrt[3]{4}$$

Solution

$$\begin{aligned}&6 + x\sqrt[3]{4} + 7 - 5x\sqrt[3]{4}\\ &= 6 + 7 + x\sqrt[3]{4} - 5x\sqrt[3]{4}\\ &= 13 - 4x\sqrt[3]{4}\end{aligned}$$

Example 15. Simplify $\sqrt{50} + \sqrt{8}$.

Solution

$$\begin{aligned}&\sqrt{50} + \sqrt{8}\\ &= \sqrt{25 \cdot 2} + \sqrt{4 \cdot 2}\\ &= 5\sqrt{2} + 2\sqrt{2}\\ &= 7\sqrt{2}\end{aligned}$$

Answers to Trial Runs

page 439 1. $4xy\sqrt{3x}$ 2. $\dfrac{3a\sqrt[3]{2a}}{5b}$ 3. $-2x\sqrt[5]{x}$ 4. $\dfrac{7x^2\sqrt{x}}{8y^3}$ 5. $(x-2)\sqrt[6]{x-2}$ 6. $x - 2$

page 442 1. $3\sqrt{3}$ 2. $5\sqrt{2} - \sqrt{5}$ 3. $-5x\sqrt{6}$ 4. $2\sqrt{2} - 9\sqrt[3]{2}$ 5. $4\sqrt{2}$ 6. $22x\sqrt{2}$ 7. $-7\sqrt[3]{3}$ 8. $\dfrac{31\sqrt{3}}{20}$

Name ______________________________ Date ______________

EXERCISE SET 7.2

Simplify. Make no assumptions about the sign of the radicand.

______ 1. $\sqrt[4]{a^4}$	______ 2. $\sqrt[6]{b^6}$
______ 3. $\sqrt[5]{y^5}$	______ 4. $\sqrt[9]{x^9}$
______ 5. $\sqrt[4]{(x - 5)^4}$	______ 6. $\sqrt[5]{(x + 2)^5}$

Simplify. Assume that any variable expression under a radical with an even index represents a nonnegative number.

______ 7. $\sqrt{12b^2}$	______ 8. $\sqrt{50a^2}$
______ 9. $\sqrt{24x^5y^7}$	______ 10. $\sqrt{45x^3y^5}$
______ 11. $9a\sqrt{16a^{12}}$	______ 12. $4b\sqrt{25b^8}$
______ 13. $\sqrt{24c^{10}d^3}$	______ 14. $\sqrt{45c^6d^5}$
______ 15. $\sqrt[3]{-40a^5b^4}$	______ 16. $\sqrt[3]{-128a^7b^8}$
______ 17. $\sqrt{\dfrac{48a^5}{9b^2}}$	______ 18. $\sqrt{\dfrac{63a^9}{16b^4}}$
______ 19. $\sqrt[3]{\dfrac{16x^4}{27y^6}}$	______ 20. $\sqrt[3]{\dfrac{54x^7}{125y^3}}$
______ 21. $\sqrt{\dfrac{144x^7}{25y^4}}$	______ 22. $\sqrt{\dfrac{121x^5}{36y^2}}$
______ 23. $\dfrac{\sqrt[5]{64x^{10}}}{\sqrt[5]{2x^3}}$	______ 24. $\dfrac{\sqrt[5]{486x^9}}{\sqrt[5]{2x^2}}$
______ 25. $\dfrac{\sqrt{147x^5}}{\sqrt{3x}}$	______ 26. $\dfrac{\sqrt{162x^7}}{\sqrt{2x}}$
______ 27. $\dfrac{\sqrt[3]{32x^5}}{\sqrt[3]{2x}}$	______ 28. $\dfrac{\sqrt[3]{162x^7}}{\sqrt[3]{3x^2}}$
______ 29. $\sqrt[4]{32x^9y^{10}}$	______ 30. $\sqrt[3]{72x^4y^3}$
______ 31. $\sqrt[5]{(x - 3)^6}$	______ 32. $\sqrt[4]{(x - 5)^5}$
______ 33. $\sqrt{x^2 - 6x + 9}$	______ 34. $\sqrt{x^2 + 14x + 49}$
______ 35. $\sqrt{4x^2 + 4x + 1}$	______ 36. $\sqrt{4x^2 + 12x + 9}$
______ 37. $\sqrt{5} + 9\sqrt{5} - 3\sqrt{5}$	______ 38. $\sqrt{3} - 6\sqrt{3} + 8\sqrt{3}$
______ 39. $\sqrt{3} - 2\sqrt{7} + 5\sqrt{3} + 8\sqrt{7}$	______ 40. $2\sqrt{2} - 5\sqrt{6} - 8\sqrt{2} + 9\sqrt{6}$

_______ 41. $3x\sqrt{2} - 4x\sqrt{2} + x\sqrt{2}$

_______ 42. $5x\sqrt{10} - 3x\sqrt{10} - 2x\sqrt{10}$

_______ 43. $9\sqrt[3]{5} - 13\sqrt[3]{5}$

_______ 44. $8\sqrt[3]{7} - 19\sqrt[3]{7}$

_______ 45. $8 - 2\sqrt{x} + 7 - 5\sqrt{x}$

_______ 46. $9 - 3\sqrt{x} - 12 + 5\sqrt{x}$

_______ 47. $3(\sqrt{5} - y) - 2(3\sqrt{5} + 4y)$

_______ 48. $6(\sqrt{5} - 2y) + 3(4\sqrt{5} - 3y)$

_______ 49. $4(\sqrt{a} - 2\sqrt[3]{b}) - 3(2\sqrt{a} + 5\sqrt[3]{b})$

_______ 50. $9(\sqrt{a} - \sqrt[3]{b}) - 4(\sqrt{a} - 2\sqrt[3]{b})$

_______ 51. $8(\sqrt[5]{2} - 3\sqrt[3]{2}) - 7(2\sqrt[5]{2} - 3\sqrt[3]{2})$

_______ 52. $7(\sqrt[4]{3} - 2\sqrt{3}) - 5(2\sqrt[4]{3} - 6\sqrt{3})$

_______ 53. $\sqrt{12} + \sqrt{3}$

_______ 54. $\sqrt{50} + \sqrt{2}$

_______ 55. $\sqrt{20} - \sqrt{80} + \sqrt{45}$

_______ 56. $\sqrt{24} - \sqrt{54} + \sqrt{600}$

_______ 57. $7x\sqrt{12} - 2\sqrt{75x^2}$

_______ 58. $9x\sqrt{20} - 3\sqrt{80x^2}$

_______ 59. $5\sqrt[3]{16} - 4\sqrt[3]{-54}$

_______ 60. $3\sqrt[3]{40} - 7\sqrt[3]{-5}$

_______ 61. $2\sqrt[5]{x^6y^7} - x\sqrt[5]{32xy^7}$

_______ 62. $3\sqrt[3]{x^5y^6} - x\sqrt[3]{27x^2y^6}$

_______ 63. $\sqrt{\frac{50}{9}} - \sqrt{\frac{98}{25}}$

_______ 64. $\sqrt{\frac{75}{16}} - \sqrt{\frac{12}{49}}$

_______ 65. $\sqrt{72} - \sqrt{75} + \sqrt{98}$

_______ 66. $\sqrt{500} - \sqrt{27} - \sqrt{48}$

_______ 67. $b\sqrt[3]{24a^5b} + 3a\sqrt[3]{81a^2b^4}$

_______ 68. $a\sqrt[3]{128a^2b^4} - 2b\sqrt[3]{54a^5b}$

☆ Stretching the Topics

Simplify.

_______ 1. $\sqrt{x^{4m+1}}$

_______ 2. $\sqrt[3]{x^{6m}y^{9n}}$

_______ 3. $3x\sqrt{2x^{-2}y^5z^{-4}} - \dfrac{2y\sqrt{50y^2z^{-2}}}{z}$

Check your answers in the back of your book.

If you can simplify the radical expressions in **Checkup 7.2,** you are ready to go on to Section 7.3.

Name ______________________________ Date ______________

✓ CHECKUP 7.2

Simplify. Assume that any variable expression under a radical with an even index represents a positive number.

_______ 1. $\sqrt{44x^5y^2}$

_______ 2. $\sqrt[3]{\dfrac{-27a^4}{8b^3}}$

_______ 3. $\sqrt[4]{(x+1)^5}$

_______ 4. $3x\sqrt{2} - 7x\sqrt{2} - 5x\sqrt{2}$

_______ 5. $3(\sqrt{7} - 5x) - 2(3x - \sqrt{7})$

_______ 6. $\sqrt{63x^2} + \sqrt{28x^2} - \sqrt{175x^2}$

_______ 7. $\sqrt[3]{80} - \sqrt[3]{270}$

_______ 8. $\sqrt[3]{-3} - \sqrt{75} + \sqrt[3]{375} + \sqrt{243}$

_______ 9. $\sqrt{\dfrac{50}{49}} - \sqrt{\dfrac{8}{9}}$

_______ 10. $6\sqrt[3]{x^7y^4} - x^2\sqrt[3]{27xy^4}$

Check your answers in the back of your book.

If You Missed Problems:	You Should Review Examples:
1, 2	5, 6
3	9, 10
4	11, 12
5	14
6–8	15, 16
9	17
10	18

7.3 Multiplying and Dividing Radical Expressions

Multiplying and Dividing Radicals with Same Index

It follows from our work with simplifying radicals that

> If $\sqrt[n]{a}$ and $\sqrt[n]{b}$ represent *real numbers,*
>
> $$\sqrt[n]{a} \cdot \sqrt[n]{b} = \sqrt[n]{ab}$$
>
> $$\frac{\sqrt[n]{a}}{\sqrt[n]{b}} = \sqrt[n]{\frac{a}{b}} \qquad (b \neq 0)$$
>
> To multiply radicals with the same index, we may multiply the radicands and keep the same index.
>
> To divide radicals with the same index, we may divide the radicands and keep the same index.

You try Example 1.

Example 1. Simplify $\sqrt{2} \cdot \sqrt{6}$.

Solution

$$\begin{aligned} \sqrt{2} \cdot \sqrt{6} &= \sqrt{2 \cdot 6} \\ &= \sqrt{12} \\ &= \sqrt{\_\_\_\_ \cdot 3} \\ &= \_\_\_\_\sqrt{\_\_\_\_} \end{aligned}$$

Check your work on page 456. ▶

Example 2. Simplify $\sqrt[3]{-2x} \cdot \sqrt[3]{32x^2}$.

Solution

$$\begin{aligned} &\sqrt[3]{-2x} \cdot \sqrt[3]{32x^2} \\ &= \sqrt[3]{-2x \cdot 32x^2} \\ &= \sqrt[3]{-64x^3} \\ &= -4x \end{aligned}$$

Example 3. Simplify $\sqrt[5]{\frac{10x}{3}} \cdot \sqrt[5]{\frac{-3x}{5}}$.

Solution

$$\sqrt[5]{\frac{10x}{3}} \cdot \sqrt[5]{\frac{-3x}{5}}$$

$= \sqrt[5]{\frac{10x}{3} \cdot \frac{-3x}{5}}$ Multiply radicands.

$= \sqrt[5]{\frac{-30x^2}{15}}$ Simplify products.

$= \sqrt[5]{-2x^2}$ Simplify quotient.

$= \sqrt[5]{-1 \cdot 2x^2}$ Notice $-1 = (-1)^5$.

$= -\sqrt[5]{2x^2}$ Simplify radical.

Now try Example 4.

Example 4. Simplify $\sqrt{3x} \cdot \sqrt{3x}$.

Solution

$$\begin{aligned} &\sqrt{3x} \cdot \sqrt{3x} \\ &= \sqrt{3x \cdot 3x} \end{aligned}$$

Check your work on page 456. ▶

From Example 4, we note that if $\sqrt{a}$ represents a *real* number, then a must be greater than or equal to zero and

$$\sqrt{a}\sqrt{a} = (\sqrt{a})^2 = a$$

Example 5. Simplify $\dfrac{\sqrt[3]{2000x^4y}}{\sqrt[3]{2x}}$.

Solution

$$\frac{\sqrt[3]{2000x^4y}}{\sqrt[3]{2x}} = \sqrt[3]{\frac{2000x^4y}{2x}}$$
$$= \sqrt[3]{1000x^3y}$$
$$= 10x\sqrt[3]{y}$$

Now complete Example 6.

Example 6. Simplify $\dfrac{\sqrt{5(x+1)^3}}{\sqrt{x+1}}$.

Solution

$$\frac{\sqrt{5(x+1)^3}}{\sqrt{x+1}} = \sqrt{\frac{5(x+1)^3}{x+1}}$$
$$= \sqrt{5(x+1)^2}$$
$$=$$

Check your work on page 456. ▶

Trial Run

Perform the indicated operation and simplify. Assume that any variable expression under a radical with an even index represents a nonnegative number.

_______ 1. $\sqrt{3} \cdot \sqrt{15}$

_______ 2. $\sqrt[3]{-10a} \cdot \sqrt[3]{4a^2}$

_______ 3. $\sqrt[4]{8x} \cdot \sqrt[4]{2x^3}$

_______ 4. $\sqrt[5]{\dfrac{12x}{5}} \cdot \sqrt[5]{\dfrac{-5x^2}{4}}$

_______ 5. $\dfrac{\sqrt{500x^5y}}{\sqrt{5x^2}}$

_______ 6. $\dfrac{\sqrt[3]{3(x-1)^5}}{\sqrt[3]{(x-1)^2}}$

Answers are on page 457.

To find products of expressions containing sums and/or differences of radicals, we must use the **distributive property** for multiplication over addition and carry out the multiplication of radicals in the usual way.

You try finding the product in Example 7.

Example 7. Simplify $\sqrt{2}(1 + \sqrt{3})$.

Solution

$$\sqrt{2}(1 + \sqrt{3})$$
$$= \sqrt{2} \cdot 1 + \sqrt{2} \cdot \sqrt{3}$$
$$= \_\_\_\_ + \_\_\_\_$$

Check your work on page 456. ▶

Example 8. Simplify $3\sqrt{11}(\sqrt{11} - 2)$.

Solution

$$3\sqrt{11}(\sqrt{11} - 2)$$
$$= 3\sqrt{11} \cdot \sqrt{11} - 2 \cdot 3\sqrt{11}$$
$$= 3 \cdot 11 - 6\sqrt{11}$$
$$= 33 - 6\sqrt{11}$$

You try Example 9.

Example 9. Simplify

$$\sqrt[3]{x}(8 + \sqrt[3]{x^2})$$

Solution

$$\sqrt[3]{x}(8 + \sqrt[3]{x^2})$$
$$= 8\sqrt[3]{x} + \sqrt[3]{x} \cdot \sqrt[3]{x^2}$$
$$= 8\sqrt[3]{x} + \sqrt[3]{\_\_\_\_\_\_}$$
$$= 8\sqrt[3]{x} + \_\_\_\_\_\_$$

Check your work on page 456. ▶

Example 10. Simplify

$$(2 - \sqrt{x})(3 + \sqrt{x})$$

Solution

$$(2 - \sqrt{x})(3 + \sqrt{x})$$
$$= 2(3 + \sqrt{x}) - \sqrt{x}(3 + \sqrt{x})$$
$$= 6 + 2\sqrt{x} - 3\sqrt{x} - \sqrt{x} \cdot \sqrt{x}$$
$$= 6 + 2\sqrt{x} - 3\sqrt{x} - x$$
$$= 6 - \sqrt{x} - x$$

It may have occurred to you that Example 10 could have been worked using the FOIL method from Chapter 3. Indeed, this is true; let's try the FOIL method on some products.

Example 11. Find

$$(\sqrt{x} + \sqrt{y})^2$$

Solution

L

F

$$(\sqrt{x} + \sqrt{y})^2 = (\sqrt{x} + \sqrt{y})(\sqrt{x} + \sqrt{y})$$

I

O

$$= \sqrt{x}\sqrt{x} + \sqrt{x}\sqrt{y} + \sqrt{y}\sqrt{x} + \sqrt{y}\sqrt{y}$$
$$= x + \sqrt{xy} + \sqrt{xy} + y$$
$$= x + 2\sqrt{xy} + y$$

Example 12. Find

$$(\sqrt{x} + 5\sqrt{3})(2\sqrt{x} - \sqrt{3})$$

Solution

$$(\sqrt{x} + 5\sqrt{3})(2\sqrt{x} - \sqrt{3})$$
$$= \sqrt{x} \cdot 2\sqrt{x} - \sqrt{x} \cdot \sqrt{3}$$
$$+ 5\sqrt{3} \cdot 2\sqrt{x} - 5\sqrt{3} \cdot \sqrt{3}$$
$$= 2x - \sqrt{3x} + 10\sqrt{3x} - 5 \cdot 3$$
$$= 2x + 9\sqrt{3x} - 15$$

Note that the product in Example 11 could have also been found using the rule for squaring a binomial:

$$(A + B)^2 = A^2 + 2AB + B^2$$
$$(\sqrt{x} + \sqrt{y})^2 = (\sqrt{x})^2 + 2\sqrt{x}\sqrt{y} + (\sqrt{y})^2$$
$$= x + 2\sqrt{xy} + y$$

Example 13. Find

$$(3 - \sqrt{x + 1})^2$$

Solution

$$(3 - \sqrt{x + 1})^2$$
$$= 3^2 - 2 \cdot 3\sqrt{x + 1} + (\sqrt{x + 1})^2$$
$$= 9 - 6\sqrt{x + 1} + x + 1$$
$$= x - 6\sqrt{x + 1} + 10$$

This time we used the rule for squaring a binomial.

Now you complete Example 14.

Example 14. Find

$$(\sqrt{7} + \sqrt{2})(\sqrt{7} - \sqrt{2})$$

Solution. We notice that we are multiplying the sum and difference of the same two terms.

$$(\sqrt{7} + \sqrt{2})(\sqrt{7} - \sqrt{2})$$
$$= (\sqrt{7})^2 - (\sqrt{2})^2$$
$$=$$
$$=$$

Check your work on page 456. ▶

In Example 14 we multiplied the irrational number $\sqrt{7} + \sqrt{2}$ times the irrational number $\sqrt{7} - \sqrt{2}$ and the product was the *rational* number 5. Such irrational numbers as $\sqrt{7} + \sqrt{2}$ and $\sqrt{7} - \sqrt{2}$ are sometimes called **real conjugates** of each other.

Real Conjugates. The irrational numbers $\sqrt{a} + \sqrt{b}$ and $\sqrt{a} - \sqrt{b}$ are called real conjugates for each other. Their product is always a rational number.

$$(\sqrt{a} + \sqrt{b})(\sqrt{a} - \sqrt{b}) = a - b$$

Trial Run

Find each product and simplify.

_______ 1. $\sqrt{3}(2 + \sqrt{5})$

_______ 2. $3\sqrt{7}(\sqrt{7} - 2)$

_______ 3. $\sqrt[3]{x}(2 - \sqrt[3]{x^2})$

_______ 4. $(1 - \sqrt{x})(3 + 2\sqrt{x})$

_______ 5. $(3\sqrt{2} - \sqrt{y})(5\sqrt{2} + 2\sqrt{y})$

_______ 6. $(\sqrt{x} - 7)(\sqrt{x} + 7)$

Answers are on page 457.

Rationalizing Denominators Containing One Term

You should recall from Chapter 1 that a number is called an **irrational number** when it cannot be expressed as a fraction of integers. Although we can *approximate* irrational numbers, we cannot find their *exact* value. For instance, from a table or calculator we could find the following approximations:

$$\sqrt{3} \doteq 1.732$$

$$\sqrt[3]{2} \doteq 1.259$$

$$1 + \sqrt{5} \doteq 1 + 2.236 \doteq 3.236$$

$$\sqrt{7} - \sqrt{2} \doteq 2.646 - 1.414 \doteq 1.232$$

The symbol $\doteq$ means "is approximately equal to."

Surely you would agree that these are not convenient numbers to work with, especially if they appear in the denominator of a fraction. For this reason (and for other reasons that will be clear to you in later mathematics courses), mathematicians usually agree that

No radical should appear in the denominator of the final form of a radical expression.

Somehow we must remove the radical from the denominator of an expression such as $\frac{1}{\sqrt{3}}$ and change the irrational number in the denominator to a *rational* number. This process is called **rationalizing the denominator.** But how can we accomplish the desired change?

We rationalize the denominator of a fraction by multiplying the numerator and denominator by an appropriate rationalizing factor.

To get rid of the $\sqrt{3}$ in the denominator of the fraction $\frac{1}{\sqrt{3}}$, we choose the rationalizing factor to be $\sqrt{3}$ (because $\sqrt{3} \cdot \sqrt{3} = 3$).

$$\frac{1}{\sqrt{3}} = \frac{1}{\sqrt{3}} \cdot \frac{\sqrt{3}}{\sqrt{3}} \qquad \text{Multiply the numerator and denominator by } \sqrt{3}.$$

$$= \frac{1\sqrt{3}}{\sqrt{3} \cdot \sqrt{3}} \qquad \text{Write the product as one fraction.}$$

$$= \frac{\sqrt{3}}{3} \qquad \text{Find the products.}$$

We have indeed rid the denominator of its irrational number; we have *rationalized the denominator*. In the examples that follow, we assume that radical expressions represent real, nonzero numbers.

You try Example 15.

Example 15. Rationalize the denominator of $\frac{5}{\sqrt{2x}}$.

Solution

$$\frac{5}{\sqrt{2x}} \cdot \frac{\sqrt{\_\_\_\_}}{\sqrt{\_\_\_\_}} = \frac{5\sqrt{\_\_\_\_}}{\sqrt{2x} \cdot \sqrt{\_\_\_\_}}$$

$$=$$

Check your work on page 456. ▶

Example 16. Rationalize the denominator of $\frac{1 + \sqrt{a}}{\sqrt{a}}$.

Solution

$$\frac{1 + \sqrt{a}}{\sqrt{a}} \cdot \frac{\sqrt{a}}{\sqrt{a}} = \frac{(1 + \sqrt{a})\sqrt{a}}{\sqrt{a} \cdot \sqrt{a}}$$

$$= \frac{\sqrt{a} + \sqrt{a} \cdot \sqrt{a}}{a}$$

$$= \frac{\sqrt{a} + a}{a}$$

Now try Example 17.

Example 17. Rationalize the denominator of $\frac{3}{\sqrt{a + b}}$.

Solution

$$\frac{3}{\sqrt{a + b}} \cdot \frac{\sqrt{a + b}}{\sqrt{a + b}}$$

$$= \frac{3\sqrt{a + b}}{\sqrt{a + b} \cdot \sqrt{a + b}}$$

$$=$$

Check your work on page 456. ▶

Example 18. Rationalize the denominator of $\frac{x + 1}{\sqrt{x^2 - 1}}$.

Solution

$$\frac{x + 1}{\sqrt{x^2 - 1}} \cdot \frac{\sqrt{x^2 - 1}}{\sqrt{x^2 - 1}} \qquad \text{Choose the rationalizing factor.}$$

$$= \frac{(x + 1)\sqrt{x^2 - 1}}{\sqrt{x^2 - 1} \cdot \sqrt{x^2 - 1}}$$

$$= \frac{(x + 1)\sqrt{x^2 - 1}}{x^2 - 1} \qquad \text{Find the product in the denominator.}$$

$$= \frac{\overset{1}{\cancel{(x + 1)}}\sqrt{x^2 - 1}}{\underset{1}{\cancel{(x + 1)}}(x - 1)} \qquad \text{Factor the denominator and reduce.}$$

$$= \frac{\sqrt{x^2 - 1}}{x - 1}$$

To rationalize the denominator of the fraction

$$\frac{5}{\sqrt[3]{2}}$$

we know that the rationalizing factor must be a cube root and that we are aiming for the *cube root of a perfect cube* as our new denominator.

$$\frac{5}{\sqrt[3]{2}} = \frac{5}{\sqrt[3]{2^1}} \cdot \frac{\sqrt[3]{(\quad)}}{\sqrt[3]{(\quad)}} = \frac{5\sqrt[3]{(\quad)}}{\sqrt[3]{2^3}}$$

Since the original radicand is 2^1, it is logical for us to aim for a radicand of 2^3. The missing radicand must be 2^2.

$$\frac{5}{\sqrt[3]{2}} = \frac{5}{\sqrt[3]{2^1}} \cdot \frac{\sqrt[3]{2^2}}{\sqrt[3]{2^2}}$$

$$= \frac{5\sqrt[3]{4}}{\sqrt[3]{2^3}}$$

$$= \frac{5\sqrt[3]{4}}{2}$$

Choosing the rationalizing factor is not difficult if you decide *first* what you are aiming for.

Example 19. Rationalize the denominator of $\frac{1}{\sqrt[3]{3x^2}}$.

Solution. We want the *cube root of a perfect cube* in the new denominator. The rationalizing factor is $\sqrt[3]{3^2x}$.

$$\frac{1}{\sqrt[3]{3x^2}} = \frac{1}{\sqrt[3]{3x^2}} \cdot \frac{\sqrt[3]{3^2x}}{\sqrt[3]{3^2x}}$$ Aim for a radicand of 3^3x^3.

$$= \frac{\sqrt[3]{9x}}{\sqrt[3]{3^3x^3}}$$ Multiply.

$$= \frac{\sqrt[3]{9x}}{3x}$$ Simplify the denominator.

Example 20. Rationalize the denominator of $\frac{a}{\sqrt[5]{4a^3}}$.

Solution. The original denominator is $\sqrt[5]{2^2a^3}$. We want the *fifth root of a perfect fifth power* in the new denominator. The rationalizing factor is $\sqrt[5]{2^3a^2}$.

$$\frac{a}{\sqrt[5]{4a^3}} = \frac{a}{\sqrt[5]{2^2a^3}} \cdot \frac{\sqrt[5]{2^3a^2}}{\sqrt[5]{2^3a^2}}$$ Aim for a radicand of 2^5a^5.

$$= \frac{a\sqrt[5]{8a^2}}{\sqrt[5]{2^5a^5}}$$ Multiply.

$$= \frac{a\sqrt[5]{8a^2}}{2a}$$ Simplify the denominator.

$$= \frac{\sqrt[5]{8a^2}}{2}$$ Reduce the fraction.

Trial Run

Rationalize each denominator. Assume that radical expressions represent real, nonzero numbers.

_______ 1. $\frac{3}{\sqrt{5}}$

_______ 2. $\frac{x}{\sqrt{2x}}$

_______ 3. $\frac{2}{\sqrt{a+3}}$

_______ 4. $\frac{x+2}{\sqrt{x^2-4}}$

_______ 5. $\dfrac{6}{\sqrt[3]{2x}}$

_______ 6. $\dfrac{1}{\sqrt[4]{x^2y^3}}$

Answers are on page 457.

To rationalize the denominator of a radical expression whose radicand is a fraction, we should use the fact that

$$\sqrt[n]{\frac{a}{b}} = \frac{\sqrt[n]{a}}{\sqrt[n]{b}}$$

before we attempt the simplifying and rationalizing processes.

Example 21. Simplify $\sqrt{\dfrac{12x^5}{y}}$.

Solution

$$\sqrt{\frac{12x^5}{y}} = \frac{\sqrt{12x^5}}{\sqrt{y}} \qquad \text{Separate radicals.}$$

$$= \frac{\sqrt{4 \cdot 3 \cdot x^2 \cdot x^2 \cdot x}}{\sqrt{y}} \qquad \text{Simplify numerator.}$$

$$= \frac{2x^2\sqrt{3x}}{\sqrt{y}} \cdot \frac{\sqrt{y}}{\sqrt{y}} \qquad \text{Choose rationalizing factor.}$$

$$= \frac{2x^2\sqrt{3xy}}{y} \qquad \text{Multiply.}$$

Example 22. Simplify $\sqrt[3]{\dfrac{16a^4b^3}{9c}}$.

Solution

$$\sqrt[3]{\frac{16a^4b^3}{9c}} = \frac{\sqrt[3]{16a^4b^3}}{\sqrt[3]{9c}}$$

$$= \frac{\sqrt[3]{8 \cdot 2 \cdot a^3 \cdot a \cdot b^3}}{\sqrt[3]{3^2c}}$$

$$= \frac{2ab\sqrt[3]{2a}}{\sqrt[3]{3^2c}} \cdot \frac{\sqrt[3]{3c^2}}{\sqrt[3]{3c^2}}$$

$$= \frac{2ab\sqrt[3]{6ac^2}}{3c}$$

Of course, **if the numerator and denominator of a fractional radicand contain common factors, the fraction should be *reduced* by the usual methods *before* we attempt the process of simplifying and rationalizing.**

You try to complete Example 23.

Example 23. Simplify $\dfrac{\sqrt{5x^5y^2}}{\sqrt{20x^2y^3}}$.

Solution

$$\frac{\sqrt{5x^5y^2}}{\sqrt{20x^2y^3}} = \sqrt{\frac{5x^5y^2}{20x^2y^3}}$$

$$= \sqrt{\frac{x^3}{4y}}$$

$$= \frac{\sqrt{x^3}}{\sqrt{4y}}$$

$$=$$

Check your work on page 457. ▶

Example 24. Simplify $\dfrac{\sqrt[3]{24a^4b^2}}{\sqrt[3]{15a^5b^4}}$.

Solution

$$\frac{\sqrt[3]{24a^4b^2}}{\sqrt[3]{15a^5b^4}} = \sqrt[3]{\frac{24a^4b^2}{15a^5b^4}} \qquad \text{Notice common factors.}$$

$$= \sqrt[3]{\frac{8}{5ab^2}} \qquad \text{Reduce the fraction.}$$

$$= \frac{\sqrt[3]{8}}{\sqrt[3]{5ab^2}} \qquad \text{Separate radicals.}$$

$$= \frac{2}{\sqrt[3]{5ab^2}} \cdot \frac{\sqrt[3]{5^2a^2b}}{\sqrt[3]{5^2a^2b}} \qquad \text{Choose rationalizing factor.}$$

$$= \frac{2\sqrt[3]{25a^2b}}{5ab} \qquad \text{Multiply.}$$

Simplifying Fractions Containing Radicals

1. Reduce the fraction if possible by treating it as a single radical.
2. Rewrite the result as a fraction of separate radicals.
3. Simplify the remaining radicals and rationalize the denominator if necessary.

Trial Run

Simplify and rationalize denominators. Assume that variables are positive.

_______ 1. $\sqrt{\dfrac{20x^3}{3y}}$

_______ 2. $\sqrt{\dfrac{26a^2x}{13ax^2}}$

_______ 3. $\dfrac{\sqrt[3]{10x}}{\sqrt[3]{9x}}$

_______ 4. $\dfrac{\sqrt{20xy^4}}{\sqrt{50x^2y}}$

Answers are on page 457.

Rationalizing Denominators Containing Two Terms

Rationalizing the denominator of a fraction such as

$$\frac{-3}{1 + \sqrt{5}}$$

requires a slightly different approach. If we multiply numerator and denominator by $1 - \sqrt{5}$, the real conjugate of the denominator, look at the result.

$$\frac{-3}{1 + \sqrt{5}} \cdot \frac{1 - \sqrt{5}}{1 - \sqrt{5}} = \frac{-3(1 - \sqrt{5})}{(1 + \sqrt{5})(1 - \sqrt{5})}$$ Multiply numerator and denominator by $1 - \sqrt{5}$.

$$= \frac{-3(1 - \sqrt{5})}{1^2 - (\sqrt{5})^2}$$ Multiply denominators.

$$= \frac{-3(1 - \sqrt{5})}{1 - 5}$$ Simplify squares.

$$= \frac{-3(1 - \sqrt{5})}{-4}$$ Simplify the denominator.

$$= \frac{3(1 - \sqrt{5})}{4}$$ Divide numerator and denominator by -1.

We have indeed rationalized the denominator.

To rationalize a denominator that is the *sum* or *difference* of two terms (one or both terms being a square root), we multiply the numerator and denominator by the *real conjugate* of the denominator.

This technique relies on the fact that when we multiply the sum and difference of the same two terms, the product is the *difference of the squares* of the two terms.

You try Example 25.

Example 25. Rationalize the denominator of $\frac{2}{\sqrt{7} - \sqrt{3}}$.

Solution. The rationalizing factor is _____ + _____ .

$$\frac{2}{\sqrt{7} - \sqrt{3}} \cdot \frac{\sqrt{7} + \sqrt{3}}{\sqrt{7} + \sqrt{3}}$$

$= \frac{2(\sqrt{7} + \sqrt{3})}{(\sqrt{7})^2 - (\sqrt{3})^2}$ Write products.

$= \frac{2(\sqrt{7} + \sqrt{3})}{\_\_\_\_\_ - \_\_\_\_\_}$ Simplify squares.

$= \frac{2(\sqrt{7} + \sqrt{3})}{\_\_\_\_\_}$ Simplify the denominator.

$= \frac{\sqrt{7} + \sqrt{3}}{\_\_\_\_\_}$ Reduce the fraction.

Check your work on page 457. ▶

Example 26. Rationalize the denominator of $\frac{\sqrt{y}}{\sqrt{x} + \sqrt{y}}$.

Solution. The rationalizing factor is $\sqrt{x} - \sqrt{y}$.

$$\frac{\sqrt{y}}{\sqrt{x} + \sqrt{y}} \cdot \frac{\sqrt{x} - \sqrt{y}}{\sqrt{x} - \sqrt{y}}$$

$$= \frac{\sqrt{y}(\sqrt{x} - \sqrt{y})}{(\sqrt{x} + \sqrt{y})(\sqrt{x} - \sqrt{y})}$$

$= \frac{\sqrt{xy} - \sqrt{y^2}}{(\sqrt{x})^2 - (\sqrt{y})^2}$ Use distributive property in numerator.

$= \frac{\sqrt{xy} - y}{x - y}$ Simplify.

Contrast Example 26 with Example 27.

Example 27. Rationalize the denominator of $\frac{\sqrt{y}}{\sqrt{x + y}}$.

Solution. Here the denominator is a *single* radical.

$\frac{\sqrt{y}}{\sqrt{x + y}} \cdot \frac{\sqrt{x + y}}{\sqrt{x + y}}$ Multiply numerator and denominator by the single radical $\sqrt{x + y}$.

$= \frac{\sqrt{y}(\sqrt{x + y})}{x + y}$ Simplify the denominator.

$= \frac{\sqrt{y(x + y)}}{x + y}$ Rewrite the numerator.

Students often confuse these last two examples. We must be careful to decide whether our denominator is a single radical term *or* a sum or difference of radical terms before we choose the correct rationalizing factor.

Trial Run

Rationalize the denominator. Assume that any variable expression under a radical with an even index represents a nonnegative number.

_______ 1. $\frac{5}{1 - \sqrt{2}}$

_______ 2. $\frac{1}{\sqrt{5} + \sqrt{2}}$

_______ 3. $\dfrac{\sqrt{a}}{\sqrt{a} + 1}$

_______ 4. $\dfrac{\sqrt{x}}{\sqrt{x} - \sqrt{2}}$

_______ 5. $\dfrac{\sqrt{x}}{\sqrt{x - 2}}$

_______ 6. $\dfrac{\sqrt{6} - 3\sqrt{2}}{2\sqrt{6} - \sqrt{2}}$

Answers are on page 457.

▶ Examples You Completed

Example 1. Simplify $\sqrt{2} \cdot \sqrt{6}$.

Solution

$$\begin{aligned} \sqrt{2} \cdot \sqrt{6} &= \sqrt{2 \cdot 6} \\ &= \sqrt{12} \\ &= \sqrt{4 \cdot 3} \\ &= 2\sqrt{3} \end{aligned}$$

Example 4. Simplify $\sqrt{3x} \cdot \sqrt{3x}$.

Solution

$$\begin{aligned} \sqrt{3x} \cdot \sqrt{3x} &= \sqrt{3x \cdot 3x} \\ &= \sqrt{9x^2} \\ &= 3x \end{aligned}$$

Example 6. Simplify $\dfrac{\sqrt{5(x + 1)^3}}{\sqrt{x + 1}}$.

Solution

$$\begin{aligned} \frac{\sqrt{5(x + 1)^3}}{\sqrt{x + 1}} &= \sqrt{\frac{5(x + 1)^3}{x + 1}} \\ &= \sqrt{5(x + 1)^2} \\ &= (x + 1)\sqrt{5} \end{aligned}$$

Example 7. Simplify $\sqrt{2}(1 + \sqrt{3})$.

Solution

$$\begin{aligned} &\sqrt{2}(1 + \sqrt{3}) \\ &= \sqrt{2} \cdot 1 + \sqrt{2} \cdot \sqrt{3} \\ &= \sqrt{2} + \sqrt{6} \end{aligned}$$

Example 9. Simplify $\sqrt[3]{x}(8 + \sqrt[3]{x^2})$.

Solution

$$\begin{aligned} &\sqrt[3]{x}(8 + \sqrt[3]{x^2}) \\ &= 8\sqrt[3]{x} + \sqrt[3]{x} \cdot \sqrt[3]{x^2} \\ &= 8\sqrt[3]{x} + \sqrt[3]{x^3} \\ &= 8\sqrt[3]{x} + x \end{aligned}$$

Example 14. Find $(\sqrt{7} + \sqrt{2})(\sqrt{7} - \sqrt{2})$.

Solution. We notice that we are multiplying the sum and difference of the same two terms.

$$\begin{aligned} &(\sqrt{7} + \sqrt{2})(\sqrt{7} - \sqrt{2}) \\ &= (\sqrt{7})^2 - (\sqrt{2})^2 \\ &= 7 - 2 \\ &= 5 \end{aligned}$$

Example 15. Rationalize the denominator of $\dfrac{5}{\sqrt{2x}}$.

Solution

$$\begin{aligned} &\frac{5}{\sqrt{2x}} \cdot \frac{\sqrt{2x}}{\sqrt{2x}} \\ &= \frac{5\sqrt{2x}}{\sqrt{2x} \cdot \sqrt{2x}} \\ &= \frac{5\sqrt{2x}}{2x} \end{aligned}$$

Example 17. Rationalize the denominator of $\dfrac{3}{\sqrt{a + b}}$.

Solution

$$\begin{aligned} &\frac{3}{\sqrt{a + b}} \cdot \frac{\sqrt{a + b}}{\sqrt{a + b}} \\ &= \frac{3\sqrt{a + b}}{\sqrt{a + b} \cdot \sqrt{a + b}} \\ &= \frac{3\sqrt{a + b}}{a + b} \end{aligned}$$

Example 23. Simplify

$$\frac{\sqrt{5x^5y^2}}{\sqrt{20x^2y^3}}.$$

Solution

$$\frac{\sqrt{5x^5y^2}}{\sqrt{20x^2y^3}} = \sqrt{\frac{5x^5y^2}{20x^2y^3}}$$
$$= \sqrt{\frac{x^3}{4y}}$$
$$= \frac{\sqrt{x^3}}{\sqrt{4y}}$$
$$= \frac{x\sqrt{x}}{2\sqrt{y}} \cdot \frac{\sqrt{y}}{\sqrt{y}}$$
$$= \frac{x\sqrt{xy}}{2y}$$

Example 25. Rationalize the denominator of $\frac{2}{\sqrt{7} - \sqrt{3}}$.

Solution. The rationalizing factor is $\sqrt{7} + \sqrt{3}$.

$$\frac{2}{\sqrt{7} - \sqrt{3}} \cdot \frac{\sqrt{7} + \sqrt{3}}{\sqrt{7} + \sqrt{3}}$$
$$= \frac{2(\sqrt{7} + \sqrt{3})}{(\sqrt{7})^2 - (\sqrt{3})^2}$$
$$= \frac{2(\sqrt{7} + \sqrt{3})}{7 - 3}$$
$$= \frac{2(\sqrt{7} + \sqrt{3})}{4}$$
$$= \frac{\sqrt{7} + \sqrt{3}}{2}$$

Answers to Trial Runs

page 448 1. $3\sqrt{5}$ 2. $-2a\sqrt[3]{5}$ 3. $2x$ 4. $-\sqrt[5]{3x^3}$ 5. $10x\sqrt{xy}$ 6. $(x - 1)\sqrt[3]{3}$

page 450 1. $2\sqrt{3} + \sqrt{15}$ 2. $21 - 6\sqrt{7}$ 3. $2\sqrt[3]{x} - x$ 4. $3 - \sqrt{x} - 2x$ 5. $30 + \sqrt{2y} - 2y$ 6. $x - 49$

page 452 1. $\frac{3\sqrt{5}}{5}$ 2. $\frac{\sqrt{2x}}{2}$ 3. $\frac{2\sqrt{a+3}}{a+3}$ 4. $\frac{\sqrt{x^2-4}}{x-2}$ 5. $\frac{3\sqrt[3]{4x^2}}{x}$ 6. $\frac{\sqrt[4]{x^2y}}{xy}$

page 454 1. $\frac{2x\sqrt{15xy}}{3y}$ 2. $\frac{\sqrt{2ax}}{x}$ 3. $\frac{\sqrt[3]{30}}{3}$ 4. $\frac{y\sqrt{10xy}}{5x}$

page 455 1. $-5 - 5\sqrt{2}$ 2. $\frac{\sqrt{5} - \sqrt{2}}{3}$ 3. $\frac{a - \sqrt{a}}{a - 1}$ 4. $\frac{x + \sqrt{2x}}{x - 2}$ 5. $\frac{\sqrt{x(x-2)}}{x - 2}$ 6. $\frac{3 - 5\sqrt{3}}{11}$

Name ______________________ Date ____________

EXERCISE SET 7.3

Perform the indicated operation and simplify, rationalizing denominators when necessary. Assume that any variable expression under a radical with an even index represents a non-negative number.

______ 1. $\sqrt{7} \cdot \sqrt{14}$	______ 2. $\sqrt{5} \cdot \sqrt{10}$
______ 3. $\sqrt{3a} \cdot \sqrt{6a^3}$	______ 4. $\sqrt{24a} \cdot \sqrt{2a^3}$
______ 5. $\sqrt[3]{-6x} \cdot \sqrt[3]{9x^2}$	______ 6. $\sqrt[3]{-10x^2} \cdot \sqrt[3]{4x}$
______ 7. $\sqrt[4]{27x^2} \cdot \sqrt[4]{3x^6}$	______ 8. $\sqrt[4]{4x^6} \cdot \sqrt[4]{4x^2}$
______ 9. $\sqrt[3]{\frac{9x^2}{2}} \cdot \sqrt[3]{\frac{-16x}{3}}$	______ 10. $\sqrt[3]{\frac{4x}{3}} \cdot \sqrt[3]{\frac{-375x^2}{2}}$
______ 11. $\sqrt[5]{\frac{24x}{5}} \cdot \sqrt[5]{\frac{20x^2}{3}}$	______ 12. $\sqrt[5]{\frac{21x^3}{2}} \cdot \sqrt[5]{\frac{6x^2}{7}}$
______ 13. $\frac{\sqrt{128x^5y}}{\sqrt{2x^2}}$	______ 14. $\frac{\sqrt{180x^3y}}{\sqrt{5x}}$
______ 15. $\sqrt{5x} \cdot \sqrt{5x}$	______ 16. $\sqrt{7y} \cdot \sqrt{7y}$
______ 17. $\sqrt{6x} \cdot \sqrt{15xy}$	______ 18. $\sqrt{10x} \cdot \sqrt{20xy}$
______ 19. $\sqrt{18x^3yz} \cdot \sqrt{8xy^4z}$	______ 20. $\sqrt{21xy^3z^5} \cdot \sqrt{14xyz^2}$
______ 21. $\sqrt{2}(\sqrt{6} + \sqrt{2})$	______ 22. $\sqrt{3}(\sqrt{15} + \sqrt{3})$
______ 23. $\sqrt{3x}(\sqrt{x} - \sqrt{3})$	______ 24. $\sqrt{5y}(\sqrt{y} - \sqrt{5})$
______ 25. $\sqrt{6xy}(\sqrt{3x} + \sqrt{2y})$	______ 26. $\sqrt{10xy}(\sqrt{5x} + \sqrt{2y})$
______ 27. $\sqrt[3]{a}(3 - \sqrt[3]{a^2})$	______ 28. $\sqrt[3]{2}(a - \sqrt[3]{4})$
______ 29. $(1 - \sqrt{x})(2 - \sqrt{x})$	______ 30. $(3 - \sqrt{x})(2 + \sqrt{x})$
______ 31. $(\sqrt{2} - 3\sqrt{y})(5\sqrt{2} + \sqrt{y})$	______ 32. $(3\sqrt{5} - 2\sqrt{y})(\sqrt{5} + \sqrt{y})$
______ 33. $(\sqrt{x} - 3)^2$	______ 34. $(\sqrt{x} - 7)^2$
______ 35. $(\sqrt{2x} - 3)(\sqrt{2x} + 3)$	______ 36. $(\sqrt{3x} - 4)(\sqrt{3x} + 4)$
______ 37. $(4\sqrt{x} + 2\sqrt{5})(4\sqrt{x} - 2\sqrt{5})$	______ 38. $(3\sqrt{7} - 8\sqrt{x})(3\sqrt{7} + 8\sqrt{x})$
______ 39. $(x - \sqrt{2y})(x + \sqrt{2y})$	______ 40. $(\sqrt{3x} - y)(\sqrt{3x} + y)$
______ 41. $(3\sqrt{x} - 5\sqrt{y})^2$	______ 42. $(4\sqrt{x} - 7\sqrt{y})^2$
______ 43. $(2 + \sqrt{x - 3})^2$	______ 44. $(3 - \sqrt{x + 4})^2$

______ 45. $\frac{4}{\sqrt{6}}$

______ 46. $\frac{3}{\sqrt{15}}$

______ 47. $\frac{\sqrt{3}}{\sqrt{6}}$

______ 48. $\frac{\sqrt{2}}{\sqrt{10}}$

______ 49. $\frac{\sqrt{2x}}{\sqrt{3x}}$

______ 50. $\frac{\sqrt{5x}}{\sqrt{2x}}$

______ 51. $\sqrt{\frac{18x^3}{5y}}$

______ 52. $\sqrt{\frac{28x^5}{3y}}$

______ 53. $\sqrt{\frac{54x^4y^2z^2}{10xz^3}}$

______ 54. $\sqrt{\frac{32x^3y^4}{12xyz}}$

______ 55. $\frac{4}{\sqrt{a+5}}$

______ 56. $\frac{3}{\sqrt{a+2}}$

______ 57. $\frac{x-3}{\sqrt{x^2-9}}$

______ 58. $\frac{x+5}{\sqrt{x^2-25}}$

______ 59. $\frac{3}{\sqrt[3]{5}}$

______ 60. $\frac{4}{\sqrt[3]{3}}$

______ 61. $\frac{64}{\sqrt[4]{3}}$

______ 62. $\frac{10x}{\sqrt[4]{2}}$

______ 63. $\frac{6}{\sqrt[3]{3x}}$

______ 64. $\frac{10}{\sqrt[3]{5x^2}}$

______ 65. $\frac{x^2}{\sqrt[4]{9x^3}}$

______ 66. $\frac{x^3}{\sqrt[5]{8x^2}}$

______ 67. $\sqrt[4]{\frac{3}{4y}}$

______ 68. $\sqrt[4]{\frac{2}{9x^3}}$

______ 69. $\sqrt[3]{\frac{5}{4y^2}}$

______ 70. $\sqrt[3]{\frac{7}{3x}}$

______ 71. $\frac{3}{2-\sqrt{3}}$

______ 72. $\frac{2}{4-\sqrt{5}}$

______ 73. $\frac{1}{\sqrt{7}+\sqrt{2}}$

______ 74. $\frac{1}{\sqrt{5}+\sqrt{3}}$

______ 75. $\frac{\sqrt{x}}{\sqrt{x}-2}$

______ 76. $\frac{\sqrt{3}}{\sqrt{3}-x}$

______ 77. $\frac{\sqrt{a}}{\sqrt{a}-3}$

______ 78. $\frac{\sqrt{a}}{\sqrt{a}+5}$

______ 79. $\frac{\sqrt{6}}{\sqrt{3}+\sqrt{2}}$

______ 80. $\frac{\sqrt{15}}{\sqrt{5}-\sqrt{3}}$

______ 81. $\frac{\sqrt{10}-2\sqrt{2}}{4\sqrt{10}-\sqrt{2}}$

______ 82. $\frac{\sqrt{3}-2\sqrt{6}}{2\sqrt{3}-\sqrt{6}}$

☆ Stretching the Topics

______ 1. Simplify $(\sqrt[3]{x}-\sqrt[3]{2})(\sqrt[3]{x^2}+\sqrt[3]{2x}+\sqrt[3]{4})$.

______ 2. Factor $9x-4y$ as the difference of two squares.

______ 3. Simplify $\frac{\sqrt{x+2}}{\sqrt{x+2}-5}$.

Check your answers in the back of your book.

If you can complete **Checkup 7.3,** you are ready to do Section 7.4.

Name ______________________ Date ____________

✓ CHECKUP 7.3

Perform the indicated operation and simplify, rationalizing denominators when necessary. Assume that any variable expression under a radical with an even index represents a nonnegative number.

______ 1. $\sqrt{6x} \cdot \sqrt{10x^2}$

______ 2. $\sqrt{\dfrac{48x^3y}{3x}}$

______ 3. $\sqrt{5}(\sqrt{10} + \sqrt{5})$

______ 4. $(2\sqrt{5} - \sqrt{10})(2\sqrt{5} + \sqrt{10})$

______ 5. $(\sqrt{x} - 6)^2$

______ 6. $\dfrac{6}{\sqrt{15}}$

______ 7. $\dfrac{\sqrt{6x}}{\sqrt{10x}}$

______ 8. $\dfrac{15x}{\sqrt[3]{-3x}}$

______ 9. $\sqrt{\dfrac{45x^3}{2y}}$

______ 10. $\dfrac{\sqrt{8}}{\sqrt{2} + \sqrt{3}}$

Check your answers in the back of your book.

If You Missed Problems:	You Should Review Examples:
1	1, 2
2	3
3–5	7–12
6, 7	15, 16
8	19
9	21
10	25

7.4 Solving Radical Equations

We have now acquired enough skill in working with radicals to enable us to solve equations containing variables within a radical expression. Such equations are called **radical equations** and some are very easily solved. For example, to solve $\sqrt{x} = 2$ we may square both sides of the equation to eliminate the radical.

$$(\sqrt{x})^2 = 2^2$$

$$x = 4$$

Is squaring both sides of an equation an acceptable manipulation? To answer that question, we must decide whether the solutions to the equation obtained after squaring are *exactly* the same as the solutions to the original equation.

Let's square both sides of another radical equation and check to see if our solution satisfies the original equation.

$\sqrt{x} = -3$	CHECK: $x = 9$?
$(\sqrt{x})^2 = (-3)^2$	$\sqrt{9} \stackrel{?}{=} -3$
$x = 9$	$3 \neq -3$

Here the solution to the squared equation does *not* satisfy the original equation. In fact, the original equation has *no* solution because the principal square root of a number is never negative.

Sometimes squaring both sides of an equation "works" and sometimes it does not. If we square both sides of an equation, we must always *check* our solutions in the original equation. If the original equation has a solution, it will always appear as a solution to the squared equation.

> We may square both sides of an equation, provided that we always check any solutions in the original equation.

A solution to a squared equation that does *not* satisfy the original equation is called an **extraneous solution.**

You try Example 1.

Example 1. Solve $\sqrt{2x + 1} = 5$.

Solution

$\sqrt{2x + 1} = 5$	CHECK: $x = \underline{\hspace{1cm}}$?
$(\sqrt{2x + 1})^2 = 5^2$	$\sqrt{2(\underline{\hspace{1cm}}) + 1} \stackrel{?}{=} 5$
$2x + 1 = \underline{\hspace{1cm}}$	$\sqrt{\underline{\hspace{1cm}} + 1} \stackrel{?}{=} 5$
$x = \underline{\hspace{1cm}}$	$\sqrt{\underline{\hspace{1cm}}} \stackrel{?}{=} 5$
	$5 = 5$

The solution is $x = \underline{\hspace{1cm}}$.

Check your work on page 467. ▶

Example 2. Solve $5 + \sqrt[3]{x} = 2$.

Solution. We isolate the radical and raise both sides to the *third* power.

$$5 + \sqrt[3]{x} = 2$$

$$\sqrt[3]{x} = -3$$

$$(\sqrt[3]{x})^3 = (-3)^3$$

$$x = -27$$

CHECK: $x = -27$?

$$5 + \sqrt[3]{-27} \stackrel{?}{=} 2$$

$$5 + (-3) \stackrel{?}{=} 2$$

$$2 = 2$$

The solution is $x = -27$.

Most of our radical equations will contain *square* roots of variable expressions. If a radical equation contains just one such radical expression, we follow these steps.

> **Solving Radical Equations Containing One Square Root**
>
> **1.** Isolate the radical expression on one side of the equation.
> **2.** Square both sides of the equation *carefully*.
> **3.** Solve the resulting equation.
> **4.** Check the solutions in the original equation.

You try Example 3.

Example 3. Solve $\sqrt{x + 2} - x = 0$.

Solution

$\sqrt{x+2} - x = 0$

$\sqrt{x+2} = x$ — Isolate the radical.

$(\sqrt{x+2})^2 = x^2$ — Square both sides.

$x + 2 = x^2$ — Note that this is a quadratic equation.

$0 = x^2 - x - 2$ — Rewrite quadratic equation in standard form.

$0 = ($________$)($________$)$ — Factor.

$x +$ ____ $= 0$ or $x -$ ____ $= 0$ — Use zero product rule.

$x =$ ____ or $x =$ ____

CHECK: $x = -1$?
Substitute -1 for x in the original equation.

$$\sqrt{-1 + 2} - (-1) \stackrel{?}{=} 0$$

$$\sqrt{1} + 1 \stackrel{?}{=} 0$$

$$1 + 1 \stackrel{?}{=} 0$$

$$2 \neq 0$$

-1 is *not* a solution.

CHECK: $x = 2$?
Substitute 2 for x in the original equation.

$$\sqrt{2 + 2} - 2 \stackrel{?}{=} 0$$

$$\sqrt{4} - 2 \stackrel{?}{=} 0$$

____ $- 2 \stackrel{?}{=} 0$

____ $= 0$

2 is a solution.

The only solution is $x =$ ____ .

Check your work on page 467. ▶

Example 4. Solve $x + \sqrt{3 - x} = 2x + 3$.

Solution

$x + \sqrt{3 - x} = 2x + 3$	
$\sqrt{3 - x} = x + 3$	Isolate the radical.
$(\sqrt{3 - x})^2 = (x + 3)^2$	Square both sides.
$3 - x = x^2 + 6x + 9$	
$0 = x^2 + 7x + 6$	Rewrite quadratic equation in standard form.
$0 = (x + 6)(x + 1)$	Factor.
$x + 6 = 0$ or $x + 1 = 0$	Use zero product rule.
$x = -6$ $\quad$ $x = -1$	

CHECK: $x = -6$?

$$-6 + \sqrt{3 - (-6)} \stackrel{?}{=} 2(-6) + 3$$
$$-6 + \sqrt{9} \stackrel{?}{=} -12 + 3$$
$$-6 + 3 \stackrel{?}{=} -9$$
$$-3 \neq -9$$

CHECK: $x = -1$?

$$-1 + \sqrt{3 - (-1)} \stackrel{?}{=} 2(-1) + 3$$
$$-1 + \sqrt{4} \stackrel{?}{=} -2 + 3$$
$$-1 + 2 \stackrel{?}{=} 1$$
$$1 = 1$$

The only solution is $x = -1$.

Trial Run

Solve each equation.

_______ 1. $\sqrt{x} = 3$

_______ 2. $\sqrt{3x + 4} = 5$

_______ 3. $9 + \sqrt{2x - 1} = 6$

_______ 4. $\sqrt[3]{x - 1} = -3$

_______ 5. $\sqrt{x + 12} - x = 0$

_______ 6. $\sqrt{7x + 14} = x + 2$

Answers are on page 468.

If a radical equation contains *two* square roots of variable expressions, it is most easily solved using the following steps.

Solving Radical Equations

1. Isolate one radical expression on one side of the equation.
2. Square both sides of the equation carefully.
3. If the equation still contains a radical expression, isolate it and square both sides again.
4. Solve the resulting equation.
5. Check the solutions in the original equation.

You try Example 5.

Example 5. Solve $\sqrt{x+1} - \sqrt{5x-3} = 0$.

Solution

$\sqrt{x+1} - \sqrt{5x-3} = 0$

$\sqrt{x+1} = \sqrt{5x-3}$ Isolate one radical.

$(\sqrt{x+1})^2 = (\sqrt{5x-3})^2$ Square both sides.

$x + 1 = 5x - 3$ This is a first-degree equation.

$1 = 4x - 3$

$\_\_\_\_ = 4x$

$\_\_\_\_ = x$

CHECK: $x = 1$?

$\sqrt{1+1} - \sqrt{5(1)-3} \stackrel{?}{=} 0$

$\sqrt{2} - \sqrt{5-3} \stackrel{?}{=} 0$

$\sqrt{2} - \sqrt{2} \stackrel{?}{=} 0$

$0 = 0$

The solution is $x =$ ____ .

Check your work on page 467. ▶

Example 6. Solve $\sqrt{2x-5} - \sqrt{x-2} = 2$.

Solution

$\sqrt{2x-5} - \sqrt{x-2} = 2$

$\sqrt{2x-5} = 2 + \sqrt{x-2}$ Isolate one radical.

$(\sqrt{2x-5})^2 = (2 + \sqrt{x-2})^2$ Square both sides.

$2x - 5 = (2 + \sqrt{x-2})(2 + \sqrt{x-2})$

$2x - 5 = 4 + 4\sqrt{x-2} + x - 2$

$2x - 5 = 2 + x + 4\sqrt{x-2}$ Combine like terms.

$x - 7 = 4\sqrt{x-2}$ Isolate the radical.

$(x-7)^2 = (4\sqrt{x-2})^2$ Square both sides again.

$x^2 - 14x - 49 = 16(x-2)$

$x^2 - 14x + 49 = 16x - 32$ Solve the quadratic equation.

$x^2 - 30x + 81 = 0$

$(x-27)(x-3) = 0$

$x - 27 = 0$ or $x - 3 = 0$

$x = 27$ $\quad$ $x = 3$

CHECK: $x = 27$?

$\sqrt{2(27)-5} - \sqrt{27-2} \stackrel{?}{=} 2$

$\sqrt{54-5} - \sqrt{25} \stackrel{?}{=} 2$

$\sqrt{49} - 5 \stackrel{?}{=} 2$

$7 - 5 \stackrel{?}{=} 2$

$2 = 2$

CHECK: $x = 3$?

$\sqrt{2(3)-5} - \sqrt{3-2} \stackrel{?}{=} 2$

$\sqrt{6-5} - \sqrt{1} \stackrel{?}{=} 2$

$\sqrt{1} - 1 \stackrel{?}{=} 2$

$1 - 1 \stackrel{?}{=} 2$

$0 \neq 2$

The only solution is $x = 27$.

As you can see, solving radical equations can be a lengthy process, but the skills required are not new ones. Skipping steps or squaring carelessly are students' usual sources of error. You must be systematic and careful as you proceed.

Example 7. One leg of a right triangle is 8 feet long. The perimeter of the triangle is 24 feet. Find the dimensions of the triangle.

Solution

We Think	***We Write***
The perimeter of a triangle is the sum of the lengths of its three sides.	$P = a + b + c$
We know the length of one leg but not the other.	Let a = length of unknown leg 8 = length of known leg
The third side is the hypotenuse. It can be found by the Pythagorean theorem.	$c^2 = a^2 + 8^2$ $c = \sqrt{a^2 + 64}$
We know the perimeter is 24 feet.	$P = a + b + c$ $24 = a + 8 + \sqrt{a^2 + 64}$

We must solve this radical equation.

$$
\begin{aligned}
24 &= a + 8 + \sqrt{a^2 + 64} \\
16 - a &= \sqrt{a^2 + 64} \\
(16 - a)^2 &= (\sqrt{a^2 + 64})^2 \\
256 - 32a + a^2 &= a^2 + 64 \\
256 - 32a &= 64 \\
-32a &= -192 \\
a &= 6
\end{aligned}
$$

The length of the unknown leg is 6 feet and the length of the hypotenuse is

$$
\begin{aligned}
\sqrt{a^2 + 64} &= \sqrt{6^2 + 64} \\
&= \sqrt{36 + 64} \\
&= \sqrt{100} \\
&= 10 \text{ ft}
\end{aligned}
$$

The triangle's dimensions are 6 feet, 8 feet, and 10 feet.

Let's solve the problem stated at the beginning of this chapter.

Example 8. One leg of a right triangle is 1 foot longer than the other. If the length of the shorter leg is increased by 4 feet and the length of the longer leg is multiplied by 6, the length of the hypotenuse of the enlarged right traingle is five times the length of the original hypotenuse. Find the dimensions of both triangles.

Solution. We can let x represent the length of the shorter leg of the original triangle and illustrate both triangles.

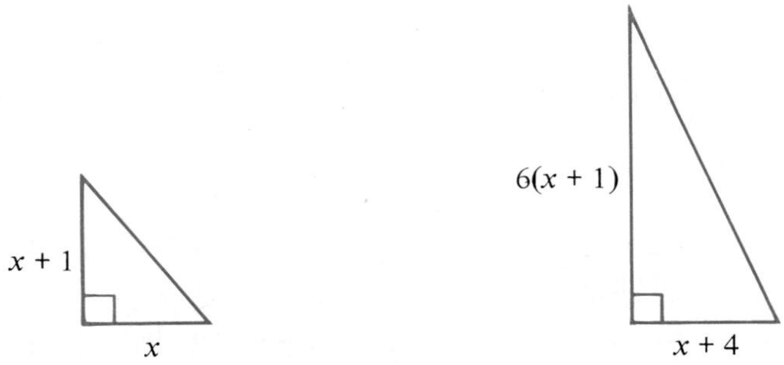

We Think / ***We Write***

In every right triangle with leg lengths a and b and hypotenuse of length c, we can use the Pythagorean theorem.

$$c^2 = a^2 + b^2$$
$$c = \sqrt{a^2 + b^2}$$

We know expressions for the lengths of the sides for both triangles, so we can write expressions for the hypotenuses.

Original hypotenuse: $\sqrt{x^2 + (x + 1)^2}$

Enlarged hypotenuse: $\sqrt{[6(x + 1)]^2 + (x + 4)^2}$

We know that the length of the enlarged hypotenuse is five times the length of the original hypotenuse.

$$\sqrt{[6(x + 1)]^2 + (x + 4)^2} = 5\sqrt{x^2 + (x + 1)^2}$$

We can solve this radical equation.

$$\sqrt{[6(x + 1)]^2 + (x + 4)^2} = 5\sqrt{x^2 + (x + 1)^2}$$
$$\sqrt{36(x^2 + 2x + 1) + x^2 + 8x + 16} = 5\sqrt{x^2 + x^2 + 2x + 1}$$
$$\sqrt{36x^2 + 72x + 36 + x^2 + 8x + 16} = 5\sqrt{2x^2 + 2x + 1}$$
$$\sqrt{37x^2 + 80x + 52} = 5\sqrt{2x^2 + 2x + 1}$$
$$(\sqrt{37x^2 + 80x + 52})^2 = (5\sqrt{2x^2 + 2x + 1})^2$$
$$37x^2 + 80x + 52 = 25(2x^2 + 2x + 1)$$
$$37x^2 + 80x + 52 = 50x^2 + 50x + 25$$
$$0 = 13x^2 - 30x - 27$$
$$0 = (13x + 9)(x - 3)$$

$$13x + 9 = 0 \quad \text{or} \quad x - 3 = 0$$
$$13x = -9 \qquad x = 3$$
$$x = -\frac{9}{13}$$

Length cannot be negative, so we reject $x = -\frac{9}{13}$ and find that $x = 3$ satisfies the equation.

Original Triangle	*Enlarged Triangle*
$x = 3$ ft	$x + 4 = 7$ ft
$x + 1 = 4$ ft	$6(x + 1) = 24$ ft
$\sqrt{x^2 + (x + 1)^2} = 5$ ft	$\sqrt{[6(x + 1)]^2 + (x + 4)^2} = 25$ ft

The dimensions of the original triangle are 3 feet, 4 feet, and 5 feet. The dimensions of the enlarged triangle are 7 feet, 24 feet, and 25 feet.

Trial Run

Solve each radical equation.

_______ 1. $\sqrt{x} = \sqrt{3x - 2}$

_______ 2. $\sqrt{2x - 3} = \sqrt{4x - 11}$

_______ 3. $\sqrt{5x - 2} + \sqrt{3x} = 0$

_______ 4. $\sqrt{x - 3} = 3 - \sqrt{x}$

Answers are on page 468.

Examples You Completed

Example 1. Solve $\sqrt{2x + 1} = 5$.

Solution

$$\sqrt{2x + 1} = 5$$
$$(\sqrt{2x + 1})^2 = 5^2$$
$$2x + 1 = 25$$
$$2x = 24$$
$$x = 12$$

CHECK: $x = 12$?

$$\sqrt{2(12) + 1} \stackrel{?}{=} 5$$
$$\sqrt{24 + 1} \stackrel{?}{=} 5$$
$$\sqrt{25} \stackrel{?}{=} 5$$
$$5 = 5$$

The solution is $x = 12$.

Example 3. Solve $\sqrt{x + 2} - x = 0$.

Solution

$$\sqrt{x + 2} - x = 0$$
$$\sqrt{x + 2} = x$$
$$(\sqrt{x + 2})^2 = x^2$$
$$x + 2 = x^2$$
$$0 = x^2 - x - 2$$
$$0 = (x + 1)(x - 2)$$
$$x + 1 = 0 \quad \text{or} \quad x - 2 = 0$$
$$x = -1 \qquad\qquad x = 2$$

CHECK: $x = -1$?

$$\sqrt{-1 + 2} - (-1) \stackrel{?}{=} 0$$
$$\sqrt{1} + 1 \stackrel{?}{=} 0$$
$$1 + 1 \stackrel{?}{=} 0$$
$$2 \neq 0$$

CHECK: $x = 2$?

$$\sqrt{2 + 2} - 2 \stackrel{?}{=} 0$$
$$\sqrt{4} - 2 \stackrel{?}{=} 0$$
$$2 - 2 \stackrel{?}{=} 0$$
$$0 = 0$$

The only solution is $x = 2$.

Example 5. Solve $\sqrt{x + 1} - \sqrt{5x - 3} = 0$.

Solution

$$\sqrt{x + 1} - \sqrt{5x - 3} = 0$$

$$\sqrt{x + 1} = \sqrt{5x - 3}$$

$$(\sqrt{x + 1})^2 = (\sqrt{5x - 3})^2$$

$$x + 1 = 5x - 3$$

$$1 = 4x - 3$$

$$4 = 4x$$

$$1 = x$$

CHECK: $x = 1$?

$$\sqrt{1 + 1} - \sqrt{5(1) - 3} \stackrel{?}{=} 0$$

$$\sqrt{2} - \sqrt{5 - 3} \stackrel{?}{=} 0$$

$$\sqrt{2} - \sqrt{2} \stackrel{?}{=} 0$$

$$0 = 0$$

The solution is $x = 1$.

Answers to Trial Runs

page 463 **1.** $x = 9$ **2.** $x = 7$ **3.** No solution **4.** $x = -26$ **5.** $x = 4$ **6.** $x = 5$ or $x = -2$

page 467 **1.** $x = 1$ **2.** $x = 4$ **3.** No solution **4.** $x = 4$

Name ______________________ Date ____________

EXERCISE SET 7.4

Solve each equation.

______ 1. $\sqrt{x} = 5$

______ 2. $\sqrt{x} = 7$

______ 3. $\sqrt{2x} - 3 = 7$

______ 4. $\sqrt{3x} + 2 = 8$

______ 5. $\sqrt{2x - 3} = 3$

______ 6. $\sqrt{3x - 5} = 7$

______ 7. $\sqrt{5x + 1} + 3 = 7$

______ 8. $\sqrt{7x + 1} + 2 = 10$

______ 9. $\sqrt{4x - 7} + 3 = 0$

______ 10. $\sqrt{3x - 5} + 2 = 0$

______ 11. $8 - \sqrt{x - 3} = 7$

______ 12. $12 - \sqrt{x - 5} = 11$

______ 13. $13 - \sqrt{2x - 3} = 10$

______ 14. $15 - \sqrt{3x + 1} = 11$

______ 15. $\sqrt[3]{x - 1} = -2$

______ 16. $\sqrt[3]{x - 1} = -4$

______ 17. $\sqrt[3]{5x - 7} = 2$

______ 18. $\sqrt[3]{2x - 5} = 1$

______ 19. $\sqrt{2x + 8} - x = 0$

______ 20. $\sqrt{4x + 21} - x = 0$

______ 21. $\sqrt{9x - 17} = x - 1$

______ 22. $\sqrt{1 - 2x} = x + 7$

______ 23. $x + \sqrt{7 - 3x} = 2x - 1$

______ 24. $2x + \sqrt{5 - x} = 3x + 1$

______ 25. $\sqrt{x} = \sqrt{5x - 8}$

______ 26. $\sqrt{3x} = \sqrt{2x + 5}$

______ 27. $\sqrt{7x - 5} + \sqrt{2x} = 0$

______ 28. $\sqrt{6x - 10} + \sqrt{4x} = 0$

______ 29. $\sqrt{7x - 1} = \sqrt{2x + 4}$

______ 30. $\sqrt{3x - 7} = \sqrt{2x + 8}$

______ 31. $\sqrt{x - 5} = 5 - \sqrt{x}$

______ 32. $\sqrt{x - 7} = 7 - \sqrt{x}$

______ 33. $\sqrt{2x + 4} = \sqrt{2x} + 1$

______ 34. $\sqrt{3x + 5} = \sqrt{3x} + 1$

______ 35. $\sqrt{3x - 2} = 5 + \sqrt{3x - 7}$

______ 36. $\sqrt{x - 5} = 4 + \sqrt{x + 3}$

______ 37. $\sqrt[3]{x^2 + x} = \sqrt[3]{20 - 7x}$

______ 38. $\sqrt[3]{x^2 + 4x} = \sqrt[3]{x + 10}$

______ 39. $\sqrt{3x - 5} + \sqrt{x - 1} = 2$

______ 40. $\sqrt{3x + 9} - \sqrt{2x + 7} = 1$

______ 41. $\sqrt{2y + 3} - \sqrt{y + 1} = 1$

______ 42. $\sqrt{4x - 3} - \sqrt{2x - 5} = 2$

Use an equation to solve each problem.

______ 43. One leg of a right triangle is 12 inches long. The perimeter of the triangle is 30 inches. Find the dimensions of the triangle.

______ 44. One leg of a right triangle is 24 centimeters long. The perimeter of the triangle is 56 centimeters. Find the dimensions of the triangle.

________ 45. The width of a rectangle is 2 feet less than its length, and its diagonal is 10 feet. Find the perimeter of the rectangle.

________ 46. The length of a rectangle is 7 inches more than its width, and its diagonal is 13 inches. Find the perimeter of the rectangle.

________ 47. One leg of a right triangle is 1 centimeter longer than the other. If the shorter leg is increased by 3 centimeters and the length of the longer leg is doubled, the length of the hypotenuse of the enlarged right triangle is twice the length of the original hypotenuse. Find the dimensions of the original triangle.

________ 48. One leg of a right triangle is 7 inches longer than the other. If the shorter leg is increased by 4 inches and the longer leg remains the same, the length of the hypotenuse of the enlarged right triangle is 2 more than the length of the original hypotenuse. Find the dimensions of the original triangle.

________ 49. The radius of a sphere is given by the formula $r = \sqrt[3]{\frac{3V}{4\pi}}$. Find the volume ($V$) if the radius of the sphere is 9 inches.

________ 50. The radius of a cone is given by the formula $r = \sqrt{\frac{3V}{\pi h}}$. Find the volume ($V$) of the cone if the radius is 6 feet and the height is 10 feet.

________ 51. Solve the formula $d = \sqrt{w^2 + l^2}$ for w (where $w > 0$).

________ 52. Solve the formula $r = \sqrt{\frac{A}{\pi}}$ for A.

________ 53. Solve the formula $t = \sqrt{\frac{s - a - k}{g}}$ for s.

________ 54. Solve the formula $v = \frac{k}{\sqrt{d}}$ for d.

☆ Stretching the Topics

________ 1. Solve $\sqrt{x + 2} = \sqrt[4]{9x - 2}$.

________ 2. Solve $\frac{\sqrt{x - 2}}{x - 2} = \frac{x - 5}{\sqrt{x - 2}}$.

________ 3. The time T for one period of a simple pendulum is given by the formula $T = 2\pi\sqrt{\frac{l}{g}}$. Solve for l, the length of the pendulum's string.

Check your answers in the back of your book.

If you can complete **Checkup 7.4,** you are ready to go on to Section 7.5.

Name ______________________ Date __________

✓ CHECKUP 7.4

Solve each equation.

______ 1. $\sqrt{3x - 5} = 4$

______ 2. $\sqrt[3]{6x} + 2 = 4$

______ 3. $\sqrt{5x + 1} + 2 = 0$

______ 4. $9 - \sqrt{x - 5} = 7$

______ 5. $7 - \sqrt{4x - 15} = 8$

______ 6. $\sqrt[3]{x + 1} = -2$

______ 7. $\sqrt{2x + 63} - x = 0$

______ 8. $\sqrt{15x - 9} = x + 3$

______ 9. $\sqrt{9x - 1} = \sqrt{3x + 11}$

______ 10. $\sqrt{x + 2} + \sqrt{x - 1} = 3$

Check your answers in the back of your book.

If You Missed Problems:	You Should Review Examples:
1	1
2	2
3–8	3, 4
9, 10	5, 6

7.5 Working with Complex Numbers

In our work with quadratic equations and square roots, we have agreed that *the square root of a negative number is not a real number*. It is time now to look at a new set of numbers in which the square root of a negative number has meaning. This new set of numbers is called the set of **complex numbers.**

Recognizing Complex Numbers

We would like to define the set of complex numbers so that *every* quadratic equation will have solutions within that set. Let's look at a few typical quadratic equations and see what kind of solutions occur.

Example 1. Solve $x^2 + 5x - 6 = 0$.

Solution

$$x^2 + 5x - 6 = 0$$
$$(x + 6)(x - 1) = 0$$
$$x + 6 = 0 \quad \text{or} \quad x - 1 = 0$$
$$x = -6 \qquad\qquad x = 1$$

Both solutions are *real* (and rational).

Example 2. Solve $x^2 - 7 = 0$.

Solution

$$x^2 - 7 = 0$$
$$x^2 = 7$$
$$x = \pm\sqrt{7}$$
$$x = \sqrt{7} \quad \text{or} \quad x = -\sqrt{7}$$

Both solutions are *real* (but irrational).

Example 3. Solve $x^2 + 5 = 0$.

Solution

$$x^2 + 5 = 0$$
$$x^2 = -5$$
$$x = \pm\sqrt{-5}$$
$$x = \sqrt{-5} \quad \text{or} \quad x = -\sqrt{-5}$$

Since $\sqrt{-5}$ is not a real number, there are *no real* solutions.

Example 4. Solve $x^2 + x + 4 = 0$.

Solution. We can use the quadratic formula with $a = 1$, $b = 1$, and $c = 4$.

$$x = \frac{-1 \pm \sqrt{1^2 - 4(1)(4)}}{2(1)}$$
$$= \frac{-1 \pm \sqrt{1 - 16}}{2}$$
$$x = \frac{-1 \pm \sqrt{-15}}{2}$$

There are *no real* solutions.

We wish to define the set of complex numbers to include solutions of the types obtained in all four of our examples. First we shall use the letter ***i*** to denote the square root of -1.

Definition of *i*

$$i = \sqrt{-1}$$

Using this notation, we may rewrite the solutions for Examples 3 and 4.

$$x = \sqrt{-5} \quad \text{or} \quad x = -\sqrt{-5}$$
$$= \sqrt{5(-1)} \qquad x = -\sqrt{5(-1)}$$
$$= \sqrt{5} \cdot \sqrt{-1} \qquad = -\sqrt{5} \cdot \sqrt{-1}$$
$$x = \sqrt{5}i \quad \text{or} \quad x = -\sqrt{5}i$$

$$x = \frac{-1 \pm \sqrt{-15}}{2}$$
$$= \frac{-1 \pm \sqrt{15(-1)}}{2}$$
$$= \frac{-1 \pm \sqrt{15} \cdot \sqrt{-1}}{2}$$
$$x = \frac{-1 \pm \sqrt{15}i}{2}$$

We call the number i the **imaginary unit** to emphasize the fact that i is *not a real number*.

Example 5. Write $\sqrt{-9}$ using the imaginary unit.

Solution

$$\sqrt{-9} = \sqrt{9(-1)}$$
$$= \sqrt{9} \cdot \sqrt{-1}$$
$$= 3i$$

Now you try Example 6.

Example 6. Write $1 + \sqrt{-3}$ using the imaginary unit.

Solution

$$1 + \sqrt{-3} = 1 + \sqrt{3(\underline{\qquad})}$$
$$= 1 + \sqrt{3} \cdot \sqrt{\underline{\qquad}}$$
$$= 1 + \sqrt{3}\,\underline{\qquad}$$

Check your work on page 483. ▶

Each of the numbers we have written is called a **complex number.**

Definition of a Complex Number. A complex number is a number that can be written in the standard form

$$a + bi$$

where a and b are *real* numbers and $i = \sqrt{-1}$. In any complex number $a + bi$,

a is called the **real part.**
b is called the **imaginary part.**
i is called the **imaginary unit.**

Now try Example 7.

Example 7. Identify the parts of $3 + i$.

Solution

$$3 + i = 3 + 1i$$

The real part is ______ .
The imaginary part is ______ .

Check your work on page 483. ▶

Example 8. Identify the parts of -6.

Solution

$$-6 = -6 + 0i$$

The real part is -6.
The imaginary part is 0.

From Example 8 we see that

Every real number belongs to the set of complex numbers. The imaginary part of a real number is always 0.

Example 9. Write $5 - \sqrt{-32}$ in standard complex form and identify its parts.

Solution

$$\begin{aligned} 5 - \sqrt{-32} &= 5 - \sqrt{32}\,\sqrt{-1} \\ &= 5 - \sqrt{16 \cdot 2}\,\sqrt{-1} \\ &= 5 - 4\sqrt{2}i \end{aligned}$$

The real part is 5.
The imaginary part is $-4\sqrt{2}$.

Now you try Example 10.

Example 10. Write $\sqrt{300}$ in standard complex form and identify its parts.

Solution

$$\begin{aligned} \sqrt{300} &= \sqrt{\_\_\_\_ \cdot 3} \\ &= \_\_\_\_\sqrt{3} \\ &= \_\_\_\_\sqrt{3} + 0i \end{aligned}$$

The real part is ____ $\sqrt{3}$.
The imaginary part is ____ .

Check your work on page 483. ▶

For two complex numbers to be equal, their real parts must be equal and their imaginary parts must be equal.

Equality of Complex Numbers. If $a + bi = c + di$, then

$$a = c \qquad \text{and} \qquad b = d$$

Example 11. Solve $2x + i = 3 - 4yi$ for real values of x and y.

Solution

$$2x + i = 3 - 4yi$$

$$2x + 1i = 3 + (-4y)i$$

Real Parts	***Imaginary Parts***
$2x = 3$	$1 = -4y$
$x = \frac{3}{2}$	$-\frac{1}{4} = y$

Now you try Example 12.

Example 12. Solve $x + yi = \sqrt{3} - \sqrt{7}i$ for real values of x and y.

Solution

$$x + yi = \sqrt{3} - \sqrt{7}i$$

$$x + yi = \sqrt{3} + (\_\_\_\_)i$$

Real Parts	***Imaginary Parts***
$x =$ ____	$y =$ ____

Check your work on page 483. ▶

To add (or subtract) complex numbers, we do what is completely natural. We must add (or subtract) the real parts, and add (or subtract) the imaginary parts and attach the imaginary unit.

Addition of Complex Numbers

$$(a + bi) + (c + di) = (a + c) + (b + d)i$$

You complete Example 13.

Example 13. Find

$$(3 + 2i) + (7 + 4i)$$

Solution. We must add the real parts and add the imaginary parts.

$$\begin{aligned} &(3 + 2i) + (7 + 4i) \\ &= (3 + \_\_) + (2 + \_\_)i \\ &= \_\_ + \_\_i \end{aligned}$$

Check your work on page 483. ▶

Example 14. Find

$$(4 - \sqrt{3}i) - (1 + \sqrt{3}i)$$

Solution. We must add the real parts and add the imaginary parts.

$$\begin{aligned} &(4 - \sqrt{3}i) - (1 + \sqrt{3}i) \\ &= (4 - 1) + (-\sqrt{3} - \sqrt{3})i \\ &= 3 + (-2\sqrt{3})i \\ &= 3 - 2\sqrt{3}i \end{aligned}$$

As you have probably noticed, the method used for simplifying sums or differences of complex numbers is similar to the methods for *combining like terms*. You may think of the i-terms as like terms.

Trial Run

Write each number in standard complex form.

________ 1. $\sqrt{-36}$

________ 2. $\sqrt{25}$

________ 3. $3 - \sqrt{-8}$

________ 4. $\sqrt{12} + \sqrt{-16}$

Find each sum or difference.

________ 5. $(7 - \sqrt{2}i) + (4 - 3\sqrt{2}i)$

________ 6. $[(3 + 2i) - (6 + i)] + (4 + i)$

Solve for real values of x and y.

________ 7. $2x - 3i = 6 + 3yi$

Answers are on page 484.

Multiplying Complex Numbers

A complex number whose real part is zero is called a **pure imaginary number.** Some pure imaginary numbers are i, $5i$, $-\sqrt{3}i$, and $\frac{1}{8}i$.

To multiply pure imaginary numbers, we may use the laws of exponents. For example,

$$i \cdot i = i^2 \qquad 5i \cdot i^2 = 5i^3 \qquad \begin{aligned} (2i)^4 &= 2^4 i^4 \\ &= 16i^4 \end{aligned}$$

But let's look more closely at the powers of i. Remembering that $i = \sqrt{-1}$, we know that $\boldsymbol{i}$ is the number whose square is -1.

> Since $i = \sqrt{-1}$, we know $i^2 = -1$

Now we can investigate some powers of i.

Power of i	Simplest Form
i^1	i
i^2	-1
i^3	$i^2 \cdot i = (-1)i = -i$
i^4	$i^2 \cdot i^2 = (-1)(-1) = 1$

The powers of i occur in cycles of four. To evaluate any power of i, we can rewrite it as a product of some of these four basic powers, using i^4 the most possible times (since $i^4 = 1$ is easy to work with).

Example 15. Find i^5.

Solution

$$i^5 = i^4 \cdot i$$
$$= 1 \cdot i$$
$$i^5 = i$$

Now you try Example 16.

Example 16. Find i^7.

Solution

$$i^7 = i^4 \cdot (\_\_\_\_)$$
$$= 1 \cdot (\_\_\_\_)$$
$$i^7 = \_\_\_\_$$

Check your work on page 483. ▶

Example 17. Find $5i^{101}$.

Solution

$$5i^{101} = 5 \cdot (i^4)^{25} \cdot i$$
$$= 5 \cdot (1)^{25} \cdot i$$
$$= 5 \cdot 1 \cdot i$$
$$= 5i$$

Now try Example 18.

Example 18. Find $\sqrt{2}i^{18}$.

Solution

$$\sqrt{2}i^{18} = \sqrt{2}\,(i^4)^4 \cdot (\_\_\_\_)$$
$$= \sqrt{2}\,(1)^4(\_\_\_\_)$$
$$= \sqrt{2}\,(1)(\_\_\_\_)$$
$$= \_\_\_\_$$

Check your work on page 484. ▶

Products of complex numbers are found using a process similar to the *distributive property* for multiplication over addition, together with the facts just learned about the *powers of i*.

You try Example 19.

Example 19. Find $-3(4 - 8i)$.

Solution

$-3(4 - 8i)$

$= -3 \cdot 4 - 3(-8i)$ Use distributive property.

$= \_\_\_\_ + \_\_\_\_i$ Simplify products.

Check your work on page 484. ▶

Example 20. Find $-6i(5 - i)$.

Solution

$-6i(5 - i)$

$= -6i \cdot 5 - 6i(-i)$ Use distributive property.

$= -30i + 6i^2$ Simplify products.

$= -30i + 6(-1)$ Substitute -1 for i^2.

$= -30i - 6$ Simplify.

$= -6 - 30i$ Put in standard form.

Sometimes a complex number does not appear in standard complex form $a + bi$. However,

> We must write complex numbers in standard form using the imaginary unit before we try to operate with them.

Example 21. Find $(3 + \sqrt{-9}) + (-2 + \sqrt{-25})$.

Solution

$(3 + \sqrt{-9}) + (-2 + \sqrt{-25})$	
$= (3 + 3i) + (-2 + 5i)$	Write each number in standard form.
$= 1 + 8i$	Add.

Example 22. Find $\sqrt{-4}(5 + \sqrt{-9})$.

Solution

$\sqrt{-4}(5 + \sqrt{-9}) = 2i(5 + 3i)$	Write each number in standard form.
$= 10i + 6i^2$	Use distributive property.
$= 10i + 6(-1)$	Substitute -1 for i^2.
$= -6 + 10i$	Write in standard form.

Make careful note of the fact that a complex number *must* be switched to standard form before it is used in performing any operation.

Right	Wrong
$\sqrt{-9} \cdot \sqrt{-4}$	$\sqrt{-9} \cdot \sqrt{-4}$
$= 3i \cdot 2i$	$\neq \sqrt{(-9)(-4)}$
$= 6i^2$	$\neq \sqrt{36}$
$= 6(-1)$	$\neq 6$
$= -6$	

Let's practice multiplying complex numbers in standard form, using methods similar to those used in the multiplication of binomials.

You complete Example 23.

Example 23. Find $(3 + 4i)(1 + 2i)$.

Solution. We can use the FOIL technique here.

$$(3 + 4i)(1 + 2i) = 3 + 6i + 4i + 8i^2$$
$$= 3 + \_\_\_\_ i + 8(\_\_\_\_)$$
$$= \_\_\_\_ + \_\_\_\_ i$$

Check your work on page 484. ▶

Example 24. Find $(5 - \sqrt{-3})(2 + \sqrt{-12})$.

Solution. We first write the numbers in standard complex form.

$$(5 - \sqrt{-3})(2 + \sqrt{-12})$$

$= (5 - \sqrt{3}i)(2 + 2\sqrt{3}i)$	Write numbers in standard form.
$= 10 + 10\sqrt{3}i - 2\sqrt{3}i - 2\sqrt{3}\sqrt{3}i^2$	Multiply.
$= 10 + 8\sqrt{3}i - 2(3)i^2$	Simplify.
$= 10 + 8\sqrt{3}i - 6(-1)$	Substitute -1 for i^2.
$= 10 + 8\sqrt{3}i + 6$	Simplify.
$= 16 + 8\sqrt{3}i$	Combine real parts.

Now try Example 25.

Example 25. Find $(3 - \sqrt{2}i)^2$.

Solution

$$(3 - \sqrt{2}i)^2$$
$$= 9 - 2 \cdot 3\sqrt{2}i + (\sqrt{2}i)^2$$

Check your work on page 484. ▶

Example 26. Find $(\sqrt{7} + 3i)(\sqrt{7} - 3i)$.

Solution

$$(\sqrt{7} + 3i)(\sqrt{7} - 3i)$$
$$= \sqrt{7} \cdot \sqrt{7} - 3\sqrt{7}i + 3\sqrt{7}i - 9i^2$$
$$= 7 - 9i^2$$
$$= 7 - 9(-1)$$
$$= 7 + 9$$
$$= 16$$
$$= 16 + 0i$$

Notice that the product of two complex numbers is always a complex number. Sometimes the product is a complex number that is *real,* as in Example 26. In Example 26, we multiplied $(\sqrt{7} + 3i)(\sqrt{7} - 3i)$ and the product was the real number 16. Indeed, whenever we multiply $(a + bi)(a - bi)$, the product will be a *real number*.

$$(a + bi)(a - bi) = a^2 - b^2i^2$$
$$= a^2 - b^2(-1)$$
$$= a^2 + b^2$$

As you can see, the imaginary unit does not appear in the product. The product is *real*.

For the complex number $a + bi$, the complex number $a - bi$ is called its (complex) **conjugate.** The product of a complex number and its conjugate is always a *real number* (namely, $a^2 + b^2$).

Complex Number	Conjugate	Product
$3 + 2i$	$3 - 2i$	$(3 + 2i)(3 - 2i)$
		$= 3^2 - 4i^2$
		$= 9 - 4(-1)$
		$= 13$
$6 - \sqrt{3}i$	$6 + \sqrt{3}i$	$(6 - \sqrt{3}i)(6 + \sqrt{3}i)$
		$= 6^2 - 3i^2$
		$= 36 - 3(-1)$
		$= 39$
$0 - i$	$0 + i$	$(0 - i)(0 + i)$
		$= 0 - i^2$
		$= -(-1)$
		$= 1$

Trial Run

Perform the indicated operation and simplify.

________ 1. i^{16}

________ 2. $7i^{65}$

________ 3. $-5(2 - 3i)$

________ 4. $4i(6 - i)$

________ 5. $\sqrt{-9}(3 + \sqrt{-25})$

________ 6. $(2 - 3i)(5 + i)$

________ 7. $(5 - \sqrt{3}i)^2$

________ 8. $(\sqrt{12} - \sqrt{-8})(\sqrt{3} + \sqrt{-2})$

________ 9. $(\sqrt{2} - 11i)(\sqrt{2} + 11i)$

________ 10. Write the conjugate of $1 + \sqrt{5}i$.

Answers are on page 484.

Dividing Complex Numbers

To divide any complex number by a *real* number, we use the same methods as we used to divide a binomial by a constant. We shall always write our answers in standard complex form.

Example 27. Find $\dfrac{3 + 2i}{6}$.

Solution

$$\frac{3 + 2i}{6} = \frac{1}{6}(3 + 2i)$$
$$= \frac{3}{6} + \frac{2}{6}i$$
$$= \frac{1}{2} + \frac{1}{3}i$$

You try Example 28.

Example 28. Find $\dfrac{4 - \sqrt{3}i}{2}$.

Solution

$$\frac{4 - \sqrt{3}i}{2} = \frac{1}{2}(4 - \sqrt{3}i)$$
$$=$$

Check your work on page 484. ▶

To divide a complex number by a *pure imaginary* number, we first eliminate the imaginary unit from the denominator. Remembering that $i = \sqrt{-1}$, we shall use a technique similar to that used for rationalizing denominators. Recalling the fact that $i \cdot i = i^2 = -1$, we simplify such quotients by multiplying the numerator and denominator by i.

Example 29. Find $\dfrac{5}{2i}$.

Solution

$$\frac{5}{2i} = \frac{5}{2i} \cdot \frac{i}{i}$$
$$= \frac{5i}{2i^2}$$
$$= \frac{5i}{2(-1)}$$
$$= -\frac{5}{2}i$$
$$= 0 - \frac{5}{2}i \quad \text{Standard complex form.}$$

Example 30. Find $\dfrac{1 - 5i}{i}$.

Solution

$$\frac{1 - 5i}{i} = \frac{1 - 5i}{i} \cdot \frac{i}{i}$$
$$= \frac{(1 - 5i)i}{i^2}$$
$$= \frac{i - 5i^2}{-1}$$
$$= \frac{i - 5(-1)}{-1}$$
$$= \frac{i + 5}{-1}$$
$$= \frac{i}{-1} + \frac{5}{-1}$$
$$= -5 - i \quad \text{Standard complex form.}$$

If a fraction contained a denominator that was a *sum* (or difference) of radicals, recall that we rationalized that denominator using the *difference* (or sum) of those radicals as the rationalizing factor. Using the same reasoning,

> We can simplify a quotient whose denominator is $a + bi$ by multiplying the numerator and denominator by the *complex conjugate*, $a - bi$.

Let's use the conjugate of the denominator to simplify some quotients, writing our answers in standard complex form.

Example 31. Simplify $\dfrac{3}{2+i}$.

Solution

$$\frac{3}{2+i} = \frac{3}{2+i} \cdot \frac{2-i}{2-i}$$
$$= \frac{3(2-i)}{(2+i)(2-i)}$$
$$= \frac{3(2-i)}{4-i^2}$$
$$= \frac{3(2-i)}{4-(-1)}$$
$$= \frac{3(2-i)}{5}$$
$$= \frac{6-3i}{5}$$
$$= \frac{6}{5} - \frac{3}{5}i$$

Example 32. Simplify $\dfrac{26i}{3-2i}$.

Solution

$$\frac{26i}{3-2i} = \frac{26i}{3-2i} \cdot \frac{3+2i}{3+2i}$$
$$= \frac{26i(3+2i)}{(3-2i)(3+2i)}$$
$$= \frac{26i(3+2i)}{9-4i^2}$$
$$= \frac{26i(3+2i)}{9-4(-1)}$$
$$= \frac{\overset{2}{\cancel{26}}i(3+2i)}{\underset{1}{\cancel{13}}}$$
$$= 2i(3+2i)$$
$$= 6i + 4i^2$$
$$= -4 + 6i$$

Usually, when simplifying quotients of complex numbers, it is wise to delay removing parentheses in the numerator just in case the new denominator contains a common factor. This did not happen in Example 31, but it did happen in Example 32.

Example 33. Simplify $\dfrac{3+\sqrt{-3}}{4+\sqrt{-3}}$.

Solution. We must write complex numbers in standard form before we begin to operate.

$\dfrac{3+\sqrt{-3}}{4+\sqrt{-3}} = \dfrac{3+\sqrt{3}i}{4+\sqrt{3}i} \cdot \dfrac{4-\sqrt{3}i}{4-\sqrt{3}i}$	Write complex numbers in standard form. Multiply by the conjugate of the denominator.
$= \dfrac{12 - 3\sqrt{3}i + 4\sqrt{3}i - 3i^2}{16 - 3i^2}$	Multiply numerators. Multiply denominators.
$= \dfrac{12 + \sqrt{3}i - 3(-1)}{16 - 3(-1)}$	Combine like terms. Substitute -1 for i^2.
$= \dfrac{12 + \sqrt{3}i + 3}{16 + 3}$	Simplify.
$= \dfrac{15 + \sqrt{3}i}{19}$	Combine like terms.
$= \dfrac{15}{19} + \dfrac{\sqrt{3}}{19}i$	Write the answer in standard form.

Trial Run

Simplify, writing answers in standard complex form.

______ 1. $\dfrac{9 - 12i}{3}$

______ 2. $\dfrac{10 - \sqrt{2}i}{5}$

______ 3. $\dfrac{-12}{\sqrt{3}i}$

______ 4. $\dfrac{\sqrt{2} - 3i}{i}$

______ 5. $\dfrac{1}{3 + i}$

______ 6. $\dfrac{17i}{3 - 5i}$

______ 7. $\dfrac{11}{3 - \sqrt{-2}}$

______ 8. $\dfrac{4 + \sqrt{5}i}{2 - \sqrt{5}i}$

Answers are on page 484.

Examples You Completed

Example 6. Write $1 + \sqrt{-3}$ using the imaginary unit.

Solution

$$\begin{aligned} 1 + \sqrt{-3} &= 1 + \sqrt{3(-1)} \\ &= 1 + \sqrt{3} \cdot \sqrt{-1} \\ &= 1 + \sqrt{3}i \end{aligned}$$

Example 7. Identify the parts of $3 + i$.

Solution

$$\begin{aligned} &3 + i \\ &= 3 + 1i \end{aligned}$$

The real part is 3.
The imaginary part is 1.

Example 10. Write $\sqrt{300}$ in standard complex form and identify its parts.

Solution

$$\begin{aligned} \sqrt{300} &= \sqrt{100 \cdot 3} \\ &= 10\sqrt{3} \\ &= 10\sqrt{3} + 0i \end{aligned}$$

The real part is $10\sqrt{3}$.
The imaginary part is 0.

Example 12. Solve $x + yi = \sqrt{3} - \sqrt{7}i$ for real values of x and y.

Solution

$$x + yi = \sqrt{3} - \sqrt{7}i$$
$$x + yi = \sqrt{3} + (-\sqrt{7})i$$

Real Parts	***Imaginary Parts***
$x = \sqrt{3}$	$y = -\sqrt{7}$

Example 13. Find $(3 + 2i) + (7 + 4i)$.

Solution

$$\begin{aligned} &(3 + 2i) + (7 + 4i) \\ &= (3 + 7) + (2 + 4)i \\ &= 10 + 6i \end{aligned}$$

Example 16. Find i^7.

Solution

$$\begin{aligned} i^7 &= i^4 \cdot i^3 \\ &= 1 \cdot (-i) \\ i^7 &= -i \end{aligned}$$

Example 18. Find $\sqrt{2}i^{18}$.

Solution

$$\begin{aligned}\sqrt{2}i^{18} &= \sqrt{2}(i^4)^4 \cdot (i^2)\\ &= \sqrt{2}(1)^4(-1)\\ &= \sqrt{2}(1)(-1)\\ &= -\sqrt{2}\end{aligned}$$

Example 19. Find $-3(4 - 8i)$.

Solution

$$\begin{aligned}&-3(4 - 8i)\\ &= -3 \cdot 4 - 3(-8i)\\ &= -12 + 24i\end{aligned}$$

Example 23. Find $(3 + 4i)(1 + 2i)$.

Solution. We can use the FOIL technique here.

$$\begin{aligned}&(3 + 4i)(1 + 2i)\\ &= 3 + 6i + 4i + 8i^2\\ &= 3 + 10i + 8(-1)\\ &= -5 + 10i\end{aligned}$$

Example 25. Find $(3 - \sqrt{2}i)^2$.

Solution

$$\begin{aligned}&(3 - \sqrt{2}i)^2\\ &= 9 - 2 \cdot 3\sqrt{2}i + (\sqrt{2}i)^2\\ &= 9 - 6\sqrt{2}i + 2i^2\\ &= 9 - 6\sqrt{2}i + 2(-1)\\ &= 7 - 6\sqrt{2}i\end{aligned}$$

Example 28. Find $\dfrac{4 - \sqrt{3}i}{2}$.

Solution

$$\begin{aligned}\frac{4 - \sqrt{3}i}{2} &= \frac{1}{2}(4 - \sqrt{3}i)\\ &= \frac{4}{2} - \frac{\sqrt{3}}{2}i\\ &= 2 - \frac{\sqrt{3}}{2}i\end{aligned}$$

Answers to Trial Runs

page 476 **1.** $0 + 6i$ **2.** $5 + 0i$ **3.** $3 - 2\sqrt{2}i$ **4.** $2\sqrt{3} + 4i$ **5.** $11 - 4\sqrt{2}i$ **6.** $1 + 2i$ **7.** $x = 3, y = -1$

page 480 **1.** 1 **2.** $7i$ **3.** $-10 + 15i$ **4.** $4 + 24i$ **5.** $-15 + 9i$ **6.** $13 - 13i$ **7.** $22 - 10\sqrt{3}i$ **8.** 10 **9.** 123 **10.** $1 - \sqrt{5}i$

page 483 **1.** $3 - 4i$ **2.** $2 - \frac{\sqrt{2}}{5}i$ **3.** $0 + 4\sqrt{3}i$ **4.** $-3 - \sqrt{2}i$ **5.** $\frac{3}{10} - \frac{1}{10}i$ **6.** $-\frac{5}{2} + \frac{3}{2}i$ **7.** $3 + \sqrt{2}i$ **8.** $\frac{1}{3} + \frac{2\sqrt{5}}{3}i$

Name ______________________ Date ____________

EXERCISE SET 7.5

Write each number in standard complex form.

_______ 1. $\sqrt{-16}$ _______ 2. $\sqrt{-64}$ _______ 3. $\sqrt{-21}$

_______ 4. $\sqrt{-15}$ _______ 5. $\sqrt{-50}$ _______ 6. $\sqrt{-45}$

_______ 7. $-\sqrt{-75}$ _______ 8. $-\sqrt{-18}$ _______ 9. $4 - \sqrt{-54}$

_______ 10. $3 - \sqrt{-60}$ _______ 11. $18 + \sqrt{-49}$ _______ 12. $32 + \sqrt{-81}$

Solve each equation for real values of x and y.

_______ 13. $4x - 2yi = 4 + 8i$ _______ 14. $5x - 3yi = 10 + 15i$

_______ 15. $2x - 4i = -8 - 12yi$ _______ 16. $3x - 9i = -9 - 18yi$

_______ 17. $4x - 5i = -2 + 3yi$ _______ 18. $2x - 7i = -7 + 5yi$

_______ 19. $14y + 3i = 21 + xi$ _______ 20. $15y + 4i = 45 + xi$

_______ 21. $(3x - 1) - i = 5 - (y + 1)i$ _______ 22. $(2x - 3) - 3i = 3 - (y + 2)i$

_______ 23. $(2x - 1) - (3y + 2)i = 7 + 2i$ _______ 24. $(x - 9) - (2y - 1)i = 8 + 9i$

Simplify the following powers.

_______ 25. i^5 _______ 26. i^9 _______ 27. i^{12} _______ 28. i^{20}

_______ 29. $-i^7$ _______ 30. $-i^{11}$ _______ 31. i^{73} _______ 32. i^{37}

Perform the indicated operation and simplify. Write answers in standard complex form.

_______ 33. $(2 + 3i) + (3 + 6i)$ _______ 34. $(5 + 4i) + (7 + 2i)$

_______ 35. $(9 - 5i) - (9 + 4i)$ _______ 36. $(9 - 6i) - (8 + 2i)$

_______ 37. $(12 - 3\sqrt{-8}) - (3 - \sqrt{-2})$ _______ 38. $(14 - \sqrt{-27}) - (11 + 5\sqrt{-3})$

_______ 39. $(-3 + \sqrt{-49}) + (8 - 7i)$ _______ 40. $(-7 + \sqrt{-36}) + (10 - 6i)$

_______ 41. $[(7 - i) - (3 + 2i)] - (6 - 4i)$ _______ 42. $[(5 - i) - (6 + 2i)] - (4 + 9i)$

_______ 43. $(7 - 4i) - [(-2 + i) - (3 + 7i)]$ _______ 44. $(8 - 3i) - [(-1 + i) - (2 + 6i)]$

_______ 45. $-6(3 + 2i)$ _______ 46. $-8(5 + 7i)$

_______ 47. $3i(7 + i)$ _______ 48. $9i(2 - i)$

_______ 49. $\sqrt{-25}(2 + \sqrt{-4})$ _______ 50. $\sqrt{-64}(3 + \sqrt{-9})$

______ 51. $(7 - 2i)(3 + 5i)$

______ 52. $(8 - 3i)(6 + 2i)$

______ 53. $(3 - \sqrt{5}i)^2$

______ 54. $(4 + \sqrt{3}i)^2$

______ 55. $(\sqrt{50} - \sqrt{-3})(\sqrt{2} - \sqrt{-48})$

______ 56. $(\sqrt{32} + \sqrt{-5})(\sqrt{2} + \sqrt{-45})$

______ 57. $(\sqrt{3} - 4i)(\sqrt{3} + 4i)$

______ 58. $(\sqrt{5} - 3i)(\sqrt{5} + 3i)$

______ 59. $(-3 - 4i)(2 - 5i)$

______ 60. $(-6 - 2i)(3 - 4i)$

______ 61. $\dfrac{6 - 15i}{3}$

______ 62. $\dfrac{10 - 25i}{5}$

______ 63. $\dfrac{4 - 10i}{5}$

______ 64. $\dfrac{3 + 14i}{7}$

______ 65. $\dfrac{12 - \sqrt{3}i}{4}$

______ 66. $\dfrac{21 - \sqrt{7}i}{3}$

______ 67. $\dfrac{5}{i}$

______ 68. $\dfrac{6}{i}$

______ 69. $\dfrac{-20}{\sqrt{5}i}$

______ 70. $\dfrac{-16}{\sqrt{2}i}$

______ 71. $\dfrac{-\sqrt{3} - i}{2i}$

______ 72. $\dfrac{-\sqrt{2} - 4i}{3i}$

______ 73. $\dfrac{39}{2 - 3i}$

______ 74. $\dfrac{10}{4 - 2i}$

______ 75. $\dfrac{2}{7 - i}$

______ 76. $\dfrac{5}{8 - i}$

______ 77. $\dfrac{13i}{7 - \sqrt{3}i}$

______ 78. $\dfrac{22i}{3 - \sqrt{2}i}$

______ 79. $\dfrac{-1}{2 - \sqrt{-25}}$

______ 80. $\dfrac{-2}{2 - \sqrt{-9}}$

______ 81. $\dfrac{5 + 4i}{3 + 6i}$

______ 82. $\dfrac{2 + i}{5 - 6i}$

______ 83. $\dfrac{2 - \sqrt{-5}}{2 + \sqrt{-5}}$

______ 84. $\dfrac{4 - \sqrt{-2}}{4 + \sqrt{-2}}$

☆ Stretching the Topics

______ 1. Find $(1 - \sqrt{3}i)^3$.

______ 2. Factor $x^2 + 9$ as the difference of two squares.

______ 3. Solve $(4 - 5i) + (x + yi) = 0$ for x and y. What is the additive inverse of $4 - 5i$?

Check your answers in the back of your book.

If you can complete **Checkup 7.5,** you are ready to go on to Section 7.6.

Name ______________________ Date __________

✓ CHECKUP 7.5

Write each number in standard complex form.

_______ 1. $\sqrt{-121}$

_______ 2. $9 - \sqrt{-28}$

_______ 3. Solve $3x - 8i = 12 + 2yi$ for real values of x and y.

_______ 4. Evaluate i^{34}.

Perform the indicated operation and simplify. Write answers in standard complex form.

_______ 5. $(8 - 3i) - (2 + 2i)$

_______ 6. $(4 - 3\sqrt{-12}) - (1 + \sqrt{-27})$

_______ 7. $(2 - 3i)(4 + 3i)$

_______ 8. $(5 - 2i)^2$

_______ 9. $\dfrac{3 - 2i}{i}$

_______ 10. $\dfrac{1 - 3i}{1 + 2i}$

Check your answers in the back of your book.

If You Missed Problems:	You Should Review Examples:
1	5
2	9
3	11, 12
4	15–18
5, 6	21
7, 8	25, 26
9	30
10	33

7.6 Solving Equations with Complex Solutions

Now that we have learned to work within the set of complex numbers, we can solve many equations that have *no real solutions*.

Example 1. Solve $x^2 + 4 = 0$.

Solution

$$x^2 + 4 = 0$$
$$x^2 = -4$$
$$x = \pm\sqrt{-4}$$
$$x = \pm 2i$$
$$x = 2i \quad \text{or} \quad x = -2i$$

Both solutions are complex numbers.

Example 2. Solve $x^2 + x + 2 = 0$.

Solution. We can use the quadratic formula. Here $a = 1$, $b = 1$, and $c = 2$.

$$x = \frac{-1 \pm \sqrt{1^2 - 4(1)(2)}}{2(1)}$$ Substitute into the quadratic formula.

$$= \frac{-1 \pm \sqrt{1 - 8}}{2}$$

$$= \frac{-1 \pm \sqrt{-7}}{2}$$ Simplify.

$$= \frac{-1 \pm \sqrt{7}i}{2}$$ Write numerator in standard form.

$$x = -\frac{1}{2} + \frac{\sqrt{7}}{2}i \quad \text{or} \quad x = -\frac{1}{2} - \frac{\sqrt{7}}{2}i$$

Both solutions are complex numbers.

Example 3. Solve $x^3 = 8$.

Solution

$$x^3 = 8$$
$$x^3 - 8 = 0$$
$$(x - 2)(x^2 + 2x + 4) = 0$$ Factor the difference of two cubes.

$$x - 2 = 0 \quad \text{or} \quad x^2 + 2x + 4 = 0$$ Use zero product rule.

$$x = 2 \quad \text{or} \quad x = \frac{-2 \pm \sqrt{2^2 - 4(1)(4)}}{2}$$

$$= \frac{-2 \pm \sqrt{4 - 16}}{2}$$

$$= \frac{-2 \pm \sqrt{-12}}{2}$$

$$= \frac{-2 \pm 2\sqrt{3}i}{2}$$ Simplify the radical.

$$= \frac{\overset{1}{\cancel{2}}(-1 \pm \sqrt{3}i)}{\underset{1}{\cancel{2}}}$$ Factor the numerator and reduce.

$$x = 2 \quad \text{or} \quad x = -1 \pm \sqrt{3}i$$

The solutions are

$$x = 2 \quad \text{or} \quad x = -1 + \sqrt{3}i \quad \text{or} \quad x = -1 - \sqrt{3}i$$

The equation has three complex solutions; one of the complex solutions is real.

You complete Example 4.

Example 4. Solve $x^4 + x^2 - 6 = 0$.

Solution

$$x^4 - x^2 - 6 = 0$$

$$(x^2 - 3)(x^2 + 2) = 0$$

$$x^2 - 3 = 0 \quad \text{or} \quad x^2 + 2 = 0$$

$$x^2 = \underline{\qquad} \qquad x^2 = \underline{\qquad}$$

$$x = \pm\sqrt{\underline{\qquad}} \qquad x = \pm\sqrt{\underline{\qquad}}$$

$$x = \pm\sqrt{\underline{\qquad}}\,i$$

The solutions are

$$x = \sqrt{\underline{\qquad}} \quad \text{or} \quad x = -\sqrt{\underline{\qquad}} \quad \text{or} \quad x = \sqrt{\underline{\qquad}}\,i \quad \text{or} \quad x = -\sqrt{\underline{\qquad}}\,i$$

The equation has ________ complex solutions. ________ of the complex solutions are real.

Check your work on page 491. ▶

From the examples we have worked, we make the observation that **nonreal solutions occur in conjugate pairs.**

> If $a + bi$ is a solution for a polynomial equation, then $a - bi$ is also a solution.

In Chapter 5 we learned that the *discriminant* ($b^2 - 4ac$ in the quadratic formula) could be used to describe the number and nature of the solutions of a quadratic equation. Now that we are permitted to use complex numbers, let's reconsider what the discriminant tells us about the solutions.

Discriminant $b^2 - 4ac$	Kind of Solutions for $ax^2 + bx + c = 0$
$b^2 - 4ac = 0$	One rational solution
$b^2 - 4ac > 0$ and perfect square	Two different rational solutions
$b^2 - 4ac > 0$ and *not* perfect square	Two different irrational solutions
$b^2 - 4ac < 0$	Two different complex (but nonreal) solutions

Trial Run

Find all complex solutions.

______ 1. $x^2 + 5 = 0$

______ 2. $x^2 - x + 1 = 0$

______ 3. $x^2 + 5x = x - 7$

______ 4. $x^4 - 4 = 0$

Answers are below.

Example You Completed

Example 4. Solve $x^4 - x^2 - 6 = 0$.

Solution

$$x^4 - x^2 - 6 = 0$$

$$(x^2 - 3)(x^2 + 2) = 0$$

$$x^2 - 3 = 0 \quad \text{or} \quad x^2 + 2 = 0$$

$$x^2 = 3 \qquad\qquad x^2 = -2$$

$$x = \pm\sqrt{3} \qquad\qquad x = \pm\sqrt{-2}$$

$$x = \pm\sqrt{2}i$$

$$x = \sqrt{3} \quad \text{or} \quad x = -\sqrt{3} \quad \text{or} \quad x = \sqrt{2}i \quad x = -\sqrt{2}i$$

The equation has four complex solutions. Two of the complex solutions are real.

Answers to Trial Run

page 491 1. $x = \pm\sqrt{5}i$ 2. $x = \frac{1}{2} \pm \frac{\sqrt{3}}{2}i$ 3. $x = -2 \pm \sqrt{3}i$ 4. $x = \pm\sqrt{2}$ or $x = \pm\sqrt{2}i$

Name ______________________ Date __________

EXERCISE SET 7.6

Find all complex solutions.

_______ 1. $x^2 + 9 = 0$

_______ 2. $x^2 + 49 = 0$

_______ 3. $4x^2 + 25 = 0$

_______ 4. $9x^2 + 16 = 0$

_______ 5. $x^2 + 3 = 0$

_______ 6. $x^2 + 5 = 0$

_______ 7. $x^2 + x + 1 = 0$

_______ 8. $x^2 - 2x + 2 = 0$

_______ 9. $3x^2 - 8x + 7 = 0$

_______ 10. $5x^2 - 10x + 7 = 0$

_______ 11. $2x^2 + 4 = 3x$

_______ 12. $6x^2 + 2 = 5x$

_______ 13. $x(x + 5) = 2x - 3$

_______ 14. $x(2x + 1) = 3x - 1$

_______ 15. $5x^2 + 6x + 4 = 0$

_______ 16. $3x^2 + 4x + 3 = 0$

_______ 17. $x^3 + 8 = 0$

_______ 18. $x^3 + 27 = 0$

_______ 19. $x^3 = 125$

_______ 20. $x^3 = 64$

_______ 21. $8x^3 - 1 = 0$

_______ 22. $27x^3 - 8 = 0$

_______ 23. $x^4 - 81 = 0$

_______ 24. $x^4 - 16 = 0$

_______ 25. $x^4 = 121$

_______ 26. $x^4 = 100$

_______ 27. $x^4 + 13x^2 + 36 = 0$

_______ 28. $x^4 + 10x^2 + 9 = 0$

_______ 29. $x^4 + 3x^2 = 10$

_______ 30. $x^4 + 4x^2 = 21$

_______ 31. $3 + \frac{4}{x} = \frac{-2}{x^2}$

_______ 32. $\frac{9}{x^2} + 1 = \frac{3}{x}$

_______ 33. $3x + \frac{1}{x} = 2$

_______ 34. $x + \frac{3}{x} = 3$

_______ 35. $\frac{1}{x} = \frac{3x}{x - 2}$

_______ 36. $\frac{x}{x - 3} = \frac{5}{x}$

☆ Stretching the Topics

Solve each equation.

_______ 1. $\sqrt{x - 1} = 3i$

_______ 2. $2ix^2 - 5x - 2i = 0$

_______ 3. Write a quadratic equation that has $2i$ and $-3i$ as its solutions.

Check your answers in the back of your book.

If you can solve the equations in **Checkup 7.6,** you are ready to do the **Review Exercises** for Chapter 7.

Name ______________________ Date ____________

✓ CHECKUP 7.6

Find all complex solutions.

______ 1. $x^2 + 25 = 0$

______ 2. $x^2 + 24 = 0$

______ 3. $x^2 - x + 25 = 0$

______ 4. $5x^2 + 9 = 12x$

______ 5. $x^3 + 125 = 0$

______ 6. $64x^3 = 27$

______ 7. $x^4 = 100$

______ 8. $x^4 + 25x^2 + 144 = 0$

______ 9. $x + \frac{7}{x} = 2$

______ 10. $\frac{2x}{x - 1} = \frac{3}{x}$

Check your answers in the back of your book.

If You Missed Problems:	You Should Review Example:
1, 2	1
3, 4	2
5, 6	3
7, 8	4
9, 10	2

Summary

In this chapter we learned to work with fractional exponents and radicals.

Fractional Exponent Form	Radical Form	Examples
$a^{\frac{1}{n}}$	$\sqrt[n]{a}$	$(-8)^{\frac{1}{3}} = \sqrt[3]{-8} = -2$
		$(x + 3)^{\frac{1}{2}} = \sqrt{x + 3}$
$a^{-\frac{1}{n}}$	$\dfrac{1}{\sqrt[n]{a}}$	$(25)^{-\frac{1}{2}} = \dfrac{1}{\sqrt{25}} = \dfrac{1}{5}$
		$(-27)^{\frac{1}{3}} = \dfrac{1}{\sqrt[3]{-27}} = -\dfrac{1}{3}$
$a^{\frac{m}{n}}$	$\sqrt[n]{a^m}$ or $(\sqrt[n]{a})^m$	$x^{\frac{2}{3}} = \sqrt[3]{x^2}$
		$32^{\frac{7}{5}} = (\sqrt[5]{32})^7 = 2^7 = 128$

If a is a real number and n is a natural number, we say the $\sqrt[n]{a} = b$ if $b^n = a$. Moreover, we noted that

1. If a is positive, then $\sqrt[n]{a}$ is real and positive.
2. If a is negative and n is *odd,* then $\sqrt[n]{a}$ is real and negative.
3. If a is negative and n is *even,* then $\sqrt[n]{a}$ is *not* a real number; instead, $\sqrt[n]{a}$ is an imaginary number.

We also discovered that

$$\sqrt[n]{a^n} = \begin{cases} a & \text{if } n \text{ is odd} \\ |a| & \text{if } n \text{ is even} \end{cases}$$

We learned to simplify radical expressions and operate with radical expressions using several techniques.

If We Want to	We Must	Examples
Simplify radical expressions	Use $\sqrt[n]{a \cdot b} = \sqrt[n]{a}\,\sqrt[n]{b}$ or Use $\sqrt[n]{\frac{a}{b}} = \frac{\sqrt[n]{a}}{\sqrt[n]{b}}$	$\sqrt{72x^3} = \sqrt{36 \cdot 2 \cdot x^2 \cdot x} = 6x\sqrt{2x}$ $\sqrt{\frac{12a^3}{25c^4}} = \frac{\sqrt{12a^3}}{\sqrt{25c^4}} = \frac{2a\sqrt{3a}}{5c^2}$
Add or subtract radical expressions	Combine like radicals	$\sqrt{12x} + \sqrt{3x} - \sqrt{75x}$ $= 2\sqrt{3x} + \sqrt{3x} - 5\sqrt{3x}$ $= -2\sqrt{3x}$
Multiply or divide radical expressions	Use $\sqrt[n]{a} \cdot \sqrt[n]{b} = \sqrt[n]{ab}$ or Use $\frac{\sqrt[n]{a}}{\sqrt[n]{b}} = \sqrt[n]{\frac{a}{b}}$	$\sqrt{7a} \cdot \sqrt{14a} = \sqrt{98a^2} = \sqrt{49 \cdot 2 \cdot a^2} = 7a\sqrt{2}$ $\frac{\sqrt[3]{24x^4}}{\sqrt[3]{3x}} = \sqrt[3]{\frac{24x^4}{3x}} = \sqrt[3]{8x^3} = 2x$
Multiply expressions containing sums or differences of radicals	Use the distributive property	$\sqrt{x}(3 + \sqrt{x}) = 3\sqrt{x} + x$ $(\sqrt{x} - \sqrt{3})(\sqrt{x} + 5\sqrt{3})$ $= x + 5\sqrt{3x} - \sqrt{3x} - 5 \cdot 3$ $= x + 4\sqrt{3x} - 15$
Rationalize the denominator of a fraction	Multiply numerator and denominator by an appropriate rationalizing factor	$\frac{7}{\sqrt{3x}} = \frac{7}{\sqrt{3x}} \cdot \frac{\sqrt{3x}}{\sqrt{3x}} = \frac{7\sqrt{3x}}{3x}$ $\frac{7}{\sqrt{3} + x} = \frac{7}{\sqrt{3} + x} \cdot \frac{\sqrt{3} - x}{\sqrt{3} - x}$ $= \frac{7(\sqrt{3} - x)}{3 - x^2}$

Radical equations are equations in which a variable expression appears in a radicand. To solve such equations, we isolate the radical and raise both sides to the appropriate power. Whenever we square both sides of a radical equation, we must remember to check our solutions in the original equation and reject **extraneous solutions.**

In this chapter we expanded our work with numbers to include square roots of negative numbers and discovered the set of **complex numbers.** Every complex number can be written in the form

$$a + bi$$

where a is real, b is real, and $i = \sqrt{-1}$. In any complex number, a is called the *real part,* b is called the *imaginary part,* and i is called the *imaginary unit*. Every *real number* is a complex number of the form $a + 0i$. A complex number of the form $0 + bi$ is called a **pure imaginary number.** The **conjugate** of a complex number $a + bi$ is the complex number $a - bi$.

	Method	Examples
Equality of complex numbers	If $a + bi = c + di$ then $a = c$ and $b = d$	If $x + yi = 3 + 2i$ then $x = 3$ and $y = 2$
Addition of complex numbers	$(a + bi) + (c + di)$ $= (a + c) + (b + d)i$	$(3 + 2i) + (7 - i) = 10 + i$
Powers of i	$i^1 = i$ $i^2 = -1$ $i^3 = -i$ $i^4 = 1$	$i^{10} = (i^4)^2 \cdot i^2$ $= (1)^2(-1)$ $= -1$
Multiplication of complex numbers	Use the distributive property	$3i(1 - i) = 3i - 3i^2 = 3 + 3i$ $(2 - 5i)(1 + 4i) = 2 + 8i - 5i - 20i^2$ $= 2 + 3i + 20$ $= 22 + 3i$
Division of complex numbers	Multiply numerator and denominator by the conjugate of the denominator	$\frac{3 + 2i}{4 - 3i} = \frac{(3 + 2i)(4 + 3i)}{(4 - 3i)(4 + 3i)}$ $= \frac{12 + 9i + 8i + 6i^2}{16 - 9i^2}$ $= \frac{12 + 17i - 6}{16 + 9}$ $= \frac{6 + 17i}{25}$ $= \frac{6}{25} + \frac{17}{25}i$

After becoming familiar with complex numbers, we used previously learned techniques to solve equations with solutions that were complex numbers.

❑ Speaking the Language of Algebra

Complete each sentence with the appropriate word or phrase.

1. The square root of a negative number is *not* a ________ number.

2. In the radical expression $\sqrt[n]{a}$, we call n the ________ and a the ________ .

3. In adding and subtracting radical expressions, we may combine ________ ________ .

4. The process by which we rid the denominator of radical expressions is called ________ ________ ________ .

5. In solving radical equations, we must check all solutions and reject any that are ________ .

6. Every number of the form $a + bi$ is called a ________ ________ , where a is ________ , b is ________ , and i = ________ .

7. For the number $a + bi$, a is called the ________ part, b is called the ________ part, and i is called the ________ ________ .

8. If two complex numbers are equal, we know their ________ ________ are equal and their ________ ________ are equal.

9. For the complex number $a + bi$, the complex number $a - bi$ is called its ________ .

10. To divide complex numbers, we multiply the numerator and denominator by the ______ of the ________ .

△ Writing About Mathematics

Write your response to each question in complete sentences.

1. Explain why $\sqrt{a^2} = |a|$.

2. Explain how to decide whether or not a quotient of two radical expressions (with the same index) should be rewritten as *one* radical expression.

3. Explain how to rationalize the denominator of a fraction when the denominator
   (a) Contains one term, and it is irrational.
   (b) Contains two terms, and one or both terms are square roots.

4. Explain why it is necessary to check every solution for a radical equation.

5. Explain why every real number is a complex number. What is the complex conjugate of every real number?

Name ________________________________ Date ______________

REVIEW EXERCISES for Chapter 7

Perform the indicated operations. Write answers with positive exponents.

________ 1. $x^{\frac{1}{2}} \cdot x^{\frac{3}{4}} \cdot x^{-\frac{1}{4}}$

________ 2. $\left(x^{\frac{1}{6}} \cdot y^{-\frac{2}{3}}\right)^{12}$

________ 3. $\left(x^{\frac{1}{2}} \cdot y^{\frac{7}{6}} \cdot z^{-\frac{2}{3}}\right)^{\frac{3}{2}}$

________ 4. $\dfrac{x^{\frac{5}{6}} \cdot y^{\frac{2}{5}}}{x^{\frac{2}{3}} \cdot y^{\frac{3}{10}}}$

Find the value of each expression.

________ 5. $\sqrt[3]{-125}$

________ 6. $\sqrt[4]{\dfrac{1}{81}}$

Write in exponential form.

________ 7. $\sqrt[5]{a^2b^3}$

________ 8. $\dfrac{1}{\sqrt[5]{(x-y)^2}}$

Write in radical form and simplify if possible.

________ 9. $27^{\frac{2}{3}}$

________ 10. $(a + b)^{\frac{3}{4}}$

Perform the indicated operation and simplify. Assume that any variable expression under an even radical represents a positive number.

________ 11. $\sqrt{18x^3y^2}$

________ 12. $\sqrt[3]{\dfrac{-24x^5}{125y^3}}$

________ 13. $4(\sqrt{7} - 2x) - (2\sqrt{7} + 3x)$

________ 14. $\sqrt{20x^2} + \sqrt{45x^2} - \sqrt{80x^2}$

________ 15. $\sqrt[3]{16} - \sqrt{75} + \sqrt[3]{54} + \sqrt{48}$

________ 16. $\sqrt{\dfrac{50}{81}} - \sqrt{\dfrac{98}{9}}$

________ 17. $\sqrt{10x} \cdot \sqrt{15x^3}$

________ 18. $\sqrt{\dfrac{27a^3b}{3a}}$

________ 19. $\sqrt{3}(\sqrt{6} + \sqrt{3})$

________ 20. $(4\sqrt{5} - \sqrt{2})(4\sqrt{5} + \sqrt{2})$

________ 21. $(\sqrt{2x} - 3)^2$

________ 22. $\dfrac{4}{\sqrt{6a}}$

________ 23. $\dfrac{3}{\sqrt[3]{2x^2}}$

________ 24. $\sqrt{\dfrac{50x^3}{3y}}$

________ 25. $\dfrac{\sqrt{3x^3y}}{\sqrt{12x^2y^4}}$

________ 26. $\dfrac{\sqrt{6}}{\sqrt{2} - \sqrt{3}}$

Find the solutions.

________ 27. $\sqrt{5x} - 7 = 3$

________ 28. Solve the formula $r = \sqrt{\dfrac{v}{\pi h}}$ for v.

________ 29. $5 - \sqrt{x - 2} = 4$

________ 30. $\sqrt[3]{2x - 1} = -3$

________ 31. $\sqrt{x - 1} = x - 3$

________ 32. $\sqrt{3x - 5} = \sqrt{3 - x}$

Write in standard complex number form.

________ 33. $\sqrt{-144}$

________ 34. $2 + \sqrt{-25}$

________ 35. $6 - \sqrt{-12}$

________ 36. $\sqrt{-4} + \sqrt{-49}$

________ 37. Solve $4x - 8i = 12 + 2yi$ for real values of x and y.

________ 38. Simplify i^{63}.

Perform the indicated operation and write answers in standard complex form.

________ 39. $(3 - 2\sqrt{-18}) - (2 + 3\sqrt{-8})$

________ 40. $(5 - 2i)(3 + 4i)$

________ 41. $\dfrac{1 - 2i}{3i}$

________ 42. $\dfrac{2 - 3i}{1 + i}$

Find the solutions.

________ 43. $x^2 + 27 = 0$

________ 44. $3x^2 - 2x + 1 = 0$

________ 45. $x^2 - 2x = -4$

________ 46. $x^3 - 64 = 0$

________ 47. $x + \dfrac{1}{x} = -1$

________ 48. $\dfrac{5x}{x + 1} = \dfrac{-2}{x}$

Check your answers in the back of your book.

If You Missed Exercises:	You Should Review Examples:	
1	SECTION 7.1	1, 2
2, 3		3, 4
4		5
5, 6		6–9
7		12, 13
8		22
9, 10		18–20
11, 12	SECTION 7.2	5–8
13–15		11–16
16		17
17	SECTION 7.3	1–3
18		5
19		7, 8
20		10
21		11
22		15
23		19, 20
24, 25		23, 24
26		25, 26
27–30	SECTION 7.4	1, 2
31		3, 4
32		5
33–36	SECTION 7.5	5–10
37		11, 12
38		15–18
39		21
40		23
41		30
42		31–33
43	SECTION 7.6	1
44–48		2–4

If you have completed the **Review Exercises** and corrected your errors, you are ready to take the **Practice Test** for Chapter 7.

Name ______________________ Date __________

PRACTICE TEST for Chapter 7

		Section	Examples
______	1. Write $\sqrt[4]{x^3y}$ in exponential form.	7.1	12, 13
______	2. Write $3(x - y)^{\frac{2}{5}}$ in radical form.	7.1	20, 21
______	3. Write $(a^2 - b^2)^{-\frac{1}{3}}$ in radical form.	7.1	22
______	4. Write $\sqrt{-49}$ in standard complex number form.	7.5	10
______	5. Write $5 - \sqrt{-75}$ in standard complex number form.	7.5	9

Perform the indicated operation and simplify. Write answers with positive exponents. Assume that any variable expression under a radical with an even index represents a positive number.

______	6. $\sqrt[3]{216}$	7.1	6, 8
______	7. $\left(\frac{8}{27}\right)^{\frac{2}{3}}$	7.1	18, 19
______	8. $x^{\frac{3}{5}} \cdot x^{\frac{1}{2}} \cdot x^{\frac{3}{10}}$	7.1	1, 2
______	9. $\left(a^{\frac{3}{5}} \cdot b^{-\frac{2}{3}}\right)^{-15}$	7.1	3
______	10. $\dfrac{x^{\frac{3}{4}}y^{\frac{7}{10}}}{x^{\frac{3}{8}}y^{\frac{1}{2}}}$	7.1	5
______	11. $\left(x^{-\frac{3}{5}} \cdot y^{\frac{6}{5}} \cdot z^{-3}\right)^{\frac{5}{12}}$	7.1	4
______	12. $\sqrt{\dfrac{63x^3}{49y^2}}$	7.2	6
______	13. $5\sqrt{8x} - \sqrt{242x} + 2\sqrt{98x}$	7.2	15, 16
______	14. $\sqrt[3]{81} - \sqrt{12} - \sqrt[3]{192} + \sqrt{108}$	7.2	13, 14
______	15. $\sqrt{\dfrac{45x^2}{4}} - \sqrt{\dfrac{5x^2}{36}}$	7.2	17
______	16. $\sqrt{12x} \cdot \sqrt{18x^4}$	7.3	2
______	17. $(4 - \sqrt{7})(3 + 2\sqrt{7})$	7.3	8, 10
______	18. $(\sqrt{3x} - 5)^2$	7.3	11
______	19. $\dfrac{\sqrt{63x^5}}{\sqrt{7x}}$	7.3	5
______	20. $\sqrt{\dfrac{200x^3}{7y}}$	7.3	21, 22

	Problem	Section	Examples
________	21. $\dfrac{\sqrt[3]{3}}{\sqrt[3]{25x^2}}$	7.3	23, 24
________	22. $\dfrac{10}{\sqrt{7} + \sqrt{3}}$	7.3	25, 26
________	23. $(3 - \sqrt{-9})(1 - \sqrt{-49})$	7.5	21
________	24. $\dfrac{4 - i}{4i}$	7.5	30
________	25. $\dfrac{5 + 2i}{3 - i}$	7.5	32, 33
________	26. Solve $4x - 9i = 8 + 21yi$ for real values of x and y.	7.5	11

Find the solutions.

	Problem	Section	Examples
________	27. $3 + \sqrt{x - 9} = 5$	7.4	1, 2
________	28. $\sqrt{8 - 7x} = x - 2$	7.4	3, 4
________	29. $3a^2 - 2a + 1 = 0$	7.6	2
________	30. Solve the formula $r = \sqrt{\dfrac{S}{4\pi}}$ for S.	7.4	1

Name ______________________ Date ____________

SHARPENING YOUR SKILLS after Chapters 1–7

Find the value of each expression.

		SECTION
______	1. $(x - 9)(-x + 7)$ when $x = 3$	1.4
______	2. $\lvert 5 - \sqrt{10} \rvert$	1.4
______	3. $-x^4y^3$ when $x = -1$ and $y = -2$	3.1
______	4. $(-8)^{-2}$	3.2
______	5. $\sqrt{\dfrac{18}{25}}$	5.2

Perform the indicated operations and simplify.

______	6. $12 - (-5)\left[\dfrac{1 + 4(11 - 6)}{-12 + 3(-3)}\right]$	1.3
______	7. $-8[2(3x - 1) + 9x] + \frac{1}{7}(14x)$	1.4
______	8. $(10x^0y^3z)(-3x^5y^7)(-\frac{1}{6}yz^4)$	3.1
______	9. $\left(\dfrac{-4x^2y^{-3}}{x^{-1}y^{-2}}\right)^{-1} \cdot \left(\dfrac{6x^4y^{-2}}{x^3y}\right)^3$	3.2
______	10. $a^2 - 9[a(a + 6) + 3a(a^2 - 2)]$	3.3
______	11. $(x + 5)(x - 1)^2$	3.4
______	12. $(21x^2 - 20x + 5) \div (3x - 2)$	3.5
______	13. $\dfrac{a^2 - 6a - 55}{5a^2 + 23a - 10}$	6.1
______	14. $\dfrac{x^2 - 169}{x^3y^8} \div \dfrac{2x + 26}{x^5y^3}$	6.2
______	15. $\dfrac{x}{x^2 - 36} - \dfrac{3}{x^2 - 6x}$	6.3
______	16. $\dfrac{6 - \dfrac{5x + 25}{x^2}}{\dfrac{3}{5} + \dfrac{1}{x}}$	6.4

Write each expression as a product of factors. SECTION

_______ 17. $6x^4yz - 12x^3y^2z - 30x^2y^3z - 4xy^4z$ 4.1

_______ 18. $3x^5y^2 + 81x^2y^5$ 4.2

_______ 19. $3a^2 - 17a + 10$ 4.3

_______ 20. $2x^4 - 14x^2 - 36$ 4.3

_______ 21. $x^4 - 2x^3 + 8x - 16$ 4.4

Solve each equation or inequality.

_______ 22. $3(x + 2) + 4(x - 6) = 3x - 22$ 2.1

_______ 23. $|7x - 2| \geq 5$ 2.4

_______ 24. $x(13x - 4) + 3 = 3x(x + 1)$ 5.1

_______ 25. $3x(3x - 5) = 1 - 9x$ 5.3

_______ 26. $x^2 - 18x + 81 \geq 0$ 5.4

_______ 27. $\dfrac{x}{x - 4} - \dfrac{12}{x - 4} = 3$ 6.5

_______ 28. $\dfrac{x}{x - 9} > 0$ 6.5

_______ 29. Two amounts of money totaling \$5000 are invested in savings certificates. One amount earns 12 percent and the other earns 9 percent. Find how much is invested at each amount if the total yearly interest is \$517.50. 2.5

_______ 30. Lee is building a 30-foot by 40-foot swimming pool bordered by a concrete walkway. If the total area allowed for the pool and walk is 1824 square feet, find the width of the walkway. 5.5

CHAPTER

8

Graphing First-Degree Equations and Inequalities

The distance required to stop a moving car varies directly as the square of the speed of the car. If the distance required to stop a car traveling 20 miles per hour is 30 feet, what distance would be required to stop a car traveling 40 miles per hour?

Until now we have solved equations and inequalities containing only one variable. In everyday algebra problems, however, it is often necessary to deal with more than one variable. In this chapter we concentrate on first-degree equations and inequalities containing two variables and learn to

1. Graph first-degree equations using the arbitrary-point method.
2. Graph first-degree equations using the intercepts method.
3. Work with distance and slope.
4. Graph first-degree equations using the slope-intercept method.
5. Find equations of lines.
6. Graph first-degree inequalities.
7. Work with function notation.
8. Work with variation.
9. Switch from word statements to functions.

8.1 Graphing First-Degree Equations

Suppose that a theater owner sells adults' tickets at \$3 each and children's tickets at \$2 each. To compute the dollars received in one day, the owner must multiply the number sold of each kind of ticket times the price of the ticket. Since we cannot assume that the number of adults' tickets sold is the same as the number of children's tickets sold, we must use *two* variables here. We can let

$$x = \text{number of adults' tickets sold}$$

$$y = \text{number of children's tickets sold}$$

Then

$$3x = \text{dollars received from adults' tickets}$$

$$2y = \text{dollars received from children's tickets}$$

and

$$3x + 2y = \text{total dollars received}$$

If the theater owner knows that a total of \$360 must be received each day in order for the theater to break even, then we may state this fact using the equation

$$3x + 2y = 360$$

This is an example of a **first-degree equation in two variables.**

> **Standard Form for First-Degree Equation in x and y**
>
> $$Ax + By = C$$
>
> where A, B, and C are constants

We turn our attention now to finding real solutions for first-degree equations in two variables.

Finding Ordered Pairs

In our example we were looking for possible values of x and y that would satisfy the equation

$$3x + 2y = 360$$

Suppose that the theater owner expects to sell 100 adults' tickets. How many children's tickets must be sold? Here we are saying that x is 100 and we find the corresponding value for y by *substituting* 100 for x in our equation.

$$3x + 2y = 360$$

$$3(100) + 2y = 360$$

$$300 + 2y = 360$$

$$2y = 60$$

$$y = 30$$

The theater owner must sell 30 children's tickets. We now know that $x = 100$ and $y = 30$ satisfies our equation. We write this solution as an **ordered pair**

$$(x, y) = (100, 30)$$

Let's find some more ordered pairs that satisfy our equation. Suppose that:

90 children's tickets are sold.

$$y = 90$$
$$3x + 2(90) = 360$$
$$3x + 180 = 360$$
$$3x = 180$$
$$x = 60$$

When $y = 90$, $x = 60$

$$(x, y) = (60, 90)$$

No adults' tickets are sold.

$$x = 0$$
$$3(0) + 2y = 360$$
$$2y = 360$$
$$y = 180$$

When $x = 0$, $y = 180$

$$(x, y) = (0, 180)$$

You should agree that there are many ordered pairs that satisfy our theater ticket equation.

Let's find some ordered pairs of numbers that satisfy the equation

$$4x + 5y = 5$$

To find such ordered pairs, we could substitute several x-values and find the corresponding y-values. Our best move, however, might be to first *isolate* y in the equation.

$$4x + 5y = 5$$
$$5y = 5 - 4x$$
$$y = \frac{5 - 4x}{5}$$

Now we may choose any x-value and find the corresponding y-value.

x	$y = \frac{5 - 4x}{5}$	(x, y)
$x = 0$	$y = \frac{5 - 4(0)}{5} = \frac{5}{5} = 1$	$(0, 1)$
$x = 2$	$y = \frac{5 - 4(2)}{5} = \frac{5 - 8}{5} = -\frac{3}{5}$	$(2, -\frac{3}{5})$
$x = -1$	$y = \frac{5 - 4(-1)}{5} = \frac{5 + 4}{5} = \frac{9}{5}$	$(-1, \frac{9}{5})$

Example 1. Find three ordered pairs of numbers that satisfy $x + 2y = 3$.

Solution. First we isolate y.

$$x + 2y = 3$$
$$2y = 3 - x$$
$$y = \frac{3 - x}{2}$$

x	$y = \frac{3 - x}{2}$	(x, y)
-3	$y = \frac{3 - (-3)}{2}$	
	$= \frac{6}{2} = 3$	$(-3, 3)$
0	$y = \frac{3 - 0}{2}$	
	$= \frac{3}{2}$	$(0, \frac{3}{2})$
3	$y = \frac{3 - 3}{2}$	
	$= \frac{0}{2} = 0$	$(3, 0)$

You try Example 2.

Example 2. Find three ordered pairs of numbers that satisfy $5x - y = 0$.

Solution. First we isolate ______ .

$$5x - y = 0$$
$$-y = -5x$$
$$y = 5x$$

x	$y = 5x$	(x, y)
-1	$y = 5(\_\_\_\_)$	
	$= \_\_\_\_$	$(-1, \_\_\_\_)$
0	$y = 5(\_\_\_\_)$	
	$= \_\_\_\_$	$(0, \_\_\_\_)$
2	$y = 5(\_\_\_\_)$	
	$= \_\_\_\_$	$(2, \_\_\_\_)$

Check your work on page 523. ▶

To find ordered pairs of real numbers that satisfy a first-degree equation in two variables, we have used the following procedure.

Finding Ordered Pairs

1. Isolate y in the original equation.
2. Choose any x-values and find the corresponding y-values.
3. Write each solution as an ordered pair (x, y).

In this method of treating first-degree equations containing x and y, we call x the **independent variable** and y the **dependent variable,** because the value of y *depends* on the value we arbitrarily choose for x. Although we usually find just a few such ordered pairs, there is always an infinite number of ordered pairs that will satisfy each equation.

Graphing First-Degree Equations Using Arbitrary Points

When we solved first-degree equations and inequalities containing one variable in Chapter 2, we often illustrated our solutions using the real number line. To illustrate ordered pairs of real numbers satisfying equations containing *two* variables, however, the real number line alone will not work. Instead, we must use *two* real number lines—a horizontal number line for the independent variable x and a vertical number line for the dependent variable y.

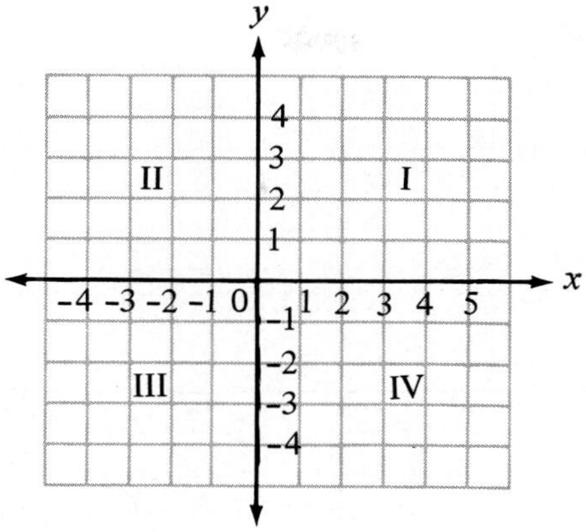

The horizontal number line is called the ***x*-axis** and the vertical number line is called the ***y*-axis.** The point where the axes intersect is called the **origin;** it is the zero point for each axis. All points to the *right* of the origin on the x-axis correspond to *positive* x-values, while points to the *left* of the origin correspond to *negative* x-values. Similarly, all points *above* the origin on the y-axis correspond to *positive* y-values, while points *below* the origin correspond to *negative* y-values.

We call the plane determined by this pair of axes the **Cartesian coordinate plane** (or the **real plane**). The axes divide this plane into four distinct regions called **quadrants,** labeled with Roman numerals in our illustration.

We can use the Cartesian coordinate plane to graph ordered pairs of real numbers. To locate a point such as (3, 2), we must realize that (3, 2) names the point where x is 3 *and* y is 2.

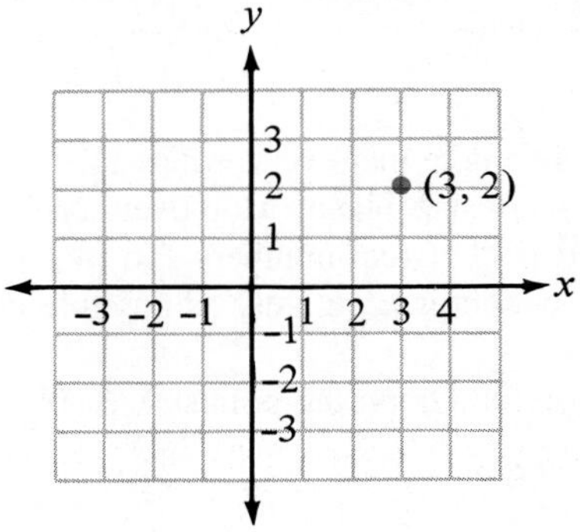

Every point in the plane has two parts.

x-coordinate (or *abscissa*): the x-value in an ordered pair
y-coordinate (or *ordinate*): the y-value in an ordered pair

Try graphing the points in Example 3.

Example 3. Graph the points $(-1, 2)$, $(2, -3)$, and $(-5, -5)$.

Solution

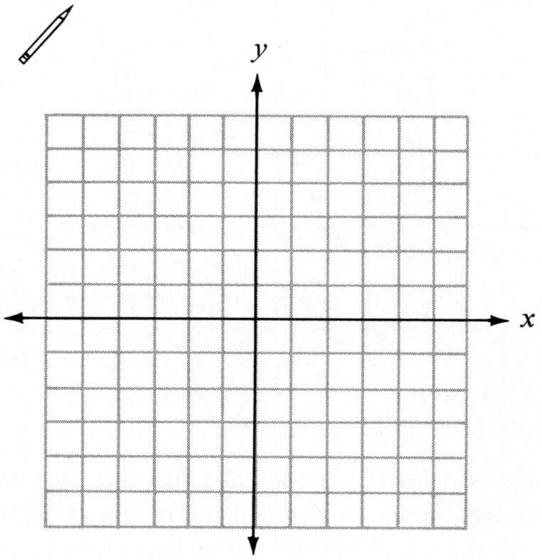

Check your work on page 523. ▶

Example 4. Graph the points $(0, -3)$, $(0, \frac{7}{2})$, $(0, 0)$, $(2, 0)$, and $(-3, 0)$.

Solution

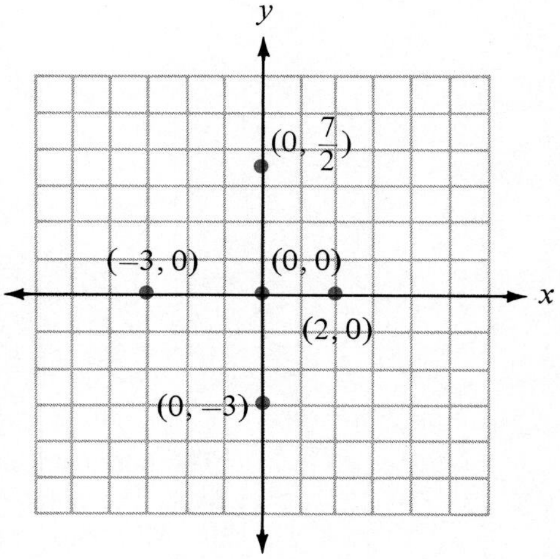

From Example 4 we note some useful facts.

> If a point has an x-coordinate of zero, that point lies on the y-axis.
> If a point has a y-coordinate of zero, that point lies on the x-axis.
> The point $(0, 0)$ is the origin.

We also note that every ordered pair of real numbers corresponds to one and only one point in the Cartesian plane. Moreover, each point in the plane corresponds to one and only one ordered pair of real numbers. To plot points with irrational coordinates, we should estimate their location as accurately as possible using a table of square roots or a calculator.

Example 5. Graph the points $(\sqrt{3}, 0)$, $(-1, \sqrt{2})$, and $(\sqrt{2}, -\sqrt{5})$.

Solution

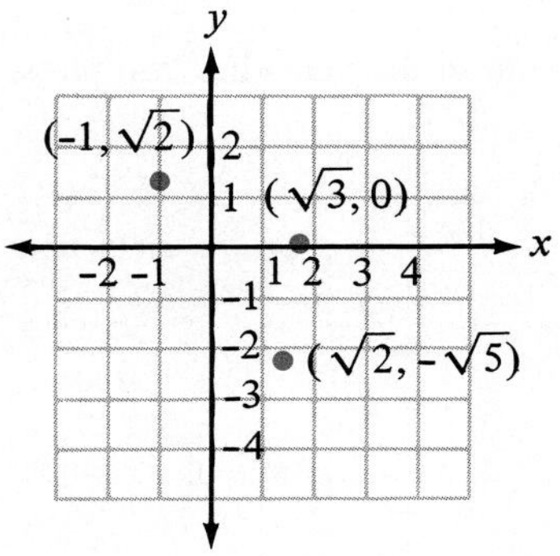

Trial Run

Find three ordered pairs of real numbers that satisfy each equation. There are lots of correct answers.

________ 1. $y = 3x - 1$ ________ 2. $2x + y = 3$ ________ 3. $3x - 5y = 1$

4. Graph and label the points.

   $A(0, -2)$

   $B(-\frac{1}{2}, -3)$

   $C(\frac{3}{2}, -2)$

   $D(4, 0)$

   $E(3, \sqrt{2})$

   $F(-\sqrt{5}, 3)$

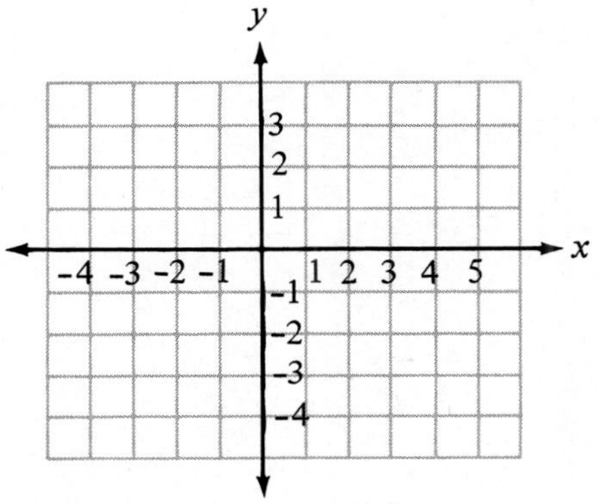

Answers are on page 524.

Earlier in this section we learned to find ordered pairs that would satisfy an equation containing two variables. Now that we have learned to plot ordered pairs in the Cartesian coordinate plane, we can graph the solutions for an equation in x and y.

For example, for the equation

$$y = 5 - x$$

we can find several ordered pairs.

x	$y = 5 - x$		(x, y)
-2	$y = 5 - (-2)$	$= 7$	$(-2, 7)$
0	$y = 5 - 0$	$= 5$	$(0, 5)$
1	$y = 5 - 1$	$= 4$	$(1, 4)$
$1\frac{1}{2}$	$y = 5 - 1\frac{1}{2}$	$= 3\frac{1}{2}$	$(1\frac{1}{2}, 3\frac{1}{2})$
$3 + \sqrt{2}$	$y = 5 - (3 + \sqrt{2})$	$= 2 - \sqrt{2}$	$(3 + \sqrt{2}, 2 - \sqrt{2})$
5	$y = 5 - 5$	$= 0$	$(5, 0)$
8	$y = 5 - 8$	$= -3$	$(8, -3)$

Now we can graph these ordered pairs in the Cartesian plane.

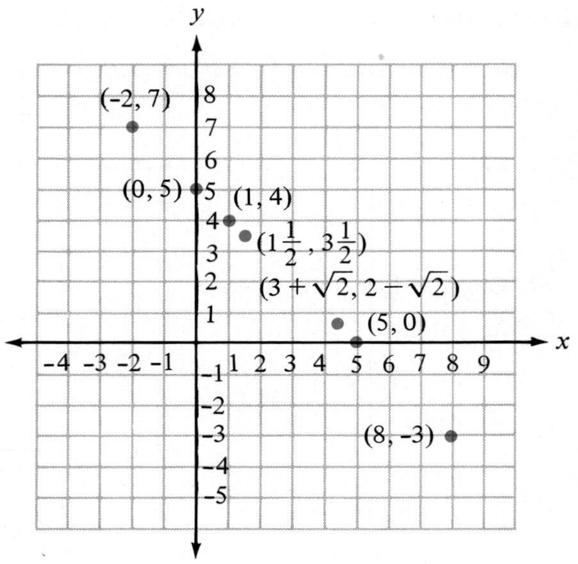

It is more than a coincidence that all these points seem to lie in a straight line. As a matter of fact, any ordered pair that satisfies the equation $y = 5 - x$ will lie on this line, and the coordinates of any point that lies on this line will satisfy that equation. In fact,

> The graph of ordered pairs satisfying a first-degree equation of the form
>
> $$Ax + By = C$$
>
> (where A and B are not *both* zero) is always a straight line.

For this reason, we sometimes call an equation of the form $Ax + By = C$ a **linear equation.**

There are several methods for graphing a linear equation. We have already discovered the method of "choose an x, find the corresponding y." This is called the **arbitrary-point method.** Of course, we know that any *two* points will determine a straight line, but to double check our accuracy, we shall use *three* points. To obtain a good spread of points, let's agree to choose one positive x-value, one negative x-value, and zero.

Example 6. Use three arbitrary points to graph $2x - y = 6$.

Solution. First we isolate y.

$$2x - y = 6$$
$$-y = -2x + 6$$
$$y = 2x - 6$$

x	$y = 2x - 6$	(x, y)
-1	$y = 2(-1) - 6$	
	$= -2 - 6 = -8$	$(-1, -8)$
0	$y = 2(0) - 6$	
	$= -6$	$(0, -6)$
3	$y = 2(3) - 6$	
	$= 6 - 6 = 0$	$(3, 0)$

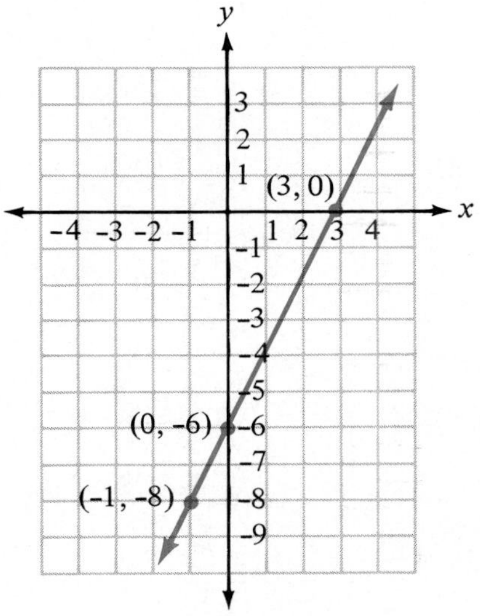

$2x - y = 6$

Notice that our points lie on a straight line that continues indefinitely in either direction.

Now try Example 7.

Example 7. Use three arbitrary points to graph $y = 3x$.

Solution. Here y is already isolated.

x	$y = 3x$	(x, y)
-2	$y = 3(\_\_\_\_) = \_\_\_\_$	$(-2, \_\_\_\_)$
0	$y = 3(\_\_\_\_) = \_\_\_\_$	$(0, \_\_\_\_)$
1	$y = 3(\_\_\_\_) = \_\_\_\_$	$(1, \_\_\_\_)$

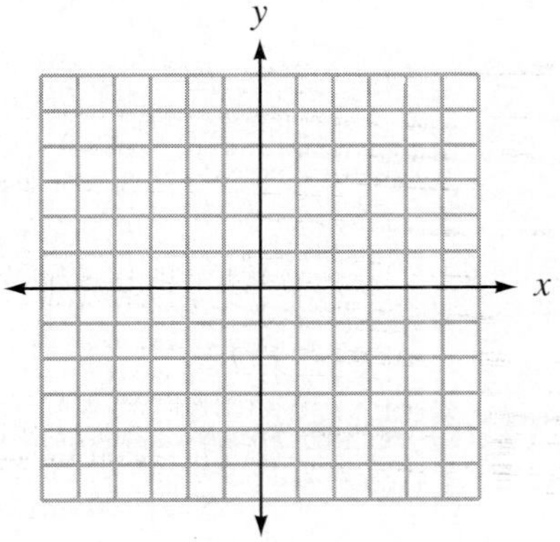

$y = 3x$

Check your work on page 523. ▶

Example 8. Use three arbitrary points to graph $3x + 2y = 6$.

Solution. First we isolate y.

$$3x + 2y = 6$$
$$2y = -3x + 6$$
$$y = \frac{-3x + 6}{2}$$

x	$y = \frac{-3x + 6}{2}$	(x, y)
-2	$y = \frac{-3(-2) + 6}{2}$	
	$= \frac{6 + 6}{2} = 6$	$(-2, 6)$
0	$y = \frac{-3(0) + 6}{2}$	
	$= \frac{6}{2} = 3$	$(0, 3)$
3	$y = \frac{-3(3) + 6}{2}$	
	$= \frac{-9 + 6}{2} = -\frac{3}{2}$	$(3, -\frac{3}{2})$

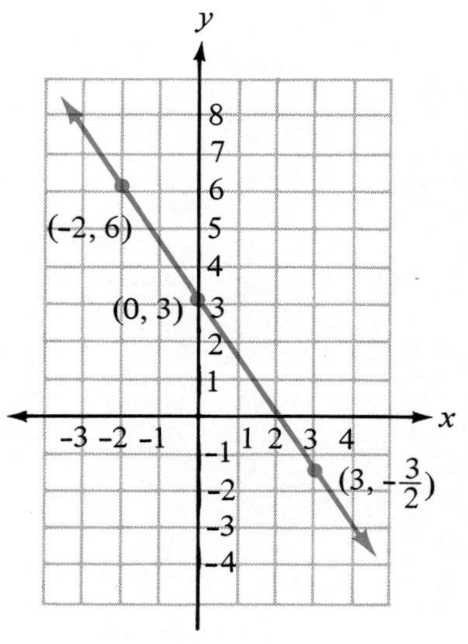

$3x + 2y = 6$

We have now developed a procedure for graphing linear equations of the form $Ax + By = C$ using the method of arbitrary points.

Graphing Lines by the Arbitrary-Point Method

1. Isolate y in the original equation.
2. Choose at least three x-values and find the corresponding y-values by substitution.
3. Write each solution as an ordered pair (x, y).
4. Plot the point in the Cartesian plane corresponding to each ordered pair.
5. Connect the points with a straight line. (If your points do not lie on a straight line, check each of your substitutions.)

Trial Run

Graph each equation using three arbitrary points.

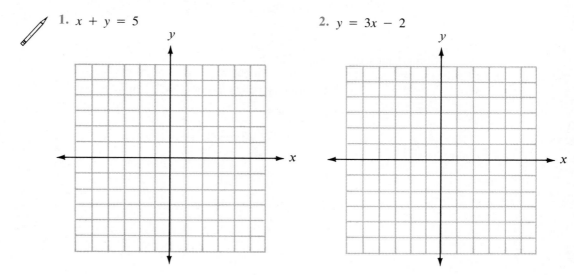

1. $x + y = 5$

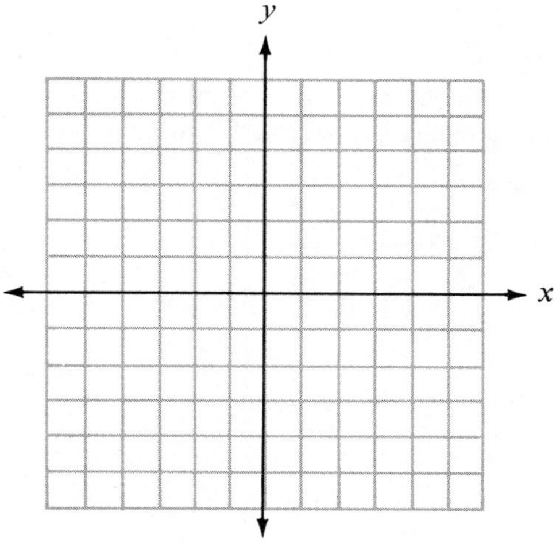

2. $y = 3x - 2$

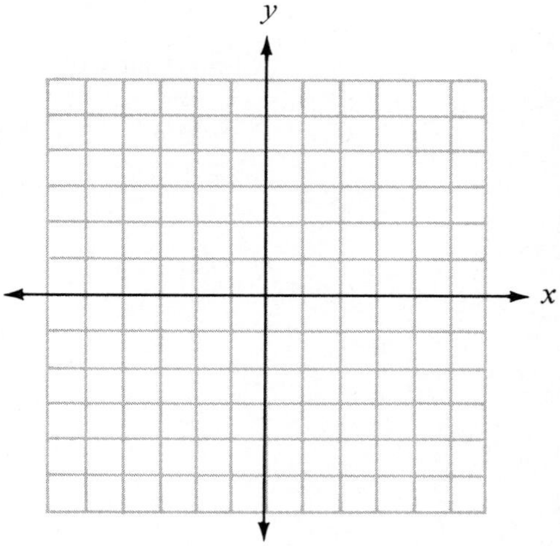

3. $2x - y = 3$

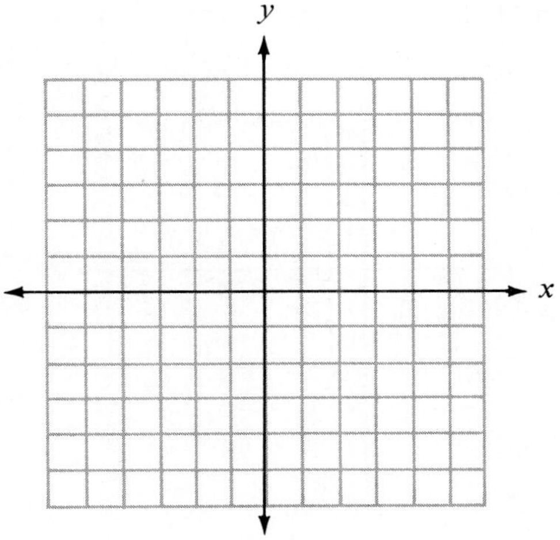

4. $y = -9x$

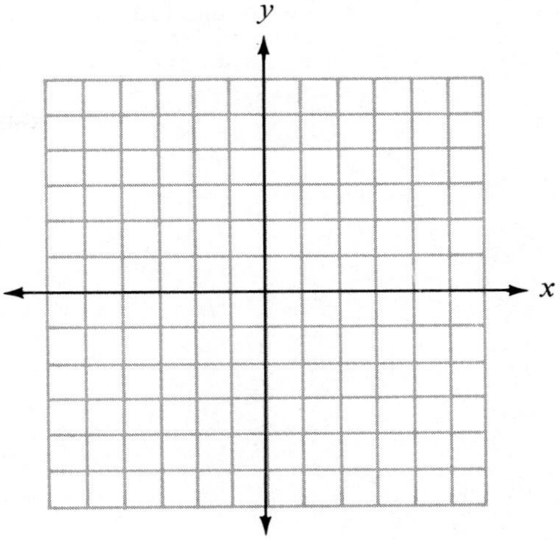

Answers are on page 525.

Graphing First-Degree Equations Using Intercepts

If you look back at the linear equations we have graphed, you will notice that each straight line crosses the x-axis at some point and also crosses the y-axis at some point. These points are called the **intercepts** for a graph.

> *x-intercept*: the x-coordinate of the point at which a graph crosses the x-axis.
> *y-intercept*: the y-coordinate of the point at which a graph crosses the y-axis.

Earlier we observed that every point on the x-axis has a y-coordinate of zero and every point on the y-axis has an x-coordinate of zero. Putting these observations together with the idea of intercepts, we conclude that

> The x-intercept for a graph is the x-coordinate of the point that has a y-coordinate of 0.
> The y-intercept for a graph is the y-coordinate of the point that has an x-coordinate of 0.

In other words, to find the x-intercept for the graph of an equation, we must let $y = 0$ in the equation. To find the y-intercept, we must let $x = 0$ in the equation. When finding intercepts, it is not necessary to begin by isolating either variable. Let's find the intercepts for the graph of

$$2x - 5y = 10$$

x-INTERCEPT: Let $y = 0$

$$2x - 5(0) = 10$$
$$2x = 10$$
$$x = 5$$

The graph crosses the x-axis at the point $(5, 0)$.

y-INTERCEPT: Let $x = 0$

$$2(0) - 5y = 10$$
$$-5y = 10$$
$$y = -2$$

The graph crosses the y-axis at the point $(0, -2)$.

Because there is less chance of error when we substitute 0 for the variable, we trust these points to be accurate and graph the line.

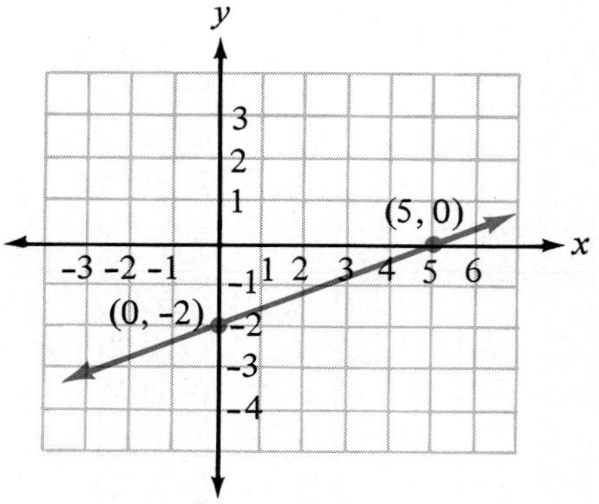

$2x - 5y = 10$

Now try Example 9.

Example 9. Use intercepts to graph $2x + 3y = 4$.

Solution

$$2x + 3y = 4$$

x-INTERCEPT: Let $y = 0$

$2x + 3(\underline{\qquad}) = 4$

$2x = 4$

$x = \underline{\qquad}$

The graph crosses the x-axis at $(\underline{\qquad}, \underline{\qquad})$.

y-INTERCEPT: Let $x = 0$

$2(\underline{\qquad}) + 3y = 4$

$3y = 4$

$y = \underline{\qquad}$

The graph crosses the y-axis at $(\underline{\qquad}, \underline{\qquad})$.

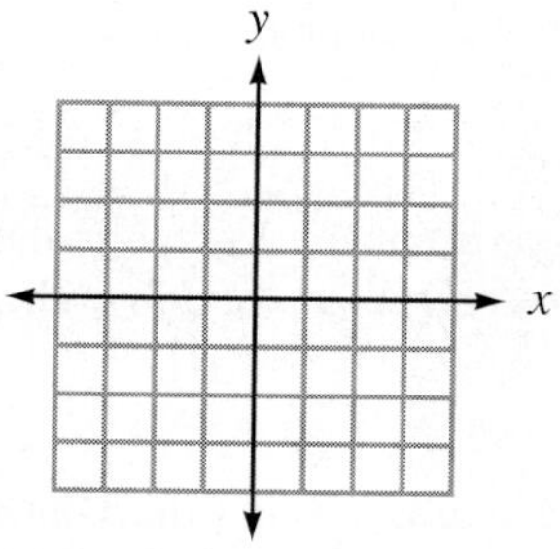

$2x + 3y = 4$

Check your work on page 524. ▶

Don't you agree that the intercepts method is a handy method to use? Will it always be the best method? Look at Example 10.

Example 10. Use intercepts to graph $x - y = 0$.

Solution

$$x - y = 0$$

x-INTERCEPT: Let $y = 0$	y-INTERCEPT: Let $x = 0$
$x - 0 = 0$	$0 - y = 0$
$x = 0$	$y = 0$
(0, 0)	(0, 0)

We seem to be in trouble here because the x-intercept and y-intercept both occur at the origin. When this happens, we must locate another point on the graph by choosing an arbitrary x-value and finding the corresponding y-value. Let $x = 2$.

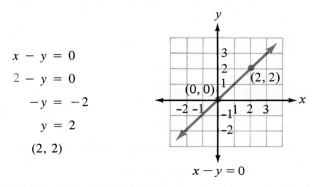

$$x - y = 0$$
$$2 - y = 0$$
$$-y = -2$$
$$y = 2$$
(2, 2)

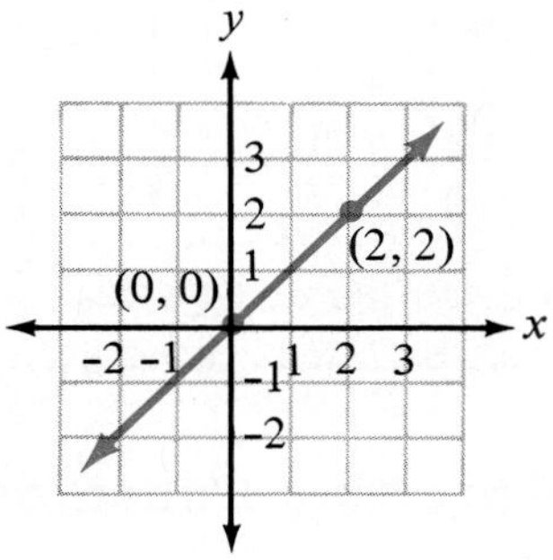

$x - y = 0$

To graph a linear equation of the form $Ax + By = C$ using the intercepts method, we have used the following procedure.

> **Graphing Lines by the Intercepts Method**
>
> **1.** Find the x-intercept by letting $y = 0$ in the original equation.
> **2.** Find the y-intercept by letting $x = 0$ in the original equation.
> **3.** Write the ordered pair corresponding to each intercept.
> **4.** Plot the intercept points and join with a straight line.

Trial Run

Graph each equation using intercepts.

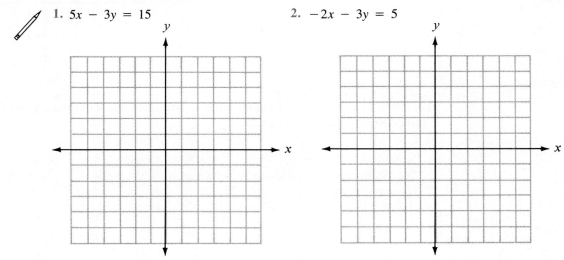

1. $5x - 3y = 15$

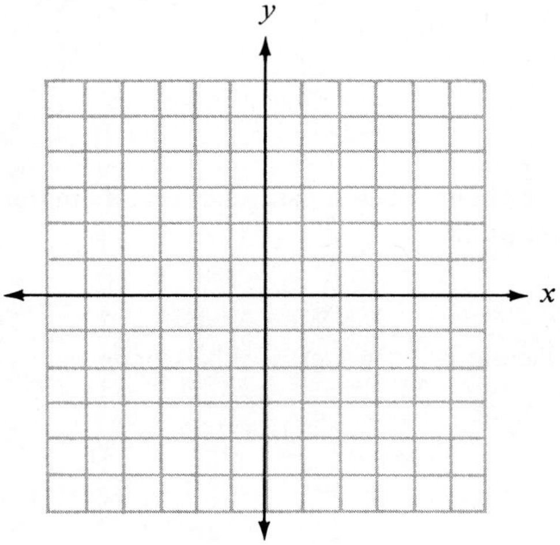

2. $-2x - 3y = 5$

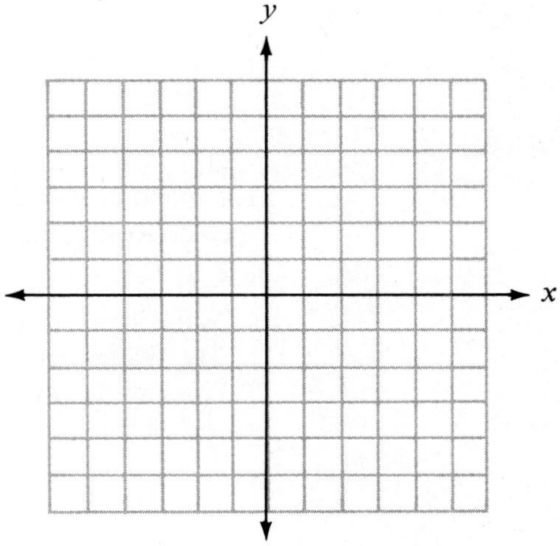

3. $-3x + 8y = 12$

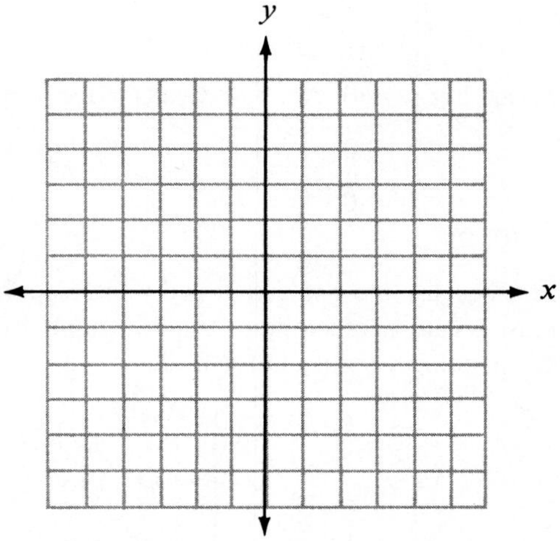

4. $4x + y = 6$

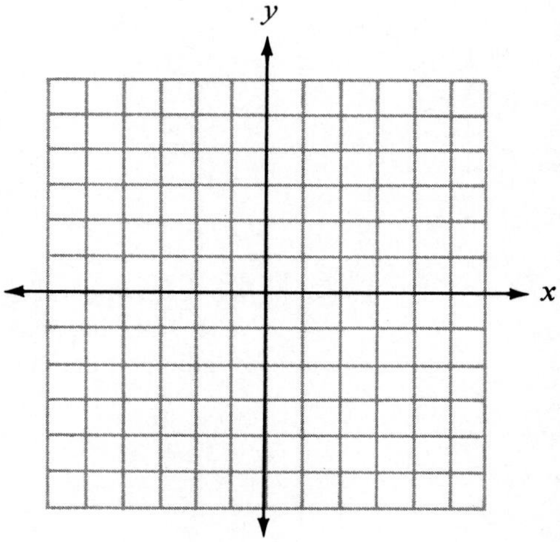

Answers are on page 525.

Graphing Constant Equations

So far we have discussed graphs of linear equations of the form $Ax + By = C$, where neither A nor B is zero, but we have not considered equations of this form in which $A = 0$ or $B = 0$. In other words, we have not discussed linear equations in which the x-term or y-term is missing.

Suppose that we wish to graph an equation such as

$$y = 2$$

which states that $y = 2$, *no matter what x is*. Let's graph some ordered pairs satisfying this equation.

x	y	(x, y)
-3	2	$(-3, 2)$
0	2	$(0, 2)$
2	2	$(2, 2)$

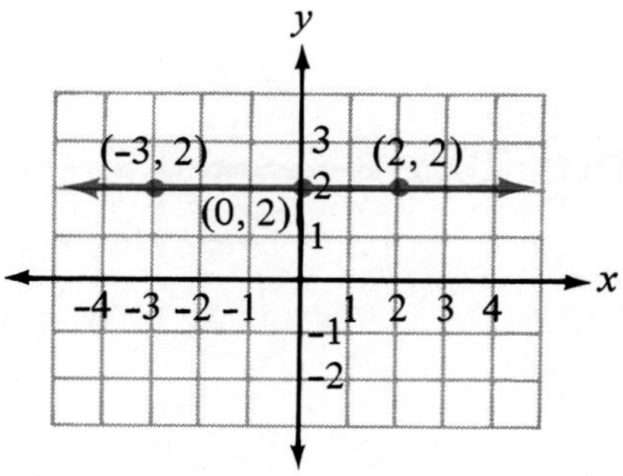

$y = 2$

The graph of $y = 2$ is a straight line, but in particular, it is a **horizontal** line (parallel to the x-axis) crossing the y-axis at (0, 2).

The graph of $y = c$ (where c is a constant) is a horizontal line (parallel to the x-axis) crossing the y-axis at $(0, c)$.

Having made this observation, we agree to graph equations of the form $y = c$ *by inspection*, without plotting several points.

Try Example 11.

Example 11. Graph by inspection

$$3y = -9$$

Solution. We isolate _____ .

$$3y = -9$$

$$y = -3$$

The graph is a ________ line, crossing the _____-axis at (0, _____).

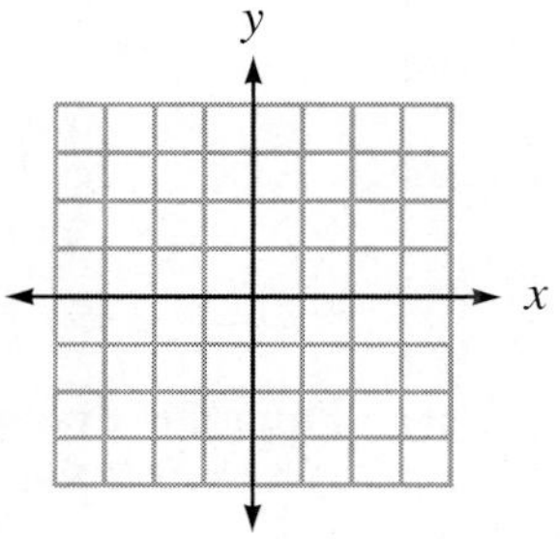

Check your work on page 524. ▶

Example 12. Graph by inspection

$$2y + 3 = 3$$

Solution. We isolate y.

$$2y + 3 = 3$$

$$2y = 0$$

$$y = 0$$

The graph is a horizontal line, crossing the y-axis at (0, 0).

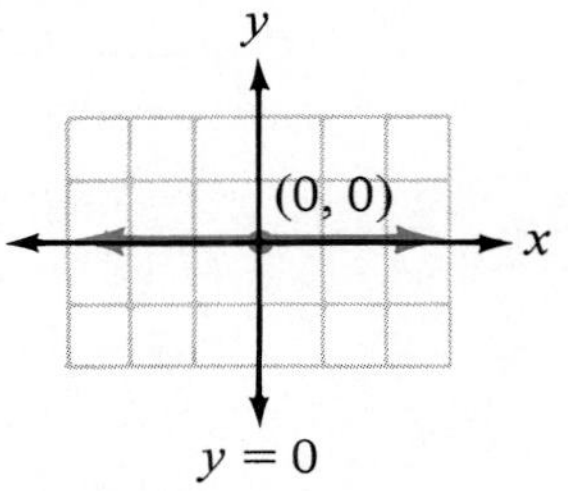

From this example we note that *the graph of* $y = 0$ *is the x-axis.*

To graph an equation such as

$$x = 1$$

we realize that $x = 1$, *no matter what y is.* We can list some ordered pairs for this equation.

(1, −2)

(1, 0)

(1, 3)

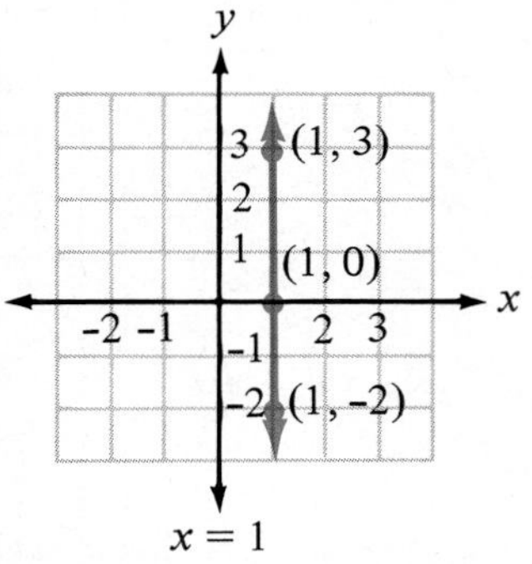

Our graph is a **vertical** line (parallel to the y-axis) crossing the x-axis at (1, 0).

> The graph of $x = c$ (where c is a constant) is a vertical line (parallel to the y-axis) crossing the x-axis at $(c, 0)$.

We may graph linear equations of the form $x = c$ by inspection without plotting several points.

Try Example 13.

Example 13. Graph $2x + 5 = 0$ by inspection.

Solution. We isolate _____ .

$$2x + 5 = 0$$
$$2x = -5$$
$$x = -\frac{5}{2}$$

The graph is a ________ line, crossing the _____ -axis at (_____ , 0).

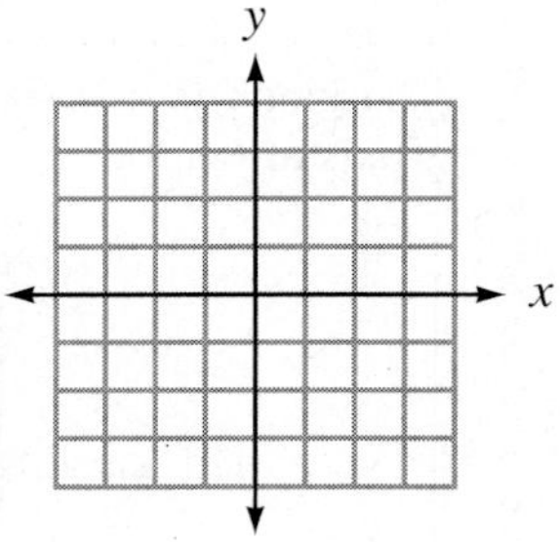

Check your work on page 524. ▶

Example 14. Graph $3(x - 1) = 5x - 3$ by inspection.

Solution. We isolate x.

$$3(x - 1) = 5x - 3$$
$$3x - 3 = 5x - 3$$
$$-2x = 0$$
$$x = 0$$

The graph is a vertical line, crossing the x-axis at (0, 0).

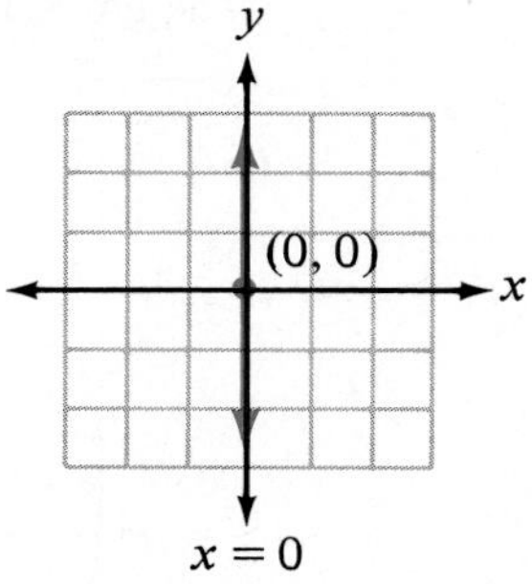

From this example, we note that *the graph of $x = 0$ is the y-axis.*

Trial Run

Graph each equation by inspection.

1. $y = 3$

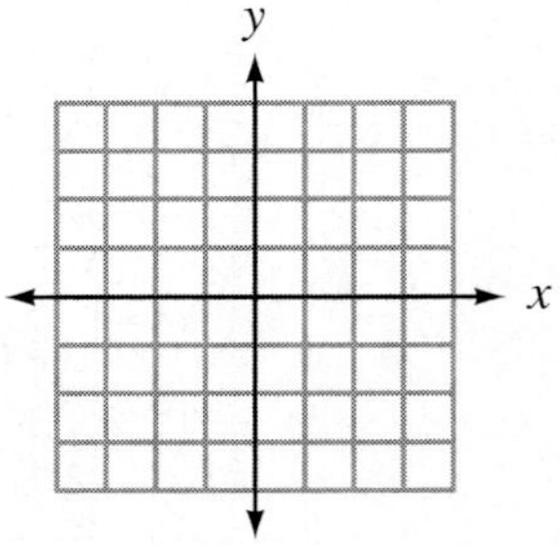

2. $2x - 3 = 0$

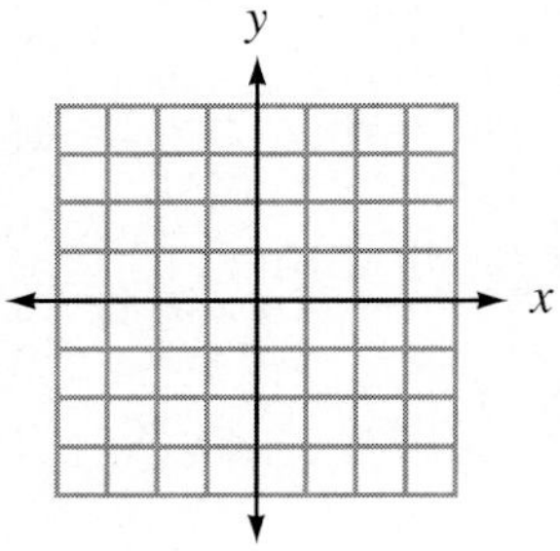

3. $3x + 13 = 10$

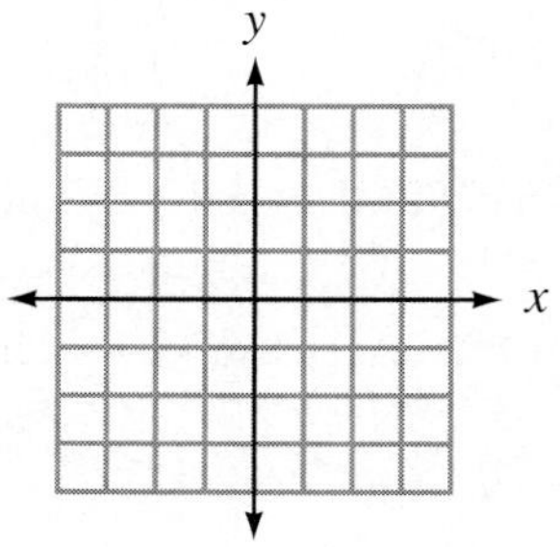

4. $9 + 4y = 1$

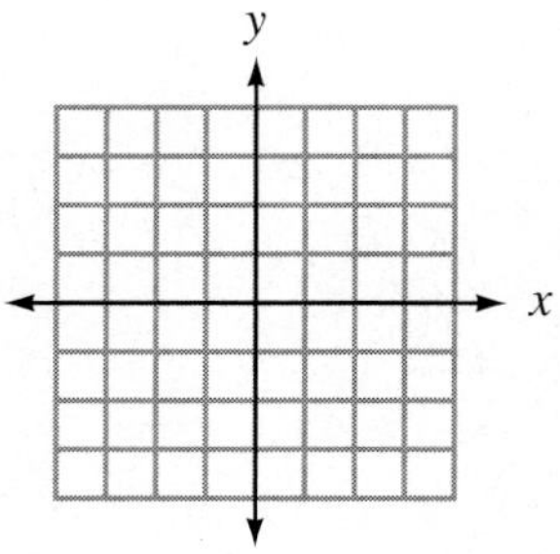

Answers are on page 525.

Let's summarize the methods we have discussed for graphing linear equations of the form $Ax + By = C$.

1. *Arbitrary-point method.* If y can be easily isolated in the given equation, we find three points by the method of "choose an x, find the corresponding y."
2. *Intercepts method.* If we do not wish to isolate y in the given equation, we can find the x-intercept (by letting $y = 0$) and find the y-intercept (by letting $x = 0$).
3. *Method for constant equations.* If the equation is of the form $y = c$ (where c is a constant), we know the graph is a horizontal line crossing the y-axis at $(0, c)$. If the equation is of the form $x = c$ (where c is a constant), we know that the graph is a vertical line crossing the x-axis at $(c, 0)$.

▶ Examples You Completed

Example 2. Find three ordered pairs of numbers that satisfy $5x - y = 0$.

Solution. First we isolate y.

$$5x - y = 0$$
$$-y = -5x$$
$$y = 5x$$

x	$y = 5x$	(x, y)
-1	$y = 5(-1)$	
	$= -5$	$(-1, -5)$
0	$y = 5(0)$	
	$= 0$	$(0, 0)$
2	$y = 5(2)$	
	$= 10$	$(2, 10)$

Example 3. Graph the points $(-1, 2)$, $(2, -3)$, and $(-5, -5)$.

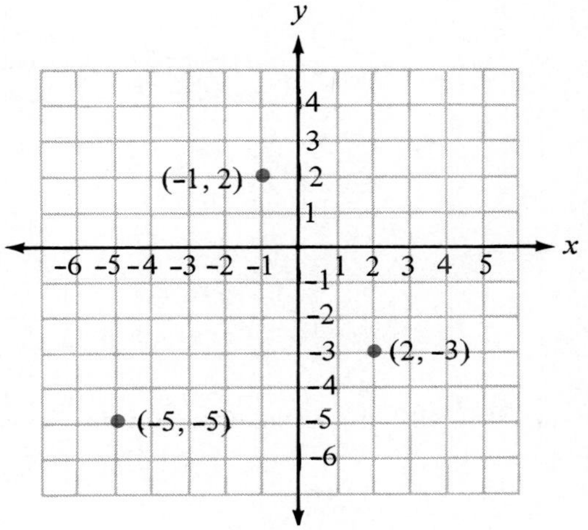

Example 7. Use three arbitrary points to graph $y = 3x$.

Solution. Here y is already isolated.

x	$y = 3x$	(x, y)
-2	$y = 3(-2) = -6$	$(-2, -6)$
0	$y = 3(0) = 0$	$(0, 0)$
1	$y = 3(1) = 3$	$(1, 3)$

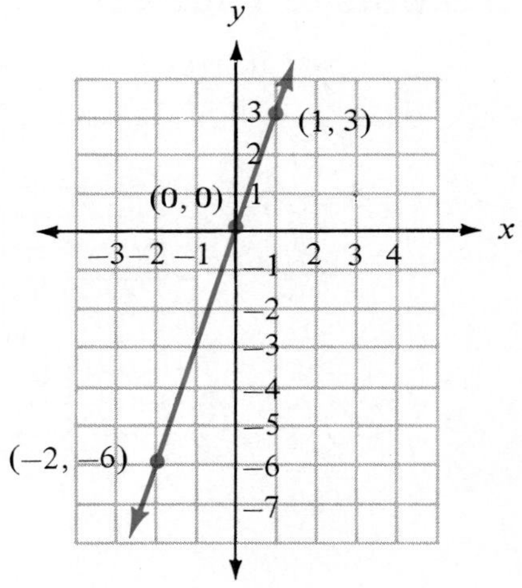

$y = 3x$

Example 9. Use intercepts to graph $2x + 3y = 4$.

Solution

$$2x + 3y = 4$$

x-INTERCEPT: Let $y = 0$

$$2x + 3(0) = 4$$
$$2x = 4$$
$$x = 2$$

The graph crosses the x-axis at $(2, 0)$.

y-INTERCEPT: Let $x = 0$

$$2(0) + 3y = 4$$
$$3y = 4$$
$$y = \frac{4}{3}$$

The graph crosses the y-axis at $(0, \frac{4}{3})$.

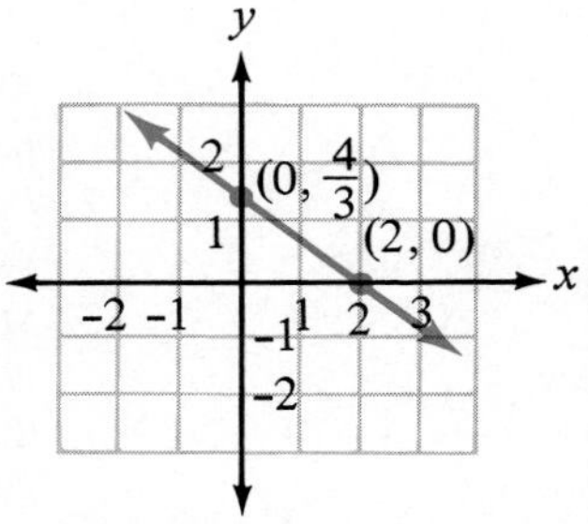

$2x + 3y = 4$

Example 11. Graph $3y = -9$ by inspection.

Solution. We isolate y.

$$3y = -9$$
$$y = -3$$

The graph is a horizontal line, crossing the y-axis at $(0, -3)$.

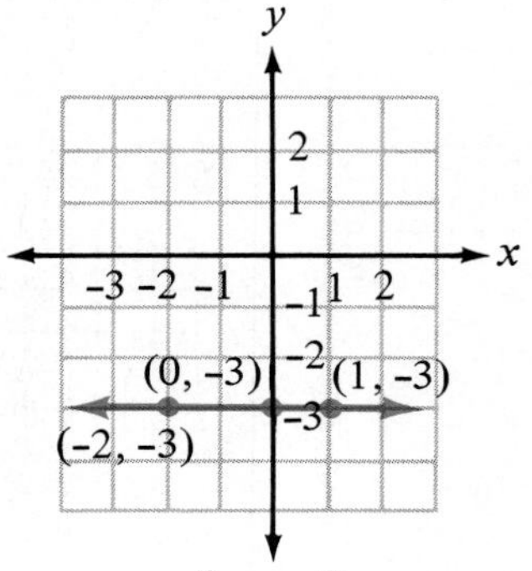

$3y = -9$

Example 13. Graph $2x + 5 = 0$ by inspection.

Solution. We isolate x.

$$2x + 5 = 0$$
$$2x = -5$$
$$x = -\frac{5}{2}$$

The graph is a vertical line, crossing the x-axis at $(-\frac{5}{2}, 0)$.

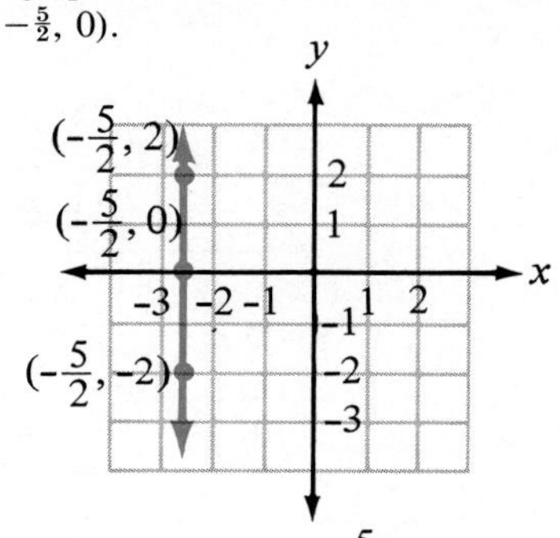

$x = -\frac{5}{2}$

Answers to Trial Runs

page 513 **4.**

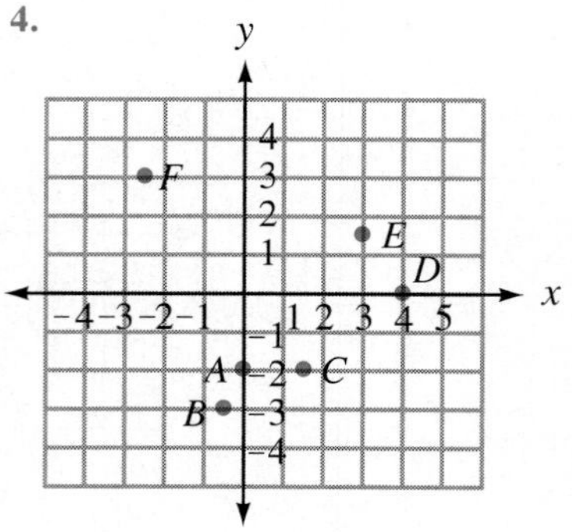

page 516

1.

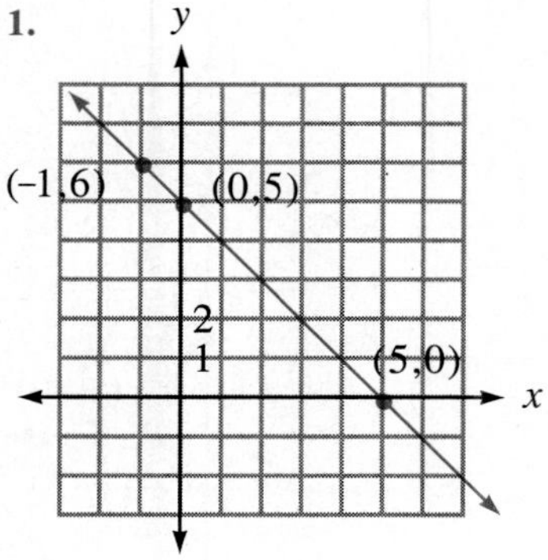

2.

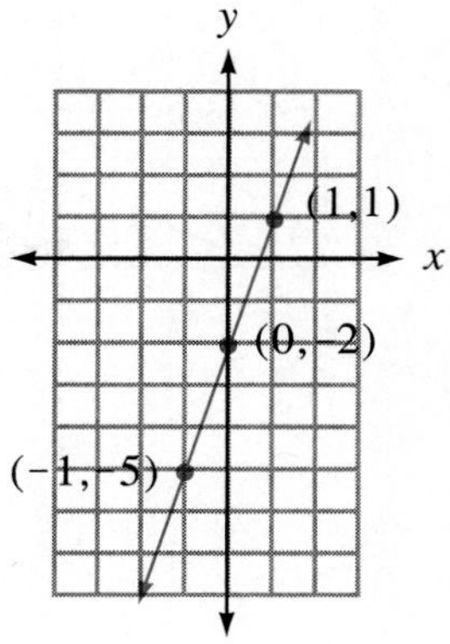

3.

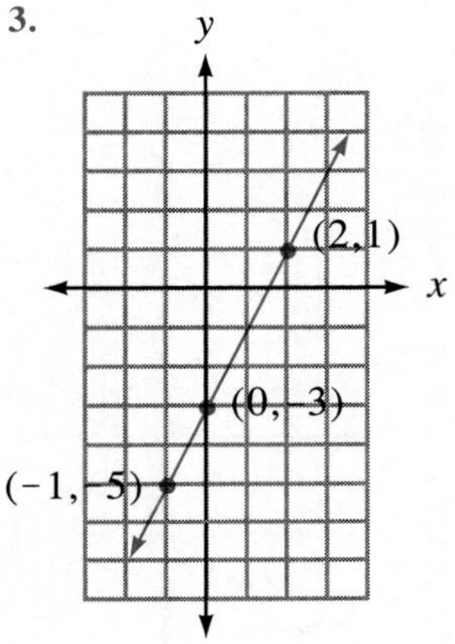

4.

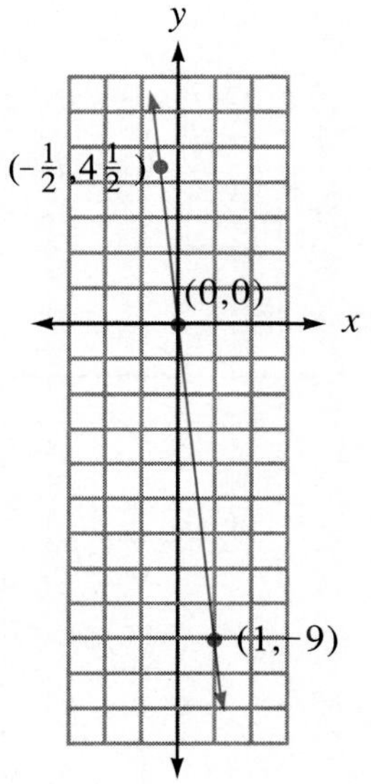

page 519

1.

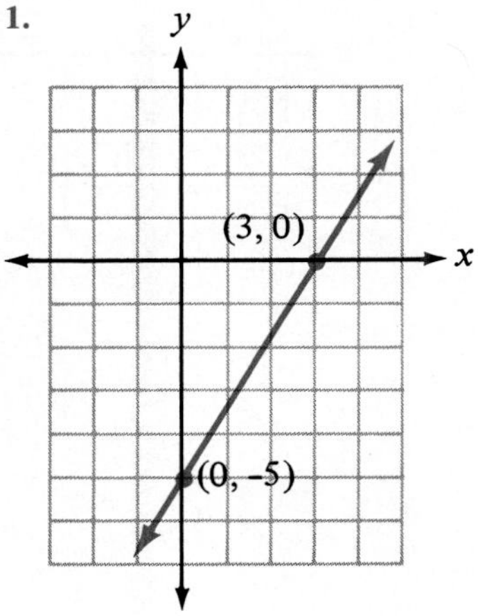

2.

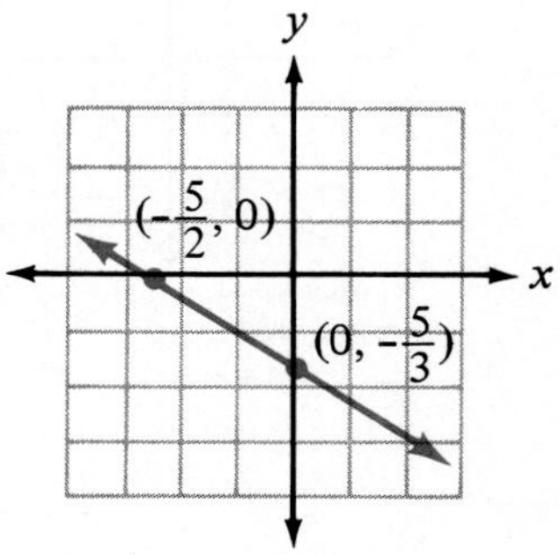

3.

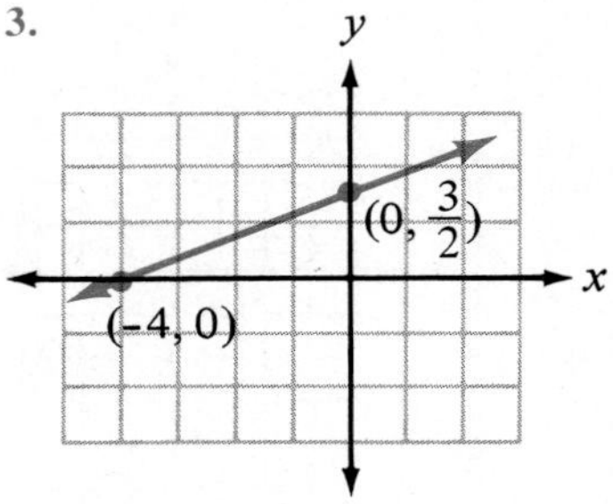

4.

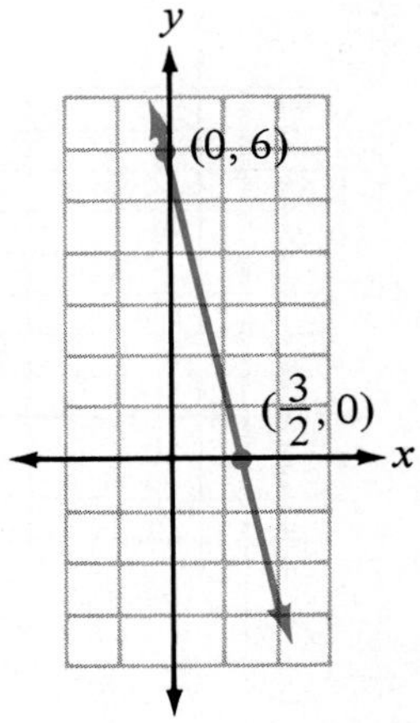

page 522 1.

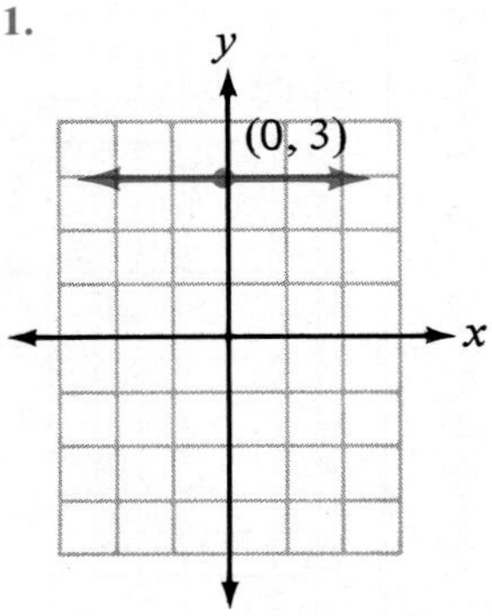

2.

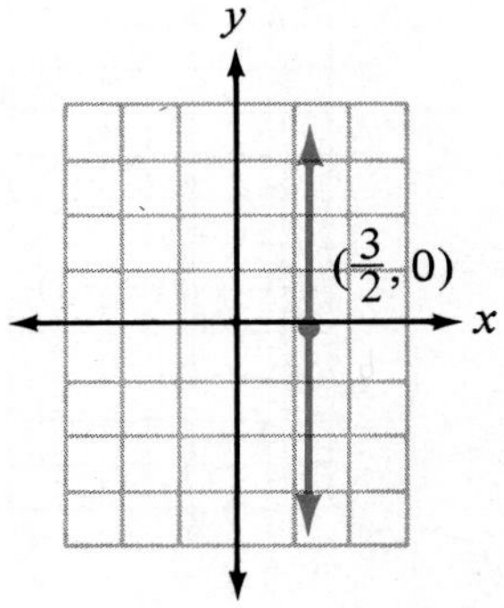

3.

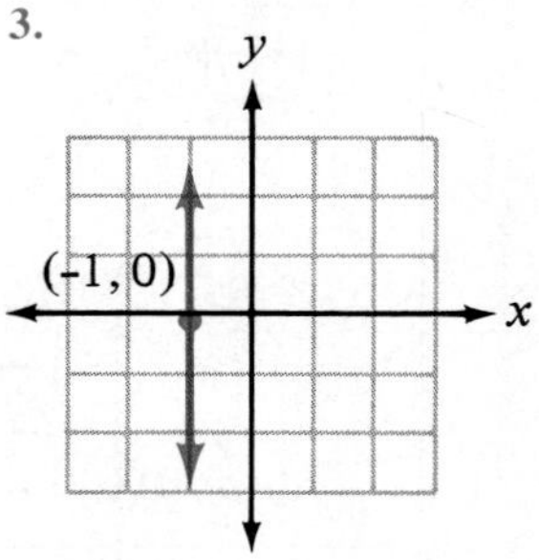

4.

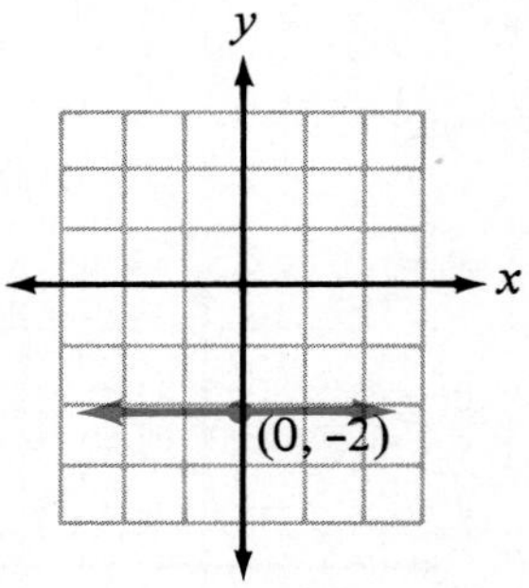

Name ______________________________ Date ______________

EXERCISE SET 8.1

Graph each equation using the arbitrary-point method.

1. $x + y = 6$
2. $x + y = 2$
3. $2y = 16x$
4. $3y = 15x$
5. $y = 5x - 2$
6. $y = 3x - 1$
7. $2x + y = 1$
8. $3x + y = 2$
9. $x + 4y = 10$
10. $x + 6y = 7$
11. $2x - y = -5$
12. $3x - y = -8$

Graph each equation using the intercepts method.

13. $x + y = -3$
14. $x + y = -2$
15. $x - y = 5$
16. $x - y = 4$
17. $x + 2y = 5$
18. $x + 4y = 6$
19. $3x - y = 7$
20. $2x - y = -3$
21. $-5x + 2y = 15$
22. $-7x - 3y = 14$
23. $-x - y = 3$
24. $-x - y = 1$

Graph each equation using any method you choose.

25. $3x + y = 6$
26. $5x + y = 5$
27. $6y - 16 = 0$
28. $8y - 28 = 0$
29. $y = 10x$
30. $y = 7x$
31. $2x - 7 = 0$
32. $3x - 5 = 0$
33. $5x + 3y = -15$
34. $3x + 4y = 12$
35. $9(x - 4) = 2x - 1$
36. $8(x - 3) = 5x - 7$
37. $3x - 4y = -14$
38. $2x - 3y = -10$
39. $3x - 5y = 0$
40. $2x - 9y = 0$

☆ Stretching the Topics

_______ 1. Graph and write the equation for the horizontal line through the point $(-2, 3)$.

_______ 2. Graph $\frac{1}{6}x + \frac{5}{2}y = \frac{5}{3}$.

_______ 3. Graph $0.2x - 0.5y = 1.5$.

Check your answers in the back of your book.

If you can complete **Checkup 8.1,** you are ready to go on to Section 8.2.

Name ________________________ Date ____________

✓ CHECKUP 8.1

Graph each equation using the arbitrary-point method.

1. $3x - y = 4$

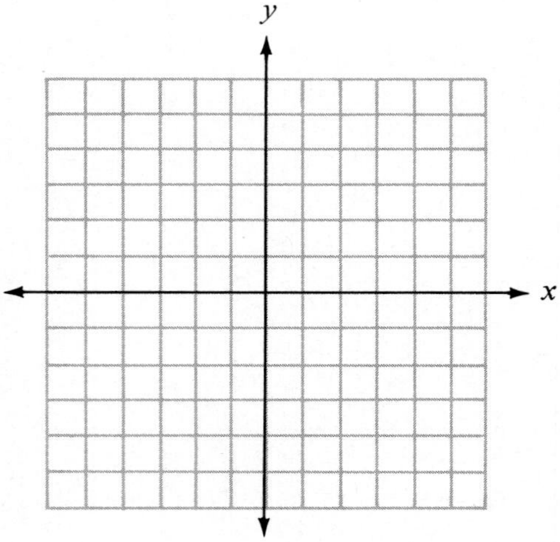

2. $4x + y = 0$

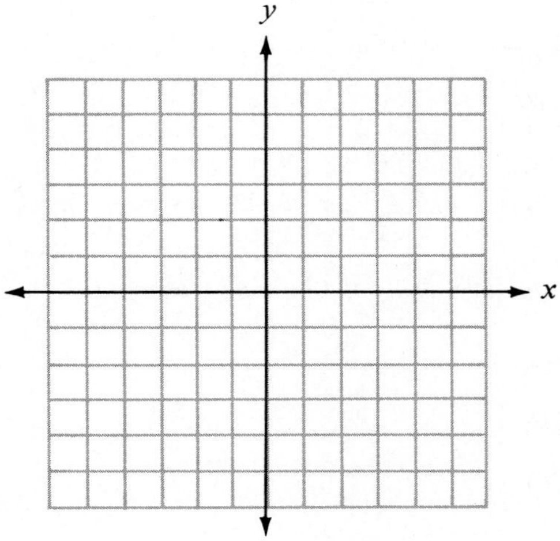

Graph each equation using the intercepts method.

3. $2x - y = 4$

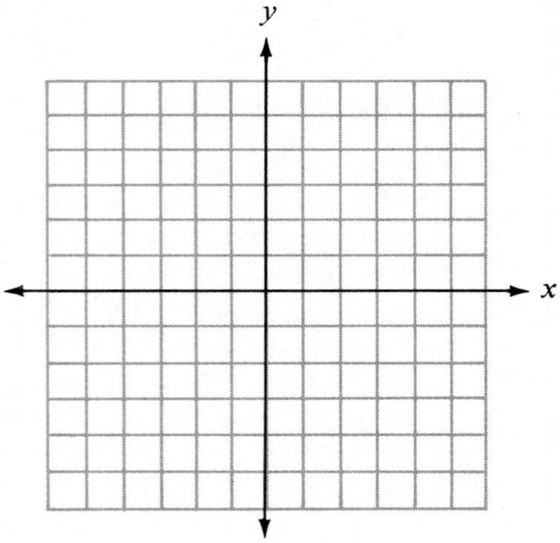

4. $3x - y = 5$

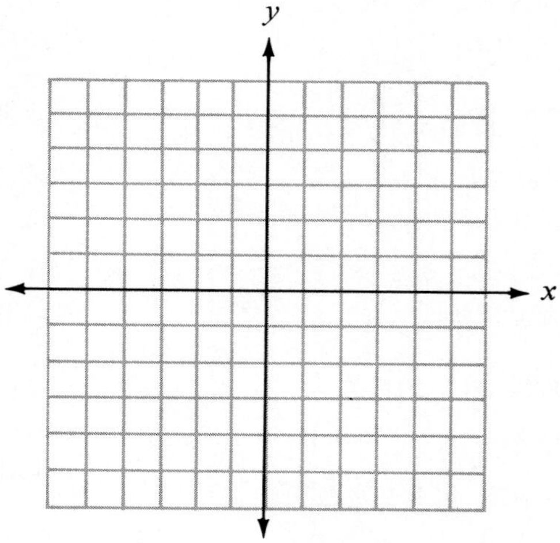

Graph each equation using any method you choose.

5. $y = \frac{2}{3}x$

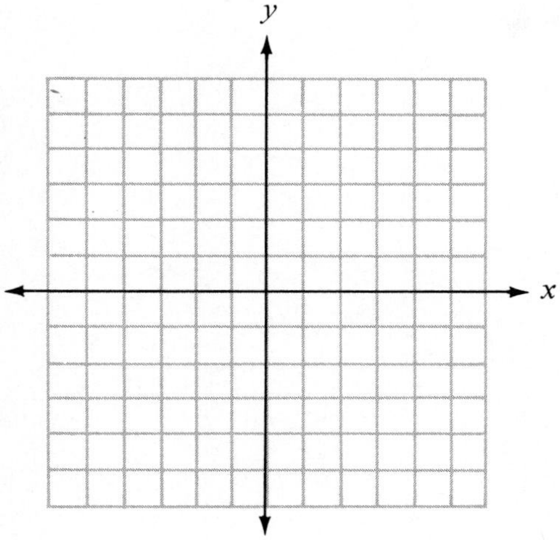

6. $3y + 7 = 0$

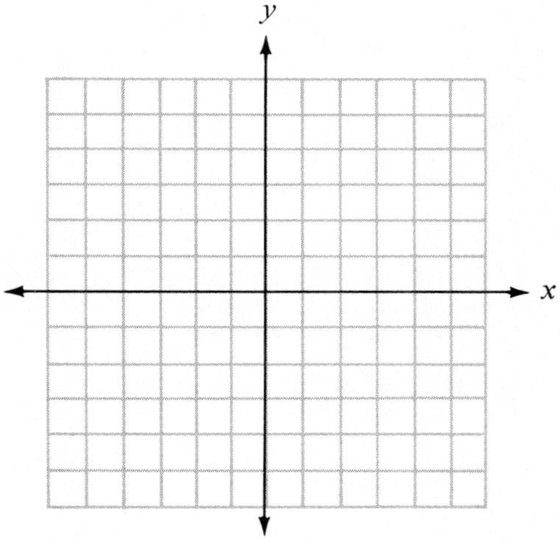

7. $2x = 8$

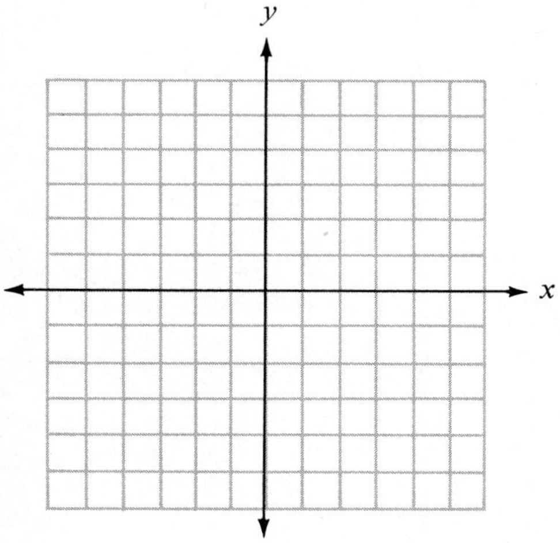

8. $-2x + 3y = -12$

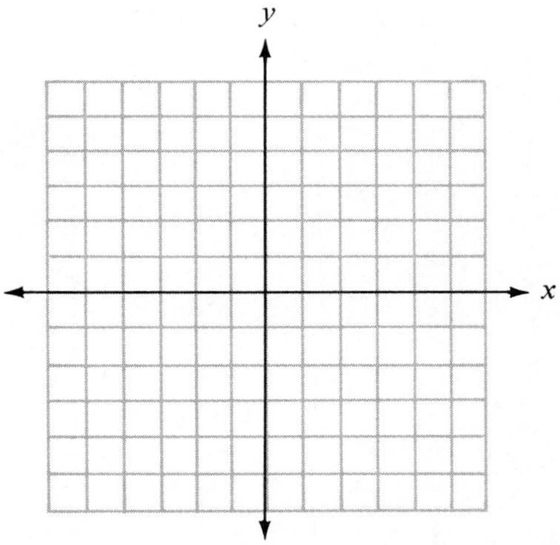

Check your answers in the back of your book.

If You Missed Problems:	You Should Review Examples:
1, 2	6–8
3, 4	9, 10
5	7
6	11
7	13
8	9

8.2 Working with Distance and Slope

Finding the Distance Between Two Points

Sometimes we wish to find the **distance between two points** on a line. For instance, suppose that we wish to find the distance between the points (1, 3) and (4, 7). First let's plot these points in the plane and treat the line segment between them as the hypotenuse of a right triangle.

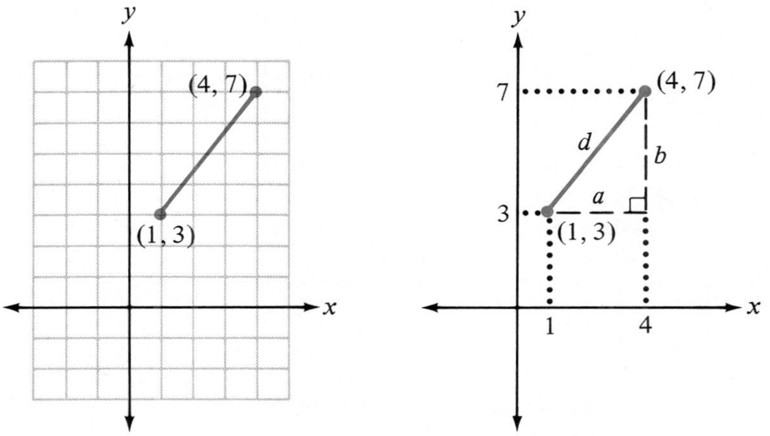

(a) (b)

Earlier we discussed the Pythagorean theorem for a right triangle, which states that

$$a^2 + b^2 = d^2$$

Here a is the difference between x-values of 4 and 1, so $a = 3$. Similarly, b is the difference between y-values of 7 and 3, so $b = 4$. Therefore, $a^2 + b^2 = d^2$ becomes

$$3^2 + 4^2 = d^2$$

$$9 + 16 = d^2$$

$$25 = d^2$$

$$d = \pm\sqrt{25}$$

$$d = \pm 5$$

Because d represents distance, it cannot be negative. Therefore, the distance between the points (1, 3) and (4, 7) is 5 units.

Having seen how to use the Pythagorean theorem to find the distance between two particular points, let's see if we can arrive at a formula for finding the distance between *any* two points.

Suppose that we represent any two points by

$$(x_1, y_1) \quad \text{and} \quad (x_2, y_2)$$

where the subscripts of 1 and 2 help us identify the coordinates of the first and second points. Let's plot these points in the plane and construct our right triangle.

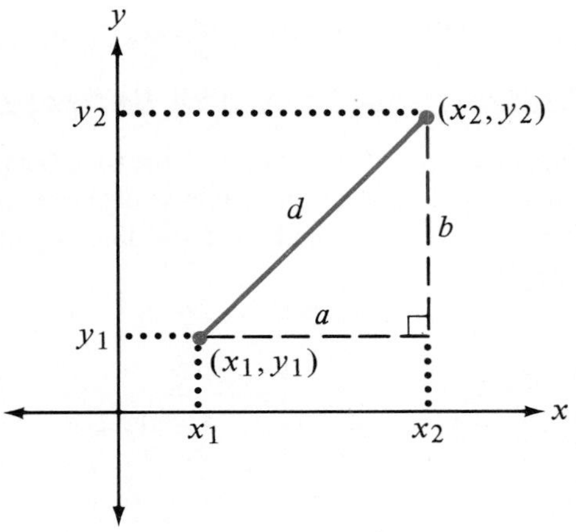

Since $x_2 > x_1$ and $y_2 > y_1$, we can see that $a = x_2 - x_1$ and $b = y_2 - y_1$. So $d^2 = a^2 + b^2$ becomes

$$d^2 = (x_2 - x_1)^2 + (y_2 - y_1)^2$$

$$d = \sqrt{(x_2 - x_1)^2 + (y_2 - y_1)^2}$$

This is the general formula for the distance between two points.

Distance Formula. If (x_1, y_1) and (x_2, y_2) are two points, then the distance between them is found by

$$d = \sqrt{(x_2 - x_1)^2 + (y_2 - y_1)^2}$$

Example 1. Find the distance between $(-2, 1)$ and $(3, -4)$.

Solution. Let $(x_1, y_1) = (-2, 1)$
$(x_2, y_2) = (3, -4)$

$$\begin{aligned} d &= \sqrt{(x_2 - x_1)^2 + (y_2 - y_1)^2} \\ &= \sqrt{[3 - (-2)]^2 + (-4 - 1)^2} \\ &= \sqrt{5^2 + (-5)^2} \\ &= \sqrt{25 + 25} \\ &= \sqrt{50} \\ &= \sqrt{25 \cdot 2} \\ &= 5\sqrt{2} \end{aligned}$$

Now you try Example 2.

Example 2. Find the distance between $(0, 3)$ and $(7, 5)$.

Solution. Let $(x_1, y_1) = (0, 3)$
$(x_2, y_2) = (7, 5)$

$$\begin{aligned} d &= \sqrt{(x_2 - x_1)^2 + (y_2 - y_1)^2} \\ &= \sqrt{(7 - 0)^2 + (5 - 3)^2} \end{aligned}$$

Check your work on page 544. ▶

You might wonder whether it matters which of the given points is labeled (x_1, y_1) and which is labeled (x_2, y_2). You can verify for yourself that the choice of labels does *not* matter; the distance between the same two points will be the same.

Finding the Midpoint of a Line Segment

The **midpoint** of a line segment joining two points (x_1, y_1) and (x_2, y_2) is the point located halfway between those two points. To find the coordinates of the midpoint, we again construct our right triangle and draw perpendicular lines from the midpoint (x_M, y_M) to both legs of the right triangle.

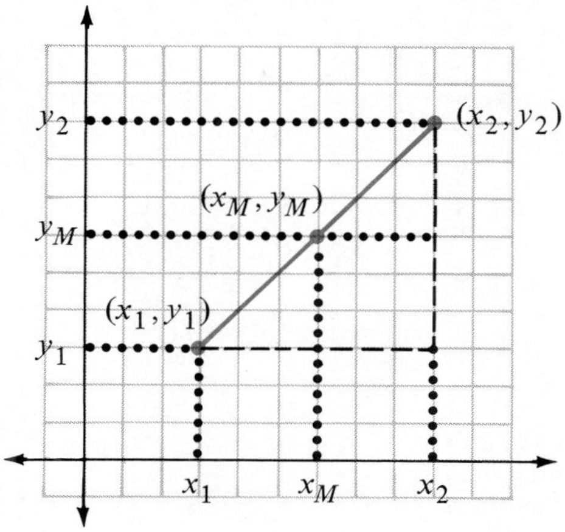

In geometry it can be proved that the perpendicular lines from the midpoint divide each leg in half, so

$$x_M - x_1 = x_2 - x_M \qquad \text{and} \qquad y_M - y_1 = y_2 - y_M$$

$$2x_M = x_1 + x_2 \qquad\qquad 2y_M = y_1 + y_2$$

$$x_M = \frac{x_1 + x_2}{2} \qquad\qquad y_M = \frac{y_1 + y_2}{2}$$

We have now discovered how to find the coordinates of the midpoint of the line segment joining two points.

> **Finding the Midpoint of a Line Segment.** The midpoint (x_M, y_M) of the line segment joining the points (x_1, y_1) and (x_2, y_2) is found by
>
> $$(x_M, y_M) = \left(\frac{x_1 + x_2}{2}, \frac{y_1 + y_2}{2}\right)$$

As you can see, the x-coordinate of the midpoint is just the *average* of the x-coordinates of the endpoints, and the y-coordinate of the midpoint is the *average* of the y-coordinates of the endpoints.

Example 3. Find the midpoint of the line segment joining (2, −3) and (6, 1).

Solution. Let $(x_1, y_1) = (2, -3)$
$(x_2, y_2) = (6, 1)$

$$x_M = \frac{x_1 + x_2}{2} \quad \text{and} \quad y_M = \frac{y_1 + y_2}{2}$$

$$= \frac{2 + 6}{2} \qquad\qquad = \frac{-3 + 1}{2}$$

$$= \frac{8}{2} \qquad\qquad = \frac{-2}{2}$$

$$x_M = 4 \qquad\qquad y_M = -1$$

$$(x_M, y_M) = (4, -1)$$

You try Example 4.

Example 4. Find the midpoint of the line segment joining (−2, 0) and (3, 0).

Solution. Let $(x_1, y_1) = (-2, 0)$
$(x_2, y_2) = (3, 0)$

$$x_M = \frac{x_1 + x_2}{2} \quad \text{and} \quad y_M = \frac{y_1 + y_2}{2}$$

$(x_M, y_M) = (\_\_\_\_, \_\_\_\_)$

Check your work on page 544. ▶

Trial Run

Find the distance between the points.

_______ 1. (0, 2) and (4, 8)

_______ 2. (−4, −1) and (1, 11)

Find the midpoint of the line segment joining the points.

_______ 3. (3, 3) and (−6, 7)

_______ 4. (0, 1) and (0, −5)

Answers are on page 545.

Defining the Slope of the Line

Let's take a look at the graphs of several linear equations.

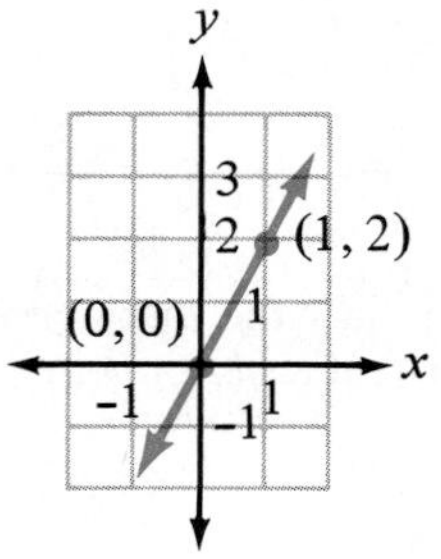

$y = 2x$

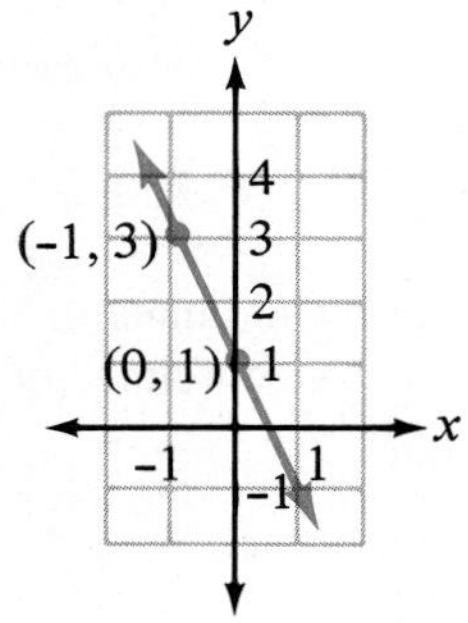

$y = 1 - 2x$

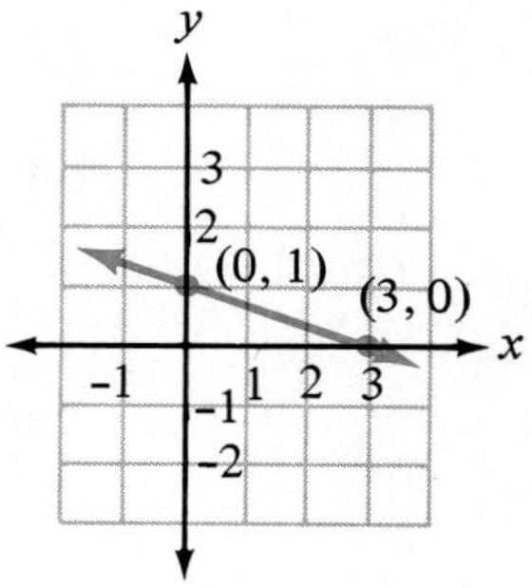

$x + 3y = 3$

If we wish to describe each of these lines in terms of their steepness, we might say that the first two lines appear to be equally steep but in opposite directions. The third line seems less steep than the other lines.

We can measure the steepness or **slope** of a line by locating any two points on that line and then comparing the units of change in the y-values (sometimes called the **rise**) to the corresponding units of change in the x-values (sometimes called the **run**).

Measuring Slope

$$\frac{\text{rise}}{\text{run}} \quad \text{or} \quad \frac{\text{vertical change}}{\text{horizontal change}} \quad \text{or} \quad \frac{\text{change in } y}{\text{change in } x}$$

To find the numerical value of the slope of a line joining two points (x_1, y_1) and (x_2, y_2), we construct our familiar triangle. The change in y-values is $y_2 - y_1$ and the corresponding change in x-values is $x_2 - x_1$.

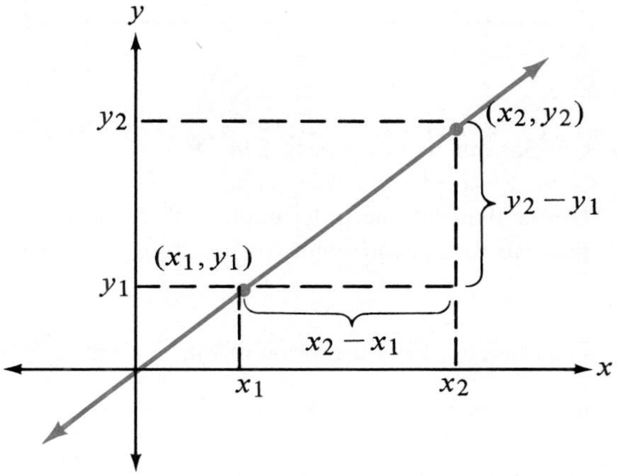

It is customary to use the letter m to represent slope.

Definition of Slope. If (x_1, y_1) and (x_2, y_2) are two points on a line, then the slope of that line is defined by

$$m = \frac{y_2 - y_1}{x_2 - x_1}$$

provided that $x_2 \neq x_1$.

Let's use the slope definition to compute the slopes of some lines. It will be helpful to sketch each line before calculating its slope. As before, it does not matter which point is labeled (x_1, y_1) and which is labeled (x_2, y_2).

You try Example 5.

Example 5. Find the slope of the line through (1, 3) and (4, 7).

Solution

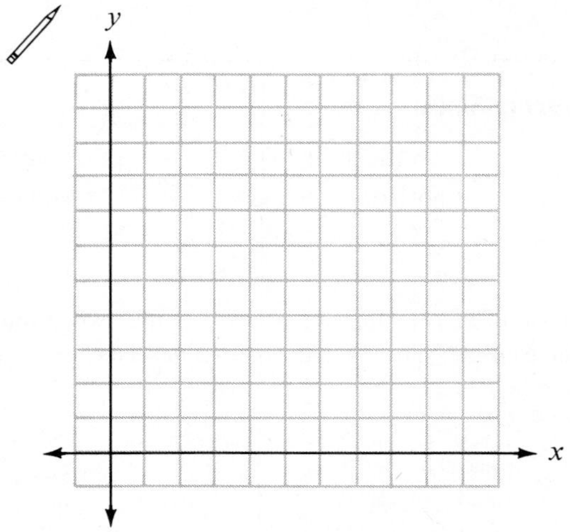

Let $(x_1, y_1) = (1, 3)$

$(x_2, y_2) = (4, 7)$

Then $m = \dfrac{y_2 - y_1}{x_2 - x_1}$

$=$

$m =$ _____

Check your work on page 544. ▶

Notice that the line in Example 5 slopes *upward* from left to right, and to move from the first point to the second point, we must go to the *right* 3 units and *up* 4 units.

Example 6. Find the slope of the line through $(-5, 1)$ and $(-3, -2)$.

Solution

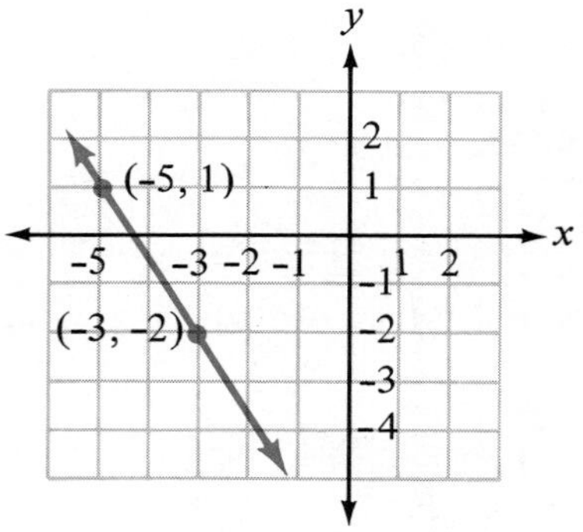

Let $(x_1, y_1) = (-5, 1)$

$(x_2, y_2) = (-3, -2)$

Then $m = \dfrac{y_2 - y_1}{x_2 - x_1}$

$= \dfrac{-2 - 1}{-3 - (-5)}$

$m = -\dfrac{3}{2}$

Notice that the line in Example 6 slopes *downward* from left to right, and to move from the first point to the second point, we must go to the *right* 2 units and *down* 3 units.

From Examples 5 and 6 we observe that

> If the slope of a line is *positive*, then that line slopes *upward* from left to right.
> If the slope of a line is *negative*, then that line slopes *downward* from left to right.

Now try Example 7.

Example 7. Find the slope of the line through (1, 4) and (−2, 4).

Solution

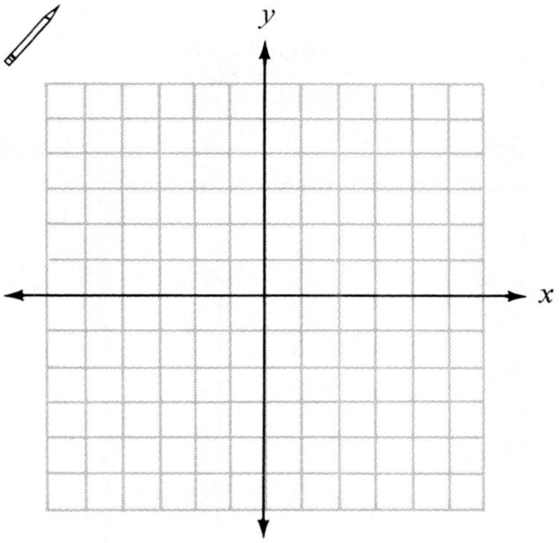

Let $(x_1, y_1) = (1, 4)$

$(x_2, y_2) = (-2, 4)$

Then $m =$

$m =$ ______

Check your work on page 544. ▶

Example 8. Find the slope of the line through (2, 3) and (2, −1).

Solution

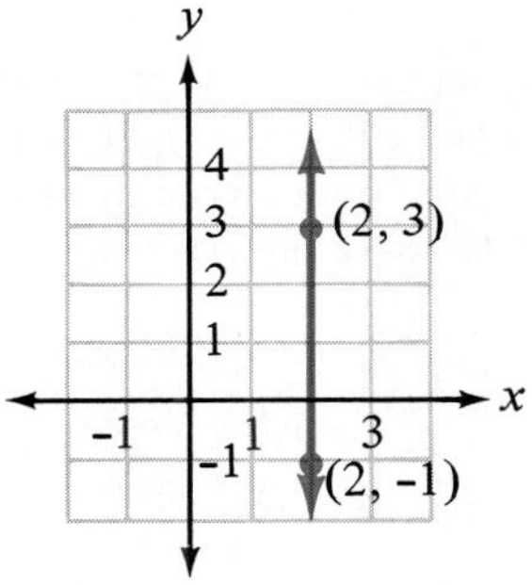

Let $(x_1, y_1) = (2, 3)$

$(x_2, y_2) = (2, -1)$

Then $m = \dfrac{-1 - 3}{2 - 2}$

$= \dfrac{-4}{0}$

m is undefined

From Examples 7 and 8 we observe that

> The slope of a horizontal line is zero.
> The slope of a vertical line is undefined.

Trial Run

Find the slope of the line through the given points.

______ **1.** (2, 1) and (3, 0)

______ **2.** (−3, 5) and (4, 5)

______ **3.** (9, −3) and (3, 1)

______ **4.** (7, −1) and (7, 5)

Answers are on page 545.

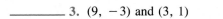

Graphing a Line Using the Slope

If we know one point on a line and the slope of that line, we can find another point on that line using the fact that

$$\text{Slope} = \frac{\text{vertical change}}{\text{horizontal change}}$$

For instance, if we know that the point (2, 1) is on a line with slope $\frac{3}{4}$, we know that we can start at the point (2, 1) and move to the *right* 4 units and *up* 3 units to arrive at another point on the same line.

$$(x_1, y_1) = (2, 1)$$

$$m = \frac{3 \uparrow}{4 \rightarrow}$$

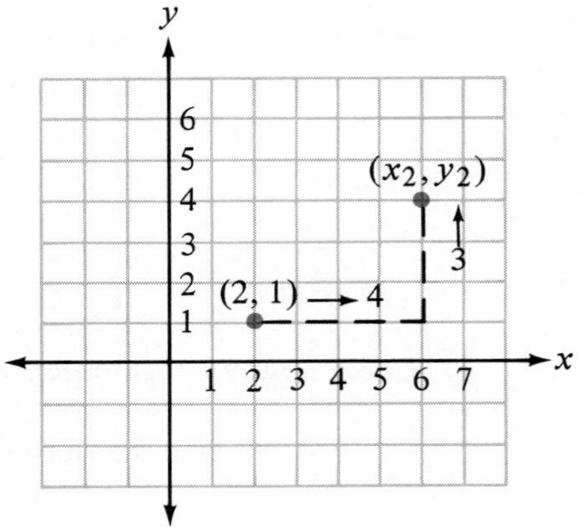

The coordinates of the point (x_2, y_2) can be found by arithmetic. If we begin at a point with x-coordinate of 2 and move 4 units to the right, we'll arrive at a point with x-coordinate of 6. Similarly, if we begin at a point with y-coordinate of 1 and move *up* 3 units, we'll arrive at a point with y-coordinate of 4. Therefore, $(x_2, y_2) = (6, 4)$, and we can graph the line.

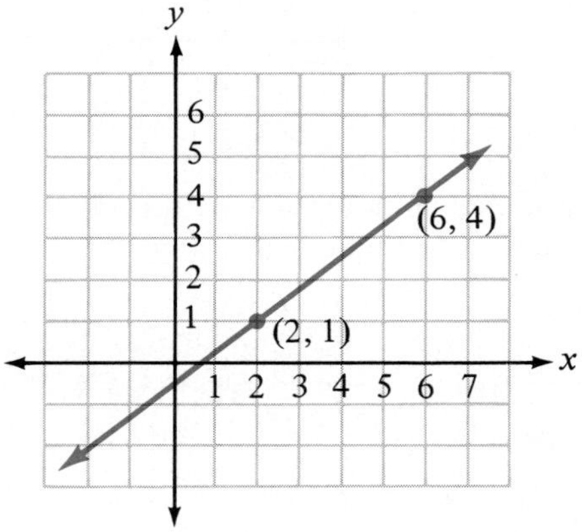

Example 9. Graph the line through $(3, -2)$ with slope of $-\frac{1}{4}$.

Solution

$$(x_1, y_1) = (3, -2)$$

$$m = -\frac{1}{4} = \frac{-1}{4}\begin{matrix}\downarrow\\ \rightarrow\end{matrix}$$

We begin at $(3, -2)$ and move to the *right* 4 units and *down* 1 unit.

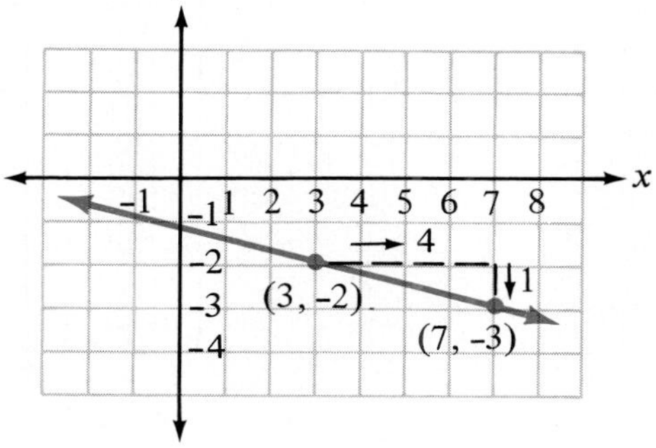

You try Example 10.

Example 10. Graph the line through the origin with slope of 3.

Solution

$$(x_1, y_1) = (0, 0)$$

$$m = 3 = \frac{3}{1}\begin{matrix}\uparrow\\ \rightarrow\end{matrix}$$

We begin at (_____ , _____) and move to the ________ 1 unit and ________ 3 units.

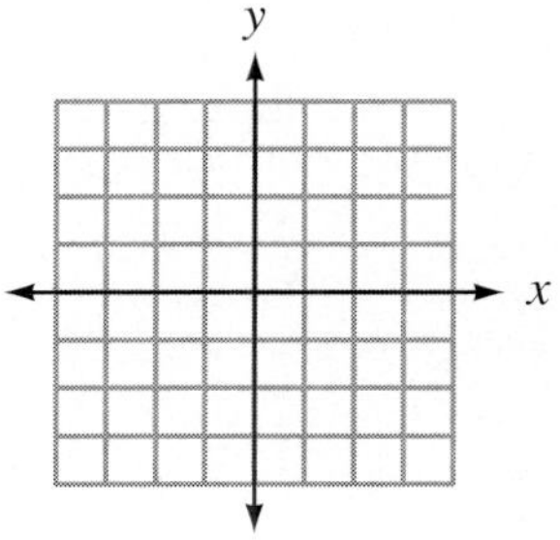

Check your work on page 545. ▶

Then try Example 11.

Example 11. Graph the line through $(5, -2)$ with slope of zero.

Solution

$$(x_1, y_1) = (5, -2)$$

$$m = 0$$

We recall that if the slope of a line is zero, that line is ________ .

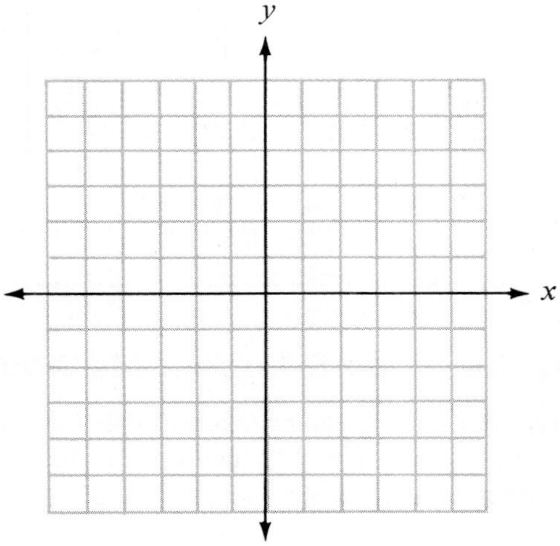

What is the *equation* for this line? Because it is a *horizontal* line, its equation must be $y =$ _____ .

Check your work on page 545. ▶

Example 12. Graph the line through $(5, -2)$ with undefined slope.

Solution

$$(x_1, y_1) = (5, -2)$$

$$m \text{ is undefined}$$

We recall that if the slope of a line is undefined, that line is vertical.

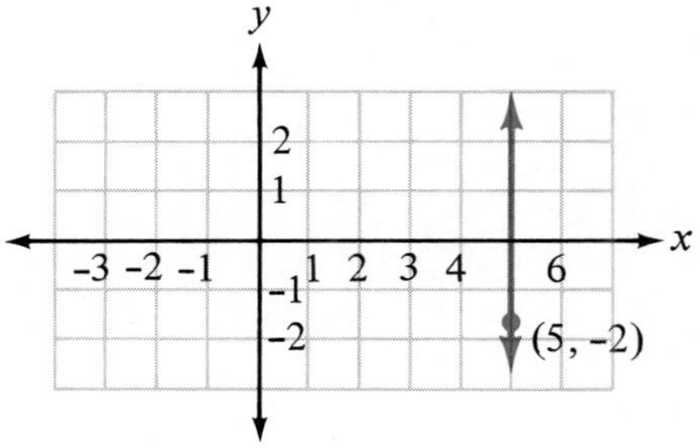

What is the *equation* for this line? Because it is a *vertical* line, its equation must be $x = 5$.

Trial Run

Graph the line through the given point with the given slope.

1. (0, 0); $m = \frac{5}{2}$

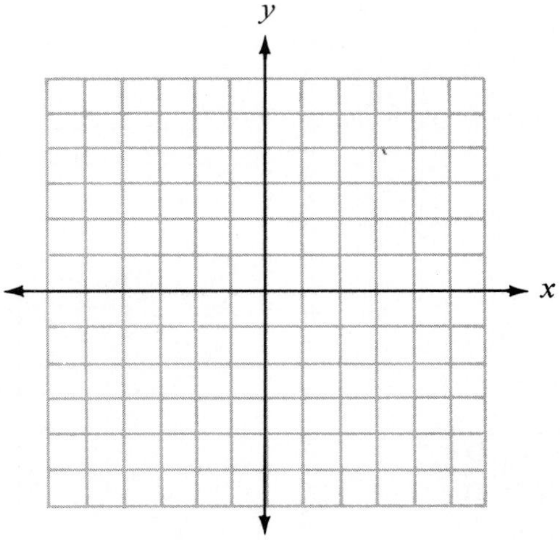

2. (4, 2); $m = 0$

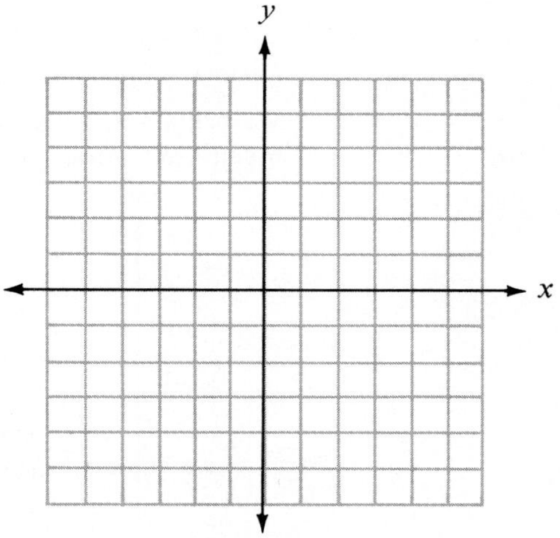

3. $(-2, -4)$; $m = -1$

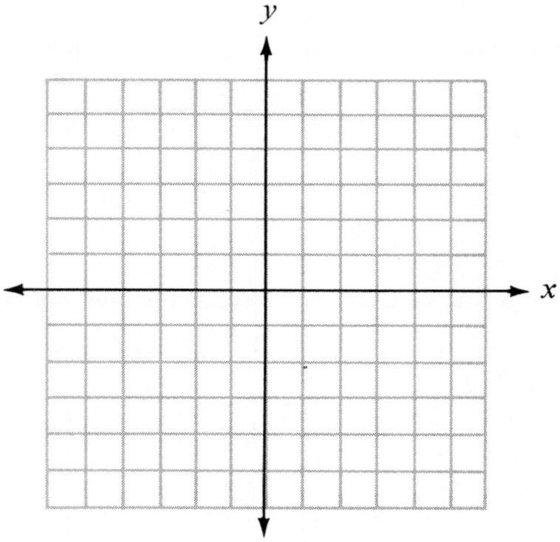

4. $(-3, 2)$; m is undefined

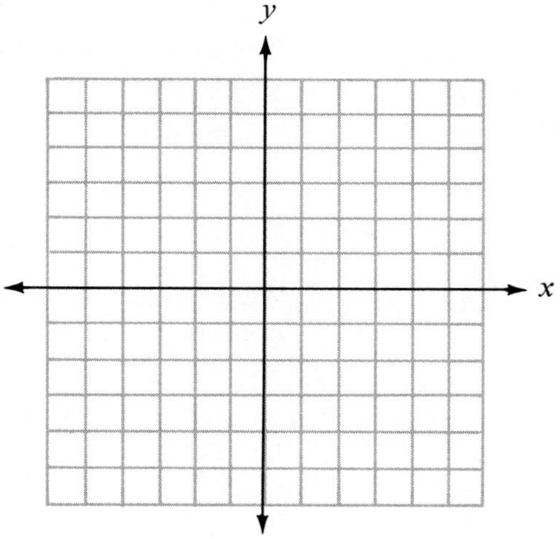

Answers are on page 545.

Graphing Lines by the Slope-Intercept Method

Take another look at the graphs of linear equations presented at the beginning of this section. For each graph, the y-intercept has been found by the usual method. Let's find the slope of each line using the points labeled on each graph.

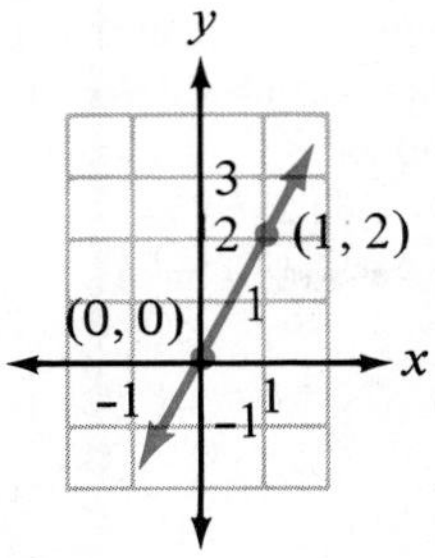

$y = 2x$

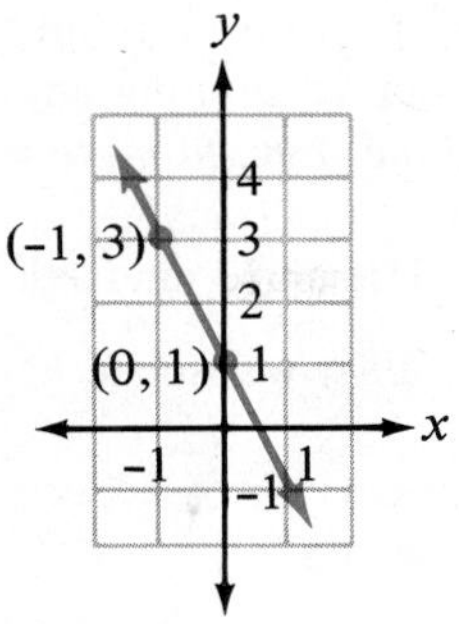

$y = 1 - 2x$

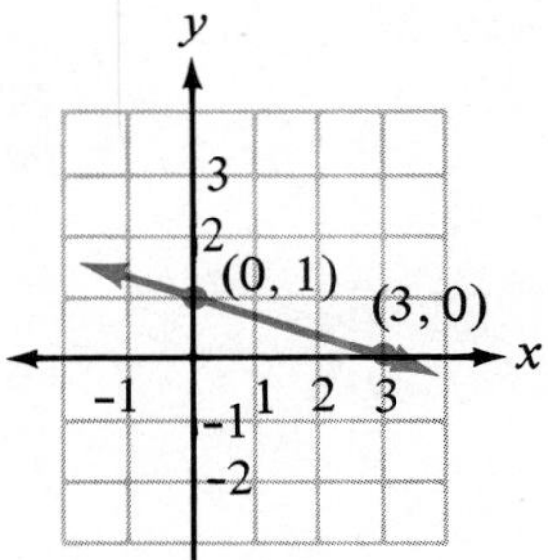

$x + 3y = 3$

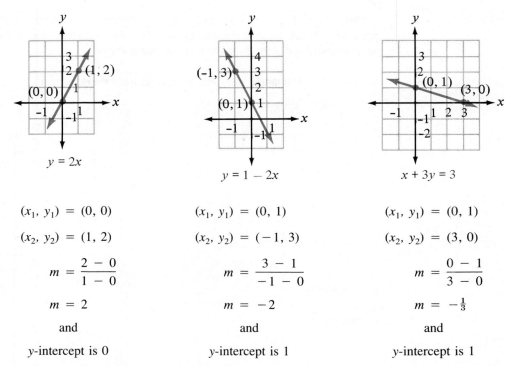

$(x_1, y_1) = (0, 0)$	$(x_1, y_1) = (0, 1)$	$(x_1, y_1) = (0, 1)$
$(x_2, y_2) = (1, 2)$	$(x_2, y_2) = (-1, 3)$	$(x_2, y_2) = (3, 0)$
$m = \frac{2 - 0}{1 - 0}$	$m = \frac{3 - 1}{-1 - 0}$	$m = \frac{0 - 1}{3 - 0}$
$m = 2$	$m = -2$	$m = -\frac{1}{3}$
and	and	and
y-intercept is 0	y-intercept is 1	y-intercept is 1

Now let's rewrite each line's equation in the same form, with y isolated on the left-hand side and the x-term and constant term on the right-hand side.

$y = 2x$	$y = 1 - 2x$	$x + 3y = 3$
becomes	becomes	becomes
$y = 2x + 0$	$y = -2x + 1$	$y = -\frac{1}{3}x + 1$

Now we summarize the information about these equations and their slopes and y-intercepts.

Equation	Slope	y-Intercept
$y = 2x + 0$	2	0
$y = -2x + 1$	-2	1
$y = -\frac{1}{3}x + 1$	$-\frac{1}{3}$	1

It is not just a coincidence that when y is isolated in a linear equation, the slope of its graph matches the coefficient of the x-term and the y-intercept matches the constant term. In fact (unless the coefficient of y is zero), every linear equation can be written in this form.

When a linear equation is written with y isolated on one side, it is said to be in **slope-intercept form** because we can read the slope and the y-intercept directly from the equation.

Slope-Intercept Form. If a linear equation is written in the form

$$y = mx + b$$

then the slope of the line is m and the y-intercept is b.

Slope-intercept form is useful because it allows us to graph a line just by looking at the equation without substituting any values for either variable. We just *begin at the y-intercept point and use the slope to move to a second point.*

Example 13. Use slope-intercept form to graph $2x - 5y = 5$.

Solution. First we isolate y.

$$2x - 5y = 5$$
$$-5y = -2x + 5$$
$$y = \frac{-2x}{-5} + \frac{5}{-5}$$
$$y = \frac{2}{5}x - 1$$

Slope: $m = \frac{2}{5} \begin{matrix}\uparrow \\ \rightarrow\end{matrix}$

y-intercept: $b = -1$

$(0, -1)$

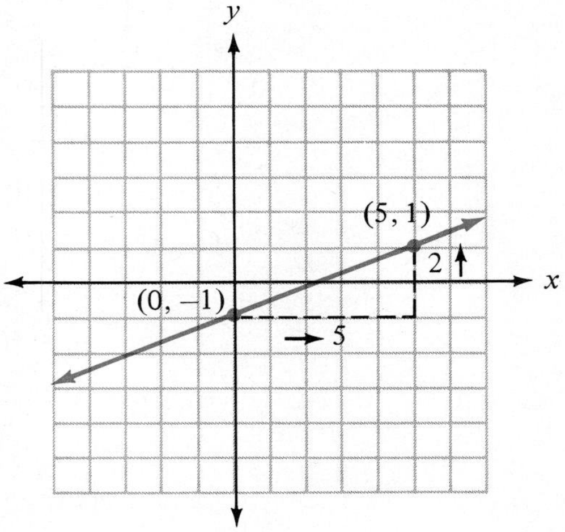

You try Example 14.

Example 14. Use slope-intercept form to graph $x + y = 0$.

Solution. First we isolate y.

$$x + y = 0$$
$$y = -x$$
$$y = -1x + 0$$

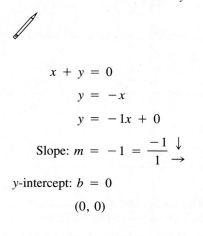

Slope: $m = -1 = \frac{-1}{1} \begin{matrix}\downarrow \\ \rightarrow\end{matrix}$

y-intercept: $b = 0$

$(0, 0)$

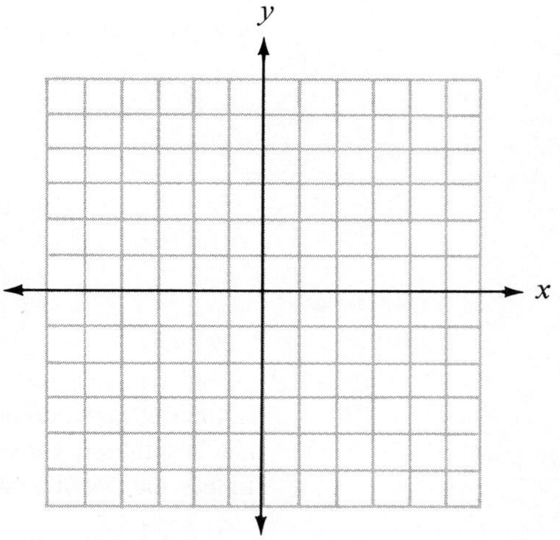

Check your graph on page 545. ▶

Trial Run

Use the slope-intercept method to graph each equation.

1. $y = -3x + 6$

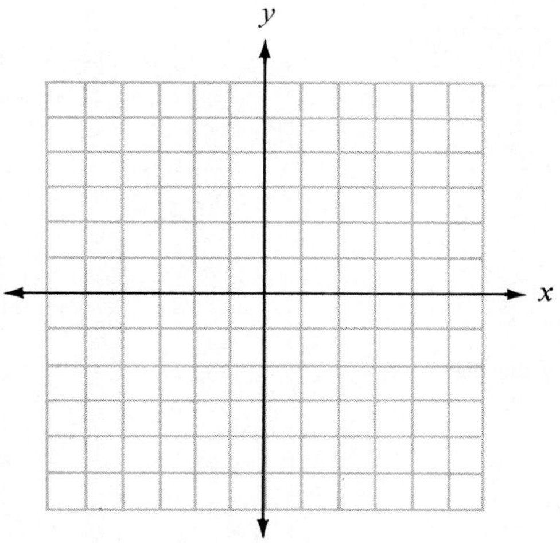

2. $x - y = 4$

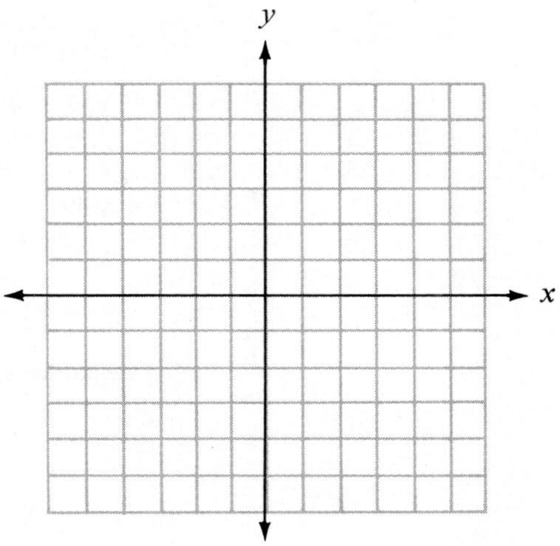

3. $x + 4y = -8$

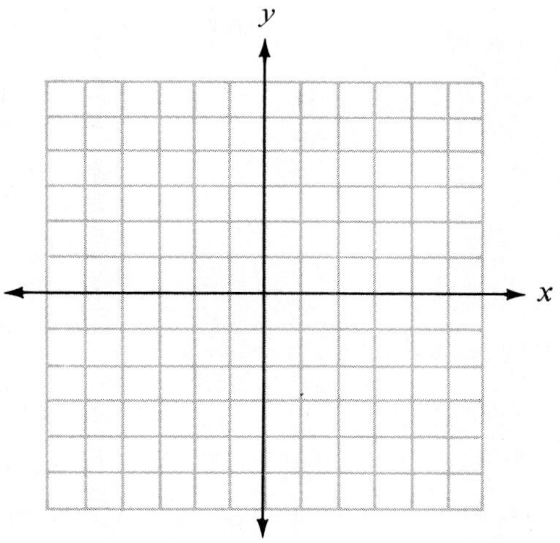

4. $3x - 4y = 12$

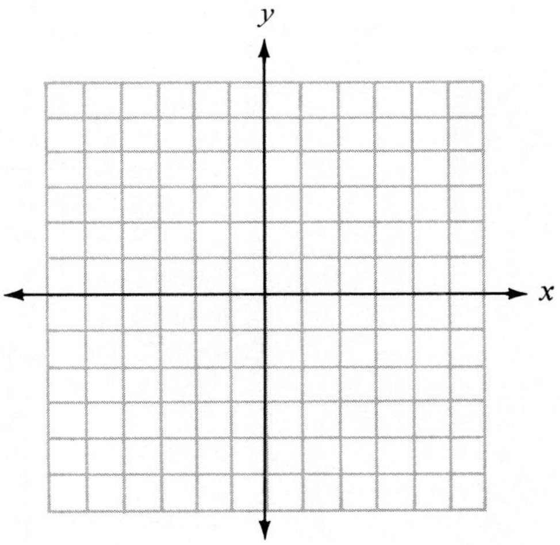

Answers are on page 546.

▶ Examples You Completed

Example 2. Find the distance between (0, 3) and (7, 5).

Solution. Let $(x_1, y_1) = (0, 3)$
$(x_2, y_2) = (7, 5)$

$$d = \sqrt{(x_2 - x_1)^2 + (y_2 - y_1)^2}$$
$$= \sqrt{(7 - 0)^2 + (5 - 3)^2}$$
$$= \sqrt{(7)^2 + (2)^2}$$
$$= \sqrt{49 + 4}$$
$$= \sqrt{53}$$

Example 4. Find the midpoint of the line segment joining (−2, 0) and (3, 0).

Solution. Let $(x_1, y_1) = (-2, 0)$
$(x_2, y_2) = (3, 0)$

$$x_M = \frac{x_1 + x_2}{2} \quad \text{and} \quad y_M = \frac{y_1 + y_2}{2}$$
$$= \frac{-2 + 3}{2} \qquad = \frac{0 + 0}{2}$$
$$= \frac{1}{2} \qquad = 0$$
$$(x_M, y_M) = (\tfrac{1}{2}, 0)$$

Example 5. Find the slope of the line through (1, 3) and (4, 7).

Solution

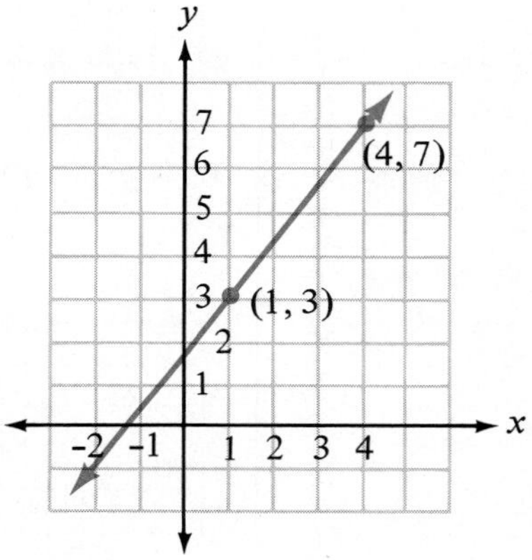

Let $(x_1, y_1) = (1, 3)$
$(x_2, y_2) = (4, 7)$

Then
$$m = \frac{y_2 - y_1}{x_2 - x_1}$$
$$= \frac{7 - 3}{4 - 1}$$
$$m = \frac{4}{3}$$

Example 7. Find the slope of the line through (1, 4) and (−2, 4).

Solution

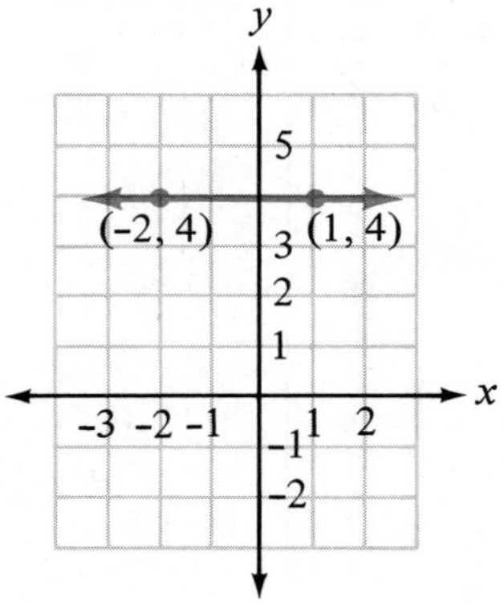

Let $(x_1, y_1) = (1, 4)$
$(x_2, y_2) = (-2, 4)$

Then
$$m = \frac{4 - 4}{-2 - 1}$$
$$= \frac{0}{-3}$$
$$m = 0$$

Example 10. Graph the line through the origin with slope 3.

Solution

$$(x_1, y_1) = (0, 0)$$

$$m = 3 = \frac{3}{1}\begin{matrix}\uparrow\\ \rightarrow\end{matrix}$$

We begin at (0, 0) and move to the right 1 unit and up 3 units.

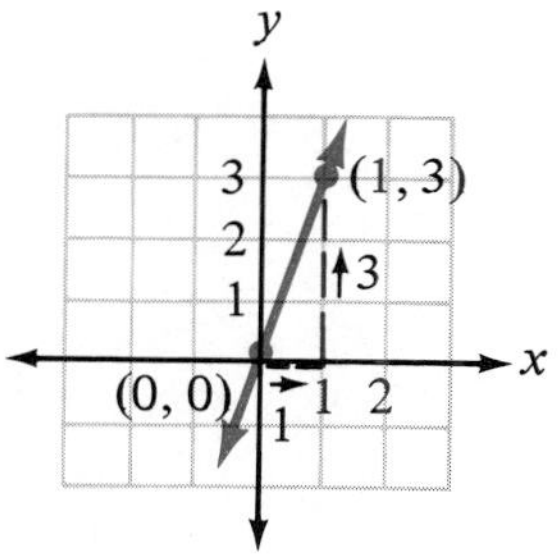

Example 11. Graph the line through (5, −2) with slope zero.

Solution

$$(x_1, y_1) = (5, -2)$$

$$m = 0$$

We recall that if the slope of a line is zero, that line is horizontal.

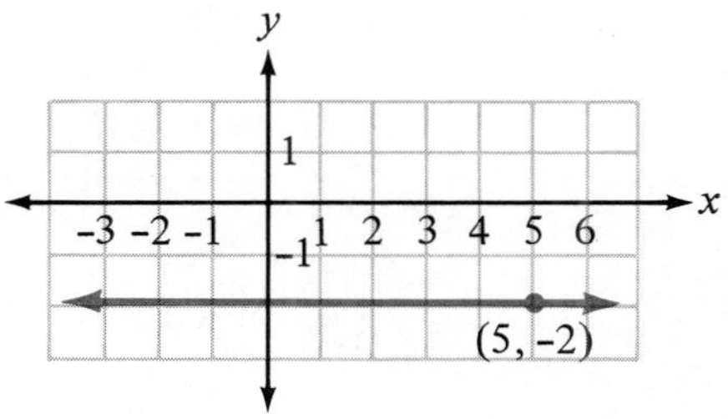

What is the *equation* for this line? Because it is a *horizontal* line, its equation must be $y = -2$.

Example 14. Use slope-intercept form to graph $x + y = 0$.

Solution. First we isolate y.

$$x + y = 0$$

$$y = -x$$

$$y = -1x + 0$$

$$\text{Slope: } m = -1 = \frac{-1}{1}\begin{matrix}\downarrow\\ \rightarrow\end{matrix}$$

$$y\text{-intercept: } b = 0$$

$$(0, 0)$$

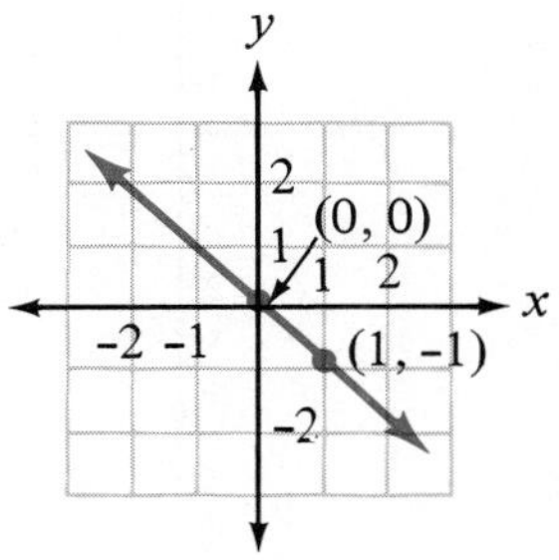

$x + y = 0$

Answers to Trial Runs

page 534 **1.** $2\sqrt{13}$ **2.** 13 **3.** $(-\frac{3}{2}, 5)$ **4.** $(0, -2)$

page 537 **1.** -1 **2.** 0 **3.** $-\frac{2}{3}$ **4.** Undefined

page 540 **1.**

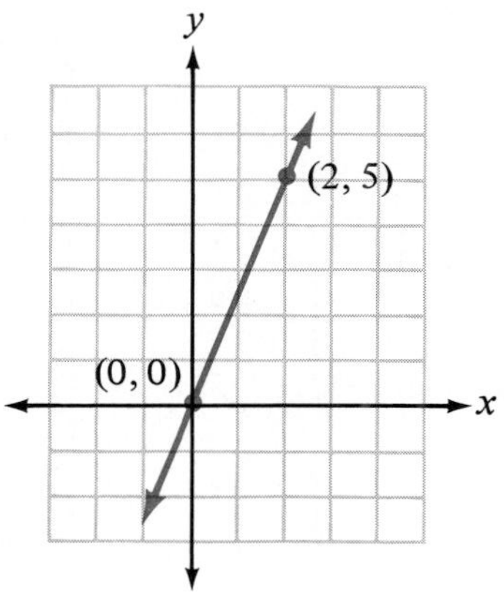

2.

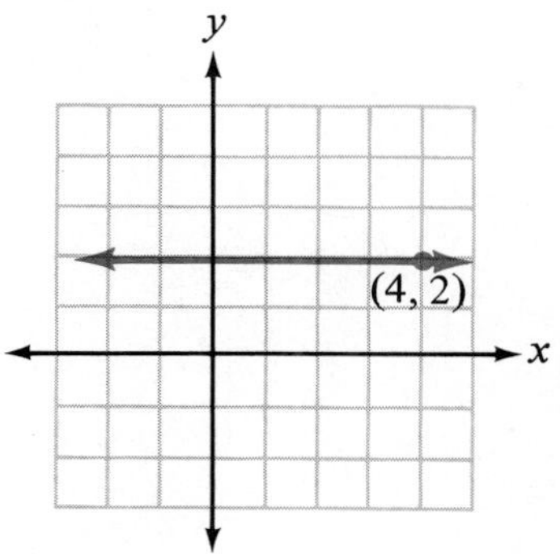

3.

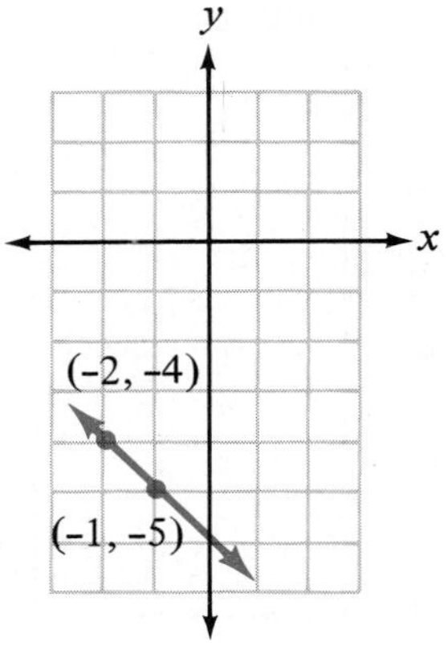

4.

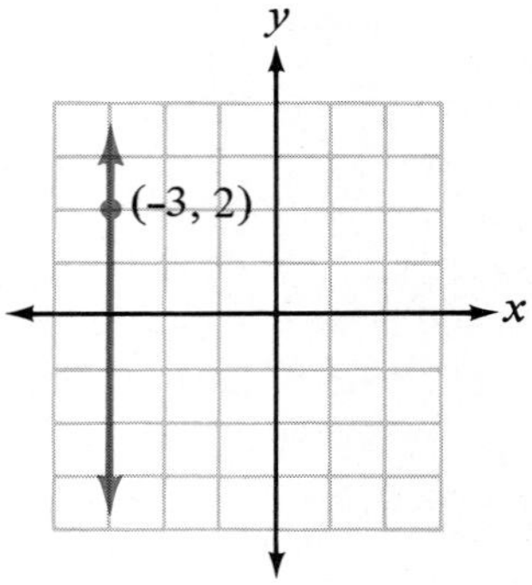

page 543 1.

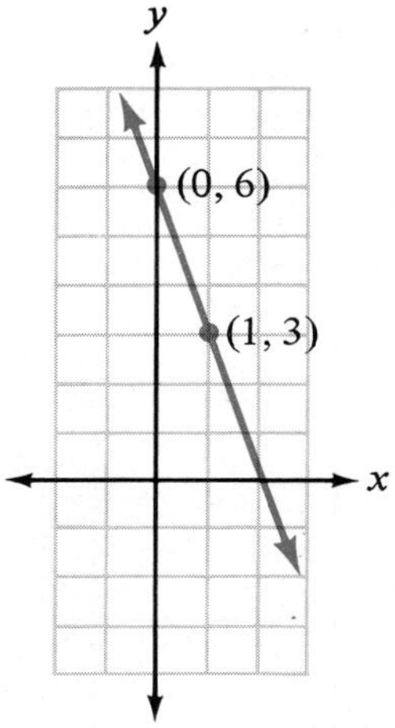

2.

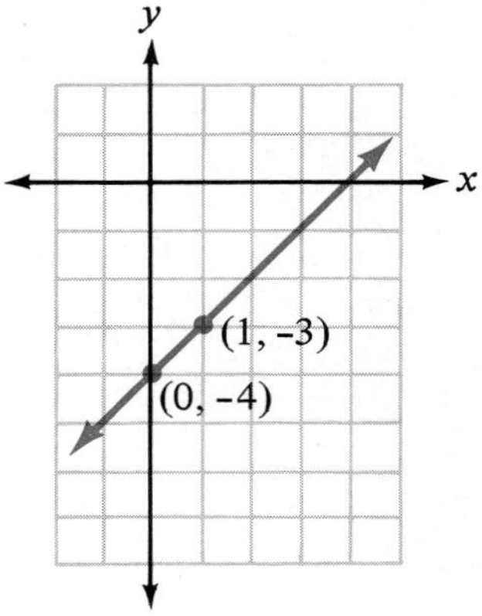

3.

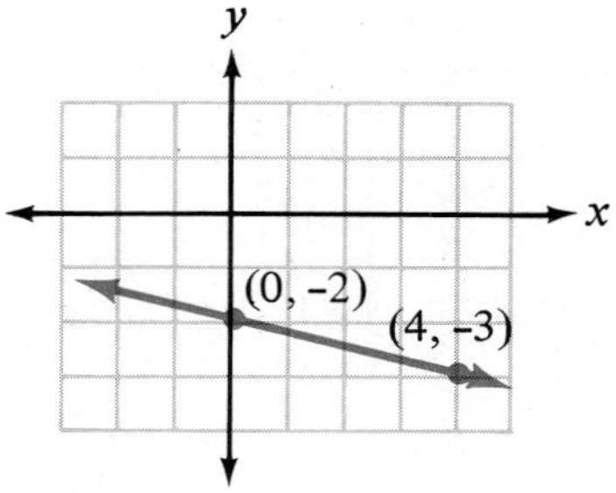

4.

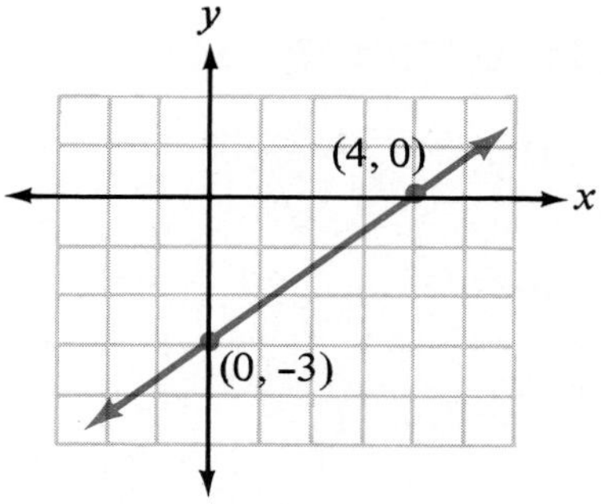

Name ______________________ Date ____________

EXERCISE SET 8.2

Find the distance between the given points.

______ 1. $(-3, 2)$ and $(0, 6)$

______ 2. $(-7, -2)$ and $(-1, 6)$

______ 3. $(2, -5)$ and $(5, 1)$

______ 4. $(4, -7)$ and $(6, -3)$

______ 5. $(-1, 2)$ and $(0, 4)$

______ 6. $(1, -2)$ and $(2, 1)$

______ 7. $(3, 6)$ and $(-2, -6)$

______ 8. $(4, -10)$ and $(-4, 5)$

______ 9. $(1\frac{1}{2}, 3\frac{1}{2})$ and $(1, 2\frac{1}{2})$

______ 10. $(\frac{1}{3}, 1\frac{1}{3})$ and $(-1, 2)$

______ 11. $(-5, 2)$ and $(-3, 7)$

______ 12. $(-4, 3)$ and $(-1, 8)$

Find the midpoint of the line segment joining the given points.

______ 13. $(3, 7)$ and $(9, 15)$

______ 14. $(1, 6)$ and $(7, 10)$

______ 15. $(-5, 3)$ and $(7, -7)$

______ 16. $(-8, 4)$ and $(12, -10)$

______ 17. $(0, -3)$ and $(-5, 9)$

______ 18. $(0, -4)$ and $(-3, 10)$

______ 19. $(-6, 8)$ and $(-6, 3)$

______ 20. $(-5, 11)$ and $(-5, 6)$

______ 21. $(12, -3)$ and $(-4, 3)$

______ 22. $(9, -4)$ and $(-7, 4)$

______ 23. $(-1\frac{1}{2}, 0)$ and $(1, 3\frac{1}{2})$

______ 24. $(-2\frac{1}{2}, 5\frac{1}{2})$ and $(0, -2)$

Find the slope of the line through the given points.

______ 25. $(0, 5)$ and $(-3, 0)$

______ 26. $(0, -7)$ and $(4, 0)$

______ 27. $(5, -1)$ and $(5, 4)$

______ 28. $(2, -7)$ and $(2, 3)$

______ 29. $(6, -3)$ and $(8, -5)$

______ 30. $(9, -7)$ and $(5, -3)$

______ 31. $(3, 5)$ and $(-2, 5)$

______ 32. $(-4, -3)$ and $(5, -3)$

______ 33. $(\frac{1}{2}, 3)$ and $(\frac{2}{3}, -\frac{1}{4})$

______ 34. $(\frac{3}{4}, \frac{5}{6})$ and $(-\frac{1}{2}, \frac{1}{3})$

______ 35. $(2, -3)$ and $(5, 6)$

______ 36. $(1, -2)$ and $(-3, 6)$

Graph the line through the given point with the given slope.

37. $(1, 3)$; $m = 2$

38. $(2, 5)$; $m = 3$

39. $(5, -1)$; $m = -3$

40. $(4, -3)$; $m = -2$

41. $(-2, 3)$; $m = -\frac{1}{2}$

42. $(-1, -1)$; $m = -\frac{1}{4}$

43. $(3, 5)$; $m = 0$

44. $(1, -4)$; $m = 0$

45. $(-8, 2)$; $m = \frac{3}{2}$

46. $(-6, 3)$; $m = \frac{4}{3}$

47. $(5, -4)$; m is undefined

48. $(3, -6)$; m is undefined

Use the slope-intercept method to graph each equation.

49. $y = -2x + 1$

50. $y = -3x + 4$

51. $x - y = 9$

52. $x - y = 6$

53. $x + 3y = 15$

54. $x + 6y = 12$

55. $3x - 2y = 6$

56. $5x - 2y = 10$

57. $-3x + 4y = -2$

58. $-2x + 9y = -6$

59. $3x + 5y = 0$

60. $4x + 3y = 0$

☆ Stretching the Topics

________ 1. Find the perimeter of a triangle with vertices $A(3, -1)$, $B(-5, 5)$, and $C(3, 5)$.

________ 2. If one endpoint of a line segment is $(\frac{1}{3}, \frac{2}{3})$ and the midpoint is $(-\frac{2}{3}, -\frac{5}{4})$, find the coordinates of the other endpoint.

________ 3. Find the slope and y-intercept for the graph of

$$4x - 3[2(5x - 1) - 4(7y - 3)] = 81$$

Check your answers in the back of your book.

If you can complete **Checkup 8.2,** you are ready to go on to Section 8.3.

Name ______________________________ Date ______________

✓ CHECKUP 8.2

________ 1. Find the distance between $(-3, 1)$ and $(0, 5)$.

________ 2. Find the midpoint of the line segment joining $(9, -9)$ and $(5, -7)$.

Find the slope of the line through the given points.

________ 3. $(16, -4)$ and $(9, 3)$

________ 4. $(1, 5)$ and $(-3, 8)$

Graph the line through the given point with the given slope.

5. $(2, -7)$; $m = 2$

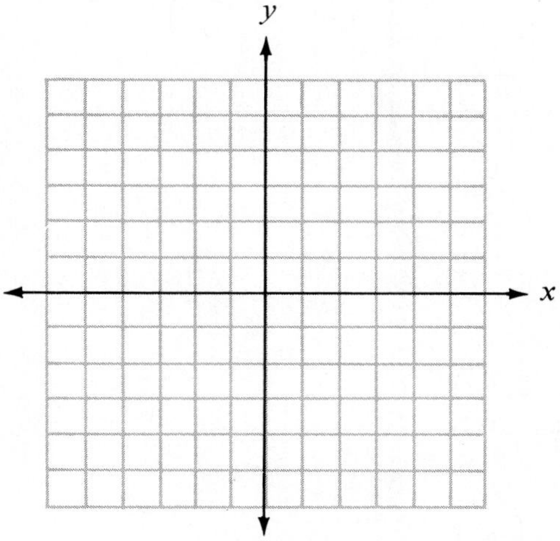

6. $(-5, -8)$; $m = -\frac{3}{2}$

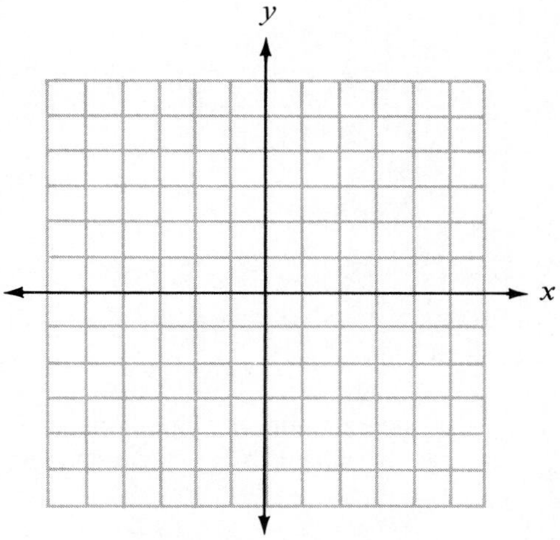

7. $(5, -1)$; $m = 0$

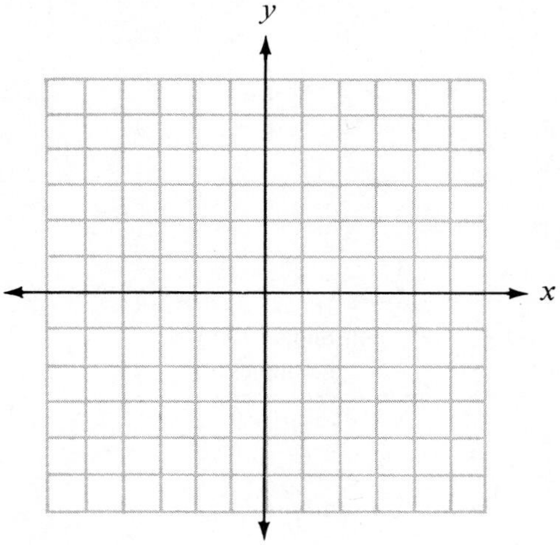

Use the slope-intercept method to graph each of the following.

8. $y = 2x - 3$

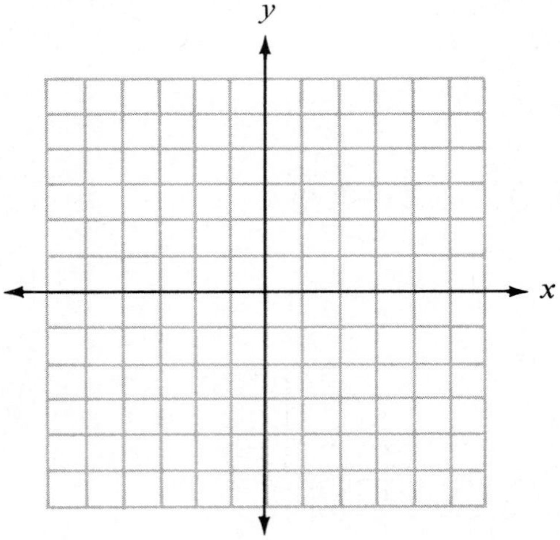

9. $3x + 5y = 10$

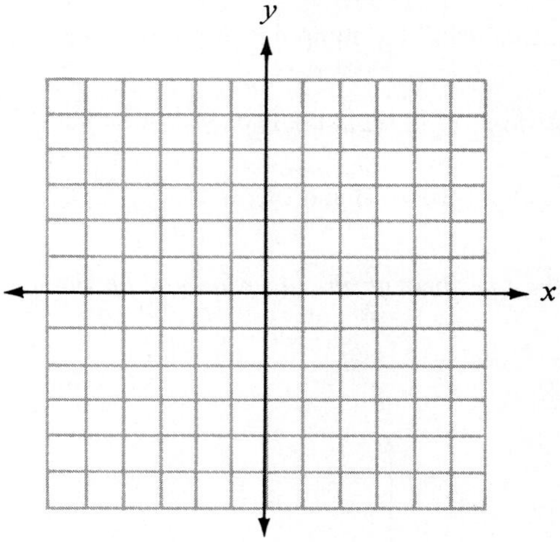

10. $3x - 4y = 0$

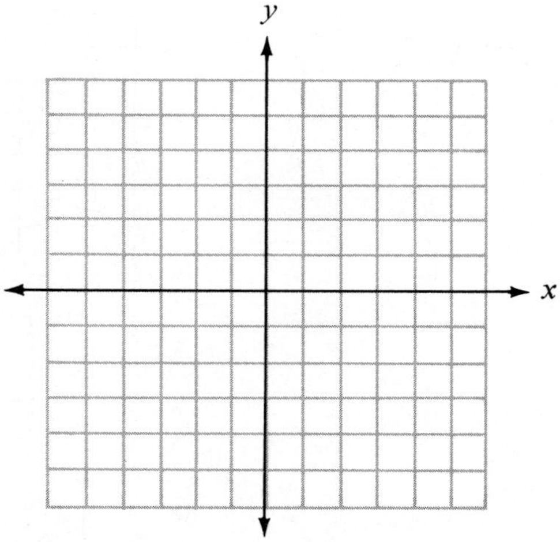

Check your answers in the back of your book.

If You Missed Problems:	You Should Review Examples:
1	1, 2
2	3, 4
3, 4	5–8
5, 6	9, 10
7	11
8–10	13, 14

8.3 Finding Equations of Lines

We have discussed several methods for graphing the solutions of first-degree equations in two variables. Now we would like to discover ways to find the equation that corresponds to a given straight line graph.

Finding Constant Equations

Suppose that you were asked to find an equation for each of these graphs.

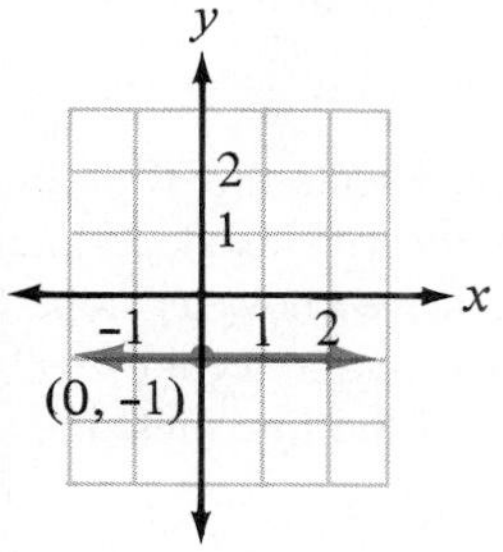

We know the equation for a *horizontal* line is always

$$y = c$$

where c is a constant. Since this line contains the point $(0, -1)$, the value of c must be the y-coordinate at that point. Thus c must be -1, and the equation of this line is

$$y = -1$$

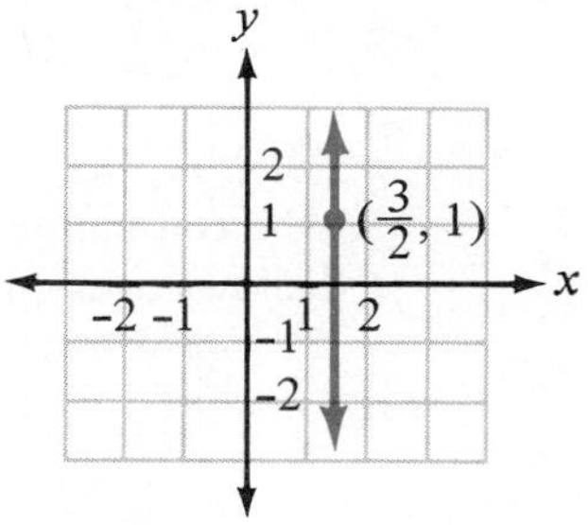

We know that the equation for a *vertical* line is always

$$x = c$$

where c is a constant. Since this line contains the point $(\frac{3}{2}, 1)$, the value of c must be the x-coordinate at that point. Thus c must be $\frac{3}{2}$, and the equation of this line is

$$x = \tfrac{3}{2}$$

You try Examples 1 and 2.

Example 1. Find the equation of the horizontal line through $(-3, 5)$.

Solution. A horizontal line has an equation ______ = ______ . For the given point, the y-coordinate is ______ , so the equation must be

______ = ______

Check your work on page 559. ▶

Example 2. Find the equation of the vertical line through $(-3, 5)$.

Solution. A vertical line has an equation ______ = ______ . For the given point, the x-coordinate is ______ , so the equation must be

______ = ______

Check your work on page 559. ▶

Trial Run

Find the equation of each line.

________ 1. Vertical line through $(4, -2)$

________ **2.** Horizontal line through $(0, -2)$

________ **3.** The x-axis ________ **4.** The y-axis

Answers are on page 560.

Finding Equations of Lines Using the Slope-Intercept Method

We learned in Section 8.2 that if a linear equation is written in the form

$$y = mx + b$$

then m, the coefficient of the x-term, represents the *slope* of the graph of the equation, and the constant term b represents the *y-intercept*. This slope-intercept form for a linear equation allows us to find the equation for a line if we know its slope (m) and its y-intercept (b).

For instance, if we know that a line has a y-intercept of -2 and a slope of 3, then we know that $m = 3$ and $b = -2$. The equation of the line must be

$$y = 3x - 2$$

Example 3. Find the equation of the line through the origin with slope $-\frac{2}{5}$.

Solution. The graph crosses the y-axis at $(0, 0)$, so $b = 0$. The slope is also given: $m = -\frac{2}{5}$. The equation of the line is

$$y = -\tfrac{2}{5}x + 0$$

$$y = -\tfrac{2}{5}x \qquad \text{Slope-intercept form.}$$

To write this equation in standard form, $Ax + By = C$, first eliminate fractions.

$$y = -\tfrac{2}{5}x$$

$$5y = -2x \qquad \text{Multiply by LCD: 5.}$$

$$2x + 5y = 0 \qquad \text{Standard form.}$$

You try Example 4.

Example 4. Find the equation of the line through $(0, -1)$ with slope $\frac{1}{2}$.

Solution. The graph crosses the y-axis at (_____ , _____), so b = _____ .The slope is also given: m = _____ .The equation of the line is

y = _____x − _____

To write this equation in standard form, we first eliminate the fractions.

$$y = \tfrac{1}{2}x - 1$$

$$2y = x - 2 \qquad \text{Multiply by LCD: 2.}$$

$$-x + 2y = -2$$

$$x - 2y = 2 \qquad \text{Standard form.}$$

Check your work on page 559. ▶

Either of the two forms we have shown is an acceptable way of writing a first-degree equation in two variables. Slope-intercept form is preferable for its graphing information, but we shall find standard form useful when solving systems of equations in Chapter 10.

Slope-intercept form:	$y = mx + b$
Standard linear form:	$Ax + By = C$

Trial Run

Find the equation of the line satisfying the given conditions. Write each equation in slope-intercept form and in standard linear form.

________ 1. Through (0, 2) with slope -4

________ 2. With y-intercept of 4 and zero slope

________ 3. With y-intercept of $-\frac{5}{6}$ and slope $\frac{2}{3}$

Answers are on page 560.

Finding Equations of Lines Using the Point-Slope Method

We now have a method for finding the equation of a line when we know its slope m and y-intercept b. Unfortunately, both those pieces of information may not be available. Let's see if we can develop methods for finding the equation for a line when we are given just the *slope and one point* on the line or when we are given *any two points* on the line.

Recall that the slope of a line is defined by

$$m = \frac{y_2 - y_1}{x_2 - x_1}$$

Suppose that we let (x, y) represent *any* point on a line that we know passes through the point (2, 5) and has a slope of 4. Here we are saying that

$$(x_1, y_1) = (2, 5)$$

$$(x_2, y_2) = (x, y)$$

$$m = 4$$

Substituting these values into the slope formula, we have

$4 = \dfrac{y - 5}{x - 2}$	Substitute.
$4(x - 2) = y - 5$	Multiply both sides by $x - 2$.
$4x - 8 = y - 5$	Remove parentheses.
$4x - 3 = y$	Isolate y.
$y = 4x - 3$	Slope-intercept form.

We have found the equation of the line through (2, 5) with slope 4.

In general, if m is the slope of a line passing through a specific point (x_1, y_1) and (x, y) represents any other point on that line, then

$$\frac{y - y_1}{x - x_1} = m$$

and multiplying both sides by $x - x_1$, we have

$$y - y_1 = m(x - x_1)$$

This is called the **point-slope form** for a linear equation.

> **Point-Slope Form.** If (x_1, y_1) is a point on a line with slope m, then the equation of that line can be found by
>
> $$y - y_1 = m(x - x_1)$$

Example 5. Find the equation of the line through $(2, -4)$ with slope $\frac{1}{3}$.

Solution. Here, $(x_1, y_1) = (2, -4)$ and $m = \frac{1}{3}$.

$$y - y_1 = m(x - x_1) \qquad \text{Point-slope formula.}$$

$$y - (-4) = \frac{1}{3}(x - 2) \qquad \text{Substitute the given values.}$$

$$y + 4 = \frac{1}{3}x - \frac{2}{3} \qquad \text{Remove parentheses.}$$

$$y = \frac{1}{3}x - \frac{2}{3} - 4 \qquad \text{Isolate } y.$$

$$y = \frac{1}{3}x - \frac{2}{3} - \frac{12}{3} \qquad \text{Combine like terms.}$$

$$y = \frac{1}{3}x - \frac{14}{3} \qquad \text{Slope-intercept form.}$$

or

$$3y = x - 14 \qquad \text{Multiply by LCD: 3.}$$

$$-x + 3y = -14 \qquad \text{Write equation in standard form.}$$

$$x - 3y = 14 \qquad \text{Multiply by } -1.$$

You complete Example 6.

Example 6. Find the equation of the line with x-intercept of -1 and slope -2.

Solution. Here, $(x_1, y_1) = (-1, 0)$ and $m = -2$.

$$y - y_1 = m(x - x_1)$$

$$y - 0 = -2[x - (-1)]$$

$$y = \_\_\_\_\_\_\_\_ \qquad \text{Slope-intercept form.}$$

or

$$\_\_\_\_\_\_\_\_ \qquad \text{Standard form.}$$

Check your work on page 559. ▶

Then try Example 7.

Example 7. Find the equation of the line through $(7, -1)$ with zero slope.

Solution. A line with zero slope is a ________ line. Therefore, the equation must be $y = c$, where c is a constant. Since the given point has a y-coordinate of _____ , the equation must be

$$y = \_\_\_\_$$

Check your work on page 560. ▶

Example 8. Find the equation of the line through $(7, -1)$ with undefined slope.

Solution. A line with undefined slope is a *vertical* line. Therefore, the equation must be $x = c$, where c is a constant. Since the given point has an x-coordinate of 7, the equation must be

$$x = 7$$

Trial Run

Find the equation of the line through the given point with the given slope.

______ 1. $(5, 1)$; $m = 3$

______ 2. $(-3, 2)$; $m = 0$

______ 3. $(-2, 3)$; $m = -\frac{1}{2}$

______ 4. $(-3, 2)$; m is undefined

Answers are on page 560.

Finding Equations of Lines from Two Points

It is very useful to be able to find the equation for a line when given just two points on the line. Actually, we already have the tools necessary for finding an equation from two points.

For instance, suppose that we wish to find the equation for the line through $(-1, 5)$ and $(2, 3)$. Both methods that we have learned for finding equations of lines require the *slope* of the line, so let's find the slope here. Letting $(x_1, y_1) = (-1, 5)$ and $(x_2, y_2) = (2, 3)$, we have

$$m = \frac{3 - 5}{2 - (-1)}$$

$$m = -\frac{2}{3}$$

Now that we know the slope, we may use it with either of the given points to find the equation using the *point-slope method*.

Let's use $(x_1, y_1) = (-1, 5)$

$$\begin{aligned}
m &= -\frac{2}{3} \\
y - y_1 &= m(x - x_1) \\
y - 5 &= -\frac{2}{3}[x - (-1)] \\
y - 5 &= -\frac{2}{3}[x + 1] \\
y - 5 &= -\frac{2}{3}x - \frac{2}{3} \\
y &= -\frac{2}{3}x - \frac{2}{3} + 5 \\
y &= -\frac{2}{3}x - \frac{2}{3} + \frac{15}{3} \\
y &= -\frac{2}{3}x + \frac{13}{3}
\end{aligned}$$

Or use $(x_2, y_2) = (2, 3)$

$$\begin{aligned}
m &= -\frac{2}{3} \\
y - y_2 &= m(x - x_2) \\
y - 3 &= -\frac{2}{3}(x - 2) \\
y - 3 &= -\frac{2}{3}x + \frac{4}{3} \\
y &= -\frac{2}{3}x + \frac{4}{3} + 3 \\
y &= -\frac{2}{3}x + \frac{4}{3} + \frac{9}{3} \\
y &= -\frac{2}{3}x + \frac{13}{3}
\end{aligned}$$

We arrive at the same equation using either of the given points. *It does not matter which point we use; the equation for the line will be the same.*

Now you try Example 9.

Example 9. Find the equation of the line through $(-4, -7)$ and $(-6, -2)$.

Solution

$$(x_1, y_1) = (-4, -7)$$

$$(x_2, y_2) = (-6, -2)$$

$$m = \frac{-2 - (-7)}{-6 - (-4)}$$

$$m = -\frac{5}{2}$$

$$y - y_1 = m(x - x_1)$$

$$y - (\_\_\_\_) = -\frac{5}{2}[x - (\_\_\_\_)]$$

Check your work on page 560. ▶

Example 10. Find the equation of the line with x-intercept of -2 and y-intercept of 6.

Solution. Remember that intercepts tell us points.

x-intercept: $(x_1, y_1) = (-2, 0)$

y-intercept: $(x_2, y_2) = (0, 6)$

$$m = \frac{6 - 0}{0 - (-2)}$$

$$= \frac{6}{2}$$

$$m = 3$$

Since the y-intercept is given, we can use slope-intercept form immediately.

$$y = mx + b$$

$$y = 3x + 6$$

Example 11. Find the equation of the line through $(5, -3)$ and $(-2, -3)$.

Solution

$$(x_1, y_1) = (5, -3)$$

$$(x_2, y_2) = (-2, -3)$$

$$m = \frac{-3 - (-3)}{-2 - 5}$$

$$= \frac{0}{-7}$$

$$= 0$$

Since the slope is zero, this is a horizontal line with equation

$$y = -3$$

You complete Example 12.

Example 12. Find the equation of the line through $(11, -1)$ and $(11, 13)$.

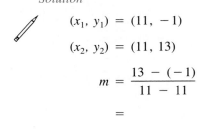

Solution

$$(x_1, y_1) = (11, -1)$$

$$(x_2, y_2) = (11, 13)$$

$$m = \frac{13 - (-1)}{11 - 11}$$

$$=$$

m is ________. Since the slope is ______, this is a ________ line with equation

Check your work on page 560. ▶

Trial Run

Find the equation of the line through the given points.

________ **1.** $(2, 3)$ and $(1, 5)$

________ **2.** $(4, 5)$ and $(-2, 5)$

________ **3.** $(-5, -8)$ and $(0, 2)$

________ **4.** $(10, 3)$ and $(10, -6)$

Answers are on page 560.

Finding Equations of Parallel and Perpendicular Lines

Let's graph the equations $y = 2x + 3$ and $y = 2x - 1$ in the same Cartesian plane, using the slope-intercept method.

(1) $y = 2x + 3$

$m = \frac{2}{1}$

$b = 3$

$(0, 3)$

(2) $y = 2x - 1$

$m = \frac{2}{1}$

$b = -1$

$(0, -1)$

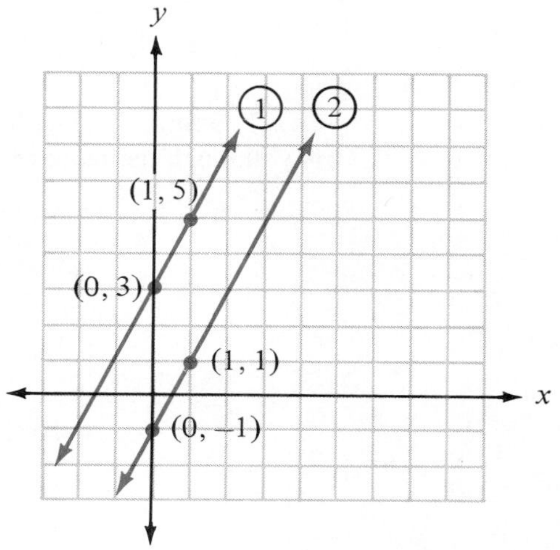

From the equations we see that these lines have the *same slope* but *different y-intercepts*. From the graph we see that the lines are **parallel.**

If two lines have the same slope but different y-intercepts, the lines are *parallel*.

If we want to find the equation of a line that is parallel to another line, we may "borrow" the slope and use it in the appropriate formula for finding the equation of the new line.

Example 13. Find the equation of the line through the point (1, 2) and parallel to the line $y = 3x + 4$.

Solution. We must "borrow" the slope from the line $y = 3x + 4$.

$$m = 3$$

$$(x_1, y_1) = (1, 2)$$

$$y - y_1 = m(x - x_1)$$

$$y - 2 = 3(x - 1)$$

$$y - 2 = 3x - 3$$

$$y = 3x - 1$$

Example 14. Find the equation of the line through the point (0, 5) and parallel to the line $2x + 3y = 1$.

Solution. We must "borrow" the slope from the line $2x + 3y = 1$. To read the slope, we must isolate y.

$$2x + 3y = 1$$

$$3y = -2x + 1$$

$$y = -\tfrac{2}{3}x + \tfrac{1}{3}$$

$$m = -\tfrac{2}{3}$$

The given point (0, 5) is the y-intercept point, so $b = 5$. The equation we need is

$$y = -\tfrac{2}{3}x + 5$$

Although the proof is beyond the scope of this course, a relationship also exists between the slopes of **perpendicular** lines.

If two lines are *perpendicular,* their slopes are *negative reciprocals* of each other. If the slope of a given line is m, then the slope of any line perpendicular to the given line is $-\frac{1}{m}$.

Example 15. Find the equation of the line perpendicular to the line $x + 2y = 3$ and passing through the point (5, 1).

Solution. We need the negative reciprocal of the slope of the given line $x + 2y = 3$.

$2y = -x + 3$	Isolate y-term.
$y = \frac{-1}{2}x + \frac{3}{2}$	Divide by 2.
$m = -\frac{1}{2}$	Read the slope of the given line.
$-\frac{1}{m} = 2$	Find the negative reciprocal of the slope.
$y - 1 = 2(x - 5)$	Use point-slope method for equation of new line.
$y - 1 = 2x - 10$	Remove parentheses.
$y = 2x - 9$	Equation of perpendicular line.

Perhaps a graph of the two equations in Example 15 will convince you that they represent perpendicular lines.

Equation (1)

$x + 2y = 3$

$y = -\frac{1}{2}x + \frac{3}{2}$

$m = -\frac{1}{2}$

$b = \frac{3}{2}$

Equation (2)

$y = 2x - 9$

$m = 2$

$b = -9$

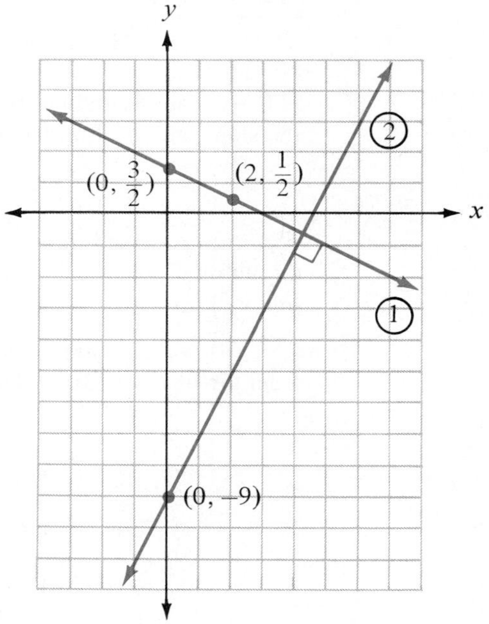

Trial Run

Find the equation of the line satisfying the given conditions.

________ 1. With y-intercept of -7 and parallel to $y = 4x + 1$

________ 2. Through (2, 4) and parallel to $x + y = 2$

________ 3. Through (3, −2) and perpendicular to $y = 3x - 1$

Answers are on page 560.

Let's summarize the methods that we have developed for finding equations of straight lines. The method to be used depends, of course, on the information we are given.

Given	Equation to Use	Name
Slope: m y-intercept: b	$y = mx + b$	Slope-intercept form
Slope: m Point: (x_1, y_1)	$y - y_1 = m(x - x_1)$	Point-slope form
Two points: (x_1, y_1) and (x_2, y_2)	$m = \dfrac{y_2 - y_1}{x_2 - x_1}$	Slope definition
	$y - y_1 = m(x - x_1)$	Point-slope form
Line is horizontal	$y = c$, a constant	Constant equation
Line is vertical	$x = c$, a constant	Constant equation

If we are told that a line is *parallel* to a given line, $y = mx + b$, we know that the slope of the parallel line is also m. If we are told that a line is *perpendicular* to a given line, $y = mx + b$ $(m \neq 0)$, we know that the slope of the perpendicular line is $-\dfrac{1}{m}$.

▶ Examples You Completed

Example 1. Find the equation of the horizontal line through (−3, 5).

Solution. A horizontal line has an equation $y = c$. For the given point, the y-coordinate is 5, so the equation must be

$$y = 5$$

Example 2. Find the equation of the vertical line through (−3, 5).

Solution. A vertical line has an equation $x = c$. For the given point, the x-coordinate is −3, so the equation must be

$$x = -3$$

Example 4. Find the equation of the line through (0, −1) with slope $\frac{1}{2}$.

Solution. The graph crosses the y-axis at (0, −1), so $b = -1$. The slope is also given: $m = \frac{1}{2}$. The equation of the line is

$$y = \tfrac{1}{2}x - 1$$

Example 6. Find the equation of the line with x-intercept of −1 and slope −2.

Solution. $(x_1, y_1) = (-1, 0)$ and $m = -2$.

$$y - y_1 = m(x - x_1)$$

$$y - 0 = -2[x - (-1)]$$

$$y = -2[x + 1]$$

$$y = -2x - 2 \qquad \text{Slope-intercept form.}$$

$$2x + y = -2 \qquad \text{Standard form.}$$

Example 7. Find the equation of the line through $(7, -1)$ with zero slope.

Solution. A line with zero slope is a horizontal line. Therefore, the equation must be $y = c$, where c is a constant. Since the given point has a y-coordinate of -1, the equation must be

$$y = -1$$

Example 12. Find the equation of the line through $(11, -1)$ and $(11, 13)$.

Solution

$$(x_1, y_1) = (11, -1)$$

$$(x_2, y_2) = (11, 13)$$

$$m = \frac{13 - (-1)}{11 - 11}$$

$$= \frac{14}{0}$$

m is undefined. Since the slope is undefined, this is a vertical line with equation

$$x = 11$$

Example 9. Find the equation of the line through $(-4, -7)$ and $(-6, -2)$.

Solution

$$(x_1, y_1) = (-4, -7)$$

$$(x_2, y_2) = (-6, -2)$$

$$m = \frac{-2 - (-7)}{-6 - (-4)}$$

$$m = -\frac{5}{2}$$

$$y - y_1 = m(x - x_1)$$

$$y - (-7) = -\frac{5}{2}[x - (-4)]$$

$$y + 7 = -\frac{5}{2}[x + 4]$$

$$y + 7 = -\frac{5}{2}x - 10$$

$$y = -\frac{5}{2}x - 17$$

Answers to Trial Runs

page 551 1. $x = 4$ 2. $y = -2$ 3. $y = 0$ 4. $x = 0$

page 553 1. $y = -4x + 2$ or $4x + y = 2$ 2. $y = 4$ 3. $y = \frac{2}{3}x - \frac{5}{6}$ or $4x - 6y = 5$

page 555 1. $y = 3x - 14$ or $3x - y = 14$ 2. $y = 2$ 3. $y = -\frac{1}{2}x + 2$ or $x + 2y = 4$ 4. $x = -3$

page 556 1. $y = -2x + 7$ or $2x + y = 7$ 2. $y = 5$ 3. $y = 2x + 2$ or $2x - y = -2$ 4. $x = 10$

page 558 1. $y = 4x - 7$ or $4x - y = 7$ 2. $y = -x + 6$ or $x + y = 6$ 3. $y = -\frac{1}{3}x - 1$ or $x + 3y = -3$

Name ______________________ Date ____________

EXERCISE SET 8.3

Find the equation of the horizontal line through the given point.

_______ 1. $(4, -5)$ _______ 2. $(8, -3)$

_______ 3. $(-1\frac{1}{2}, \frac{1}{2})$ _______ 4. $(2\frac{1}{2}, \frac{3}{4})$

Find the equation of the vertical line through the given point.

_______ 5. $(2, -6)$ _______ 6. $(3, -4)$

_______ 7. $(-\frac{2}{3}, 4)$ _______ 8. $(-1\frac{1}{4}, 2)$

Find the equation of the line with the given slope and y-intercept.

_______ 9. $m = 4, b = 2$ _______ 10. $m = 3, b = 5$

_______ 11. $m = -2, b = 1$ _______ 12. $m = -3, b = 4$

_______ 13. $m = \frac{1}{3}, b = -\frac{2}{3}$ _______ 14. $m = \frac{1}{4}, b = -\frac{3}{4}$

_______ 15. $m = 0, b = 10$ _______ 16. $m = 0, b = 8$

_______ 17. $m = \frac{1}{2}, b = -\frac{5}{4}$ _______ 18. $m = \frac{3}{2}, b = -\frac{1}{10}$

_______ 19. $m = 7, b = -2$ _______ 20. $m = 5, b = -1$

Find the equation of the line through the given point with the given slope.

_______ 21. $(-1, 5); m = 8$ _______ 22. $(-2, 3); m = 6$

_______ 23. $(2, 7); m = -1$ _______ 24. $(3, 5); m = -2$

_______ 25. $(-6, -3); m = 0$ _______ 26. $(5, 2); m = 0$

_______ 27. $(-1, 4); m = \frac{3}{4}$ _______ 28. $(6, -3); m = \frac{2}{5}$

_______ 29. $(-2, 1)$; m is undefined _______ 30. $(-6, 4)$; m is undefined

_______ 31. $(-\frac{1}{4}, \frac{3}{4}); m = -\frac{2}{3}$ _______ 32. $(\frac{1}{7}, -\frac{5}{7}); m = -\frac{1}{2}$

Find the equation of the line through the given points.

_______ 33. $(0, 2), (1, 0)$ _______ 34. $(0, 4), (-8, 0)$

_______ 35. $(-1, 7), (-4, 12)$ _______ 36. $(4, 6), (7, -3)$

_______ 37. $(9, -8), (-1, -8)$ _______ 38. $(6, 8), (-3, 8)$

_______ 39. $(-3, 7), (-1, 2)$ _______ 40. $(1, 11), (-2, 4)$

_______ 41. $(6, -5), (6, -10)$

_______ 42. $(-3, 4), (-3, 9)$

_______ 43. $(\frac{5}{2}, \frac{7}{5}), (\frac{1}{2}, -\frac{3}{5})$

_______ 44. $(\frac{7}{3}, -\frac{11}{8}), (\frac{1}{3}, -\frac{7}{8})$

Find the equation of the line through the given point and parallel to the given line.

_______ 45. $(-3, 2)$; $y = -2x + 5$

_______ 46. $(-2, 1)$; $y = 3x + 4$

_______ 47. $(-1, 1)$; $2x - y = 4$

_______ 48. $(-3, 3)$; $3x + y = -2$

_______ 49. $(0, 4)$; $x + 3y = -4$

_______ 50. $(2, 0)$; $x + 5y = 1$

Find the equation of the line through the given point and perpendicular to the given line.

_______ 51. $(4, 1)$; $y = 3x$

_______ 52. $(3, 2)$; $y = 5x$

_______ 53. $(-2, 0)$; $x + 2y = 1$

_______ 54. $(-4, 0)$; $x - 3y = 2$

_______ 55. $(3, -2)$; $4x - 2y = 5$

_______ 56. $(5, -1)$; $3x - 5y = 1$

_______ 57. Find the equation of the line with x-intercept of 3 and y-intercept of -5.

_______ 58. Find the equation of the line with x-intercept of -2 and y-intercept of -7.

_______ 59. Find the equation of the line with x-intercept of 2 and y-intercept of 4.

_______ 60. Find the equation of the line with x-intercept of 5 and y-intercept of 1.

☆ Stretching the Topics

_______ 1. Find the equation of the line through $(-4, 10)$ and parallel to the line through the points $(0, 5)$ and $(-8, -8)$.

_______ 2. Find the equation of the line through the midpoint of and perpendicular to the line segment with endpoints $(5, -1)$ and $(-3, 1)$.

_______ 3. Find the equations of the sides of a triangle whose vertices are $(1, -8)$, $(4, 2)$, and $(-3, 5)$.

Check your answers in the back of your book.

If you can complete **Checkup 8.3,** you are ready to go on to Section 8.4.

Name ______________________ Date ____________

✓ CHECKUP 8.3

________ **1.** Find the equation of the horizontal line through $(-3, 7)$.

________ **2.** Find the equation of the vertical line through $(\frac{1}{2}, -\frac{2}{3})$.

Find the equation of the line with the given slope and y-intercept.

________ **3.** $m = -4, b = 2$

________ **4.** $m = -\frac{2}{3}, b = 0$

Find the equation of the line through the given point with the given slope.

________ **5.** $(-1, 1); m = -4$

________ **6.** $(-3, 5)$; m is undefined

Find the equation of the line through the given points.

________ **7.** $(1, -1)$ and $(3, 9)$

________ **8.** $(\frac{4}{5}, -\frac{1}{2})$ and $(-\frac{6}{5}, \frac{3}{2})$

________ **9.** Find the equation of the line through the point $(-1, 2)$ and parallel to $3x - y = 5$.

________ **10.** Find the equation of the line through the point $(0, -5)$ and perpendicular to $x + 4y = 0$.

Check your answers in the back of your book.

If You Missed Problems:	You Should Review Examples:
1, 2	1, 2
3, 4	3, 4
5, 6	5–8
7, 8	9–12
9	13, 14
10	15

8.4 Graphing First-Degree Inequalities

Graphing a first-degree equation in two variables allows us to "picture" the set of all ordered pairs satisfying that equation. It is also handy to be able to picture the set of all ordered pairs that satisfy a first-degree **inequality** in two variables.

In each case we will graph the *equation* corresponding to the given *inequality* before we graph the inequality itself.

For example, to find all points that satisfy the inequality $y \leq 2$, we must first graph the equation $y = 2$. We know the graph of $y = 2$ is a horizontal line crossing the y-axis at (0, 2). To satisfy the inequality $y \leq 2$, we want all points in the plane whose y-coordinates are *less than 2 or equal 2*. All such points lie *on* the line $y = 2$ or *below* that line. We shade the appropriate region.

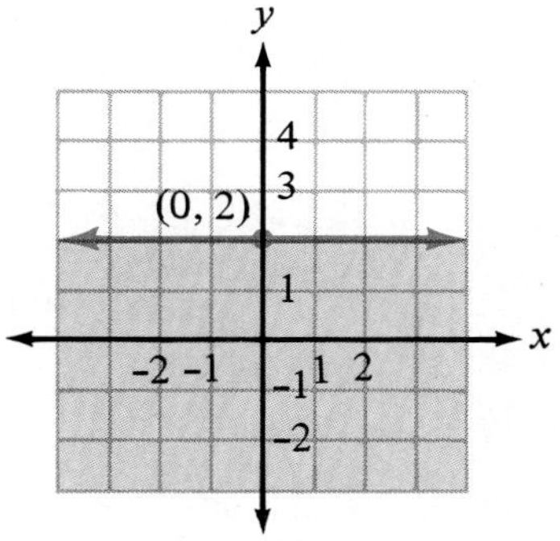

$y \leqslant 2$

Example 1. Graph $x > -1$.

Solution. We want all points in the plane whose x-coordinates are greater than -1. First we graph the vertical line $x = -1$. All x-values that are greater than -1 lie to the *right* of this line, but *not* on the line itself. Just as we used an *open* dot on a number line to show that a point was acting as a boundary but was not included in the graph itself, we can use a *dashed* line here to indicate that the vertical line $x = -1$ is acting as a boundary but is not included. Then we shade the region to the right of that line.

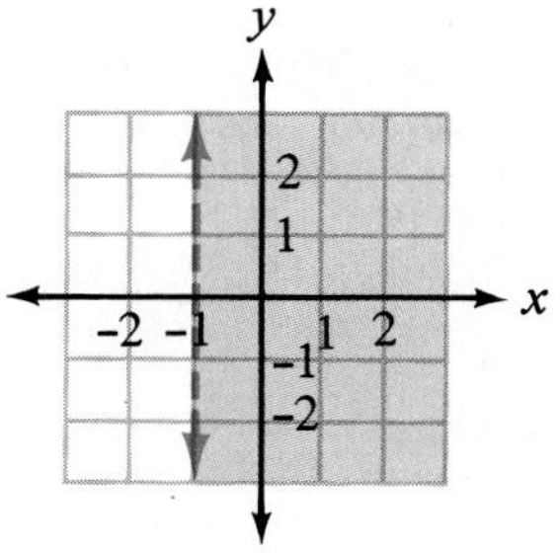

$x > -1$

In graphing linear inequalities, it is wise to decide right away whether the graph of the corresponding boundary equation should be solid or dashed.

> If the given inequality is of the $\leq$ or $\geq$ type, the graph of the corresponding boundary equation is a *solid* line. If the given inequality is of the $<$ or $>$ type, the graph of the corresponding boundary equation is a *dashed* line.

To graph the inequality

$$y < 4x - 1$$

we must first graph the boundary equation

$$y = 4x - 1$$

using a *dashed* line. This equation is in slope-intercept form, so we know that the line has a slope of 4 and a y-intercept of -1.

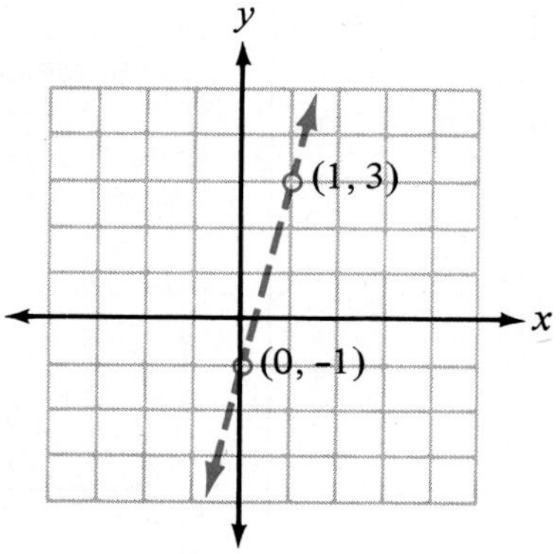

To determine the region in which $y < 4x - 1$ we can look at some points on either side of the line and see where ordered pairs satisfying the inequality must lie.

Above the line?			***Below the line?***		
(0, 2)	$2 \overset{?}{<} 4(0) - 1$		(3, 0)	$0 \overset{?}{<} 4(3) - 1$	
	$2 \overset{?}{<} -1$	No		$0 \overset{?}{<} 11$	Yes
(−2, 1)	$1 \overset{?}{<} 4(-2) - 1$		(0, −2)	$-2 \overset{?}{<} 4(0) - 1$	
	$1 \overset{?}{<} -9$	No		$-2 \overset{?}{<} -1$	Yes

Points *below* the line $y = 4x - 1$ satisfy the less-than inequality $y < 4x - 1$, but points above the line do not. Our graph is the shaded region *below* the line.

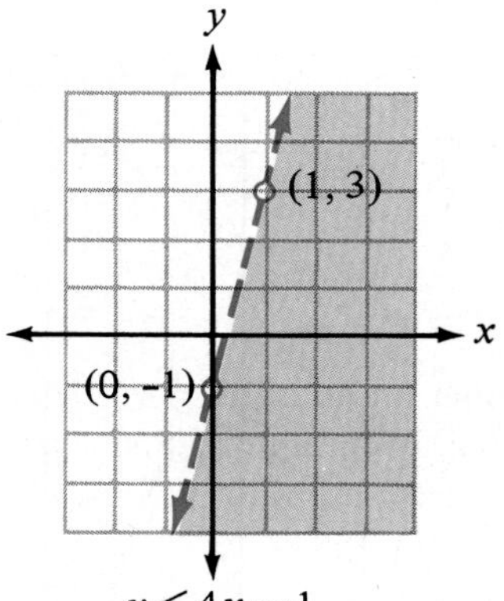

$y < 4x - 1$

Example 2. Graph $x + 3y > 0$.

Solution. We isolate y.

$$x + 3y > 0$$
$$3y > -x$$
$$y > -\tfrac{1}{3}x$$

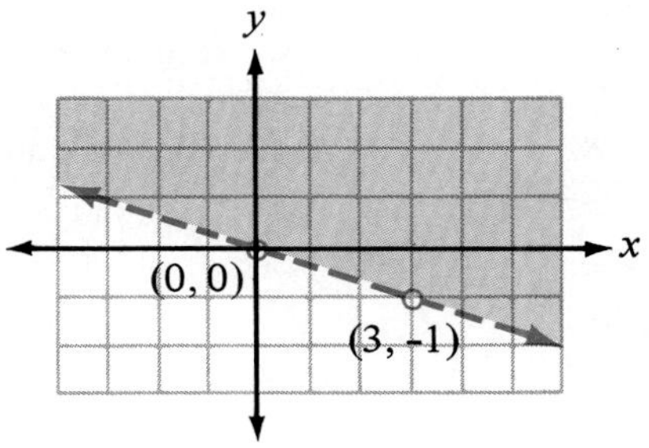

$x + 3y > 0$

Now we graph the corresponding boundary equation $y = -\frac{1}{3}x$, which is a dashed line with slope of $-\frac{1}{3}$, crossing the y-axis at (0, 0). Testing points above and below the line, we discover that this greater-than inequality is satisfied by points *above* the line.

In general, we may use the following steps to graph linear inequalities.

Graphing Linear Inequalities

1. Isolate y in the given inequality.
2. Write the corresponding boundary equation and decide whether the boundary line is solid ($\leq$ or $\geq$) or dashed ($<$ or $>$).
3. By testing points above and below the line, shade the appropriate region.

From the examples we have worked, we can make some observations that will allow us to shade the appropriate region without using test points. After isolating y in the given inequality we note that

> If $y < mx + b$, the line is *dashed* and we shade *below* the line.
>
> If $y \leq mx + b$, the line is *solid* and we shade *below* the line.
>
> If $y > mx + b$, the line is *dashed* and we shade *above* the line.
>
> If $y \geq mx + b$, the line is *solid* and we shade *above* the line.

Example 3. Graph the inequality $3x - 5y < 10$.

Solution. We isolate y.

$$3x - 5y < 10$$
$$-5y < -3x + 10$$
$$y > \frac{3}{5}x - 2$$ Remember to reverse the direction of inequality. Why?

The corresponding boundary equation is

$$y = \frac{3}{5}x - 2$$

so we draw a *dashed* line with slope $\frac{3}{5}$ and y-intercept of -2. Then we shade *above* the line, since the inequality is $y > \frac{3}{5}x - 2$.

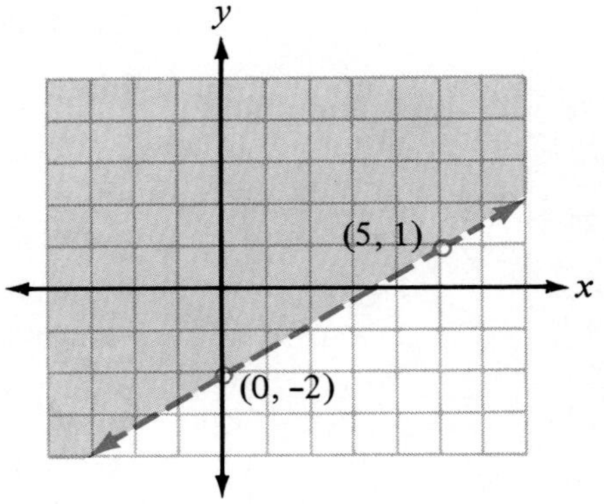

$3x - 5y < 10$

You try Example 4.

Example 4. Graph the inequality $y \leq 2 - 3x$.

Solution

$$y \leq 2 - 3x$$

$$y \leq -3x + 2$$

The corresponding boundary equation is $y = -3x + 2$ so we draw a ________ line with slope _____ and y-intercept of _____. Then we shade ________ the line.

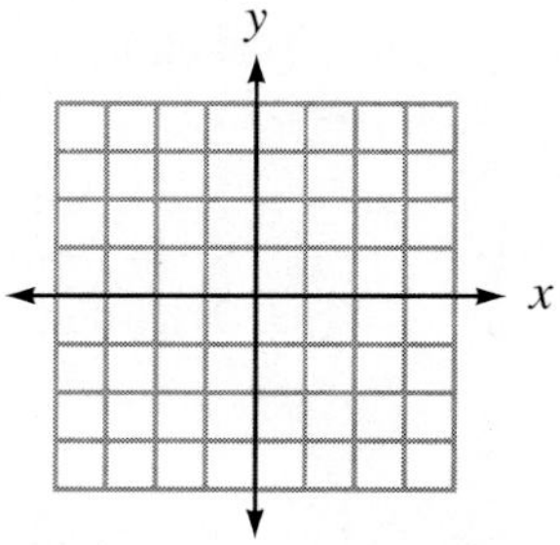

Check your work on page 569. ▶

Trial Run

Graph the inequalities.

1. $y < -2$

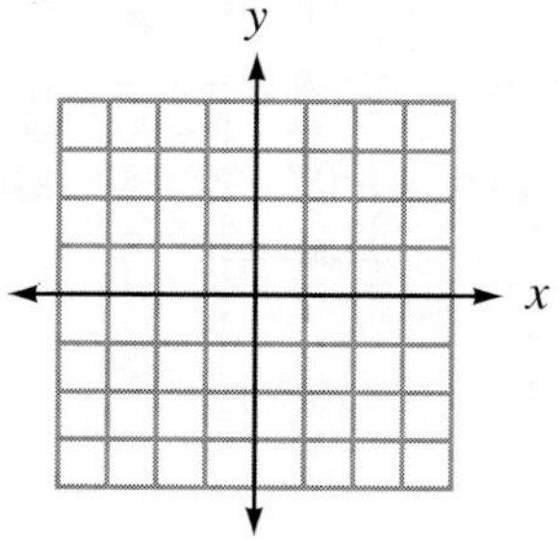

2. $x \geq -\frac{1}{2}$

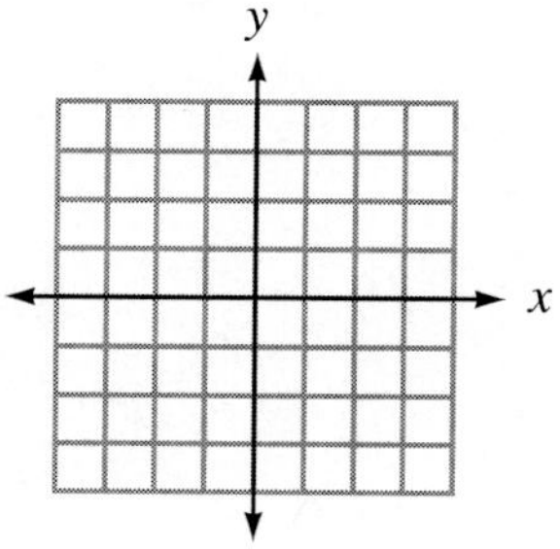

3. $y \geq 2x - 1$

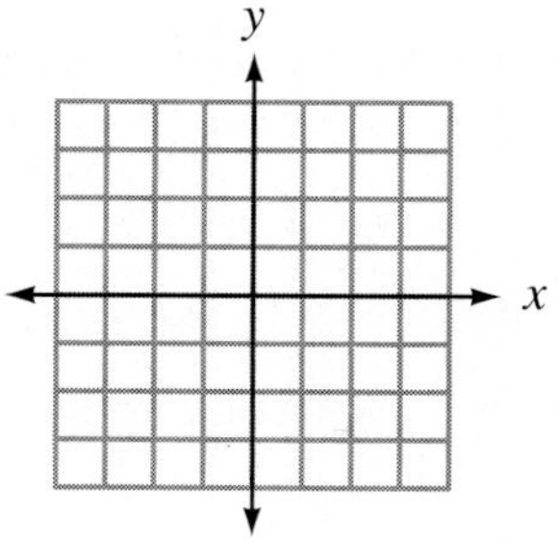

4. $2x - 3y > 6$

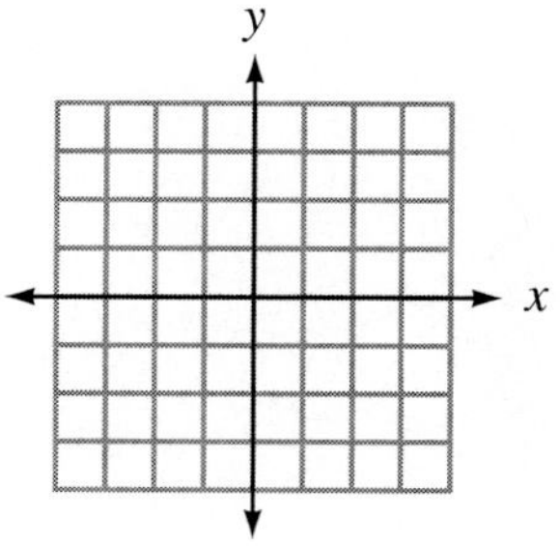

Answers are below.

▶ Example You Completed

Example 4. Graph the inequality $y \leq 2 - 3x$.

Solution

$$y \leq 2 - 3x$$

$$y \leq -3x + 2$$

The corresponding boundary equation is $y = -3x + 2$, so we draw a solid line with slope of -3 and y-intercept of 2. Then we shade below the line.

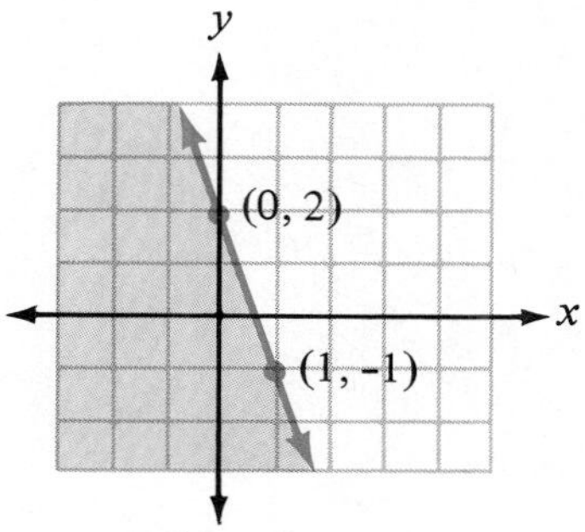

$y \leq 2 - 3x$

Answers to Trial Run

page 568 **1.**

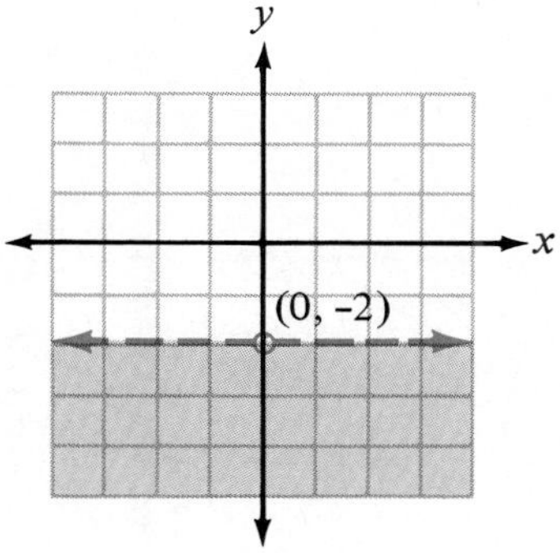

2.

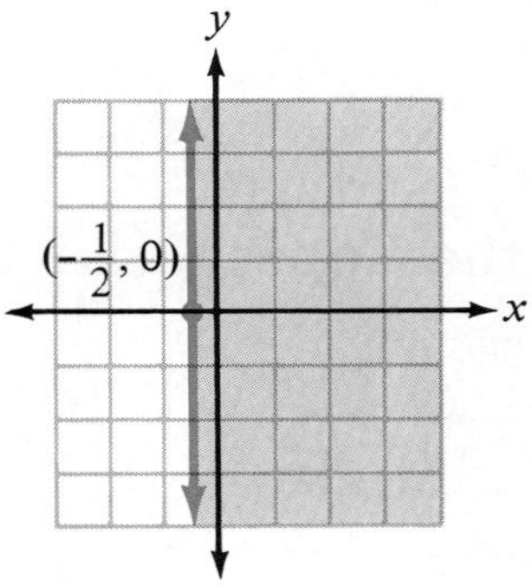

3.

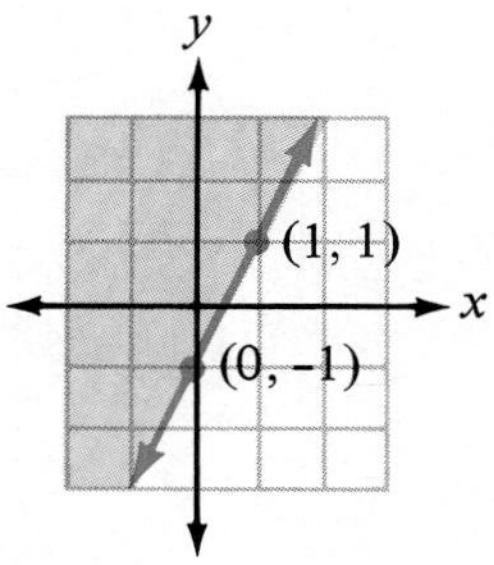

4.

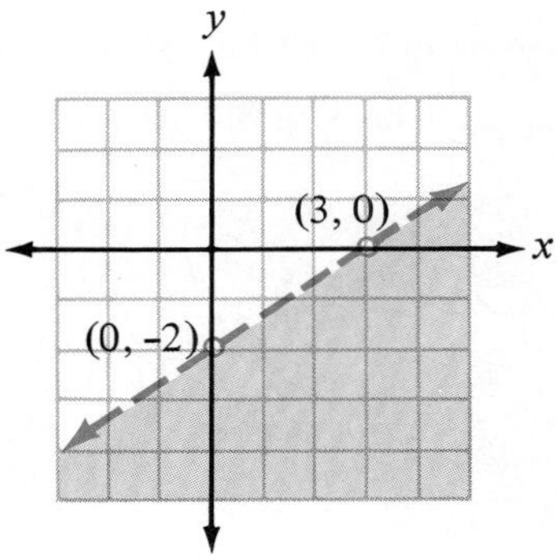

Name ______________________ Date ____________

EXERCISE SET 8.4

Graph each inequality.

1. $y > 5$
2. $y < -1$
3. $x \le 2$
4. $x \ge -4$
5. $y \ge 3x - 2$
6. $y \ge -4x + 1$
7. $x + y < -2$
8. $x + y < 4$
9. $y \le \frac{2}{3}x$
10. $y \le \frac{3}{4}x$
11. $2x - y > 1$
12. $3x - y > -2$
13. $x > 3$
14. $x \ge -2$
15. $x + 2y > -4$
16. $x + 3y > -9$
17. $3x - 2y > 8$
18. $2x - 3y > 6$
19. $3x + 5y \ge 7$
20. $2x + 3y \ge 8$

☆ Stretching the Topics

Graph each inequality.

1. $7 - [5 - (4 - y)] > 2 - x$
2. $-5 \le x < 8$
3. $|y| > 5$

Check your answers in the back of your book.

If you can complete **Checkup 8.4,** you are ready to go on to Section 8.5.

Name ______________________ Date ______________

✓ CHECKUP 8.4

Graph each inequality.

1. $x \leq 4$

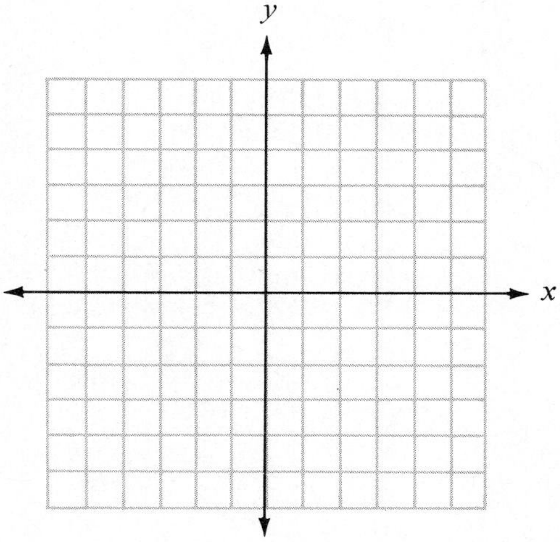

2. $x + 2y > -2$

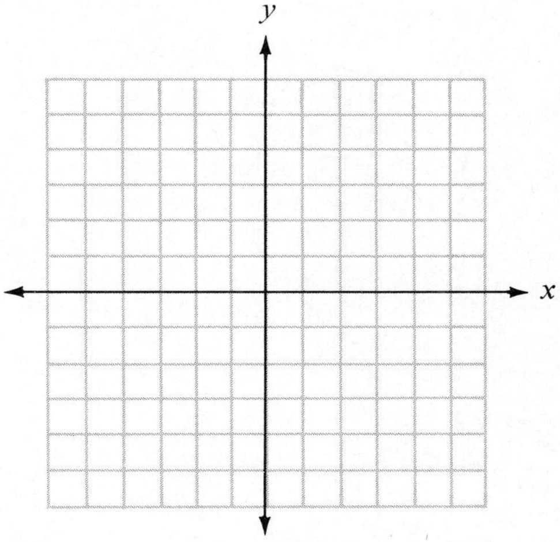

3. $2x - 3y > -3$

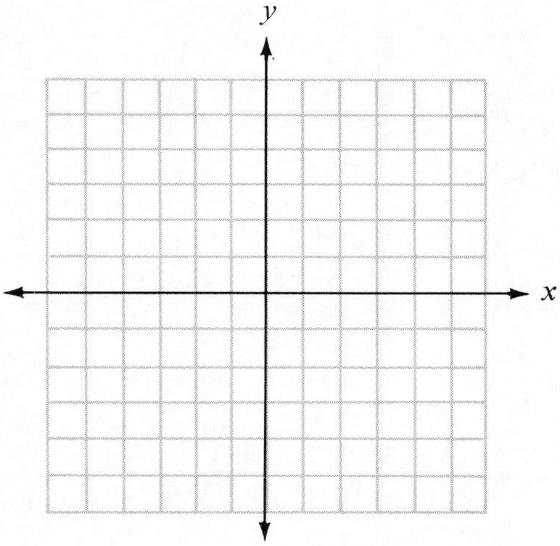

4. $y \leq 4 - 2x$

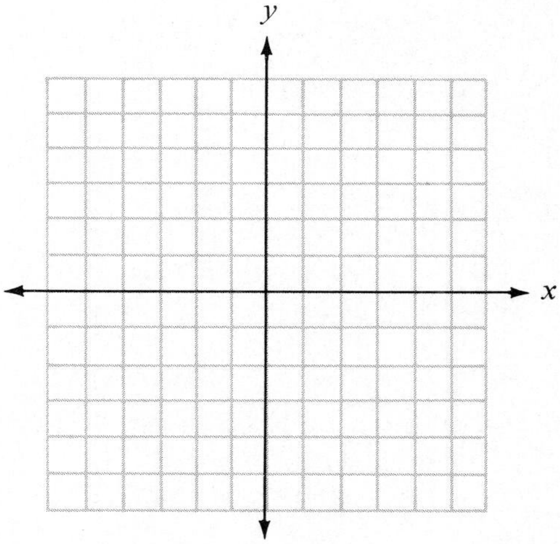

Check your answers in the back of your book.

If You Missed Problem:	You Should Review Example:
1	1
2	2
3	3
4	4

8.5 Working with Relations and Functions

Recognizing Functions

In this chapter we have been working with sets of ordered pairs and their graphs. A set of ordered pairs is sometimes called a **relation.** We have already noted that in each ordered pair (x, y), x is called the **independent variable** and y is called the **dependent variable.** The set of all permissible values for the independent variable is called the **domain** for the relation, and the set of all permissible values for the dependent variable is called the **range** for the relation.

Consider the relation described by this finite set of ordered pairs:

$$\{(-2, 8), (0, 0), (1, 5), (2, -1)\}$$

Here, the domain is the set of all the first coordinates used in the ordered pairs and the range is the set of all the second coordinates.

$$\text{Domain: } \{-2, 0, 1, 2\}$$

$$\text{Range: } \{-1, 0, 5, 8\}$$

Most often in algebra, a relation represents an infinite number of ordered pairs and we are given an *equation* to describe the ordered pairs. For instance, in the relation described by

$$y = 2x + 1$$

some ordered pairs are

$$\{(-2, -3), (0, 1), (\tfrac{1}{2}, 2), (1, 3), (4.3, 9.6), (\sqrt{3}, 2\sqrt{3} + 1), \ldots\}$$

The independent variable is x; the dependent variable is y. The domain and range are the set of all real numbers. Or, using notation discussed in Chapter 1, we may write

$$\text{Domain: } x \in R$$

$$\text{Range: } y \in R$$

We can use a graph to represent the ordered pairs in this relation in yet another way. Every ordered pair in the relation appears as a point on the line graphed here.

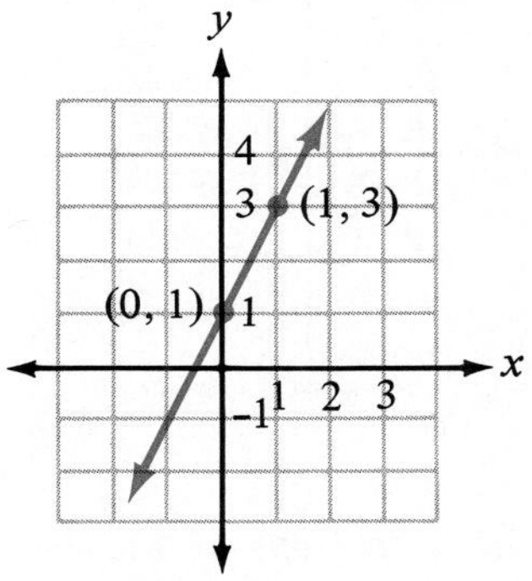

$y = 2x + 1$

In this relation, the equation $y = 2x + 1$ provides a rule for arriving at the ordered pairs. To write the relation correctly, we must write $\{(x, y) : y = 2x + 1\}$. However, we shall usually refer to such a set of ordered pairs by stating only its equation.

Every set of ordered pairs graphed in this chapter has been a relation. In some of the relations, each x in the domain has been paired with *one and only one* y in the range. When that occurs, we say that the relation is a **function.**

Relation: any set of ordered pairs.

Domain: the set of all replacement values for the independent variable in a relation.

Range: the set of all replacement values for the dependent variable in a relation.

Function: a set of ordered pairs in which each element in the domain is paired with one and only one element in the range.

Is the set of ordered pairs described by $y = 2x + 1$ a function? Certainly, because if we substitute any x-value into this equation, we arrive at one and only one y-value.

Example 1. Discuss the relation $y = 3$.

Solution. Some ordered pairs in this relation are

$$\{(-2, 3), (0, 3), (\tfrac{1}{2}, 3), (\sqrt{2}, 3), \ldots\}$$

No matter what real number we substitute for x, the y-value is always 3.

Domain: $x \in R$

Range: $y = 3$

Is this a function? Is any x-value paired with more than one y-value? No. Therefore,

$y = 3$ is a function

and we can easily graph it.

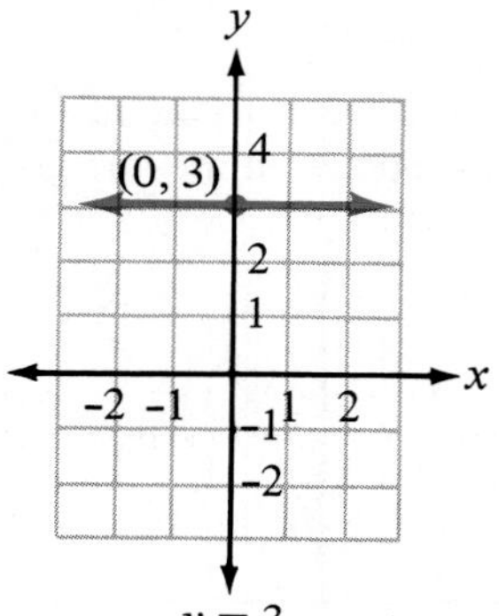

$y = 3$

Example 2. Discuss the relation $x = 2$.

Solution. Some ordered pairs in this relation are

$$\{(2, -5), (2, -\tfrac{1}{3}), (2, 0), (2, \sqrt{6}), \ldots\}$$

No matter what real number we substitute for y, the x-value is always 2.

Domain: $x = 2$

Range: $y \in R$

Is this a function? The only x-value in the domain is 2, and it is paired with infinitely many different y-values. Therefore,

$x = 2$ is *not* a function

but we can graph it.

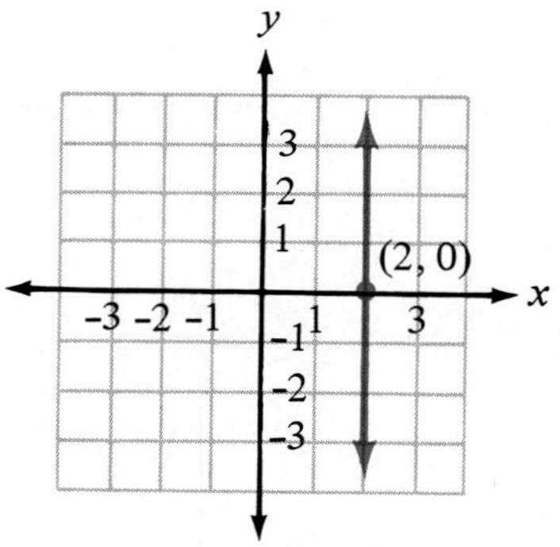

$x = 2$

Notice how the *graph* of a relation can help us decide whether that relation is a function. We may locate on the x-axis any x-value in the domain. Then we may see how many points on our graph correspond to that x-value. If any x-value corresponds to more than one point on the graph, that x is being paired with more than one y, and our relation is *not* a function. This technique is called the **vertical-line test.**

Vertical-Line Test. If a vertical line drawn through any x-value in the domain of a relation crosses the graph of the relation in more than one point, the relation is *not* a function.

Example 3. Use the vertical-line test to decide whether each graph represents a function. Give the domain and range for each.

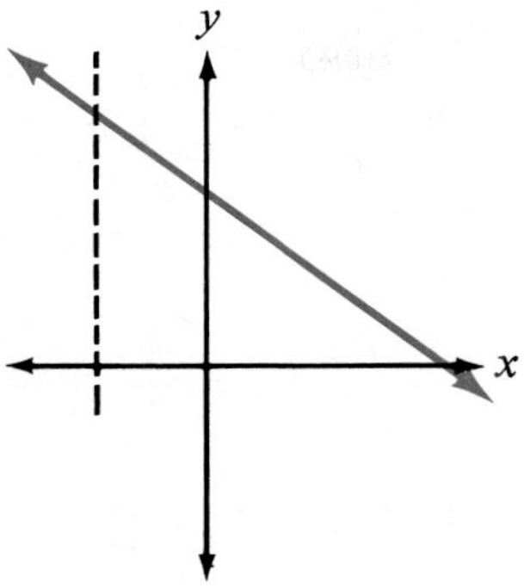

Function? Yes
Domain: $x \in R$
Range: $y \in R$

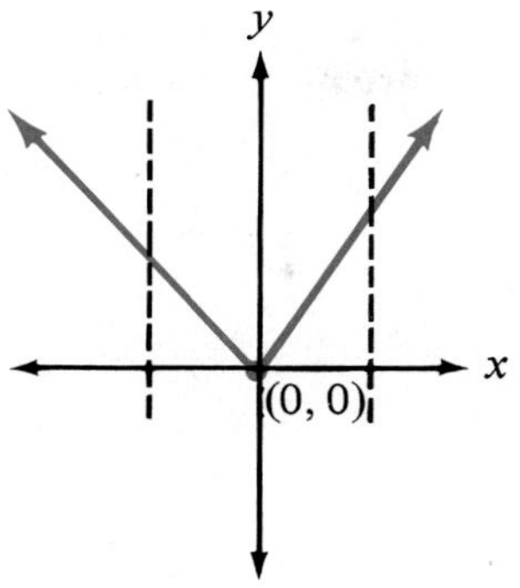

Function? Yes
Domain: $x \in R$
Range: $y \in R,\ y \geq 0$

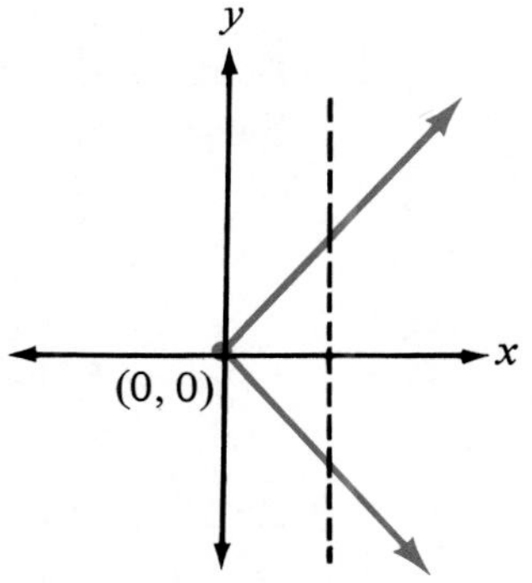

Function? No
Domain: $x \in R,\ x \geq 0$
Range: $y \in R$

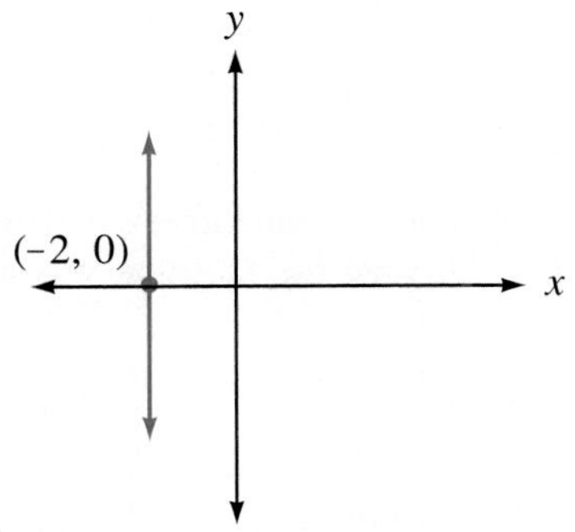

Function? No
Domain: $x = -2$
Range: $y \in R$

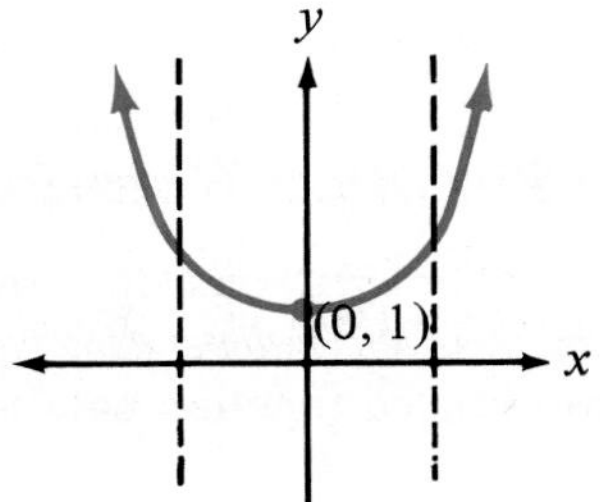

Function? Yes
Domain: $x \in R$
Range: $y \in R,\ y \geq 1$

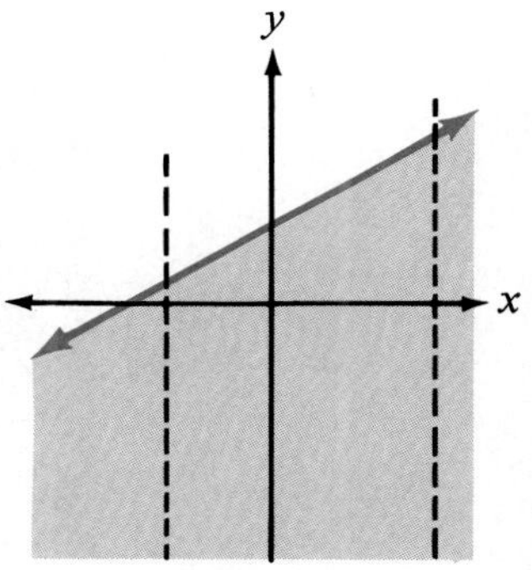

Function? No
Domain: $x \in R$
Range: $y \in R$

Trial Run

For each relation, give the domain and range and tell whether it is a function.

1. $\{(-1, 3), (1, 5), (0, 0), (1, 4), (-1, 2)\}$
Domain: _____ Range: _____ Function? _____

2. $\{(-2, 4), (-1, 0), (0, 4)\}$
Domain: _____ Range: _____ Function? _____

3. $\{(x, y) : y = x - 1\}$
Domain: _____ Range: _____ Function? _____

Use the vertical-line test to tell which graphs represent functions. Also give the domain and range.

______ 4.
Domain: ____
Range: ____

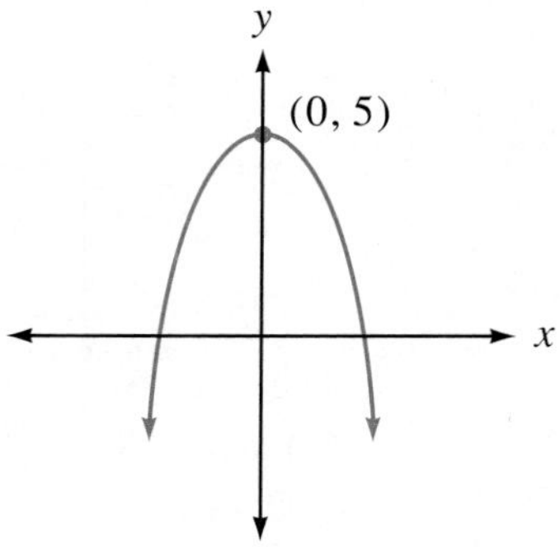

______ 5.
Domain: ____
Range: ____

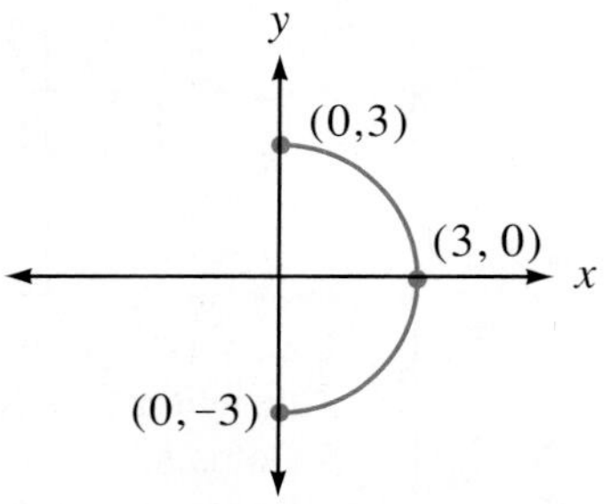

______ 6.
Domain: ____
Range: ____

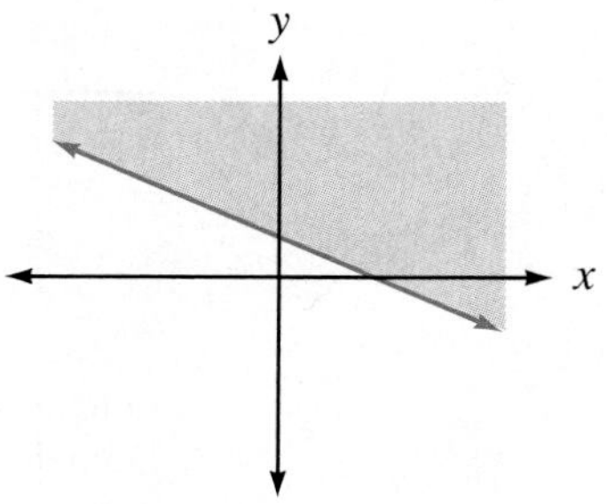

Answers are on page 579.

Using Function Notation

When a set of ordered pairs represents a function, we sometimes say that *the dependent variable is a function of the independent variable*. To state the fact that y is a function of x, mathematicians invented **function notation.**

> **Function Notation**
> $y = f(x)$ is read
>
> "y is a function of x"
>
> or
>
> "y equals f of x"

As a matter of fact, any letter can be used to name a function. When dealing with several functions of x in the same discussion, we must use different letters to name the different functions.

Function notation can be used whenever we wish to describe a function. For example,

$y = 2x + 1$	can be written	$f(x) = 2x + 1$
$y = 3$	can be written	$g(x) = 3$
$y = x^2 - 2$	can be written	$F(x) = x^2 - 2$

Function notation provides us with a quick way of finding ordered pairs. For example, if $f(x) = 7 - 3x$, then to find $f(2)$ we must evaluate $7 - 3x$ when x is 2.

$f(x) = 7 - 3x$	(x, y)
$f(2) = 7 - 3(2) = 1$	$(2, 1)$
$f(0) = 7 - 3(0) = 7$	$(0, 7)$
$f(\sqrt{2}) = 7 - 3\sqrt{2}$	$(\sqrt{2}, 7 - 3\sqrt{2})$

Function notation is also useful in finding function values when the independent variable x is replaced by other *variable* expressions. For example, if

$$F(x) = 5x - 2$$

$F(3)$	$F(a)$	$F(2x)$	$F(x + 3)$	$F(x + h)$
$5(3) - 2$	$5a - 2$	$5(2x) - 2$	$5(x + 3) - 2$	$5(x + h) - 2$
$= 13$		$= 10x - 2$	$= 5x + 13$	$= 5x + 5h - 2$

In each case, the expression replacing x was substituted for x in $5x - 2$.

You complete Example 4.

Example 4. If $f(x) = x^2 - 3x + 1$, find $f(0)$, $f(-2)$, $f(3a)$, and $f(x + h)$.

Solution

$$f(x) = x^2 - 3x + 1$$

$f(0)$	$f(-2)$	$f(3a)$	$f(x + h)$
$(0)^2 - 3(0) + 1$	$(-2)^2 - 3(-2) + 1$	$(3a)^2 - 3(3a) + 1$	$(\_\_\_\_)^2 - 3(\_\_\_\_) + 1$
$= 0 - 0 + 1$	$= \_\_\_\_ + \_\_\_\_ + 1$	$= \_\_\_\_ - \_\_\_\_ + 1$	$= \_\_\_\_\_\_\_\_$
$= \_\_\_\_$	$= \_\_\_\_$		

Check your table on page 579. ▶

Restricting Domains of Functions

When dealing with fractions and square roots, we learned to make *restrictions* on the domain of the variable to avoid expressions that were undefined in the set of real numbers. Recall that

> The denominator of a fraction *cannot* equal zero.
> The even root of a negative number is *not* real.

We can use these same ideas to restrict domains for functions.

Try Example 5.

Example 5. Give the domain for $f(x) = \frac{1}{x}$.

Solution. We know that $\frac{1}{x}$ is *not* defined when $x =$ _____, so

Domain: $x \in R,\ x \neq$ _____

Check your work on page 579. ▶

Example 6. Give the domain for $h(x) = \frac{x}{x^2 + 2x - 3}$.

Solution. We rewrite the denominator in factored form.

$$h(x) = \frac{x}{(x + 3)(x - 1)}$$

The denominator cannot be zero, so

Domain: $x \in R,\ x \neq -3,\ x \neq 1$

Example 7. Give the domain for $g(x) = \sqrt{x}$.

Solution. We know that $\sqrt{x}$ is *not* a real number when x is *negative*. So

Domain: $x \in R,\ x \geq 0$

Example 8. Give the domain for $F(x) = \sqrt{2 - x}$.

Solution. We know that $2 - x$ cannot be negative.

$$2 - x \geq 0$$
$$-x \geq -2$$
$$x \leq 2$$

Domain: $x \in R,\ x \leq 2$

Trial Run

If $f(x) = 3x - 2$, find

________ **1.** $f(4)$

________ **2.** $f(x + h)$

If $g(x) = x^2 - 2x + 3$, find

________ **3.** $g(0)$

________ **4.** $g(-1)$

Give the domain for each function.

________ **5.** $f(x) = \frac{1}{x - 1}$

________ **6.** $h(x) = \frac{x}{x^2 - x - 12}$

________ **7.** $F(x) = \sqrt{3 - x}$

________ **8.** $g(x) = \frac{1}{\sqrt{x}}$

Answers are on page 579.

▶ Examples You Completed

Example 4. If $f(x) = x^2 - 3x + 1$, find $f(0)$, $f(-2)$, $f(3a)$, and $f(x+h)$.

Solution

$$f(x) = x^2 - 3x + 1$$

$f(0)$	$f(-2)$	$f(3a)$	$f(x+h)$
$(0)^2 - 3(0) + 1$	$(-2)^2 - 3(-2) + 1$	$(3a)^2 - 3(3a) + 1$	$(x+h)^2 - 3(x+h) + 1$
$= 0 - 0 + 1$	$= 4 + 6 + 1$	$= 9a^2 - 9a + 1$	$= x^2 + 2xh + h^2 - 3x - 3h + 1$
$= 1$	$= 11$		

Example 5. Give the domain for $f(x) = \frac{1}{x}$.

Solution. We know that $\frac{1}{x}$ is *not* defined when $x = 0$, so

$$\text{Domain: } x \in R,\ x \neq 0$$

Answers to Trial Runs

page 575 **1.** Domain: $\{-1, 0, 1\}$; Range: $\{0, 2, 3, 4, 5\}$; not a function
2. Domain: $\{-2, -1, 0\}$; Range: $\{0, 4\}$; function **3.** Domain: $x \in R$; Range: $y \in R$; function
4. Function; Domain: $x \in R$; Range: $y \in R$, $y \le 5$
5. Not function; Domain: $x \in R$, $0 \le x \le 3$; Range: $y \in R$, $-3 \le y \le 3$
6. Not function; Domain: $x \in R$; Range: $y \in R$

page 578 **1.** 10 **2.** $3x + 3h - 2$ **3.** 3 **4.** 6 **5.** $x \in R$, $x \neq 1$ **6.** $x \in R$, $x \neq 4$, $x \neq -3$
7. $x \in R$, $x \le 3$ **8.** $x \in R$, $x > 0$

Name ______________________________ Date ______________

EXERCISE SET 8.5

For each of the following relations, give the domain and the range and state which are functions.

________ 1. $\{(1, 0), (2, 3), (3, 5), (2, -3)\}$

________ 2. $\{(-5, 0), (-1, 4), (0, 6), (-1, -4)\}$

________ 3. $\{(x, y) : y = 2x\}$

________ 4. $\{(x, y) : y = x + 2\}$

________ 5. $\{(-6, 3), (-2, 3), (0, 3), (3, 3)\}$

________ 6. $\{(-1, -4), (0, -4), (2, -4), (5, -4)\}$

________ 7. $\{(x, y) : x = 3\}$

________ 8. $\{(x, y) : x = 7\}$

________ 9. $\{(1, 3), (-1, 3), (2, 5), (-2, 5)\}$

________ 10. $\{(2, -3), (-2, -3), (4, -6), (-4, -6)\}$

Use the vertical-line test to tell which of the following graphs represent functions.

________ 11.

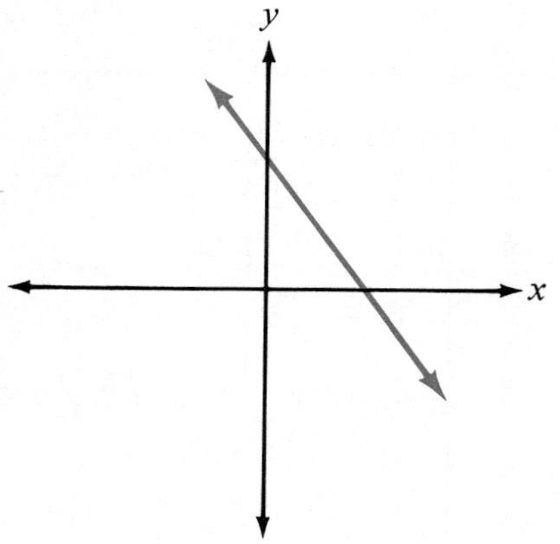

________ 12.

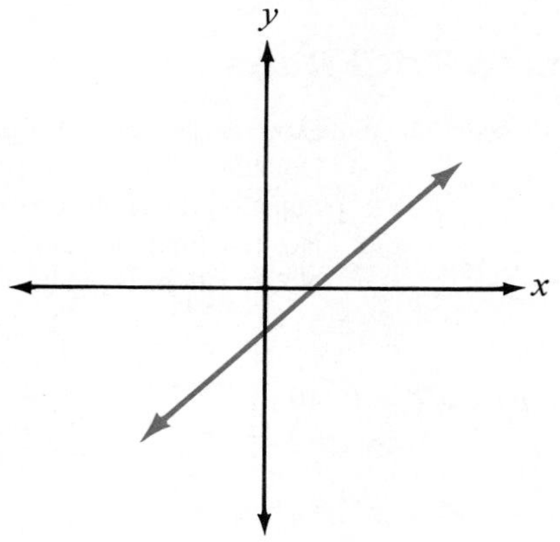

________ 13.

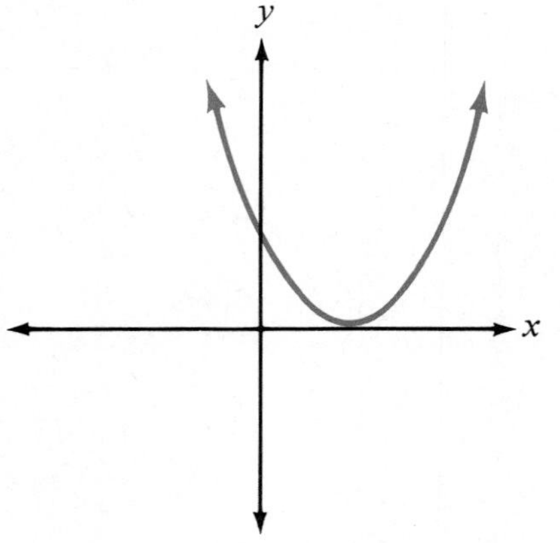

________ 14.

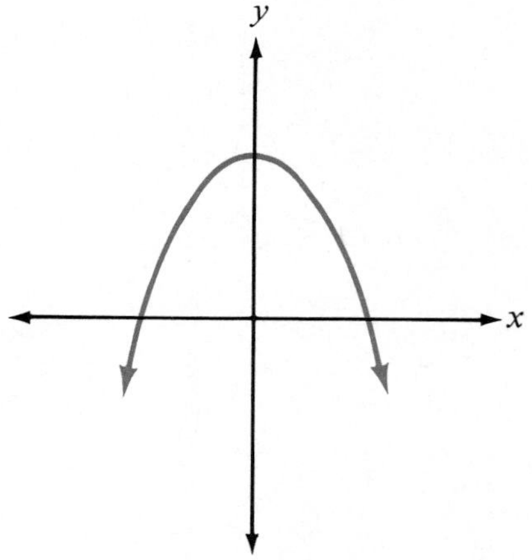

_______ 15.

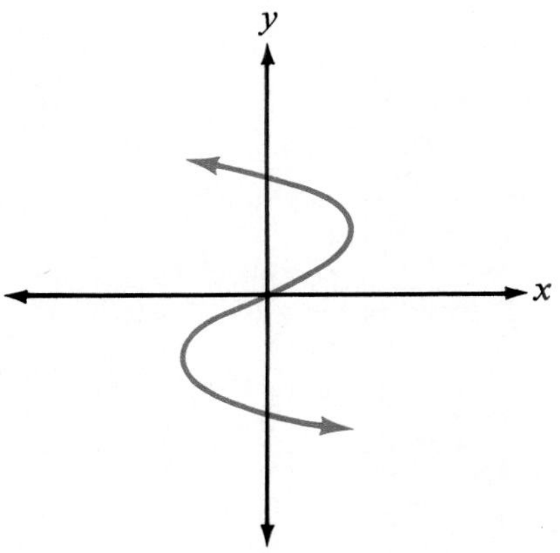

_______ 16.

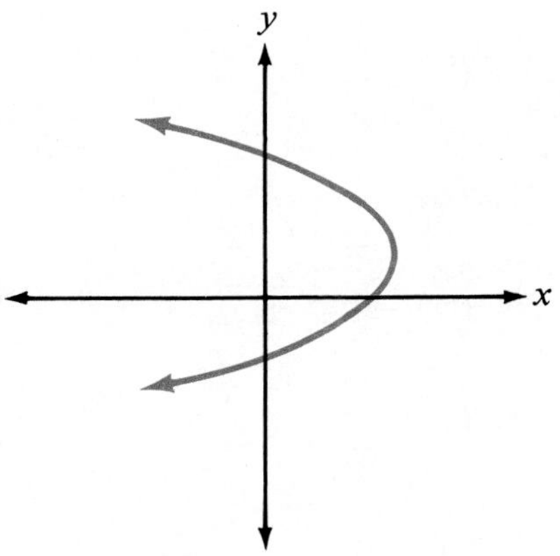

_______ 17.

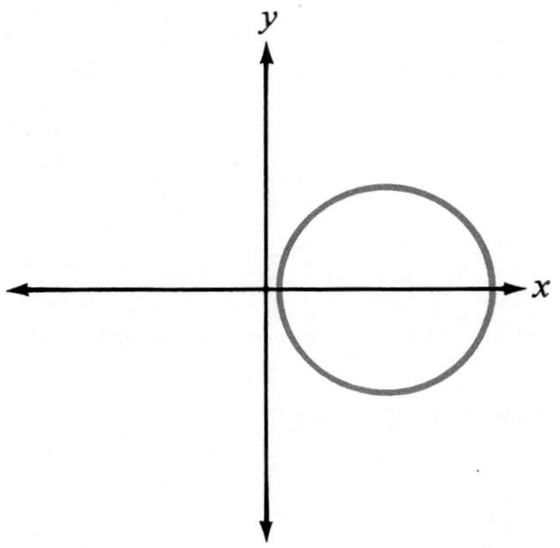

_______ 18.

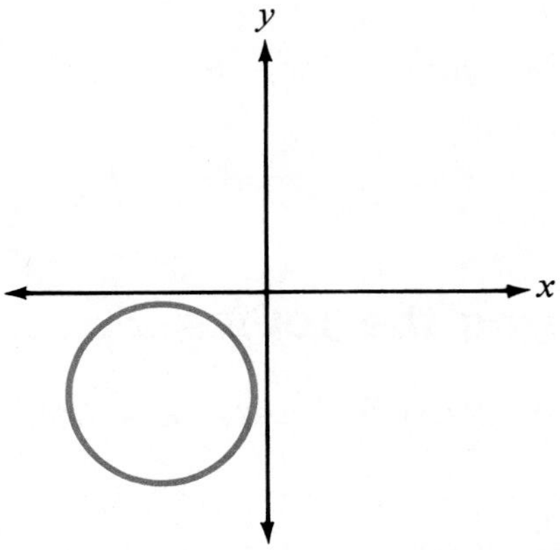

If $f(x) = 3x + 1$ *and* $g(x) = x^2 + 2x - 8$, *find each of the following.*

_______ 19. $f(2)$

_______ 20. $f(4)$

_______ 21. $g(-1)$

_______ 22. $g(-3)$

_______ 23. $f(x + h)$

_______ 24. $f(x - h)$

_______ 25. $f(-3) + g(0)$

_______ 26. $f(-4) - g(0)$

If $f(x) = 2x - 3$ *and* $g(x) = x^2 - x - 1$, *find each of the following.*

_______ 27. $f(x + 1)$

_______ 28. $f(x + 2)$

_______ 29. $g(a)$

_______ 30. $g(c)$

_______ 31. $g(1) - f(2)$

_______ 32. $g(3) - f(0)$

_______ 33. $g(a + 1)$

_______ 34. $g(a + 3)$

Give the domain for each of the following functions.

_______ 35. $f(x) = \dfrac{1}{x + 4}$

_______ 36. $f(x) = \dfrac{1}{x - 6}$

_______ 37. $f(x) = \dfrac{3}{2x - 1}$

_______ 38. $f(x) = \dfrac{5}{3x - 2}$

_______ 39. $g(x) = \sqrt{x + 1}$

_______ 40. $g(x) = \sqrt{x + 3}$

_______ 41. $f(x) = \dfrac{x - 1}{x - 3}$

_______ 42. $f(x) = \dfrac{x - 4}{x - 5}$

_______ 43. $h(x) = \sqrt{5 - x}$

_______ 44. $h(x) = \sqrt{1 - x}$

_______ 45. $G(x) = \dfrac{x}{x^2 - 16}$

_______ 46. $G(x) = \dfrac{2x}{x^2 - 25}$

_______ 47. $f(x) = \sqrt{2x - 6}$

_______ 48. $f(x) = \sqrt{3x - 9}$

_______ 49. $F(x) = \dfrac{-1}{2x^2 - 9x + 7}$

_______ 50. $F(x) = \dfrac{-5}{3x^2 + 14x - 5}$

☆ Stretching the Topics

_______ 1. If $f(x) = 2x^2 - 3x + 4$, find $\dfrac{f(a + h) - f(a)}{h}$, where $h \neq 0$.

_______ 2. If $g(x) = \dfrac{x^2}{x^2 + 1}$, find $g(\sqrt{a})$.

_______ 3. Give the domain for $f(x) = \dfrac{\sqrt{x}}{2x^3 - 11x^2 + 12x}$.

Check your answers in the back of your book.

If you can complete **Checkup 8.5,** you are ready to go on to Section 8.6.

Name ______________________ Date ____________

✓ CHECKUP 8.5

______ 1. Give the domain and range for the relation $\{(x, y) : x = 5\}$. Is this relation a function?

Use the vertical-line test to tell which of the following graphs represent functions.

______ 2.

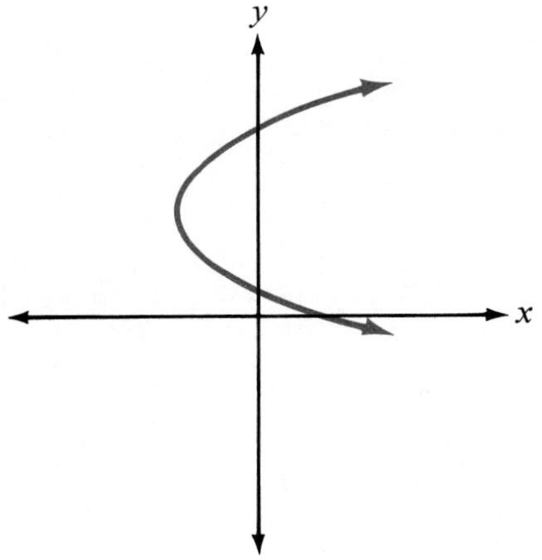

______ 3.

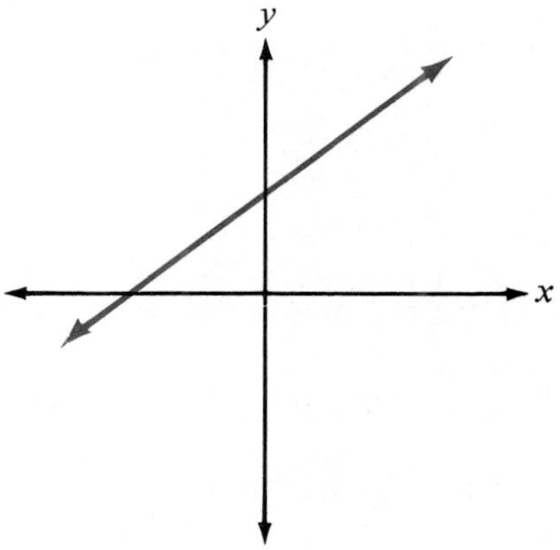

If $f(x) = 2x + 1$ *and* $g(x) = x^2 + x - 1$, *find each of the following.*

______ 4. $f(3)$

______ 5. $g(-2)$

______ 6. $f(-5) + g(0)$

Give the domain for each of the following functions.

______ 7. $f(x) = \dfrac{x + 1}{2x - 3}$

______ 8. $g(x) = \sqrt{x - 4}$

______ 9. $f(x) = \dfrac{x}{x^2 - 3x - 28}$

Check your answers in the back of your book.

If You Missed Problems:	You Should Review Examples:
1	2
2, 3	3
4–6	4
7–9	5–8

8.6 Switching from Word Statements to Functions

Many relationships between two variables can be expressed as functions or relations.

Working with Variation

If a dealer sells calculators for $29 each, you should agree that the dealer's revenue from the calculators depends on the number of calculators sold. In other words, the revenue *is a function of* the number of calculators sold. If we let

x = number of calculators sold

r = revenue from sale of calculators

then we can express the revenue function as

$$r = 29x$$

Of course, for this function to make sense, the domain must be the set of whole numbers. Why?

In the revenue function $r = 29x$, you should agree that the revenue will *increase* as the number of calculators sold *increases*. In such a function, the dependent variable (r) is said to **vary directly** with the independent variable (x). This function is an example of **direct variation.**

> **Direct Variation.** If y varies directly as x, then
>
> $$y = k \cdot x$$
>
> and k is called the constant of variation.

Example 1. Write a function to describe the fact that the number of freshmen at a university varies directly with the total enrollment.

Solution. Let F = number of freshmen
t = total enrollment
then $F = kt$

Example 2. If the number of freshmen is 2500 when total enrollment is 9000, find the constant of variation k. Then find the freshman enrollment when the total enrollment is 13,500.

Solution. If $F = 2500$ when $t = 9000$, we can use substitution to find k.

$$F = kt$$

$$2500 = k(9000) \quad \text{Substitute the given values.}$$

$$\frac{2500}{9000} = k \quad \text{Divide by 9000 to isolate } k.$$

$$k = \frac{5}{18} \quad \text{Reduce the fraction.}$$

Now we know that $F = \frac{5}{18}t$ and we can substitute to find F when $t = 13{,}500$.

$$F = \frac{5}{18}t \qquad \text{Substitute } \frac{5}{18} \text{ for } k.$$

$$= \frac{5}{18}(13{,}500) \qquad \text{Substitute 13,500 for } t.$$

$$= 3750 \qquad \text{Multiply.}$$

When total enrollment is 13,500, freshman enrollment will be 3750.

Now we consider the problem stated at the beginning of this chapter.

Example 3. The distance required to stop a moving car varies directly as the square of the speed of the car. If the distance required to stop a car traveling 20 miles per hour is 30 feet, what distance would be required to stop the same car traveling 40 miles per hour?

Solution. Let d = distance (feet)
s = speed of car (miles per hour)
then $d = k \cdot s^2$

To find k, we substitute 30 for d and 20 for s.

$$30 = k(20)^2$$

$$30 = k(400)$$

$$\frac{30}{400} = k$$

$$k = \frac{3}{40}$$

Now we know that $d = \frac{3}{40}s^2$ and we can find d when $s = 40$.

$$d = \frac{3}{40}s^2$$

$$= \frac{3}{40}(40)^2$$

$$= 120 \text{ ft}$$

If a car is traveling 40 miles per hour, then 120 feet is the distance required to stop.

Now consider a relationship between two variables in which the variation is *not* direct. If a rectangle with length l and width w has an area of 148 square feet, then we know that

$$l \cdot w = 148$$

We can express the length of the rectangle as a function of the width by writing

$$l = \frac{148}{w}$$

This is a function whose domain is the set of positive real numbers. You should see that the length must *decrease* as the width *increases*. In such a function, the dependent variable (l) is said to **vary inversely** with the independent variable (w). This function is an example of **inverse variation.**

Inverse Variation. If y varies inversely as x, then

$$y = \frac{k}{x}$$

and k is called the constant of variation.

Try Example 4.

Example 4. Write a function to describe the fact that the time (t) for a person to walk a fixed distance varies inversely as the rate (r) at which the person walks.

Solution

$$t = \_\_\_\_$$

Check your function on page 590. ▶

Example 5. If, in Example 4, we know that $t = 3$ hours when $r = 2$ mph, find the constant of variation. Then find t when $r = 1.5$ mph and when $r = \frac{2}{3}$ mph.

Solution

$$t = \frac{k}{r}$$

$$3 = \frac{k}{2} \qquad \text{Substitute 3 for } t \text{ and 2 for } r.$$

$$6 = k \qquad \text{Solve for } k.$$

Therefore $t = \dfrac{6}{r}$.

When $r = 1.5$ mph

$$t = \frac{6}{1.5} = 4 \text{ hr}$$

When $r = \frac{2}{3}$ mph

$$t = \frac{6}{\frac{2}{3}} = 6 \cdot \tfrac{3}{2} = 9 \text{ hr}$$

If one variable varies *directly* with the *product* of two other variables, we call such a relationship **joint variation**.

> **Joint Variation.** If y varies jointly with x and z, then
>
> $$y = k \cdot x \cdot z$$
>
> where k is the constant of variation.

Example 6. Write a function that states that the volume (V) of a cone varies jointly with the height (h) and the square of the radius (r) of the base.

Solution

$$V = k \cdot h \cdot r^2$$

You try Example 7.

Example 7. Write a function stating that the simple interest (I) from a certain investment varies jointly with the interest rate (r) and the length of time (t).

Solution

$$I = \_\_\_\_\_\_\_\_$$

Check your function on page 590. ▶

When a variable varies directly with some variable and inversely with another variable, such a relationship is called **mixed variation.**

> **Mixed Variation.** If y varies directly as x and inversely as z, then
>
> $$y = \frac{kx}{z}$$
>
> where k is the constant of variation.

Example 8. When a customer agrees to pay for a purchase in n installments at an advertised rate of interest r, then the true rate of interest (i) varies jointly as n and r, and inversely as the sum $n + 1$. Write a function to describe this relationship.

Solution

$$i = \frac{knr}{n + 1}$$

Example 9. A certain variable V varies directly with b and inversely with a. Write the function describing this relationship. Then, if $V = -3$ when $b = 12$ and $a = 2$, find k. Finally, find V when $b = 3$ and $a = -1$.

Solution

$$V = \frac{k \cdot b}{a}$$

We know that $V = -3$ when $b = 12$ and $a = 2$. We can substitute to find k.

$-3 = \frac{k(12)}{2}$	Substitute -3 for V, 12 for b, and 2 for a.
$-3 = k \cdot 6$	Simplify.
$\frac{-3}{6} = k$	Divide both sides by 6 to isolate k.
$k = -\frac{1}{2}$	Simplify the fraction.

and the function can be written $V = -\frac{1}{2} \cdot \frac{b}{a}$. Now find V when $b = 3$ and $a = -1$.

$$V = -\frac{1}{2} \cdot \frac{3}{-1}$$

$$= \frac{3}{2}$$

Trial Run

Write the function that describes the variation.

________ 1. The volume (V) of a sphere varies directly as the cube of its radius (r).

________ 2. The average fixed costs ($\bar{c}$) for a company vary inversely as the number of units (n) produced.

________ 3. The area (A) of a trapezoid varies jointly as its height (h) and the sum of its bases (a and b).

________ 4. The strength (s) of a string varies directly as its thickness (t) and inversely as its length (l).

Answers are on page 590.

Working with Other Functions and Relations

Let's consider some more functions and relations involving two variables that might arise in everyday situations.

Example 10. A car rental agency charges its customers $20 per day plus 32 cents per mile driven. Write an equation to describe one day's total charge (C) as a function of the number of miles driven.

Solution

We Think	*We Write*
Total daily charge is the sum of the flat charge and the mileage charge.	$C = 20 +$ mileage charge
The mileage charge depends on the miles driven.	Let $m =$ number of miles driven
The mileage rate is 32 cents per mile.	$0.32m =$ mileage charge
We write the equation for the total daily charge, C.	$C = 20 + 0.32m$

You try Example 11.

Example 11. If the length of a rectangular garden is 10 feet more than its width, write an equation for the perimeter (P) of the garden as a function of its width.

Solution

We Think	*We Write*
The perimeter of a rectangle is the sum of 2 lengths and 2 widths.	$P = 2$ lengths $+ 2$ widths
We wish the width to be the independent variable.	Let $w =$ width

We Think	***We Write***
The garden is 10 feet longer than it is wide.	_____ + _____ = length
We can write an equation for perimeter.	P = 2 lengths + 2 widths = 2(_____ + _____) + 2(_____)
We can simplify the right side.	P =

The perimeter can be expressed as a function of the width by P = _________ .

Check your work below. ▶

Example 12. If a company sells chairs for \$90 each and tables for \$130 each, write a relation that states that the company wishes its total revenue from chairs and tables to be at least \$5850 per week.

Solution

We Think	***We Write***
The company's total revenue comes from selling chairs and tables.	Total revenue = chair revenue + table revenue
Chair revenue depends on how many chairs are sold. Table revenue depends on how many tables are sold.	Let x = number of chairs sold y = number of tables sold
Revenue is found by multiplying price per unit times number of units sold.	$90x$ = chair revenue $130y$ = table revenue
The company wants total revenue to be at least \$5850.	$90x + 130y \geq 5850$

The inequality $90x + 130y \geq 5850$ describes the fact that total revenue must be at least \$5850. This is a relation whose domain and range are the set of whole numbers.

▶ Examples You Completed

Example 4 (*Solution*)

$$t = \frac{k}{r}$$

Example 7 (*Solution*)

$$I = k \cdot r \cdot t$$

Example 11 (*Solution*)

We Write

P = 2 lengths + 2 widths
Let w = width
$w + 10$ = length
P = 2 lengths + 2 widths
$= 2(w + 10) + 2(w)$
$P = 2w + 20 + 2w$
$= 4w + 20$

The perimeter can be expressed as a function of the width by $P = 4w + 20$.

Answers to Trial Run

page 589 1. $V = kr^3$ 2. $\bar{c} = \dfrac{k}{n}$ 3. $A = kh(a + b)$ 4. $s = \dfrac{kt}{l}$

Name ______________________ Date ____________

EXERCISE SET 8.6

______ 1. Write an equation of variation and find k if y varies directly as x, and $y = 48$ when $x = 4$.

______ 2. Write an equation of variation and find k if y varies directly as x, and $y = 35$ when $x = 7$.

______ 3. Write an equation of variation and find k if y varies directly as x, and $y = 16$ when $x = 80$.

______ 4. Write an equation of variation and find k if y varies directly as x, and $y = 42$ when $x = 63$.

______ 5. Write an equation of variation and find k if y varies inversely as x, and $y = 6$ when $x = 10$.

______ 6. Write an equation of variation and find k if y varies inversely as x, and $y = 7$ when $x = 5$.

______ 7. Write an equation of variation and find k if y varies inversely as x, and $y = 0.3$ when $x = 15$.

______ 8. Write an equation of variation and find k if y varies inversely as x, and $y = 4.5$ when $x = 0.9$.

______ 9. Write an equation of variation and find k if z varies jointly as x and y, and $z = 6$ when $x = 4$ and $y = 2$.

______ 10. Write an equation of variation and find k if w varies jointly as u and v, and $w = 10$ when $u = 5$ and $v = 3$.

______ 11. Write an equation of variation and find k if y varies directly as x and inversely as the square of z, and $y = 12$ when $x = 9$ and $z = 2$.

______ 12. Write an equation of variation and find k if y varies directly as the square of z and inversely as x, and $y = 50$ when $x = 2$ and $z = 5$.

______ 13. The number of rubber bands produced by a machine varies directly as the amount of time the machine is operating. If 1200 rubber bands are produced in 3 hours, how many can be produced in 8 hours?

______ 14. The cost of producing pencils varies directly with the number of pencils produced. If the cost of producing 2000 pencils is \$300, how much will it cost to produce 5000 pencils?

______ 15. The time required to empty a tank varies inversely as the rate of the pump. If the tank can be emptied in 25 minutes by a pump that is pumping at the rate of 400 gallons per minute, how long will it take to empty the tank with a pump that will pump at the rate of 500 gallons per minute?

______ 16. The number of hours required to clean the arena after a ball game varies inversely as the number of students hired to do the cleaning. If it takes 9 hours for 5 students to do the job, how many hours will it take for 15 students to do the job?

________ 17. The length a spring stretches varies directly as the force applied. If the length is 15 inches when the force is 9 pounds, find the force needed to stretch the spring to 20 inches.

________ 18. The water pressure at a point under water varies directly as the height of the water above the point. If the pressure is 312 pounds per square foot when the height is 10 feet, find the height of the water if the pressure being exerted at a point is 1248 pounds per square foot.

________ 19. The average fixed cost for a manufacturing firm varies inversely as the number of items in daily output. If the average fixed cost is $35 when the daily output is 120 items, how many items must be produced to reduce the average fixed cost to $21?

________ 20. The volume of a gas varies inversely as the pressure of the gas on its container. If the volume is 96 cubic feet when the pressure is 25 pounds, what pressure is needed for the volume to be 100 cubic feet?

________ 21. The distance a body falls varies directly as the square of the time it has been falling. If an object falls 144 feet in 3 seconds, how long will it take it to fall 400 feet?

________ 22. The brightness of light at a given distance from its source varies inversely as the square of this distance. If the brightness of a certain light is 50 (candlepower) at a distance of 20 feet from its source, find the brightness when the distance is 12 feet.

________ 23. The weight of a round metal rod varies jointly as the length and the square of the diameter. If the weight of a rod 1 inch in diameter and 8 inches long is 1.5 pounds, find the weight of a rod 2.5 inches in diameter and 16 feet long.

________ 24. The weight of a circular disk varies jointly as its thickness and the square of its radius. A disk that is 2 inches thick with a radius of 5 inches weighs 3.1 pounds. What is the weight of a disk of the same material if it is 3.2 inches thick with a radius of 8.6 inches? (Round answer to the tenths place.)

________ 25. A car rental agency charges a fee of $18 per day, plus 30 cents per mile. Write an equation to describe one day's total charge as a function of the number of miles driven.

________ 26. A local restaurant charges a fee of $75 plus $7 a plate for serving a banquet. Write an equation to describe the cost of a banquet as a function of the number of people attending.

________ 27. Mr. Gillenwater sells vacuum cleaners. He earns $100 a week plus $75 commission for each vacuum cleaner he sells. Write an equation to express that his weekly salary is a function of the number of vacuum cleaners he sells.

________ 28. Ms. Jolly has a weekly salary of $340 and she earns $10.50 for each hour she works overtime. Write an equation to express that her pay each week is a function of the number of hours she works overtime.

_______ 29. For traveling on the toll road, Pat needs four times as many quarters as dimes. Write an equation to describe the total amount needed as a function of the number of dimes.

_______ 30. Mrs. Abell's recipe for punch calls for twice as many cans of frozen orange juice as frozen lemonade. If a can of orange juice costs 53 cents and a can of lemonade costs 62 cents, write an equation to describe the total cost as a function of the number of cans of lemonade.

_______ 31. If the length of a rectangle is 12 feet longer than the width, write an equation that expresses the perimeter as a function of the width.

_______ 32. If the width of a rectangle is 3 feet less than the length, write an equation that expresses the perimeter as a function of the length.

_______ 33. If the width of a rectangle is $\frac{1}{3}$ of the length, write an equation that expresses the area as a function of the length.

_______ 34. If the length of a rectangle is five times the width, write an equation that expresses the area as a function of the width.

_______ 35. The base of a triangle is 4 more than the altitude. Write an equation that expresses the area as a function of the altitude.

_______ 36. The altitude of a triangle is 3 less than the base. Write an equation that expresses the area as a function of the base.

_______ 37. Beth has decided that each day she will walk twice as long as she jogs. If she walks at 4 miles per hour and jogs at 6 miles per hour, write an equation that expresses the total distance as a function of the time she jogs.

_______ 38. Wayne can jog 2 miles an hour faster than he can walk. If he decides to do each for a half-hour each day, write an equation that expresses the total distance as a function of the rate he can walk.

_______ 39. If Simon has lost 5 pounds, 3 pounds, and 4 pounds, during the first three weeks of his diet, write an equation that expresses his average weight loss as a function of the number of pounds he loses during the fourth week.

_______ 40. If Amanda scored 85, 72, and 90 on the first three tests during the semester, write an equation that expresses her test average for the semester as a function of the score she makes on the fourth test.

_______ 41. Venton buys a bedroom suite for $150 down and $55 a month. Write an equation that expresses the total cost as a function of the number of months he must make payments.

_______ 42. A consumer service advertises a stereo system for $33 down and only 10 monthly payments. Write an equation that expresses the total cost as a function of the amount of each monthly payment.

_______ 43. If the Fountain Players sells an adult's ticket for $5 and a child's ticket for $3, write a relation to state that the theater group wishes its total receipts from one night's performance to be at least $2000.

_______ **44.** Mr. Shouse invested a certain amount of money at 8.5 percent and another amount at 9.75 percent. Write a relation that states Mr. Shouse wishes his income from the interest to be at least $1500.

☆ Stretching the Topics

_______ **1.** If y varies directly as x^2, what is the effect on y when x is multiplied by 3? What is the effect on x when y is multiplied by 4?

_______ **2.** The weight that can be safely supported by a beam with rectangular ends varies jointly as the width and the square of the length of the rectangle and inversely as the length of the beam. If a 4-inch by 6-inch beam that is 15 feet long can support a 1200-pound load, what weight can be supported by a 2-inch by 5-inch beam that is 20 feet long?

_______ **3.** The profit (in dollars) for manufacturing a certain item is given by $P(x) = -500 + 0.5x + x^2$, where x is the number of items manufactured. Find $P(20)$ and interpret your answer.

Check your answers in the back of your book.

If you can complete **Checkup 8.6,** you are ready to do the **Review Exercises** for Chapter 8.

Name ______________________ Date ____________

✓ CHECKUP 8.6

______ 1. The amount of stretching of a spiral spring varies directly as the force applied. If 6 pounds of force stretched the spring $\frac{2}{3}$ inch, how many inches will 18 pounds of force stretch it?

______ 2. The rate in traveling a certain distance varies inversely as the time. If a car whose rate is 40 miles per hour travels a certain distance in 7 hours, how long will it take a car traveling 50 miles per hour to travel the same distance?

______ 3. Write an equation of variation and find k if w varies jointly as x and the square of y, and $w = 12$ when $x = 3$ and $y = 2$.

______ 4. Exhibitors at the county fair must pay a fee of $6 plus $1.75 for each item exhibited. Write an equation to describe an exhibitor's fee as a function of the number of items exhibited.

______ 5. The length of a rectangle is 5 feet less than five times its width. Write an equation that expresses the perimeter as a function of the width.

Check your answers in the back of your book.

If You Missed Problem:	You Should Review Examples:
1	1, 2
2	4
3	6
4	10
5	11

Summary

In this chapter we discovered that solutions for equations and inequalities containing two variables are **ordered pairs** of real numbers. We called the set of replacements for the independent variable the **domain** and the set of replacements for the dependent variable the **range.** Then we learned to distinguish between a **relation** and a **function.**

> **Relation:** any set of ordered pairs.
> **Function:** a set of ordered pairs in which each element of the domain is paired with *one and only one* element of the range.

Using the Cartesian coordinate plane, we learned to graph ordered pairs of real numbers as points and to find certain information about two points (x_1, y_1) and (x_2, y_2) in the plane.

To Find	Use the Formula
Distance between the points	$d = \sqrt{(x_2 - x_1)^2 + (y_2 - y_1)^2}$
Midpoint of the line segment joining the points	$(x_M, y_M) = \left(\frac{x_1 + x_2}{2}, \frac{y_1 + y_2}{2}\right)$
Slope of the line through the points	$m = \frac{y_2 - y_1}{x_2 - x_1}$

We learned that the graph of a **linear equation** of the form $Ax + By = C$ (where A and B are not both zero) is always a straight line and we developed several techniques for graphing lines.

Method	Procedure
Arbitrary points	Choose at least three x-values and find the corresponding y-values. Plot the three points and draw the line.
Intercepts	Let $y = 0$ to find the x-intercept. Let $x = 0$ to find the y-intercept. Plot the two intercept points and draw the line.
Slope-intercept	Rewrite the equation in the form $y = mx + b$. Then use the y-intercept, b, and the slope, m, to graph the line.
Inspection	For $y = c$, a constant, the graph is a horizontal line crossing the y-axis at $(0, c)$. For $x = c$, a constant, the graph is a vertical line crossing the x-axis at $(c, 0)$.

After learning to graph straight lines from given equations, we discovered how to find the equation for a line when given certain information about that line.

Given	Equation to Use	Name
Slope: m y-intercept: b	$y = mx + b$	Slope-intercept equation
Slope: m Point: (x_1, y_1)	$y - y_1 = m(x - x_1)$	Point-slope equation
Two points: (x_1, y_1) and (x_2, y_2)	$m = \frac{y_2 - y_1}{x_2 - x_1} \quad x_1 \neq x_2$	Slope definition
	$y - y_1 = m(x - x_1)$	Point-slope equation
Line is horizontal	$y = c$	Constant function
Line is vertical	$x = c$	Constant relation

To graph **linear inequalities,** we agreed to *isolate* the dependent variable y, graph the corresponding equation as a solid or dashed line, and then shade the appropriate region.

If $y < mx + b$, the line is *dashed* and we shade *below* the line.
If $y \leq mx + b$, the line is *solid* and we shade *below* the line.
If $y > mx + b$, the line is *dashed* and we shade *above* the line.
If $y \geq mx + b$, the line is *solid* and we shade *above* the line.

Finally, we discussed function notation and learned to express verbal relationships between variables using functions and relations containing two variables. In particular, we learned to write equations describing *variation*.

Type of Variation	Statement of Variation	Equation
Direct	y varies directly as x	$y = kx$
Inverse	y varies inversely as x	$y = \frac{k}{x}$
Joint	y varies directly as x and z	$y = kxz$
Mixed	y varies directly as x and inversely as z	$y = \frac{kx}{z}$

❑ Speaking the Language of Algebra

Complete each sentence with the appropriate word or phrase.

1. Solutions for equations and inequalities containing two variables are ________ ________ .

2. In the ordered pair (x, y) we call x the ________ variable and y the ________ variable.

3. The set of permissible replacement values for the independent variable is called the ________ and the set of replacement values for the dependent variable is called the ________ .

4. The graph of $y = mx + b$ is a ________ ________ , for which m is the ________ and b is the ________ .

5. The graph of $y = c$, where c is a constant, is a ________ ________ . The graph of $x = c$, where c is a constant, is a ________ ________ .

6. To graph the inequality $y > 3x - 1$, we first graph $y = 3x - 1$ as a ________ line and then we shade ________ that line.

7. To find the equation of a line through two points, we must first find the ________ of the line.

8. For a linear equation, $Ax + By = C$ is called ________ form and $y = mx + b$ is called ________ form.

9. If $f(x) = 5x + 2$, then $f(3) =$ ________ .

10. The equation $y = kx$ describes ________ variation, while the equation $y = \frac{k}{x}$ describes ________ variation.

Writing About Mathematics

Write your response to each question in complete sentences.

1. Describe the Cartesian coordinate system to someone who is unfamiliar with it. Explain how to locate in the Cartesian plane a point corresponding to an ordered pair of numbers.

2. Explain how to graph the line $6x + 2y = 4$ using
   (a) The intercepts method.
   (b) The slope-intercept method.

3. Explain why
   (a) The slope of a horizontal line is zero.
   (b) The slope of a vertical line is undefined.

4. Explain the steps you would use to find the equation of the line through two given points.

5. Explain how to decide (from their equations in standard form) whether two lines are parallel, perpendicular, or neither.

6. In graphing linear inequalities in two variables, how do you decide whether the boundary line is solid or dashed?

7. Explain why the vertical-line test can be used to tell whether a graph represents a function.

Writing About Mathematics

Name ______________________________ Date ______________

REVIEW EXERCISES for Chapter 8

Graph using the arbitrary-point method.

1. $2x - y = -5$

2. $-3x + y = 0$

Graph using the intercept method.

3. $2x + 5y = 10$

4. $-4x + y = 2$

________ 5. Find the distance between the points $(-3, 2)$ and $(1, -1)$.

________ 6. Find the midpoint of the line segment joining $(-3, 8)$ and $(7, -5)$.

________ 7. Find the slope of the line through $(5, -3)$ and $(0, 7)$.

Graph the line through the given point with the given slope.

8. $(1, -2)$; $m = \frac{2}{3}$

9. $(-4, 2)$; $m = 0$

Use the slope-intercept method to graph each of the following.

10. $y = 3x + 2$

11. $2x - 4y = 1$

Graph using any method you choose.

12. $3x + 4y = 0$

13. $x - 2y = -6$

14. $2x + 4 = 0$

15. $5 - y = 0$

Find the equation of the line satisfying the given conditions.

________ 16. With $m = -3$, $b = -1$

________ 17. With $m = \frac{1}{2}$, $b = 0$

________ 18. Through $(-2, 2)$; m is undefined

________ 19. Through $(-1, 3)$; $m = \frac{3}{4}$

________ 20. Through $(-7, 2)$ and $(-5, 0)$

________ 21. Through $(\frac{2}{3}, -\frac{1}{2})$, and $(\frac{5}{3}, \frac{3}{2})$

________ 22. Through $(3, 2)$ and $(3, -5)$

________ 23. Through $(4, -5)$ and $(-2, -5)$

________ 24. Through $(-2, 1)$ and parallel to $2x - y = 6$

________ 25. Through $(6, 0)$ and perpendicular to $x + 2y = -1$

Graph each inequality.

26. $x \geq -3$

27. $y < -\frac{2}{3}x$

28. $2x - y \leq 6$

29. $3x - 5y < 15$

For each of these relations give the domain and range.

_________ **30.** $\{(-2, 2), (-1, 1), (0, 0), (1, -1), (2, -2)\}$

_________ **31.** $\{(x, y) : y = 3\}$

Use the vertical-line test to tell which of the following graphs represent functions.

_________ **32.**

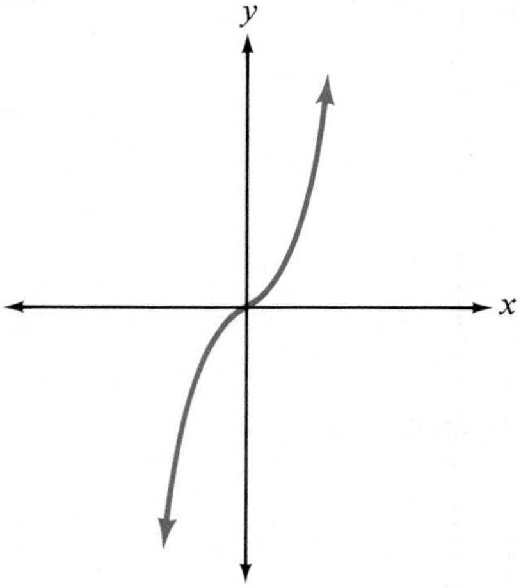

_________ **33.**

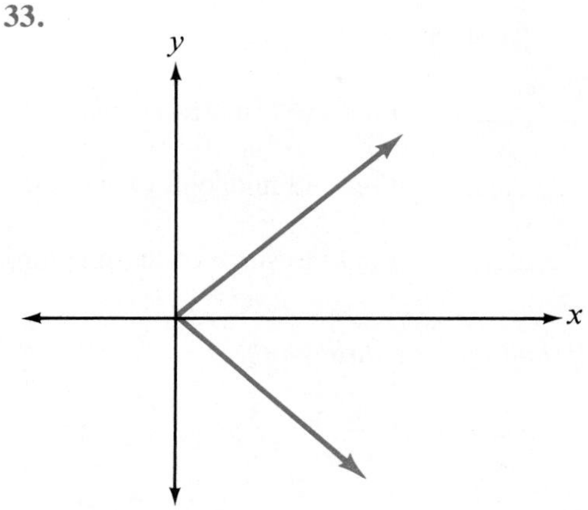

If $f(x) = 2x - 1$ *and* $g(x) = x^2 - 9$, *find each of the following.*

_________ **34.** $f(-3)$

_________ **35.** $f(2) - g(-1)$

In Exercises 36 and 37 give the domain.

_________ **36.** $f(x) = \dfrac{x - 1}{x^2 - 4}$

_________ **37.** $f(x) = \sqrt{3 - x}$

_________ **38.** Write an equation of variation and find k if y varies directly as x, and $y = 56$ when $x = 8$.

_________ **39.** Write an equation of variation and find k if y varies inversely as x, and $y = \frac{1}{2}$ when $x = 6$.

_________ **40.** Write an equation of variation and find k if z varies jointly as x and the square of y, and $z = \frac{1}{4}$ when $x = 9$ and $y = 4$.

_________ **41.** A certain variable T varies directly as D and inversely as R. Write a function describing this relationship and find k if $T = 300$ when $D = 20$ and $R = 4$. Then find T when $D = 60$ and $R = 9$.

_________ **42.** The cost of copying a pamphlet varies directly with the number of pages in it. If the cost of copying a pamphlet of 50 pages is \$3.50, how much will it cost to copy a pamphlet of 80 pages?

________ **43.** The distance a body falls varies directly as the square of the time it has been falling. If an object falls 400 feet in 5 seconds, how far will it fall in 12 seconds?

________ **44.** If the length of a rectangle is 3 feet less than twice the width, write an equation that expresses the area as a function of the width.

Check your answers in the back of your book.

If You Missed Exercises:	You Should Review Examples:	
1, 2	SECTION 8.1	6–8
3, 4		9, 10
5	SECTION 8.2	1, 2
6		3, 4
7		5–8
8, 9		9–12
10, 11		13–15
12–15	SECTIONS 8.1 and 8.2	
16, 17	SECTION 8.3	1–4
18, 19		5–8
20–23		9–12
24		13, 14
25		15
26–29	SECTION 8.4	1–4
30, 31	SECTION 8.5	1, 2
32, 33		3
34, 35		4
36, 37		5–8
38	SECTION 8.6	2
39		4
40		6
41		9
42		2
43		3
44		11

If you have completed the **Review Exercises** and corrected your errors, you are ready to take the **Practice Test** for Chapter 8.

Name ________________________________ Date ______________

PRACTICE TEST for Chapter 8

	Section	Examples
1. Use the arbitrary-point method to graph $2x + y = 5$.	8.1	1, 2

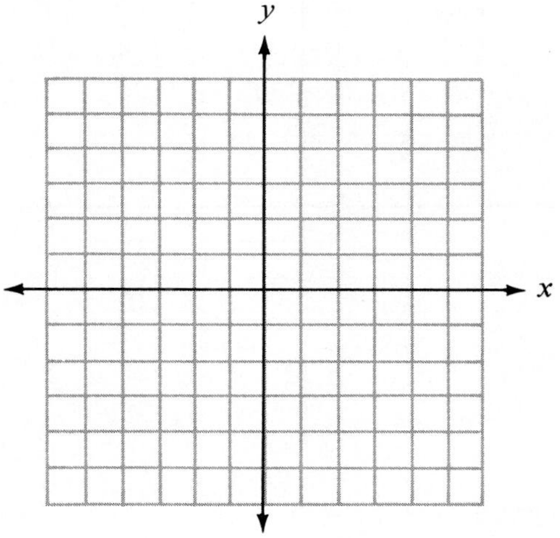

	Section	Examples
2. Use the intercepts method to graph $4x - 3y = 9$.	8.1	9

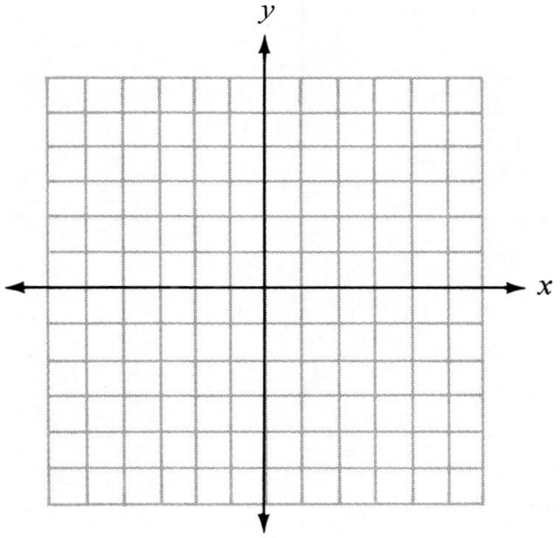

	Section	Examples
________ 3. Find the distance between the points $(-2, 3)$ and $(4, 5)$.	8.2	1, 2
________ 4. Find the midpoint of the line segment joining $(0, 5)$ and $(-3, 2)$.	8.2	3, 4
________ 5. Find the slope of the line through the points $(3, -1)$ and $(7, -5)$.	8.2	6

	Section	Example
6. Graph the line through (1, -3) with slope $m = \frac{4}{3}$.	8.2	9

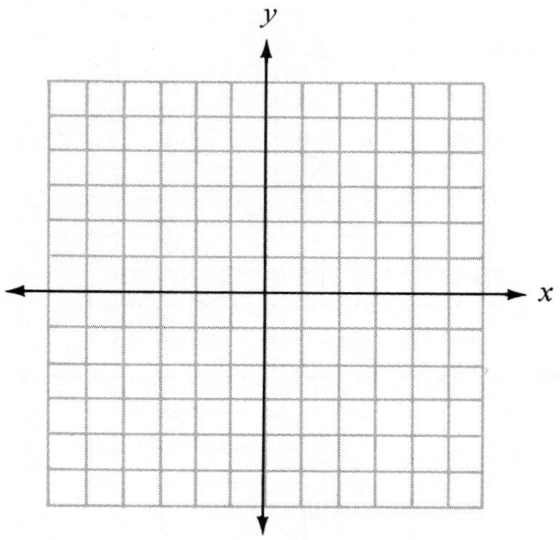

7. Use the slope-intercept method to graph $2x + 5y = 5$.	8.2	13

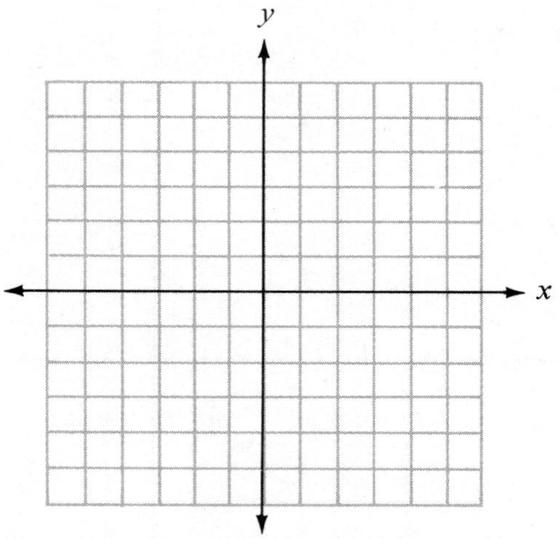

8. Graph $y - 3x = 0$ using any method you choose.	8.1	7

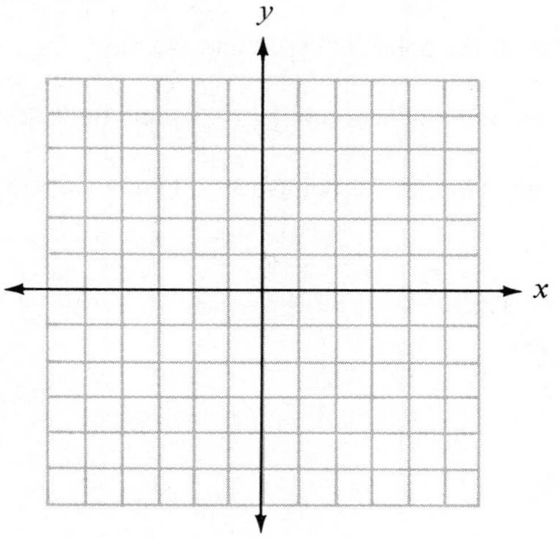

	SECTION	EXAMPLE
9. Graph $2y + 6 = 0$ using any method you choose.	8.2	15

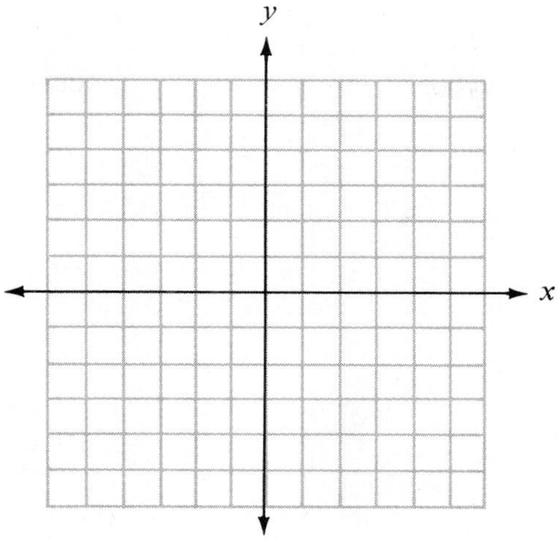

Find the equation of the line satisfying the given conditions.

	SECTION	EXAMPLE
________ **10.** Through the origin with slope $m = -2$	8.3	3
________ **11.** Through the point $(-6, 1)$ with slope $m = -\frac{2}{3}$	8.3	6
________ **12.** Through the points $(5, -3)$ and $(5, 1)$	8.3	12
________ **13.** Through the points $(2, -4)$ and $(5, -2)$	8.3	9
________ **14.** Through $(3, -5)$ and parallel to $x - y = 1$	8.3	13
________ **15.** Through $(-1, 4)$ and perpendicular to $2x + y = -1$	8.3	15
16. Graph $x \leq 1$.	8.4	1

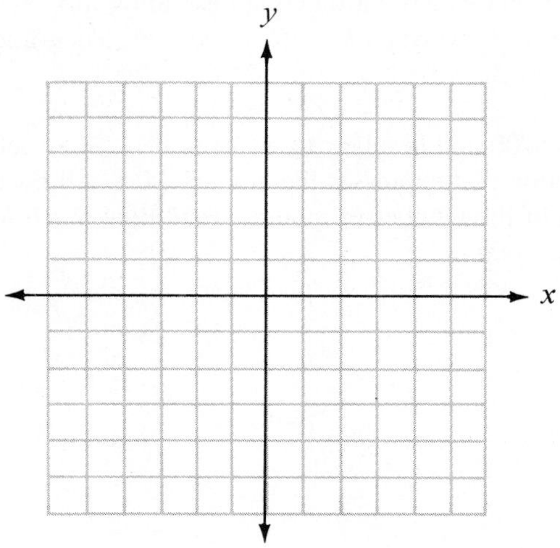

		Section	Examples
	17. Graph $4x - y < -2$.	8.4	2

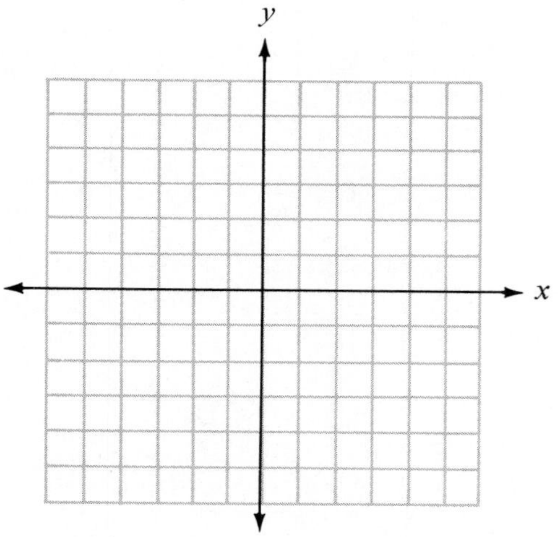

		Section	Examples
________	18. For the relation $\{(x, y) : y = 8\}$ give the domain and range. Is this a function?	8.5	1, 2
________	19. If $f(x) = -4x + 1$, find $f(-\frac{3}{2})$.	8.5	4
________	20. If $g(x) = x^2 - 2x - 3$, find $g(-1)$.	8.5	4
________	21. Give the domain for $f(x) = \sqrt{2x - 1}$.	8.5	8
________	22. Give the domain for $g(x) = \frac{x}{4x^2 + 3x}$.	8.5	6
________	23. Write an equation of variation and find k if y varies directly with the square of x, and $y = 18$ when $x = -3$.	8.6	3
________	24. If A varies jointly as b and h, write a function describing this relationship. Find k if $A = 10$ when $b = 5$ and $h = 4$. Then find A when $b = 7$ and $h = 5$.	8.6	5
________	25. If Max owes Charles \$700 and he plans to pay him back \$25 each month, write an equation that expresses the amount Max still owes Charles as a function of the number of months. Find how much Max owes Charles after one year.	8.6	10

Name ______________________ Date ______________

SHARPENING YOUR SKILLS after Chapters 1–8

		Section
________	1. Find $(-3)^{-3}$	3.2
________	2. Write $7 - \sqrt{-12}$ in standard complex form.	7.5
________	3. Find i^{36}.	7.5
________	4. Use the discriminant to describe the solutions for $x^2 - 2x + 5 = 0$.	5.3
________	5. Write $8a^{\frac{1}{3}}$ in radical form.	7.1
________	6. Write $\sqrt[3]{(a - b)^4}$ in exponential form.	7.1

Perform the indicated operations.

________	7. $5\left[-10 + \dfrac{-18 + 3}{-4 + 9}\right] - 6$	1.3
________	8. $x - 9[x - 7(2x - 1)]$	1.4
________	9. $\left(\dfrac{13x^2y^7}{9z^3}\right)^0$	3.1
________	10. $(a^{-10}b^7c^{-3})(a^{12}b^{-9}c^4)$	3.2
________	11. $3ab(a^{-1}b^2 - 5a^2b^{-1}) + 4a^7b^{-2}(a^{-4}b^5)$	3.3
________	12. $(x^2 - 7x + 12)(x^2 - x - 12)$	3.4
________	13. $\dfrac{2x^2 - 13x - 24}{x^2 - 64} \div \dfrac{4x^2 - 9}{x + 8}$	6.2
________	14. $\dfrac{x + 8}{x^2 + 3x - 10} + \dfrac{4}{x + 5}$	6.3
________	15. $\dfrac{1 + x^{-1}}{1 - x^{-2}}$	6.4
________	16. $\sqrt{48x} - \sqrt{27x} + \sqrt{300x}$	7.2
________	17. $(\sqrt{3x} + 5)(\sqrt{3x} - 5)$	7.3
________	18. $\dfrac{2 - 3i}{4 - 5i}$	7.5

Factor completely.

		Section
________	19. $x^3 + 125$	4.2
________	20. $9x^4 - 30x^2 + 25$	4.3
________	21. $63a^2 - 14ab - 45a + 10b$	4.3

Solve each equation or inequality.

________	22. $4(3x - 1) + 1 = 3(9 - x)$	2.1
________	23. $\left\lvert \frac{x}{3} + 5 \right\rvert \geq 10$	2.4
________	24. $11x^3 - 25x^2 + 6x = 0$	5.1
________	25. $\frac{3}{x - 9} - \frac{4}{x + 5} = \frac{2}{2x + 10}$	6.5
________	26. $\sqrt{5x - 21} + \sqrt{2x} = 0$	7.4
________	27. $x^2 + 20 \leq 12x$	5.4
________	28. Solve and graph $-2 \leq \frac{4a - 6}{2} < 0$.	2.3
________	29. George has doubled the lengths of two opposite sides of his old square garden and has increased the other two sides by 4 feet. If the difference between the areas of the old garden and new garden is 560 square feet, find the dimensions of George's new garden.	5.5

CHAPTER

9

Graphing More Relations and Functions

If a person throws a ball directly upward at a speed of 96 feet per second, the ball's height y feet above the ground after x seconds is given by the function $y = -16x^2 + 96x$. Use the graph of this function to find the maximum height of the ball.

Although we have glanced briefly at a few graphs that were not straight lines, until now we have concentrated mainly on linear functions and relations. In this chapter we graph sets of ordered pairs satisfying functions and relations that are *not* linear. In particular, we learn to

1. Graph any function or relation using arbitrary points.
2. Graph quadratic functions.
3. Graph the conic sections.
4. Graph second-degree inequalities.

9.1 Graphing Functions by Arbitrary Points

Recall that our first method for graphing linear functions required that we find several ordered pairs that satisfied the given function. This method of arbitrary points will be even more important to us as we attempt to graph nonlinear functions and relations. Although we used at most three points to graph a straight line, more points are necessary for graphs of most other relations, especially if we have no idea about the appearance of the graph. No matter what relation we are graphing, we shall always be interested in finding the *intercepts* for the graph whenever possible.

Suppose that we try to graph several sets of ordered pairs, using tables to organize our work and labeling the points we find.

Example 1. Graph $y = x^2$.

Solution. Choose several x-values and find corresponding y-values.

x	$y = x^2$	(x, y)
-3	$(-3)^2 = 9$	$(-3, 9)$
-2	$(-2)^2 = 4$	$(-2, 4)$
-1	$(-1)^2 = 1$	$(-1, 1)$
0	$0^2 = 0$	$(0, 0)$
1	$1^2 = 1$	$(1, 1)$
2	$2^2 = 4$	$(2, 4)$
3	$3^2 = 9$	$(3, 9)$

Plotting these points and connecting them with a smooth curve, we have our graph.

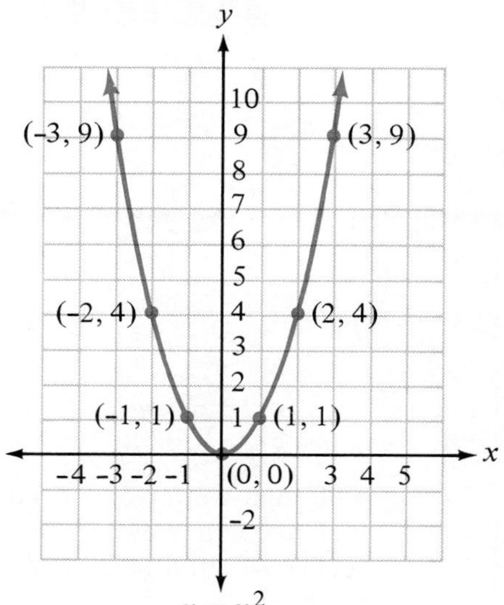

$y = x^2$

A graph with this shape is called a **parabola.** Is this a function? Since each x-value corresponds to one and only one y-value (remember the vertical-line test), we agree that $y = x^2$ describes a function. In particular, it is a second-degree or **quadratic function.**

From the graph of $y = x^2$ we can determine the domain and range for the function.

Domain: $x \in R$
Range: $y \in R$ and $y \geq 0$

Example 2. Graph $y = 2x - x^2$. Find x- and y-intercepts, construct a table of ordered pairs, and decide whether this is a function. Then state its domain and range.

Solution

$$y = 2x - x^2$$

y-INTERCEPT: Let $x = 0$	x-INTERCEPTS: Let $y = 0$
$y = 2(0) - 0^2$	$0 = 2x - x^2$
$y = 0$	$0 = x(2 - x)$
$(0, 0)$	$x = 0$ or $2 - x = 0$
	$x = 2$
	$(0, 0)$ $(2, 0)$

x	$y = 2x - x^2$	(x, y)
-1	$y = 2(-1) - (-1)^2$ $= -2 - 1 = -3$	$(-1, -3)$
0	$y = 0$	$(0, 0)$
1	$y = 2(1) - (1)^2$ $= 2 - 1 = 1$	$(1, 1)$
2	$y = 0$	$(2, 0)$
3	$y = 2(3) - (3)^2$ $= 6 - 9 = -3$	$(3, -3)$

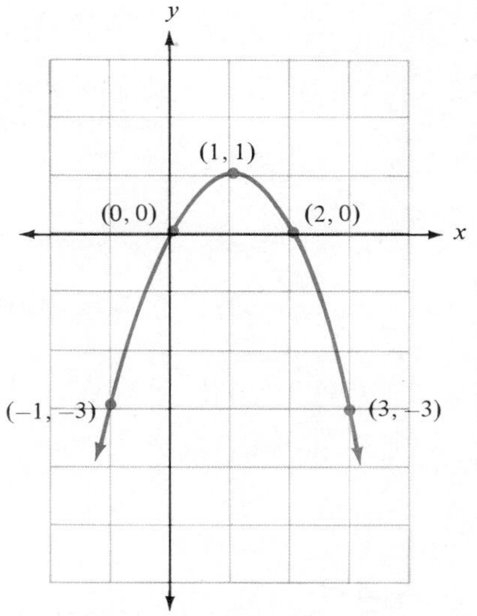

Once again our graph is a parabola (opening *down* in this example). This is certainly a function; in particular, it is another quadratic function. From the graph we see

Domain: $x \in R$
Range: $y \in R$ and $y \leq 1$

Example 3. Graph $y = x^3$ using a table of ordered pairs. Tell whether it is a function, and state its domain and range.

Solution

$$y = x^3$$

y-INTERCEPT: Let $x = 0$

$$y = 0^3 = 0$$

$$(0, 0)$$

x-INTERCEPT: Let $y = 0$

$$0 = x^3$$

$$0 = x$$

$$(0, 0)$$

x	$y = x^3$	(x, y)
-2	$y = (-2)^3 = -8$	$(-2, -8)$
-1	$y = (-1)^3 = -1$	$(-1, -1)$
0	$y = 0$	$(0, 0)$
1	$y = 1^3 = 1$	$(1, 1)$
2	$y = 2^3 = 8$	$(2, 8)$

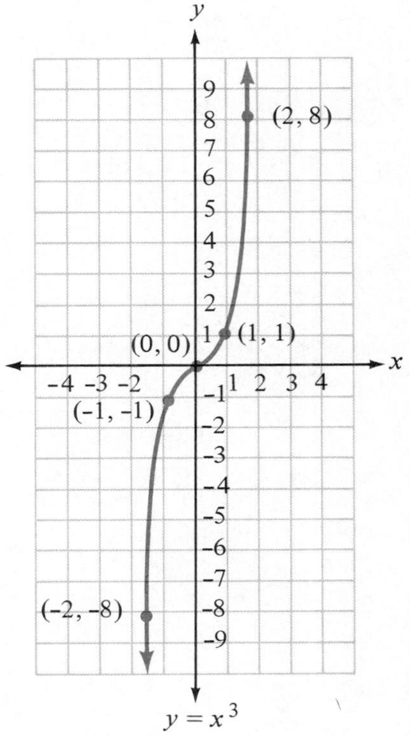

Is $y = x^3$ a function? Yes; it is a third-degree or **cubic function** and its graph is sort of S-shaped. From the graph we state

Domain: $x \in R$
Range: $y \in R$

Now you try Example 4.

Example 4. Graph $y = x^3 - x$. Find the intercepts, construct a table of ordered pairs, and decide whether this is a function. Then state its domain and range.

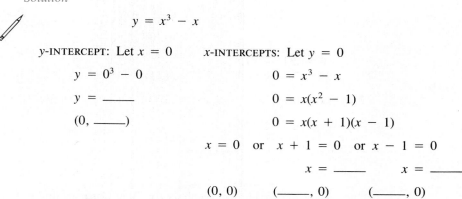

Solution

$$y = x^3 - x$$

y-INTERCEPT: Let $x = 0$

$y = 0^3 - 0$

$y =$ ____

$(0,$ ____$)$

x-INTERCEPTS: Let $y = 0$

$0 = x^3 - x$

$0 = x(x^2 - 1)$

$0 = x(x + 1)(x - 1)$

$x = 0$ or $x + 1 = 0$ or $x - 1 = 0$

$x =$ ____ $\quad x =$ ____

$(0, 0) \quad ($____$, 0) \quad ($____$, 0)$

x	$y = x^3 - x$	(x, y)
-2	$y = (-2)^3 - (-2)$ $= -8 + 2 =$ ____	$(-2,$ ____ $)$
-1	$y = 0$	$(-1, 0)$
$-\frac{1}{2}$	$y = (-\frac{1}{2})^3 - (-\frac{1}{2})$ $= -\frac{1}{8} + \frac{4}{8} =$ ____	$(-\frac{1}{2},$ ____$)$
0	$y = 0$	$(0, 0)$
$\frac{1}{2}$	$y = (\frac{1}{2})^3 - \frac{1}{2}$ $= \frac{1}{8} - \frac{4}{8} =$ ____	$(\frac{1}{2},$ ____$)$
1	$y = 0$	$(1, 0)$
2	$y = 2^3 - 2$ $= 8 - 2 =$ ____	$(2,$ ____$)$

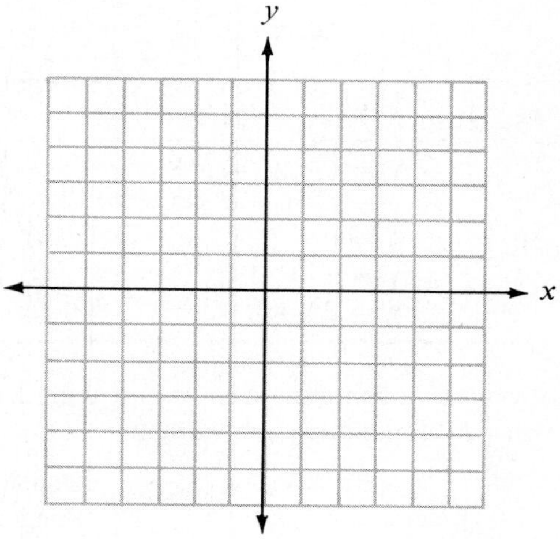

Is $y = x^3 - x$ a function? ________ . From the graph we state

Domain: ________
Range: ________

Check your work on page 617. ▶

Example 5. Graph $y = \frac{1}{x}$.

Solution. We immediately notice the restriction on the domain and agree that $x \neq 0$. Our graph will *not* cross the y-axis.

y-INTERCEPT: None, because $x \neq 0$.

x-INTERCEPT: Let $y = 0$

$$0 = \frac{1}{x}$$

$$0 \cdot x = 1$$

$$0 = 1$$

This false statement tells us there is *no* x-intercept. Our graph will *not* cross either axis.

x	$y = \frac{1}{x}$	(x, y)
-3	$y = \frac{1}{-3} = -\frac{1}{3}$	$(-3, -\frac{1}{3})$
-1	$y = \frac{1}{-1} = -1$	$(-1, -1)$
$-\frac{1}{2}$	$y = \frac{1}{-\frac{1}{2}} = 1(-\frac{2}{1}) = -2$	$(-\frac{1}{2}, -2)$
$\frac{1}{2}$	$y = \frac{1}{\frac{1}{2}} = 1 \cdot \frac{2}{1} = 2$	$(\frac{1}{2}, 2)$
1	$y = \frac{1}{1} = 1$	$(1, 1)$
3	$y = \frac{1}{3}$	$(3, \frac{1}{3})$

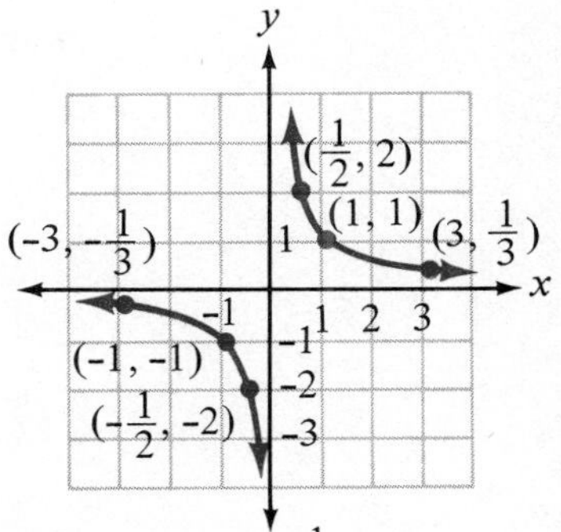

$y = \frac{1}{x}$

We see that $y = \frac{1}{x}$ is a function; in particular, it is called a **rational function.** From the graph, we can state

Domain: $x \in R$ and $x \neq 0$
Range: $y \in R$ and $y \neq 0$

Trial Run

Graph each function. Then state its domain and range.

Domain: ________ **1.** $y = x^2 - 4$
Range: ________

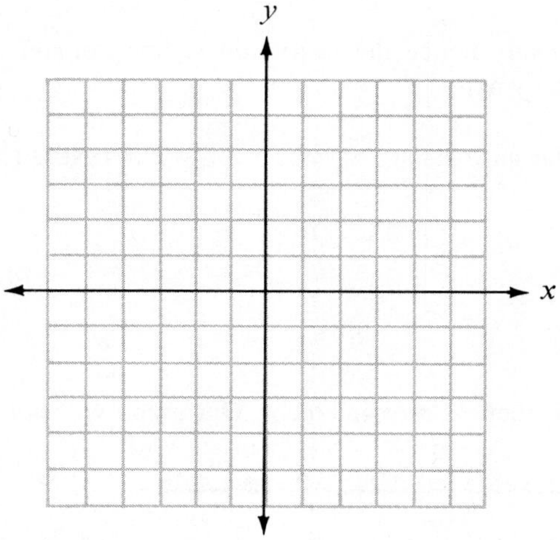

Domain: ________ **2.** $y = 1 - x^2$
Range: ________

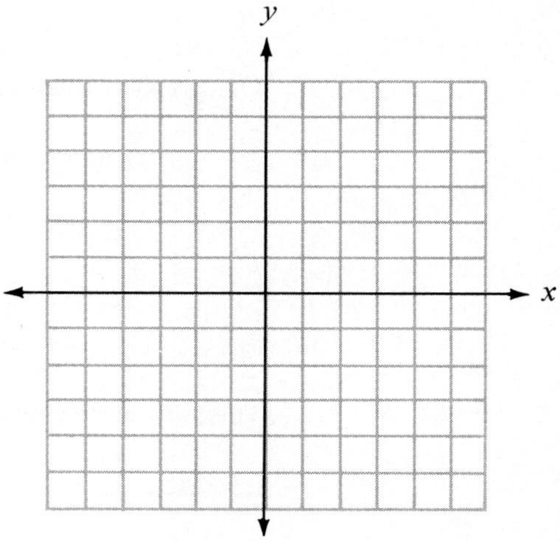

Domain: ________ **3.** $y = x^3 - 1$
Range: ________

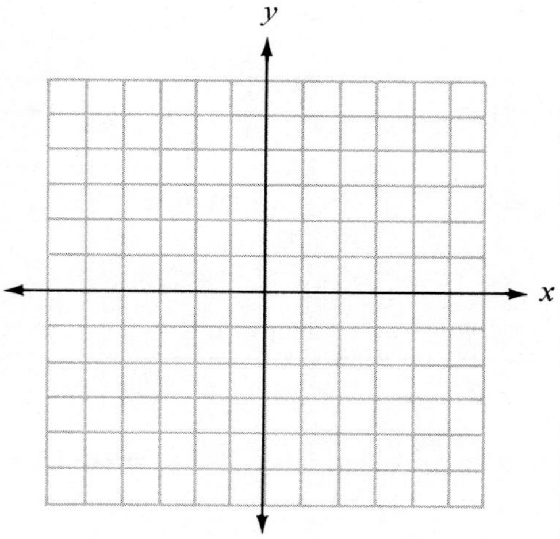

Domain: ________ **4.** $y = \frac{2}{x}$
Range: ________

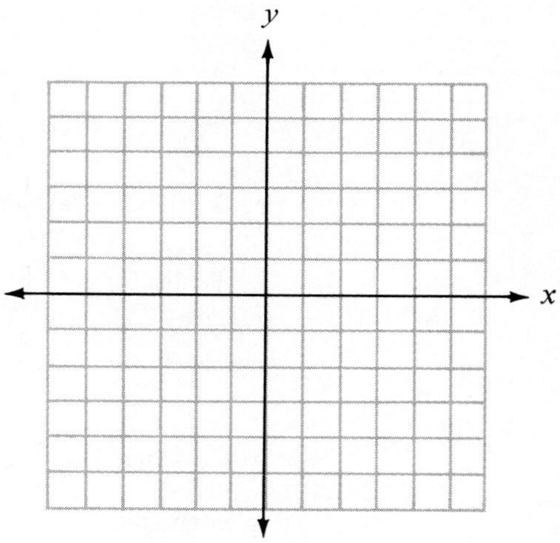

Answers are on page 618.

▶ Example You Completed

Example 4. Graph $y = x^3 - x$. Find the intercepts, construct a table of ordered pairs, and decide whether this is a function. Then state its domain and range.

Solution

$$y = x^3 - x$$

y-INTERCEPT: Let $x = 0$

$$y = 0^3 - 0$$
$$y = 0$$
$$(0, 0)$$

x-INTERCEPTS: Let $y = 0$

$$0 = x^3 - x$$
$$0 = x(x^2 - 1)$$
$$0 = x(x + 1)(x - 1)$$
$$x = 0 \quad \text{or} \quad x + 1 = 0 \quad \text{or} \quad x - 1 = 0$$
$$x = -1 \qquad x = 1$$
$$(0, 0) \qquad (-1, 0) \qquad (1, 0)$$

x	$y = x^3 - x$	(x, y)
-2	$y = (-2)^3 - (-2)$ $= -8 + 2 = -6$	$(-2, -6)$
-1	$y = 0$	$(-1, 0)$
$-\frac{1}{2}$	$y = (-\frac{1}{2})^3 - (-\frac{1}{2})$ $= -\frac{1}{8} + \frac{4}{8} = \frac{3}{8}$	$(-\frac{1}{2}, \frac{3}{8})$
0	$y = 0$	$(0, 0)$
$\frac{1}{2}$	$y = (\frac{1}{2})^3 - \frac{1}{2}$ $= \frac{1}{8} - \frac{4}{8} = -\frac{3}{8}$	$(\frac{1}{2}, -\frac{3}{8})$
1	$y = 0$	$(1, 0)$
2	$y = 2^3 - 2$ $= 8 - 2 = 6$	$(2, 6)$

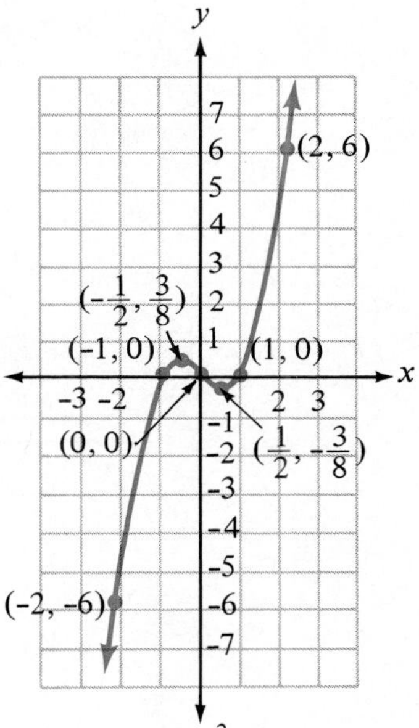

$y = x^3 - x$

Is $y = x^3 - x$ a function? Yes. From the graph we state

Domain: $x \in R$
Range: $y \in R$

Answers to Trial Run

page 616 **1.**

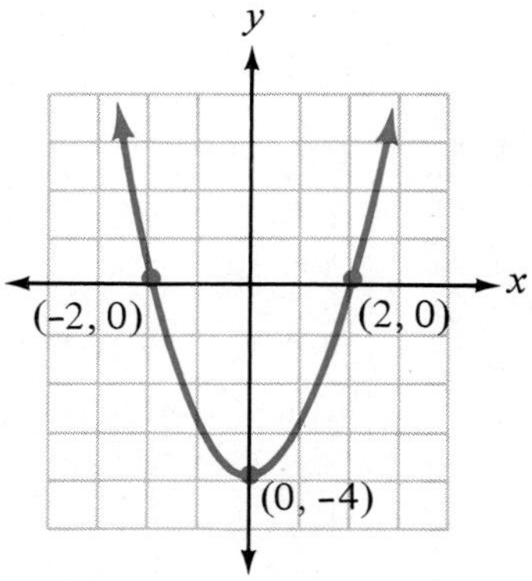

Domain: $x \in R$
Range: $y \in R$ and $y \geq -4$

2.

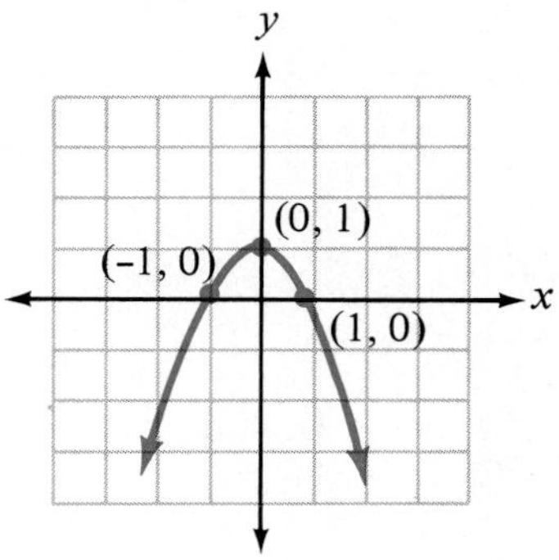

Domain: $x \in R$
Range: $y \in R$ and $y \leq 1$

3.

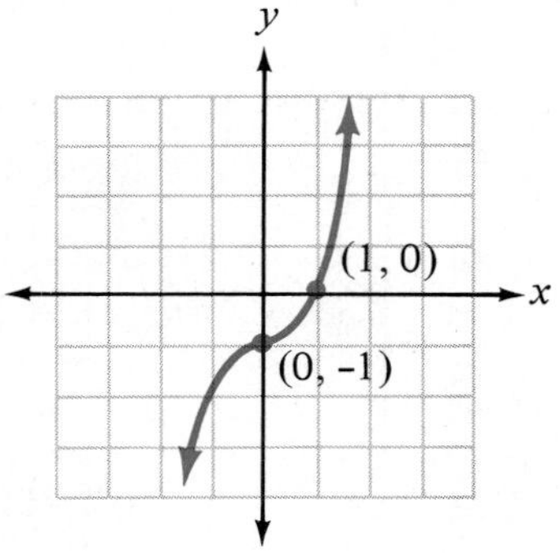

Domain: $x \in R$
Range: $y \in R$

4.

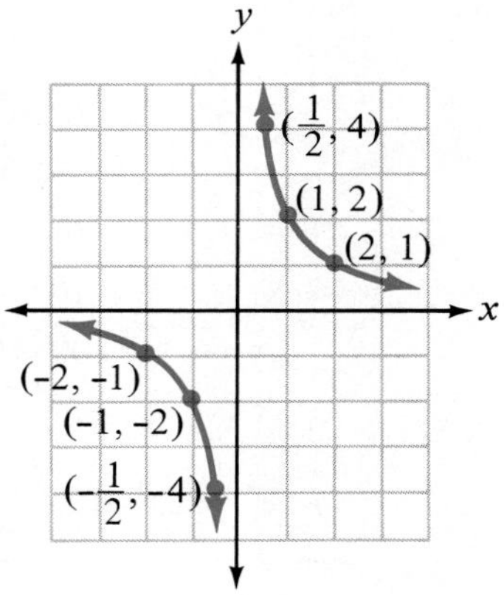

Domain: $x \in R$ and $x \neq 0$
Range: $y \in R$ and $y \neq 0$

Name ______________________ Date ____________

EXERCISE SET 9.1

Graph each function. State the domain and range of each.

Domain: ________ 1. $y = x^2 - 9$
Range: ________

Domain: ________ 2. $y = x^2 - 16$
Range: ________

Domain: ________ 3. $y = -x^2 + 1$
Range: ________

Domain: ________ 4. $y = -x^2 + 4$
Range: ________

Domain: ________ 5. $y = x^2 - 3$
Range: ________

Domain: ________ 6. $y = x^2 - 5$
Range: ________

Domain: ________ 7. $y = x^2 - 2x$
Range: ________

Domain: ________ 8. $y = x^2 + 4x$
Range: ________

Domain: ________ 9. $y = x^3 + 1$
Range: ________

Domain: ________ 10. $y = x^3 - 8$
Range: ________

Domain: ________ 11. $y = -x^3 - 1$
Range: ________

Domain: ________ 12. $y = -x^3 + 8$
Range: ________

Domain: ________ 13. $y = x^3 - 16x$
Range: ________

Domain: ________ 14. $y = x^3 - 4x$
Range: ________

Domain: ________ 15. $y = \frac{3}{x}$
Range: ________

Domain: ________ 16. $y = \frac{4}{x}$
Range: ________

Domain: ________ 17. $y = \frac{-2}{x}$
Range: ________

Domain: ________ 18. $y = \frac{-1}{x}$
Range: ________

☆ Stretching the Topics

________ 1. Graph $y = \frac{2}{x - 2}$.

________ 2. Graph $y = x^3 - x^2 - 6x$.

Check your answers in the back of your book.

If you can graph the functions in **Checkup 9.1,** you are ready to go on to Section 9.2.

Name ______________________ Date ____________

✓ CHECKUP 9.1

Graph each function. State the domain and range of each.

Domain: ________ **1.** $y = x^2 - 5$
Range: ________

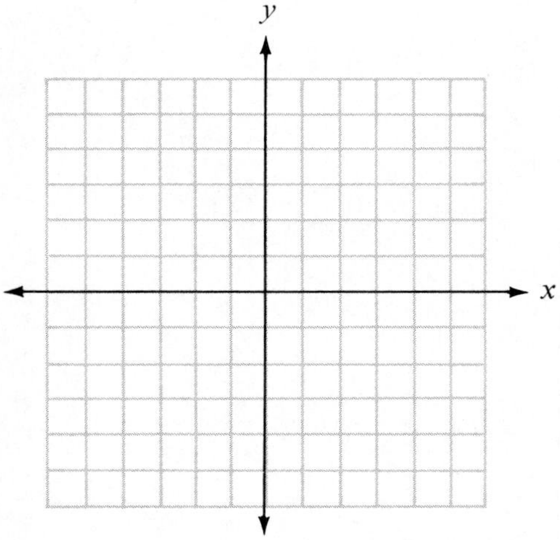

Domain: ________ **2.** $y = -2x^2 + 8$
Range: ________

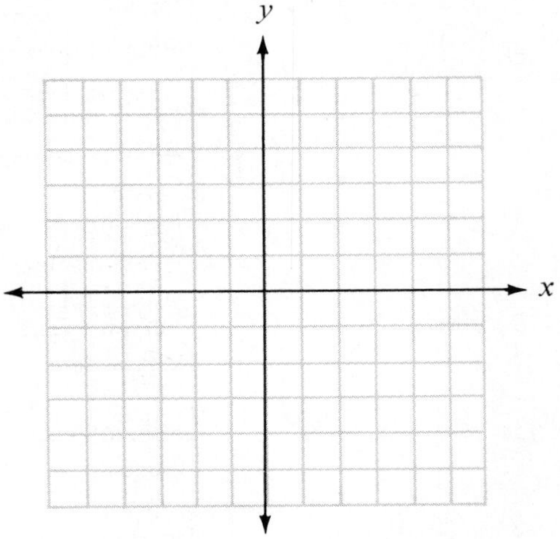

Domain: ________ **3.** $y = -x^3$
Range: ________

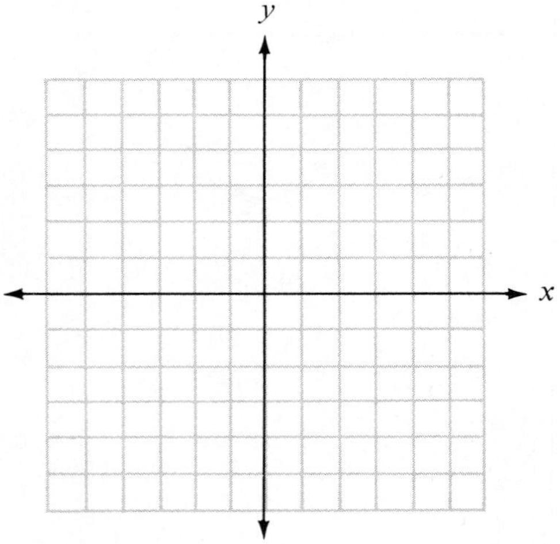

Domain: ________ **4.** $y = x^3 - 16x$
Range: ________

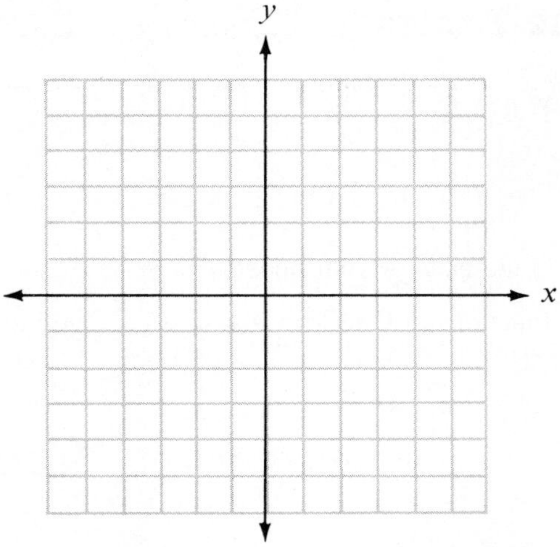

Domain: ________ 5. $y = \dfrac{-4}{x}$
Range: ________

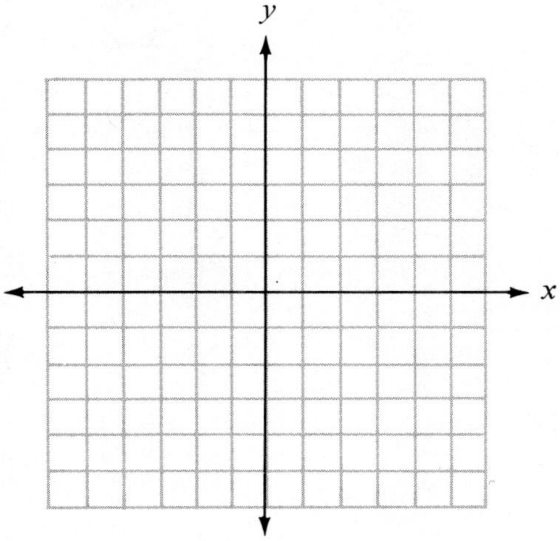

Check your answers in the back of your book.

If You Missed Problems:	You Should Review Examples:
1, 2	1, 2
3	3
4	4
5	5

9.2 Graphing Quadratic Functions

In earlier chapters we learned to solve quadratic equations of standard form

$$ax^2 + bx + c = 0$$

Now we turn our attention to graphing sets of ordered pairs satisfying equations of the form

$$y = ax^2 + bx + c$$

Because each x-value substituted into such an equation will yield one and only one y-value, you should agree that the resulting ordered pairs represent a *function*.

Quadratic Function

$$y = ax^2 + bx + c \qquad (a \neq 0)$$

We note that $a \neq 0$ in such a function or else the polynomial would contain no x^2-term and the function would be *linear* rather than quadratic.

Graphing Quadratic Functions: $y = ax^2 + bx + c$

In Examples 1 and 2 of Section 9.1 we discovered that the graph of a quadratic function is a *parabola*. Let's use those examples to make some general observations about parabolas that are graphs of quadratic functions.

Characteristics of Graphs of Quadratic Functions (Parabolas)

1. Every parabola opens either upward or downward.
2. There is a *vertical* line, called the **axis of symmetry,** that passes down the middle of a parabola. If the Cartesian plane were folded along this axis of symmetry, the left-hand portion of the parabola would lie exactly on top of the right-hand portion.
3. Every parabola has either a low point (if it opens upward) or a high point (if it opens downward). This point is called the **vertex** of the parabola, and it *always* lies on the axis of symmetry.

Example 1. Sketch the graphs of $y = x^2$ and $y = 2 - x^2$ and observe their characteristics.

Solution

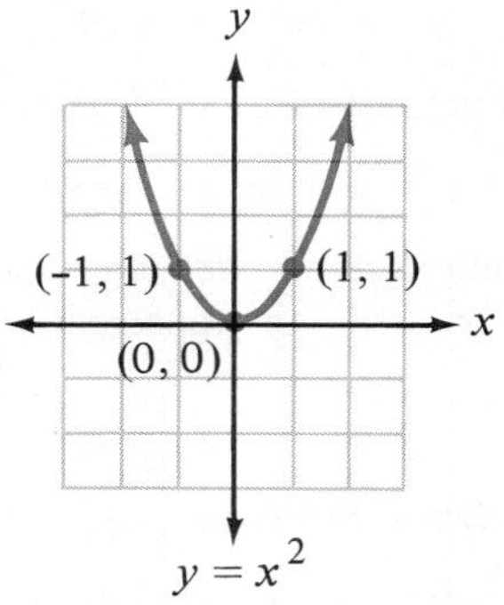

(a)

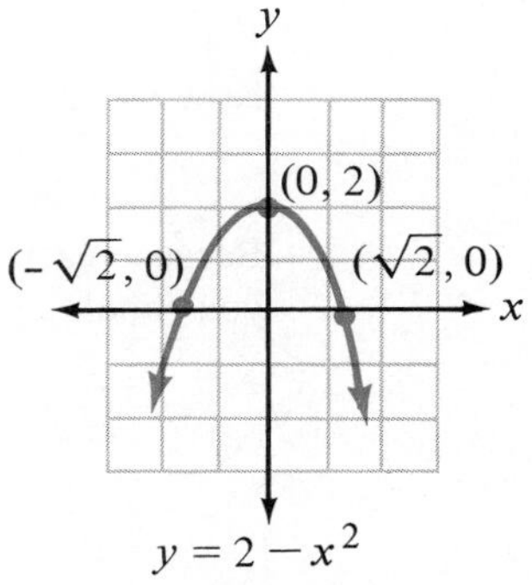

(b)

For $y = x^2$

1. The parabola opens upward.
2. The axis of symmetry is the vertical line $x = 0$ (the y-axis).
3. The vertex is the point (0, 0). It is the function's **minimum** (low) point.

For $y = 2 - x^2$

1. The parabola opens downward.
2. The axis of symmetry is the vertical line $x = 0$ (the y-axis).
3. The vertex is the point (0, 2). It is the function's **maximum** (high) point.

Having graphed several parabolas, we observe that whether a parabola opens upward or downward can be determined by the *sign* of a, the coefficient of the x^2-term in the equation $y = ax^2 + bx + c$.

> If a is *positive,* the parabola opens *upward.* (See $y = x^2$.)
> If a is *negative,* the parabola opens *downward.* (See $y = 2 - x^2$.)

Now that we have made several observations about the nature of parabolas, let's develop a procedure for graphing any quadratic function of the form

$$y = ax^2 + bx + c$$

Our approach will be very visual and will rely heavily on the observations we have made.

Let's try to graph the function

$$y = x^2 - 4x$$

We immediately note that this parabola opens *upward* because the coefficient of the x^2-term is positive. Let's find the intercepts.

y-INTERCEPT: Let $x = 0$

$$y = 0^2 - 4(0)$$
$$y = 0$$
$$(0, 0)$$

x-INTERCEPTS: Let $y = 0$

$$0 = x^2 - 4x$$
$$0 = x(x - 4)$$
$$x = 0 \quad \text{or} \quad x - 4 = 0$$
$$x = 4$$
$$(0, 0) \qquad (4, 0)$$

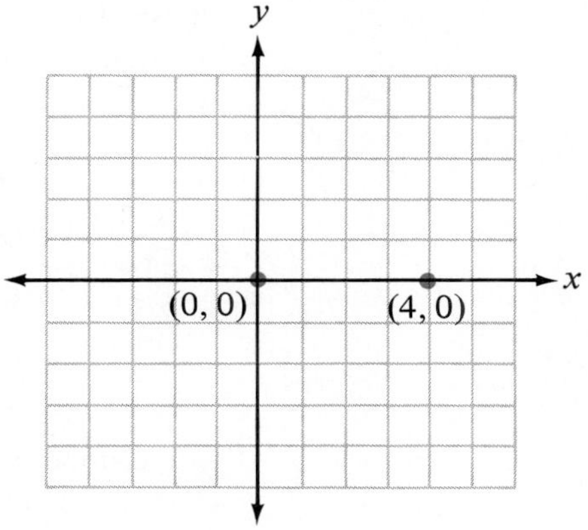

From our understanding of the *axis of symmetry,* we know it is a vertical line halfway between the two x-intercepts. The x-coordinate of the point halfway between (0, 0) and (4, 0) is 2, so the axis of symmetry must be the vertical line $x = 2$.

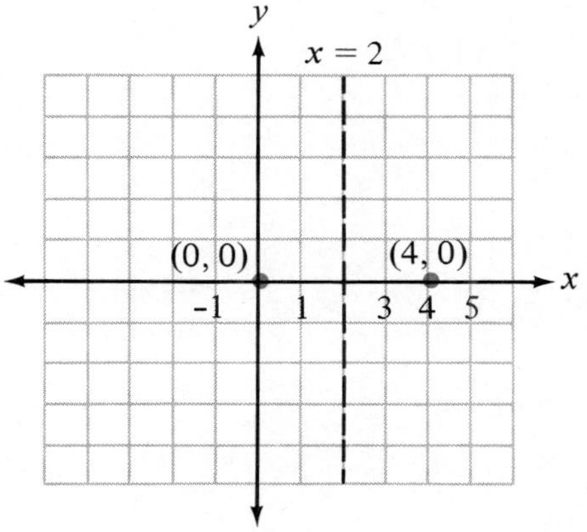

Moreover, we know that the *vertex* of the parabola must lie on the axis of symmetry, $x = 2$. Therefore, the x-coordinate for the vertex must be 2 and we can find the y-coordinate by substituting 2 for x in the equation for the parabola: $y = x^2 - 4x$.

$$y = 2^2 - 4(2)$$
$$= 4 - 8$$
$$= -4$$

The vertex is the point $(2, -4)$.

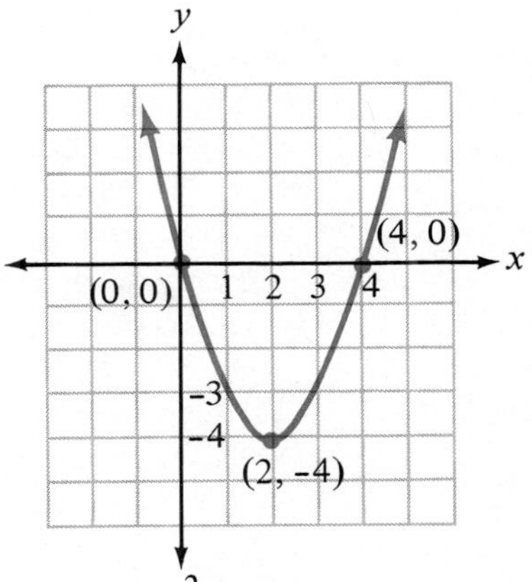

$y = x^2 - 4x$

Domain: $x \in R$
Range: $y \in R$ and $y \geq -4$

It is worth mentioning here that we can use part of the *midpoint formula* to find the x-coordinate of the point halfway between two known points on a parabola.

Midpoint Formula

$$x_M = \frac{x_1 + x_2}{2}$$

If you can locate the halfway point by looking, then by all means do so. We shall reserve the midpoint formula for situations in which the halfway point is difficult to locate on sight.

As you can see, we need only a few key pieces of information to graph any parabola.

Graphing Parabolas

1. Find the y-intercept.
2. Find the x-intercepts.
3. Find the axis of symmetry.
4. Find the vertex.

Example 2. Graph $y = 2 + x - x^2$. State its domain and range.

Solution. $y = 2 + x - x^2$. This parabola will open downward.

y-INTERCEPT: Let $x = 0$

$$y = 2 + 0 - 0^2$$
$$= 2$$
$$(0, 2)$$

x-INTERCEPT: Let $y = 0$

$$0 = 2 + x - x^2$$
$$x^2 - x - 2 = 0$$
$$(x - 2)(x + 1) = 0$$
$$x - 2 = 0 \quad \text{or} \quad x + 1 = 0$$
$$x = 2 \qquad x = -1$$
$$(2, 0) \qquad (-1, 0)$$

AXIS OF SYMMETRY

$$x_M = \frac{2 + (-1)}{2}$$
$$= \frac{1}{2}$$

$x = \frac{1}{2}$ is the axis of symmetry

VERTEX: Let $x = \frac{1}{2}$

$$y = 2 + x - x^2$$
$$y = 2 + \tfrac{1}{2} - (\tfrac{1}{2})^2$$
$$= 2 + \tfrac{1}{2} - \tfrac{1}{4}$$
$$= \tfrac{8}{4} + \tfrac{2}{4} - \tfrac{1}{4}$$
$$= \tfrac{9}{4}$$

$(\frac{1}{2}, \frac{9}{4})$ is the vertex

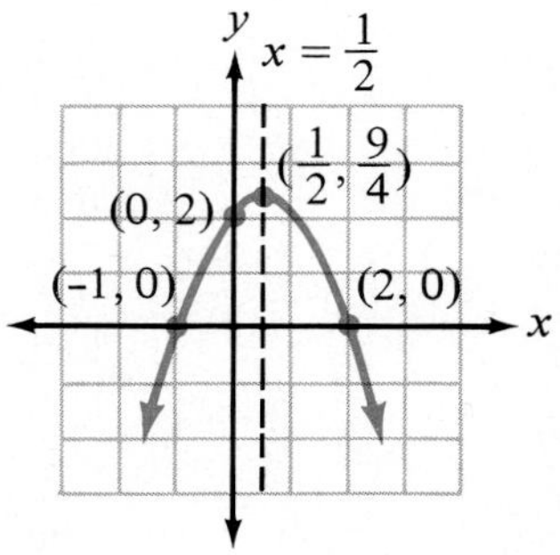

$y = 2 + x - x^2$

Domain: $x \in R$
Range: $y \in R$ and $y \leq \frac{9}{4}$

Example 3. Graph $y = x^2 + x - 1$. State its domain and range.

Solution. This parabola will open upward.

y-INTERCEPT: Let $x = 0$

$$y = 0^2 + 0 - 1$$
$$= -1$$
$$(0, -1)$$

x-INTERCEPTS: Let $y = 0$

$$0 = x^2 + x - 1$$

Since the quadratic expression does not factor, we must use the quadratic formula with $a = 1$, $b = 1$, and $c = -1$.

$$x = \frac{-b \pm \sqrt{b^2 - 4ac}}{2a}$$
$$x = \frac{-1 \pm \sqrt{1^2 - 4(1)(-1)}}{2}$$
$$= \frac{-1 \pm \sqrt{1 + 4}}{2}$$
$$x = \frac{-1 \pm \sqrt{5}}{2}$$
$$\left(\frac{-1 + \sqrt{5}}{2}, 0\right) \quad \text{and} \quad \left(\frac{-1 - \sqrt{5}}{2}, 0\right)$$

AXIS OF SYMMETRY

$$x_M = \frac{\frac{-1 + \sqrt{5}}{2} + \frac{-1 - \sqrt{5}}{2}}{2}$$
$$= \frac{\frac{-1 + \sqrt{5} - 1 - \sqrt{5}}{2}}{2}$$
$$= \frac{\frac{-2}{2}}{2}$$
$$x_M = \frac{-1}{2}$$

$x = -\frac{1}{2}$ is the axis of symmetry

VERTEX: Let $x = -\frac{1}{2}$.

$$y = x^2 + x - 1$$
$$y = \left(-\frac{1}{2}\right)^2 + \left(-\frac{1}{2}\right) - 1$$
$$= \frac{1}{4} - \frac{1}{2} - 1$$
$$= \frac{1}{4} - \frac{2}{4} - \frac{4}{4}$$
$$y = -\frac{5}{4}$$

$(-\frac{1}{2}, -\frac{5}{4})$ is the vertex

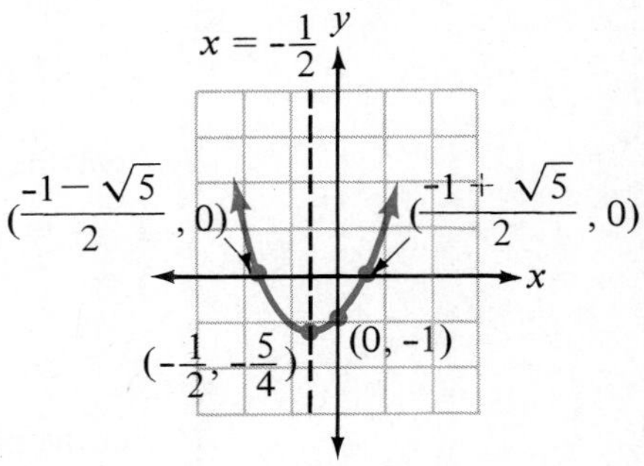

$y = x^2 + x - 1$

Domain: $x \in R$
Range: $y \in R$ and $y \geq -\frac{5}{4}$

From Example 3 we note that our four-step procedure will work even when the x-intercepts are irrational numbers that must be found using the quadratic formula. Because the arithmetic involved in finding the axis of symmetry when the intercepts are irrational numbers is rather tedious, we shall now develop a method for locating the axis of symmetry for *any* parabola.

For a parabola with the equation $y = ax^2 + bx + c$, the x-intercepts can always be found using the quadratic formula.

$$x\text{-INTERCEPTS:} \quad \text{Let } y = 0$$

$$ax^2 + bx + c = 0$$

$$x = \frac{-b \pm \sqrt{b^2 - 4ac}}{2a}$$

$$\left(\frac{-b + \sqrt{b^2 - 4ac}}{2a}, 0\right) \text{ and } \left(\frac{-b - \sqrt{b^2 - 4ac}}{2a}, 0\right)$$

Now we can use the midpoint formula to locate the axis of symmetry halfway between these intercepts.

AXIS OF SYMMETRY

$$x_M = \frac{\dfrac{-b + \sqrt{b^2 - 4ac}}{2a} + \dfrac{-b - \sqrt{b^2 - 4ac}}{2a}}{2}$$

$$= \left[\frac{-b + \sqrt{b^2 - 4ac} + (-b - \sqrt{b^2 - 4ac})}{2a}\right] \div 2$$

$$= \frac{-b + \sqrt{b^2 - 4ac} - b - \sqrt{b^2 - 4ac}}{2a} \cdot \frac{1}{2}$$

$$= \frac{-2b}{2a} \cdot \frac{1}{2}$$

$$= \frac{-b}{a} \cdot \frac{1}{2}$$

$$x_M = -\frac{b}{2a}$$

$$x = -\frac{b}{2a} \text{ is the axis of symmetry}$$

To find the y-coordinate of the vertex, we substitute $-\dfrac{b}{2a}$ for x in the original equation $y = ax^2 + bx + c$. This discussion leads us to the following general conclusion.

> For the parabola with equation $y = ax^2 + bx + c$, the axis of symmetry is the vertical line
>
> $$x = -\frac{b}{2a}$$

Although this formula for the axis of symmetry was derived by assuming that the parabola had real x-intercepts, it can be proved that this formula "works" even when that is not so. We shall not attempt such a proof in this course. Let's practice our new formula for finding the axis of symmetry and vertex of a parabola.

Example 4. Graph $y = x^2 - 3x - 10$. Note that this parabola will open upward. Note also that $a = 1$ and $b = -3$.

y-INTERCEPT: Let $x = 0$

$$y = 0^2 - 3(0) - 10$$
$$= -10$$
$$(0, -10)$$

x-INTERCEPTS: Let $y = 0$

$$0 = x^2 - 3x - 10$$
$$0 = (x - 5)(x + 2)$$
$$x - 5 = 0 \text{ or } x + 2 = 0$$
$$x = 5 \qquad x = -2$$
$$(5, 0) \qquad (-2, 0)$$

AXIS OF SYMMETRY

$$x = -\frac{b}{2a}$$
$$x = -\frac{-3}{2(1)}$$
$$x = \frac{3}{2} \text{ is the axis of symmetry}$$

VERTEX: Let $x = \frac{3}{2}$

$$y = x^2 - 3x - 10$$
$$y = \left(\frac{3}{2}\right)^2 - 3\left(\frac{3}{2}\right) - 10$$
$$= \frac{9}{4} - \frac{9}{2} - 10$$
$$= \frac{9}{4} - \frac{18}{4} - \frac{40}{4}$$
$$y = -\frac{49}{4}$$

$(\frac{3}{2}, -\frac{49}{4})$ is the vertex

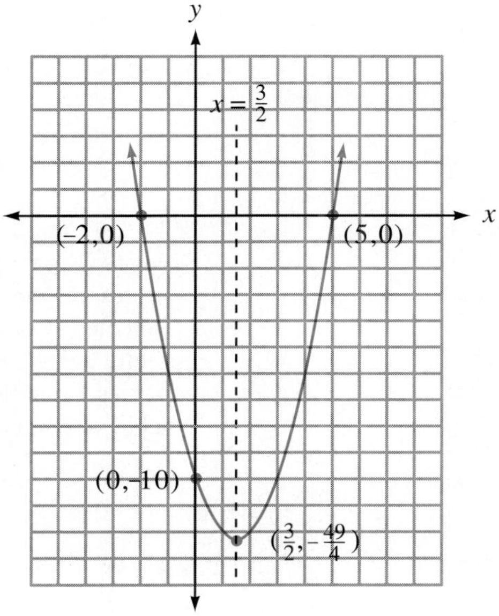

Now you try the problem stated at the beginning of the chapter.

Example 5. If a person throws a ball directly upward at a speed of 96 feet per second, the ball's height y above the ground after x seconds is given by the function $y = -16x^2 + 96x$. Use the graph of this function to find the maximum height of the ball.

Solution. The graph of $y = -16x^2 + 96x$ is a ________ that opens ________ . We note that $a =$ ________ and $b =$ ________.

y-INTERCEPT: Let $x = 0$.

$$y = -16(\_\_\_)^2 + 96(\_\_\_)$$

$$y = \_\_\_$$

$$(0, \_\_\_)$$

x-INTERCEPTS: Let $y = 0$

$$0 = -16x^2 + 96x$$

$$0 = -16x(x - \_\_\_)$$

$$-16x = 0 \quad \text{or} \quad x - \_\_\_ = 0$$

$$x = 0 \qquad\qquad x = \_\_\_$$

$$(0, 0) \qquad\qquad (\_\_\_, 0)$$

AXIS OF SYMMETRY

$$x = -\frac{b}{2a}$$

$$= -\frac{96}{2(-16)}$$

$$= \_\_\_$$

$x =$ _____ is the axis of symmetry.

VERTEX: Let $x = 3$

$$y = -16x^2 + 96x$$

$$y = -16(3)^2 + 96(3)$$

$$= -16(9) + 288$$

$$= -144 + 288$$

$$y = \_\_\_$$

$(3, \_\_\_)$ is the vertex.

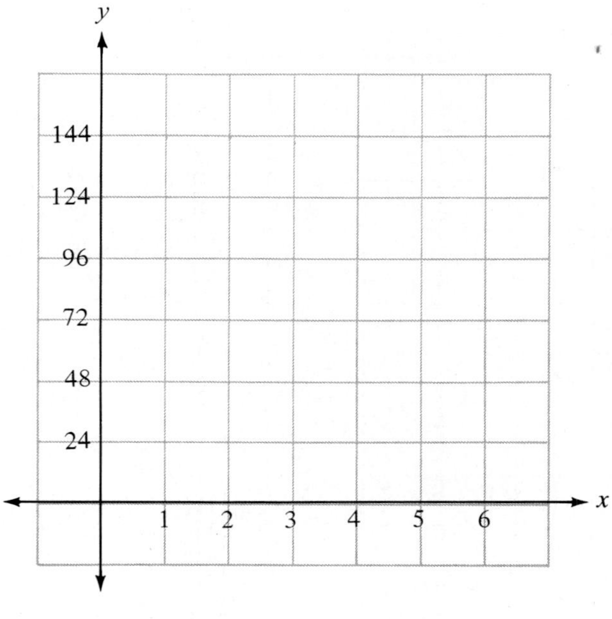

[*Note:* Here we chose to use different scales on the two axes. As long as you clearly label the units, this is perfectly acceptable and is useful when values for either (or both) of the variables are very large or very small.]

Since this parabola opens downward, its vertex is its ________ point. The maximum height of the ball is _____ feet.

Check your work on page 635. ▶

Trial Run

Graph each quadratic function. State the domain and range.

Domain: ________ 1. $y = x^2 - 6x$

Range: ________

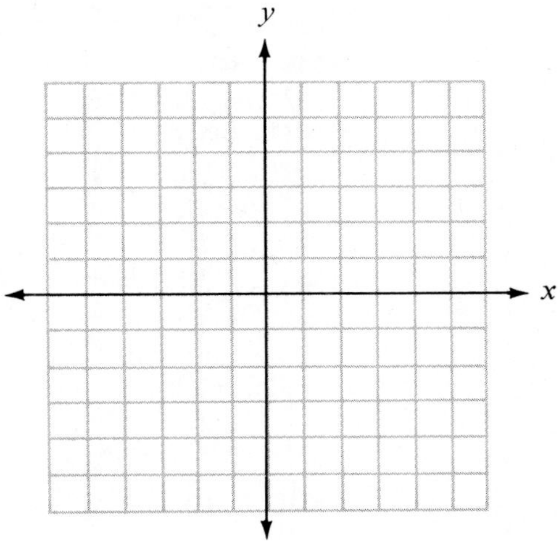

Domain: ________ 2. $y = -x^2 + 8$

Range: ________

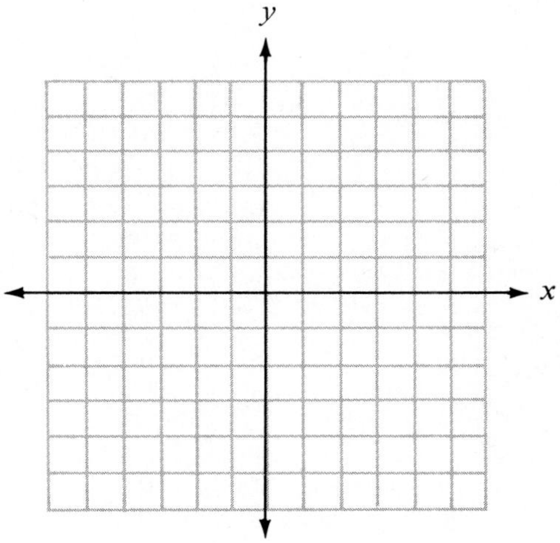

Domain: ________ 3. $y = -x^2 - x + 6$

Range: ________

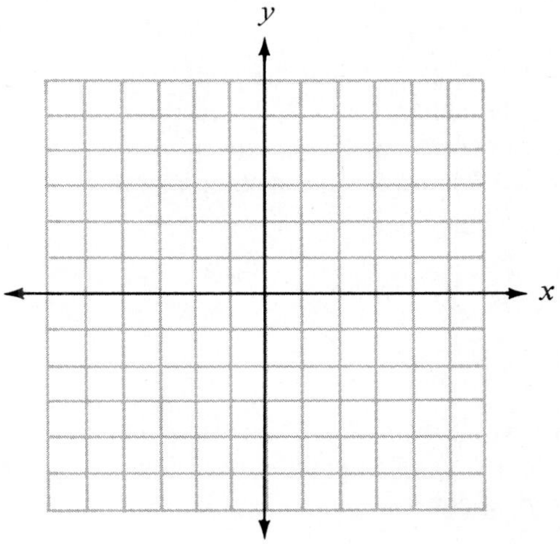

Answers are on page 636.

Graphing Quadratic Functions with No x-Intercepts

Let's graph the quadratic function

$$y = x^2 + x + 1$$

We note that the graph will be a parabola opening upward. We note also that $a = 1$ and $b = 1$.

y-INTERCEPT: Let $x = 0$

$$y = 0^2 + 0 + 1$$
$$y = 1$$
$$(0, 1)$$

x-INTERCEPTS: Let $y = 0$

$$x = \frac{-1 \pm \sqrt{1^2 - 4(1)(1)}}{2}$$
$$= \frac{-1 \pm \sqrt{-3}}{2}$$

Since $\sqrt{-3}$ is not a real number, there are no x-intercepts.

AXIS OF SYMMETRY

$$x = \frac{-b}{2a}$$
$$= \frac{-1}{2(1)}$$
$$x = -\frac{1}{2}$$

VERTEX: Let $x = -\frac{1}{2}$

$$y = x^2 + x + 1$$
$$y = \left(-\frac{1}{2}\right)^2 + \left(-\frac{1}{2}\right) + 1$$
$$= \frac{1}{4} - \frac{1}{2} + 1$$
$$= \frac{1}{4} - \frac{2}{4} + \frac{4}{4}$$
$$= \frac{3}{4}$$

$(-\frac{1}{2}, \frac{3}{4})$ is the vertex

So far we have found the y-intercept (0, 1) and the vertex $(-\frac{1}{2}, \frac{3}{4})$. Since there are *no* x-intercepts, our graph never crosses the x-axis. It must look something like this.

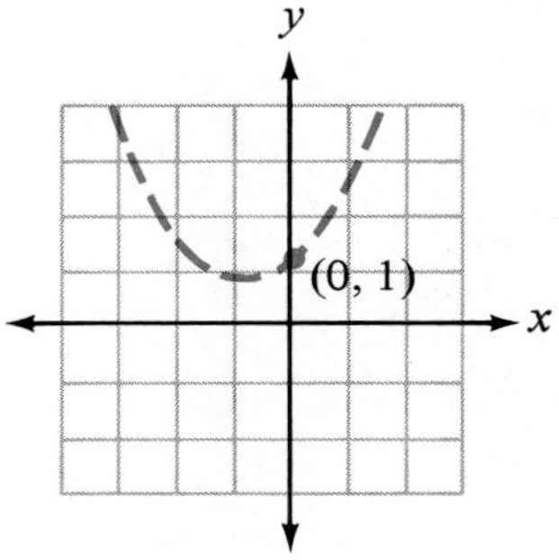

We would like to find another point on the parabola. Let's find the point symmetric to the point (0, 1). That point must also have a y-coordinate of 1. To find it we can let $y = 1$ in the original equation.

$$y = x^2 + x + 1$$
$$1 = x^2 + x + 1$$
$$0 = x^2 + x$$
$$0 = x(x + 1)$$
$$x = 0 \quad \text{or} \quad x + 1 = 0$$
$$x = -1$$
$$(0, 1) \qquad (-1, 1)$$

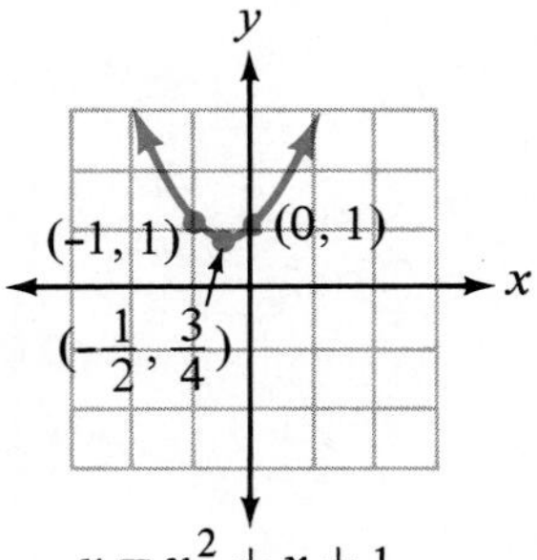

$y = x^2 + x + 1$

Domain: $x \in R$ Range: $y \in R$ and $y \geq \frac{3}{4}$

Example 6. Graph $y = -x^2 - 1$. Then state its domain and range.

Solution. First we note that the parabola opens downward and that $a = 1$ and $b = 0$.

y-INTERCEPT: Let $x = 0$

$$y = -0^2 - 1$$
$$= -1$$

$(0, -1)$

x-INTERCEPTS: Let $y = 0$

$$0 = -x^2 - 1$$
$$x^2 = -1$$
$$x = \pm\sqrt{-1} \quad \text{Not real numbers}$$

No x-intercepts

AXIS OF SYMMETRY

$$x = -\frac{b}{2a}$$
$$= -\frac{0}{2(-1)}$$
$$= 0$$

$x = 0$ is the axis of symmetry

VERTEX: Let $x = 0$

$$y = -0^2 - 1$$
$$= -1$$

$(0, -1)$ is the vertex

Since there are no x-intercepts and the y-intercept is the vertex, we shall look for two more points, one to the left of the vertex and one to the right of the vertex. Let's find the points corresponding to x-coordinates of 1 and -1.

Let $x = -1$

$$y = -(-1)^2 - 1$$
$$= -1 - 1$$
$$= -2$$

$(-1, -2)$

Let $x = 1$

$$y = -(1)^2 - 1$$
$$= -1 - 1$$
$$= -2$$

$(1, -2)$

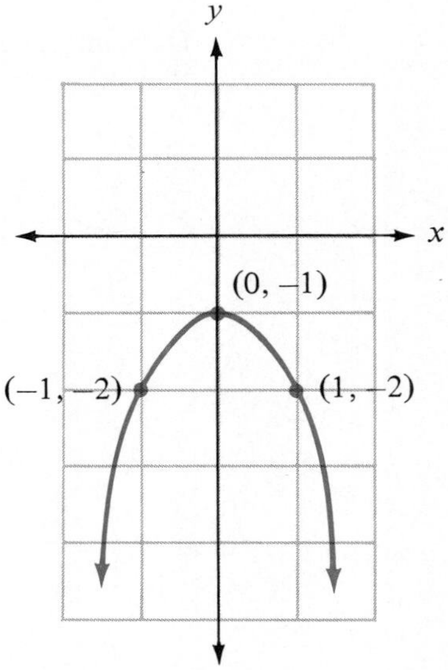

$y = -x^2 - 1$

Domain: $x \in R$
Range: $y \in R$ and $y \leq -1$

Trial Run

Graph each quadratic function. State the domain and range.

Domain: ________ **1.** $y = x^2 + 2x + 4$
Range: ________

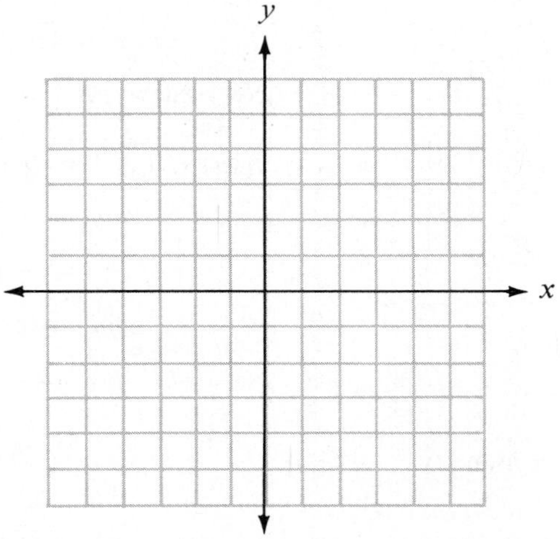

Domain: ________ **2.** $y = -x^2 + 3x - 3$
Range: ________

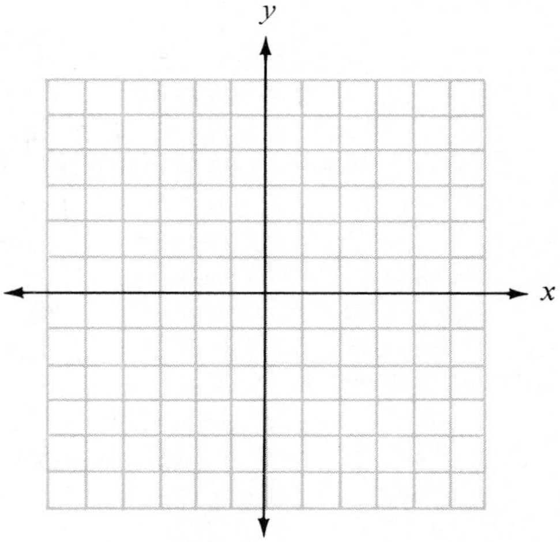

Domain: ________ **3.** $y = -3x^2 - 6$
Range: ________

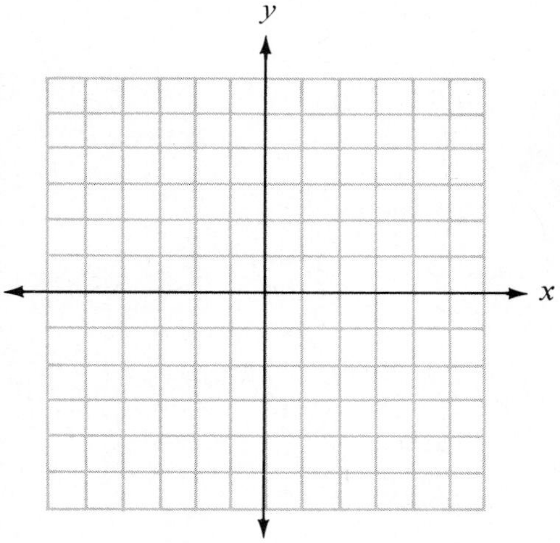

Answers are on page 636.

Now we can summarize our steps for graphing quadratic functions.

Graphing Quadratic Functions: $y = ax^2 + bx + c$

1. Find the y-intercept (by letting $x = 0$).
2. Find the x-intercepts (by letting $y = 0$).
3. Find the axis of symmetry $\left(\text{using the midpoint formula or } x = -\dfrac{b}{2a}\right)$.
4. Find the vertex (by substituting the x-value from the axis of symmetry into the original equation).
5. If you do not have three different points on the parabola after steps 1 through 4, use appropriate substitutions of x-values or y-values to obtain the needed points.
6. Sketch the graph.

▶ Example You Completed

Example 5. If a person throws a ball directly upward at a speed of 96 feet per second, then the ball's height y above the ground after x seconds is given by the function $y = -16x^2 + 96x$. Use the graph of this function to find the maximum height of the ball.

Solution. The graph of $y = -16x^2 + 96x$ is a parabola that opens downward. We note that $a = -16$ and $b = 96$.

y-INTERCEPT: Let $x = 0$

$$y = -16(0)^2 + 96(0)$$
$$y = 0$$
$(0, 0)$

AXIS OF SYMMETRY

$$x = -\frac{b}{2a}$$
$$= -\frac{96}{2(-16)}$$
$$= 3$$

$x = 3$ is the axis of symmetry

x-INTERCEPTS: Let $y = 0$

$$0 = -16x^2 + 96x$$
$$0 = -16x(x - 6)$$
$$-16x = 0 \quad \text{or} \quad x - 6 = 0$$
$$x = 0 \qquad\qquad x = 6$$
$$(0, 0) \qquad\qquad (6, 0)$$

VERTEX: Let $x = 3$

$$y = -16x^2 + 96x$$
$$y = -16(3)^2 + 96(3)$$
$$= -16(9) + 288$$
$$= -144 + 288$$
$$y = 144$$

$(3, 144)$ is the vertex

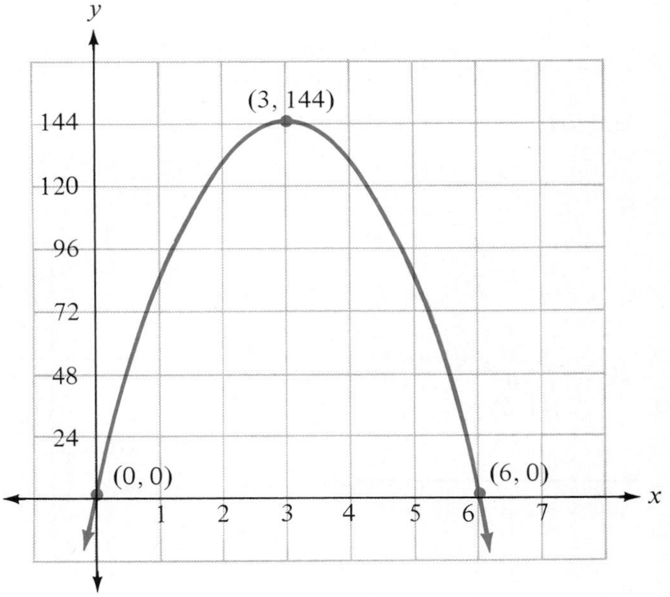

Since this parabola opens downward, its vertex is its maximum point. The maximum height of the ball is 144 feet.

Answers to Trial Runs

page 631

1.

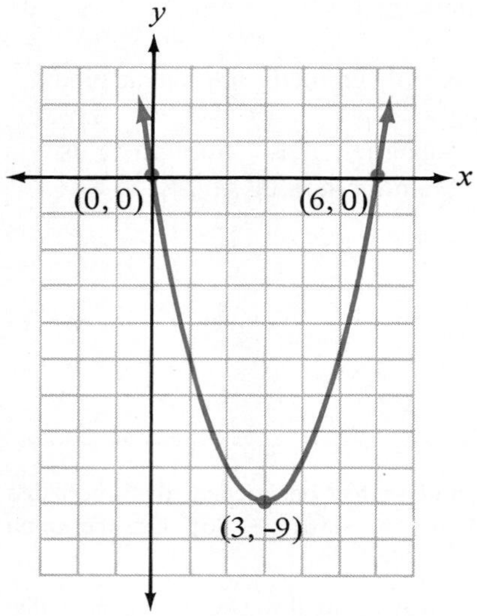

Domain: $x \in R$
Range: $y \in R$ and $y \geq -9$

2.

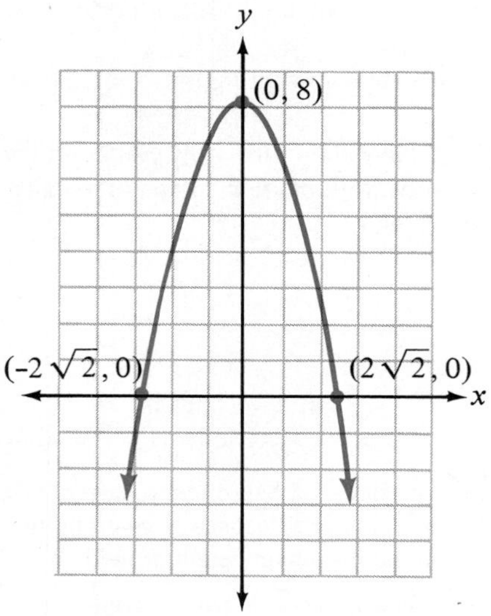

Domain: $x \in R$
Range: $y \in R$ and $y \leq 8$

3.

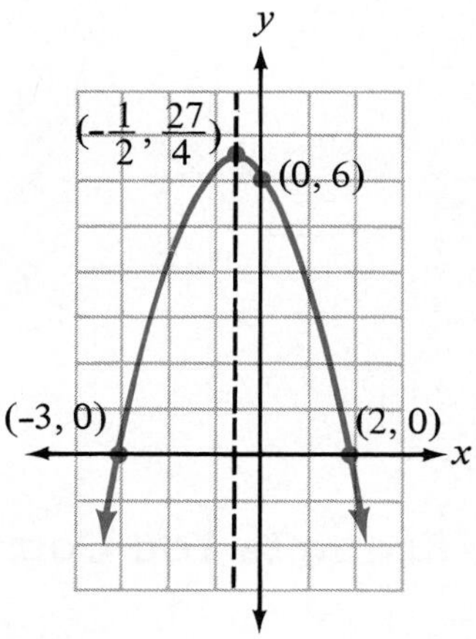

Domain: $x \in R$
Range: $y \in R$ and $y \leq \frac{27}{4}$

page 634

1.

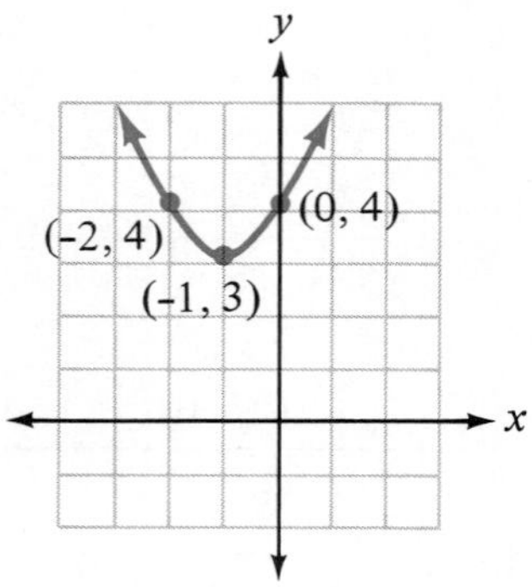

Domain: $x \in R$
Range: $y \in R$ and $y \geq 3$

2.

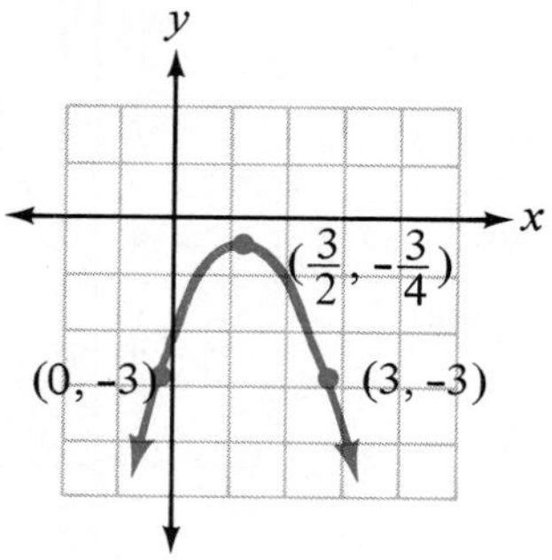

Domain: $x \in R$
Range: $y \in R$ and $y \leq -\frac{3}{4}$

3.

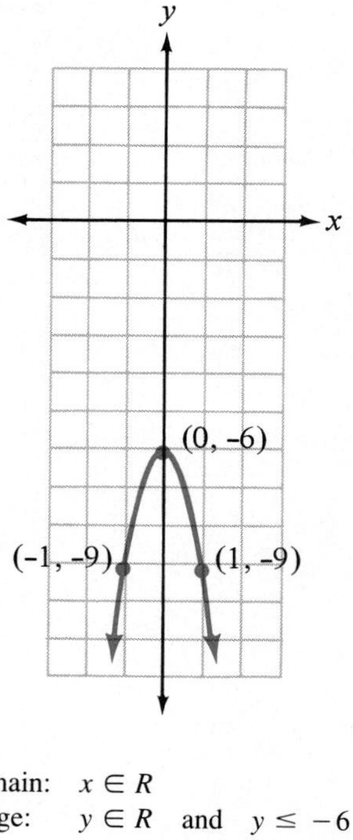

Domain: $x \in R$
Range: $y \in R$ and $y \leq -6$

Name ______________________ Date ____________

EXERCISE SET 9.2

Graph each function. State the domain and range.

Domain: ________ 1. $y = -x^2 + 2x + 15$
Range: ________

Domain: ________ 2. $y = -x^2 - 3x + 4$
Range: ________

Domain: ________ 3. $y = x^2 + 4x$
Range: ________

Domain: ________ 4. $y = x^2 - 2x$
Range: ________

Domain: ________ 5. $y = x^2 - 9$
Range: ________

Domain: ________ 6. $y = x^2 - 16$
Range: ________

Domain: ________ 7. $y = x^2 - 10x + 25$
Range: ________

Domain: ________ 8. $y = x^2 - 6x + 9$
Range: ________

Domain: ________ 9. $y = 3x^2$
Range: ________

Domain: ________ 10. $y = -4x^2$
Range: ________

Domain: ________ 11. $y = -x^2 + 6$
Range: ________

Domain: ________ 12. $y = x^2 + 12$
Range: ________

Domain: ________ 13. $y = -5x^2 - 10$
Range: ________

Domain: ________ 14. $y = -2x^2 - 8$
Range: ________

Domain: ________ 15. $y = -2x^2 - x + 6$
Range: ________

Domain: ________ 16. $y = -3x^2 + 2x + 5$
Range: ________

Domain: ________ 17. $y = 3x^2 - 3x + 1$
Range: ________

Domain: ________ 18. $y = 5x^2 - 2x + 1$
Range: ________

Domain: ________ 19. $y = x^2 + x + 2$
Range: ________

Domain: ________ 20. $y = x^2 + 3x + 5$
Range: ________

Domain: ________ 21. $y = -x^2 - 2x - 4$
Range: ________

Domain: ________ 22. $y = -x^2 + 4x - 6$
Range: ________

Domain: ________ 23. $y = -2x^2 + 5x - 5$
Range: ________

Domain: ________ 24. $y = -3x^2 + 8x - 6$
Range: ________

________ 25. Estelle runs a television repair shop. The formula she uses to approximate her weekly profits is $y = -3x^2 + 108x$, where y is the profit from repairing x television sets. Use a graph to determine how many sets she must repair to make a maximum profit. What is the maximum profit?

________ 26. A refrigerator manufacturer uses the formula $y = -5x^2 + 600x$ to determine the profit y when x refrigerators are sold. Use a graph to determine how many refrigerators must be sold to make a maximum profit. What is the maximum profit?

_______ 27. The function for the cost y of building x picnic tables is $y = x^2 - 12x + 72$. Use a graph to determine how many picnic tables Mr. Denison should build to have a minimum cost. What is the minimum cost?

_______ 28. The cost y of running a steak house is related to the number of cartons of steaks x that are sold by the function $y = x^2 - 18x + 1000$. Use a graph to determine how many cartons of steaks must be sold to keep costs at a minimum. What is the minimum cost?

☆ Stretching the Topics

_______ 1. Estimate to the nearest tenth the x-intercepts of the parabola determined by $y = x^2 - 3x - 5$.

_______ 2. The sum of the width and length of a rectangle is 18 feet. Write an equation that expresses the fact that the area is a function of the width. Use a graph to determine what the width should be so that the rectangle will have a maximum area. What is the maximum area?

Check your answers in the back of your book.

If you can graph the functions in **Checkup 9.2,** you are ready to go on to Section 9.3.

Name ______________________ Date ____________

✓ CHECKUP 9.2

Graph each function. State the domain and range.

Domain: ________ 1. $y = -2x^2$
Range: ________

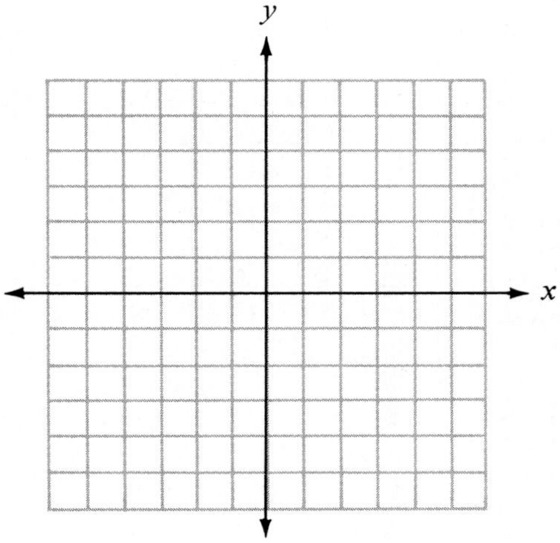

Domain: ________ 2. $y = x^2 - 1$
Range: ________

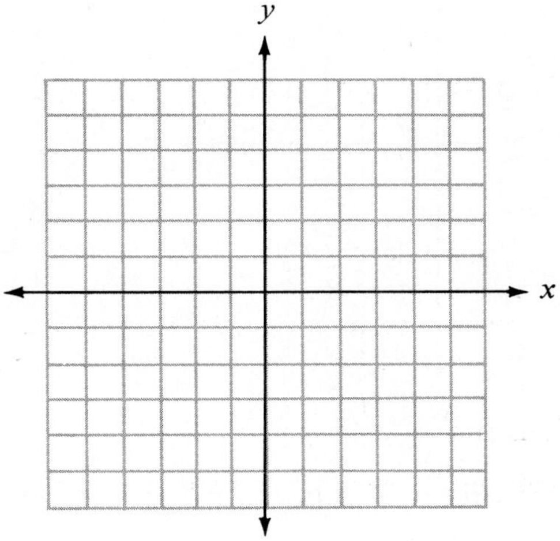

Domain: ________ 3. $y = x^2 - x - 2$
Range: ________

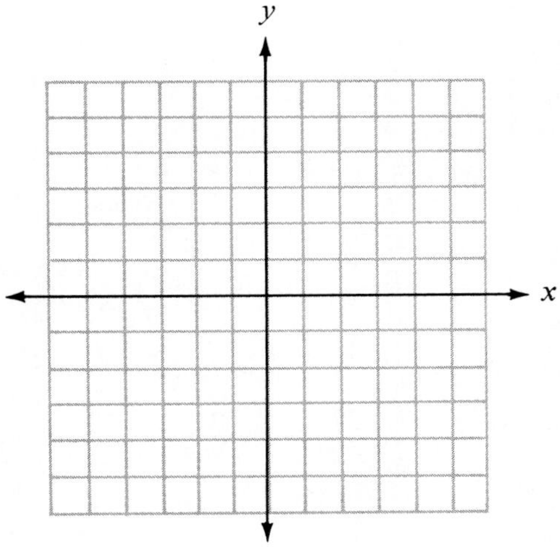

Domain: ________ 4. $y = -x^2 + 5x$
Range: ________

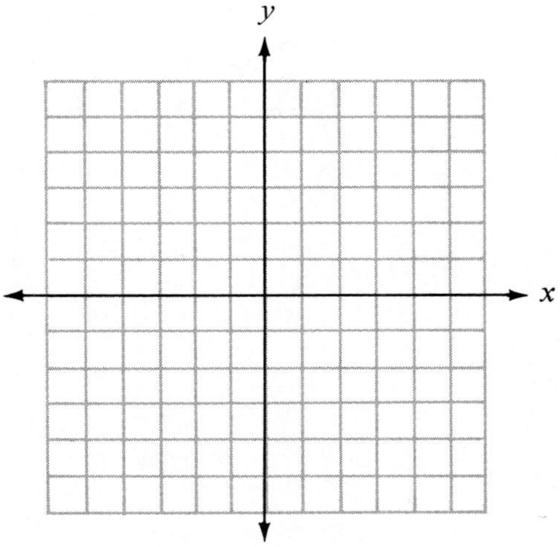

Domain: ________ 5. $y = x^2 + 2x - 2$
Range: ________

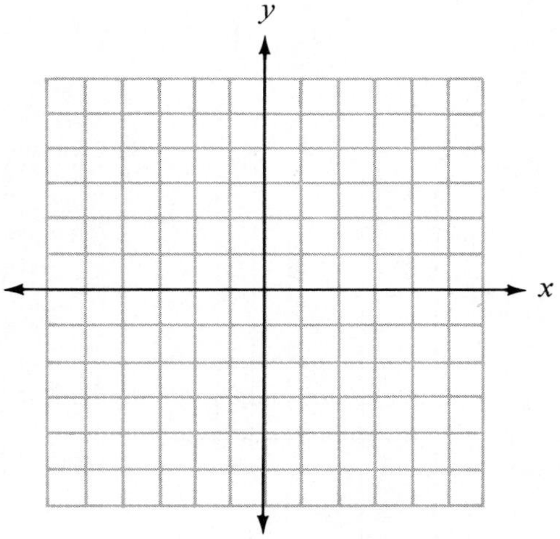

Domain: ________ 6. $y = -x^2 + 3x - 5$
Range: ________

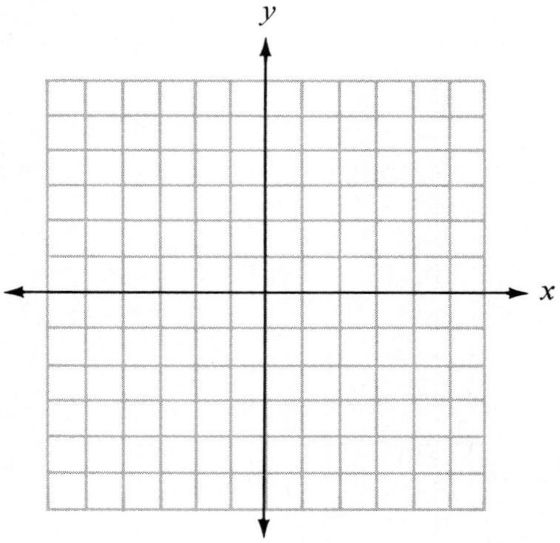

Check your answers in the back of your book.

If You Missed Problems:	You Should Review Examples:
1	1
2, 3	2, 4
4	5
5	3
6	6

9.3 Graphing the Conic Sections

An interesting family of curves results when the surface of a cone is cut by a plane.

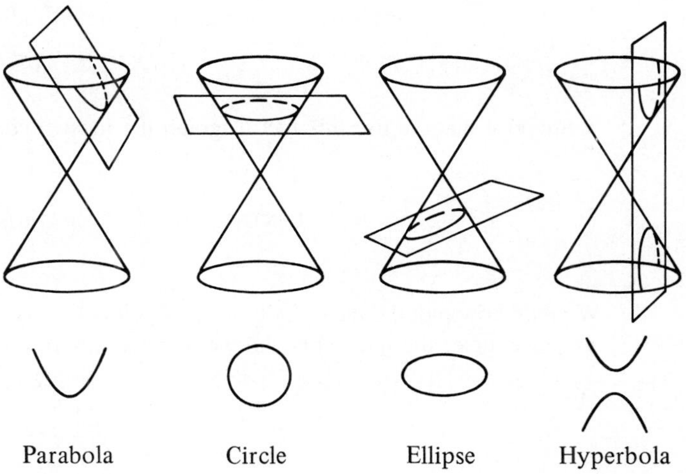

Because of their relationship to the cone, these curves are called the **conic sections.** The first conic section, the **parabola,** was discussed in detail in Section 9.2. Now we must see what kinds of equations would result in the remaining conic sections.

Graphing Circles

Suppose we look at the second conic section—the **circle.** Consider the graph of a circle with center at the origin and radius, r. We know the distance between any point on our circle and the circle's center must be r units.

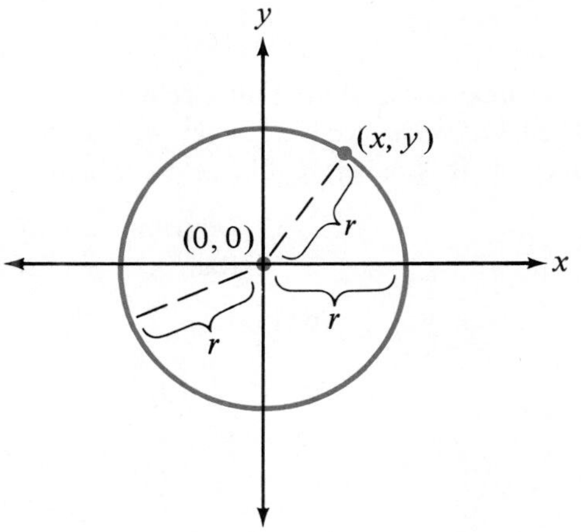

Letting (x, y) represent the coordinates of any point on the circle and (0, 0) represent the origin, we can use the distance formula from Chapter 8 and state that

$$r = \sqrt{(x - 0)^2 + (y - 0)^2}$$
$$= \sqrt{x^2 + y^2}$$
$$r^2 = x^2 + y^2$$

The graph of the equation

$$x^2 + y^2 = r^2$$

is a *circle* with center (0, 0) and radius r (where $r > 0$).

Suppose that we use this fact to graph the most fundamental circle:

$$x^2 + y^2 = 1$$

CENTER: (0, 0) RADIUS: $r^2 = 1$

$$r = 1$$

We may now plot the most easily located points lying $r = 1$ unit from the origin. Those points are, of course, the intercepts and we join them with a smooth curve.

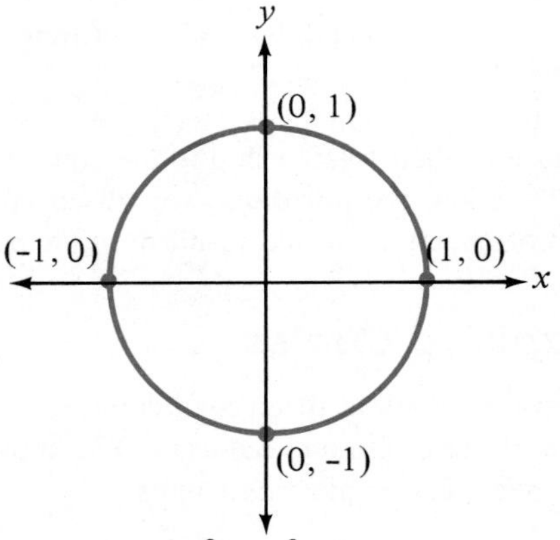

$x^2 + y^2 = 1$

This circle is called the **unit circle** and it is of great importance in the branch of mathematics called *trigonometry*. Is $x^2 + y^2 = 1$ a function? The vertical line test tells us that it is *not* a function. It is merely a relation for which

Domain: $x \in R$ and $-1 \leq x \leq 1$
Range: $y \in R$ and $-1 \leq y \leq 1$

You complete Example 1.

Example 1. Graph $2x^2 + 2y^2 = 4$. State its domain and range.

Solution

$$2x^2 + 2y^2 = 4$$

Dividing both sides by 2 yields

$$x^2 + y^2 = 2$$

The graph will be a ________ .

CENTER: (____, ____)

RADIUS: $r^2 = 2$

$r = \sqrt{\_\_\_\_}$

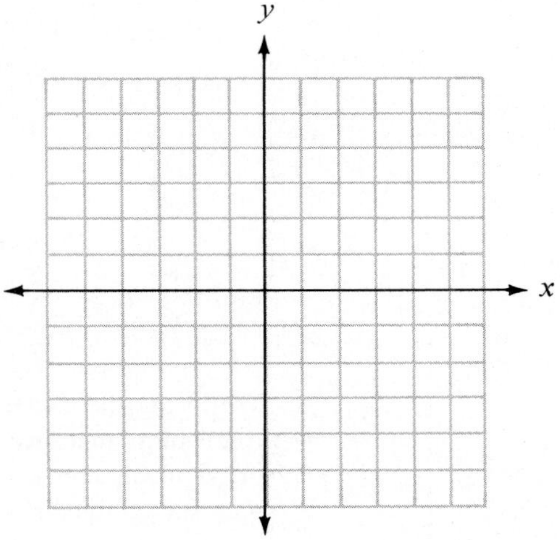

Domain: $x \in R$ and ____ $\leq x \leq$ ____

Range: $y \in R$ and ____ $\leq y \leq$ ____

Check your work on page 652. ▶

Graphing circles is quite systematic, but we cannot expect circles always to be centered at the origin. Let's investigate what happens when a circle is centered at some other point, say (h, k). The definition of a circle still demands that the distance from the center (h, k) to any point (x, y) on the circle must be r.

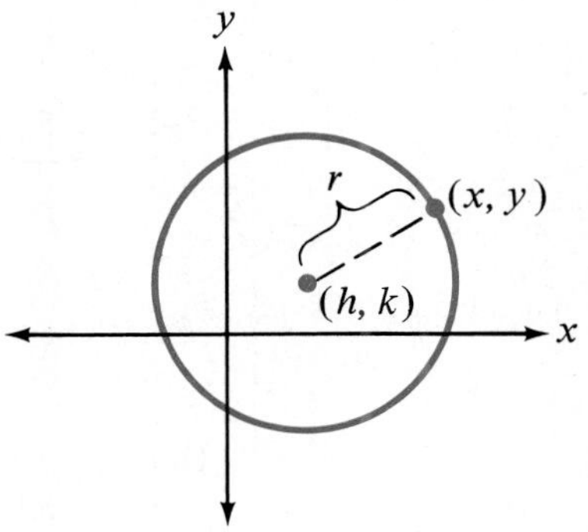

$$r = \sqrt{(x - h)^2 + (y - k)^2}$$

$$r^2 = (x - h)^2 + (y - k)^2$$

> The graph of the equation
>
> $$(x - h)^2 + (y - k)^2 = r^2$$
>
> is a *circle* with center (h, k) and radius r (where h, k, and r are constants, and $r > 0$).

Example 2. Graph $(x - 3)^2 + (y - 5)^2 = 4$. State its domain and range.

Solution

$$(x - 3)^2 + (y - 5)^2 = 4$$

CENTER

$x - h = x - 3 \qquad y - k = y - 5$

$h = 3 \qquad k = 5$

$$(h, k) = (3, 5)$$

RADIUS: $r^2 = 4$

$r = 2$

We locate the center at (3, 5) and then locate four points that are 2 units directly to the left of, right of, above, and below the center. We then sketch the circle.

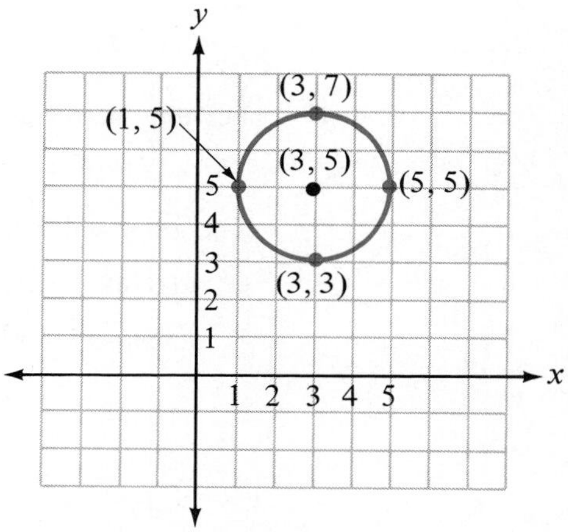

$(x - 3)^2 + (y - 5)^2 = 4$

Domain: $x \in R$ and $1 \leq x \leq 5$
Range: $y \in R$ and $3 \leq y \leq 7$

Example 3. Graph $(x + 2)^2 + y^2 = 5$. State its domain and range.

Solution

$$(x + 2)^2 + y^2 = 5$$

$$[x - (-2)]^2 + (y - 0)^2 = 5$$

CENTER

$x - h = x - (-2) \qquad y - k = y - 0$

$h = -2 \qquad k = 0$

$$(h, k) = (-2, 0)$$

RADIUS: $r^2 = 5$

$r = \sqrt{5}$

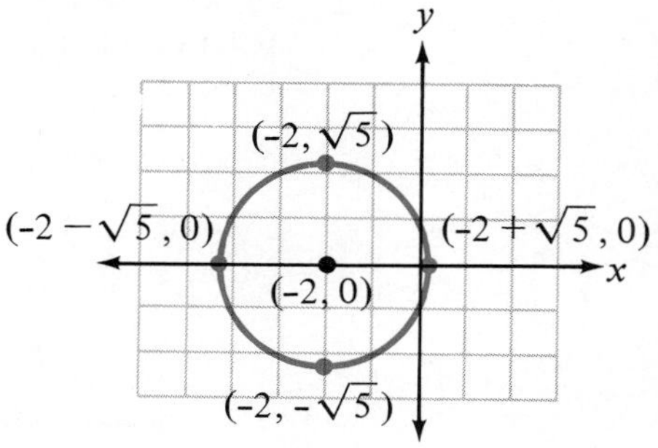

Domain: $x \in R$ and $-2 - \sqrt{5} \leq x \leq -2 + \sqrt{5}$
Range: $y \in R$ and $-\sqrt{5} \leq y \leq \sqrt{5}$

When a coordinate of a point is an irrational number, we may find its approximate value using a calculator or the table inside the back cover of this book. Then we approximate the point's location in the plane and label it with its exact irrational coordinates.

Trial Run

Graph each relation. State the domain and range.

1. $x^2 + y^2 = 9$

Domain: ________
Range: ________

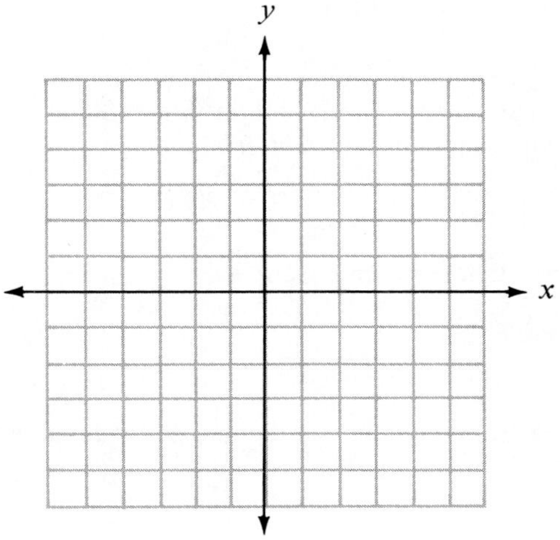

2. $3x^2 + 3y^2 = 6$

Domain: ________
Range: ________

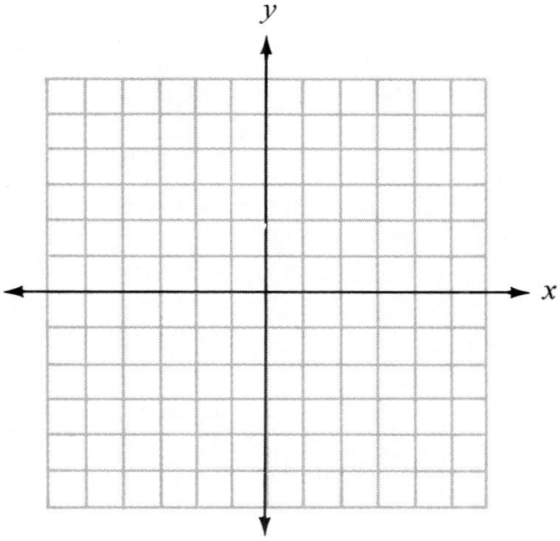

3. $(x - 3)^2 + (y - 1)^2 = 1$

Domain: ________
Range: ________

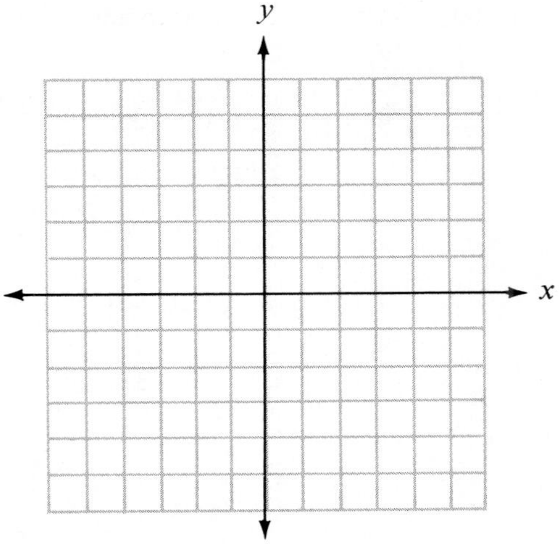

4. $(x + 1)^2 + (y + 2)^2 = 25$

Domain: ________
Range: ________

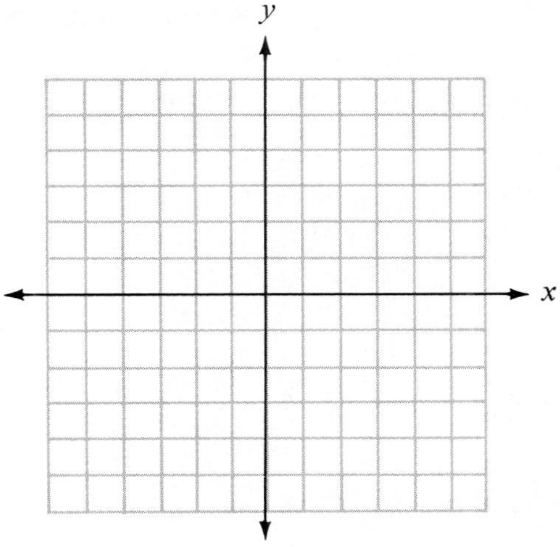

Answers are on page 653.

Graphing Ellipses

The third kind of conic section is the **ellipse.** We concentrate here only on ellipses centered at the origin. Since an ellipse is sort of an elongated circle, we expect its equation to be similar to that of a circle.

> The graph of the equation
>
> $$Ax^2 + By^2 = C$$
>
> is an *ellipse* centered at the origin (where A, B, and C are *positive* constants and $A \neq B$).

Note that if $A = B$, this equation represents a circle.

To graph an equation such as

$$9x^2 + 4y^2 = 36$$

we rely heavily on knowing from the start that the graph will be an ellipse. First we find the *intercepts*.

x-INTERCEPTS: Let $y = 0$

$$9x^2 + 4(0)^2 = 36$$
$$9x^2 = 36$$
$$x^2 = 4$$
$$x = \pm 2$$

$(2, 0)$ and $(-2, 0)$

y-INTERCEPTS: Let $x = 0$

$$9(0)^2 + 4y^2 = 36$$
$$4y^2 = 36$$
$$y^2 = 9$$
$$y = \pm 3$$

$(0, 3)$ and $(0, -3)$

Since this is the first ellipse we have graphed, let's plot a few more points.

Let $x = 1$

$$9(1)^2 + 4y^2 = 36$$
$$9 + 4y^2 = 36$$
$$4y^2 = 27$$
$$y^2 = \frac{27}{4}$$
$$y = \frac{\pm\sqrt{27}}{2}$$
$$y = \frac{\pm 3\sqrt{3}}{2}$$

$\left(1, \frac{3\sqrt{3}}{2}\right)$ and $\left(1, -\frac{3\sqrt{3}}{2}\right)$

Let $x = -1$

$$9(-1)^2 + 4y^2 = 36$$
$$9 + 4y^2 = 36$$
$$4y^2 = 27$$
$$y^2 = \frac{27}{4}$$
$$y = \frac{\pm\sqrt{27}}{2}$$
$$y = \frac{\pm 3\sqrt{3}}{2}$$

$\left(-1, \frac{3\sqrt{3}}{2}\right)$ and $\left(-1, -\frac{3\sqrt{3}}{2}\right)$

Notice the symmetry of the four points. In order to plot these points, we recall that $\sqrt{3} \doteq 1.732$, so we may *approximate* the locations of the points by

$(1, 2.6)$ and $(1, -2.6)$ $\quad$ $(-1, 2.6)$ and $(-1, -2.6)$

but we label the points with their *exact* coordinates.

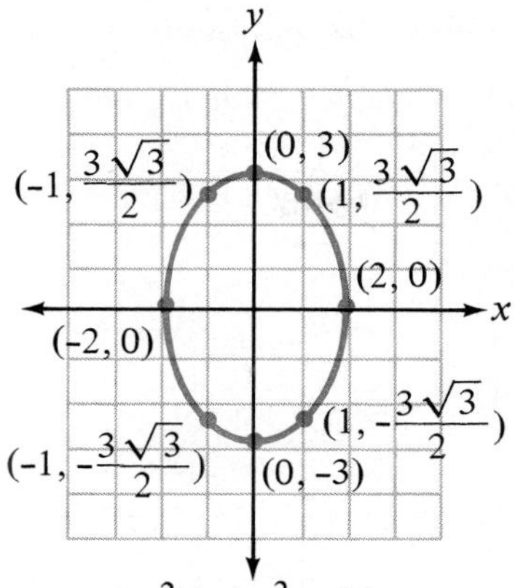

$9x^2 + 4y^2 = 36$

Does this graph represent a function? No, it is merely a relation for which

Domain: $x \in R$ and $-2 \leq x \leq 2$
Range: $y \in R$ and $-3 \leq y \leq 3$

Now that we have seen that the graph of $Ax^2 + By^2 + C$ (where A, B, $C > 0$, and $A \neq B$) is indeed an ellipse centered at the origin, we shall plot only the four intercept points and connect them with a smooth curve.

You complete Example 4.

Example 4. Graph $x^2 + 2y^2 = 4$. State its domain and range.

Solution

$$x^2 + 2y^2 = 4$$

The graph will be an ________ .

x-INTERCEPTS: Let $y = 0$

$$x^2 + 2(0)^2 = 4$$
$$x^2 = 4$$
$$x = \pm\_\_\_\_$$

(____, 0) and (____, 0)

y-INTERCEPTS: Let $x = 0$

$$0^2 + 2y^2 = 4$$
$$2y^2 = 4$$
$$y^2 = 2$$
$$y = \pm\sqrt{\_\_\_\_}$$

(0, ____) and (0, ____)

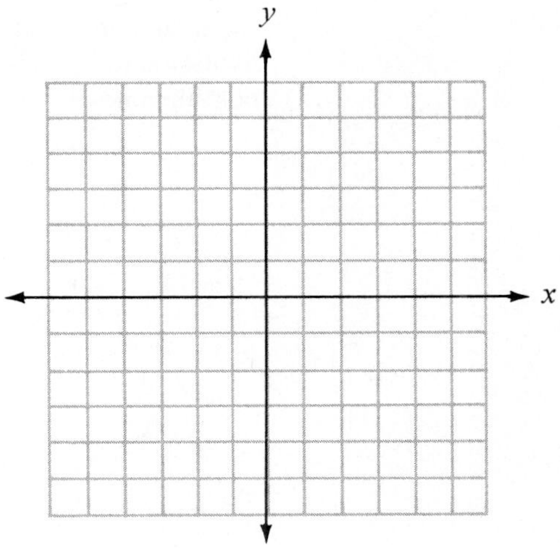

Domain: $x \in R$ and ____ $\leq x \leq$ ____
Range: $y \in R$ and ____ $\leq y \leq$ ____

Check your work on page 652. ▶

Trial Run

Graph each relation. State the domain and range.

Domain: ________ 1. $9x^2 + 4y^2 = 36$
Range: ________

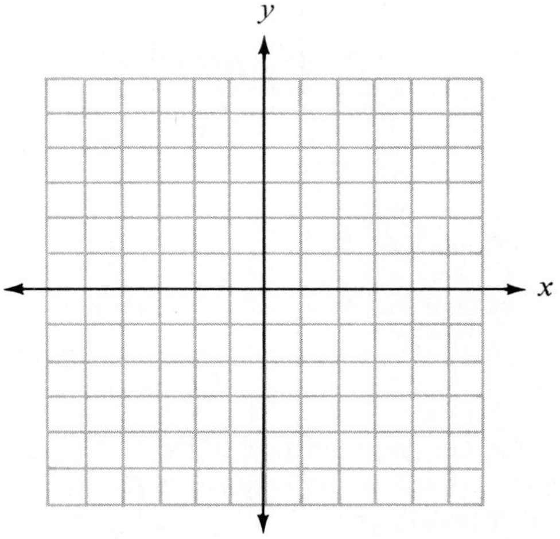

Domain: ________ 2. $2x^2 + 3y^2 = 6$
Range: ________

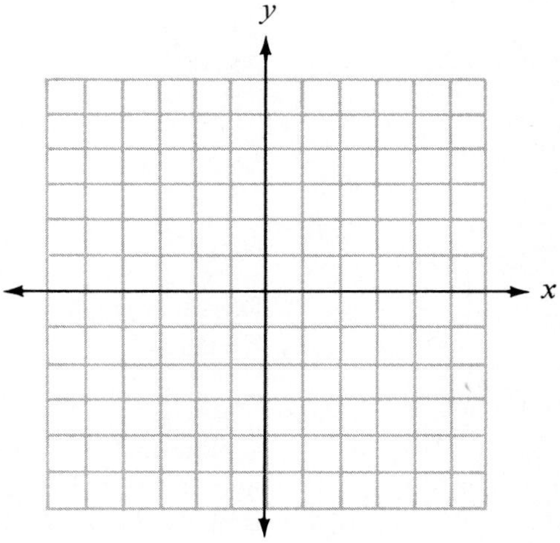

Answers are on page 654.

Graphing Hyperbolas

The final conic section that we must investigate is the **hyperbola.** Once again we restrict our discussion to hyperbolas that are centered at the origin. Hyperbolas centered at the origin will look like one of these figures.

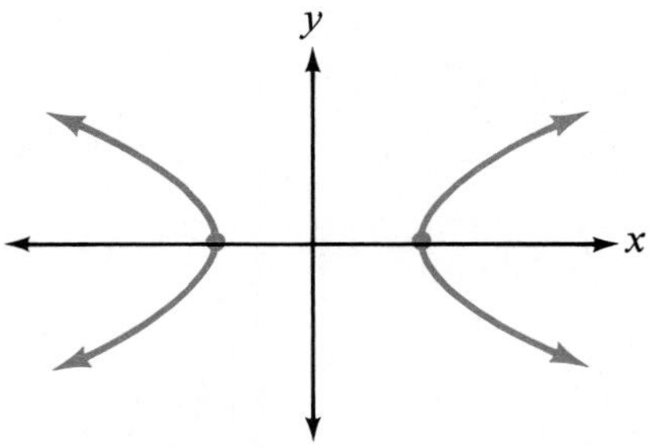

(1)

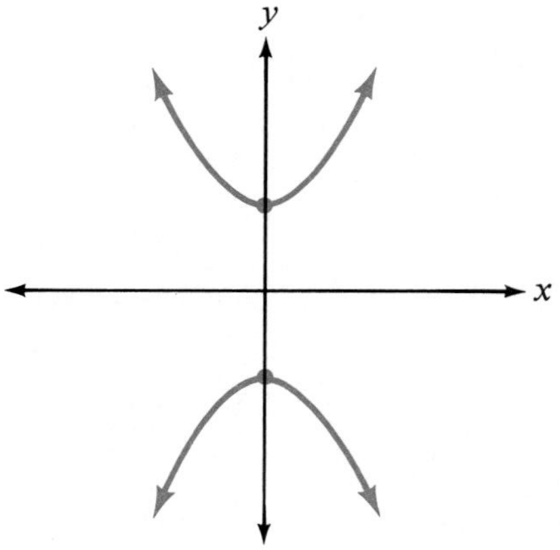

(2)

The general equation for a hyperbola is similar to the equation for an ellipse, except that *A and B must have opposite signs.*

> The graph of the equation
>
> $$Ax^2 + By^2 = C$$
>
> is a *hyperbola* (where A, B, and C are constants, and A and B have *opposite* signs).

Example 5. Graph $x^2 - 4y^2 = 16$. State its domain and range.

Solution. Let's find intercepts.

$$x^2 - 4y^2 = 16$$

x-INTERCEPTS: Let $y = 0$	y-INTERCEPTS: Let $x = 0$
$x^2 - 4(0)^2 = 16$	$0^2 - 4y^2 = 16$
$x^2 = 16$	$-4y^2 = 16$
$x = \pm 4$	$y^2 = -4$
$(4, 0)$ and $(-4, 0)$	There are no y-intercepts.

Since our graph does *not* cross the y-axis, it must resemble Figure 1. We need a few more points to be sure of its shape. From Figure 1 we know that only x-values to the left of $(-4, 0)$ or to the right of $(4, 0)$ will correspond to real y-values. To take advantage of the symmetry of the hyperbola, let's substitute two x-values equally spaced to the left of -4 and to the right of 4.

Let $x = -5$	Let $x = 5$
$x^2 - 4y^2 = 16$	$x^2 - 4y^2 = 16$
$(-5)^2 - 4y^2 = 16$	$(5)^2 - 4y^2 = 16$
$25 - 4y^2 = 16$	$25 - 4y^2 = 16$
$-4y^2 = -9$	$-4y^2 = -9$
$y^2 = \frac{9}{4}$	$y^2 = \frac{9}{4}$
$y = \pm\frac{3}{2}$	$y = \pm\frac{3}{2}$
$(-5, \frac{3}{2})$ and $(-5, -\frac{3}{2})$	$(5, \frac{3}{2})$ and $(5, -\frac{3}{2})$

Notice that our symmetrically chosen x-values yielded symmetric y-values. This will always occur if you choose x-values that are opposites. Now we sketch the graph.

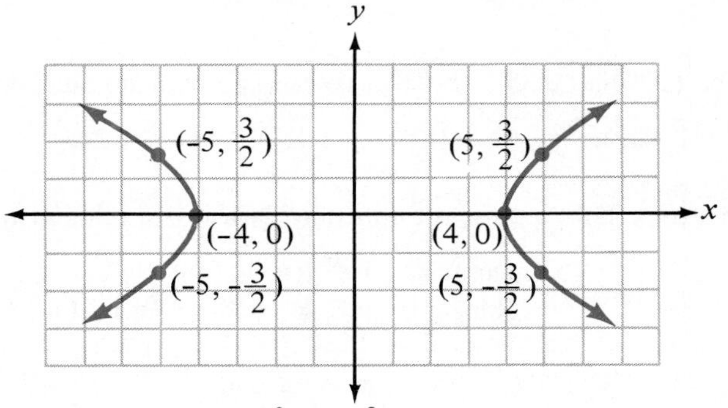

$x^2 - 4y^2 = 16$

Does this hyperbola represent a function? No, it is merely a relation for which

Domain: $x \in R$ and $x \geq 4$ or $x \leq -4$
Range: $y \in R$

Example 6. Graph $5y^2 - 2x^2 = 5$. State its domain and range.

Solution. The graph will be a hyperbola.

x-INTERCEPTS: Let $y = 0$

$$5(0)^2 - 2x^2 = 5$$
$$-2x^2 = 5$$
$$x^2 = -\frac{5}{2}$$

There are no x-intercepts.

y-INTERCEPTS: Let $x = 0$

$$5y^2 - 2(0)^2 = 5$$
$$5y^2 = 5$$
$$y^2 = 1$$
$$y = \pm 1$$

$(0, 1)$ and $(0, -1)$

This graph will resemble Figure 2, so we choose x-values equally spaced to the left and right of $(0, 1)$ and $(0, -1)$. Let's substitute $x = -1$ and $x = 1$ into the original equation.

Let $x = -1$

$$5y^2 - 2x^2 = 5$$
$$5y^2 - 2(-1)^2 = 5$$
$$5y^2 - 2 = 5$$
$$5y^2 = 7$$
$$y^2 = \frac{7}{5}$$
$$y = \pm\sqrt{\frac{7}{5}}$$

Let $x = 1$

$$5y^2 - 2x^2 = 5$$
$$5y^2 - 2(1)^2 = 5$$
$$5y^2 - 2 = 5$$
$$5y^2 = 7$$
$$y^2 = \frac{7}{5}$$
$$y = \pm\sqrt{\frac{7}{5}}$$

We should rationalize the denominators.

$$y = \frac{\pm\sqrt{7}}{\sqrt{5}} \cdot \frac{\sqrt{5}}{\sqrt{5}}$$
$$= \frac{\pm\sqrt{35}}{5}$$

The points on the hyperbola are

$$\left(-1, \frac{\sqrt{35}}{5}\right), \left(-1, -\frac{\sqrt{35}}{5}\right), \left(1, \frac{\sqrt{35}}{5}\right), \left(1, -\frac{\sqrt{35}}{5}\right)$$

Using the fact that $\sqrt{35} \doteq 5.916$, we can approximate our points by

$(-1, 1.2)$, $(-1, -1.2)$, $(1, 1.2)$, $(1, -1.2)$

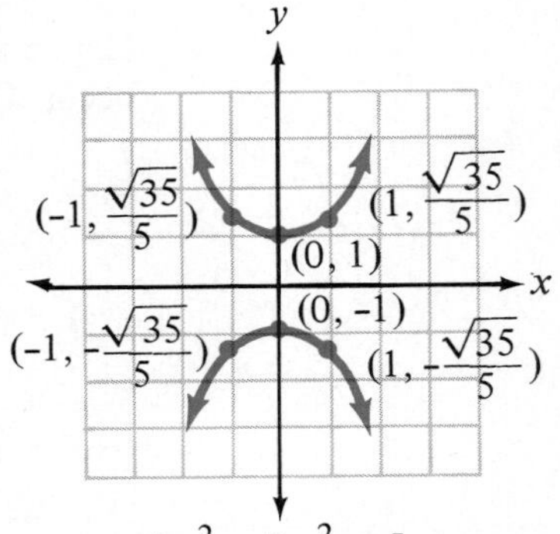

$5y^2 - 2x^2 = 5$

Does this hyperbola represent a function? No, it is a relation for which

Domain: $x \in R$
Range: $y \in R$ and $y \geq 1$ or $y \leq -1$

Trial Run

Graph each relation. State the domain and range.

Domain: ________ **1.** $4x^2 - 2y^2 = 16$
Range: ________

Domain: ________ **2.** $4y^2 - x^2 = 8$
Range: ________

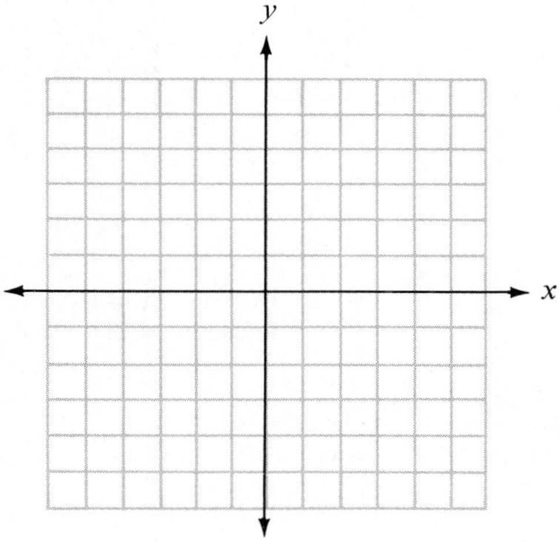

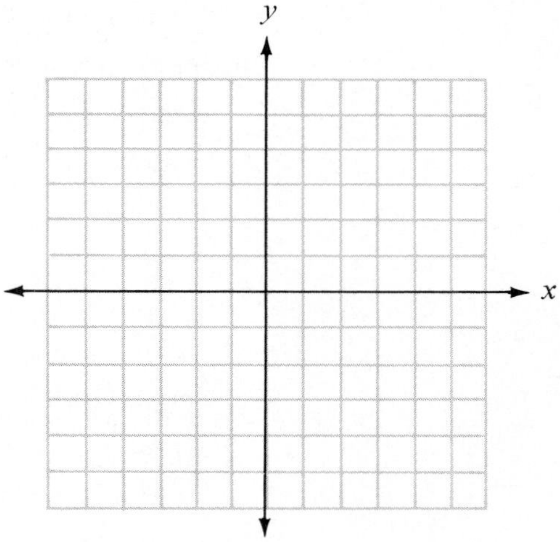

Answers are on page 654.

In discussing the conic sections we have learned to identify each graph from its equation.

Conic Section	Equation	Graphing Method
Parabola	$y = ax^2 + bx + c$ $(a \neq 0)$	Find intercepts Find axis of symmetry Find vertex
Circle	$x^2 + y^2 = r^2$ $(r > 0)$	Center at $(0, 0)$ Radius r
	$(x - h)^2 + (y - k)^2 = r^2$ $(r > 0)$	Center at (h, k) Radius r
Ellipse	$Ax^2 + By^2 = C$ $(A, B, C > 0; A \neq B)$	Find x-intercepts Find y-intercepts
Hyperbola	$Ax^2 + By^2 = C$ (A and B opposite in sign)	Find intercepts Find four symmetric points

We discovered that the parabola is the only conic section that represents a function.

▶ Examples You Completed

Example 1. Graph $2x^2 + 2y^2 = 4$. State its domain and range.

Solution

$$2x^2 + 2y^2 = 4$$

Dividing both sides by 2 yields

$$x^2 + y^2 = 2$$

The graph will be a circle.

CENTER: $(0, 0)$

RADIUS: $r^2 = 2$

$$r = \sqrt{2}$$

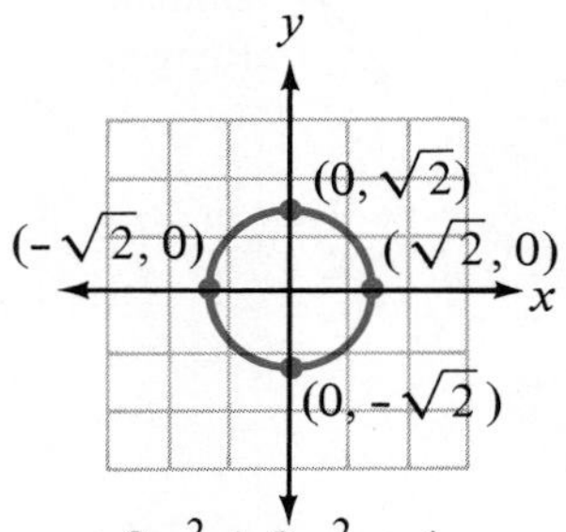

Domain: $x \in R$ and $-\sqrt{2} \le x \le \sqrt{2}$
Range: $y \in R$ and $-\sqrt{2} \le y \le \sqrt{2}$

Example 4. Graph $x^2 + 2y^2 = 4$. State its domain and range.

Solution

$$x^2 + 2y^2 = 4$$

The graph will be an ellipse.

x-INTERCEPTS: Let $y = 0$

$$x^2 + 2(0)^2 = 4$$

$$x^2 = 4$$

$$x = \pm 2$$

$(2, 0)$ and $(-2, 0)$

y-INTERCEPTS: Let $x = 0$

$$0^2 + 2y^2 = 4$$

$$2y^2 = 4$$

$$y^2 = 2$$

$$y = \pm\sqrt{2}$$

$(0, \sqrt{2})$ and $(0, -\sqrt{2})$

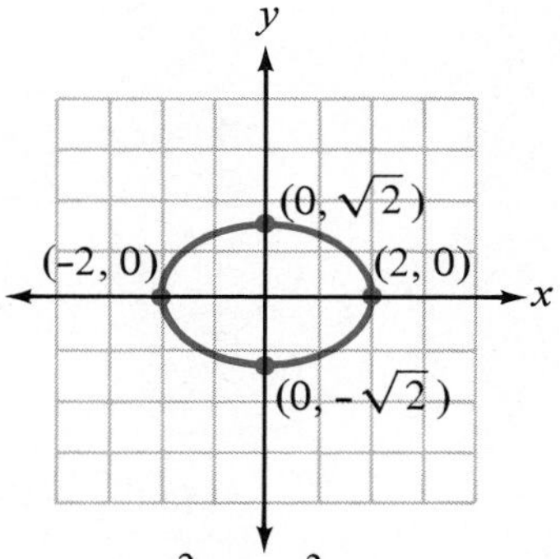

Domain: $x \in R$ and $-2 \le x \le 2$
Range: $y \in R$ and $-\sqrt{2} \le y \le \sqrt{2}$

Answers to Trial Runs

page 645 1.

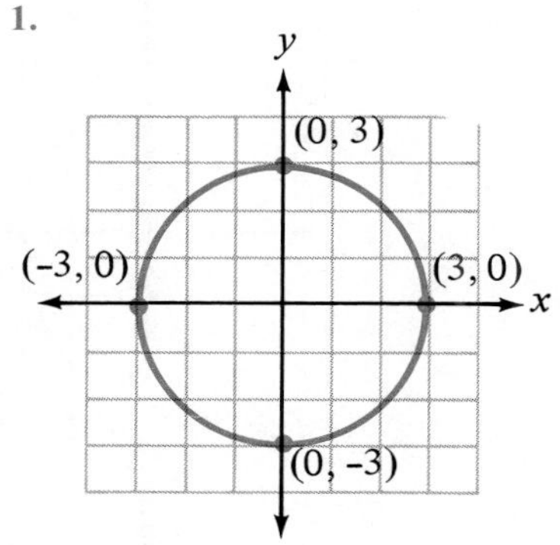

Domain: $x \in R$ and $-3 \leq x \leq 3$
Range: $y \in R$ and $-3 \leq y \leq 3$

2.

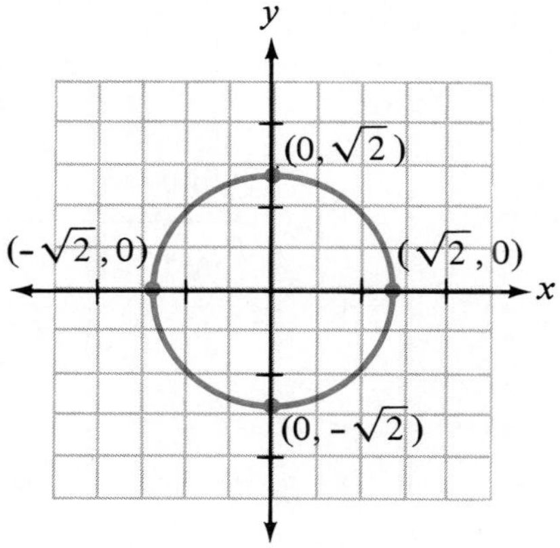

Domain: $x \in R$ and $-\sqrt{2} \leq x \leq \sqrt{2}$
Range: $y \in R$ and $-\sqrt{2} \leq y \leq \sqrt{2}$

3.

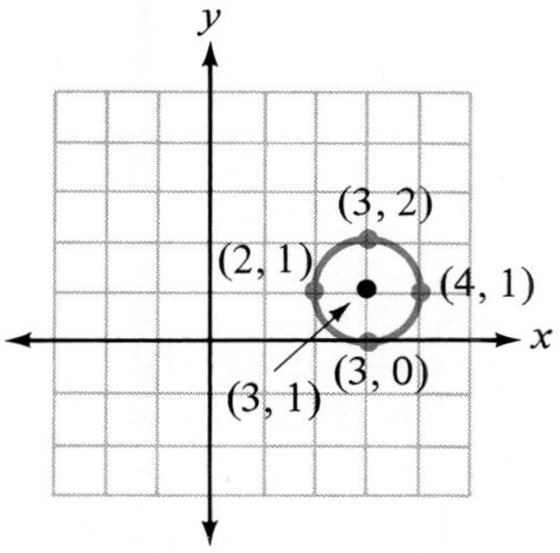

Domain: $x \in R$ and $2 \leq x \leq 4$
Range: $y \in R$ and $0 \leq y \leq 2$

4.

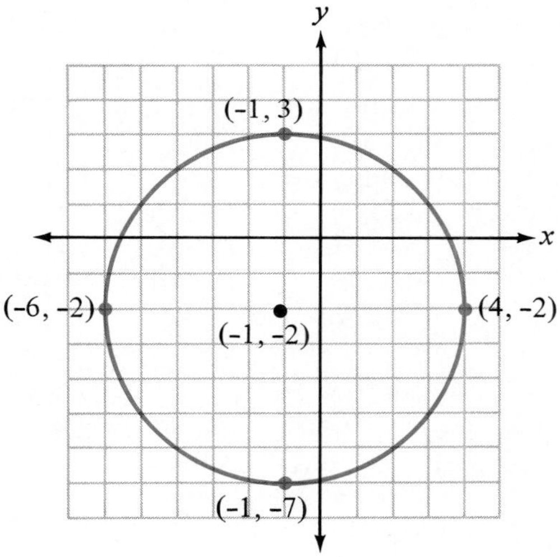

Domain: $x \in R$ and $-6 \leq x \leq 4$
Range: $y \in R$ and $-7 \leq y \leq 3$

page 648 **1.**

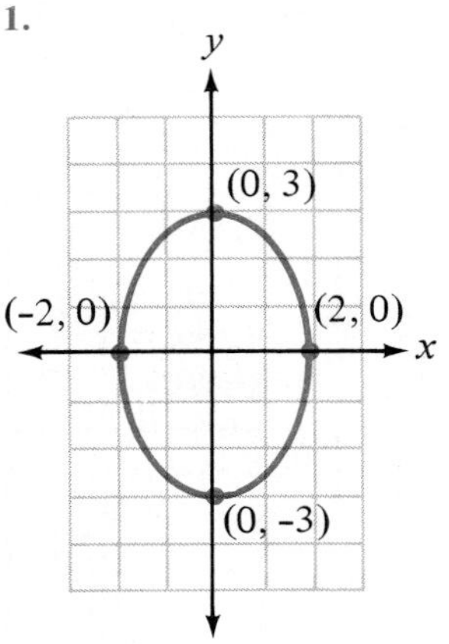

Domain: $x \in R$ and $-2 \leq x \leq 2$
Range: $y \in R$ and $-3 \leq y \leq 3$

2.

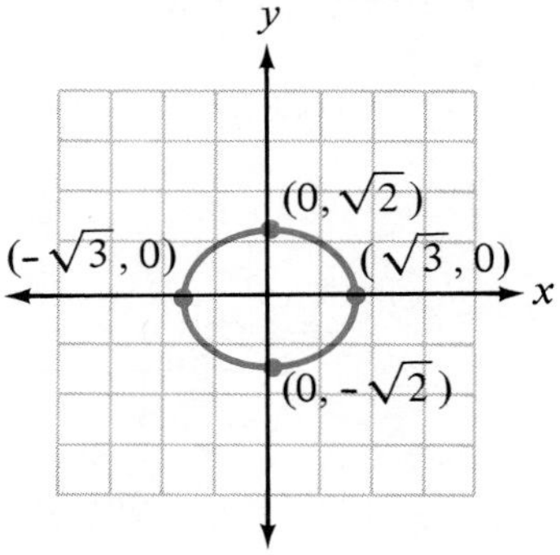

Domain: $x \in R$ and $-\sqrt{3} \leq x \leq \sqrt{3}$
Range: $y \in R$ and $-\sqrt{2} \leq y \leq \sqrt{2}$

page 651 **1.**

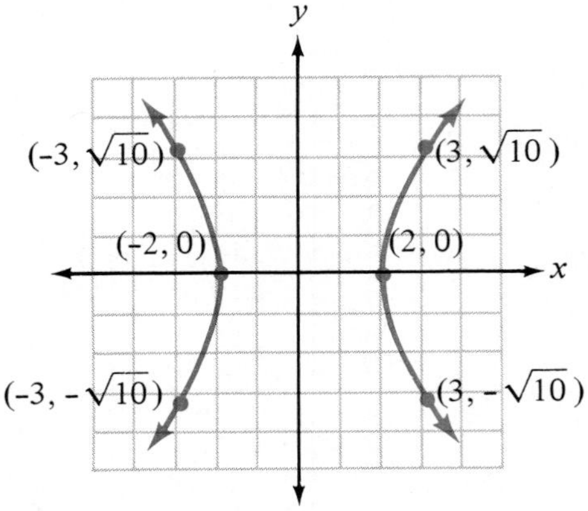

Domain: $x \in R$ and
$x \leq -2$ or $x \geq 2$
Range: $y \in R$

2.

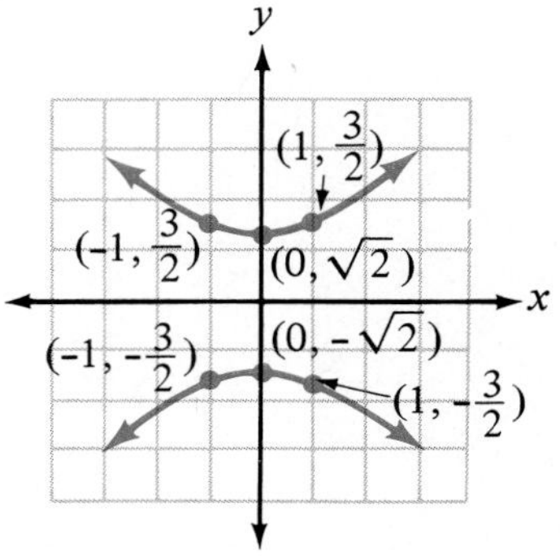

Domain: $x \in R$
Range: $y \in R$ and
$y \leq -\sqrt{2}$ or $y \geq \sqrt{2}$

Name ______________________ Date ____________

EXERCISE SET 9.3

Graph each relation. State the domain and range.

Domain: ________ 1. $x^2 + y^2 = 16$
Range: ________

Domain: ________ 2. $x^2 + y^2 = 25$
Range: ________

Domain: ________ 3. $x^2 + y^2 = 6$
Range: ________

Domain: ________ 4. $x^2 + y^2 = 10$
Range: ________

Domain: ________ 5. $5x^2 + 5y^2 = 45$
Range: ________

Domain: ________ 6. $4x^2 + 4y^2 = 36$
Range: ________

Domain: ________ 7. $y^2 = 16 - x^2$
Range: ________

Domain: ________ 8. $y^2 = 36 - x^2$
Range: ________

Domain: ________ 9. $(x - 1)^2 + (y - 3)^2 = 25$
Range: ________

Domain: ________ 10. $(x - 2)^2 + (y - 4)^2 = 9$
Range: ________

Domain: ________ 11. $(x + 3)^2 + (y - 1)^2 = 1$
Range: ________

Domain: ________ 12. $(x + 9)^2 + (y - 3)^2 = 81$
Range: ________

Domain: ________ 13. $x^2 + (y - 6)^2 = 4$
Range: ________

Domain: ________ 14. $x^2 + (y - 8)^2 = 49$
Range: ________

Domain: ________ 15. $(x + 2)^2 + (y + 7)^2 = 64$
Range: ________

Domain: ________ 16. $(x + 3)^2 + (y + 5)^2 = 144$
Range: ________

Domain: ________ 17. $9x^2 + 25y^2 = 225$
Range: ________

Domain: ________ 18. $4x^2 + 9y^2 = 36$
Range: ________

Domain: ________ 19. $16x^2 + y^2 = 16$
Range: ________

Domain: ________ 20. $9x^2 + y^2 = 9$
Range: ________

Domain: ________ 21. $12x^2 + y^2 = 36$
Range: ________

Domain: ________ 22. $x^2 + 8y^2 = 16$
Range: ________

Domain: ________ 23. $3x^2 + 5y^2 = 15$
Range: ________

Domain: ________ 24. $7x^2 + 5y^2 = 35$
Range: ________

Domain: ________ 25. $x^2 - 4y^2 = 4$
Range: ________

Domain: ________ 26. $x^2 - 16y^2 = 16$
Range: ________

Domain: ________ 27. $4y^2 - 25x^2 = 100$
Range: ________

Domain: ________ 28. $16y^2 - 9x^2 = 144$
Range: ________

Domain: ________ 29. $3x^2 - y^2 = 6$
Range: ________

Domain: ________ 30. $5x^2 - y^2 = 15$
Range: ________

Domain: ________ 31. $y^2 - x^2 = 100$
Range: ________

Domain: ________ 32. $y^2 - x^2 = 25$
Range: ________

☆ Stretching the Topics

________ 1. Write the equation of a circle with the center at $(-3, 4)$ and a radius of 5 units.

________ 2. Estimate to the nearest tenth the x- and y-intercepts of $6x^2 + 9y^2 = 27$.

________ 3. Graph $y = \frac{1}{3}\sqrt{81 - x^2}$. State its domain and range.

Check your answers in the back of your book.

If you can complete **Checkup 9.3**, you are ready to go on to Section 9.4.

Name ______________________ Date ____________

✓ CHECKUP 9.3

Graph each relation. State the domain and range.

Domain: ________ **1.** $x^2 + y^2 = 64$
Range: ________

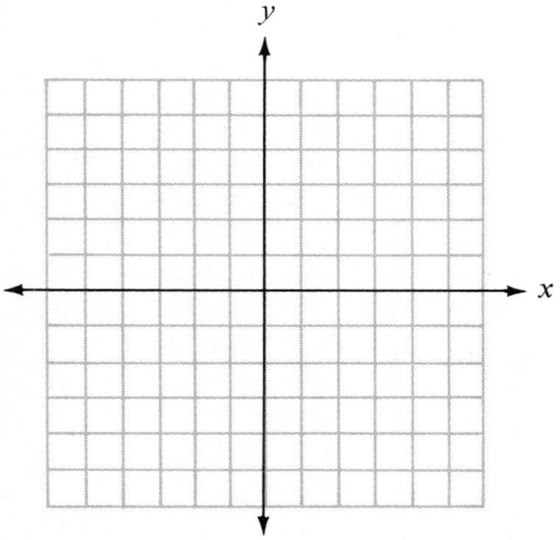

Domain: ________ **2.** $3x^2 + 3y^2 = 27$
Range: ________

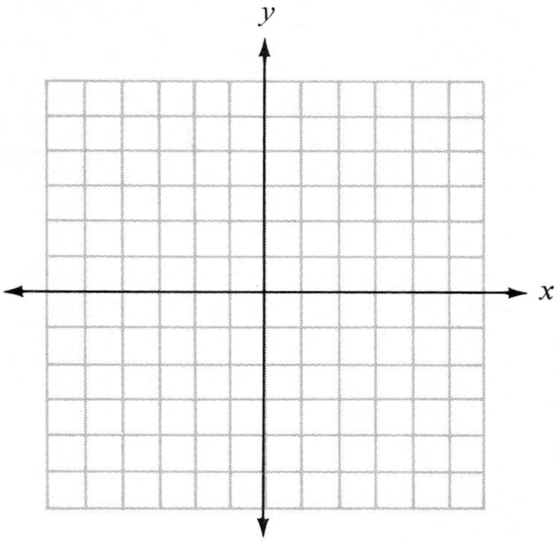

Domain: ________ **3.** $(x + 4)^2 + (y - 9)^2 = 81$
Range: ________

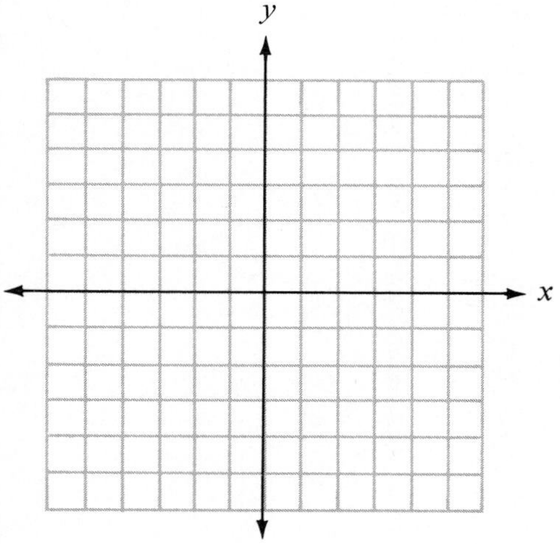

Domain: ________ **4.** $9x^2 + 16y^2 = 144$
Range: ________

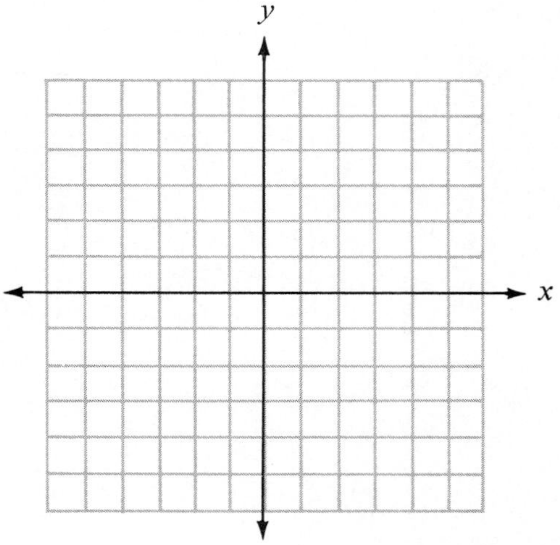

Domain: ________ 5. $16y^2 - 4x^2 = 64$
Range: ________

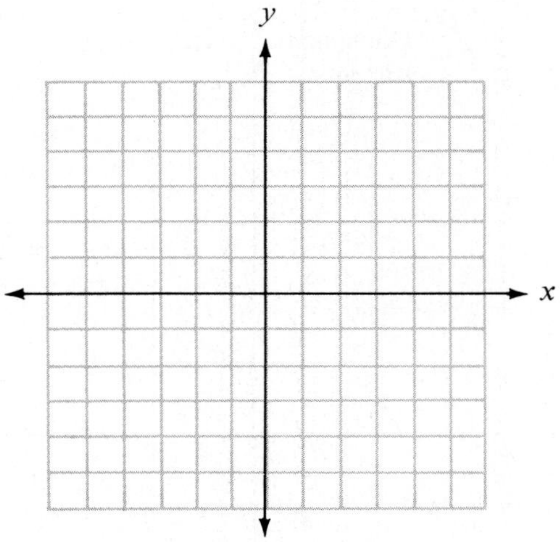

Check your answers in the back of your book.

If You Missed Problems:	You Should Review Examples:
1, 2	1
3	2
4	4
5	5, 6

9.4 Graphing Second-Degree Inequalities

Now that we have learned to graph second-degree equations, we turn our attention to graphing **second-degree inequalities.** Recall from Chapter 8 that we learned to graph a *first-degree inequality* by graphing the corresponding first-degree equation (straight line) and then shading the region containing the points that satisfied the inequality. We shall use the same approach with second-degree inequalities.

Example 1. Graph $y < x^2 - 2x - 8$.

Solution. First we must graph the corresponding boundary equation

$$y = x^2 - 2x - 8$$

which we recognize as a parabola.

y-INTERCEPT: Let $x = 0$

$$y = -8$$

$$(0, -8)$$

x-INTERCEPTS: Let $y = 0$

$$x^2 - 2x - 8 = 0$$

$$(x - 4)(x + 2) = 0$$

$$x = 4 \quad \text{or} \quad x = -2$$

$$(4, 0) \quad \text{and} \quad (-2, 0)$$

AXIS OF SYMMETRY: $x = -\dfrac{b}{2a}$

$$x = -\frac{-2}{2(1)}$$

$$x = 1$$

VERTEX: Let $x = 1$

$$y = 1^2 - 2(1) - 8$$

$$y = -9$$

$$(1, -9)$$

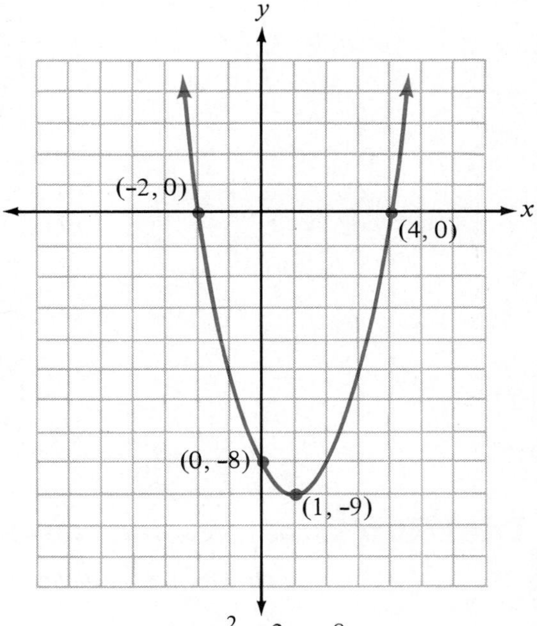

$y = x^2 - 2x - 8$

Notice that the parabola divides the plane into two regions: the region above the parabola and the region below the parabola. To find the region in which $y < x^2 - 2x - 8$, we may choose any point in the plane that does not satisfy the equation for the parabola. The origin (0, 0) is an easy point to use, so let's test it.

$y < x^2 - 2x - 8$	Original inequality.
$0 \overset{?}{<} 0^2 - 2(0) - 8$	Substitute 0 for x and 0 for y.
$0 \overset{?}{<} -8$	See if the inequality is true.

No; therefore, the region containing (0, 0) is *not* the correct region. The points above the parabola do *not* satisfy the inequality, so the points *below* the parabola must be correct.

We might try a point in that region, say (0, −10).

$$y \overset{?}{<} x^2 - 2x - 8$$

$$-10 \overset{?}{<} 0^2 - 2(0) - 8$$

$$-10 \overset{?}{<} -8$$

The inequality is satisfied, so we graph our inequality using a *dashed* curve (to indicate that the parabola itself is not in the solution) and shading *below* the dashed parabola.

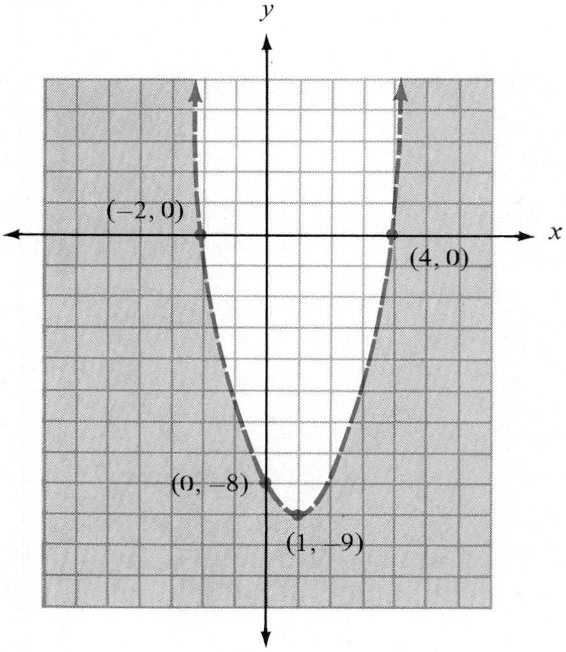

Trial Run

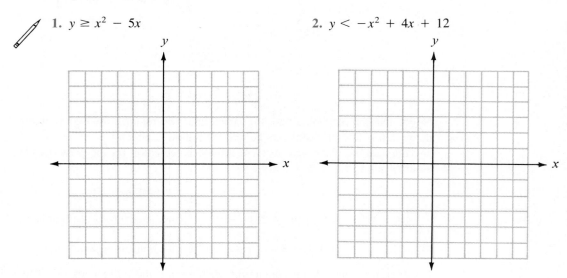

Graph each inequality.

1. $y \geq x^2 - 5x$

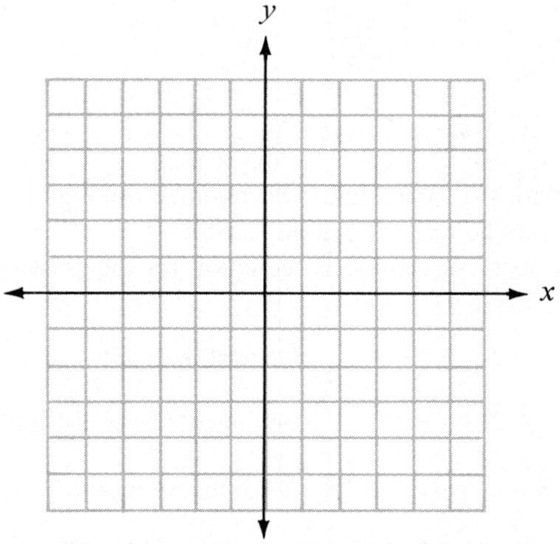

2. $y < -x^2 + 4x + 12$

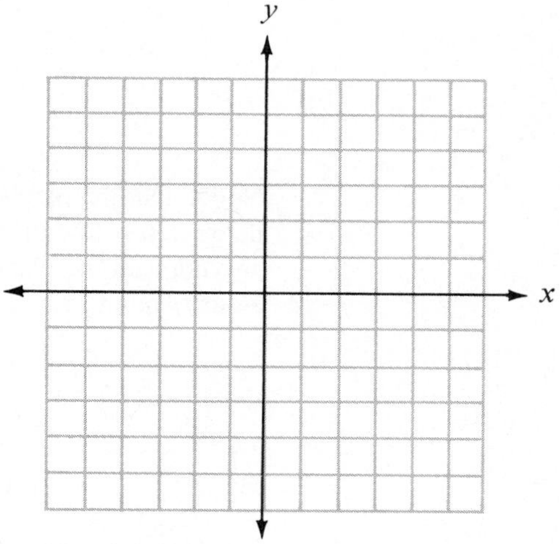

Answers are on page 663.

Similar techniques can be used to graph inequalities whose corresponding equations yield graphs that are circles, ellipses, or hyperbolas. We first graph the ordered pairs satisfying the boundary equation and then use a test point (often the origin) to determine the correct region in the plane.

Example 2. Graph the inequality $x^2 + y^2 \leq 4$.

Solution. First we graph the points satisfying the boundary equation

$$x^2 + y^2 = 4$$

which we recognize as a circle.

CENTER: (0, 0) RADIUS: $r = 2$

Now we test the point (0, 0) that lies *inside* the circle.

$$x^2 + y^2 \leq 4$$

$$0^2 + 0^2 \stackrel{?}{\leq} 4$$

$$0 \stackrel{?}{\leq} 4$$

Yes, points *inside or on* the circle satisfy the inequality.

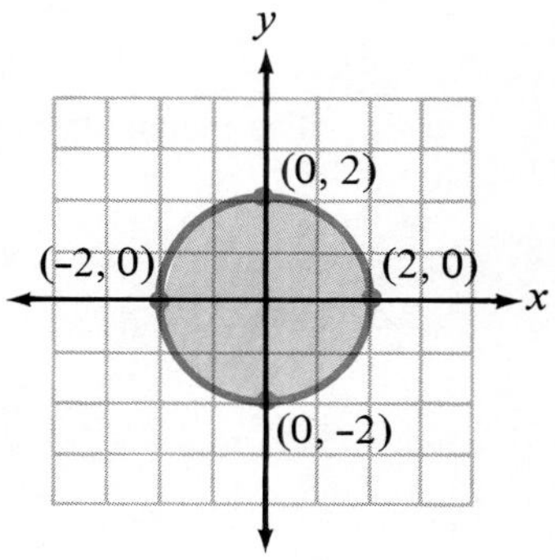

Example 3. Graph the inequality $3x^2 - y^2 \leq 12$.

Solution. First we graph the points satisfying the boundary equation

$$3x^2 - y^2 = 12$$

which we recognize as a hyperbola. Graphing by the usual methods, we have

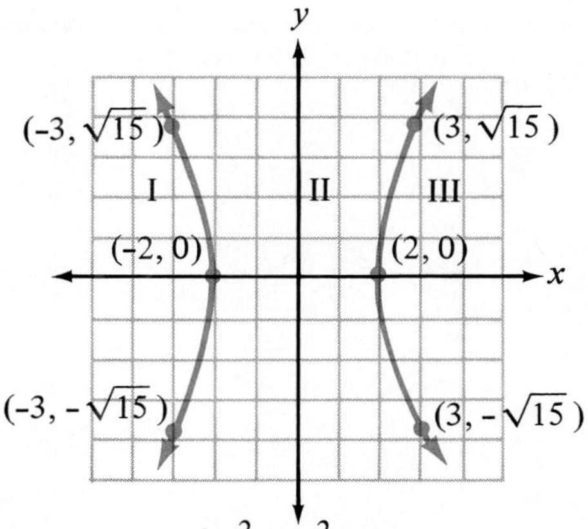

$3x^2 - y^2 = 12$

The hyperbola seems to divide the plane into *three* regions, each of which must be tested to see whether the inequality $3x^2 - y^2 \leq 12$ is satisfied.

Region I	Region II	Region III
Test point: $(-3, 0)$	Test point: $(0, 0)$	Test Point: $(3, 0)$
$3x^2 - y^2 \leq 12$	$3x^2 - y^2 \leq 12$	$3x^2 - y^2 \leq 12$
$3(-3)^2 - 0^2 \overset{?}{\leq} 12$	$3(0)^2 - 0^2 \overset{?}{\leq} 12$	$3(3)^2 - 0^2 \overset{?}{\leq} 12$
$3(9) - 0 \overset{?}{\leq} 12$	$0 - 0 \overset{?}{\leq} 12$	$3(9) - 0 \overset{?}{\leq} 12$
$27 \overset{?}{\leq} 12$	$0 \overset{?}{\leq} 12$	$27 \overset{?}{\leq} 12$
No	Yes	No

The inequality is satisfied only in region II *and* on the hyperbola itself. The graph becomes

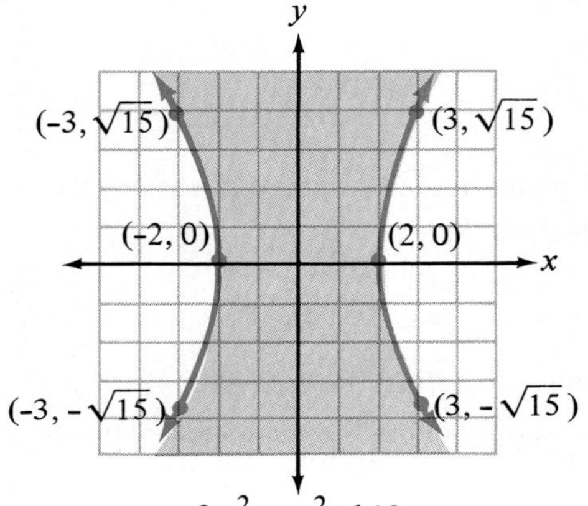

Trial Run

Graph each inequality.

1. $x^2 + y^2 \leq 9$

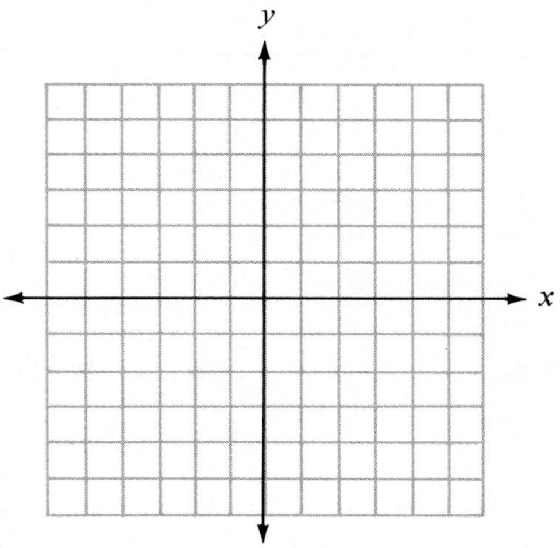

2. $x^2 + 4y^2 > 16$

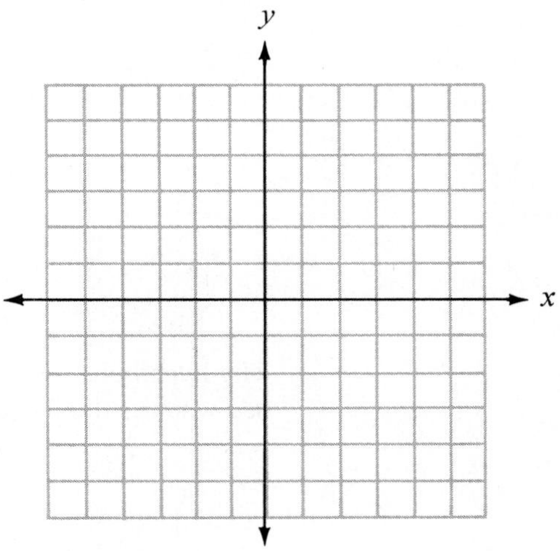

Answers are on page 663.

We have discovered that graphing second-degree inequalities requires a few short steps.

> **Graphing Second-Degree Inequalities**
>
> 1. Graph the boundary equation corresponding to the given inequality using a solid or dashed curve.
> 2. Test a point (often the origin) not on the graph of the boundary equation to determine whether it satisfies the inequality.
> 3. Decide on the appropriate region and shade it.

Notice that all inequalities represent relations that are *not* functions.

Answers to Trial Runs

page 660 1.

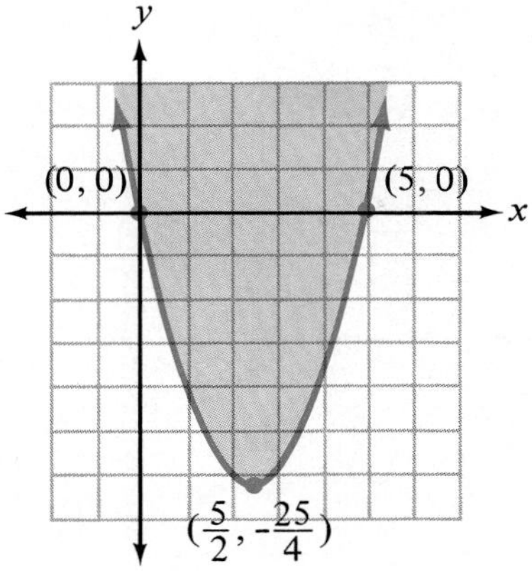

2.

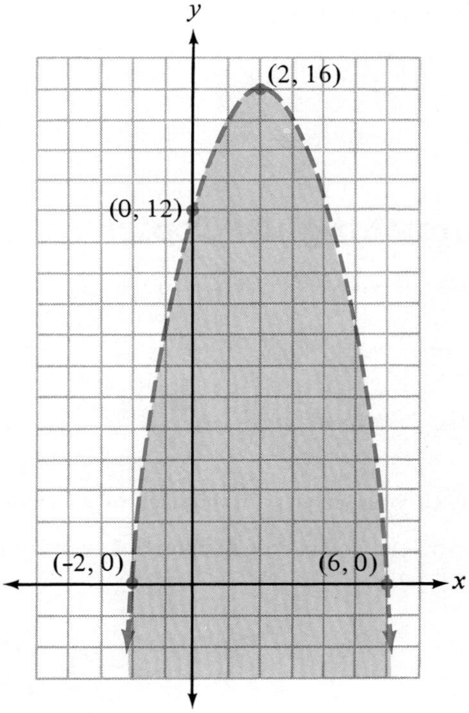

page 662 1.

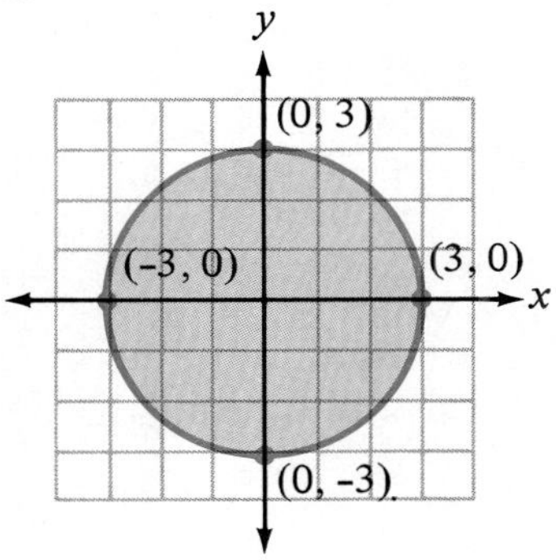

2.

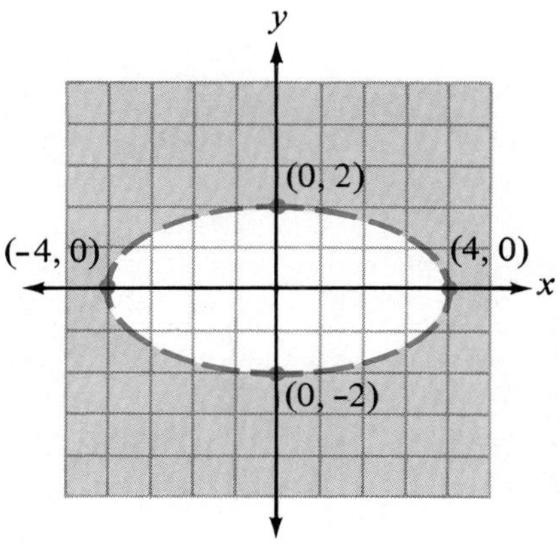

Name ______________________ Date ____________

EXERCISE SET 9.4

Graph each relation.

1. $y \geq 4x^2$
2. $y \geq x^2 - 2$
3. $y > 6x - x^2$
4. $y > 4x + x^2$
5. $y \leq x^2 - x - 20$
6. $y \leq x^2 + 5x - 14$
7. $y < -x^2 + 4x - 1$
8. $y < -x^2 + 2x + 5$
9. $x^2 + y^2 \geq 16$
10. $x^2 + y^2 \geq 25$
11. $3x^2 + 3y^2 < 9$
12. $5x^2 + 5y^2 < 10$
13. $(x - 1)^2 + (y + 3)^2 \leq 49$
14. $(x + 3)^2 + (y - 5)^2 \leq 36$
15. $4x^2 + 9y^2 < 36$
16. $25x^2 + y^2 < 25$
17. $9x^2 - y^2 \leq 36$
18. $4x^2 - y^2 \leq 16$
19. $y^2 - 4x^2 > 36$
20. $y^2 - x^2 > 25$

☆ Stretching the Topics

________ 1. Graph $y > x^3 - x$.

________ 2. Graph $xy < 4$.

________ 3. Graph $\frac{x^2}{9} + \frac{y^2}{4} \leq 1$.

Check your answers in the back of your book.

If you can complete **Checkup 9.4,** you are ready to do the **Review Exercises** for Chapter 9.

Name ______________________ Date ____________

✓ CHECKUP 9.4

Graph each relation.

1. $y \leq -x^2 + 4$

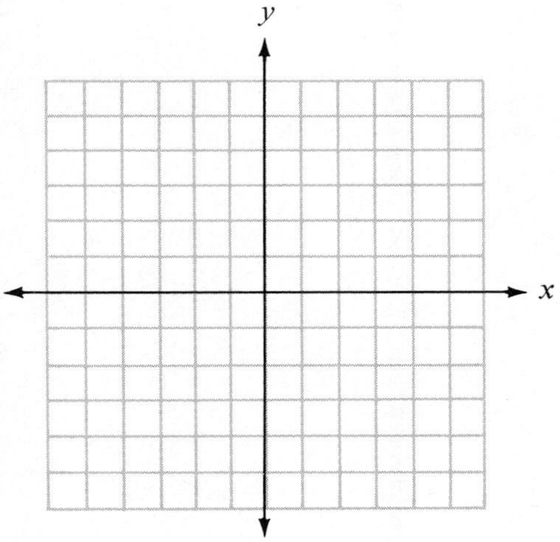

2. $x^2 + y^2 > 25$

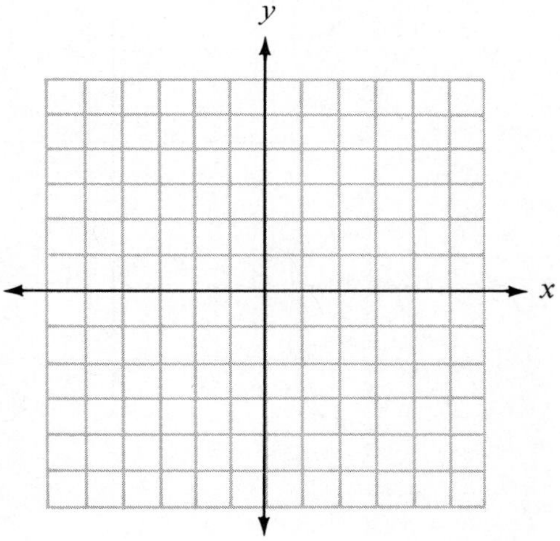

3. $(x - 3)^2 + (y + 1)^2 < 25$

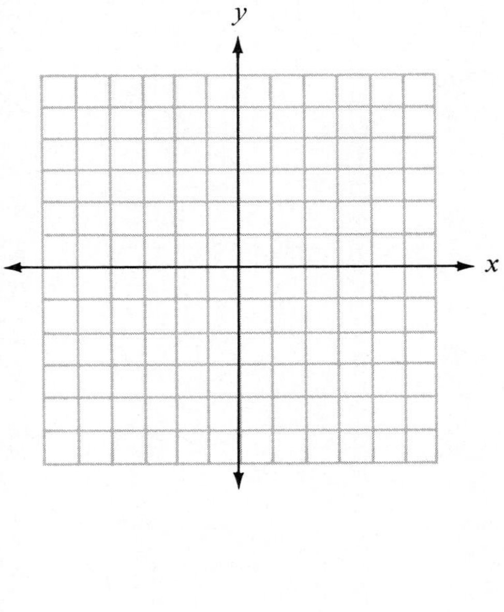

4. $4x^2 + 9y^2 \leq 36$

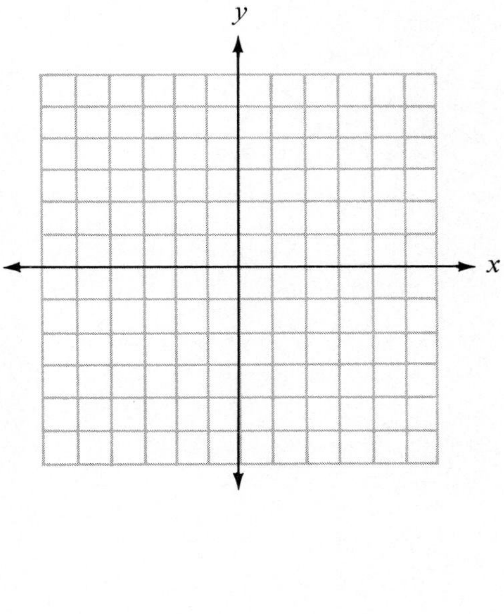

5. $9x^2 - y^2 \leq 81$

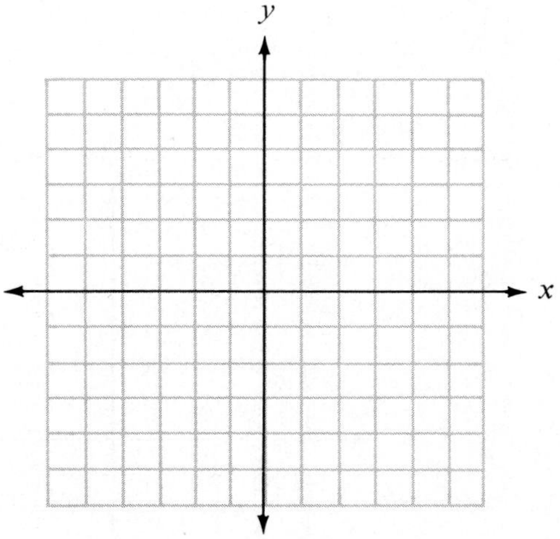

Check your answers in the back of your book.

If You Missed Problems:	You Should Review Example:
1	1
2–4	2
5	3

Summary

In this chapter we have extended our understanding of graphing solution sets to include ordered pairs satisfying nonlinear equations and inequalities. After using arbitrary points to determine several unfamiliar graphs, we made some observations about the **conic sections** and generalized the techniques for graphing them.

Conic Section	Equation	Graphing Method
Parabola	$y = ax^2 + bx + c$ $(a \neq 0)$	Find x- and y-intercepts Find axis of symmetry Find vertex
Circle	$(x - h)^2 + (y - k)^2 = r^2$ $(r > 0)$	Center at (h, k) Radius r
Ellipse	$Ax^2 + By^2 = C$ $(A, B, C > 0; A \neq B)$	Find x-intercepts Find y-intercepts
Hyperbola	$Ax^2 + By^2 = C$ (A and B opposite in sign)	Find x- and y-intercepts Find four symmetric points

To graph **second-degree inequalities,** we learned to graph the corresponding boundary equation and then choose the appropriate region using test points.

Throughout this chapter we agreed that the graph of ordered pairs satisfying an equation or inequality helped us determine whether those ordered pairs represented a function or a relation. We also found that it was possible to determine domain and range from the appearance of the graph.

❑ Speaking the Language of Algebra

Complete each sentence with the appropriate word or phrase.

1. The graph of ordered pairs satisfying the equation $y = ax^2 + bx + c$ is a ________ . If a is positive, we know that the graph opens ________ .

2. For the graph of $y = ax^2 + bx + c$, we know the axis of symmetry is $x =$ ________ and the graph's maximum or minimum point is called its ________ .

3. The graph of $3x^2 + 3y^2 = 9$ is a(n) ________ .
   The graph of $3x^2 + 2y^2 = 9$ is a(n) ________ .
   The graph of $3x^2 - 2y^2 = 9$ is a(n) ________ .

4. The graph of $x^2 + y^2 = 1$ is called the ________ circle. The graph of $x^2 + y^2 < 1$ is the region ________ the circle, while the graph of $x^2 + y^2 > 1$ is the region ________ the circle.

Writing About Mathematics

Write your response to each question in complete sentences.

1. What general observations can you make about the graph of $y = ax^2 + c$?

2. Explain how to use the distance formula to find the equation of a circle with center $(2, -1)$ and radius 3.

3. Tell how you would recognize the equation for each of the four conic sections and tell which of them represents a function.

4. Describe the process you would use to graph a second-degree inequality in two variables. Tell how you would decide whether the boundary curve is solid or dashed.

Name ________________________________ Date ______________

REVIEW EXERCISES for Chapter 9

Graph each relation. State the domain and range.

Domain: ________ **1.** $y = x^2 - 9$
Range: ________

Domain: ________ **2.** $y = -3x^2 + 10$
Range: ________

Domain: ________ **3.** $y = x^3 - 2$
Range: ________

Domain: ________ **4.** $y = x^3 - 4x$
Range: ________

Domain: ________ **5.** $y = \frac{5}{x}$
Range: ________

Domain: ________ **6.** $y = -x^2 + 6x$
Range: ________

Domain: ________ **7.** $y = 9x^2 - 12x + 14$
Range: ________

Domain: ________ **8.** $y = x^2 - 4x - 5$
Range: ________

Domain: ________ **9.** $x^2 + y^2 = 121$
Range: ________

Domain: ________ **10.** $x^2 + y^2 = 5$
Range: ________

Domain: ________ **11.** $(x - 3)^2 + y^2 = 49$
Range: ________

Domain: ________ **12.** $4x^2 + 25y^2 = 100$
Range: ________

Domain: ________ **13.** $x^2 - y^2 = 9$
Range: ________

Domain: ________ **14.** $9y^2 - 16x^2 = 144$
Range: ________

Graph each relation.

15. $y \geq x^2 - 4x$

16. $(x - 5)^2 + (y + 1)^2 > 25$

________ **17.** Deal on Wheels Tire Company finds its profit (y) is directly related to the number of sets of tires sold (x) by the function $y = -x^2 + 200x$. Use a graph to determine the maximum profit.

Check your answers in the back of your book.

If You Missed Problems:	You Should Review Examples:	
1, 2	SECTION 9.2	2–4
3, 4	SECTION 9.1	3, 4
5		5
6–8	SECTION 9.2	2–4, 6
9, 10	SECTION 9.3	1, 3
11		2
12		4
13, 14		5, 6
15	SECTION 9.4	1
16		2
17	SECTION 9.2	5

If you have completed the **Review Exercises** and corrected your errors, you are ready to take the **Practice Test** for Chapter 9.

Name ______________________ Date ______________

PRACTICE TEST for Chapter 9

Graph each relation. State the domain and range.

	SECTION	EXAMPLES
1. $y = 4x - 2x^2$	9.2	2

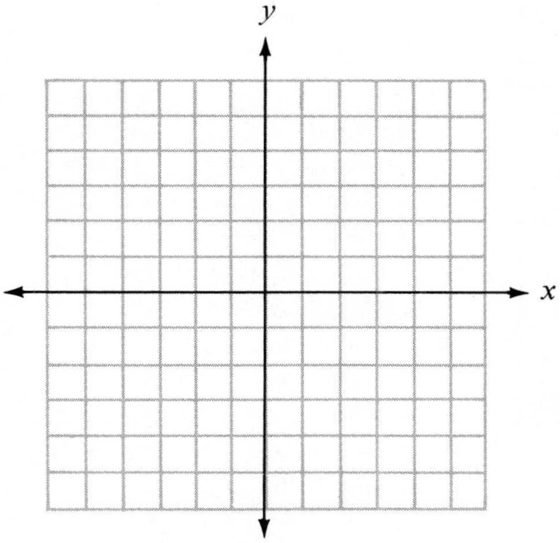

Domain: ________
Range: ________

2. $y = x^3 - 1$	9.1	3, 4

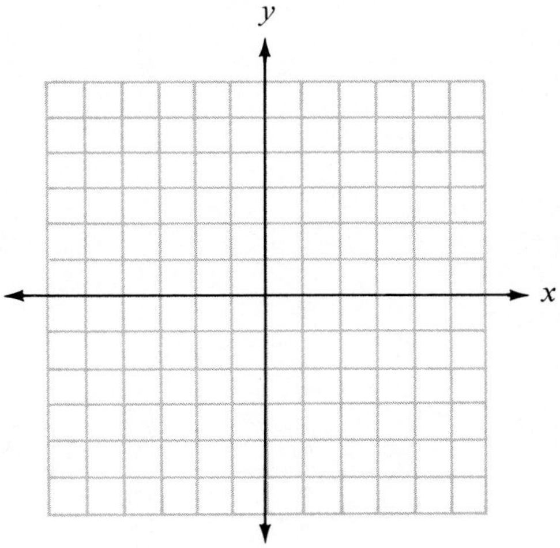

Domain: ________
Range: ________

	Section	Example
3. $y = \frac{4}{x}$	9.1	5

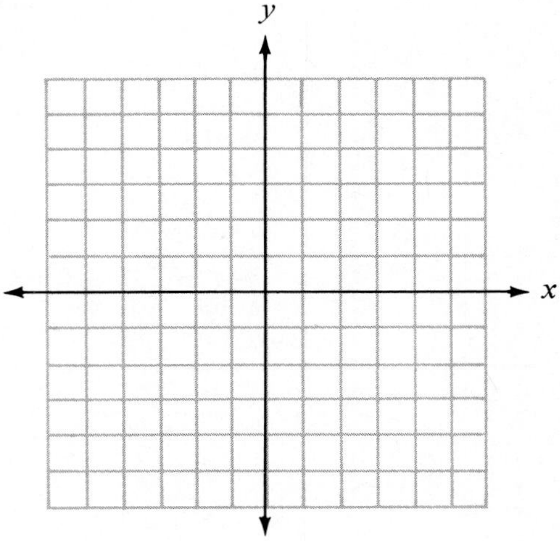

Domain: ________
Range: ________

4. $y = \frac{1}{2}x^2$	9.2	1

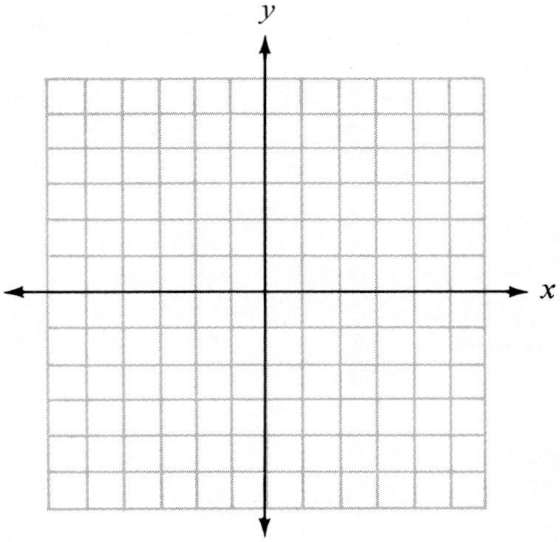

Domain: ________
Range: ________

	SECTION	EXAMPLE
5. $y = x^2 - x - 6$	9.2	4

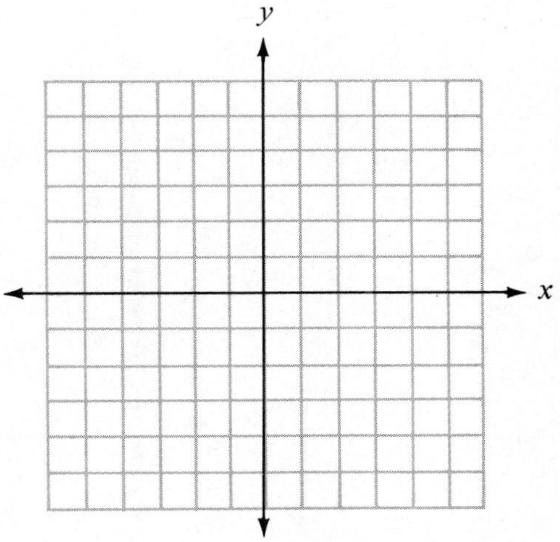

Domain: ________
Range: ________

6. $(x - 2)^2 + (y + 1)^2 = 4$	9.3	2

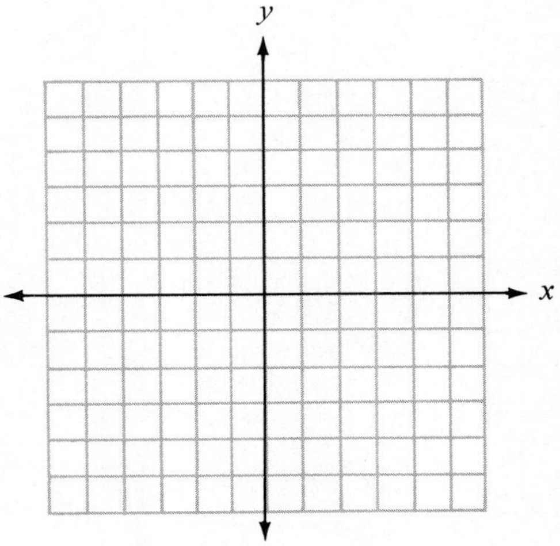

Domain: ________
Range: ________

	Section	Example
7. $x^2 + 5y^2 = 25$	9.3	4

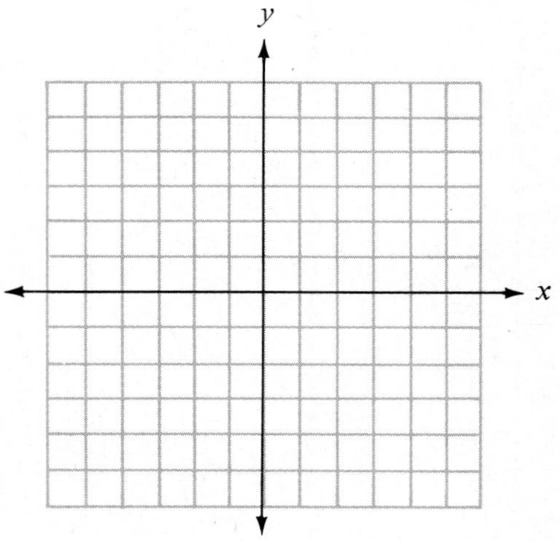

Domain: ________
Range: ________

	Section	Example
8. $3y^2 - x^2 = 12$	9.3	6

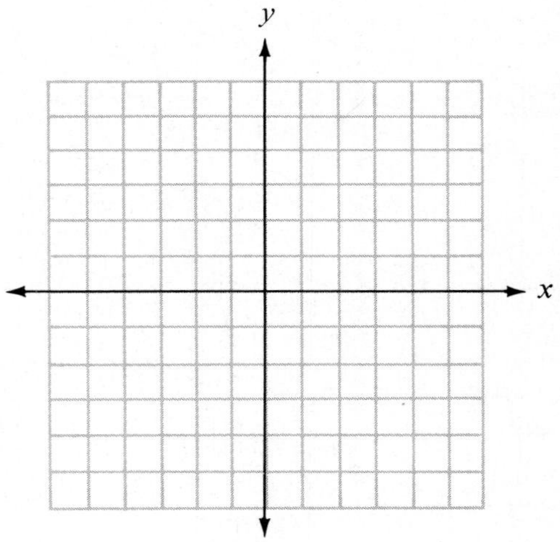

Domain: ________
Range: ________

Graph each relation. SECTION EXAMPLE

9. $y \geq x^2 - 2$ 9.4 1

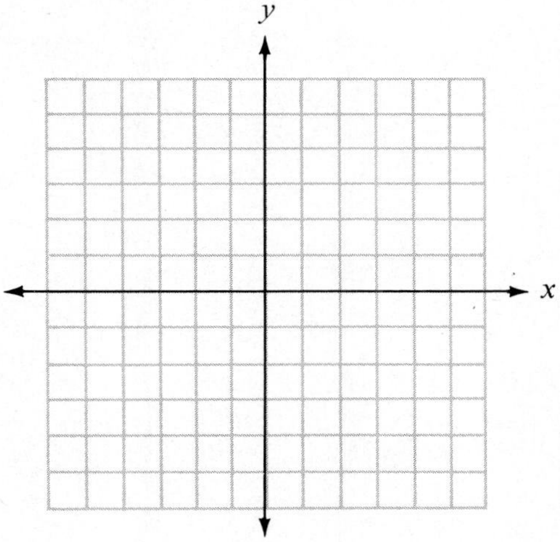

10. $x^2 + y^2 > 1$ 9.4 2

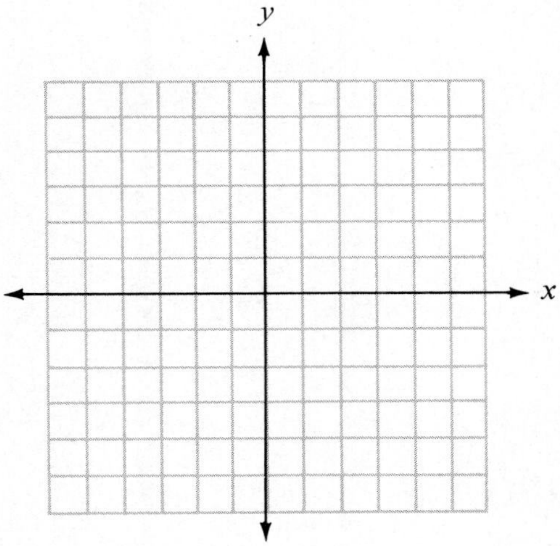

Name ______________________ Date ____________

SHARPENING YOUR SKILLS after Chapters 1–9

		SECTION
______	1. Evaluate $3a^2 - 2ab + b^2$ when $a = -2$ and $b = 3$.	3.1
______	2. Write $7 - \sqrt{-72}$ in standard complex number form.	7.5
______	3. Find the equation of the line through the points $(-\frac{2}{3}, \frac{7}{4})$ and $(\frac{13}{3}, -\frac{1}{4})$.	8.3
______	4. Find the distance between $(-3, -1)$ and $(5, -16)$.	8.2
______	5. Write a quadratic equation with the solutions $x = -1$ and $x = 7$.	5.1

Perform the indicated operation and simplify.

______	6. $-[3(a + 7b) + 7(2a - 3b)]$	1.4
______	7. $10a(-3a^7b^4)^2(ab^3)^3$	3.1
______	8. $(x^2y^{-9}z^{-3})(x^{-5}y^9z^5)$	3.2
______	9. $\dfrac{-7x^4y^{-3}}{x^{-2}y^{-2}}$	3.2
______	10. $a(a - 9)(a + 9) - 12a(a - 3)$	3.4
______	11. $(3a^3 - 8a^2 - 20a + 16) \div (3a - 2)$	3.5
______	12. $\dfrac{4x^2 - 11x + 6}{x^3 - 125} \div \dfrac{4x^2 + 13x - 12}{4x^3 + 20x^2 + 100x}$	6.3
______	13. $\dfrac{5}{x - 4} - \dfrac{3}{x + 3} - \dfrac{35}{x^2 - x - 12}$	6.3
______	14. $\dfrac{x^{-1} + 2y^{-1}}{x^{-2} - 4y^{-2}}$	6.4
______	15. $\left(\dfrac{x^{\frac{3}{4}}y^{-\frac{5}{12}}}{z^{\frac{1}{4}}}\right)^{\frac{4}{9}}$	7.1
______	16. $5\sqrt[4]{x^5y^8} - x\sqrt[4]{81xy^8}$	7.2
______	17. $\dfrac{1}{\sqrt{2} + \sqrt{5}}$	7.3

Solve each equation or inequality. SECTIONS

________ 18. $|7x - 2| > -8$ 2.4

________ 19. $2x(x + 1) + 5(x^2 - 4) = 5(2x - 4)$ 5.1

________ 20. $5x^2 - 6x - 2 = 0$ 5.3

________ 21. $5(x + 2) \geq 6(x + 1)$ 5.4

________ 22. $\dfrac{3}{3x + 2} - \dfrac{4}{3x - 2} = \dfrac{1}{9x^2 - 4}$ 6.5

________ 23. $\dfrac{x - 1}{2x - 3} < 1$ 6.5

________ 24. $\sqrt[3]{x^2 + 2x} = \sqrt[3]{x + 12}$ 7.4

________ 25. $125x^3 - 27 = 0$ 7.6

Graph using any method you choose.

26. $5x - y = 6$ 8.1, 8.2

27. $y = \dfrac{2}{3}x$ 8.1, 8.2

28. $x + 5y > -10$ 8.4

________ 29. The velocity v of a falling body varies directly as the time t during which it falls. Write an equation of variation and find k if at the end of 6 seconds a falling body has a velocity of 192 feet per second. Under the same conditions, what will be the velocity at the end of 8 seconds? 8.6

________ 30. Solve $h = \dfrac{2v^2}{g}$ for v. 5.2, 7.3

C H A P T E R

10

Solving Systems of Equations and Inequalities

During the Fourth of July Festival, Marcia sold hot dogs and hamburgers at her refreshment stand. On Friday she made a profit of $130 by selling 200 hot dogs and 150 hamburgers. On Saturday she made $225 by selling 300 hot dogs and 300 hamburgers. What profit did Marcia make on each hot dog and each hamburger during the two days?

In Chapter 2 we learned how to find all values of *one* variable that satisfied two inequalities at the same time. Now that we have discussed solutions for equations and inequalities containing *two* variables, we must investigate ways to find ordered pairs that satisfy more than one condition at the same time. A set of equations or inequalities that must be satisfied at the same time is called a system of simultaneous equations or inequalities, or simply a **system of equations** or inequalities. Finding the solutions that satisfy all equations in a given system is called **solving a system of equations.** In this chapter we learn how to

1. Solve systems of linear equations by graphing.
2. Solve systems of linear equations by substitution.
3. Solve systems of linear equations by addition.
4. Solve systems of linear equations in three variables.
5. Solve nonlinear systems of equations.
6. Solve systems of inequalities.
7. Switch from words to systems of equations.

10.1 Solving Systems of Linear Equations by Graphing and Substitution

Solving Systems by Graphing

If you were asked to find all the ordered pairs that satisfy *both* equations in the system

$$(1)\quad x + y = 7$$
$$(2)\quad 2x - y = -10$$

you might be able to come up with an ordered pair by trial and error, but that would be a tedious process.

Considered separately, the set of ordered pairs satisfying each of these equations can be graphed as a straight line. Let's graph those straight lines and look for any points that the two lines have in common. After all, any *common* points would represent ordered pairs satisfying *both* equations, and that is what we are looking for.

(1) $x + y = 7$

x-INTERCEPT: Let $y = 0$

$x + 0 = 7$

$x = 7$ $\quad (7, 0)$

y-INTERCEPT: Let $x = 0$

$0 + y = 7$

$y = 7$ $\quad (0, 7)$

(2) $2x - y = -10$

x-INTERCEPT: Let $y = 0$

$2x - 0 = -10$

$2x = -10$

$x = -5$ $\quad (-5, 0)$

y-INTERCEPT: Let $x = 0$

$2(0) - y = -10$

$-y = -10$

$y = 10$ $\quad (0, 10)$

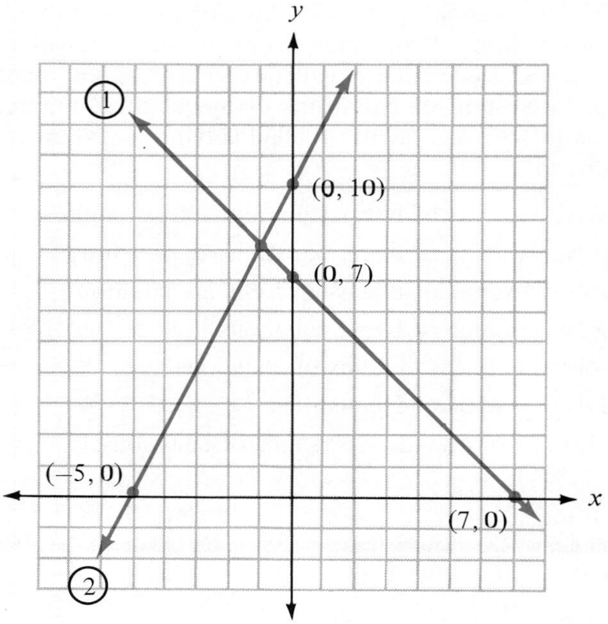

(1) $x + y = 7$

(2) $2x - y = -10$

You should agree that there is just one point common to both lines, and it appears that the coordinates of that point are $(-1, 8)$. But we must check to be sure that point satisfies both equations in the system.

(1)		(2)	
	$x + y = 7$		$2x - y = -10$
	$-1 + 8 \stackrel{?}{=} 7$		$2(-1) - 8 \stackrel{?}{=} -10$
	$7 = 7$		$-10 = -10$

Indeed, the ordered pair $(-1, 8)$ is a solution for our system. Are there any other possible solutions? From the graph we see that these lines have exactly one point in common. Since we have found that point, we know that $(-1, 8)$ is the *only* solution for this system. You try to complete Example 1.

Example 1. Use graphing to solve the system.

$$(1)\ y = 3$$
$$(2)\ y = 5x - 2$$

Solution. First we graph each equation.

(1) $y = 3$

Horizontal line

y-intercept: 3

(2) $y = 5x - 2$

Slope: $m =$ _____

y-intercept: $b =$ _____

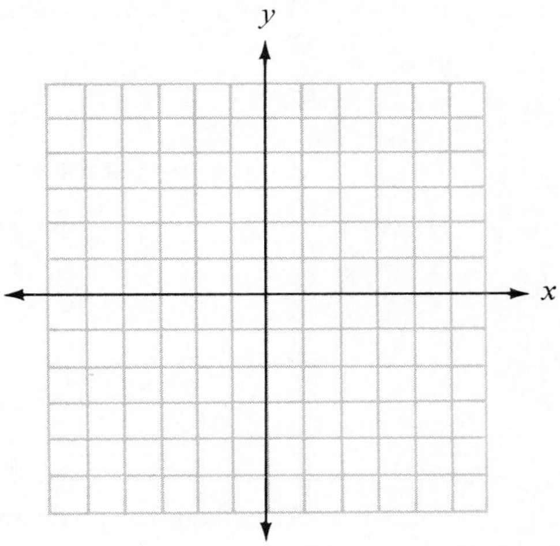

It appears that (_____ , _____) is the common point for the lines. We can check this ordered pair in each equation.

(1)		(2)	
	$y = 3$		$y = 5x - 2$
	_____ $= 3$		_____ $\stackrel{?}{=} 5($_____$) - 2$
			_____ $=$ _____

Check your work on page 688. ▶

Example 2. Use graphing to solve the system.

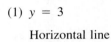

$$(1)\ y = -2x$$
$$(2)\ 6x + 3y = 9$$

Solution. First we graph each equation.

(1) $y = -2x$

Slope: $m = -2$

y-intercept: $b = 0$

(2) $6x + 3y = 9$

$3y = -6x + 9$

$y = -2x + 3$

Slope: $m = -2$

y-intercept: $b = 3$

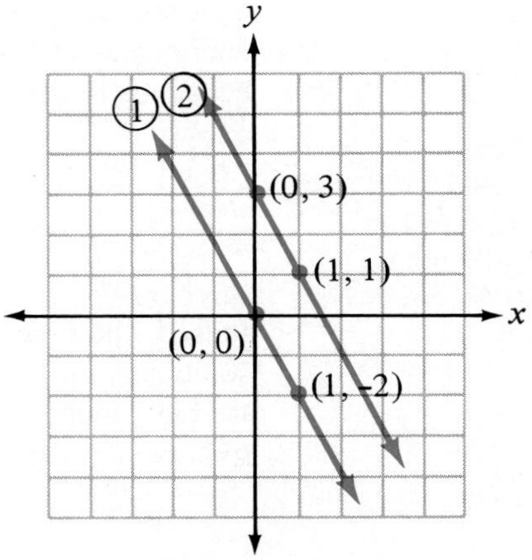

(1) $y = -2x$

(2) $6x + 3y = 9$

The fact that these lines are parallel should not be surprising, since they have the same slope and different y-intercepts. They share *no* common point, so there is *no solution* for this system of equations.

Example 3. Use graphing to solve the system.

(1) $2x - 3y = 9$

(2) $6y = 4x - 18$

Solution. First we graph each equation.

(1) $2x - 3y = 9$

$-3y = -2x + 9$

$y = \frac{2}{3}x - 3$

Slope: $m = \frac{2}{3}$

y-intercept: $b = -3$

(2) $6y = 4x - 18$

$y = \frac{4}{6}x - 3$

$y = \frac{2}{3}x - 3$

Slope: $m = \frac{2}{3}$

y-intercept: $b = -3$

Since both lines have the same slope and y-intercept, their graphs must be the same straight line.

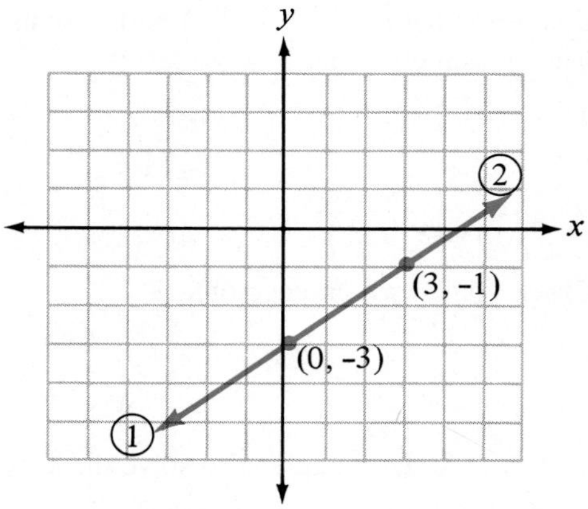

(1) $2x - 3y = 9$

(2) $6y = 4x - 18$

What solutions do the two equations share? Any ordered pair that satisfies one equation will satisfy the other. We may write the solution to the system as:

All ordered pairs satisfying $2x - 3y = 9$

or

All ordered pairs satisfying $6y = 4x - 18$

Systems of linear equations can be classified according to the number of solutions they have.

Independent and consistent system. The system has *one* solution (intersecting lines).
Inconsistent system. The system has *no* solutions (parallel lines).
Dependent system. The system has an *infinite number* of solutions (coinciding lines).

Now try Example 4.

Example 4. Solve and classify the system.

$$(1)\ 3x + y = 5$$
$$(2)\ y = -\tfrac{1}{2}x$$

Solution. First we graph each equation.

(1) $3x + y = 5$

$y = -3x + 5$

Slope: $m =$ _____

y-intercept: $b =$ _____

(2) $y = -\frac{1}{2}x$

Slope: $m =$ _____

y-intercept: $b =$ _____

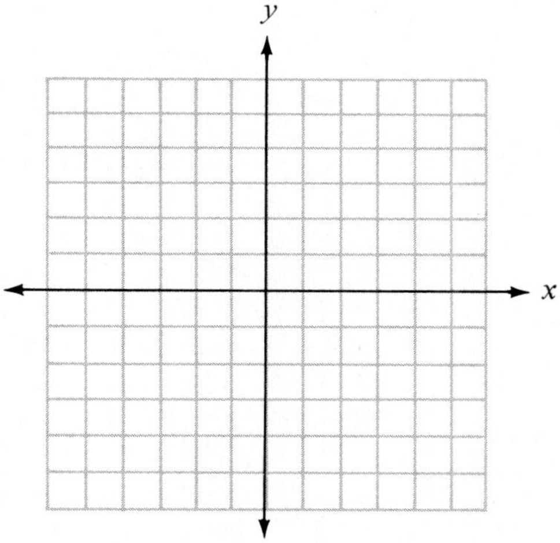

(1) $3x + y = 5$

(2) $y = -\frac{1}{2}x$

The common point seems to be (_____ , _____). Let's check in each equation.

(1) $3x + y = 5$

$3(\_\_\_\_) + (\_\_\_\_) \stackrel{?}{=} 5$

$\_\_\_\_ - \_\_\_\_ \stackrel{?}{=} 5$

$\_\_\_\_ = 5$

(2) $y = -\dfrac{1}{2}x$

$(\_\_\_\_) \stackrel{?}{=} -\dfrac{1}{2}(\_\_\_\_)$

$\_\_\_\_ = \_\_\_\_$

The solution for the system is (_____ , _____) and the system is _________ and _________ .

Check your work on page 689. ▶

Trial Run

Solve by graphing and classify the system.

_______ 1. $3x - y = 4$
$2x + y = 11$

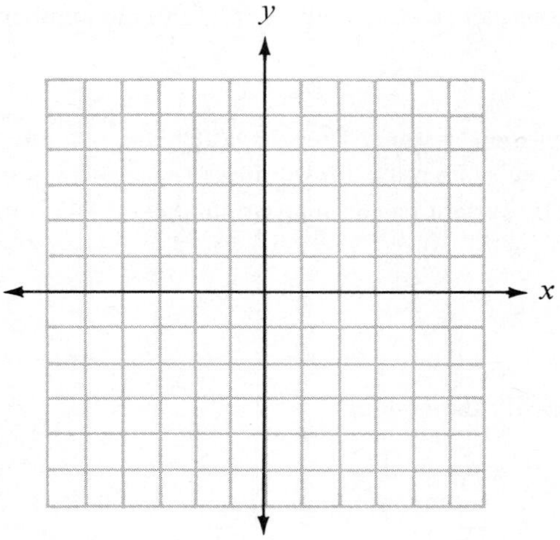

_______ 2. $y = 5x - 2$
$10x - 2y = 4$

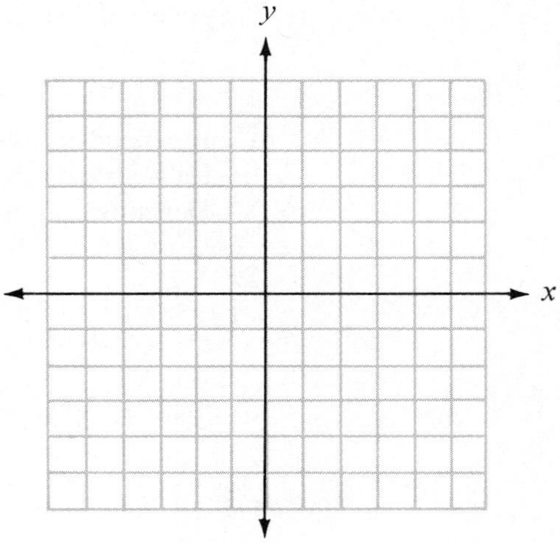

_______ 3. $2x - 3y = 3$
$y = 3$

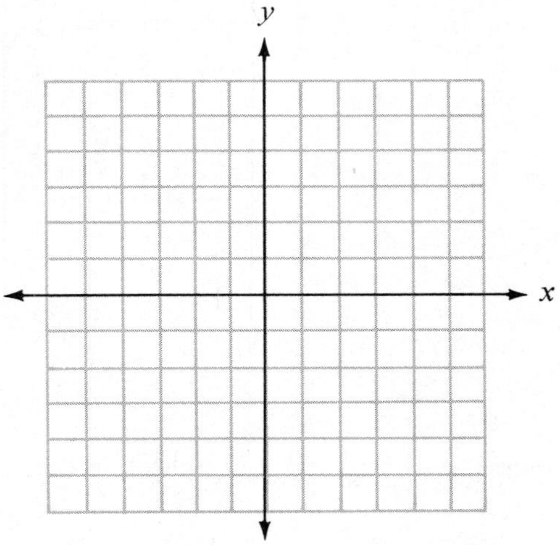

_______ 4. $x + 2y = 6$
$y = -\frac{1}{2}x + 1$

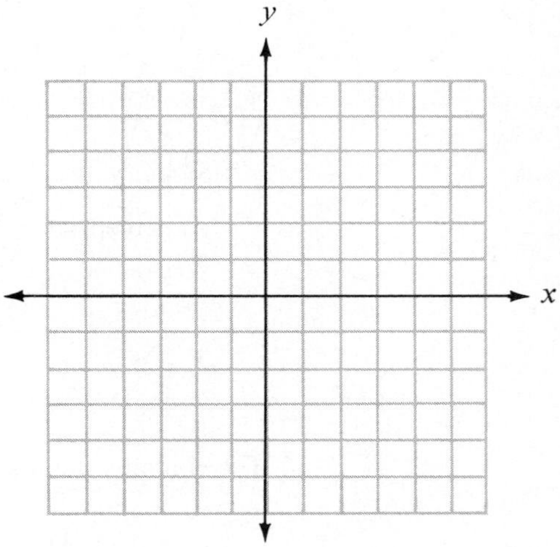

Answers are on page 690.

Solving Systems by Substitution

It has probably occurred to you that the method of graphing is useful in solving linear systems only when it is easy to determine the point where two lines intersect. If two lines intersected at a point such as $(\frac{2}{17}, -\frac{4}{17})$, it would be almost impossible to locate that point with any accuracy using only a graph. Or if the lines intersected at a point such as $(124, -67)$, graphing would be a very awkward way of locating that point.

Therefore, we would like to find an algebraic method for solving systems of equations without graphing. In fact, there are two algebraic methods that can be used; we first discuss the **method of substitution.**

In solving a system such as

$$(1)\ x + y = 7$$

$$(2)\ y = 2x + 10$$

remember that we are looking for the ordered pair that will satisfy equation (1) and equation (2) at the same time. Because we are looking for a *common* (x, y), we may "borrow" the expression for y from equation (2) and use it in equation (1).

$x + y = 7$	Equation (1).
$y = 2x + 10$	Equation (2).
$x + (2x + 10) = 7$	Substitute $2x + 10$ for y in equation (1).
$3x + 10 = 7$	Combine like terms.
$3x = -3$	Isolate the x-term.
$x = -1$	Isolate x.

We have found the x-coordinate for the common point. To find the corresponding y-coordinate, we can substitute -1 for x in either of the original equations.

(1)	$x + y = 7$	or	(2) $y = 3x + 10$
	$-1 + y = 7$		$y = 2(-1) + 10$
	$y = 8$		$y = 8$

The solution for the system is $(-1, 8)$, which matches the solution obtained earlier by graphing.

The method of substitution is especially useful if in either of the original equations, one of the variables is already isolated or has a coefficient of 1 and therefore can be easily isolated.

Solving a System of Equations by Substitution

1. Look for a variable appearing in an equation with a coefficient of 1, and isolate that variable in its equation.
2. Substitute the expression obtained for the isolated variable into the other equation.
3. Solve for the value of the remaining variable.
4. Substitute that value into either of the original equations and solve for the other variable.
5. Write the solution as an ordered pair.

Example 5. Solve the system by substitution.

$$(1)\quad 3y - 2x = 10$$

$$(2)\quad 2y + x = 9$$

Solution. In equation (2) we can isolate x.

$$2y + x = 9$$

$$x = 9 - 2y$$

Now we substitute $9 - 2y$ for x in equation (1) and solve for y.

$$3y - 2(9 - 2y) = 10$$

$$3y - 18 + 4y = 10$$

$$7y - 18 = 10$$

$$7y = 28$$

$$y = 4$$

We find the corresponding x-value by substituting 4 for y in either equation. The easiest way to find x is by using equation (2) with x isolated.

$$x = 9 - 2y$$

$$x = 9 - 2(4)$$

$$x = 1$$

The ordered pair (1, 4) is the solution for the system. You should check to be sure that the solution satisfies equation (1).

You try Example 6.

Example 6. Solve the system by substitution.

$$(1)\quad y - 5 = 0$$

$$(2)\quad 3x + 2y = 8$$

Solution. We can isolate ____ in equation (1).

$$y - 5 = 0$$

$$y = \underline{\qquad}$$

Now we know the y-value for our solution, so we can substitute ____ for y in equation (2) and solve for x.

$$3x + 2y = 8$$

$$3x + 2(\underline{\qquad}) = 8$$

$$3x + \underline{\qquad} = 8$$

$$3x = \underline{\qquad}$$

$$x = \underline{\qquad}$$

The ordered pair (____ , ____) is the solution for the system.

Check your work on page 690. ▶

Example 7. Solve the system by substitution.

$$(1)\ 4x - y = 3$$
$$(2)\ 2y - 8x = 1$$

Solution. We can isolate y in equation (1).

$$4x - y = 3$$
$$4x - 3 = y$$

Now substitute $4x - 3$ for y in equation (2) and solve for x.

$$2y - 8x = 1$$
$$2(4x - 3) - 8x = 1$$
$$8x - 6 - 8x = 1$$
$$-6 = 1$$

What does this false statement imply? We know that $-6 \neq 1$. Let's return to our original equations and write each one in slope-intercept form.

(1) $4x - y = 3$	(2) $2y - 8x = 1$
$4x - 3 = y$	$2y = 8x + 1$
	$y = 4x + \frac{1}{2}$
Slope: $m = 4$	Slope: $m = 4$
y-intercept: $b = -3$	y-intercept: $b = \frac{1}{2}$

These lines have the same slope but different y-intercepts, so they must be *parallel*. The system is inconsistent and has *no solution*.

> If, in solving a system of equations algebraically, we arrive at a statement that is *false*, we know that the system is *inconsistent* and has *no solution*.

Example 8. Solve the system by substitution.

$$(1)\ y = 2x - 3$$
$$(2)\ 5y + 15 = 10x$$

Solution. Since y is isolated in equation (1) we can substitute $2x - 3$ for y in equation (2) and solve for x.

$$5y + 15 = 10x$$
$$5(2x - 3) + 15 = 10x$$
$$10x - 15 + 15 = 10x$$
$$10x - 10x = 0$$
$$0 = 0$$

Once again we seem to have lost our variable, but this time we are left with the *true* statement $0 = 0$. To interpret this result, let's discuss the graphs of the original equations.

(1) $y = 2x - 3$

(2) $5y + 15 = 10x$

$5y = 10x - 15$

$y = 2x - 3$

Slope: $m = 2$ — Slope: $m = 2$

y-intercept: $b = -3$ — y-intercept: $b = -3$

Since the graphs have the same slope and the same y-intercept, we know they represent the *same straight line*. The system is dependent and the solutions are all ordered pairs satisfying $y = 2x - 3$ (or all ordered pairs satisfying $5y + 15 = 10x$).

> If, in solving a system of equations algebraically, we arrive at a statement that is *always true*, we know that the system is *dependent*, and any ordered pair that satisfies one equation will satisfy the other.

Trial Run

Solve each system by substitution.

______ 1. $x + y = 7$
$2x - y = 5$

______ 2. $5x - y = 2$
$y = 3$

______ 3. $3x - y = 1$
$y = 3x + 2$

______ 4. $5x + 2y = 9$
$x + 3y = 4$

______ 5. $y = 3 - x$
$2x + 2y = 6$

Answers are on page 691.

Examples You Completed

Example 1. Use graphing to solve the system.

(1) $y = 3$

(2) $y = 5x - 2$

Solution. First we graph each equation.

(1) $y = 3$

Horizontal line

y-intercept: 3

(2) $y = 5x - 2$

Slope: $m = \frac{5}{1}$

y-intercept: $b = -2$

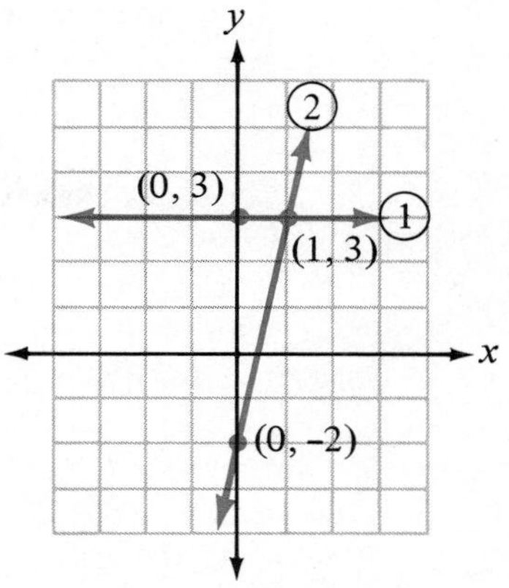

(1) $y = 3$

(2) $y = 5x - 2$

It appears that (1, 3) is the common point for the lines. We can check this ordered pair in each equation.

$$(1)\ y = 3 \qquad\qquad (2)\ y = 5x - 2$$
$$3 = 3 \qquad\qquad 3 \stackrel{?}{=} 5(1) - 2$$
$$3 = 3$$

Example 4. Solve and classify the system.

$$(1)\ 3x + y = 5$$
$$(2)\ y = -\tfrac{1}{2}x$$

Solution. First we graph each equation.

(1) $3x + y = 5$

$y = -3x + 5$

Slope: $m = \frac{-3}{1}$

y-intercept: $b = 5$

(2) $y = -\frac{1}{2}x$

Slope: $m = -\frac{1}{2} = \frac{-1}{2}$

y-intercept: $b = 0$

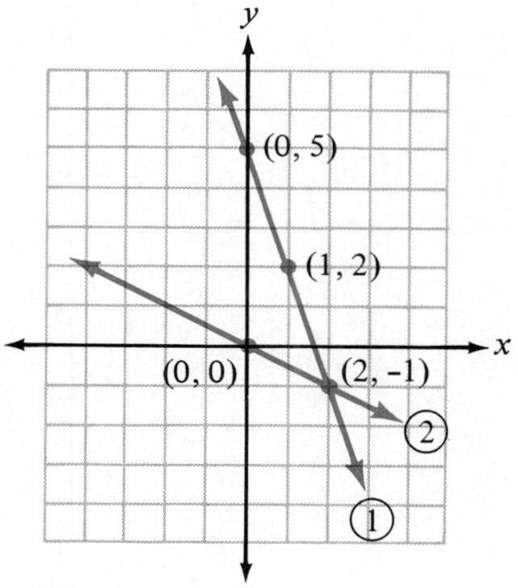

(1) $3x + y = 5$

(2) $y = -\frac{1}{2}x$

The common point seems to be $(2, -1)$. Let's check in each equation.

$$(1)\quad 3x + y = 5 \qquad\qquad (2)\quad y = -\frac{1}{2}x$$
$$3(2) + (-1) \stackrel{?}{=} 5$$
$$6 - 1 \stackrel{?}{=} 5 \qquad\qquad (-1) \stackrel{?}{=} -\frac{1}{2}(2)$$
$$5 = 5 \qquad\qquad -1 = -1$$

The solution for the system is $(2, -1)$ and the system is independent and consistent.

Example 6. Solve the system by substitution.

$$(1)\quad y - 5 = 0$$
$$(2)\quad 3x + 2y = 8$$

Solution. We can isolate y in equation (1).

$$y - 5 = 0$$
$$y = 5$$

Now we know the y-value for our solution, so we can substitute 5 for y in equation (2) and solve for x.

$$3x + 2y = 8$$
$$3x + 2(5) = 8$$
$$3x + 10 = 8$$
$$3x = -2$$
$$x = -\frac{2}{3}$$

The ordered pair $(-\frac{2}{3}, 5)$ is the solution for the system.

Answers to Trial Runs

page 684 **1.** (3, 5)

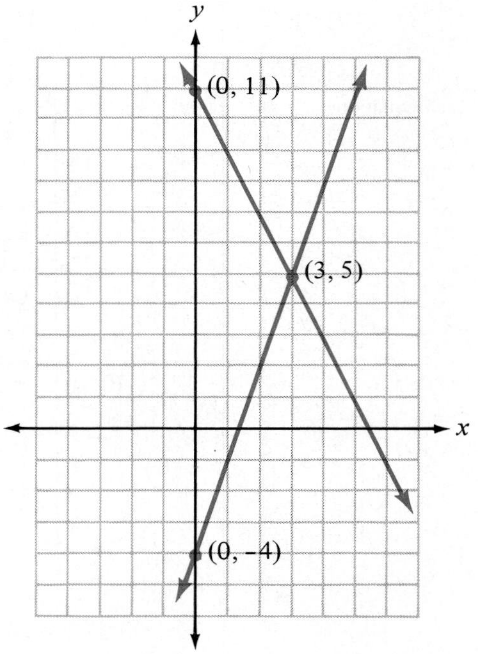

Independent, consistent

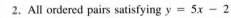

2. All ordered pairs satisfying $y = 5x - 2$

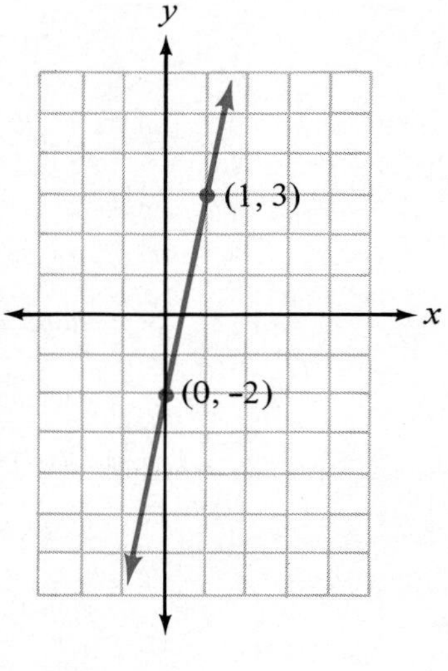

Dependent

3. (6, 3)

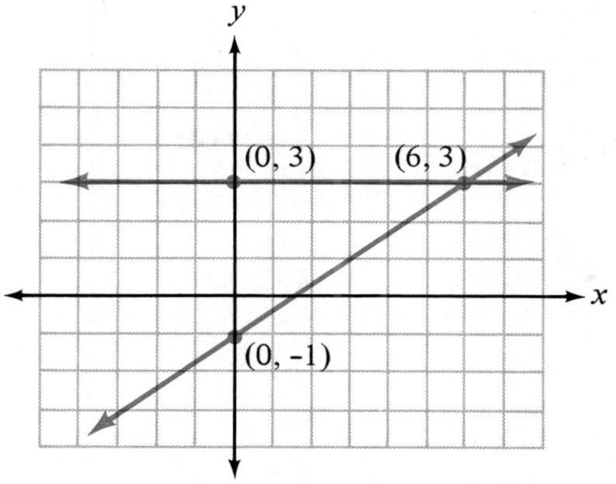

Independent, consistent

4. No solution

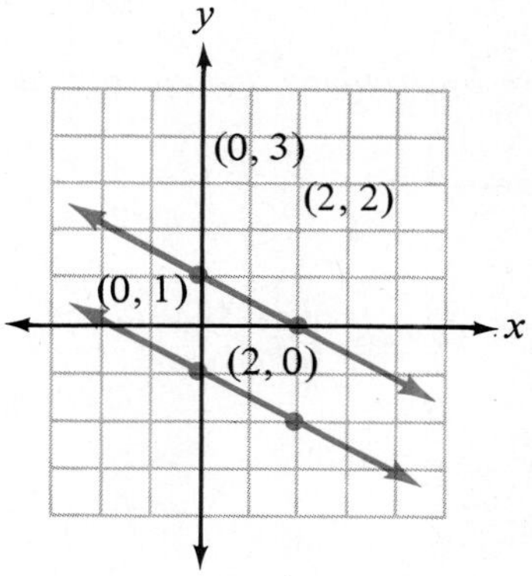

Inconsistent

page 688 1. (4, 3) 2. (1, 3) 3. No solution (inconsistent) 4. $(\frac{19}{13}, \frac{11}{13})$
5. All ordered pairs satisfying $y = 3 - x$ (dependent)

Name ______________________ Date ____________

EXERCISE SET 10.1

Solve by graphing and classify each system.

______ 1. $2x - y = 5$
$x + y = 7$

______ 2. $3x - y = 4$
$x - y = 2$

______ 3. $y = 3x + 7$
$2y - 6x = 5$

______ 4. $y = -2x + 3$
$3y + 6x = 6$

______ 5. $3x - y = 6$
$y = 6x$

______ 6. $-2x + y = 8$
$y = 4x$

______ 7. $2x - y = 10$
$y = 4$

______ 8. $3x + y = 9$
$y = -3$

______ 9. $x + 4y = 1$
$8y = 2 - 2x$

______ 10. $x - 3y = 2$
$6y = 2x - 4$

______ 11. $3x - 2y = 0$
$x - y = -1$

______ 12. $4x + 3y = 0$
$x + y = -1$

______ 13. $3x + 2y = 6$
$2x + y = 2$

______ 14. $5x + 4y = 20$
$x + y = 6$

______ 15. $4x - 2y = 16$
$-2x + y = -4$

______ 16. $-3x + 2y = -6$
$6x - 4y = 8$

Solve each system by substitution.

______ 17. $3x - 2y = 12$
$x - y = 5$

______ 18. $2x - 3y = 11$
$x + y = 3$

______ 19. $3x + y = 2$
$4x - 2y = -3$

______ 20. $5x + y = -1$
$13x + 3y = 2$

______ 21. $8x - 3y = 7$
$x = 2$

______ 22. $7x - 11y = -1$
$x = 3$

______ 23. $3x - 8y = 13$
$y = 3x$

______ 24. $5x - 3y = 28$
$y = -4x$

______ 25. $14x - 2y = 10$
$7x - y = 2$

______ 26. $10x + 2y = 3$
$5x + y = -1$

______ 27. $4x - y = 8$
$2x + 3y = 1$

______ 28. $5x + y = 7$
$4x - 2y = -3$

______ 29. $10x - 5y = 20$
$y = 2x - 4$

______ 30. $3x - 15y = 15$
$x = 5y + 5$

______ 31. $x - y = -2$
$11x - 3y = 2$

______ 32. $x - y = -4$
$7x + 5y = -4$

☆ Stretching the Topics

Solve the system by substitution.

_______ 1. $0.3x - 0.2y = 0.7$
$0.4x - 0.1y = 0.6$

_______ 2. Find the value(s) of k for which the system

$$y = kx + 5$$
$$y = 3x - 2$$

will be inconsistent.

Check your answers in the back of your book.

If you can complete **Checkup 10.1,** you are ready to go on to Section 10.2.

Name ______________________ **Date** __________

✓ CHECKUP 10.1

Solve by graphing and classify each system.

________ 1. $x - 2y = 8$
$3x + 2y = 0$

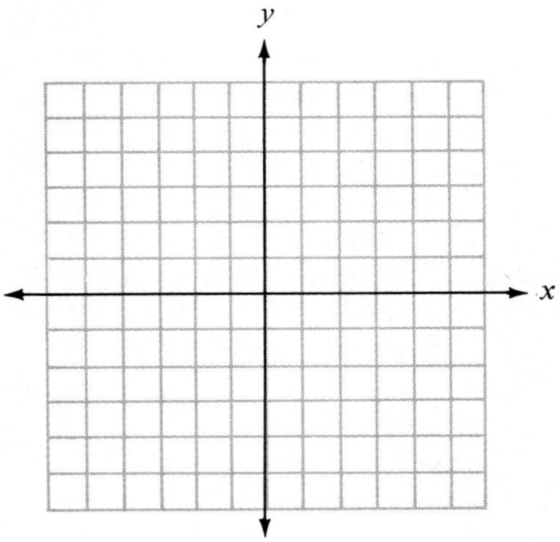

________ 2. $-x + 2y = 4$
$5x - 10y = -20$

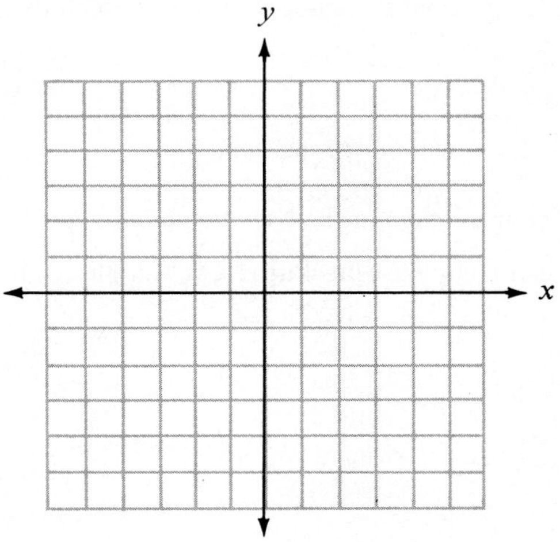

Solve each system by substitution.

________ 3. $9x - 4y = 10$
$2x + y = 6$

________ 4. $x - y = -3$
$-4x + 4y = 1$

________ 5. $4x + 7y = 16$
$x + 4y = 7$

Check your answers in the back of your book.

If You Missed Problem:	You Should Review Examples:
1	1, 4
2	3
3	5
4	7
5	5

10.2 Solving Systems of Linear Equations by Addition (Elimination)

What happens when we want to solve a system of equations algebraically but neither variable can easily be isolated in either equation? To solve a system such as

$$(1)\ 5x - 2y = 9$$
$$(2)\ 3x + 2y = 7$$

by the method of substitution would be a tedious process because neither equation contains a variable with a coefficient of 1.

In this section we make use of two important facts from earlier chapters.

> We may multiply both sides of an equation by any nonzero constant.
> We may add equal quantities to both sides of an equation.

Using the Method of Addition

Consider the system of equations introduced above.

$$(1)\ 5x - 2y = 9$$
$$(2)\ 3x + 2y = 7$$

Since the quantities on the left and right sides of these equations are equal, we may *add* the left-hand quantities and *add* the right-hand quantities.

$$\begin{array}{r} 5x - 2y = 9 \\ \underline{3x + 2y = 7} \\ 8x + 0y = 16 \end{array}$$

We have eliminated the y-terms, and now we can solve for x.

$$8x = 16$$
$$x = 2$$

We have found the x-value of the common solution. To find the corresponding y-value, we can substitute 2 for x in either of the original equations.

$$\begin{array}{rrcrr} (1) & 5x - 2y = 9 & \text{or} & (2) & 3x + 2y = 7 \\ & 5(2) - 2y = 9 & & & 3(2) + 2y = 7 \\ & 10 - 2y = 9 & & & 6 + 2y = 7 \\ & -2y = -1 & & & 2y = 1 \\ & y = \dfrac{-1}{-2} & & & y = \dfrac{1}{2} \\ & y = \dfrac{1}{2} & & & \end{array}$$

The ordered pair $(2, \frac{1}{2})$ is the solution for the system.

Now let's discuss how we arrived at the solution. We began by noting that the coefficients of the y-terms were *opposites* of each other. Then, when we added the equations together, the y-terms were eliminated. For obvious reasons, this method is called the **method of addition** (or the **method of elimination**).

Example 1. Solve the system by addition.

$$\begin{aligned} (1)\ 6x - 10y &= 5 \\ (2)\ 2x - 3y &= 1 \end{aligned}$$

Solution. We note that neither variable has opposite coefficients, but we can make the x-coefficients opposites by multiplying equation (2) by -3.

$6x - 10y = 5$	Equation (1).
$-6x + 9y = -3$	Multiply equation (2) by -3.
$0x - y = 2$	Add the equations.
$-y = 2$	Simplify.
$y = -2$	Solve for y.

We find the corresponding x-value by substituting -2 for y in either of the original equations. Let's use equation (2).

$$\begin{aligned} 2x - 3y &= 1 \\ 2x - 3(-2) &= 1 \\ 2x + 6 &= 1 \\ 2x &= -5 \\ x &= -\frac{5}{2} \end{aligned}$$

The solution for the system is $(-\frac{5}{2}, -2)$ and you should check this ordered pair in equation (1).

Solving a System by Addition (Elimination)

1. Write both equations in the form $ax + by = c$.
2. Look to see if the coefficient of one of the variables in one equation is the opposite of the coefficient of the same variable in the other equation.
3. If the coefficients are not opposites, multiply one or both equations by constants that will make the coefficients opposites.
4. Add the equations together to eliminate one variable.
5. Solve for the value of the remaining variable.
6. Substitute that value into either of the original equations and solve for the other variable; *or* repeat the addition process, eliminating the other variable.
7. Write the solution as an ordered pair.

Now you try Example 2.

Example 2. Solve the system by addition.

$$\begin{aligned} (1)\ x + 5y &= 3 \\ (2)\ 10y &= 4 - 2x \end{aligned}$$

Solution. First we rewrite equation (2) in standard form so that like variables are lined up beneath each other in the system.

$$\begin{aligned} (1)\ x + 5y &= 3 \\ (2)\ 2x + 10y &= 4 \end{aligned}$$

To eliminate the x-terms, we can multiply equation (1) by -2.

$\underline{\qquad}x - \underline{\qquad}y = \underline{\qquad}$	Multiply equation (1) by -2.
$2x + 10y = 4$	Equation (2).
$\underline{\qquad}x + \underline{\qquad}y = \underline{\qquad}$	Add the equations.
$0 = -2$	Simplify.

In attempting to eliminate the x-terms, we have also eliminated the y-terms! Since $0 = -2$ is a *false* statement, this system must be ________ ; it has ________ solution. Graphing the original equation would yield a pair of ________ lines.

Check your work on page 701. ▶

Example 3. Solve the system by addition.

$$(1)\ 5x + 2y = 1$$
$$(2)\ 3x - 3y = 2$$

Solution. The coefficients of the y-terms are opposite in sign, so let's try to eliminate the y-terms. Unfortunately (unless we use fractions), we cannot make the coefficients match by multiplying just one equation by a constant. But if we multiply equation (1) by 3 and multiply equation (2) by 2, the coefficients of the y-terms will be opposites.

$5x + 2y = 1$	Equation (1).
$3x - 3y = 2$	Equation (2).
$15x + 6y = 3$	Multiply equation (1) by 3.
$6x - 6y = 4$	Multiply equation (2) by 2.
$21x + 0y = 7$	Add the equations.
$21x = 7$	Simplify.
$x = \dfrac{7}{21}$	Solve for x.
$x = \dfrac{1}{3}$	

Now we substitute $\frac{1}{3}$ for x in either equation. Let's use equation (2).

$$\begin{aligned} 3x - 3y &= 2 \\ 3\left(\frac{1}{3}\right) - 3y &= 2 \\ 1 - 3y &= 2 \\ -3y &= 1 \\ y &= -\frac{1}{3} \end{aligned}$$

The ordered pair $(\frac{1}{3}, -\frac{1}{3})$ is the solution for the system.

Now you try Example 4.

Example 4. Solve the system by addition.

$$(1)\ 3x - 3y = 6$$
$$(2)\ 5x - 10 = 5y$$

Solution. Line up variables beneath each other and try to eliminate the x-terms.

$3x - 3y = 6$	Equation (1).
$5x - 5y = 10$	Equation (2).
$15x - 15y = 30$	Multiply equation (1) by 5.
$-15x + 15y = -30$	Multiply equation (2) by -3.
$\_\_x + \_\_y = \_\_\_\_$	Add the equations.
$\_\_\_\_ = \_\_\_\_$	Simplify.

Since this statement is *always true,* the system is ________ . The solutions are all ordered pairs satisfying ________ . Graphing the original equations yields the ________ line.

Check your work on page 702. ▶

Trial Run

Solve each system by addition.

________ 1. $5x - y = 7$
$3x + 4y = -5$

________ 2. $2x + 3y = 7$
$x + 5y = 4$

________ 3. $4x - 3y = -5$
$2x - 5y = 1$

________ 4. $3x - 2y = 2$
$-6x + 4y = -1$

________ 5. $7x - 3y = 9$
$2x + 5y = -4$

________ 6. $4x - 5y = 2$
$12x = 15y + 6$

Answers are on page 702.

Solving Systems of Linear Equations in Three Variables

The solutions for an equation containing three variables x, y, and z are **ordered triples** rather than ordered pairs. For instance, the ordered triple $(x, y, z) = (1, 3, 2)$ is one solution for the equation

$$x + 2y - 3z = 1$$

There are infinitely many other ordered triples that satisfy this equation.

Graphing equations in three variables requires a three-dimensional coordinate system. We shall not investigate such graphs in this book. Instead, we shall use algebraic methods to find ordered triples that satisfy three equations at the same time. We shall use the methods of *addition* and/or *substitution* to solve systems of three equations in three variables.

Our method is basically the same method used for solving systems of two equations in two variables. Of course, more steps will be required and we must keep careful track of what we are doing.

Example 5. Solve the system by addition.

$$\begin{aligned} &(1)\quad x - y + 2z = 6 \\ &(2)\quad 2x + y + 2z = 5 \\ &(3)\quad x + 3y - z = -4 \end{aligned}$$

Solution

We can eliminate y-terms from equation (1) and equation (2).

$$\begin{array}{rl} (1) & x - y + 2z = 6 \\ (2) & 2x + y + 2z = 5 \\ \hline & 3x + 0y + 4z = 11 \\ & 3x + 4z = 11 \end{array}$$

Let's label this new equation (a).

$$\text{(a)}\quad 3x + 4z = 11$$

We must bring equation (3) into the discussion and use another equation with it to eliminate y-terms again. Multiply equation (1) by 3.

$$\begin{array}{r} 3x - 3y + 6z = 18 \\ x + 3y - z = -4 \\ \hline 4x + 0y + 5z = 14 \\ 4x + 5z = 14 \end{array}$$

Label this new equation (b).

$$\text{(b)}\quad 4x + 5z = 14$$

Now equations (a) and (b) represent a system of two equations in two variables, x and z.

$$\begin{aligned} &\text{(a)}\quad 3x + 4z = 11 \\ &\text{(b)}\quad 4x + 5z = 14 \end{aligned}$$

We proceed as before to eliminate x.

$12x + 16z = 44$	Multiply equation (a) by 4.
$-12x - 15z = -42$	Multiply equation (b) by -3.
$z = 2$	Add the equations.

Substitute 2 for z in equation (a) or equation (b) and solve for x.

$$\begin{aligned} \text{(a)}\quad 3x + 4z &= 11 \\ 3x + 4(2) &= 11 \\ 3x &= 3 \\ x &= 1 \end{aligned}$$

Substitute 2 for z and 1 for x in equation (1) or (2) or (3) and solve for y.

$$\begin{aligned} (1)\quad x - y + 2z &= 6 \\ 1 - y + 2(2) &= 6 \\ -y &= 1 \\ y &= -1 \end{aligned}$$

The ordered triple is $(1, -1, 2)$. You should verify that this ordered triple also satisfies equations (2) and (3).

Example 6. Solve the system by addition.

$$\begin{aligned} &(1)\quad 3x + z = 7 \\ &(2)\quad x + 5y = 2 \\ &(3)\quad 3y - z = -1 \end{aligned}$$

Solution. We line up like terms beneath each other.

$$\begin{array}{lrrrrl} (1) & 3x & & + z & = & 7 \\ (2) & x & + 5y & & = & 2 \\ (3) & & 3y & - z & = & -1 \end{array}$$

First use equations (1) and (3) to eliminate z.

$$\begin{array}{lrcrcr} (1) & 3x & & +z & = & 7 \\ (3) & & 3y & -z & = & -1 \\ \hline (a) & 3x & +3y & & = & 6 \end{array}$$

Now use equation (a) with equation (2) to eliminate x. Multiply equation (2) by -3.

$$\begin{array}{lrcrcr} (a) & 3x & + & 3y & = & 6 \\ (2) & -3x & - & 15y & = & -6 \\ \hline & & & -12y & = & 0 \\ & & & y & = & 0 \end{array}$$

Substitute 0 for y in equation (a) or equation (2) to find x.

$$\begin{aligned} (2)\quad x + 5y &= 2 \\ x + 5(0) &= 2 \\ x &= 2 \end{aligned}$$

Substitute 2 for x in equation (1) to find z.

$$\begin{aligned} (1)\quad 3x + z &= 7 \\ 3(2) + z &= 7 \\ z &= 1 \end{aligned}$$

The solution is (2, 0, 1). Verify that this solution satisfies each of the original equations.

Example 7. Solve the system by addition.

$$\begin{aligned} (1)\quad x - \tfrac{1}{2}y - \tfrac{1}{2}z &= 4 \\ (2)\quad x - \tfrac{3}{2}y - 2z &= 3 \\ (3)\quad \tfrac{1}{4}x + \tfrac{1}{4}y - \tfrac{1}{4}z &= 0 \end{aligned}$$

Solution. First we clear the fractions from each equation.

(1) $2x - y - z = 8$	Multiply equation (1) by LCD: 2.
(2) $2x - 3y - 4z = 6$	Multiply equation (2) by LCD: 2.
(3) $x + y - z = 0$	Multiply equation (3) by LCD: 4.

Eliminate y from equations (1) and (3).

$$\begin{array}{lrcrcr} (1) & 2x - y & - & z & = & 8 \\ (3) & x + y & - & z & = & 0 \\ \hline (a) & 3x & - & 2z & = & 8 \end{array}$$

Eliminate y from equations (2) and (3). Multiply equation (3) by 3.

$$\begin{array}{lrcrcr} & 2x - 3y & - & 4z & = & 6 \\ & 3x + 3y & - & 3z & = & 0 \\ \hline (b) & 5x & - & 7z & = & 6 \end{array}$$

Eliminate x from equations (a) and (b).

$$\begin{aligned} (a)\quad 3x - 2z &= 8 \\ (b)\quad 5x - 7z &= 6 \end{aligned}$$

$15x - 10z = 40$	Multiply equation (a) by 5.
$-15x + 21z = -18$	Multiply equation (b) by -3.
$11z = 22$	Add the equations.
$z = 2$	Solve for z.

Substitute 2 for z in equation (a) to find x.

$$\begin{aligned} (a)\quad 3x - 2z &= 8 \\ 3x - 2(2) &= 8 \\ 3x &= 12 \\ x &= 4 \end{aligned}$$

Substitute 4 for x and 2 for z in equation (3) to find y.

$$\begin{aligned} (3)\quad x + y - z &= 0 \\ 4 + y - 2 &= 0 \\ y + 2 &= 0 \\ y &= -2 \end{aligned}$$

The solution for the system is (4, -2, 2).

Trial Run

Solve each system by addition.

_______ 1. $\begin{aligned} x + y - z &= 4 \\ x - 2y + z &= -1 \\ 3x - y + z &= 4 \end{aligned}$

_______ 2. $\begin{aligned} 2x + z &= 4 \\ x - 5y &= 3 \\ 4y - z &= 2 \end{aligned}$

_______ 3. $\begin{aligned} x + \tfrac{1}{2}y - \tfrac{1}{4}z &= 0 \\ \tfrac{1}{2}x + y + \tfrac{1}{2}z &= 3 \\ \tfrac{3}{2}x + \tfrac{1}{4}y + z &= -2 \end{aligned}$

Answers are on page 702.

As was the case with a system of two equations in two variables, a system of three equations in three variables may not yield exactly one ordered triple. When solving such a system algebraically, we must watch for the same signals as before.

> If we arrive at a *false* statement, the system is *inconsistent*. There is no solution for the system.
>
> If we arrive at a statement that is *always true*, the system is *dependent*. There is not just one solution for the system.

Examples You Completed

Example 2. Solve the system by addition.

$$(1)\ x + 5y = 3$$
$$(2)\ 10y = 4 - 2x$$

Solution. First we rewrite equation (2) in standard form so that like variables are lined up beneath each other in the system.

$$(1)\ x + 5y = 3$$
$$(2)\ 2x + 10y = 4$$

To eliminate the x-terms, we can multiply equation (1) by -2.

$-2x - 10y = -6$	Multiply equation (1) by -2.
$2x + 10y = 4$	Equation (2).
$0x + 0y = -2$	Add the equations.
$0 = -2$	Simplify.

In attempting to eliminate the x-terms, we have also eliminated the y-terms! Since $0 = -2$ is a *false* statement, this system must be inconsistent; it has no solution. Graphing the original equations would yield a pair of parallel lines.

Example 4. Solve the system by addition.

$$(1)\ 3x - 3y = 6$$

$$(2)\ 5x - 10 = 5y$$

Solution. Line up variables beneath each other and try to eliminate the x-terms.

$3x - 3y = 6$	Equation (1).
$5x - 5y = 10$	Equation (2).
$15x - 15y = 30$	Multiply equation (1) by 5.
$-15x + 15y = -30$	Multiply equation (2) by -3.
$0x + 0y = 0$	Add the equations.
$0 = 0$	Simplify.

Since this is a statement that is *always true,* the system is dependent. The solutions are all ordered pairs satisfying $3x - 3y = 6$. Graphing the original equations yields the same line.

Answers to Trial Runs

page 698 **1.** $(1, -2)$ **2.** $(\frac{23}{7}, \frac{1}{7})$ **3.** $(-2, -1)$ **4.** No solution **5.** $(\frac{33}{41}, -\frac{46}{41})$
6. All ordered pairs satisfying $4x - 5y = 2$

page 701 **1.** $(2, 1, -1)$ **2.** $(3, 0, -2)$ **3.** $(-2, 4, 0)$

Name ______________________ Date __________

EXERCISE SET 10.2

Solve each system by addition.

______ 1. $3x - y = 4$
$2x + y = 1$

______ 2. $4x - y = 5$
$3x + y = 9$

______ 3. $3x - 2y = 0$
$5x + 2y = 6$

______ 4. $-4x + 5y = 0$
$4x + y = 2$

______ 5. $7x - 10y = 4$
$x + 5y = 7$

______ 6. $8x - 4y = 6$
$3x + 2y = 4$

______ 7. $2x - 8y = 9$
$-x + 4y = -6$

______ 8. $-3x + 3y = 1$
$x - y = 8$

______ 9. $5x - 4y = 1$
$2x + 3y = 5$

______ 10. $7x - 6y = 20$
$3x + 5y = 1$

______ 11. $7x - 5y = 2$
$2x - 3y = -1$

______ 12. $2x - 3y = 5$
$5x - 2y = -4$

______ 13. $3x - 6y = 9$
$-2x + 4y = -6$

______ 14. $8x - 4y = -4$
$-10x + 5y = 5$

______ 15. $y = \frac{5}{2}x - \frac{5}{2}$
$4x - 6y = 15$

______ 16. $10x - 9y = 12$
$y = \frac{2}{3}x - \frac{4}{3}$

______ 17. $2x - 7y = 3$
$3x + 2y = -1$

______ 18. $5x - 2y = -2$
$4x + 3y = 5$

______ 19. $\frac{1}{2}x - \frac{1}{9}y = 0$
$\frac{1}{5}x - \frac{1}{4}y = \frac{3}{20}$

______ 20. $\frac{1}{3}x - \frac{1}{7}y = 0$
$\frac{1}{2}x - \frac{1}{4}y = -\frac{3}{8}$

______ 21. $\frac{1}{3}x - \frac{2}{3}y = -2$
$\frac{1}{2}x + \frac{1}{6}y = 2$

______ 22. $\frac{2}{3}x + \frac{1}{2}y = -1$
$x - \frac{1}{4}y = -\frac{7}{2}$

______ 23. $0.5x - y = -1$
$2.5x + 6y = 28$

______ 24. $0.25x - 2y = -4$
$2x + 7y = -9$

______ 25. $0.2x + 0.7y = 15$
$0.8x + 2.1y = 11$

______ 26. $0.3x + 0.8y = 3$
$0.9x + 3.4y = 12$

______ 27. $x - 2y + z = 3$
$2x + 3y - z = -4$
$-x + y + z = 0$

______ 28. $2x - y + z = 3$
$x + 3y - 2z = 5$
$3x - 4y + z = 2$

______ 29. $2x - y + 4z = 17$
$3x + 4y - z = -1$
$4x + 3y + 2z = 11$

______ 30. $4x - 3y - z = 16$
$2x + 2y + 3z = 8$
$x + 4y + 5z = 3$

______ 31. $2x - y - z = 6$
$2x - 2y + z = 10$
$x + y - 3z = -2$

______ 32. $x + y + z = 6$
$3x - 2y - z = 13$
$2x - y + 3z = 26$

_______ 33. $\begin{aligned} x + y &= 0 \\ 3y + z &= 1 \\ 2x - 3z &= -10 \end{aligned}$

_______ 34. $\begin{aligned} 2x - y &= 0 \\ 3y - z &= 5 \\ x + z &= 2 \end{aligned}$

_______ 35. $\begin{aligned} \tfrac{1}{3}x + \tfrac{1}{6}y + z &= -2 \\ x + \tfrac{1}{2}y + z &= 0 \\ \tfrac{2}{3}x + y - \tfrac{1}{3}z &= 5 \end{aligned}$

_______ 36. $\begin{aligned} \tfrac{1}{3}x + y + \tfrac{1}{2}z &= 7 \\ \tfrac{1}{2}x + \tfrac{3}{2}y + z &= 11 \\ \tfrac{1}{3}x + \tfrac{1}{3}y - z &= 0 \end{aligned}$

☆ Stretching the Topics

Solve each system.

_______ 1. $\begin{aligned} 3(2x - 3) + 4y - 20 &= 0 \\ 4(y + 3) - 3x - 10 &= 0 \end{aligned}$

_______ 2. $\begin{aligned} 2(x + 1) - 3(y - 4) &= 2 \\ 4(x - 1) + 2(y + 1) &= 6 \end{aligned}$

_______ 3. Find the values of m and b that will make the line $y = mx + b$ pass through the points (3, 2) and (5, 4).

Check your answers in the back of your book.

If you can complete **Checkup 10.2,** you are ready to go on to Section 10.3.

Name ______________________ Date ____________

✓ CHECKUP 10.2

Solve each system by addition.

______ 1. $\begin{aligned} x - 3y &= 2 \\ 2x + y &= -3 \end{aligned}$

______ 2. $\begin{aligned} 2x + 3y &= 5 \\ x + y &= 2 \end{aligned}$

______ 3. $\begin{aligned} 2x - 3y &= 4 \\ 5x + 11y &= 47 \end{aligned}$

______ 4. $\begin{aligned} 3x - 6y &= 3 \\ 5x - 10y &= 5 \end{aligned}$

______ 5. $\begin{aligned} x + y - z &= 4 \\ 2x - 3y + z &= -4 \\ 3x - y - z &= 2 \end{aligned}$

Check your answers in the back of your book.

If You Missed Problems:	You Should Review Examples:
1–3	1, 3
4	4
5	5

10.3 Using Determinants to Solve Systems (Optional)

Mathematicians invented the **matrix** as a convenient way to organize a pattern of numbers.

> **Definition of a Matrix.** A matrix is a rectangular array of numbers.

A matrix has horizontal rows and vertical columns, enclosed in brackets.

$$\begin{bmatrix} 1 & 2 \\ 3 & 0 \\ 4 & -1 \end{bmatrix}$$

This matrix has 3 rows and 2 columns. It is called a 3 × 2 ("three by two") matrix.

$$\begin{bmatrix} 5 & -2 \\ 3 & 1 \end{bmatrix}$$

This matrix has 2 rows and 2 columns. It is a 2 × 2 matrix.

If a matrix has m rows and n columns, it is called an $m \times n$ matrix. If a matrix has the same number of rows as columns, $(m = n)$, it is called a **square matrix.**

Finding Determinants

Associated with every *square* matrix is a unique real number called its **determinant.** Although every $n \times n$ matrix has a determinant, we shall discuss here the method for finding the determinant of a 2 × 2 matrix only. We denote the determinant of a matrix by enclosing the array with vertical bars rather than brackets.

For the matrix

$$\begin{bmatrix} a_1 & b_1 \\ a_2 & b_2 \end{bmatrix} \quad \text{the determinant is written} \quad \begin{vmatrix} a_1 & b_1 \\ a_2 & b_2 \end{vmatrix}$$

Now we must find the real number value of this determinant using the definition of a 2 × 2 determinant.

> **A 2 × 2 Determinant**
>
> $$\begin{vmatrix} a_1 & b_1 \\ a_2 & b_2 \end{vmatrix} = a_1b_2 - a_2b_1$$

Example 1. Find $\begin{vmatrix} 3 & 5 \\ -1 & 2 \end{vmatrix}$.

Solution

$$\begin{vmatrix} 3 & 5 \\ -1 & 2 \end{vmatrix} = 3(2) - (-1)(5)$$

$$= 6 + 5$$

$$= 11$$

You complete Example 2.

Example 2. Find $\begin{vmatrix} 2 & -3 \\ -4 & 6 \end{vmatrix}$.

Solution

$$\begin{vmatrix} 2 & -3 \\ -4 & 6 \end{vmatrix} =$$

Check your work on page 711. ▶

Trial Run

Find the value of each determinant.

_______ 1. $\begin{vmatrix} 2 & 3 \\ 1 & -4 \end{vmatrix}$ _______ 2. $\begin{vmatrix} -\frac{1}{2} & 0 \\ \frac{7}{3} & 4 \end{vmatrix}$ _______ 3. $\begin{vmatrix} 0.7 & 0.6 \\ 0.1 & 0.2 \end{vmatrix}$

Answers are on page 711.

Using Cramer's Rule

Determinants can be used to solve a system of linear equations according to a theorem called **Cramer's rule**. We shall discuss the use of that rule in solving systems of *two* equations in *two* variables only. Consider such a system.

$$(1)\quad a_1x + b_1y = c_1$$

$$(2)\quad a_2x + b_2y = c_2$$

Here we are using a, b, and c with subscripts to indicate *any* constants.

To solve this system by addition, we can eliminate the y-terms by multiplying equation (1) by b_2 and multiplying equation (2) by $-b_1$.

$b_2[a_1x + b_1y = c_1]$	Multiply equation (1) by b_2.
$-b_1[a_2x + b_2y = c_2]$	Multiply equation (2) by $-b_1$.
$a_1b_2x + b_1b_2y = c_1b_2$ $-a_2b_1x - b_1b_2y = -c_2b_1$	Remove parentheses.
$a_1b_2x - a_2b_1x = c_1b_2 - c_2b_1$	Add the equations.
$x(a_1b_2 - a_2b_1) = c_1b_2 - c_2b_1$	Factor out x on left side.
$x = \dfrac{c_1b_2 - c_2b_1}{a_1b_2 - a_2b_1}$	Isolate x.

Notice that both the numerator and denominator on the right can be rewritten as determinants.

$$x = \frac{\begin{vmatrix} c_1 & b_1 \\ c_2 & b_2 \end{vmatrix}}{\begin{vmatrix} a_1 & b_1 \\ a_2 & b_2 \end{vmatrix}}$$

Observe that the denominator is the determinant of the matrix made up of the coefficients of the variables in the original system. The determinant of that coefficient matrix is often represented by $\boldsymbol{D}$.

Observe that the numerator is a similar determinant in which the column of x-coefficients in D has been replaced by the column of constants from the original system. This determinant is represented by $\boldsymbol{D_x}$. Another determinant, $\boldsymbol{D_y}$, is formed when the column of y-coefficients in D is replaced by the column of constants.

For the system

$$a_1x + b_1y = c_1$$

$$a_2x + b_2y = c_2$$

we define

$$D = \begin{vmatrix} a_1 & b_1 \\ a_2 & b_2 \end{vmatrix} \qquad D_x = \begin{vmatrix} c_1 & b_1 \\ c_2 & b_2 \end{vmatrix} \qquad D_y = \begin{vmatrix} a_1 & c_1 \\ a_2 & c_2 \end{vmatrix}$$

In our system we discovered that $x = \frac{D_x}{D}$. A similar procedure (eliminating x-terms instead) would result in the statement that $y = \frac{D_y}{D}$. These facts are summarized in **Cramer's rule**.

Cramer's Rule. For the system

$$a_1x + b_1y = c_1$$
$$a_2x + b_2y = c_2$$

the solutions are found by

$$x = \frac{D_x}{D} \quad \text{and} \quad y = \frac{D_y}{D}$$

where $D \neq 0$ and

$$D = \begin{vmatrix} a_1 & b_1 \\ a_2 & b_2 \end{vmatrix} \qquad D_x = \begin{vmatrix} c_1 & b_1 \\ c_2 & b_2 \end{vmatrix} \qquad D_y = \begin{vmatrix} a_1 & c_1 \\ a_2 & c_2 \end{vmatrix}$$

Example 3. Use Cramer's rule to solve the system.

$$3x + 4y = 2$$
$$7x - 2y = 9$$

Solution. Here

$$a_1x + b_1y = c_1 \quad \text{is} \quad 3x + 4y = 2$$
$$a_2x + b_2y = c_2 \quad \text{is} \quad 7x - 2y = 9$$

We must find each determinant.

$$D = \begin{vmatrix} a_1 & b_1 \\ a_2 & b_2 \end{vmatrix} = \begin{vmatrix} 3 & 4 \\ 7 & -2 \end{vmatrix} = 3(-2) - 7(4) = -6 - 28$$
$$D = -34$$

$$D_x = \begin{vmatrix} c_1 & b_1 \\ c_2 & b_2 \end{vmatrix} = \begin{vmatrix} 2 & 4 \\ 9 & -2 \end{vmatrix} = 2(-2) - 9(4) = -4 - 36$$
$$D_x = -40$$

$$D_y = \begin{vmatrix} a_1 & c_1 \\ a_2 & c_2 \end{vmatrix} = \begin{vmatrix} 3 & 2 \\ 7 & 9 \end{vmatrix} = 3(9) - 7(2) = 27 - 14$$
$$D_y = 13$$

Now we can find x and y.

$$x = \frac{D_x}{D} = \frac{-40}{-34} \qquad y = \frac{D_y}{D} = \frac{13}{-34}$$
$$x = \frac{20}{17} \qquad y = -\frac{13}{34}$$

The solution is $(\frac{20}{17}, -\frac{13}{34})$.

Example 4. Use Cramer's rule to solve the system.

$$x + y = 1$$
$$3x + 3y = 4$$

Solution. Here $a_1x + b_1y = c_1$ is $1x + 1y = 1$

and $a_2x + b_2y = c_2$ is $3x + 3y = 4$

$$D = \begin{vmatrix} a_1 & b_1 \\ a_2 & b_2 \end{vmatrix} = \begin{vmatrix} 1 & 1 \\ 3 & 3 \end{vmatrix} = 1(3) - 3(1) = 3 - 3 = 0$$

Since $D = 0$, we cannot use Cramer's rule. The system does not have one unique solution; it is either inconsistent or dependent. We must solve by some other method to decide.

Using the addition method, we multiply the first equation by -3 to eliminate x.

$-3x - 3y = -3$	Multiply equation (1) by -3.
$3x + 3y = 4$	Equation (2).
$0 = 1$	Add equations.

This system is inconsistent; there is *no* solution.

As this example indicates, Cramer's rule provides a quick procedure for determining whether a system has a unique solution. If $D = 0$, we know that the system is inconsistent or dependent.

Example 5. Use Cramer's rule to solve the system.

$$y = 3x + 2$$
$$5x + 4y = 8$$

Solution. Our system is

$$-3x + 1y = 2$$
$$5x + 4y = 8$$

$$D = \begin{vmatrix} -3 & 1 \\ 5 & 4 \end{vmatrix} = -3(4) - 5(1) = -12 - 5 = -17$$

$$D_x = \begin{vmatrix} 2 & 1 \\ 8 & 4 \end{vmatrix} = 2(4) - 8(1) = 8 - 8 = 0$$

$$D_y = \begin{vmatrix} -3 & 2 \\ 5 & 8 \end{vmatrix} = -3(8) - 5(2) = -24 - 10 = -34$$

Now we can find x and y.

$$x = \frac{D_x}{D} = \frac{0}{-17}, \quad x = 0 \qquad \text{and} \qquad y = \frac{D_y}{D} = \frac{-34}{-17}, \quad y = 2$$

The solution is (0, 2).

Cramer's rule can also be extended to solving systems of more equations in more variables, but that is a topic for your next algebra course.

Trial Run

Use Cramer's rule to solve each system.

______ 1. $3x - y = 1$
$x + 3y = 2$

______ 2. $4x + 3y = -8$
$-x + y = 2$

______ 3. $y = 5 - 2x$
$4x + 5y = 7$

______ 4. $-2x + 3y = 5$
$5x + 7y = 1$

Answers are below.

Example You Completed

Example 2. Find $\begin{vmatrix} 2 & -3 \\ -4 & 6 \end{vmatrix}$.

Solution

$$\begin{vmatrix} 2 & -3 \\ -4 & 6 \end{vmatrix} = 2(6) - (-4)(-3) = 12 - 12 = 0$$

Answers to Trial Runs

page 708 **1.** -11 **2.** -2 **3.** 0.08

page 711 **1.** $(\frac{1}{2}, \frac{1}{2})$ **2.** $(-2, 0)$ **3.** $(3, -1)$ **4.** $(-\frac{32}{29}, \frac{27}{29})$

Name ______________________ Date ____________

EXERCISE SET 10.3

Find the value of each determinant.

______ 1. $\begin{vmatrix} 2 & 3 \\ -2 & 5 \end{vmatrix}$

______ 2. $\begin{vmatrix} -3 & -2 \\ 5 & 3 \end{vmatrix}$

______ 3. $\begin{vmatrix} 0 & -3 \\ 4 & 0 \end{vmatrix}$

______ 4. $\begin{vmatrix} -7 & 0 \\ 0 & 9 \end{vmatrix}$

______ 5. $\begin{vmatrix} \frac{3}{4} & \frac{1}{6} \\ -\frac{1}{2} & \frac{1}{3} \end{vmatrix}$

______ 6. $\begin{vmatrix} -\frac{1}{3} & \frac{5}{6} \\ -\frac{3}{5} & \frac{7}{10} \end{vmatrix}$

Use Cramer's rule to solve each system.

______ 7. $\begin{aligned} 2x + 3y &= 7 \\ x + 2y &= 3 \end{aligned}$

______ 8. $\begin{aligned} 5x - 12y &= 4 \\ 4x - 7y &= -2 \end{aligned}$

______ 9. $\begin{aligned} 3x + 2y &= 4 \\ x + \tfrac{2}{3}y &= -2 \end{aligned}$

______ 10. $\begin{aligned} 4x - y &= 5 \\ -x + \tfrac{1}{4}y &= -2 \end{aligned}$

______ 11. $\begin{aligned} -x + y &= 1 \\ 3x - 4y &= -3 \end{aligned}$

______ 12. $\begin{aligned} 2x + 3y &= 4 \\ 4x + y &= -2 \end{aligned}$

______ 13. $\begin{aligned} 3x + 2y &= 1 \\ 3x - 2y &= -5 \end{aligned}$

______ 14. $\begin{aligned} 2x + 3y &= 5 \\ x + 5y &= -2 \end{aligned}$

______ 15. $\begin{aligned} \tfrac{1}{2}x + \tfrac{1}{3}y &= \tfrac{3}{2} \\ \tfrac{2}{3}x - \tfrac{1}{4}y &= \tfrac{5}{3} \end{aligned}$

______ 16. $\begin{aligned} \tfrac{1}{4}x + \tfrac{1}{2}y &= \tfrac{5}{4} \\ \tfrac{1}{2}x - \tfrac{3}{4}y &= \tfrac{3}{2} \end{aligned}$

______ 17. $\begin{aligned} 0.3x + 1.5y &= 6 \\ 1.5x - 0.5y &= 6 \end{aligned}$

______ 18. $\begin{aligned} 0.8x + 0.3y &= 3 \\ 0.3x + 0.5y &= 1.9 \end{aligned}$

☆ Stretching the Topics

______ 1. If $\begin{vmatrix} a + 2 & -3 \\ a + 10 & 4 \end{vmatrix} = 0$, find the value of a.

______ 2. Use Cramer's rule to solve for x and y in terms of a and b.

$$\begin{aligned} ax - 2y &= 3 \\ 2x + by &= -2 \end{aligned}$$

Check your answers in the back of your book.

If you can complete **Checkup 10.3,** you are ready to go on to Section 10.4.

Name ____________________ Date ____________

✓ CHECKUP 10.3

Find the value of each determinant.

_______ 1. $\begin{vmatrix} -4 & 5 \\ 3 & 0 \end{vmatrix}$

_______ 2. $\begin{vmatrix} -12 & -3 \\ 9 & 2 \end{vmatrix}$

Use Cramer's rule to solve each system.

_______ 3. $2x - y = 5$
$-x + y = -3$

_______ 4. $\frac{2}{3}x - y = -2$
$2x - 3y = 3$

_______ 5. $3x + 2y = 5$
$4x - 5y = 22$

_______ 6. $4x + 9y = 3$
$9x - 4y = 2$

Check your answers in the back of your book.

If You Missed Problems:	You Should Review Examples:
1, 2	1, 2
3	3
4	4
5, 6	3–5

10.4 Solving Nonlinear Systems of Equations and Systems of Inequalities

Solving Nonlinear Systems of Equations

As you might imagine, a **nonlinear system of equations** is a system in which at least one of the equations is *not* linear. The graph of a nonlinear system might contain:

1. A line and a parabola, circle, ellipse, or hyperbola.

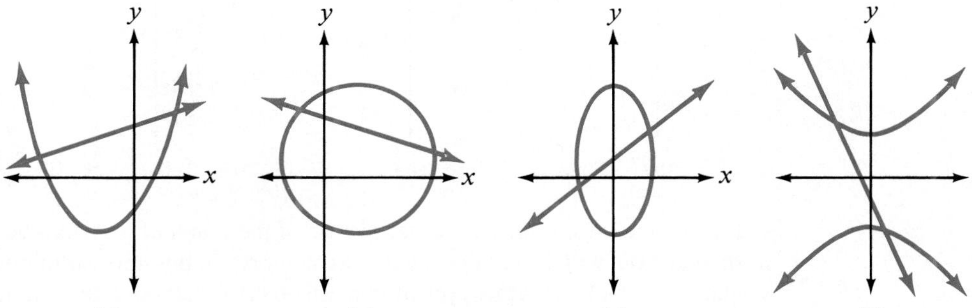

(a) (b) (c) (d)

2. A parabola and a circle, ellipse, or hyperbola.

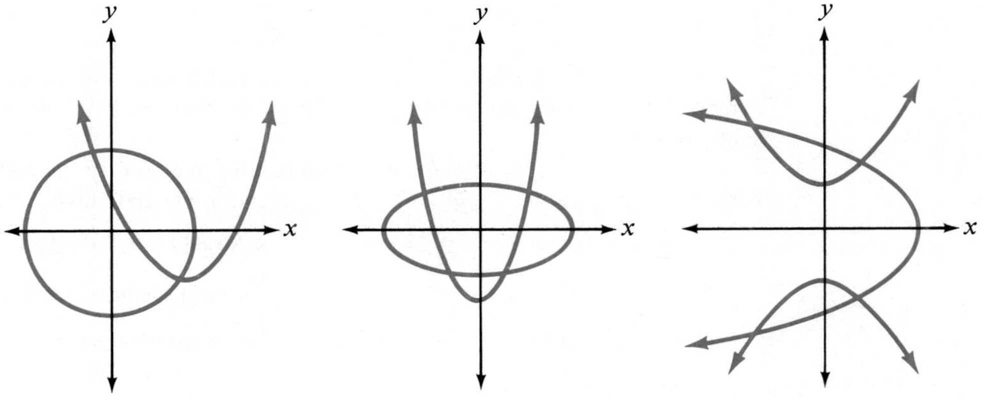

(a) (b) (c)

3. A circle and an ellipse or hyperbola.

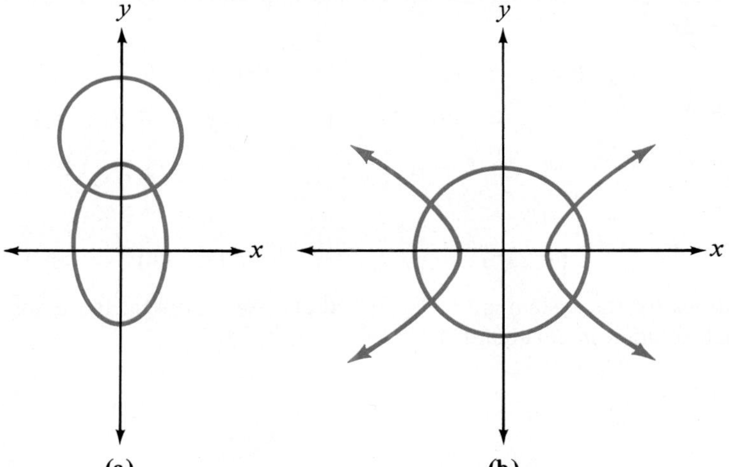

(a) (b)

4. An ellipse and a hyperbola.

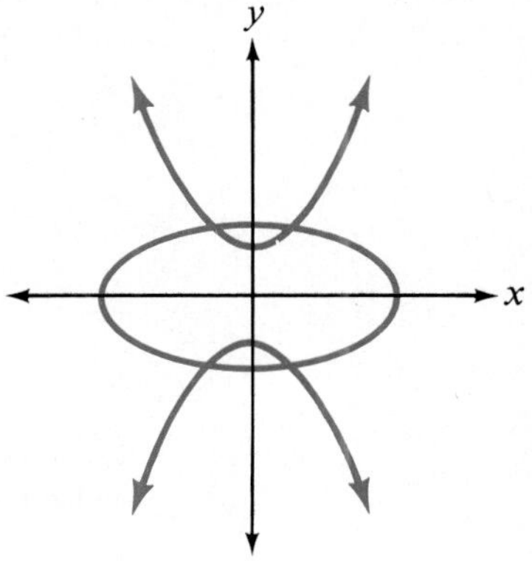

5. A pair of parabolas, circles, ellipses, or hyperbolas.

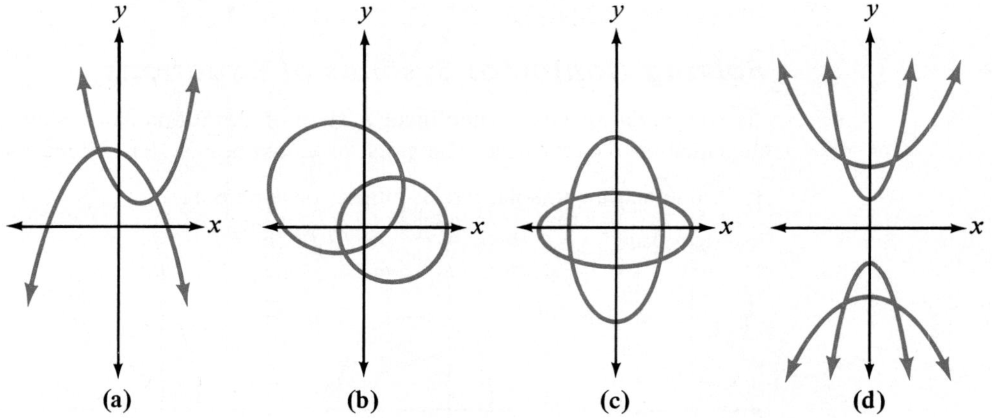

It is a good idea to have a mental picture of the graph of a system before attempting to solve it so that you will know the maximum number of possible solutions. In solving nonlinear systems we shall use whichever of our established methods seems most appropriate.

Example 1. Solve the system.

$$(1)\ y = -3x - 1$$

$$(2)\ y = x^2 - 2x - 3$$

Solution. The graph of equation (1) is a straight line, and the graph of equation (2) is a parabola. We seek the points of intersection of the line and parabola; there can be at most *two* such points.

Since y is isolated in at least one equation, the method of *substitution* seems appropriate here. We can substitute $-3x - 1$ for y in equation (2) and solve for x.

$-3x - 1 = x^2 - 2x - 3$ Substitute $-3x - 1$ for y in equation (2).

$0 = x^2 + x - 2$ Set the quadratic expression equal to zero.

$0 = (x + 2)(x - 1)$ Factor the trinomial.

$x + 2 = 0$ or $x - 1 = 0$

$x = -2$ $\quad$ $x = 1$

The graphs intersect at the points with x-coordinates of -2 and 1. To find the corresponding y-coordinates, we substitute each x-value into equation (1) or (2). Let's use equation (1) $y = -3x - 1$.

For $x = -2$

$$y = -3(-2) - 1$$
$$= 6 - 1$$
$$y = 5$$
$$(-2, 5)$$

For $x = 1$

$$y = -3(1) - 1$$
$$= -3 - 1$$
$$y = -4$$
$$(1, -4)$$

The solutions for the system are $(-2, 5)$ and $(1, -4)$. Look at the graph of the system and check each solution in equation (2) $y = x^2 - 2x - 3$.

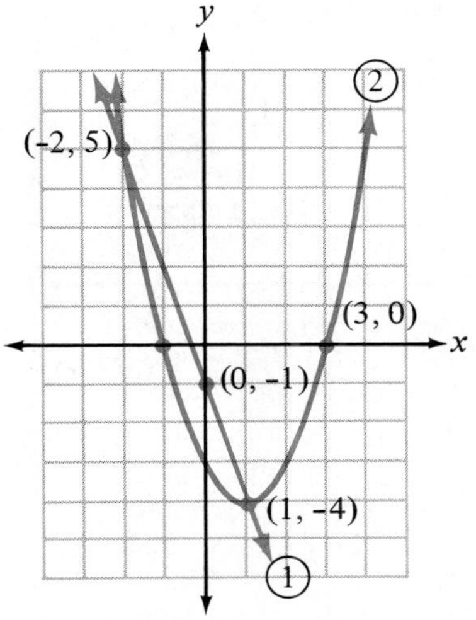

(1) $y = -3x - 1$

(2) $y = x^2 - 2x - 3$

CHECK: $x = -2$ and $y = 5$

$5 \stackrel{?}{=} (-2)^2 - 2(-2) - 3$

$5 \stackrel{?}{=} 4 + 4 - 3$

$5 = 5$

CHECK: $x = 1$ and $y = -4$

$-4 \stackrel{?}{=} (1)^2 - 2(1) - 3$

$-4 \stackrel{?}{=} 1 - 2 - 3$

$-4 = -4$

Example 2. Solve the system.

$$(1)\ x^2 + y^2 = 4$$
$$(2)\ x^2 + 2y^2 = 6$$

Solution. Equation (1) represents a circle and equation (2) represents an ellipse. There can be at most *four* points of intersection.

Since neither variable is easily isolated in either equation, we shall try *elimination* rather than substitution here. We can eliminate the x^2-terms by multiplying equation (1) by -1.

$-x^2 - y^2 = -4$	Multiply equation (1) by -1.
$x^2 + 2y^2 = 6$	Equation (2).
$y^2 = 2$	Add the equations.
$y = \pm\sqrt{2}$	Solve for y.

$y = \sqrt{2}$ or $y = -\sqrt{2}$

To find corresponding x-values, substitute these y-values into equation (1).

For $y = \sqrt{2}$	For $y = -\sqrt{2}$
$x^2 + y^2 = 4$	$x^2 + y^2 = 4$
$x^2 + (\sqrt{2})^2 = 4$	$x^2 + (-\sqrt{2})^2 = 4$
$x^2 + 2 = 4$	$x^2 + 2 = 4$
$x^2 = 2$	$x^2 = 2$
$x = \pm\sqrt{2}$	$x = \pm\sqrt{2}$
$(\sqrt{2}, \sqrt{2})$ and $(-\sqrt{2}, \sqrt{2})$	$(\sqrt{2}, -\sqrt{2})$ and $(-\sqrt{2}, -\sqrt{2})$

This system has four solutions. Let's sketch the graph.

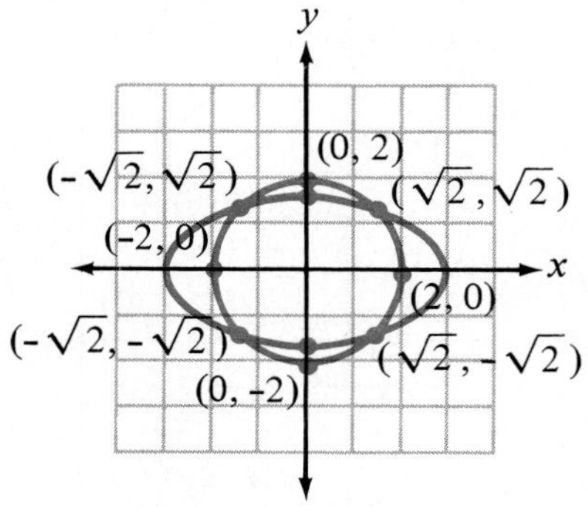

(1) $x^2 + y^2 = 4$
(2) $x^2 + 2y^2 = 6$

Example 3. Solve the system.

$$(1)\quad x^2 + y^2 = 1$$
$$(2)\quad 2x^2 - 3y^2 = 8$$

Solution. The graph of equation (1) is a circle and equation (2) is a hyperbola. There can be at most *four* points of intersection. Let's solve the system by eliminating x^2-terms.

$-2x^2 - 2y^2 = -2$	Multiply equation (1) by -2.
$2x^2 - 3y^2 = 8$	Equation (2).
$-5y^2 = 6$	Add the equations.
$y^2 = -\frac{6}{5}$	Isolate y^2.
$y = \pm\sqrt{-\frac{6}{5}}$	Solve for y.

There are no real y-values. The system has *no real number solutions*; the circle and hyperbola do *not* intersect. Look at the graph.

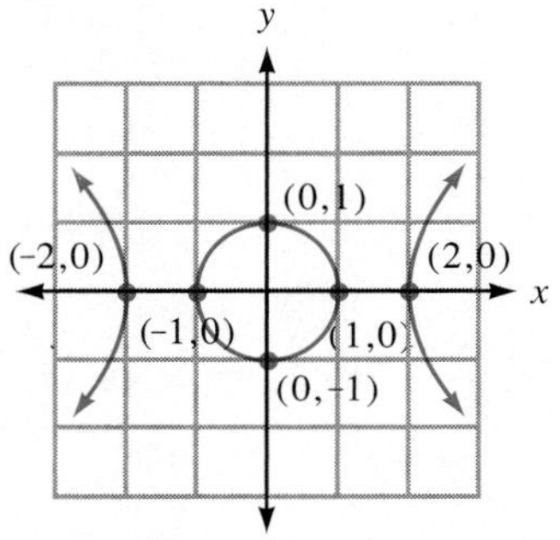

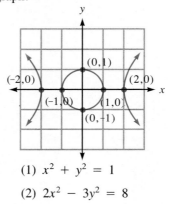

(1) $x^2 + y^2 = 1$
(2) $2x^2 - 3y^2 = 8$

Now you complete Example 4.

Example 4. Solve the system.

$$(1)\quad y = x^2 + 2$$
$$(2)\quad x^2 + y^2 = 4$$

Solution. The graph of equation (1) is a ________ and the graph of equation (2) is a ________ . There can be at most _____ points of intersection.

We could solve this equation by substitution, since y is isolated in equation (1), but such a substitution would change equation (2) into a fourth-degree equation. Instead, let's line up the variables and eliminate the x^2-terms.

$$\begin{array}{rrcrcl} (1) & -x^2 & & + y & = & 2 \\ (2) & x^2 & + y^2 & & = & 4 \\ \hline & & y^2 & + y & = & 6 \\ & & y^2 + y & - 6 & = & 0 \end{array}$$

$$(y + \_\_\_\_)(y - \_\_\_\_) = 0$$

$$y + \_\_\_\_ = 0 \qquad \text{or} \qquad y - \_\_\_\_ = 0$$

$$y = \_\_\_\_ \qquad\qquad y = \_\_\_\_$$

We substitute these y-values into equation (1).

Let $y = -3$	Let $y = 2$
$y = x^2 + 2$	$y = x^2 + 2$
$-3 = x^2 + 2$	$2 = x^2 + 2$
$-5 = x^2$	$0 = x^2$
$\pm\sqrt{-5} = x$	$x = \_\_\_\_$
No real solutions	$(\_\_\_\_, 2)$

The only solution is $(\_\_\_\_, 2)$. The parabola and circle intersect in just one point. Let's see if this result is verified by the graph.

Check your work on page 722. ▶

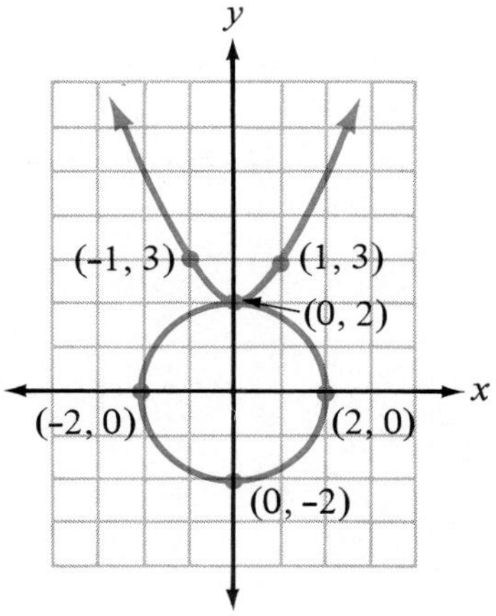

(1) $y = x^2 + 2$

(2) $x^2 + y^2 = 4$

Trial Run

Solve each system.

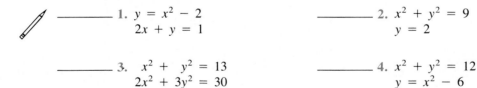

______ 1. $y = x^2 - 2$
$2x + y = 1$

______ 2. $x^2 + y^2 = 9$
$y = 2$

______ 3. $x^2 + y^2 = 13$
$2x^2 + 3y^2 = 30$

______ 4. $x^2 + y^2 = 12$
$y = x^2 - 6$

Answers are on page 723.

Solving Systems of Inequalities

In Chapter 8 we learned to graph inequalities as shaded regions in the plane. If we wish to find all points satisfying more than one inequality, we must graph each inequality and look for shaded regions *common* to all the given inequalities.

Example 5. Solve the system of inequalities.

$$(1)\ y > 2x - 1$$

$$(2)\ y \le 3 - x$$

Solution. We graph each inequality by the usual methods of Chapter 8.

(1) $y > 2x - 1$

Consider the boundary equation

$$y = 2x - 1$$

Slope: $m = 2$

y-intercept: $b = -1$

The line is *dashed* and we use vertical shading for the region *above* the line.

(2) $y \leq 3 - x$

Consider the boundary equation

$$y = -x + 3$$

Slope: $m = -1$

y-intercept: $b = 3$

The line is *solid* and we use horizontal shading for the region *below* the line.

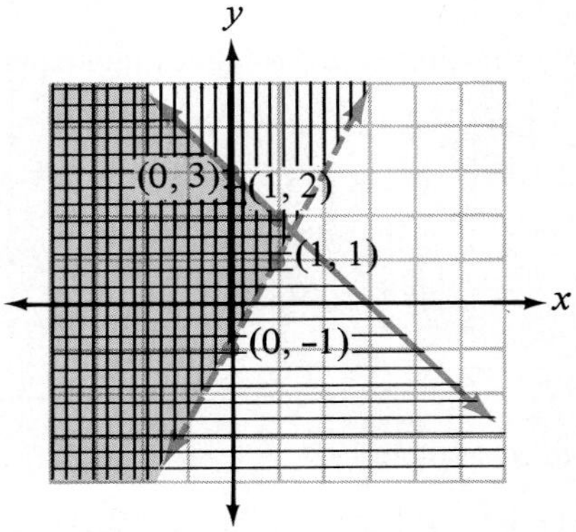

(1) $y > 2x - 1$

(2) $y \leq 3 - x$

The region in which both inequalities are satisfied is the region containing both kinds of shading. It is the region shaded in red in the figure.

You try Example 6.

Example 6. Solve the system of inequalities.

$$(1)\ (x - 2)^2 + (y + 1)^2 < 4$$

$$(2)\ y + x \geq 0$$

Solution. We graph each inequality.

(1) $(x - 2)^2 + (y + 1)^2 < 4$

Consider the boundary equation

$$(x - 2)^2 + (y + 1)^2 = 4$$

Center: (_____, _____)

Radius: $r =$ _____

The graph is a dashed ________ and we shade the region ________ the circle.

(2) $y + x \geq 0$

$$y \geq -x$$

Consider the boundary equation

$$y = -x$$

Slope: $m =$ _____

y-intercept: $b =$ _____

The graph is a ________ line and we shade the region ________ the line.

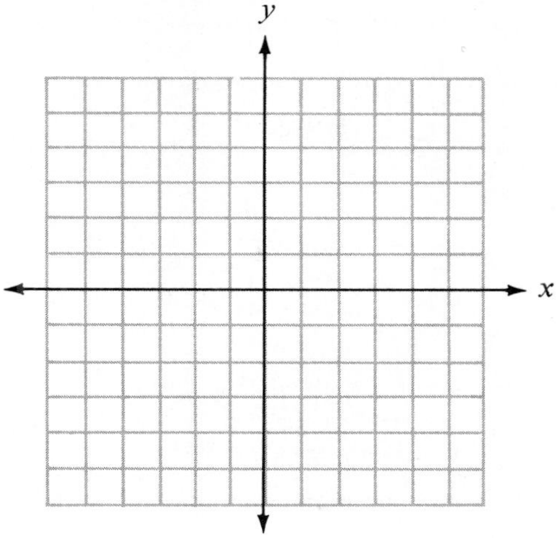

Check your work on page 723. ▶

Example 7. Solve the system of inequalities.

(1) $y \leq 5 - x^2$

(2) $x < 1$

(3) $y \geq 2$

Solution. We graph each inequality.

(1) $y \leq 5 - x^2$

Consider the boundary equation

$$y = 5 - x^2$$

This is a *solid* parabola that opens down and is symmetric to the y-axis. The vertex is (0, 5). We shade *below* the parabola.

(2) $x < 1$

Consider the boundary equation

$$x = 1$$

This is a *dashed* vertical line and we shade to the *left* of the line.

(3) $y \geq 2$

Consider the boundary equation

$$y = 2$$

This is a *solid* horizontal line and we shade *above* the line.

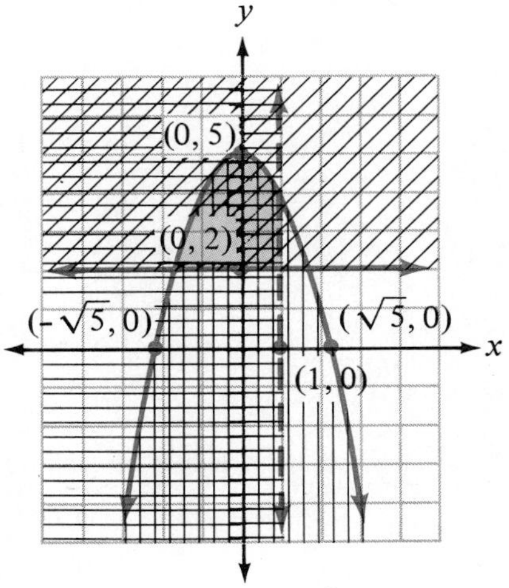

(1) $y \leq 5 - x^2$

(2) $x < 1$

(3) $y \geq 2$

The region in which all *three* kinds of shading appear is the region shaded in red in the graph.

Trial Run

Solve each system of inequalities.

1. $5x - 2y \le 6$
   $y < x$

2. $x^2 + y^2 < 25$
   $y \ge -x + 5$

3. $y < -x^2 + 9$
   $y > x + 1$
   $x \ge 0$

Answers are on page 723.

Examples You Completed

Example 4. Solve the system.

$$(1)\ y = x^2 + 2$$

$$(2)\ x^2 + y^2 = 4$$

Solution. The graph of equation (1) is a parabola and the graph of equation (2) is a circle. There can be at most four points of intersection.

$$\begin{array}{lrcl} (1) & -x^2 \qquad + y & = & 2 \\ (2) & x^2 + y^2 \qquad & = & 4 \\ \hline & y^2 + y & = & 6 \\ & y^2 + y - 6 & = & 0 \\ & (y + 3)(y - 2) & = & 0 \end{array}$$

$$y + 3 = 0 \quad \text{or} \quad y - 2 = 0$$

$$y = -3 \qquad\qquad y = 2$$

We substitute these y-values into equation (1).

Let $y = -3$	Let $y = 2$
$y = x^2 + 2$	$y = x^2 + 2$
$-3 = x^2 + 2$	$2 = x^2 + 2$
$-5 = x^2$	$0 = x^2$
$\pm\sqrt{-5} = x$	$x = 0$
No real solutions	$(0, 2)$

The solution is $(0, 2)$.

Example 6. Solve the system of inequalities.

$$(1)\ (x - 2)^2 + (y + 1)^2 < 4$$

$$(2)\ y + x \ge 0$$

Solution. We graph each inequality.

(1) $(x - 2)^2 + (y + 1)^2 < 4$

Consider the boundary equation

$$(x - 2)^2 + (y + 1)^2 = 4$$

Center: $(2, -1)$

Radius: $r = 2$

The graph is a dashed circle and we shade the region inside the circle.

(2) $y + x \geq 0$

$$y \geq -x$$

Consider the boundary equation

$$y = -x$$

Slope: $m = \frac{-1}{1}$

y-intercept: $b = 0$

The graph is a solid line and we shade the region above the line.

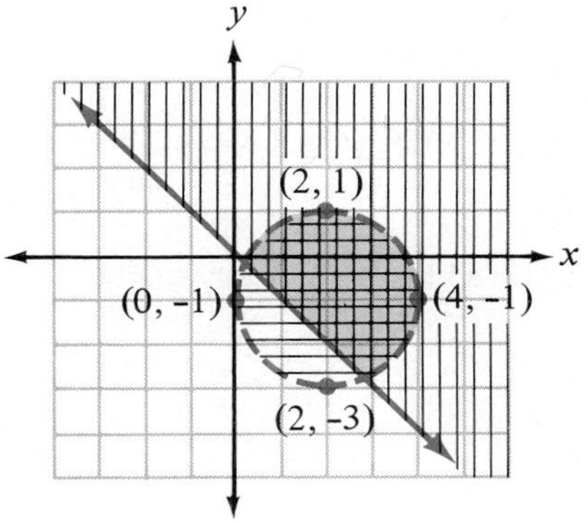

Answers to Trial Runs

page 719 **1.** $(1, -1)$ and $(-3, 7)$ **2.** $(\sqrt{5}, 2)$ and $(-\sqrt{5}, 2)$
3. $(-3, 2)$, $(3, 2)$, $(-3, -2)$, and $(3, -2)$
4. $(\sqrt{3}, -3)$, $(-\sqrt{3}, -3)$, $(2\sqrt{2}, 2)$, and $(-2\sqrt{2}, 2)$

page 722 **1.**

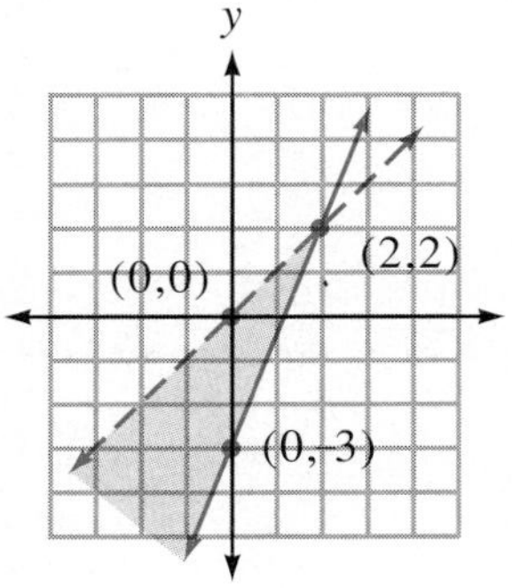

2.

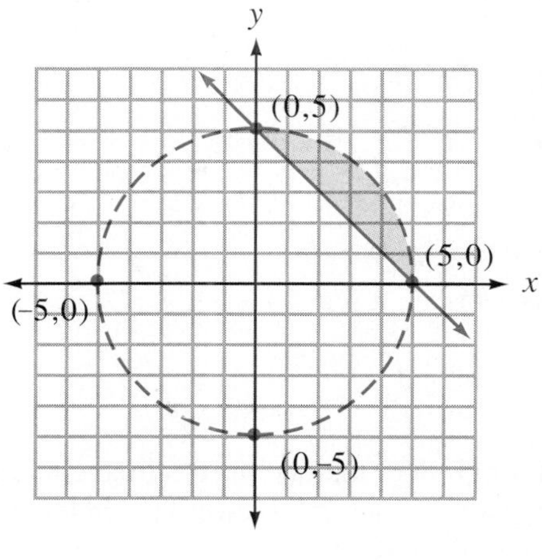

3.

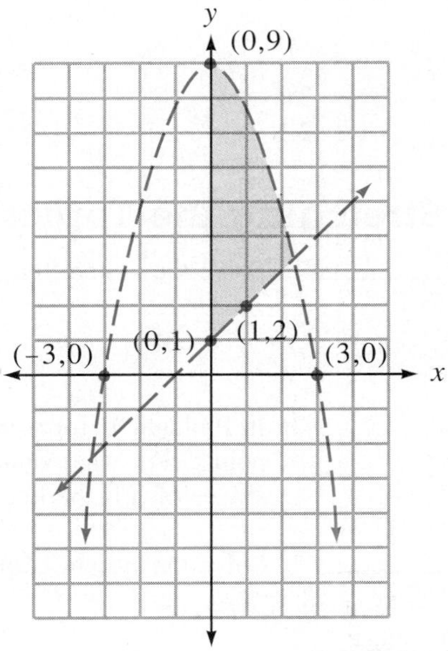

Name ______________________________ Date ______________

EXERCISE SET 10.4

Solve each system of equations.

______ 1. $y = x^2 - 5$
$x + y = -3$

______ 2. $y = x^2 + 3$
$x + y = 9$

______ 3. $x^2 + y^2 = 20$
$x - y = 2$

______ 4. $x^2 + y^2 = 25$
$x + y = -1$

______ 5. $x^2 + y^2 = 25$
$y = -x^2 + 13$

______ 6. $x^2 + y^2 = 16$
$y = x^2 - 10$

______ 7. $4x^2 - 9y^2 = 16$
$y = \frac{2}{5}x$

______ 8. $9x^2 - 27y^2 = 9$
$y = \frac{1}{2}x$

______ 9. $x^2 + y^2 = 8$
$3x^2 + 4y^2 = 28$

______ 10. $x^2 + y^2 = 25$
$2x^2 + 5y^2 = 98$

______ 11. $4x^2 + 9y^2 = 36$
$x^2 - 4y^2 = 49$

______ 12. $4x^2 + 25y^2 = 100$
$2x^2 - 9y^2 = 72$

______ 13. $y = x + 3$
$xy = 10$

______ 14. $y = 2x - 3$
$xy = -1$

______ 15. $x = y^2 - 3$
$x = 1$

______ 16. $x = y^2 - 5$
$x = 4$

Solve each system of inequalities.

17. $2x + 4y \leq 20$
$3x - 2y < 6$

18. $2x - y < 7$
$3x + 4y \geq -6$

19. $x^2 + y^2 \leq 16$
$3x - y \geq 1$

20. $x^2 + y^2 \geq 4$
$x - y \geq 0$

21. $-3x + 2y \leq -6$
$6x - 4y \geq 8$

22. $4x - 2y \geq 16$
$-2x + y \leq -4$

23. $y > x^2 - 4x$
$3y < 2x$
$x > 2$

24. $y > x^2 + 6x$
$y < -x$
$y > 1$

☆ Stretching the Topics

______ 1. Solve the system for y in terms of a.

$$x^2 + y^2 = 16$$
$$x + y = a$$

______ 2. In Problem 1, for what values of a does the line intersect the circle in two points? For what values of a does the line fail to meet the circle? For what values of a is the line tangent to the circle?

______ 3. Solve the system graphically.

$$x^2 + y^2 \geq 4$$
$$4x^2 + 9y^2 \leq 36$$

Check your answers in the back of your book.

If you can complete **Checkup 10.4,** you are ready to go on to Section 10.5.

Name ______________________ Date ____________

✓ CHECKUP 10.4

Solve each system of equations.

______ 1. $y = x^2 - 8$
$3x + y = -4$

______ 2. $x^2 + y^2 = 9$
$y = -x^2 + 3$

______ 3. $x^2 - y^2 = 16$
$4x^2 + y^2 = 9$

4. Graph the system.

$y > 5x - 2$

$2x - 3y > -3$

$y \geq 1$

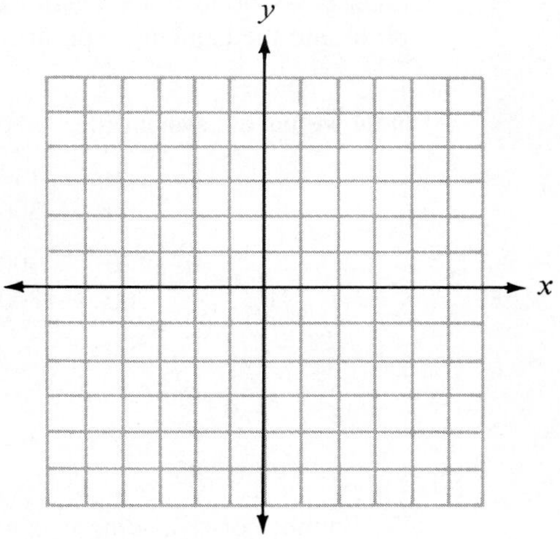

Check your answers in the back of your book.

If You Missed Problem:	You Should Review Example:
1	1
2	4
3	2
4	7

10.5 Switching from Words to Systems of Equations

Systems of equations are useful in solving word problems involving more than one unknown quantity that must meet more than one condition at the same time. Let's look at the problem stated at the beginning of this chapter.

Example 1. During the Fourth of July Festival, Marcia sold hot dogs and hamburgers at her refreshment stand. On Friday she made a profit of \$130 by selling 200 hot dogs and 150 hamburgers. On Saturday she made \$225 by selling 300 hot dogs and 300 hamburgers. What profit did Marcia make on each hot dog and each hamburger during the two days?

Solution

We Think	***We Write***
We do not know the profit on each hot dog and each hamburger.	Let x = profit on each hot dog (dollars) y = profit on each hamburger (dollars)
We can find the profit for each kind of sandwich by *multiplying* the number sold times the profit per sandwich.	(see table below)
Each day's profit is the *sum* of the hot dog profit and the hamburger profit.	Friday: $200x + 150y = 130$ Saturday: $300x + 300y = 225$

	Hot Dog Profit	Hamburger Profit
Friday	$200x$	$150y$
Saturday	$300x$	$300y$

Now we have a system of two equations to solve by elimination.

$$(1)\quad 200x + 150y = 130$$
$$(2)\quad 300x + 300y = 225$$

$$-400x - 300y = -260 \qquad \text{Multiply equation (1) by } -2.$$
$$300x + 300y = 225 \qquad \text{Equation (2).}$$
$$-100x = -35 \qquad \text{Add the equations.}$$
$$x = \frac{-35}{-100} \qquad \text{Isolate } x.$$
$$x = 0.35 \qquad \text{Simplify.}$$

We find the corresponding y-value by substituting 0.35 for x in either of the original equations. Let's use equation (1).

$$200x + 150y = 130$$
$$200(0.35) + 150y = 130$$
$$70 + 150y = 130$$
$$150y = 60$$
$$y = \frac{60}{150}$$
$$y = \frac{2}{5}$$
$$y = 0.40$$

Marcia made a profit of \$0.35 on each hot dog and \$0.40 on each hamburger.

Example 2. A sail in the shape of a right triangle has a perimeter of 30 feet. If the longest side (hypotenuse) of the sail measures 13 feet, what are the lengths of the other two sides (legs)?

Solution

We Think	***We Write***
We do not know the lengths of the two legs.	Let x = length of one leg (feet) y = length of other leg (feet)
An illustration is always useful in a geometry problem.	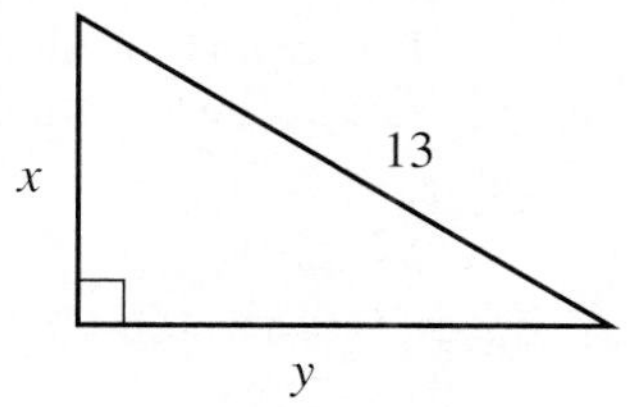
The perimeter of a triangle is the *sum* of the lengths of the three sides.	Perimeter: $x + y + 13 = 30$ $x + y = 17$
We need another equation. Since the sail is a *right* triangle, we can use the Pythagorean theorem.	Pythagorean theorem: $x^2 + y^2 = 13^2$ $x^2 + y^2 = 169$

We now have a system of two equations to solve.

$$(1)\quad x + y = 17$$
$$(2)\quad x^2 + y^2 = 169$$

Elimination will not work here, so we use the method of substitution. Solve equation (1) for y.

$y = 17 - x$	Isolate y in equation (1).
$x^2 + (17 - x)^2 = 169$	Substitute $17 - x$ for y in equation (2).
$x^2 + 289 - 34x + x^2 = 169$	Remove parentheses.
$2x^2 - 34x + 289 = 169$	Combine like terms.
$2x^2 - 34x + 120 = 0$	Set the quadratic expression equal to zero.
$2(x^2 - 17x + 60) = 0$	Factor out the common 2.
$2(x - 5)(x - 12) = 0$	Factor the trinomial.

$$x - 5 = 0 \quad \text{or} \quad x - 12 = 0$$
$$x = 5 \qquad\qquad x = 12$$

Substitute 5 for x in equation (1).	Substitute 12 for x in equation (1).
(1) $x + y = 17$	(1) $x + y = 17$
$5 + y = 17$	$12 + y = 17$
$y = 12$	$y = 5$
$(5, 12)$	$(12, 5)$

The sides of the sail must measure 5 feet and 12 feet.

Example 3. Anna and Beth are planning to drive to Canada, sharing the time at the wheel. Anna always drives at one constant rate and Beth always drives at another constant rate. If Anna drives for 7 hours and Beth for 4 hours, they can travel 570 miles. If Anna drives for 4 hours and Beth for 7 hours, they can travel 585 miles. How fast does each one drive?

Solution

We Think	***We Write***
We do not know either driver's rate.	Let a = Anna's rate (mph) b = Beth's rate (mph)
Distance is found by *multiplying* rate × time.	(see table below)
The total distance for either plan is the *sum* of the drivers' distances.	First plan: $7a + 4b = 570$ Second plan: $4a + 7b = 585$

	Anna's Distance	Beth's Distance
First plan	$7a$	$4b$
Second plan	$4a$	$7b$

Now we have our system of two equations to solve by elimination. Let's eliminate b.

$$\begin{aligned} (1)\quad 7a + 4b &= 570 \\ (2)\quad 4a + 7b &= 585 \end{aligned}$$

$$\begin{aligned} 49a + 28b &= 3990 && \text{Multiply equation (1) by 7.} \\ -16a - 28b &= -2340 && \text{Multiply equation (2) by } -4. \\ \hline 33a &= 1650 && \text{Add the equations.} \\ a &= \frac{1650}{33} && \text{Isolate } a. \\ a &= 50 && \text{Simplify.} \end{aligned}$$

We can find the corresponding b-value by substituting 50 for a in either equation.

$$\begin{aligned} (1)\quad 7a + 4b &= 570 \\ 7(50) + 4b &= 570 \\ 350 + 4b &= 570 \\ 4b &= 220 \\ b &= \frac{220}{4} \\ &= 55 \end{aligned}$$

Anna's rate is 50 miles per hour and Beth's rate is 55 miles per hour.

Example 4. Beside his house, Carlo wishes to plant a rectangular garden which is to be enclosed on the two shorter sides and one of the longer sides by 38 feet of fencing. If he wishes the area of the garden to be 176 square feet, what should be its dimensions?

Solution

We Think	***We Write***
We do not know the dimensions of the garden.	Let x = length of shorter side (feet) y = length of longer side (feet)

An illustration is helpful in a geometry problem.

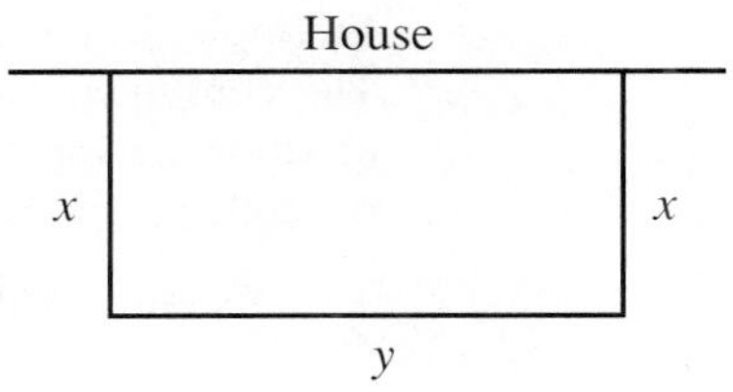

The fencing will be used on three sides.

Fencing: $x + x + y = 38$
$2x + y = 38$

The area of a rectangle is found by *multiplying* length times width.

Area: $x \cdot y = 176$

We now have a nonlinear system that can be solved by substitution.

$$\begin{aligned} (1)\quad 2x + y &= 38 \\ (2)\quad xy &= 176 \end{aligned}$$

$y = 38 - 2x$ — Isolate y in equation (1).

$x(38 - 2x) = 176$ — Substitute $38 - 2x$ for y in equation (2).

$38x - 2x^2 = 176$ — Remove parentheses.

$0 = 2x^2 - 38x + 176$ — Set the quadratic expression equal to zero.

$0 = 2(x^2 - 19x + 88)$ — Factor out the common 2.

$0 = 2(x - 8)(x - 11)$ — Factor the trinomial.

$x - 8 = 0$ or $x - 11 = 0$

$x = 8$ $\quad x = 11$

Substitute 8 for x in equation (2).

$$(2)\quad xy = 176$$
$$8y = 176$$
$$y = \frac{176}{8}$$
$$y = 22$$
$$(8, 22)$$

Substitute 11 for x in equation (2).

$$(2)\quad xy = 176$$
$$11y = 176$$
$$y = \frac{176}{11}$$
$$y = 16$$
$$(11, 16)$$

In each case the y-value is larger than the x-value, so either pair of dimensions satisfies the condition of our problem. Carlo's garden can measure 8 feet by 22 feet *or* 11 feet by 16 feet.

When a company puts a product on the market, the number of units (or quantity) the company is willing to **supply** depends on the price (per unit) of that product. The higher the price, the more units the company will supply. The quantity of the product that consumers will **demand** also depends on the price of the product. For consumers, however, the higher the price, the fewer units they will demand.

The point at which the producer's quantity supplied is equal to the consumers' quantity demanded is called the **equilibrium point**. If we are given a system of two equations that relate price and quantity for the producer and the consumers, we may find the equilibrium price and quantity by solving that system.

Example 5. For a producer of calculators, the monthly supply equation is $8p - 3x = 64$. The consumers' monthly demand equation for the calculators is $2p + 5x = 62$. In both equations, x represents the quantity (in hundreds of units) and p represents the price per unit (in dollars). Find the equilibrium quantity and price.

Solution. We must solve the following system.

$$\begin{aligned} \text{Supply: } 8p - 3x &= 64 \\ \text{Demand: } 2p + 5x &= 62 \end{aligned}$$

Let's use the method of addition.

$$\begin{array}{lrcrl} \text{Supply:} & 8p - 3x & = & 64 & \\ \text{Demand:} & -8p - 20x & = & -248 & \text{Multiply the Demand equation by } -4. \\ \hline & -23x & = & -184 & \text{Add the equations.} \\ & x & = & \dfrac{-184}{-23} & \text{Isolate } x. \\ & x & = & 8 & \text{Simplify.} \end{array}$$

To find the corresponding value of p, we substitute 8 for x in either equation.

$$\begin{aligned} \text{Demand: } 2p + 5x &= 62 \\ 2p + 5(8) &= 62 \\ 2p + 40 &= 62 \\ 2p &= 22 \\ p &= 11 \end{aligned}$$

At a price of \$11 per calculator, the producer will supply 8 hundred calculators per month and consumers will demand 8 hundred calculators per month. Look at the graph of this company's supply and demand system.

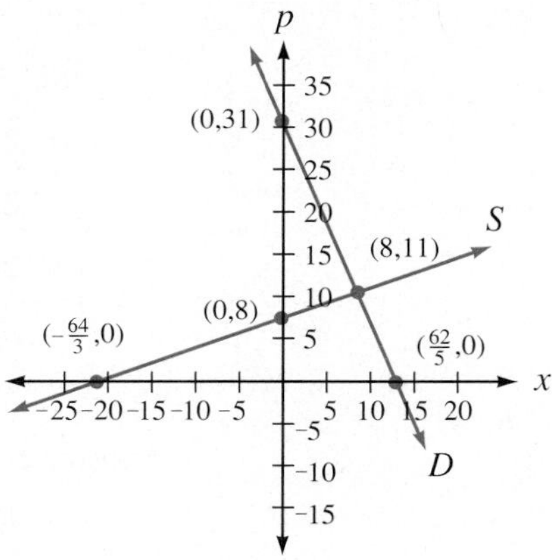

Name ______________________________ Date ______________

EXERCISE SET 10.5

Use a system of equations to solve each problem.

________ 1. For a United Way fund-raising dinner, a total of 245 tickets were sold. Adults' tickets sold for \$6.50 each and children's tickets sold for \$3. If \$1410.50 worth of tickets were sold, how many adults' tickets and how many children's tickets were sold?

________ 2. A total of 2825 tickets were sold for a concert and total ticket receipts were \$25,898. If advance tickets sold for \$8 and tickets at the door sold for \$10, how many tickets of each kind were sold?

________ 3. Harry works part-time mowing lawns and lifeguarding at the pool. When he mows for 5 hours and guards for 9 hours, he earns \$61. When he mows 3 hours and guards 12 hours, he earns \$63. How much does he earn per hour at each job?

________ 4. Five pounds of cashews and 8 pounds of peanuts cost \$26.27, and 7 pounds of cashews and 4 pounds of peanuts cost \$23.89. What is the price per pound of each kind of nut?

________ 5. Art sold a box of books at a yard sale for \$56. He sold some of the books for \$2 each and some for \$1.75 each. If there were 30 books in the box, how many did he sell at each price?

________ 6. At Garvin's Department Store, Martha and Laura found skirts on sale at one price and shoes on sale at another price. Martha bought 3 skirts and 4 pairs of shoes for \$85. Laura bought 5 skirts and 2 pairs of shoes for \$81. Find the sale price for the skirts and shoes.

________ 7. J. K. invested part of his money at 10 percent and the rest at 12 percent. The income from both investments was \$2700. If the amount invested at 10 percent is \$5000 less than twice the amount invested at 12 percent, find the amount that he has invested at each rate.

________ 8. Mr. Van Hooser has \$20,000 to invest. He invests part of it at 9 percent and the rest at 13 percent. If he earns \$2120 interest after 1 year, how much does he have invested at each rate?

________ 9. Sam and his roommate are planning to drive to Florida during spring break. Sam averages one rate of speed when he drives and his roommate averages another rate. If Sam drives for 5 hours and his roommate for 5 hours, they can travel 600 miles, but if Sam drives for only 3 hours and his roommate drives for 7 hours, they can travel 620 miles. How fast does each one drive?

________ 10. When Craig went to interview for a job, he made part of the trip by bus and the other part by taxi. The time he traveled by bus was 10 minutes longer than the amount of time by taxi. He averaged 40 miles per hour by bus and 30 miles per hour by taxi. If the total trip was 30 miles, how long did he spend traveling by bus and how long by taxi?

________ 11. The length of a carpet is 20 feet less than three times the width. The difference between the length and the width is 4 feet. Find the dimensions of the carpet.

________ 12. The width of a patio is 5 feet less than twice its length. The difference between the length and width is 1 foot. Find the dimensions of the patio.

________ 13. A community organization distributed 500 pounds of mixed candy in its Christmas cheer boxes. Part of the candy was bought at \$1.10 a pound and part at \$0.95 a pound. How many pounds were bought at each price if the total cost was \$501.25?

________ 14. A horse breeder used a mixture of shelled corn and oats to feed his horses. He found 100 bushels of this mixture weighed 4160 pounds. How many bushels of each kind did he use if a bushel of shelled corn weighs 56 pounds and a bushel of oats weighs 32 pounds?

________ 15. A pharmacist wishes to put 500 grains of a certain drug into 3-grain and 2-grain capsules. If he fills 200 capsules, how many capsules of each kind will he fill?

________ 16. Find the capacity of each of Mr. Ratley's two trucks if 5 loads of the smaller truck and 2 loads of the larger will carry 10 tons, while 3 loads of the smaller truck and 6 loads of the larger will carry 24 tons.

________ 17. A piece of wire 60 inches long is bent in the form of a right triangle whose hypotenuse is 25 inches. Find the other two sides of the triangle.

________ 18. In the corner of his yard Nelson has a flower bed that is a right triangle. If the flower bed requires 30 feet of fencing and the hypotenuse is 13 feet, find the lengths of the other two sides.

________ 19. Agnes is redecorating her sewing room. She will need 34 feet of a decorative ceiling border and 72 square feet of new carpeting. Find the dimensions of the sewing room.

________ 20. Find the dimensions of Bridget's rectangular garden if she needs 260 feet of fencing to enclose an area of 4200 square feet.

________ 21. The area of a rectangular piece of material is 12 square feet. The material is cut along the diagonal forming two triangles. If the length of the diagonal is 5 feet, find the dimensions of the rectangle.

________ 22. The area of a rectangular piece of wooden paneling is 120 square feet. The panel is sawed along the diagonal forming two triangles. If the length of the diagonal is 17 feet, find the dimensions of the rectangle.

________ 23. The House of Pottery determines that the monthly supply equation for pitchers is $5p - 7x = 89$. The consumers' monthly demand equation for the pitchers is $3p + 2x = 72$. Find the equilibrium quantity (in hundreds of units) and price (in dollars).

________ 24. For a producer of weed-trimmers, the monthly supply equation is $2p - 3x = 169$. The consumers' monthly demand equation for the weed-trimmers is $3p + 5x = 320$. Find the equilibrium quantity (in hundreds of units) and price (in dollars).

☆ Stretching the Topics

________ 1. The sum of the perimeters of two squares is 52 inches and the sum of the areas is 97 square inches. Find the length of each side of both squares.

________ 2. A farmer pressed 3000 pounds of tobacco into bales. If he had made each bale 15 pounds heavier, he would have had 10 fewer bales. Find the number of bales that he pressed and also the weight of each bale.

________ 3. Find the dimensions of a rectangle whose area remains unchanged if the length is decreased 2 feet and the width is increased 1 foot, and whose area is decreased by 44 square feet if the length is decreased by 4 and the width is decreased by 3 feet.

Check your answers in the back of your book.

If you can solve the problems in **Checkup 10.5,** you are ready to do the **Review Exercises** for Chapter 10.

Name ________________________________ Date ______________

✓ CHECKUP 10.5

Use a system of equations to solve each problem.

_______ 1. A grocer bought 4 crates of lettuce and 3 crates of cabbage. Later he bought 2 crates of lettuce and 5 crates of cabbage. His first bill was $34.20 and the second bill was $31.80. Find the cost of each crate of lettuce and each crate of cabbage.

_______ 2. The color flags hanging in the stadium at State University are right triangles. The perimeter of each flag is 36 feet and the longest side measures 15 feet. Find the lengths of the other two sides.

_______ 3. Willie drove his car 289 miles in 6 hours. Part of the time he averaged 45 miles per hour and the rest of the time he averaged 53 miles per hour. How much time did he drive at each speed?

_______ 4. Acme Trucking Company has a rectangular parking lot which is enclosed on the two shorter sides and one of the longer sides by 260 feet of fencing. If the area of the parking lot is 8000 square feet, what are its dimensions?

Check your answers in the back of your book.

If You Missed Problem:	You Should Review Example:
1	1
2	2
3	3
4	4

Summary

In this chapter we learned to find ordered pairs of numbers that satisfy two equations at the same time. We discussed several methods for solving such systems of equations and established ways of determining which method to use in different circumstances.

	Graphing	Substitution	Addition
When to use	When common points are easily located	When either variable is easily isolated in either equation	When neither variable is easily isolated
How to use	**1.** Graph the original equations. **2.** Locate the point of intersection. **3.** Write the solution as an ordered pair.	**1.** Isolate one variable in one equation. **2.** Substitute the found expression into the other equation, and solve for the value of the remaining variable. **3.** Substitute that value into either original equation and solve for the other variable. **4.** Write the solution as an ordered pair.	**1.** Write both equations in form $ax + by = c$. **2.** Multiply one or both equations by constants to make the coefficients of one variable *opposites* in the two equations. **3.** Add equations to eliminate one variable, and solve for the value of the remaining variable. **4.** Substitute that value into either original equation and solve for the other variable. **5.** Write the solution as an ordered pair.

We became alerted to the possibility that a system of linear equations might have *no* solutions in common or *all* solutions in common.

Label	Common Solutions	Signal when Solving	Graph
Consistent and independent	One ordered pair	Statement is sometimes true	Intersecting lines 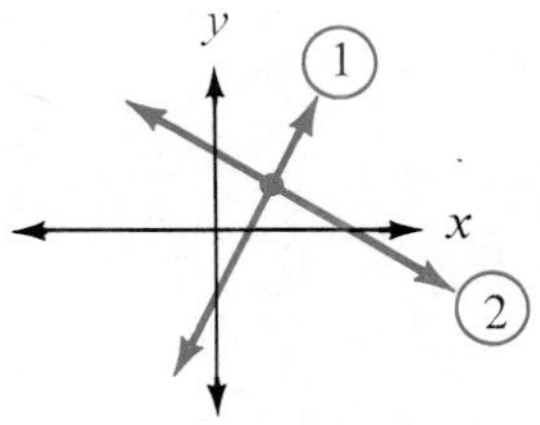
Inconsistent	No ordered pairs	Statement is never true	Parallel lines 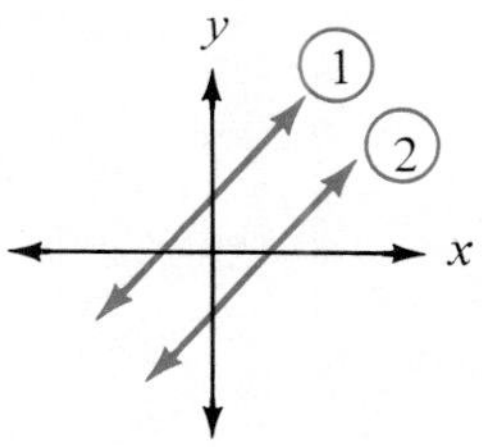

Label	Common Solutions	Signal when Solving	Graph
Dependent	All ordered pairs	Statement is always true	Coinciding lines 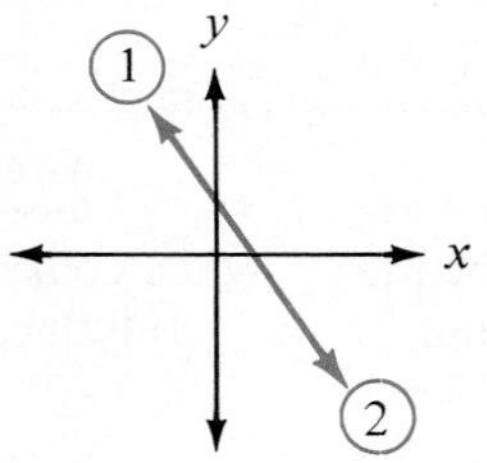

We discovered that we could use the same methods of substitution and/or addition to solve *systems of three equations in three variables* and to solve *nonlinear systems of equations*.

To solve *systems of inequalities in two variables,* we used the graphing method, looking for the shaded regions where all the given inequalities were satisfied.

Finally, we discussed the usefulness of systems of equations in finding solutions to problems stated in words.

❑ Speaking the Language of Algebra

Complete each statement with the appropriate word or phrase.

1. In solving a system of linear equations, we must find the ________ ________ that satisfies both equations.

2. If it is easy to isolate one variable in one of the equations, we may solve the system by the method of ________ .

3. In using the method of addition, we wish the coefficient of one variable in one equation to be the ________ of the coefficient of the same variable in the other equation.

4. If we arrive at a statement such as ''0 = 3'' in solving a linear system algebraically, we know the system has ________ solution and the graph of the system is a pair of ________ lines.

5. If the graph of one equation is a circle and the graph of another equation is a parabola, the system of these two equations will have *at most* ________ solutions.

Writing About Mathematics

Write your response to each question in complete sentences.

1. Explain what you are looking for when you solve a system of equations.

2. Explain why the method of graphing is often not a very efficient method for solving a system of equations.

3. In solving a system of two first-degree equations algebraically, tell when you would conclude that the system is
   (a) Inconsistent.
   (b) Dependent.
   (c) Independent and consistent.
   Describe the graphs for situations **(a)**, **(b)**, and **(c)**.

4. Write a word problem that would lead you to write this system of equations.

$$5x + 3y = 53$$
$$2x + 7y = 56$$

Name ________________________________ Date ______________

REVIEW EXERCISES for Chapter 10

Solve by graphing and classify each system.

________ 1. $3x - y = 5$
$2x + y = 0$

________ 2. $2x - 3y = -1$
$-4x + 6y = 12$

Solve each system by substitution.

________ 3. $2x + y = -5$
$3x + 2y = -6$

________ 4. $5x + 2y = 4$
$x + 2y = -4$

Solve each system by addition.

________ 5. $2x + 3y = -1$
$-x + y = 13$

________ 6. $4x - 4y = 9$
$-3x + 3y = 2$

________ 7. $4x + 5y = 25$
$3x + 7y = 9$

________ 8. $x + y + z = 1$
$5x - 2y - z = -3$
$2x + y - 3z = 5$

Solve each system.

________ 9. $y = x^2 - 15$
$5x - y = -9$

________ 10. $x^2 + y^2 = 20$
$2x^2 - y^2 = 28$

11. $x^2 + y^2 < 25$
$y > 3x - 1$

12. $y < 3x + 4$
$x \leq 3$
$x + y > -1$

Use a system of equations to solve each problem.

________ 13. During the month of May, the Union County Sheriff's Department issued 117 traffic tickets for speeding or failing to obey traffic signals. The fine for speeding was \$50 and the fine for failure to obey traffic signals was \$35. If the Sheriff's Department collected \$5340 in fines from these two offenses, how many of each kind of ticket were issued?

________ 14. From Henshaw, Henry bicycled due east to Grove Center. He then bicycled due north to Boxville. Finally, he returned 35 miles to Henshaw by completing the triangle. Henry traveled a total of 84 miles. If the shortest stretch was between Henshaw and Grove Center, how far apart are those two towns?

________ 15. The area of a rectangle is 60 square meters and the diagonal is 13 meters. Find the dimensions of the rectangle.

________ 16. Homefolks Furniture Fashions determines that the monthly supply equation for decorator lamps is $2p - x = 63$. The consumers' monthly demand equation for the lamps is $p + 3x = 84$. Find the equilibrium quantity (in hundreds of units) and the price (in dollars).

Check your answers in the back of your book.

If You Missed Exercises:	You Should Review Examples:	
1, 2	SECTION 10.1	1–4
3, 4		5–8
5–7	SECTION 10.2	1–4
8		5
9	SECTION 10.4	1
10		3
11		6
12		7
13	SECTION 10.5	1
14		2
15		4
16		5

If you have completed the **Review Exercises** and corrected your errors, you are ready to take the **Practice Test** for Chapter 10.

Name ____________________ Date ____________

PRACTICE TEST for Chapter 10

	Section	Example
1. Solve the system by graphing.	10.1	4

$$\begin{aligned} -3x + 2y &= -4 \\ x - y &= 3 \end{aligned}$$

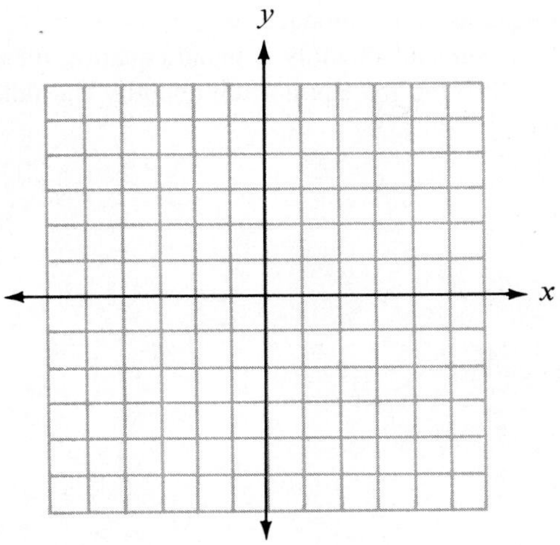

	Section	Example
______ 2. Solve the system by substitution. $6x - 2y = -9$, $4x + y = 1$	10.1	5
______ 3. Solve the system by addition. $2x + 5y = 4$, $3x + 6y = 1$	10.2	1

2. $$\begin{aligned} 6x - 2y &= -9 \\ 4x + y &= 1 \end{aligned}$$

3. $$\begin{aligned} 2x + 5y &= 4 \\ 3x + 6y &= 1 \end{aligned}$$

Solve each system.

	Section	Example
______ 4. $6x - 4y = -20$; $2y = 3x + 10$	10.2	4
______ 5. $7 + y = 0$; $x - 4y = 13$	10.1	6
______ 6. $x + 2y = 3$; $3x + 6y = 6$	10.2	2
______ 7. $x + 2y + z = -3$; $-x + y + 3z = -3$; $2x - 3y + 4z = 8$	10.2	5
______ 8. $y = x^2 - 2x$; $2x - y = 3$	10.4	1
9. $x + y < 0$; $y \geq 2x + 3$	10.4	5
10. $x^2 + y^2 \leq 9$; $2x - y > 3$	10.4	6

		Section	Example
________	11. During one week, Barbara ate 5 apples and 7 peaches for a total of 820 calories. Sam ate 10 apples and 3 peaches for a total of 980 calories. Find the number of calories in one apple and in one peach.	10.5	1
________	12. The monthly supply equation for a producer of car stereo systems is $2p - 5x = 35$. The consumers' monthly demand equation for the car stereos is $p + 2x = 130$. Find the equilibrium quantity (in hundreds of units) and the price (in dollars).	10.5	5

Name ______________________ Date ____________

SHARPENING YOUR SKILLS after Chapters 1–10

		Section
______	1. State the restrictions on the variable and reduce the fraction $\dfrac{a^2 - 6a - 55}{5a^2 + 23a - 10}$.	6.1
______	2. Find the real values for x and y so that $(3x - 5) - (5y + 6)i = 2x + i$.	7.5
______	3. Find the equation of the horizontal line through $(-2, 1)$.	8.3
______	4. Find the midpoint of the line segment joining $(3, -2)$ and $(5, 9)$.	8.2
______	5. Find the equation of a line through $(-4, 1)$ parallel to $2x - y = 3$.	8.3
______	6. If $g(x) = -x^2 + 2x - 1$, find $g(-2)$.	8.5
______	7. Give the domain for $f(x) = \sqrt{5 - 2x}$.	8.5
______	8. Use synthetic division to find the remainder when $x^4 + x^3 + 5x - 2$ is divided by $x + 2$.	3.5
______	9. Solve by graphing: $\dfrac{3x - 1}{7} < 1 - x$ or $\dfrac{x}{6} - 1 < -2$.	2.3

Perform the indicated operation and simplify.

______	10. $\left(\dfrac{3a}{2b}\right)^{-1}\left(\dfrac{a^2b}{a^{-3}b^4}\right)^2\left(\dfrac{7a^{-11}b^7}{9a^{-1}b^{-9}}\right)^0$	3.2
______	11. $7x - x[3x^2 - 4x(x + 2)]$	3.3
______	12. $(x - 1)^4$	3.4
______	13. $\dfrac{6x^2 + 19x - 20}{5x^2 - 17x + 6} \div \dfrac{20 - 9x - 18x^2}{3x^2 - 5x - 12}$	6.2
______	14. $\dfrac{3x - 1}{x^2 - 3x + 2} + \dfrac{2x + 1}{x^2 + 2x - 8} - \dfrac{x - 3}{x^2 + 3x - 4}$	6.3
______	15. $\dfrac{x - \dfrac{15}{x + 2}}{x + 10 + \dfrac{40}{x - 3}}$	6.4
______	16. $\left(\dfrac{8}{27}\right)^{-\frac{2}{3}}$	7.1
______	17. $\left(x^{\frac{4}{5}}y^{-\frac{2}{3}}z^{\frac{8}{3}}\right)^{\frac{1}{16}}$	7.2
______	18. $\sqrt[3]{\dfrac{14x^2}{5}} \cdot \sqrt[3]{\dfrac{-40x}{7}}$	7.3

SECTION

_______ 19. $(\sqrt{3} - 2\sqrt{y})(4\sqrt{3} + \sqrt{y})$ 7.3

_______ 20. $\dfrac{1 - \sqrt{-6}}{1 + \sqrt{-6}}$ 7.5

Solve each equation or inequality.

_______ 21. $(5x - 1)(x + 2) = -5(x^2 + 1) - 8x$ 5.1

_______ 22. $16x^3 - 48x^2 - x + 3 = 0$ 5.1

_______ 23. $8x^2 + 10x + 1 = 0$ 5.3

_______ 24. $144 - x^2 > 0$ 5.4

_______ 25. $\dfrac{x}{x - 6} + \dfrac{1}{2x - 1} = \dfrac{2x^2 - 18}{2x^2 - 13x + 6}$ 6.5

_______ 26. $\dfrac{x - 10}{x} \geq \dfrac{1}{3}$ 6.5

_______ 27. $\sqrt{x - 11} = 11 - \sqrt{x}$ 7.4

_______ 28. $\dfrac{3}{x^2} + 7 = \dfrac{8}{x}$ 7.6

Graph using any method you choose.

29. $8x - 3y = -15$ 8.1

30. $3x - 8 = 0$ 8.1

31. $x + y < -4$ 8.4

Graph each relation. State the domain and range.

32. $y = x^3 - 9x$ 9.1

33. $y = -x^2 - x + 12$ 9.2

34. $(x + 5)^2 + (y - 8)^2 = 64$ 9.3

35. $y > 8x - x^2$ 9.4

_______ 36. Lela can rake all the leaves from the Harpers' yard in 5 hours. If her sister Linda helps her, they can finish the yard in 3 hours. How long would it take Linda by herself to rake the yard? 6.6

CHAPTER

11

Working with Exponential and Logarithmic Functions

When \$100 is deposited in a savings account paying compound interest at the rate of 2 percent per quarter (every 3 months), the total amount, A, in the account after n quarters is given by the function $A = 100(1.02)^n$. Use a calculator to find the amount in the account after 7 quarters; after 5 years.

We have worked with many functions in the preceding chapters. Most functions we have looked at have been polynomial functions. Now it is time to investigate two functions of different types. In this chapter we

1. Work with exponential functions.
2. Work with logarithmic functions.
3. Use the properties of logarithms.
4. Use common logarithms.
5. Solve exponential and logarithmic equations.

11.1 Working with Exponential Functions

In earlier discussions of polynomial functions such as

$$y = 2x + 3 \qquad y = x^2 + 7x + 1 \qquad y = 1 - x^3$$

we were dealing with expressions containing constant powers of a variable base. Now we turn our attention to expressions in which the base is a constant and the variable appears in the exponent. We wish to investigate expressions such as

$$2^x \qquad (\tfrac{1}{4})^x \qquad (-3)^x$$

Evaluating Exponential Expressions

To evaluate exponential expressions for different values of the variable, we substitute the given x-value wherever x appears in the exponent. If you need to refresh your memory about exponents, see Chapters 3 and 7.

x	2^x	$(\frac{1}{4})^x$	$(-3)^x$
3	$2^3 = 8$	$(\frac{1}{4})^3 = \frac{1}{64}$	$(-3)^3 = -27$
0	$2^0 = 1$	$(\frac{1}{4})^0 = 1$	$(-3)^0 = 1$
-1	$2^{-1} = \frac{1}{2}$	$(\frac{1}{4})^{-1} = 4$	$(-3)^{-1} = -\frac{1}{3}$
$\frac{1}{2}$	$2^{\frac{1}{2}} = \sqrt{2}$	$(\frac{1}{4})^{\frac{1}{2}} = \sqrt{\frac{1}{4}} = \frac{1}{2}$	$(-3)^{\frac{1}{2}} = \sqrt{-3} = \sqrt{3}i$ (not a real number)

As you can see from the last entry in our table, an exponential expression with a *negative* base will have a *real* value only if we avoid x-values in the exponent that cause us to take an *even* root. Therefore, we must avoid substituting x-values that are fractions with *even* denominators. Such a restriction on x-values can become a real headache, because there are infinitely many fractions with even denominators that must be avoided. We shall discuss this dilemma in more detail shortly.

For now, let's practice evaluating some more exponential expressions.

Example 1. Complete the table of values for 10^x and 5^{x+1}.

x	10^x	5^{x+1}
-2	$10^{-2} = \frac{1}{10^2} =$ ____	$5^{-2+1} = 5^{-1} =$ ____
-1	$10^{-1} =$ ____	$5^{-1+1} = 5^0 =$ ____
0	$10^0 =$ ____	$5^{0+1} = 5^1 =$ ____
1	$10^1 =$ ____	$5^{1+1} = 5^2 =$ ____
2	$10^2 =$ ____	$5^{2+1} = 5^3 =$ ____

Check your table on page 754. ▶

Example 2. Complete the table of values for -4^{2x} and $(-4)^{2x}$.

Solution. Recall that an exponent belongs only to the base to which it is attached unless parentheses indicate otherwise. Thus $-4^{2x} = -1 \cdot 4^{2x}$.

x	-4^{2x}	$(-4)^{2x}$
-1	$-4^{2(-1)} = -1 \cdot 4^{-2} = \frac{-1}{4^2} = -\frac{1}{16}$	$(-4)^{2(-1)} = (-4)^{-2} = \frac{1}{(-4)^2} = \frac{1}{16}$
0	$-4^0 = -1 \cdot 4^0 = -1 \cdot 1 = -1$	$(-4)^0 = 1$
$\frac{1}{4}$	$-4^{2(\frac{1}{4})} = -1 \cdot 4^{\frac{1}{2}} = -1\sqrt{4} = -2$	$(-4)^{2(\frac{1}{4})} = (-4)^{\frac{1}{2}} = \sqrt{-4} = 2i$ (not real)
$\frac{1}{2}$	$-4^{2(\frac{1}{2})} = -1 \cdot 4^1 = -1 \cdot 4 = -4$	$(-4)^{2(\frac{1}{2})} = (-4)^1 = -4$

Trial Run

Evaluate each expression for the given x-values.

_______ 1. 5^x when $x = -2$; when $x = 0$; when $x = 2$.

_______ 2. 2^{x-1} when $x = -1$; when $x = 0$; when $x = 2$.

_______ 3. -3^{4x} when $x = -\frac{1}{2}$; when $x = 0$; when $x = \frac{1}{8}$; when $x = \frac{1}{2}$.

_______ 4. $(-3)^{4x}$ when $x = -\frac{1}{2}$; when $x = 0$; when $x = \frac{1}{8}$; when $x = \frac{1}{2}$.

Answers are on page 754.

Defining the Exponential Function

Now that we have practiced evaluating exponential expressions, it is time to investigate sets of ordered pairs of real numbers defined by relations such as

$$y = 2^x \qquad \text{and} \qquad y = (\tfrac{1}{2})^x$$

Notice that we shall avoid expressions such as $(-2)^x$ and $(-3)^x$ because such expressions may represent numbers that are *not real*. Therefore, exponential expressions with negative bases will *not* be considered in this discussion of ordered pairs of real numbers.

We shall restrict our discussion to relations of the form

$$y = a^x \qquad \text{where} \qquad a > 0$$

Before deciding whether such relations are *functions*, let's try to graph two examples, using the method "choose an x, find a y."

Example 3. Use arbitrary points to graph $y = 2^x$.

Solution

x	$y = 2^x$	(x, y)
-2	$2^{-2} = \frac{1}{4}$	$(-2, \frac{1}{4})$
-1	$2^{-1} = \frac{1}{2}$	$(-1, \frac{1}{2})$
0	$2^0 = 1$	$(0, 1)$
$\frac{1}{2}$	$2^{\frac{1}{2}} = \sqrt{2}$	$(\frac{1}{2}, \sqrt{2})$
1	$2^1 = 2$	$(1, 2)$
2	$2^2 = 4$	$(2, 4)$

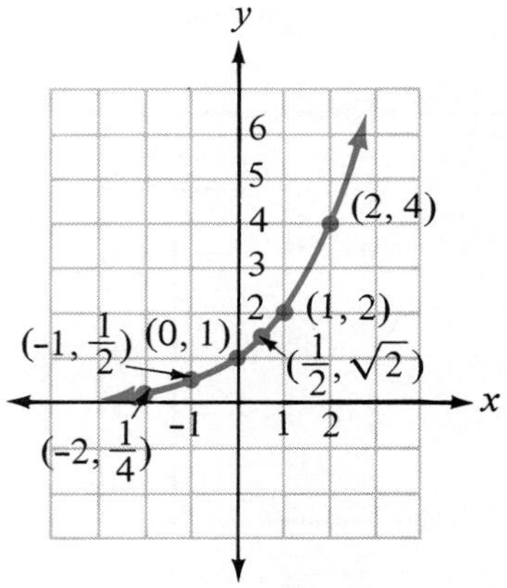

$y = 2^x$

In joining our points with a smooth curve, we assumed that we could find a real y-value corresponding to every real x-value, whether it be rational or irrational. This is indeed the case. For instance, to evaluate 2^x when $x = \sqrt{3}$, we must realize that since

$$1 < \sqrt{3} < 2$$

we know that

$$2^1 < 2^{\sqrt{3}} < 2^2$$

$$2 < 2^{\sqrt{3}} < 4$$

so that $2^{\sqrt{3}}$ is some irrational number between 2 and 4, probably closer to 4. The relation $y = 2^x$ is defined for all real values of x.

Example 4. Use arbitrary points to graph $y = 2^{-x}$.

Solution

$$y = 2^{-x}$$
$$= (2^{-1})^x$$
$$y = \left(\frac{1}{2}\right)^x$$

x	$y = (\frac{1}{2})^x$	(x, y)
-2	$(\frac{1}{2})^{-2} = \frac{1}{2^{-2}} = 4$	$(-2, 4)$
-1	$(\frac{1}{2})^{-1} = 2$	$(-1, 2)$
0	$(\frac{1}{2})^0 = 1$	$(0, 1)$
$\frac{1}{2}$	$(\frac{1}{2})^{\frac{1}{2}} = \frac{1}{\sqrt{2}} = \frac{\sqrt{2}}{2}$	$\left(\frac{1}{2}, \frac{\sqrt{2}}{2}\right)$
1	$(\frac{1}{2})^1 = \frac{1}{2}$	$(1, \frac{1}{2})$
2	$(\frac{1}{2})^2 = \frac{1}{4}$	$(2, \frac{1}{4})$

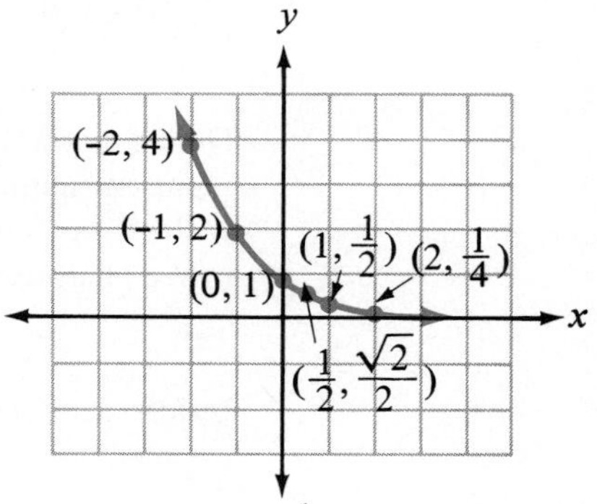

$y = (\frac{1}{2})^x$

Notice that the graph of $y = (\frac{1}{2})^x$ looks like the graph of $y = 2^x$, flipped across the y-axis.

Before summarizing our observations about the exponential function, let's consider the graph of $y = 1^x$.

Example 5. Graph $y = 1^x$.

Solution

x	$y = 1^x$	(x, y)
-2	$1^{-2} = \frac{1}{1^2} = 1$	$(-2, 1)$
-1	$1^{-1} = 1$	$(-1, 1)$
0	$1^0 = 1$	$(0, 1)$
$\frac{1}{2}$	$1^{\frac{1}{2}} = \sqrt{1} = 1$	$(\frac{1}{2}, 1)$
1	$1^1 = 1$	$(1, 1)$
2	$1^2 = 1$	$(2, 1)$

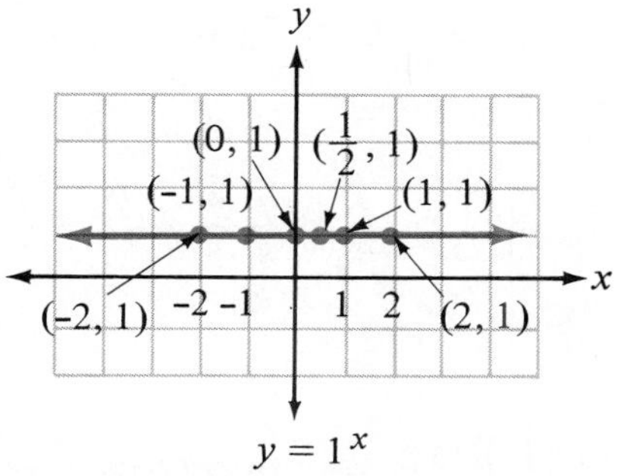

These ordered pairs are indeed boring and the graph does not resemble the other examples of exponential relations. In fact, this graph looks more like the graph of a constant function. For this reason, we exclude the base 1 from our discussion of exponential relations.

From our examples we see that the exponential relation is indeed a function because each x-value corresponds to one and only one y-value. We also note that the exponent can be *any* real number, but the resulting y-value will always be a *positive* real number. Now we are ready to summarize our findings regarding **exponential functions.**

Definition of Exponential Function

$$y = a^x$$

where $a > 0$ and $a \neq 1$ is called an exponential function.

Domain: $x \in R$
Range: $y \in R,\ y > 0$

For our own convenience, we make note of the fact that all exponential functions of the form $y = a^x$ contain the point $(0, 1)$. Moreover, we can determine whether $y = a^x$ is increasing or decreasing from left to right by looking at the base a.

If $0 < a < 1$, then $y = a^x$ is decreasing (Example 4).
If $a > 1$, then $y = a^x$ is increasing (Example 3).

To graph exponential functions that are variations on the basic function $y = a^x$, we continue to use arbitrary points.

Example 6. Graph $y = 5^{x+1}$.

Solution

x	$y = 5^{x+1}$	(x, y)
-2	$5^{-2+1} = 5^{-1} = \frac{1}{5}$	$(-2, \frac{1}{5})$
-1	$5^{-1+1} = 5^0 = 1$	$(-1, 1)$
$-\frac{1}{2}$	$5^{-\frac{1}{2}+1} = 5^{\frac{1}{2}} = \sqrt{5}$	$(-\frac{1}{2}, \sqrt{5})$
0	$5^{0+1} = 5^1 = 5$	$(0, 5)$
1	$5^{1+1} = 5^2 = 25$	$(1, 25)$

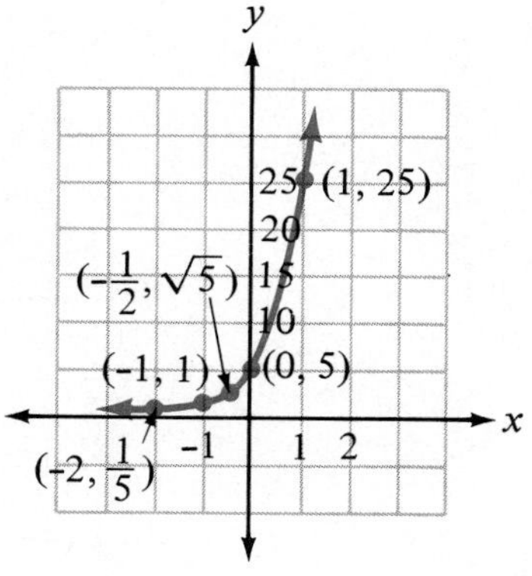

$y = 5^{x+1}$

Trial Run

Graph each function.

1. $y = 3^x$

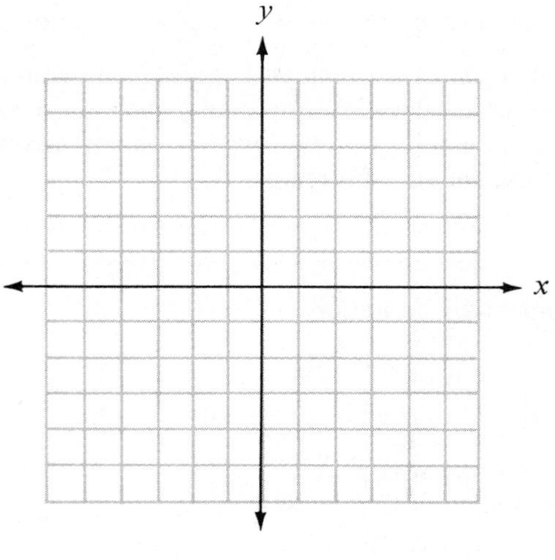

2. $y = \left(\frac{1}{3}\right)^x$

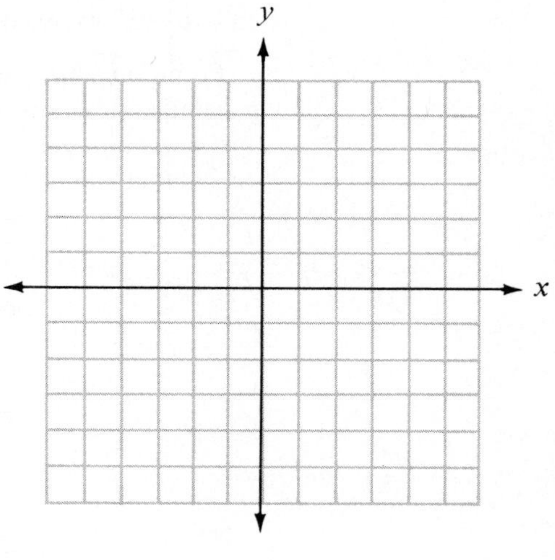

3. $y = 2^{x+1}$

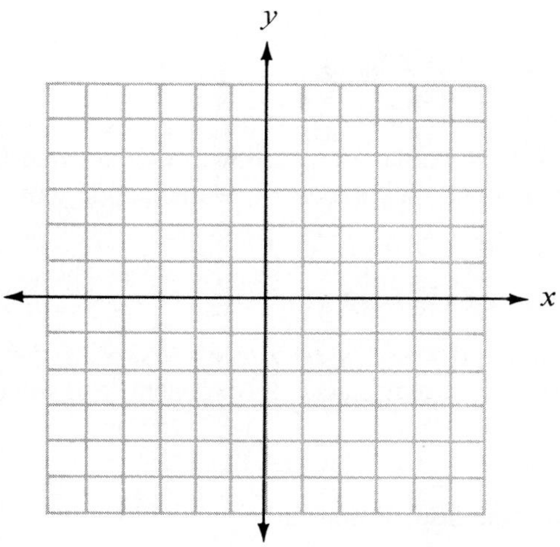

4. $y = \left(\frac{1}{2}\right)^{2x}$

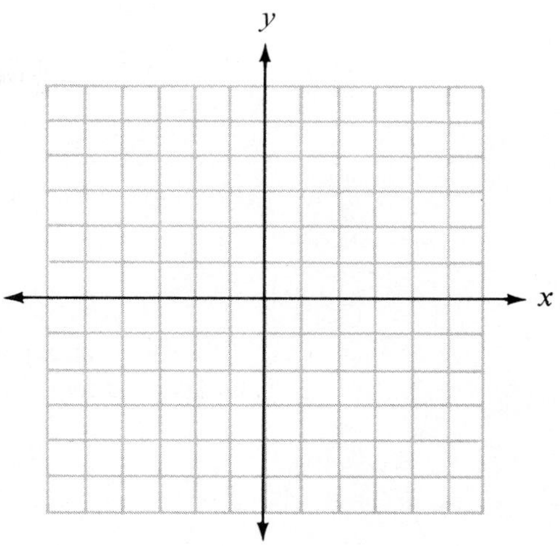

Answers are on page 754.

Exponential functions have many applications in the real world. Population growth, chemical decay, and compound interest problems can all be solved using exponential functions. Consider the problem stated at the beginning of this chapter.

Example 7. When \$100 is deposited in a savings account paying compound interest at the rate of 2 percent per quarter (every 3 months), the total amount, A, in the account after n quarters is given by the function $A = 100(1.02)^n$. Use a calculator to find the amount in the account after 7 quarters; after 5 years.

Solution. We can use a calculator to evaluate the function

$$A = 100(1.02)^n$$

After 7 quarters

$$\begin{aligned} n &= 7 \text{ quarters} \\ A &= 100(1.02)^7 \\ &\doteq 100(1.1486) \\ &\doteq 114.86 \end{aligned}$$

There will be approximately \$114.86 in the account after 7 quarters.

After 5 years

$$\begin{aligned} n &= 5(4) = 20 \text{ quarters} \\ A &= 100(1.02)^{20} \\ &\doteq 100(1.4859) \\ &\doteq 148.59 \end{aligned}$$

There will be approximately \$148.59 in the account after 5 years (or 20 quarters).

Solving Exponential Equations

Suppose that we wish to find x-values corresponding to specific y-values for an exponential function. For instance, suppose that we wish to find the value of x for which

$$2^x = 8$$

Looking back at the graph of $y = 2^x$, you should agree that there will be only one x-value corresponding to the y-value of 8. Since we can rewrite 8 as a power of 2, our equation becomes

$$2^x = 8$$
$$2^x = 2^3$$

Since each side of this equation has been expressed as a power of the same base and since each power of 2 corresponds to one and only one exponent, we conclude that

$$\text{if} \quad 2^x = 2^3$$
$$\text{then} \quad x = 3$$

To solve an exponential equation, we first *try to rewrite each side of the equation as a power of the same base*. If this is possible, we may equate the exponents and solve the equation.

Exponential Equations

If $a^m = a^n$

then $m = n$

Example 8. Solve $3^x = \frac{1}{3}$.

Solution

$$3^x = \frac{1}{3}$$
$$3^x = 3^{-1}$$
$$x = -1$$

You try Example 9.

Example 9. Solve $5^{-x} = 25$.

Solution

$$5^{-x} = 25$$
$$5^{-x} = 5^2$$

Check your work on page 754. ▶
Then try Example 10.

Example 10. Solve $2^{x+1} = \frac{1}{8}$.

Solution

$$2^{x+1} = \frac{1}{8}$$
$$2^{x+1} = \frac{1}{2^3}$$
$$2^{x+1} = 2^{-3}$$

Check your work on page 754. ▶

Example 11. Solve $9^x = 3^{x-5}$.

Solution

$$9^x = 3^{x-5}$$
$$(3^2)^x = 3^{x-5}$$
$$3^{2x} = 3^{x-5}$$
$$2x = x - 5$$
$$x = -5$$

Example 12. Solve $(\frac{1}{2})^{3x} = 16^{x+1}$.

Solution

$$\begin{aligned}
\left(\frac{1}{2}\right)^{3x} &= 16^{x+1} && \\
(2^{-1})^{3x} &= (2^4)^{x+1} && \text{Rewrite each base as a power of 2.} \\
2^{-3x} &= 2^{4(x+1)} && \text{Raise each power to its power.} \\
2^{-3x} &= 2^{4x+4} && \text{Remove parentheses.} \\
-3x &= 4x + 4 && \text{Equate exponents.} \\
-7x &= 4 && \text{Isolate } x\text{-term.} \\
x &= -\frac{4}{7} && \text{Isolate } x.
\end{aligned}$$

Example 13. Solve $2 \cdot 2^x = 8^{1-x}$.

Solution

$$\begin{aligned}
2 \cdot 2^x &= 8^{1-x} && \\
2^1 \cdot 2^x &= 8^{1-x} && \text{Rewrite 2 as } 2^1. \\
2^{1+x} &= 8^{1-x} && \text{Multiply powers of 2 on left by adding exponents.} \\
2^{1+x} &= (2^3)^{1-x} && \text{Rewrite 8 as power of 2.} \\
2^{1+x} &= 2^{3(1-x)} && \text{Raise the power to the power.} \\
2^{1+x} &= 2^{3-3x} && \text{Remove parentheses.} \\
1 + x &= 3 - 3x && \text{Equate exponents.} \\
4x &= 2 && \text{Isolate } x\text{-term.} \\
x &= \frac{1}{2} && \text{Isolate } x.
\end{aligned}$$

If it is not possible to rewrite both sides of an exponential equation as powers of the same base, the equation cannot be solved using our techniques. Equations such as

$$5^x = 7 \qquad \text{and} \qquad 2^{x+1} = 3^x$$

will be discussed in Section 11.5.

Trial Run

Solve each exponential equation.

________ 1. $10^x = \frac{1}{10}$

________ 2. $2^{-x} = \frac{1}{4}$

________ 3. $3^{x+1} = \frac{1}{27}$

________ 4. $4^x = 2^{x+3}$

________ 5. $(\frac{1}{5})^{2x} = 25^{x+2}$

________ 6. $3^2 \cdot 3^x = 9^{3-x}$

Answers are on page 755.

▶ Examples You Completed

Example 1. Complete the table of values for 10^x and 5^{x+1}.

x	10^x	5^{x+1}
-2	$10^{-2} = \frac{1}{10^2} = \frac{1}{100}$	$5^{-2+1} = 5^{-1} = \frac{1}{5}$
-1	$10^{-1} = \frac{1}{10}$	$5^{-1+1} = 5^0 = 1$
0	$10^0 = 1$	$5^{0+1} = 5^1 = 5$
1	$10^1 = 10$	$5^{1+1} = 5^2 = 25$
2	$10^2 = 100$	$5^{2+1} = 5^3 = 125$

Example 9. Solve $5^{-x} = 25$.

Solution

$$5^{-x} = 25$$
$$5^{-x} = 5^2$$
$$-x = 2$$
$$x = -2$$

Example 10. Solve $2^{x+1} = \frac{1}{8}$.

Solution

$$2^{x+1} = \frac{1}{8}$$
$$2^{x+1} = \frac{1}{2^3}$$
$$2^{x+1} = 2^{-3}$$
$$x + 1 = -3$$
$$x = -4$$

Answers to Trial Runs

page 747 **1.** $\frac{1}{25}$; 1; 25 **2.** $\frac{1}{4}$; $\frac{1}{2}$; 2 **3.** $-\frac{1}{9}$; -1; $-\sqrt{3}$; -9 **4.** $\frac{1}{9}$; 1; $\sqrt{3}i$ (not real); 9

page 750 **1.**

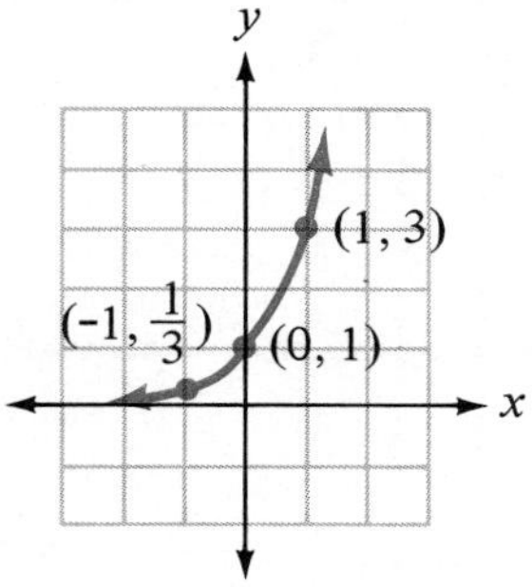

2.

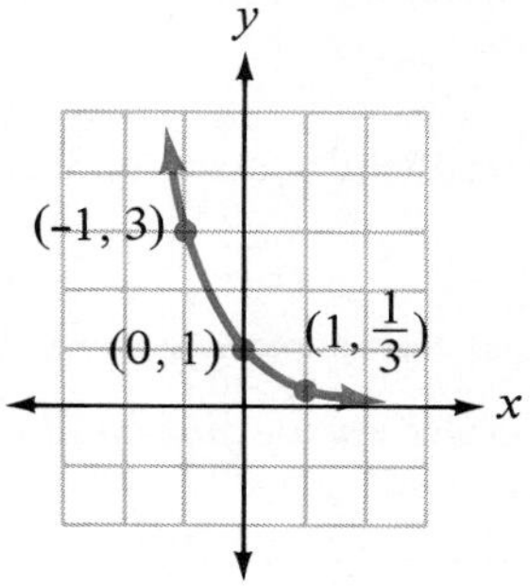

3.

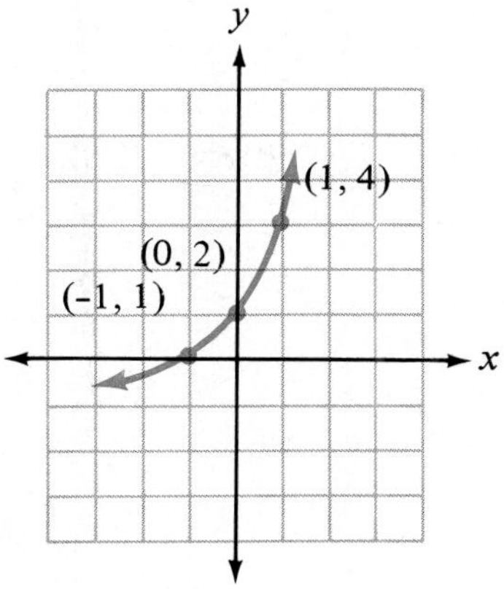

4.

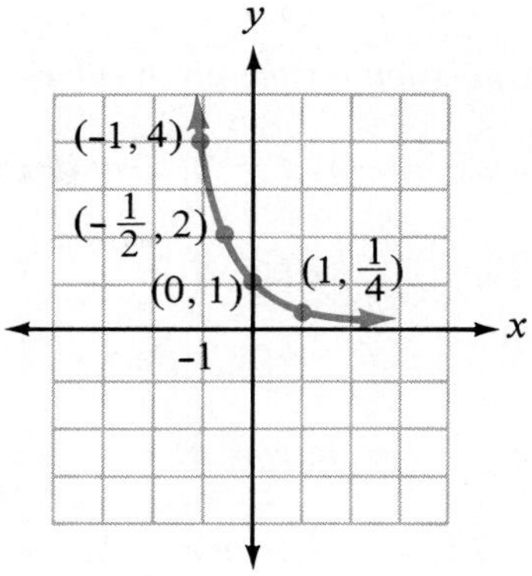

page 753 1. $x = -1$ 2. $x = 2$ 3. $x = -4$ 4. $x = 3$ 5. $x = -1$ 6. $x = \frac{4}{3}$

Name ______________________________ Date ______________

EXERCISE SET 11.1

Evaluate each expression for the given values of x.

________ 1. 4^x; $x = -\frac{1}{2}, x = 0, x = \frac{1}{2}, x = 2$

________ 2. 9^x; $x = -\frac{1}{2}, x = 0, x = \frac{1}{2}, x = 2$

________ 3. 5^{x-1}; $x = -1, x = \frac{1}{2}, x = \frac{3}{2}, x = 3$

________ 4. 6^{x-2}; $x = 0, x = \frac{3}{2}, x = 2, x = 4$

________ 5. -2^{3x}; $x = -1, x = -\frac{1}{3}, x = 0, x = \frac{2}{3}$

________ 6. -3^{5x}; $x = -\frac{3}{5}, x = 0, x = \frac{2}{5}, x = 1$

________ 7. $(-2)^{4x}$; $x = -\frac{1}{2}, x = 0, x = \frac{1}{8}, x = 1$

________ 8. $(-3)^{2x}$; $x = -\frac{1}{4}, x = \frac{1}{2}, x = 0, x = 2$

________ 9. $\left(\frac{1}{2}\right)^{2-x}$; $x = 1, x = 0, x = \frac{3}{2}, x = 4$

________ 10. $\left(\frac{1}{3}\right)^{1-x}$; $x = 1, x = 0, x = \frac{1}{2}, x = 3$

Graph each exponential function.

11. $y = 6^x$

12. $y = 4^x$

13. $y = \left(\frac{1}{4}\right)^x$

14. $y = \left(\frac{1}{6}\right)^x$

15. $y = 3^{x+1}$

16. $y = 2^{x-3}$

17. $y = \left(\frac{1}{3}\right)^{2x}$

18. $y = \left(\frac{1}{2}\right)^{3x}$

19. $y = 9^{\frac{x}{2}}$

20. $y = 10^{\frac{x}{2}}$

21. $y = 5^{3-x}$

22. $y = 6^{2-x}$

23. $y = \left(\frac{1}{2}\right)^{2x-1}$

24. $y = \left(\frac{1}{3}\right)^{2x+1}$

When \$1000 is deposited in a savings account paying compound interest at the rate of 2 percent per quarter, the total amount, A, in the account after n quarters is given by the function $A = 1000(1.02)^n$. Use a calculator to find the amounts in the account at the times given in problems 25–28.

________ 25. After 5 quarters

________ 26. After 9 quarters

________ 27. After 3 years

________ 28. After 4 years

Solve each exponential equation.

_______ 29. $5^x = \frac{1}{5}$

_______ 30. $3^x = \frac{1}{3}$

_______ 31. $4^{-x} = \frac{1}{16}$

_______ 32. $5^{-x} = \frac{1}{25}$

_______ 33. $10^{x+1} = \frac{1}{1000}$

_______ 34. $8^{x-1} = \frac{1}{64}$

_______ 35. $\left(\frac{1}{9}\right)^{2x} = 81^{x+2}$

_______ 36. $\left(\frac{1}{4}\right)^{3x} = 64^{x+1}$

_______ 37. $16^x = 2^{2x-1}$

_______ 38. $25^x = 5^{3x+1}$

_______ 39. $\left(\frac{1}{25}\right)^{\frac{1}{x}} = 5$

_______ 40. $\left(\frac{1}{81}\right)^{\frac{2}{x}} = 9$

_______ 41. $2^3 \cdot 2^x = 8^{-x}$

_______ 42. $5^2 \cdot 5^x = 25^{x-1}$

☆ Stretching the Topics

_______ 1. Graph $y = 2 + \left(\frac{1}{2}\right)^{x-1}$.

_______ 2. Solve $(2^x)^{x+1} = 16^{x+1}$ for x.

_______ 3. A bond paying 8 percent yearly interest is bought for \$4000. How much will the bond be worth 10 years from now if the interest is compounded quarterly?

Check your answers in the back of your book.

If you can complete **Checkup 11.1,** you are ready to go on to Section 11.2.

Name ______________________ Date ____________

✓ CHECKUP 11.1

_______ 1. Evaluate 3^{x-1} when $x = -1$, $x = 0$, $x = \frac{1}{2}$, $x = 2$, $x = \frac{3}{2}$.

Graph each exponential function.

2. $y = 4^{x-1}$

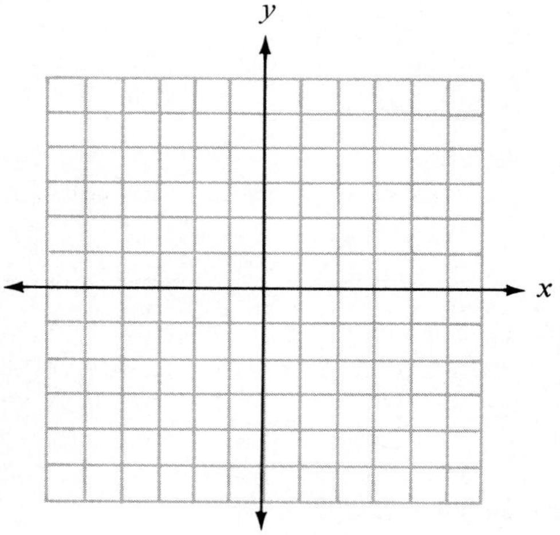

3. $y = 2^{-2x}$

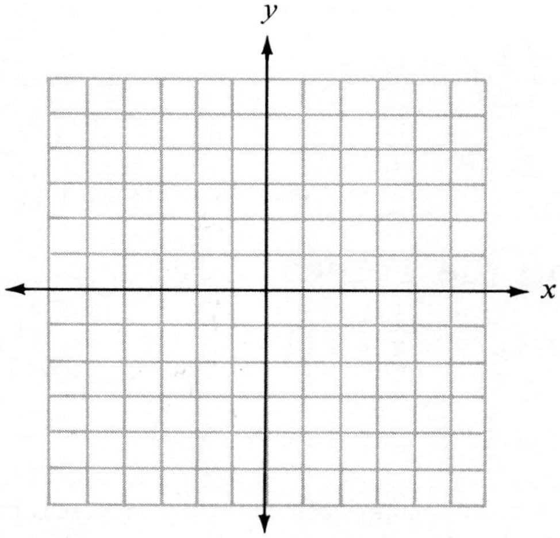

Solve each exponential equation.

_______ 4. $3^{2x-1} = \dfrac{1}{243}$

_______ 5. $\left(\dfrac{1}{32}\right)^x = 4^{x-1}$

Check your answers in the back of your book.

If You Missed Problem:	You Should Review Examples:
1	1, 2
2	6
3	4
4	10
5	12

11.2 Working with Logarithmic Functions

In an exponential statement such as $a^x = b$, we would like to find a way to express the exponent x in terms of a and b.

Writing Logarithmic Expressions

When solving equations such as $2^x = 8$, we were looking for the exponent to which we could raise the base 2 to arrive at the number 8. In other words, we could say

$$x = \text{the exponent to which we must raise the base 2 to get 8}$$

As is typical, mathematicians have invented a way of symbolizing such statements, and their invention is called a **logarithm.** We may rewrite our statement as

$$x = \text{the logarithm, base 2, of 8}$$

or simply

$$x = \log_2 8$$

The subscript 2 identifies the base, the x represents the exponent on the 2, and the number 8 tells the result of raising 2 to the x power. In other words,

$$x = \log_2 8 \qquad \text{means} \qquad 2^x = 8$$

These two statements say *exactly the same thing*. The logarithmic form of the statement merely allows us to isolate the exponent in the statement. Every logarithmic statement corresponds to an exponential statement, and vice versa. The sooner you realize this fact, the better off you will be in your understanding of logarithms.

Logarithmic Notation

$$x = \log_a b \qquad \text{means} \qquad a^x = b$$

Because these statements are equivalent, you should see that *a logarithm is just an exponent*.

Because of the close relationship between logarithms and the exponential functions discussed in Section 11.1, we shall continue to restrict ourselves to *positive bases* different from 1. Let's practice switching back and forth between exponential statements and logarithmic statements.

Example 1. Write each exponential statement as a logarithmic statement.

$$3^2 = 9$$

$$2^{-2} = \tfrac{1}{4}$$

$$5^{\frac{1}{2}} = \sqrt{5}$$

Solution

$3^2 = 9$ means $\log_3 9 = 2$

$2^{-2} = \frac{1}{4}$ means $\log_2 \frac{1}{4} = -2$

$5^{\frac{1}{2}} = \sqrt{5}$ means $\log_5 \sqrt{5} = \frac{1}{2}$

You complete Example 2.

Example 2. Write each logarithmic statement as an exponential statement.

$$\log_5 125 = 3$$

$$\log_7 1 = 0$$

$$\log_8 2 = \tfrac{1}{3}$$

Solution

$\log_5 125 = 3$ means _____ = _____

$\log_7 1 = 0$ means _____ = _____

$\log_8 2 = \frac{1}{3}$ means _____ = _____

Check your work on page 766. ▶

Trial Run

Write each exponential statement as a logarithmic statement.

______ 1. $2^2 = 4$ ______ 2. $5^{-2} = \frac{1}{25}$ ______ 3. $9^{\frac{1}{2}} = 3$

Write each logarithmic statement as an exponential statement.

______ 4. $\log_3 81 = 4$ ______ 5. $\log_2 \frac{1}{8} = -3$ ______ 6. $\log_4 2 = \frac{1}{2}$

Answers are on page 766.

Simplifying Logarithmic Expressions

Most of us feel more comfortable working with exponential statements than with logarithmic statements because we have been dealing with exponents throughout algebra. To simplify logarithmic expressions, it is probably wise to assign a variable name to the logarithm and then switch to exponential form. For example, let's find $\log_4 16$.

$\log_4 16 = x$ Call the logarithm "x."

$4^x = 16$ Switch to exponential form.

$4^x = 4^2$ Rewrite both sides as powers of 4.

$x = 2$ Solve the exponential equation.

Therefore, $\log_4 16 = 2$.

You try Example 3.

Example 3. Find $\log_2 32$.

Solution

$$\log_2 32 = x$$

$$2^x = 32$$

$$2^x = 2^{\_\_}$$

$$x = \_\_\_\_$$

Therefore, $\log_2 32 =$ ______ .

Check your work on page 766. ▶

Then complete Example 4.

Example 4. Find $\log_3 \frac{1}{3}$.

Solution

$$\log_3 \frac{1}{3} = x$$

$$3^x = \frac{1}{3}$$

$$3^x = 3^{\_\_}$$

$$x = \_\_\_\_$$

Therefore, $\log_3 \frac{1}{3} =$ ______ .

Check your work on page 766. ▶

Example 5. Find $\log_9 1$.

Solution

$$\log_9 1 = x$$

$$9^x = 1$$

$$9^x = 9^0$$

$$x = 0$$

Therefore, $\log_9 1 = 0$.

Example 6. Find $\log_6 6^5$.

Solution

$$\log_6 6^5 = x$$

$$6^x = 6^5$$

$$x = 5$$

Therefore, $\log_6 6^5 = 5$.

Example 7. Find $\log_3 3 + \log_2 \frac{1}{8}$.

Solution. We assign a variable name to each logarithm.

$$\begin{aligned} \log_3 3 &= x & \text{and} \quad \log_2 \tfrac{1}{8} &= y \\ 3^x &= 3 & 2^y &= \tfrac{1}{8} \\ 3^x &= 3^1 & 2^y &= 2^{-3} \\ x &= 1 & y &= -3 \\ \log_3 3 &= 1 & \text{and} \quad \log_2 \tfrac{1}{8} &= -3 \end{aligned}$$

Therefore,

$$\begin{aligned} \log_3 3 + \log_2 \tfrac{1}{8} &= 1 + (-3) \\ &= -2 \end{aligned}$$

Example 8. Find $\log_5 0$.

Solution

$$\log_5 0 = x$$

$$5^x = 0$$

There is *no* exponent to which we can raise 5 to arrive at the answer 0. Therefore, $\log_5 0$ does *not* exist.

Example 9. Find $\log_2 (-2)$.

Solution

$$\log_2 (-2) = x$$

$$2^x = -2$$

There is *no* exponent to which we can raise 2 to arrive at the answer -2. Therefore, $\log_2 (-2)$ does *not* exist.

From these examples we can make some useful observations.

For $a > 0$, $a \neq 1$

$\log_a 1 = 0$ (because $a^0 = 1$)

$\log_a a = 1$ (because $a^1 = a$)

$\log_a a^n = n$ (because $a^n = a^n$)

$\log_a b$ exists only when $b > 0$.

Our last observation deserves some emphasis.

The logarithm of a negative number *or* zero does not exist.

Trial Run

Find the value of each expression.

_______ 1. $\log_5 \frac{1}{5}$

_______ 2. $\log_4 1$

_______ 3. $\log_3 27$

_______ 4. $\log_7 7^3$

_______ 5. $\log_2 32 + \log_3 \frac{1}{81}$

_______ 6. $\log_3 (-9)$

Answers are on page 766.

Solving Logarithmic Equations

Now that we have learned to treat logarithmic statements as exponential statements, we should be able to find any unknown quantity in a logarithmic statement.

You try Example 10.

Example 10. Solve $\log_4 \frac{1}{4} = x$.

Solution

$$\log_4 \frac{1}{4} = x$$

$$4^x = \frac{1}{4}$$

$$4^x = 4^{\_\_}$$

$$x = \_\_\_\_$$

Check your work on page 766. ▶

Example 11. Solve $\log_a 25 = 2$.

Solution

$$\log_a 25 = 2$$

$$a^2 = 25$$

$$a = \pm 5$$

But remember that the base, a, must be positive. So

$$a = 5$$

Example 12. Solve $\log_6 b = -2$.

Solution

$$\log_6 b = -2$$

$$6^{-2} = b$$

$$\frac{1}{36} = b$$

Now try Example 13.

Example 13. Solve $\log_2 (x + 1) = 3$.

Solution

$$\log_2 (x + 1) = 3$$

$$2^3 = x + 1$$

$$\_\_\_\_ = x + 1$$

$$\_\_\_\_ = x$$

Check your work on page 766. ▶

Example 14. Solve $\log_4 64 = 2 - 3x$.

Solution

$\log_4 64 = 2 - 3x$	
$4^{2-3x} = 64$	Switch to exponential form.
$4^{2-3x} = 4^3$	Write 64 as a power of base 4.
$2 - 3x = 3$	Equate exponents.
$-3x = 1$	Isolate x-term.
$x = -\frac{1}{3}$	Isolate x.

Example 15. Solve $\log_x (x + 2) = 2$.

Solution

$\log_x (x + 2) = 2$	
$x^2 = x + 2$	Switch to exponential form.
$x^2 - x - 2 = 0$	Set quadratic expression equal to zero.
$(x - 2)(x + 1) = 0$	Factor the trinomial.
$x - 2 = 0 \quad \text{or} \quad x + 1 = 0$	Set each factor equal to zero.
$x = 2 \qquad\qquad x = -1$	Solve for x.

Since x appears as the base in the original logarithmic statement, x cannot be negative. The solution is $x = 2$.

Trial Run

Solve each logarithmic equation.

_______ 1. $\log_{10} \frac{1}{10} = x$

_______ 2. $\log_a 36 = 2$

_______ 3. $\log_3 b = -4$

_______ 4. $\log_5 (x - 1) = 2$

_______ 5. $\log_2 32 = 3 - 2x$

_______ 6. $\log_x (2x + 3) = 2$

Answers are on page 766.

Graphing the Logarithmic Function

Suppose that we look at ordered pairs (x, y) satisfying the equation

$$y = \log_2 x$$

We could come up with some ordered pairs by choosing x and finding y, but we must be careful to choose x-values whose logarithms are easily found. It is easier to rewrite this statement in exponential form before finding ordered pairs.

$$y = \log_2 x \qquad \text{means} \qquad 2^y = x$$

Now we can find ordered pairs more easily by substituting values for y and finding corresponding x-values. Let's try that; but remember that our ordered pairs should be written (x, y).

Example 16. Graph $y = \log_2 x$.

Solution

$$y = \log_2 x \quad \text{means} \quad 2^y = x$$

y	$x = 2^y$	(x, y)
-2	$2^{-2} = \frac{1}{4}$	$(\frac{1}{4}, -2)$
-1	$2^{-1} = \frac{1}{2}$	$(\frac{1}{2}, -1)$
0	$2^0 = 1$	$(1, 0)$
$\frac{1}{2}$	$2^{\frac{1}{2}} = \sqrt{2}$	$(\sqrt{2}, \frac{1}{2})$
1	$2^1 = 2$	$(2, 1)$
2	$2^2 = 4$	$(4, 2)$

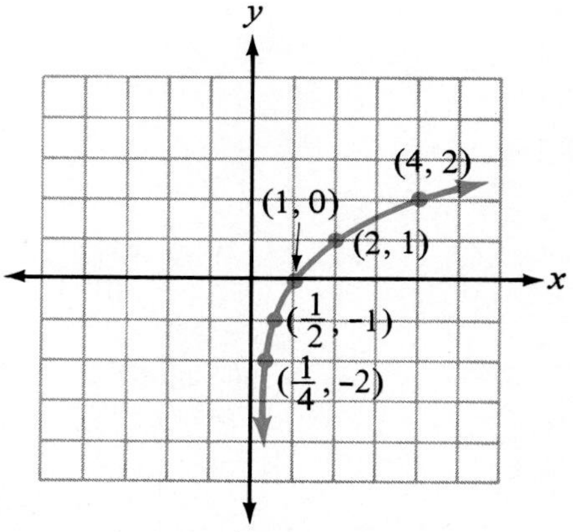

$y = \log_2 x$

Notice that our y-values could be positive, negative, or zero, but the x-values always turned out to be *positive*.

Example 17. Graph $y = \log_{\frac{1}{3}} x$.

Solution

$$y = \log_{\frac{1}{3}} x \quad \text{means} \quad \left(\frac{1}{3}\right)^y = x$$

y	$x = \left(\frac{1}{3}\right)^y$	(x, y)
-1	$\left(\frac{1}{3}\right)^{-1} = 3$	$(3, -1)$
0	$\left(\frac{1}{3}\right)^0 = 1$	$(1, 0)$
$\frac{1}{2}$	$\left(\frac{1}{3}\right)^{\frac{1}{2}} = \sqrt{\frac{1}{3}} = \frac{\sqrt{3}}{3}$	$\left(\frac{\sqrt{3}}{3}, \frac{1}{2}\right)$
1	$\left(\frac{1}{3}\right)^1 = \frac{1}{3}$	$\left(\frac{1}{3}, 1\right)$

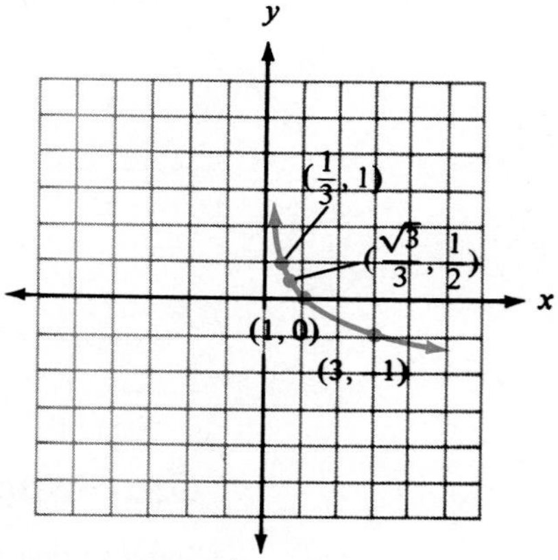

$y = \log_{1/3} x$

From the graphs of $y = \log_2 x$ and $y = \log_{\frac{1}{3}} x$, you should agree that the sets of ordered pairs satisfying such equations are indeed *functions*. Each x in the domain corresponds to one and only one y in the range. We notice that x is always *positive,* even though y can be any real number.

Let's summarize our findings for **logarithmic functions.**

Definition of Logarithmic Function

$$y = \log_a x$$

where $a > 0$, $a \neq 1$, is called a logarithmic function.

Domain: $x \in R$ and $x > 0$

Range: $y \in R$

To graph a logarithmic function more easily, we have agreed to follow these steps.

Graphing a Logarithmic Function

1. Rewrite the logarithmic statement as an exponential statement with x isolated.
2. Substitute y-values and find the corresponding x-values.
3. Write the ordered pairs (x, y).
4. Use the ordered pairs to sketch the graph.

Trial Run

Graph each logarithmic function.

1. $y = \log_{10} x$

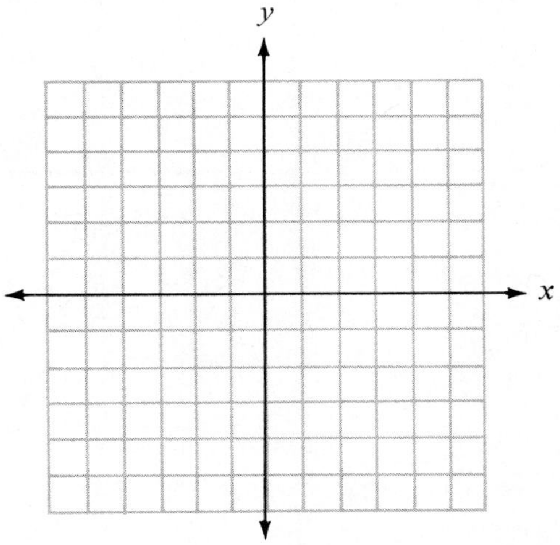

2. $y = \log_{\frac{1}{2}} x$

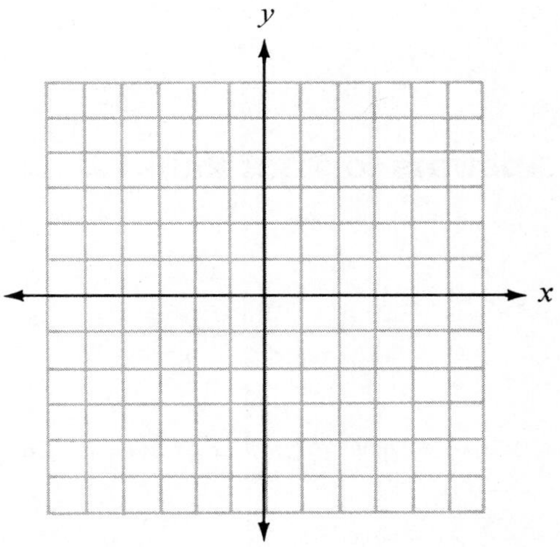

Answers are on page 766.

▶ Examples You Completed

Example 2 *(Solution)*

$\log_5 125 = 3$ means $5^3 = 125$

$\log_7 1 = 0$ means $7^0 = 1$

$\log_8 2 = \frac{1}{3}$ means $8^{\frac{1}{3}} = 2$

Example 3. Find $\log_2 32$.

Solution

$$\log_2 32 = x$$
$$2^x = 32$$
$$2^x = 2^5$$
$$x = 5$$

Therefore, $\log_2 32 = 5$.

Example 4. Find $\log_3 \frac{1}{3}$.

Solution

$$\log_3 \frac{1}{3} = x$$
$$3^x = \frac{1}{3}$$
$$3^x = 3^{-1}$$
$$x = -1$$

Therefore, $\log_3 \frac{1}{3} = -1$.

Example 10. Solve $\log_4 \frac{1}{4} = x$.

Solution

$$\log_4 \frac{1}{4} = x$$
$$4^x = \frac{1}{4}$$
$$4^x = 4^{-1}$$
$$x = -1$$

Example 13. Solve $\log_2 (x + 1) = 3$.

Solution

$$\log_2 (x + 1) = 3$$
$$2^3 = x + 1$$
$$8 = x + 1$$
$$7 = x$$

Answers to Trial Runs

page 760 1. $\log_2 4 = 2$ 2. $\log_5 \frac{1}{25} = -2$ 3. $\log_9 3 = \frac{1}{2}$ 4. $3^4 = 81$ 5. $2^{-3} = \frac{1}{8}$ 6. $4^{\frac{1}{2}} = 2$

page 762 1. -1 2. 0 3. 3 4. 3 5. 1 6. Does not exist

page 763 1. $x = -1$ 2. $a = 6$ 3. $b = \frac{1}{81}$ 4. $x = 26$ 5. $x = -1$ 6. $x = 3$

page 765 1.

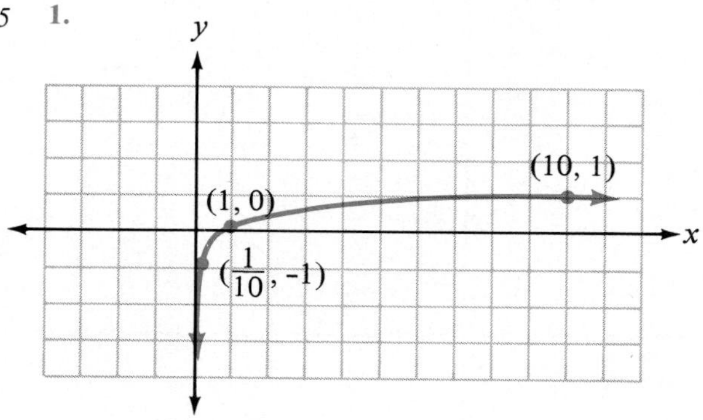

2.

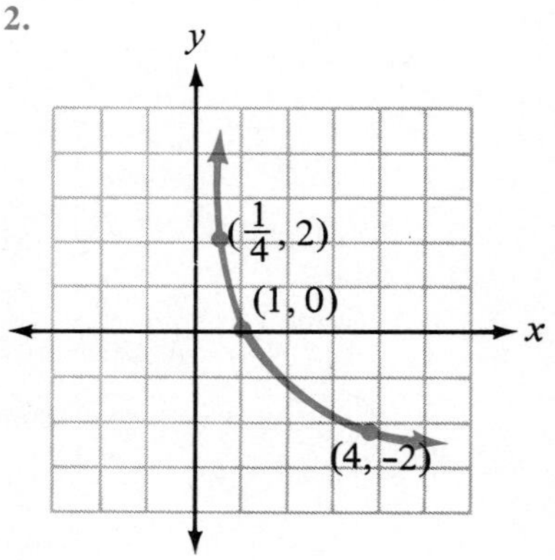

Name ______________________ Date __________

EXERCISE SET 11.2

Write each exponential statement as a logarithmic statement.

______ 1. $3^2 = 9$

______ 2. $2^4 = 16$

______ 3. $4^{-2} = \frac{1}{16}$

______ 4. $6^{-2} = \frac{1}{36}$

______ 5. $16^{\frac{1}{2}} = 4$

______ 6. $27^{\frac{1}{3}} = 3$

______ 7. $9^{-\frac{1}{2}} = \frac{1}{3}$

______ 8. $25^{-\frac{1}{2}} = \frac{1}{5}$

Write each logarithmic statement as an exponential statement.

______ 9. $\log_5 125 = 3$

______ 10. $\log_4 64 = 3$

______ 11. $\log_2 \frac{1}{32} = -5$

______ 12. $\log_3 \frac{1}{27} = -3$

______ 13. $\log_{16} 4 = \frac{1}{2}$

______ 14. $\log_{27} 3 = \frac{1}{3}$

______ 15. $\log_{36} \frac{1}{6} = -\frac{1}{2}$

______ 16. $\log_{64} \frac{1}{8} = -\frac{1}{2}$

Find the value of each expression.

______ 17. $\log_8 \frac{1}{8}$

______ 18. $\log_7 \frac{1}{7}$

______ 19. $\log_{10} 1$

______ 20. $\log_{12} 1$

______ 21. $\log_5 125$

______ 22. $\log_6 36$

______ 23. $\log_2 2^7$

______ 24. $\log_3 3^5$

______ 25. $\log_5 0$

______ 26. $\log_7 (-49)$

______ 27. $\log_5 25 - \log_2 \frac{1}{8}$

______ 28. $\log_3 \frac{1}{27} + \log_2 2$

______ 29. $\log_{25} \frac{1}{5} + \log_9 3$

______ 30. $\log_{49} \frac{1}{7} + \log_4 2$

Solve each logarithmic equation.

______ 31. $\log_{12} \frac{1}{12} = x$

______ 32. $\log_2 \frac{1}{2} = x$

______ 33. $\log_a 49 = 2$

______ 34. $\log_a 121 = 2$

______ 35. $\log_2 b = -5$

______ 36. $\log_3 b = -2$

______ 37. $\log_4 (x + 1) = 3$

______ 38. $\log_3 (2x + 1) = 3$

______ 39. $\log_3 81 = 5 + 2x$

______ 40. $\log_2 64 = 4 - 3x$

______ 41. $\log_x (x + 6) = 2$

______ 42. $\log_x (2x + 8) = 2$

Graph each logarithmic function.

43. $y = \log_3 x$

44. $y = \log_5 x$

45. $y = \log_{\frac{1}{3}} x$

46. $y = \log_{\frac{1}{10}} x$

☆ Stretching the Topics

_______ 1. Solve $\log_2 (x^2 - 5x + 4) = 3$ for x.

_______ 2. Graph $y = \log_3 (2 + x)$.

_______ 3. Solve $\dfrac{x + 5}{\log_3 9} = \dfrac{\log_2 8}{x}$ for x.

Check your answers in the back of your book.

If you can complete **Checkup 11.2,** you are ready to go on to Section 11.3.

Name ______________________ Date ____________

✓ CHECKUP 11.2

________ 1. Write $5^2 = 25$ as a logarithmic statement.

________ 2. Write $64^{-\frac{1}{2}} = \frac{1}{8}$ as a logarithmic statement.

________ 3. Write $\log_3 81 = 4$ as an exponential statement.

________ 4. Write $\log_5 \frac{1}{25} = -2$ as an exponential statement.

________ 5. Find $\log_2 \frac{1}{32}$.

________ 6. Find $\log_8 2$.

________ 7. Solve $\log_a 125 = 3$.

________ 8. Solve $\log_4 (x - 3) = \frac{1}{2}$.

9. Graph $y = \log_6 x$.

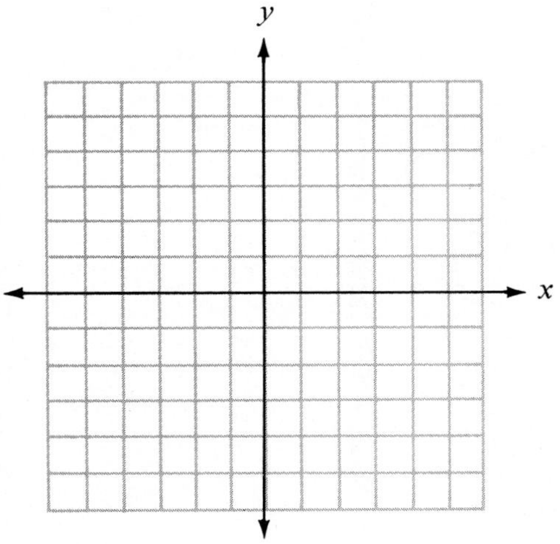

10. Graph $y = \log_{\frac{1}{5}} x$.

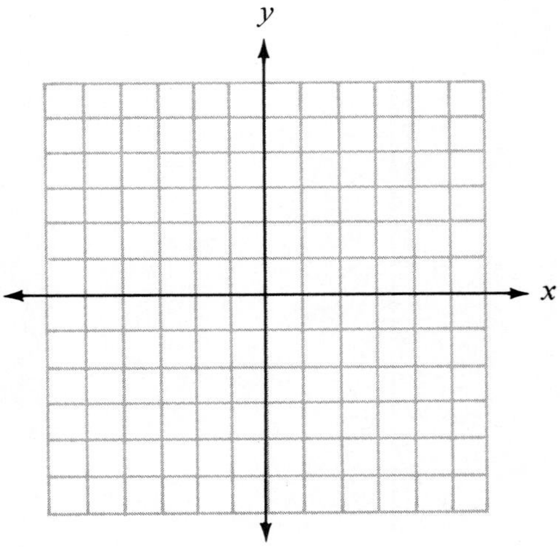

Check your answers in the back of your book.

If You Missed Problems:	You Should Review Examples:
1, 2	1
3, 4	2
5, 6	3–7
7	11, 15
8	13
9, 10	16, 17

11.3 Using the Properties of Logarithms

Since *logarithms are exponents,* you should suspect that the rules for working with logarithms will be similar to the laws of exponents. Let's recall a few laws of exponents.

First Law of Exponents:	$a^m \cdot a^n$	$= a^{m+n}$
Second Law of Exponents:	$(a^m)^n$	$= a^{m \cdot n}$
Fifth Law of Exponents:	$\dfrac{a^m}{a^n}$	$= a^{m-n}$

Using the Laws of Logarithms

First let's consider the logarithm of a *product.* The logarithm of a product of factors is the *sum* of the logarithms of each of those factors.

First Law of Logarithms

$$\log_a (M \cdot N) = \log_a M + \log_a N$$

Let's verify this law with a numerical example.

$$\log_2 (16 \cdot 8) \stackrel{?}{=} \log_2 16 + \log_2 8$$

$$\log_2 128 \stackrel{?}{=} \log_2 16 + \log_2 8$$

$$7 \stackrel{?}{=} 4 + 3 \qquad \text{Remember: } 2^7 = 128,\ 2^4 = 16, \text{ and } 2^3 = 8.$$

$$7 = 7$$

We have seen that this law "works" for our numerical example, but let's prove it for *any* base a. Let

$$\log_a M = m \quad \text{and} \quad \log_a N = n$$

$$a^m = M \quad \text{and} \quad a^n = N \qquad \text{Switch to exponential form.}$$

$$M \cdot N = a^m \cdot a^n \qquad \text{Consider the product } M \cdot N.$$

$$M \cdot N = a^{m+n} \qquad \text{Use First Law of Exponents.}$$

$$\log_a (M \cdot N) = m + n \qquad \text{Switch to logarithmic form.}$$

$$\log_a (M \cdot N) = \log_a M + \log_a N \qquad \text{Substitute: } m = \log_a M,\ n = \log_a N$$

We have proved the First Law of Logarithms.

Example 1. Find $\log_3 (27 \cdot 81)$.

Solution

$$\log_3 (27 \cdot 81) = \log_3 27 + \log_3 81 \qquad \text{Use the First Law of Logarithms.}$$

$$= 3 + 4 \qquad \text{Remember: } 3^3 = 27 \text{ and } 3^4 = 81.$$

$$= 7$$

Let's consider a rule for finding the logarithm of a *quotient*. The logarithm of a quotient is the *difference* between the logarithm of the numerator and the logarithm of the denominator.

Second Law of Logarithms

$$\log_a \frac{M}{N} = \log_a M - \log_a N$$

Once again, we shall verify this law with a numerical example before proving it in general.

$$\log_3 \frac{243}{27} \stackrel{?}{=} \log_3 243 - \log_3 27$$

$$\log_3 9 \stackrel{?}{=} \log_3 243 - \log_3 27$$

$$2 \stackrel{?}{=} 5 - 3 \qquad \text{Remember: } 3^2 = 9,\ 3^5 = 243, \text{ and } 3^3 = 27.$$

$$2 = 2$$

Now we can prove the Second Law of Logarithms for *any* base a. Let

$$\log_a M = m \quad \text{and} \quad \log_a N = n$$

$$a^m = M \quad \text{and} \quad a^n = N \qquad \text{Switch to exponential form.}$$

$$\frac{M}{N} = \frac{a^m}{a^n} \qquad \text{Consider the quotient } \frac{M}{N}.$$

$$\frac{M}{N} = a^{m-n} \qquad \text{Use Fifth Law of Exponents.}$$

$$\log_a \frac{M}{N} = m - n \qquad \text{Switch to logarithmic form.}$$

$$\log_a \frac{M}{N} = \log_a M - \log_a N \qquad \text{Substitute: } m = \log_a M,\ n = \log_a N$$

We have proved the Second Law of Logarithms.

Example 2. Find $\log_2 \frac{32}{256}$.

Solution

$$\log_2 \frac{32}{256} = \log_2 32 - \log_2 256 \qquad \text{Use the Second Law of Logarithms.}$$

$$= 5 - 8 \qquad \text{Remember: } 2^5 = 32 \text{ and } 2^8 = 256.$$

$$= -3$$

Finally, we seek a way to simplify the logarithm of a *power*. The logarithm of a power is the *product* of the exponent and the logarithm of the base.

Third Law of Logarithms

$$\log_a M^c = c \log_a M$$

We again verify this law with a numerical example before proving it in general.

$$\log_2 4^3 \stackrel{?}{=} 3 \log_2 4$$

$$\log_2 64 \stackrel{?}{=} 3 \log_2 4$$

$$6 \stackrel{?}{=} 3 \cdot 2 \qquad \text{Remember: } 2^6 = 64 \text{ and } 2^2 = 4.$$

$$6 = 6$$

Now let's prove the Third Law of Logarithms for *any* base a. Let

$$\log_a M = m$$

$a^m = M$	Switch to exponential form.
$M^c = (a^m)^c$	Consider the power M^c.
$M^c = a^{cm}$	Use Second Law of Exponents.
$\log_a M^c = cm$	Switch to logarithmic form.
$\log_a M^c = c \log_a M$	Substitute: $m = \log_a M$

Our proof is complete.

You finish Example 3.

Example 3. Find $\log_4 16^7$.

Solution

$$\log_4 16^7 = 7 \log_4 16$$

$$= 7(\underline{\qquad})$$

$$= \underline{\qquad}$$

Check your work on page 777. ▶

Example 4. Find $\log_5 \sqrt{125}$.

Solution

$$\log_5 \sqrt{125} = \log_5 (125)^{\frac{1}{2}}$$

$$= \tfrac{1}{2} \log_5 125$$

$$= \tfrac{1}{2} (3)$$

$$= \tfrac{3}{2}$$

The three laws of logarithms are very useful in changing logarithms of complicated expressions into sums, differences, or constant multiples of simpler logarithmic expressions. Often it will be necessary to use more than one law in the simplifying process.

You complete Example 5.

Example 5. Find $\log_2 (4^7 \cdot 2^5)$.

Solution

$\log_2 (4^7 \cdot 2^5) = \log_2 4^7 + \log_2 2^5$	Use First Law of Logarithms.
$= 7 \log_2 4 + 5 \log_2 2$	Use Third Law of Logarithms.
$= 7(\underline{\qquad}) + 5(\underline{\qquad})$	Simplify the logarithms.
$= \underline{\qquad} + \underline{\qquad}$	Find each product.
$= \underline{\qquad}$	Find the sum.

Check your work on page 777. ▶

Example 6. Find $\log_3 \dfrac{3^5 \cdot 81^2}{27^{10}}$.

Solution

$$\log_3 \frac{3^5 \cdot 81^2}{27^{10}}$$

$= \log_3 3^5 + \log_3 81^2 - \log_3 27^{10}$	Use First and Second Laws of Logarithms.
$= 5 \log_3 3 + 2 \log_3 81 - 10 \log_3 27$	Use Third Law of Logarithms.
$= 5(1) + 2(4) - 10(3)$	Remember: $3^1 = 3$, $3^4 = 81$, and $3^3 = 27$.
$= 5 + 8 - 30$	Find each product.
$= -17$	Simplify.

In Examples 7 through 10 we will use the fact that $\log_{10} 2 \doteq 0.3010$ and $\log_{10} 3 \doteq 0.4771$ to find approximate values for each logarithm.

Example 7. Find $\log_{10} 6$.

Solution

$\log_{10} 6 = \log_{10} (3 \cdot 2)$	Rewrite 6 as $3 \cdot 2$.
$= \log_{10} 3 + \log_{10} 2$	Use First Law of Logarithms.
$\doteq 0.3010 + 0.4771$	Substitute given values.
$\log_{10} 6 \doteq 0.7781$	Find the sum.

You complete Example 8.

Example 8. Find $\log_{10} \frac{8}{9}$.

Solution

$\log_{10} \dfrac{8}{9} = \log_{10} 8 - \log_{10} 9$	Use Second Law of Logarithms.
$= \log_{10} 2^3 - \log_{10} 3^2$	Rewrite 8 as 2^3 and 9 as 3^2.
$= 3 \log_{10} 2 - 2 \log_{10} 3$	Use Third Law of Logarithms.
$\doteq 3(\_\_\_\_\_\_) - 2(\_\_\_\_\_\_)$	Substitute given values.
$\doteq \_\_\_\_\_\_ - \_\_\_\_\_\_$	Find each product.
$\log_{10} \dfrac{8}{9} \doteq \_\_\_\_\_\_$	Simplify.

Check your work on page 777. ▶

Example 9. Find $\log_{10} 600$.

Solution

$\log_{10} 600 = \log_{10} (2 \cdot 3 \cdot 100)$	Rewrite 600 as $2 \cdot 3 \cdot 100$.
$= \log_{10} 2 + \log_{10} 3 + \log_{10} 100$	Use First Law of Logarithms.
$\doteq 0.3010 + 0.4771 + 2$	Remember: $10^2 = 100$.
$\log_{10} 600 \doteq 2.7781$	

Example 10. Find $\log_{10} 1500$.

Solution

$\log_{10} 1500 = \log_{10} \dfrac{3000}{2}$	Rewrite 1500 as $\dfrac{3000}{2}$.
$= \log_{10} \dfrac{3 \cdot 1000}{2}$	Rewrite 3000 as $3 \cdot 1000$.
$= \log_{10} 3 + \log_{10} 1000 - \log_{10} 2$	Use laws of logarithms.
$\doteq 0.4771 + 3 - 0.3010$	Remember: $10^3 = 1000$.
$\doteq 3.4771 - 0.3010$	
$\log_{10} 1500 \doteq 3.1761$	

Trial Run

Find each logarithm.

_______ 1. $\log_3 (9 \cdot 27)$

_______ 2. $\log_2 \frac{8}{128}$

_______ 3. $\log_5 25^3$

_______ 4. $\log_6 \sqrt{36}$

_______ 5. $\log_3 (9^4 \cdot 3^2)$

If $\log_{10} 2 \doteq 0.3010$ *and* $\log_{10} 3 \doteq 0.4771$, *approximate each logarithm.*

_______ 6. $\log_{10} 18$

_______ 7. $\log_{10} 400$

_______ 8. $\log_{10} \frac{16}{27}$

Answers are on page 777.

Rewriting Logarithmic Expressions Containing Variables

In the examples above we used the laws of logarithms to simplify logarithmic expressions to a single number. Sometimes it is not possible or even necessary to reduce a logarithmic expression to a numerical value. In such cases we may merely wish to rewrite the original expression in another form, using the laws of logarithms.

Example 11. Use the laws of logarithms to rewrite $\log_a \dfrac{x^2(x + 2)}{x - 3}$.

Solution

$$\log_a \frac{x^2(x + 2)}{x - 3}$$

$= \log_a x^2(x + 2) - \log_a(x - 3)$	Use Second Law of Logarithms.
$= \log_a x^2 + \log_a (x + 2) - \log_a (x - 3)$	Use First Law of Logarithms.
$= 2 \log_a x + \log_a (x + 2) - \log_a (x - 3)$	Use Third Law of Logarithms.

You complete Example 12.

Example 12. Use the laws of logarithms to rewrite $\log_a \frac{(x+5)^3}{\sqrt{x-4}}$.

Solution

$$\log_a \frac{(x+5)^3}{\sqrt{x-4}}$$

$$= \log_a \frac{(x+5)^3}{(x-4)^{\frac{1}{2}}}$$ Rewrite $\sqrt{x-4}$ as $(x-4)^{\frac{1}{2}}$.

$$= \log_a (x+5)^3 - \log_a (x-4)^{\frac{1}{2}}$$ Use Second Law of Logarithms.

$$= \_\_\_\_ \log_a (\_\_\_\_\_\_) - \_\_\_\_ \log_a (\_\_\_\_\_\_)$$ Use Third Law of Logarithms.

Check your work on page 777. ▶

On the other hand, sometimes we may wish to rewrite a sum or difference of several logarithmic expressions as a *single* logarithmic expression. To accomplish this, we must reverse the procedure used above.

Example 13. Rewrite as a single logarithm $\log_a (2x+3) - 5 \log_a x$.

Solution

$$\log_a (2x+3) - 5 \log_a x$$

$$= \log_a (2x+3) - \log_a x^5$$ Use Third Law of Logarithms.

$$= \log_a \frac{2x+3}{x^5}$$ Use Second Law of Logarithms.

Example 14. Rewrite as a single logarithm.

$$3 \log_a x + \tfrac{1}{2} \log_a (x+7) - \tfrac{1}{3} \log_a (x^2+5)$$

Solution

$$3 \log_a x + \tfrac{1}{2} \log_a (x+7) - \tfrac{1}{3} \log_a (x^2+5)$$

$$= \log_a x^3 + \log_a (x+7)^{\frac{1}{2}} - \log_a (x^2+5)^{\frac{1}{3}}$$

$$= \log_a x^3 + \log_a \sqrt{x+7} - \log_a \sqrt[3]{x^2+5}$$

$$= \log_a x^3 \sqrt{x+7} - \log_a \sqrt[3]{x^2+5}$$

$$= \log_a \frac{x^3 \sqrt{x+7}}{\sqrt[3]{x^2+5}}$$

Trial Run

Rewrite, using the laws of logarithms.

______ 1. $\log_a (x+1)(2x-5)^2$

______ 2. $\log_a \frac{x^2+1}{\sqrt[3]{x+9}}$

Rewrite as a single logarithm.

________ 3. $3 \log_a x + \frac{1}{2} \log_a (x^2 + 2)$

________ 4. $2 \log_a 5x + 4 \log_a (x^2 + 9) - \frac{1}{5} \log_a 7x - 3 \log_a (x - 1)$

Answers are below.

As you may have noticed from our work with logarithmic expressions, there is no rule for simplifying the logarithm of a sum or difference.

$$\log_a (M + N) \neq \log_a M + \log_a N$$

$$\log_a (M - N) \neq \log_a M - \log_a N$$

▶ Examples You Completed

Example 3. Find $\log_4 16^7$.

Solution

$$\begin{aligned} \log_4 16^7 &= 7 \log_4 16 \\ &= 7(2) \\ &= 14 \end{aligned}$$

Example 5. Find $\log_2 (4^7 \cdot 2^5)$.

Solution

$$\begin{aligned} \log_2 (4^7 \cdot 2^5) &= \log_2 4^7 + \log_2 2^5 \\ &= 7 \log_2 4 + 5 \log_2 2 \\ &= 7(2) + 5(1) \\ &= 14 + 5 \\ &= 19 \end{aligned}$$

Example 8. Find $\log_{10} \frac{8}{9}$.

Solution

$$\begin{aligned} \log_{10} \frac{8}{9} &= \log_{10} 8 - \log_{10} 9 \\ &= \log_{10} 2^3 - \log_{10} 3^2 \\ &= 3 \log_{10} 2 - 2 \log_{10} 3 \\ &\doteq 3(0.3010) - 2(0.4771) \\ &\doteq 0.9030 - 0.9542 \\ &\doteq -0.0512 \end{aligned}$$

Example 12. Use the laws of logarithms to rewrite $\log_a \frac{(x + 5)^3}{\sqrt{x - 4}}$.

Solution

$$\begin{aligned} &\log_a \frac{(x + 5)^3}{\sqrt{x - 4}} \\ &= \log_a \frac{(x + 5)^3}{(x - 4)^{\frac{1}{2}}} \\ &= \log_a (x + 5)^3 - \log_a (x - 4)^{\frac{1}{2}} \\ &= 3 \log_a (x + 5) - \tfrac{1}{2} \log_a (x - 4) \end{aligned}$$

Answers to Trial Runs

page 775 1. 5 2. -4 3. 6 4. 1 5. 10 6. 1.2552 7. 2.6020 8. -0.2273

page 776 1. $\log_a (x + 1) + 2 \log_a (2x - 5)$ 2. $\log_a (x^2 + 1) - \frac{1}{3} \log_a (x + 9)$ 3. $\log_a x^3 \sqrt{x^2 + 2}$
4. $\log_a \frac{(5x)^2(x^2 + 9)^4}{\sqrt[5]{7x}\,(x - 1)^3}$

Name ______________________ Date ____________

EXERCISE SET 11.3

Find the following logarithms.

_______ 1. $\log_2 (16 \cdot 8)$

_______ 2. $\log_2 (4 \cdot 32)$

_______ 3. $\log_3 \frac{27}{81}$

_______ 4. $\log_3 \frac{9}{243}$

_______ 5. $\log_7 49^2$

_______ 6. $\log_6 36^3$

_______ 7. $\log_3 \sqrt{27}$

_______ 8. $\log_2 \sqrt[3]{16}$

_______ 9. $\log_5 (5^3 \cdot 25^2)$

_______ 10. $\log_2 (8^3 \cdot 16^2)$

_______ 11. $\log_3 \frac{27^2}{9^3}$

_______ 12. $\log_5 \frac{25^3}{125^2}$

_______ 13. $\log_2 \frac{2^5 \cdot 8^2}{16^3}$

_______ 14. $\log_3 \frac{9^2 \cdot 27^3}{81^2}$

Approximate each of the following logarithms, given that $\log_{10} 2 \doteq 0.3010$, $\log_{10} 3 \doteq 0.4771$, $\log_{10} 5 \doteq 0.6990$, *and* $\log_{10} 7 \doteq 0.8451$.

_______ 15. $\log_{10} 15$

_______ 16. $\log_{10} 14$

_______ 17. $\log_{10} \sqrt{200}$

_______ 18. $\log_{10} \sqrt{300}$

_______ 19. $\log_{10} 350$

_______ 20. $\log_{10} 210$

_______ 21. $\log_{10} \frac{49}{6}$

_______ 22. $\log_{10} \frac{25}{14}$

_______ 23. $\log_{10} 45$

_______ 24. $\log_{10} 63$

Rewrite using the laws of logarithms.

_______ 25. $\log_a x^3(x + 5)^2$

_______ 26. $\log_a x^2(x + 1)^5$

_______ 27. $\log_a \frac{(x - 3)^5}{\sqrt{3x}}$

_______ 28. $\log_a \frac{7x - 5}{\sqrt[3]{2x}}$

_______ 29. $\log_a \frac{\sqrt[3]{x^2}}{x^2(x - 9)^5}$

_______ 30. $\log_a \frac{\sqrt{x^5}}{x^4(x^2 + 1)}$

_______ 31. $\log_a \frac{(x - 3)\sqrt{2 - x}}{\sqrt{x}(x + 1)^5}$

_______ 32. $\log_a \frac{(x - 2)\sqrt{3 - x}}{\sqrt[3]{x}(x + 2)^4}$

Rewrite as a single logarithm.

_______ 33. $2 \log_a (x - 1) + \frac{1}{3} \log_a x$

_______ 34. $3 \log_a x + \frac{1}{5} \log_a (x + 1)$

_______ 35. $\frac{1}{2} \log_a x - 4 \log_a (x + 2)$

_______ 36. $\frac{1}{3} \log_a x - 2 \log_a (x^2 + 2)$

_______ 37. $\frac{1}{2} \log_a 3x + 3 \log_a x - \log_a (x^2 + 1)$

_______ 38. $\frac{1}{3} \log_a 2x + 4 \log_a x - \log_a (2x - 1)$

_______ 39. $3 \log_a (x + 1) + \frac{2}{3} \log_a (x + 1) - \log_a x - 5 \log_a (x^2 + 1)$

_______ 40. $2 \log_a (x - 1) + \frac{3}{2} \log_a (x - 1) - \log_a 2x - 3 \log_a (x - 2)$

☆ Stretching the Topics

_______ 1. For the formula $V = \frac{1}{3}\pi r^2 h$, write $\log_a r$ in expanded form.

_______ 2. Show that $5 \log_a \sqrt[3]{a} - 7 \log_a \sqrt{a} + 10 \log_a \sqrt[3]{a} - \frac{3}{2} \log_a a = 0$.

Check your answers in the back of your book.

If you can complete **Checkup 11.3,** you are ready to go on to Section 11.4.

Name ______________________________ Date ______________

✓ CHECKUP 11.3

Find the following logarithms.

______ 1. $\log_2 (16 \cdot 8)$

______ 2. $\log_3 \dfrac{9}{\sqrt{27}}$

Find the following logarithms given that $\log_{10} 2 \doteq 0.3010$ *and* $\log_{10} 3 \doteq 0.4771$.

______ 3. $\log_{10} 72$

______ 4. $\log_{10} \sqrt{80}$

______ 5. $\log_{10} 360$

______ 6. $\log_{10} \dfrac{90}{8}$

Rewrite using the laws of logarithms.

______ 7. $\log_a (x^2 \sqrt{x^2 + 1})$

______ 8. $\log_a \dfrac{\sqrt[3]{x^2}\,(x^2 + 1)}{(x^2 + 4)^5}$

Rewrite as a single logarithm.

______ 9. $2 \log_a (x + 1) - \frac{2}{3} \log_a x$

______ 10. $\frac{1}{2} \log_a (3x) + \log_a (x + 5) - 4 \log_a x - \log_a (2x + 1)$

Check your answers in the back of your book.

If You Missed Problems:	You Should Review Examples:
1, 2	1–6
3–6	7–10
7, 8	11, 12
9, 10	13, 14

11.4 Using Common Logarithms

Before the advent of the inexpensive hand-held calculator, logarithms provided the tool for scientific calculations. Scientists did not relish lengthy computation with decimal numbers any more than college students do, and they used logarithms to make the arithmetic of such numbers less tedious.

Using Scientific Notation

The key to using logarithms for computation is our ability to write every decimal number as a product of some number between 1 and 10, and an appropriate power of 10. Let us spend a moment reviewing some integer powers of 10.

$$10^0 = 1$$

$$10^1 = 10 \qquad 10^{-1} = \frac{1}{10^1} = \frac{1}{10} = 0.1$$

$$10^2 = 100 \qquad 10^{-2} = \frac{1}{10^2} = \frac{1}{100} = 0.01$$

$$10^3 = 1000 \qquad 10^{-3} = \frac{1}{10^3} = \frac{1}{1000} = 0.001$$

$$10^4 = 10{,}000 \qquad 10^{-4} = \frac{1}{10^4} = \frac{1}{10{,}000} = 0.0001$$

$$10^5 = 100{,}000 \qquad 10^{-5} = \frac{1}{10^5} = \frac{1}{100{,}000} = 0.00001$$

Suppose we assume that n represents a natural number. To write a power of 10 without using exponents, we first write down the numeral 1. Then if the power is n, we add zeros to the *right* of that 1 and stop and place a decimal point when we have moved n places right. If the power is $-n$, we add zeros to the left of that 1 and stop and place a decimal point when we have moved n places left. If the power is 0, we leave the 1 as is.

Example 1. Write 10^9.

Solution

$$10^9 = 1{,}000{,}000{,}000$$

9 places to the right of 1

Example 2. Write 10^{-7}.

Solution

$$10^{-7} = 0.0000001$$

7 places to the left of 1

You should agree that each of the numbers below has been rewritten in an equivalent form.

$$325 = 3.25 \times 100 = 3.25 \times 10^2$$

$$69{,}000 = 6.9 \times 10{,}000 = 6.9 \times 10^4$$

$$87 = 8.7 \times 10 = 8.7 \times 10^1$$

$$0.325 = 3.25 \times 0.1 = 3.25 \times 10^{-1}$$

$$0.0069 = 6.9 \times 0.001 = 6.9 \times 10^{-3}$$

$$1.3 = 1.3 \times 1 = 1.3 \times 10^0$$

The third version of each number is called its **scientific notation** form, and it is not difficult to rewrite a number using scientific notation if you have understood our discussion of powers of 10.

Scientific Notation. To write a number using scientific notation, we must express it as a *product* of some factor between 1 and 10, and an appropriate power of 10.

We summarize here the steps needed to write a number using scientific notation.

1. Insert a caret ($\wedge$) in the given number to obtain the necessary factor between 1 and 10.
2. Count the places from your caret to the actual position of the decimal point in the given number.
3. If the actual decimal point is n places to the *right* of your caret, the appropriate power of 10 is 10^n. If the actual decimal point is n places to the *left* of your caret, the appropriate power of 10 is 10^{-n}.
4. Write the given number as the product of your factor between 1 and 10, and the appropriate power of 10.

For instance, to write 578 using scientific notation, we proceed as follows:

$$578 = 5_\wedge 78 \quad \text{(2 places right)}$$
$$= 5.78 \times 10^2$$

Similarly, we may write 0.00578 using scientific notation by reasoning as follows:

$$0.00578 = 0.005_\wedge 78 \quad \text{(3 places left)}$$
$$= 5.78 \times 10^{-3}$$

Now you complete Examples 3 and 4.

Example 3. Use scientific notation to write 76,300.

Solution

$$76{,}300 = 7_\wedge 6300 \quad \text{(4 places right)}$$
$$= 7.63 \times \_\_\_\_$$

Check your work on page 789. ▶

Example 4. Use scientific notation to write 0.000012.

Solution

$$0.000012 = 0.00001_\wedge 2 \quad \text{(5 places left)}$$
$$= \_\_\_\_ \times \_\_\_\_$$

Check your work on page 789. ▶

Example 5. Use scientific notation to write 123; 123,000; 0.123; and 0.000123.

Solution

$$123 = 1_\wedge 23 = 1.23 \times 10^2$$
$$123{,}000 = 1_\wedge 23000 = 1.23 \times 10^5$$
$$0.123 = 0.1_\wedge 23 = 1.23 \times 10^{-1}$$
$$0.000123 = 0.0001_\wedge 23 = 1.23 \times 10^{-4}$$

Notice in Example 5 that all the numbers shared the same factor between 1 and 10. The numbers differed only in their powers of 10. We could think of these numbers as belonging to the same "family," perhaps called the 1.23 family. We shall make use of this observation shortly.

We can reverse our steps to change a number from scientific notation to standard form.

Example 6. Write 6.4×10^5 in standard form.

Solution

$$6.4 \times 10^5 = 6{}_\wedge 40000$$

5 places right

$$= 640{,}000$$

Example 7. Write 3×10^{-8} in standard form.

Solution

$$3 \times 10^{-8} = 3.0 \times 10^{-8}$$

$$= 00000003{}_\wedge$$

8 places left

$$= 0.00000003$$

Trial Run

Use scientific notation to write each number.

_______ 1. 899

_______ 2. 0.00054

_______ 3. 2,010,000

Write each number in standard form.

_______ 4. 10^{-5}

_______ 5. 7.1×10^3

_______ 6. 1.05×10^{-3}

Answers are on page 789.

Finding Common Logarithms

Suppose that we wish to find numerical values for *logarithms* of numbers written in scientific notation. Since every such number is expressed as a product, we use the First Law of Logarithms to write

$$\log_a (1.23 \times 10^2) = \log_a 1.23 + \log_a 10^2$$

$$\log_a (1.23 \times 10^5) = \log_a 1.23 + \log_a 10^5$$

$$\log_a (1.23 \times 10^{-1}) = \log_a 1.23 + \log_a 10^{-1}$$

$$\log_a (1.23 \times 10^{-4}) = \log_a 1.23 + \log_a 10^{-4}$$

What would be a good choice for the base, a, in computing these logarithms? Since the second term on the right of each logarithmic statement requires that we find the logarithm of a power of 10, it seems that a base of 10 might be a wise choice. Why? Because from earlier work with logarithms, we recall that $\log_a a^n = n$, so $\log_{10} 10^n = n$.

Therefore, using a base of 10 in our logarithmic statements we may say that

$$\log_{10} (1.23 \times 10^2) = \log_{10} 1.23 + \log_{10} 10^2 = \log_{10} (1.23) + 2$$

$$\log_{10} (1.23 \times 10^5) = \log_{10} 1.23 + \log_{10} 10^5 = \log_{10} (1.23) + 5$$

$$\log_{10} (1.23 \times 10^{-1}) = \log_{10} 1.23 + \log_{10} 10^{-1} = \log_{10} (1.23) + (-1)$$

$$\log_{10} (1.23 \times 10^{-4}) = \log_{10} 1.23 + \log_{10} 10^{-4} = \log_{10} (1.23) + (-4)$$

To finish finding the logarithm of each of these numbers, we only need to know $\log_{10} 1.23$, and we shall return to that matter in a minute. First, however, let's notice how the logarithms of the given numbers differ from each other.

$$\log_{10}(1.23 \times 10^2) = 2 + \log_{10} 1.23$$

$$\log_{10}(1.23 \times 10^5) = 5 + \log_{10} 1.23$$

$$\log_{10}(1.23 \times 10^{-1}) = -1 + \log_{10} 1.23$$

$$\log_{10}(1.23 \times 10^{-4}) = -4 + \log_{10} 1.23$$

The part of each logarithm that we have already computed *exactly matches the power of 10* in the scientific notation form of each of the given numbers. This part is called the **characteristic** of the logarithm because it characterizes the given number by indicating its decimal-point location.

For each of our logarithms, the $\log_{10} 1.23$ is called the **mantissa** and it is the same for *all* numbers in the 1.23 family. For example,

$$\log_{10}(1230) = \log_{10}(1.23 \times 10^3) = \overbrace{3}^{\text{characteristic}} + \underbrace{\log_{10} 1.23}_{\text{mantissa}}$$

Let's finish finding each of these logarithms by finding $\log_{10} 1.23$. Fortunately, approximations for base 10 logarithms of numbers between 1 and 10 have already been found by mathematicians. Also called **common logarithms,** they appear in the Table of Common Logarithms inside the back cover of your book.

To locate $\log_{10} 1.23$ in the table, we first locate 1.2 in the leftmost column. Then we move across the top row to find the 3 and look for the entry in the body of the table where the 1.2 row meets the 3 column. That entry is 0.0899 and we conclude that $\log_{10} 1.23 \doteq 0.0899$.

We may now write the common logarithms of all our original numbers.

$$\log_{10}(1.23 \times 10^2) \doteq 2 + 0.0899$$

$$\log_{10}(1.23 \times 10^4) \doteq 4 + 0.0899$$

$$\log_{10}(1.23 \times 10^{-1}) \doteq -1 + 0.0899$$

$$\log_{10}(1.23 \times 10^{-5}) \doteq -5 + 0.0899$$

To find the common logarithm of any number, then, we may use the following steps.

Finding the Common Logarithm of a Number

1. Rewrite the given number in scientific notation.
2. Use the power of 10 as the characteristic of the common logarithm of the given number.
3. Use the common logarithm of the factor between 1 and 10 as the mantissa of the common logarithm of the given number.
4. Find the mantissa in a table of common logarithms.
5. Write the common logarithm of the given number as the sum of its characteristic and its mantissa.

If the characteristic of a logarithm is positive, we are permitted to write the characteristic and mantissa as a single number. For example,

$$\log_{10} (1.23 \times 10^2) = 2 + 0.0899 = 2.0899$$

$$\log_{10} (1.23 \times 10^4) = 4 + 0.899 = 4.0899$$

It is understood that the number preceding the decimal point in each of these logarithms represents the characteristic and the number after the decimal point is the mantissa.

However, when the characteristic is negative, we are not permitted to write the characteristic and mantissa together as a single number. Learning the method for combining characteristic and mantissa in such cases serves no useful purpose here. For our work with logarithms, we shall agree to write a negative characteristic separate from its mantissa.

By popular agreement, mathematicians decided to use the notation **log N**, with no written base, to represent $\log_{10} N$.

$$\log N = \log_{10} N$$

Example 8. Find log 639,000.

Solution

$$\begin{aligned} \log 639{,}000 &= \log (6.39 \times 10^5) \\ &= 5 + \log 6.39 \\ &= 5 + 0.8055 \\ &= 5.8055 \end{aligned}$$

Example 9. Find log 0.0048.

Solution

$$\begin{aligned} \log 0.0048 &= \log (4.80 \times 10^{-3}) \\ &= -3 + \log 4.80 \\ &= -3 + 0.6812 \end{aligned}$$

If we are given the logarithm of a number, we may find that number merely by reversing our steps. For example, suppose we know that

$$\begin{aligned} \log N &= 2.5514 \\ &= 2 + 0.5514 \end{aligned}$$

We know that the characteristic is 2 and the mantissa is 0.5514. Locating 0.5514 in the *body* of our table, we see that it is the logarithm of 3.56. Of course, we know that 2 is the logarithm of 10^2. We may write

$$\log N = 2 + 0.5514$$

$$= \log 10^2 + \log 3.56 \qquad \text{Rewrite numbers as logarithms.}$$

$$\log N = \log (3.56 \times 10^2) \qquad \text{Use First Law of Logarithms.}$$

We conclude that

$$\begin{aligned} N &= 3.56 \times 10^2 \\ &= 356 \end{aligned}$$

Finding the number with a given logarithm is sometimes called finding the **antilogarithm** of the given logarithm. In our last example,

$$356 = \text{antilog}_{10} (2.5514)$$

Definition of Antilogarithm

$$N = \text{antilog}_a L$$

means

"N is the number whose logarithm is L"

Therefore,

$$\text{If } N = \text{antilog } L$$
$$\text{then } L = \log N$$

Example 10. Find antilog $(-2 + 0.9557)$.

Solution. We use the table to find the number whose logarithm is 0.9557.

$$\begin{aligned}\text{antilog } (-2 + 0.9557) &= 9.03 \times 10^{-2} \\ &= 0.0903\end{aligned}$$

Example 11. Find antilog_{10} (7.7619).

Solution. We use the table to find the number whose logarithm is 0.7619.

$$\begin{aligned}\text{antilog } (7 + 0.7619) &= 5.78 \times 10^{7} \\ &= 57{,}800{,}000\end{aligned}$$

Trial Run

Find each common logarithm.

________ 1. log 6.45 ________ 2. log 180

________ 3. log 478,000 ________ 4. log 0.00759

Find each antilogarithm.

________ 5. antilog 3.9504 ________ 6. antilog $(-2 + 0.5416)$

Answers are on page 789.

Computing with Common Logarithms

Let's turn our attention to using common logarithms as a computational tool. Keep in mind that logarithms were available to students of mathematics long before calculators came into existence. The primary reason for the inclusion of this section is to reinforce your understanding of the laws of logarithms, *not* to suggest that computation using logarithms is more efficient than computation using a calculator.

Since $y = \log x$ is a function, we know that if $A = B$, then $\log A = \log B$. This fact allows us to "take the logarithm of both sides of an equation."

Suppose that you wish to calculate a power such as $(2.9)^5$ without tedious multiplication.

$N = (2.9)^5$	
$\log N = \log (2.9)^5$	Take logarithms of both sides.
$= 5 \log 2.9$	Use Third Law of Logarithms.
$= 5 \log (2.9 \times 10^0)$	Rewrite 2.9 in scientific notation.
$= 5(0.4624)$	Find the logarithm in the table.
$\log N = 2.3120$	Remove parentheses. Note the characteristic is 2 and the mantissa is 0.3120.
$N = \text{antilog } 2.3120$	Isolate N.
$\doteq 2.05 \times 10^2$	Use table to find number whose logarithm is closest to 0.3120.
$\doteq 205$	Simplify.

Therefore, $(2.9)^5 \doteq 205$.

Let's see whether this answer is reasonable. We know that

$$2 < 2.9 < 3$$

so

$$2^5 < (2.9)^5 < 3^5$$

$$32 < (2.9)^5 < 243$$

Since $32 < 205 < 243$ our answer seems appropriate, especially since 2.9 is much closer to 3 than to 2. It is not a bad idea to roughly check your answers to see whether they make sense.

Example 12. Use common logarithms to compute (6280)(0.000132).

Solution

$N = (6280)(0.000132)$	
$\log N = \log (6280)(0.000132)$	Take logarithms of both sides.
$= \log (6280) + \log (0.000132)$	Use First Law of Logarithms.
$= \log (6.28 \times 10^3) + \log (1.32 \times 10^{-4})$	Use scientific notation.
$= (3 + 0.7980) + (-4 + 0.1206)$	Find logarithms in the table.
$= 3 + (-4) + 0.7980 + 0.1206$	Combine characteristics and mantissas.
$\log N = -1 + 0.9186$	
$N = \text{antilog } (-1 + 0.9186)$	Isolate N.
$\doteq 8.29 \times 10^{-1}$	Approximate N from the table.
$\doteq 0.829$	Write N in standard form.

Therefore, $(6280)(0.000132) \doteq 0.829$.

Example 13. Use common logarithms to find $\dfrac{(753)(0.054)^3}{2200}$.

Solution

$$N = \frac{(753)(0.054)^3}{2200}$$

$$\log N = \log \frac{(753)(0.054)^3}{2200}$$ Take logarithms of both sides.

$$= \log 753 + \log (0.054)^3 - \log 2200$$ Use First and Second Laws of Logarithms.

$$= \log 753 + 3 \log 0.054 - \log 2200$$ Use Third Law of Logarithms.

$$= \log (7.53 \times 10^2) + 3 \log (5.4 \times 10^{-2}) - \log (2.2 \times 10^3)$$

$$= (2 + 0.8768) + 3(-2 + 0.7324) - (3 + 0.3424)$$ Find each logarithm.

$$= 2 + 0.8768 - 6 + 2.1972 - 3 - 0.3424$$ Remove parentheses.

$$= 2 + 0.8768 - 6 + 2 + 0.1972 - 3 - 0.3424$$ Separate mantissas and characteristics.

$$= (2 - 6 + 2 - 3) + (0.8768 + 0.1972 - 0.3424)$$ Rearrange terms.

$$\log N = -5 + 0.7316$$ Combine characteristics and mantissas.

$$N = \text{antilog}\,(-5 + 0.7316)$$ Isolate N.

$$\doteq 5.39 \times 10^{-5}$$ Approximate N from the table.

$$\doteq 0.0000539$$ Write N in standard form.

Therefore, $\dfrac{(753)(0.054)^3}{2200} \doteq 0.0000539$.

This may seem like a tedious process, but think of the arithmetic necessary in computing such a quotient without logarithms. The use of logarithms reduces multiplication and/or division of very large or very small numbers to addition and/or subtraction of decimal numbers. It reduces finding powers and/or roots of very large or very small numbers to multiplication by a rational number.

Example 14. Use logarithms to find $N = \sqrt{0.0753}$.

Solution

$$N = \sqrt{0.0753}$$

$$\log N = \log \sqrt{0.0753}$$ Take logarithms of both sides.

$$= \log (0.0753)^{\frac{1}{2}}$$ Rewrite root using exponent.

$$= \tfrac{1}{2} \log (0.0753)$$ Use Third Law of Logarithms.

$$= \tfrac{1}{2} \log (7.53 \times 10^{-2})$$ Use scientific notation.

$= \frac{1}{2}(-2 + 0.8768)$	Find logarithm.
$= \frac{1}{2}(-2) + \frac{1}{2}(0.8768)$	Use distributive property.
$\log N = -1 + 0.4384$	Simplify products.
$N = \text{antilog}(-1 + 0.4384)$	Isolate N.
$\doteq 2.74 \times 10^{-1}$	Approximate N from table.
$\doteq 0.274$	Write N in standard form.

Therefore, $\sqrt{0.0753} \doteq 0.274$.

To those of us who are accustomed to doing calculations with calculators, the benefits of using logarithms in computations may seem questionable. However, such practice does reinforce understanding of the laws of logarithms and the process for finding antilogarithms.

Trial Run

Use logarithms to approximate each N.

______ 1. $N = (271)(0.0136)$

______ 2. $N = \dfrac{5730}{28}$

______ 3. $N = (0.0619)^4$

______ 4. $N = \sqrt[3]{32}$

Answers are below.

Examples You Completed

Example 3. Use scientific notation to write 76,300.

Solution

$$76{,}300 = 7_{\wedge}6300_{\curvearrowright}$$

4 places right

$$= 7.63 \times 10^4$$

Example 4. Use scientific notation to write 0.000012.

Solution

$$0.000012 = 0.00001_{\wedge}2$$

5 places left

$$= 1.2 \times 10^{-5}$$

Answers to Trial Runs

page 783 1. 8.99×10^2 2. 5.4×10^{-4} 3. 2.01×10^6 4. 0.00001 5. 7100 6. 0.00105

page 786 1. 0.8096 2. 2.2553 3. 5.6794 4. $-3 + 0.8802$ 5. 8920 6. 0.0348

page 789 1. $N \doteq 3.69$ 2. $N \doteq 205$ 3. $N \doteq 1.47 \times 10^{-5} = 0.0000147$ 4. $N \doteq 3.17$

Name ______________________ Date ____________

EXERCISE SET 11.4

Use scientific notation to rewrite each number.

_______ 1. 0.000542

_______ 2. 0.0000361

_______ 3. 23.5

_______ 4. 46.7

_______ 5. 89,000,000

_______ 6. 73,000,000

_______ 7. 0.00071

_______ 8. 0.00523

_______ 9. 78,300

_______ 10. 67,400

Write each number in standard form.

_______ 11. 1.62×10^{-4}

_______ 12. 2.31×10^{-5}

_______ 13. 3.82×10^{5}

_______ 14. 7.16×10^{4}

_______ 15. 3.01×10^{1}

_______ 16. 9.23×10^{2}

_______ 17. 4.21×10^{-2}

_______ 18. 8.37×10^{-1}

_______ 19. 1.50×10^{7}

_______ 20. 2.40×10^{6}

Find each common logarithm.

_______ 21. log 87,500

_______ 22. log 293,000

_______ 23. log 0.000139

_______ 24. log 0.00245

_______ 25. log 18.7

_______ 26. log 23.6

_______ 27. log 85,000,000

_______ 28. log 9,310,000

_______ 29. log 0.00000000621

_______ 30. log 0.000000000725

Find each antilogarithm.

_______ 31. antilog $(-2 + 0.8791)$

_______ 32. antilog $(-3 + 0.7818)$

_______ 33. antilog 5.1931

_______ 34. antilog 7.5551

_______ 35. antilog $(-8 + 0.6474)$

_______ 36. antilog $(-9 + 0.7782)$

_______ 37. antilog 0.3560

_______ 38. antilog 0.9355

_______ 39. antilog 4.9474

_______ 40. antilog 3.8228

Use common logarithms to approximate each N.

_______ 41. $N = (0.316)(9.72)$

_______ 42. $N = (0.813)(7.46)$

_______ 43. $N = (0.871)^4$

_______ 44. $N = (0.0316)^3$

_______ 45. $N = \dfrac{83.6}{28.7}$

_______ 46. $N = \dfrac{42.8}{17.5}$

_______ 47. $N = \dfrac{(34.7)^2}{80.3}$

_______ 48. $N = \dfrac{(7.31)^3}{62.3}$

_______ 49. $N = \dfrac{(86)(12.3)}{217}$

_______ 50. $N = \dfrac{(59)(15.7)}{328}$

_______ 51. $N = \sqrt[3]{(48.5)(3.13)}$

_______ 52. $N = \sqrt[3]{(56.7)(9.36)}$

_______ 53. $N = \dfrac{\sqrt{87.2}\,(96)}{8.09}$

_______ 54. $N = \dfrac{\sqrt{96.3}\,(28)}{7.63}$

☆ Stretching the Topics

_______ 1. Use logarithms to approximate $N = \dfrac{\sqrt{451}\,\sqrt[3]{73.4}}{(1.65)^5(2.01)^2}$.

_______ 2. The time T for one period of a simple pendulum is given by the formula $T = 2\pi\sqrt{\dfrac{l}{g}}$, where l is the length of the pendulum and g is the acceleration due to gravity. Find l if $T = 6.5$ seconds, $g = 98$ meters per second, and $\pi \doteq 3.14$. Use logarithms.

Check your answers in the back of your book.

If you can complete **Checkup 11.4,** you are ready to go on to Section 11.5.

Name ______________________ Date ____________

✓ CHECKUP 11.4

_______ 1. Use scientific notation to write 0.00231.

_______ 2. Write 1.21×10^4 in standard form.

Find each common logarithm.

_______ 3. log 726

_______ 4. log 0.00983

Find each antilogarithm.

_______ 5. antilog $(-3 + 0.0899)$

_______ 6. antilog (4.9106)

Use common logarithms to find each N.

_______ 7. $N = (5.37)(0.00325)$

_______ 8. $N = \frac{75.9}{36.3}$

_______ 9. $N = \sqrt[3]{8360}$

_______ 10. $N = (0.0725)^4$

Check your answers in the back of your book.

If You Missed Problems:	You Should Review Examples:
1	4
2	6
3, 4	8, 9
5, 6	10, 11
7–10	12–14

11.5 Solving More Logarithmic and Exponential Equations

In Section 11.2 we learned to solve logarithmic equations using the definition of a logarithm. Recall that for $a > 0$, $a \neq 1$, $b > 0$ and $n \in R$,

$$\log_a b = n \qquad \text{means} \qquad a^n = b$$

If a logarithmic equation contains a single logarithmic expression, we shall continue to use this definition to solve the equation. Recalling that *the logarithm of zero or a negative number does not exist,* we shall always *check solutions* for logarithmic equations.

Example 1. Solve $\log_2 (3x + 1) = 4$.

Solution

$\log_2 (3x + 1) = 4$

$2^4 = 3x + 1$	Switch to exponential form.
$16 = 3x + 1$	Simplify the power.
$15 = 3x$	Isolate the x-term.
$5 = x$	Isolate x.

CHECK: $x = 5$

$\log_2 (3 \cdot 5 + 1) \stackrel{?}{=} 4$

$\log_2 16 \stackrel{?}{=} 4$

$4 = 4$

The solution is $x = 5$.

Example 2. Solve $\log_5 (x + 1)(x - 3) = 1$.

Solution

$\log_5 (x + 1)(x - 3) = 1$

$5^1 = (x + 1)(x - 3)$	Switch to exponential form.
$5 = x^2 - 2x - 3$	Remove parentheses.
$0 = x^2 - 2x - 8$	Set quadratic expression equal to zero.
$0 = (x - 4)(x + 2)$	Factor the trinomial.
$x - 4 = 0$ or $x + 2 = 0$	Use the zero product rule.
$x = 4$ $\quad$ $x = -2$	Solve for x.

CHECK: $x = 4$	CHECK: $x = -2$
$\log_5 (4 + 1)(4 - 3) \stackrel{?}{=} 1$	$\log_5 (-2 + 1)(-2 - 3) \stackrel{?}{=} 1$
$\log_5 (5)(1) \stackrel{?}{=} 1$	$\log_5 (-1)(-5) \stackrel{?}{=} 1$
$\log_5 5 \stackrel{?}{=} 1$	$\log_5 5 \stackrel{?}{=} 1$
$1 = 1$	$1 = 1$

The solutions are $x = 4$ or $x = -2$.

Solving Equations Containing Only Logarithmic Expressions

From the graph of the logarithmic function, you should agree that each y in the range corresponds to one and only one x in the domain. This observation is very important because it implies that if we know that the logarithm of one quantity equals the logarithm of another quantity, we may conclude that those two quantities are equal.

$$\text{If } \log_a M = \log_a N$$
$$\text{then } \quad M = N$$

With this fact in mind, we can solve logarithmic equations that contain several logarithmic expressions and no other terms.

Example 3. Solve $\log_2 (x + 5) = \log_2 (4x - 1)$.

Solution

$$\log_2 (x + 5) = \log_2 (4x - 1)$$
$$x + 5 = 4x - 1$$
$$6 = 3x$$
$$2 = x$$

CHECK: $x = 2$

$$\log_2 (2 + 5) \stackrel{?}{=} \log_2 (4 \cdot 2 - 1)$$
$$\log_2 7 = \log_2 7$$

You complete Example 4.

Example 4. Solve $\log_3 (5 - x) = \log_3 (x^2 - 1)$.

Solution

$$\log_3 (5 - x) = \log_3 (x^2 - 1)$$
$$5 - x = x^2 - 1$$
$$0 = x^2 + x - 6$$

$$x = \underline{\qquad} \quad \text{or} \quad x = \underline{\qquad}$$

CHECK: $x = -3$

$$\log_3 [5 - (-3)] \stackrel{?}{=} \log_3 [(-3)^2 - 1]$$
$$\log_3 [5 + \underline{\qquad}] \stackrel{?}{=} \log_3 [\underline{\qquad} - 1]$$
$$\log_3 \underline{\qquad} = \log_3 \underline{\qquad}$$

CHECK: $x = 2$

$$\log_3 (5 - 2) \stackrel{?}{=} \log_3 (2^2 - 1)$$
$$\log_3 \underline{\qquad} = \log_3 \underline{\qquad}$$

The solutions are $x = \underline{\qquad}$ or $x = \underline{\qquad}$.

Check your work on page 803. ▶

When an equation contains several logarithmic expressions, it is best to combine those expressions so that there is just one logarithmic expression on each side of the equation. The laws of logarithms provide us with the key for such manipulation.

Solving Equations Containing Only Logarithmic Expressions

1. Use the laws of logarithms to rewrite each side of the equation as the logarithm of a single quantity.
2. Set those quantities equal to each other and solve for the variable.
3. Check each solution in the original equation to be sure that the logarithm of a negative number does not appear.

Example 5. Solve $\log_a x + \log_a (x + 2) = \log_a (3x + 2)$.

Solution

$\log_a x + \log_a (x + 2) = \log_a (3x + 2)$

$\log_a x(x + 2) = \log_a (3x + 2)$ — Use First Law of Logarithms.

$x(x + 2) = 3x + 2$ — If $\log_a M = \log_a N$, then $M = N$.

$x^2 + 2x = 3x + 2$ — Remove parentheses.

$x^2 - x - 2 = 0$ — Set quadratic expression equal to zero.

$(x - 2)(x + 1) = 0$ — Factor the trinomial.

$x - 2 = 0$ or $x + 1 = 0$ — Set each factor equal to 0.

$x = 2$ $\quad x = -1$ — Solve.

We must check our solutions. Since our base, a, is not a specific number, our check will only inform us whether either solution puts us in the position of writing the logarithm of a *negative* number. If so, we must discard that solution.

$$\log_a x + \log_a (x + 2) = \log_a (3x + 2)$$

CHECK: $x = 2$

$\log_a 2 + \log_a (2 + 2) \stackrel{?}{=} \log_a (3 \cdot 2 + 2)$

$\log_a 2 + \log_a 4 \stackrel{?}{=} \log_a 8$

CHECK: $x = -1$

$\log_a (-1) + \log_a (-1 + 2) \stackrel{?}{=} \log_a [3(-1) + 2]$

$\log_a (-1) + \log_a (1) \stackrel{?}{=} \log_a (-1)$

In checking -1 in the original equation, logarithms of negative numbers appeared. Therefore, we must reject that solution. Our only solution is $x = 2$.

Example 6. Solve $\log_a (x + 1) - \log_a 3 = \log_a (1 - 2x)$.

Solution

$\log_a (x + 1) - \log_a 3 = \log_a (1 - 2x)$

$\log_a \dfrac{x + 1}{3} = \log_a (1 - 2x)$ — Use Second Law of Logarithms.

$\dfrac{x + 1}{3} = 1 - 2x$ — If $\log_a M = \log_a N$, then $M = N$.

$\dfrac{3}{1}\left(\dfrac{x + 1}{3}\right) = 3(1 - 2x)$ — Multiply by LCD: 3.

$x + 1 = 3 - 6x$ — Remove parentheses.

$7x = 2$ — Isolate x-term.

$x = \dfrac{2}{7}$ — Isolate x.

CHECK: $x = \frac{2}{7}$

$$\log_a \left(\tfrac{2}{7} + 1\right) - \log_a 3 \stackrel{?}{=} \log_a \left[1 - 2\left(\tfrac{2}{7}\right)\right]$$

$$\log_a \left(\tfrac{2}{7} + \tfrac{7}{7}\right) - \log_a 3 \stackrel{?}{=} \log_a \left(\tfrac{7}{7} - \tfrac{4}{7}\right)$$

$$\log_a \tfrac{9}{7} - \log_a 3 \stackrel{?}{=} \log_a \tfrac{3}{7}$$

Since no logarithms of negative numbers appear, we accept our solution, $x = \frac{2}{7}$.

Example 7. Solve $\log_a (x + 5) + \log_a (x - 3) = \log_a (2x + 1)$.

Solution

$$\log_a (x + 5) + \log_a (x - 3) = \log_a (2x + 1)$$
$$\log_a (x + 5)(x - 3) = \log_a (2x + 1)$$
$$(x + 5)(x - 3) = 2x + 1$$
$$x^2 + 2x - 15 = 2x + 1$$
$$x^2 - 16 = 0$$
$$(x - 4)(x + 4) = 0$$
$$x - 4 = 0 \quad \text{or} \quad x + 4 = 0$$
$$x = 4 \qquad\qquad x = -4$$

CHECK: $x = 4$

$$\log_a (4 + 5) + \log_a (4 - 3) \stackrel{?}{=} \log_a [2(4) + 1]$$
$$\log_a 9 + \log_a 1 \stackrel{?}{=} \log_a 9$$

CHECK: $x = -4$

$$\log_a (-4 + 5) + \log_a (-4 - 3) \stackrel{?}{=} \log_a [2(-4) + 1]$$
$$\log_a 1 + \log_a (-7) \stackrel{?}{=} \log_a (-7)$$

Since logarithms of negative numbers appear in the check for $x = -4$, we must reject that solution. The only solution is $x = 4$.

Trial Run

Solve each equation.

_______ 1. $\log_3 (2x - 1) = 2$

_______ 2. $\log_2 (x - 2)(x + 5) = 3$

_______ 3. $\log_5 (8 - 5x) = \log_5 (x^2 - 6)$

_______ 4. $\log_a (2x - 1) - \log_a 5 = \log_a (4 - x)$

_______ 5. $\log_3 x + \log_3 (x + 4) = \log_3 (x + 10)$

_______ 6. $\log_{10} \left(\dfrac{x + 3}{2}\right) + \log_{10} (x + 1) = \log_{10} (x + 9)$

Answers are on page 804.

Solving Equations Containing Logarithmic Expressions and Constants

If a logarithmic equation contains several logarithmic expressions together with one or more constants, we must use the laws of logarithms and the definition of a logarithm to solve it. In general, we shall use the following procedure.

Solving Equations Containing Logarithmic Expressions and Constants

1. Write all logarithmic expressions on one side of the equation and combine all constants on the other side.
2. Use the laws of logarithms to rewrite the logarithmic expressions as a single logarithmic expression.
3. Use the definition of a logarithm to rewrite the equation in exponential form.
4. Simplify the exponential equation and solve for the variable.
5. Check each solution in the original equation to be sure that the logarithm of a negative number does not appear.

Example 8. Solve $\log_3 (2x - 1) = 2 + \log_3 4$.

Solution

$$\log_3 (2x - 1) = 2 + \log_3 4$$

$$\log_3 (2x - 1) - \log_3 4 = 2$$ Write logarithmic expressions on one side.

$$\log_3 \left(\frac{2x - 1}{4}\right) = 2$$ Rewrite logarithms as a single logarithm.

$$3^2 = \frac{2x - 1}{4}$$ Switch to exponential form.

$$9 = \frac{2x - 1}{4}$$ Simplify power.

$$36 = 2x - 1$$ Multiply both sides by 4.

$$37 = 2x$$ Isolate x-term.

$$\frac{37}{2} = x$$ Isolate x.

CHECK: $x = \frac{37}{2}$

$$\log_3 [2(\tfrac{37}{2}) - 1] \stackrel{?}{=} 2 + \log_3 4$$

$$\log_3 (37 - 1) \stackrel{?}{=} 2 + \log_3 4$$

$$\log_3 36 \stackrel{?}{=} 2 + \log_3 4$$

Since no logarithms of negative numbers appear, we accept the solution $x = \frac{37}{2}$.

You complete Example 9.

Example 9. Solve $\log_2 5 + \log_2 (x + 3) = 3$.

Solution

$$\log_2 5 + \log_2 (x + 3) = 3$$

$$\log_2 5(x + 3) = 3$$ Rewrite logarithms as a single logarithm.

Switch to exponential form.

Simplify both sides.

Isolate x-term.

$$\_\_\_\_\_ = x$$ Isolate x.

CHECK: $x =$ _____

$$\log_2 5 + \log_2\left(-\tfrac{7}{5} + 3\right) \stackrel{?}{=} 3$$

$$\log_2 5 + \log_2\left(-\tfrac{7}{5} + \tfrac{15}{5}\right) \stackrel{?}{=} 3$$

$$\log_2 5 + \log_2\left(\tfrac{8}{5}\right) \stackrel{?}{=} 3$$

Since no logarithms of negative numbers appear, we accept the solution $x = -\frac{7}{5}$.

Check your work on page 803. ▶

Example 10. Solve $\log_{10}(x - 2) + 4 = 5 - \log_{10}(x + 1)$.

Solution

$$\log_{10}(x - 2) + 4 = 5 - \log_{10}(x + 1)$$

$$\log_{10}(x - 2) + \log_{10}(x + 1) = 5 - 4$$ Write logarithms on one side, constants on other side.

$$\log_{10}(x - 2)(x + 1) = 1$$ Rewrite logarithms as a single logarithm. Combine constants.

$$10^1 = (x - 2)(x + 1)$$ Switch to exponential form.

$$10 = x^2 - x - 2$$ Remove parentheses.

$$0 = x^2 - x - 12$$ Set quadratic expression equal to zero.

$$0 = (x - 4)(x + 3)$$ Factor the trinomial.

$$x - 4 = 0 \quad \text{or} \quad x + 3 = 0$$ Set each factor equal to zero.

$$x = 4 \quad \text{or} \quad x = -3$$ Solve for x.

CHECK: $x = 4$

$$\log_{10}(4 - 2) + 4 \stackrel{?}{=} 5 - \log_{10}(4 + 1)$$

$$\log_{10} 2 + 4 \stackrel{?}{=} 5 - \log_{10}(5)$$

CHECK: $x = -3$

$$\log_{10}(-3 - 2) + 4 \stackrel{?}{=} 5 - \log_{10}(-3 + 1)$$

$$\log_{10}(-5) + 4 \stackrel{?}{=} 5 - \log_{10}(-2)$$

Since logarithms of negative numbers appear in the check of $x = -3$, we must reject that solution. The only solution is $x = 4$.

Example 11. Solve $\log_3(2x - 1) + \log_3(2x + 7) = 2 + \log_3 x$.

Solution

$$\log_3(2x - 1) + \log_3(2x + 7) = 2 + \log_3 x$$

$$\log_3(2x - 1) + \log_3(2x + 7) - \log_3 x = 2$$ Write logarithms on one side.

$$\log_3 \frac{(2x - 1)(2x + 7)}{x} = 2$$ Rewrite logarithms as a single logarithm.

$$3^2 = \frac{(2x - 1)(2x + 7)}{x}$$ Switch to exponential form.

$$9 = \frac{(2x - 1)(2x + 7)}{x}$$ Simplify power.

$$9x = (2x - 1)(2x + 7)$$ Multiply by LCD: x.

$$9x = 4x^2 + 12x - 7$$ Remove parentheses.

$$0 = 4x^2 + 3x - 7 \qquad \text{Set quadratic expression equal to zero.}$$

$$0 = (4x + 7)(x - 1) \qquad \text{Factor trinomial.}$$

$$4x + 7 = 0 \quad \text{or} \quad x - 1 = 0$$

$$4x = -7$$

$$x = -\frac{7}{4} \qquad\qquad x = 1$$

CHECK: $x = -\frac{7}{4}$

$$\log_3 [2(-\tfrac{7}{4}) - 1] + \log_3 [2(-\tfrac{7}{4}) + 7] = 2 + \log_3 (-\tfrac{7}{4})$$

We see the logarithm of a negative number on the right side and *reject* the solution $x = -\frac{7}{4}$.

CHECK: $x = 1$

$$\log_3 (2 \cdot 1 - 1) + \log_3 (2 \cdot 1 + 7) \stackrel{?}{=} 2 + \log_3 1$$

$$\log_3 1 + \log_3 (9) \stackrel{?}{=} 2 + \log_3 1$$

No logarithms of negative numbers appear, so we accept the solution $x = 1$.

Trial Run

Solve each equation.

_______ 1. $\log_2 x + \log_2 6 = 1$

_______ 2. $\log_3 (x + 2) - \log_3 5 = 2$

_______ 3. $\log_6 x + \log_6 (x + 5) = 2$

_______ 4. $\log_2 (x - 1) = \log_2 (2x + 3) + 1$

_______ 5. $\log_{10} (x + 3) + 3 = 4 - \log_{10} x$

Answers are on page 804.

Solving More Exponential Equations

In Section 11.1 we learned to solve exponential equations by rewriting both sides of the equation as powers of the same base. At that time we agreed that equations such as $5^x = 7$ and $2^{x+1} = 3^x$ could not be solved using that technique. Instead, we can now solve them by taking the common logarithm of each side.

$$5^x = 7$$

$$\log 5^x = \log 7 \qquad \text{If } M = N, \text{ then } \log M = \log N.$$

$$x \log 5 = \log 7 \qquad \text{Use Third Law of Logarithms.}$$

$$x = \frac{\log 7}{\log 5} \qquad \text{Isolate } x.$$

Keep in mind that log 7 and log 5 are constants; their values can be *approximated* using a table of common logarithms.

$$x = \frac{\log 7}{\log 5} \doteq \frac{0.8451}{0.6990}$$

$$x \doteq 1.209$$

Remember, the *exact* value for x is $\frac{\log 7}{\log 5}$, but the *approximate* value is 1.209.

Example 12. Solve $3^{x+2} = 4$. Then approximate the solution to the thousandths place.

Solution

$$3^{x+2} = 4$$

$$\log 3^{x+2} = \log 4 \qquad \text{If } M = N, \text{ then } \log M = \log N.$$

$$(x + 2) \log 3 = \log 4 \qquad \text{Use Third Law of Logarithms.}$$

$$x \log 3 + 2 \log 3 = \log 4 \qquad \text{Use distributive property.}$$

$$x \log 3 = \log 4 - 2 \log 3 \qquad \text{Isolate } x\text{-term.}$$

$$x = \frac{\log 4 - 2 \log 3}{\log 3} \qquad \text{Isolate } x.$$

$$x \doteq \frac{0.6021 - 2(0.4771)}{0.4771} \qquad \text{Substitute values from table.}$$

$$x \doteq \frac{0.6021 - 0.9542}{0.4771} \qquad \text{Remove parentheses.}$$

$$x \doteq \frac{-0.3521}{0.4771} \qquad \text{Simplify the numerator.}$$

$$x \doteq -0.738 \qquad \text{Find the quotient.}$$

Example 13. Solve $2^{x+1} = 3^x$. Then approximate the solution to the thousandths place.

Solution

$$2^{x+1} = 3^x$$

$$\log 2^{x+1} = \log 3^x \qquad \text{If } M = N, \text{ then } \log M = \log N.$$

$$(x + 1) \log 2 = x \log 3 \qquad \text{Use Third Law of Logarithms.}$$

$$x \log 2 + \log 2 = x \log 3 \qquad \text{Use distributive property.}$$

$$\log 2 = x \log 3 - x \log 2 \qquad \text{Isolate } x\text{-terms.}$$

$$\log 2 = x(\log 3 - \log 2) \qquad \text{Factor out } x \text{ on right.}$$

$$\frac{\log 2}{\log 3 - \log 2} = x \qquad \text{Isolate } x.$$

$$\frac{0.3010}{0.4771 - 0.3010} \doteq x \qquad \text{Substitute values from table.}$$

$$\frac{0.3010}{0.1761} \doteq x \qquad \text{Simplify the denominator.}$$

$$1.709 \doteq x \qquad \text{Find the quotient.}$$

We used logarithms to solve these equations because we could not rewrite both sides as powers of the same base. However, do not forget to check that possibility before beginning to solve an exponential equation. An equation such as $2^{x+1} = 4^x$ can be solved more efficiently by writing each side as a power of 2 than by using common logarithms (although either method can be used).

Trial Run

Solve for the variable. Approximate your solution to the thousandths place.

_______ 1. $6^x = 10$ _______ 2. $4^{x-1} = 5$

_______ 3. $(1.03)^r = 2$ _______ 4. $7^x = 9^{x+2}$

Answers are on page 804.

When money is deposited in an account that pays **compound interest,** the depositor earns interest on the original amount (called the **principal,** P) and also on previously earned interest. If the interest rate is r per payment period, the total amount, A, in the account after n payment periods is found by the formula

$$A = P(1 + r)^n$$

To evaluate the right side without a calculator we may use common logarithms.

Example 14. If \$2000 is deposited in an account paying 2 percent compound interest per quarter, find the total amount after 10 years.

Solution. Since there are 4 quarters in one year, there are 40 quarters in 10 years. So $n = 40$, $P = 2000$, and $r = 2\% = 0.02$.

$$\begin{aligned}
A &= P(1 + r)^n && \text{Formula for compound amount.}\\
&= 2000(1 + 0.02)^{40} && \text{Substitute into formula.}\\
&= 2000(1.02)^{40} && \\
\log A &= \log[2000(1.02)^{40}] && \text{If } M = N, \text{ then } \log M = \log N.\\
&= \log 2000 + \log (1.02)^{40} && \text{Use First Law of Logarithms.}\\
&= \log (2000) + 40 \log (1.02) && \text{Use Third Law of Logarithms.}\\
&\doteq 3.3010 + 40(0.0086) && \text{Substitute logarithms from table.}\\
&\doteq 3.3010 + 0.344 && \text{Remove parentheses.}\\
\log A &\doteq 3.6450 && \text{Simplify.}\\
A &= \text{antilog } 3.6450 && \text{Isolate } A.\\
&\doteq 4.42 \times 10^3 && \text{Approximate } A \text{ from the table.}\\
&\doteq 4420 && \text{Write } A \text{ in standard form.}
\end{aligned}$$

After 10 years in an account paying 2 percent per quarter, a principal of \$2000 will have increased to approximately \$4420.

Example 15. If Hannah deposits $1500 today in an account paying 7 percent compound interest per year, in how many years will the amount in her account have doubled?

Solution. We want the final amount A to be twice the original principal, so $A = 2(1500) = 3000$. Here we know that $P = 1500$, $A = 3000$, and $r = 7\% = 0.07$, but we must find n.

$$A = P(1 + r)^n \qquad \text{Formula for compound amount.}$$

$$3000 = 1500(1 + 0.07)^n \qquad \text{Substitute into formula.}$$

$$\frac{3000}{1500} = (1.07)^n \qquad \text{Divide by 1500.}$$

$$2 = (1.07)^n \qquad \text{Simplify left side.}$$

We must solve an exponential equation.

$$\log 2 = \log (1.07)^n \qquad \text{If } M = N, \text{ then } \log M = \log N.$$

$$\log 2 = n \log 1.07 \qquad \text{Use Third Law of Logarithms.}$$

$$\frac{\log 2}{\log 1.07} = n \qquad \text{Isolate } n.$$

$$\frac{0.3010}{0.0294} \doteq n \qquad \text{Substitute values from table.}$$

$$10.2 \doteq n \qquad \text{Find the quotient.}$$

The amount in the account paying 7 percent per year will double in about 10.2 years.

In biology and chemistry, patterns of growth and decay are often described by exponential functions.

Example 16. If the number of bacteria in a certain culture triples every hour and there are 100 bacteria present initially, then the number of bacteria present after t hours is found by the formula

$$B = 100 \cdot 3^t$$

Find the approximate number of bacteria present after 30 minutes.

Solution. Here $t = 30$ minutes or $\frac{1}{2}$ hour.

$$B = 100 \cdot 3^t$$

$$= 100 \cdot 3^{\frac{1}{2}} \qquad \text{Substitute } \tfrac{1}{2} \text{ for } t.$$

$$= 100\sqrt{3} \qquad \text{Interpret fractional exponent.}$$

$$\doteq 100(1.732) \qquad \text{Substitute value for } \sqrt{3}.$$

$$\doteq 173.2 \qquad \text{Simplify.}$$

After 30 minutes, there will be approximately 173 bacteria.

You complete Example 17.

Example 17. Under the conditions of Example 16, in how many hours would the number of bacteria reach 5000?

Solution. Here we know $B = 5000$, but we must find t.

$$\begin{aligned} B &= 100 \cdot 3^t \\ 5000 &= 100 \cdot 3^t \\ \frac{5000}{100} &= 3^t \\ 50 &= 3^t \\ \log 50 &= \log 3^t \\ \log 50 &= t \log 3 \\ &= t \end{aligned}$$

The number of bacteria will reach 5000 after approximately ________ hours.

Check your work on page 804. ▶

▶ Examples You Completed

Example 4. Solve $\log_3 (5 - x) = \log_3 (x^2 - 1)$.

Solution

$$\begin{aligned} \log_3 (5 - x) &= \log_3 (x^2 - 1) \\ 5 - x &= x^2 - 1 \\ 0 &= x^2 + x - 6 \\ 0 &= (x + 3)(x - 2) \end{aligned}$$

$$x + 3 = 0 \quad \text{or} \quad x - 2 = 0$$

$$x = -3 \qquad\qquad x = 2$$

CHECK: $x = -3$

$$\begin{aligned} \log_3 [5 - (-3)] &\stackrel{?}{=} \log_3 [(-3)^2 - 1] \\ \log_3 [5 + 3] &\stackrel{?}{=} \log_3 (9 - 1) \\ \log_3 8 &= \log_3 8 \end{aligned}$$

CHECK: $x = 2$

$$\begin{aligned} \log_3 (5 - 2) &\stackrel{?}{=} \log_3 (2^2 - 1) \\ \log_3 3 &= \log_3 3 \end{aligned}$$

The solutions are $x = -3$ or $x = 2$.

Example 9. Solve $\log_2 5 + \log_2 (x + 3) = 3$.

Solution

$$\begin{aligned} \log_2 5 + \log_2 (x + 3) &= 3 \\ \log_2 5(x + 3) &= 3 \\ 2^3 &= 5(x + 3) \\ 8 &= 5x + 15 \\ -7 &= 5x \\ -\frac{7}{5} &= x \end{aligned}$$

Example 17. Under the conditions of Example 16, in how many hours would the number of bacteria reach 5000?

Solution. Here we know $B = 5000$, but we must find t.

$$B = 100 \cdot 3^t$$

$$5000 = 100 \cdot 3^t$$

$$\frac{5000}{100} = 3^t$$

$$50 = 3^t$$

$$\log 50 = \log 3^t$$

$$\log 50 = t \log 3$$

$$\frac{\log 50}{\log 3} = t$$

$$\frac{1.6990}{0.4771} \doteq t$$

$$3.56 \doteq t$$

The number of bacteria will reach 5000 after approximately 3.56 hours (or about 3 hours 34 minutes).

Answers to Trial Runs

page 796 1. $x = 5$ 2. $x = -6, x = 3$ 3. $x = -7$ 4. $x = 3$ 5. $x = 2$ 6. $x = 3$

page 799 1. $x = \frac{1}{3}$ 2. $x = 43$ 3. $x = 4$ 4. No solution 5. $x = 2$

page 801 1. $x \doteq 1.285$ 2. $x \doteq 2.161$ 3. $r \doteq 23.516$ 4. $x \doteq -17.492$

Name ______________________ Date __________

EXERCISE SET 11.5

Solve each equation.

______ 1. $\log_5 (3x + 4) = 2$

______ 2. $\log_3 (5x + 2) = 3$

______ 3. $\log_4 x(3x - 8) = 2$

______ 4. $\log_6 2x(x + 2) = 1$

______ 5. $\log_3 (x + 7)(x - 1) = 2$

______ 6. $\log_2 (x - 5)(x - 1) = 5$

______ 7. $\log_2 (5x - 17) = \log_2 (x + 3)$

______ 8. $\log_4 (3x - 2) = \log_4 (12 + x)$

______ 9. $\log_5 (9 - 2x) = \log_5 (x - 6)$

______ 10. $\log_3 (8 - 5x) = \log_3 (2x - 13)$

______ 11. $\log_a 3 + \log_a (x - 2) = \log_a (x + 1)$

______ 12. $\log_a 5 + \log_a (2x - 1) = \log_a (4x + 7)$

______ 13. $\log_3 x + \log_3 (x + 9) = \log_3 (8x + 12)$

______ 14. $\log_a x + \log_a (2x - 1) = \log_a (4x - 3)$

______ 15. $\log_{10} (3x + 2) - \log_{10} 4 = \log_{10} (x - 9)$

______ 16. $\log_{10} (2x - 5) - \log_{10} 3 = \log_{10} (x - 5)$

______ 17. $\log_2 (x - 3) = \log_2 (3x - 8) - \log_2 x$

______ 18. $\log_2 (2x + 1) = \log_2 (2x + 3) - \log_2 x$

______ 19. $\log_a x + \log_a (x - 7) = \log_a 2 + \log_a (x + 11)$

______ 20. $\log_a (x - 3) + \log_a (x + 1) = \log_a (7x - 23)$

______ 21. $\log_3 x + \log_3 4 = 2$

______ 22. $\log_3 x + \log_3 2 = 2$

______ 23. $\log_4 x - \log_4 3 = 1$

______ 24. $\log_4 x - \log_4 5 = 1$

______ 25. $\log_2 (x - 1) - \log_2 4 = 3$

______ 26. $\log_2 (2x - 3) - \log_2 5 = 2$

______ 27. $\log_4 x + \log_4 (x - 3) = 1$

______ 28. $\log_2 x + \log_2 (x - 4) = 5$

______ 29. $\log_5 (2x + 7) - \log_5 x = 1$

______ 30. $\log_3 (2x + 3) - \log_3 x = 2$

______ 31. $\log_2 (x - 3) = \log_2 (x - 2) + 2$

______ 32. $\log_3 (x - 5) = \log_3 (x + 2) + 1$

______ 33. $\log_5 (2x - 9) + 2 = 3 - \log_5 x$

_______ 34. $\log_6 (x - 1) + 3 = 4 - \log_6 x$

_______ 35. $\log_2 x + \log_2 (x - 3) = 2 + \log_2 (x + 2)$

_______ 36. $\log_2 (x - 4) + \log_2 (x - 6) = 2 + \log_2 x$

Solve each equation and approximate each solution to the thousandths place.

_______ 37. $4^x = 25$

_______ 38. $3^x = 28$

_______ 39. $3^{2x} = 16$

_______ 40. $2^{3x} = 12$

_______ 41. $5^{x-1} = 17$

_______ 42. $6^{x+2} = 37$

_______ 43. $9^{2x+1} = 78.3$

_______ 44. $8^{3x-1} = 94.5$

_______ 45. $1.76^x = 23.4$

_______ 46. $1.37^x = 42.3$

_______ 47. $8 = 14^{x-1}$

_______ 48. $9 = 15^{1-x}$

_______ 49. $5^{x-1} = 6^x$

_______ 50. $7^{x+1} = 10^{2x}$

_______ 51. $3^{2x+1} = 7^{x-2}$

_______ 52. $4^{2x+3} = 5^{x-2}$

_______ 53. If \$3000 is deposited in an account paying 3 percent per quarter, find the total amount after 6 years.

_______ 54. If \$4000 is deposited in an account paying 2 percent per quarter, find the total amount after 5 years.

_______ 55. Find the total amount in an account after 8 years if \$10,000 is invested at 10 percent per year compounded semiannually.

_______ 56. Find the total amount in an account after 6 years if \$8000 is invested at 12 percent per year compounded semiannually.

_______ 57. If \$2000 is invested today in an account paying 8 percent per year compounded yearly, in how many years will the amount have tripled?

_______ 58. If \$5000 is invested today in an account paying 10 percent per year compounded semiannually, in how many years will the amount have doubled?

_______ 59. If the number of bacteria in a culture triples every hour and there are 200 bacteria present initially, find the number of bacteria present after 4 hours. In how many hours would the number of bacteria reach 10,000?

_______ 60. If the number of bacteria in a culture doubles every hour and there are 50 bacteria present initially, find the number of bacteria present after 30 minutes. In how many hours would the number of bacteria reach 400?

☆ Stretching the Topics

________ 1. Solve $3 \cdot 7^{x+1} = 7 \cdot 3^{x+2}$ for x.

________ 2. Find the total amount on deposit at the end of 4 years if \$68,900 is invested at a yearly rate of 13.2 percent and the interest is compounded monthly.

________ 3. Under certain conditions the number of bacteria in a culture is given by $B = n \cdot 2^{t/20}$, where n is the original number of bacteria and t is the number of minutes after the experiment begins. If the original number of bacteria is 250,000, find how many minutes it would take for the number of bacteria to reach 1,000,000.

Check your answers in the back of your book.

If you can complete **Checkup 11.5,** you are ready to do the **Review Exercises** for Chapter 11.

Name ______________________ Date ____________

✓ CHECKUP 11.5

Solve each equation.

______ 1. $\log_a x + \log_a (x + 10) = \log_a (13x + 28)$

______ 2. $\log_{10} (3x - 1) - \log_{10} 3 = \log_{10} (2x + 1)$

______ 3. $\log_3 (x + 12) - \log_3 2 = 2$

______ 4. $\log_2 (3x + 2) - 4 = \log_2 (x - 2) - 2$

______ 5. $5^{x+1} = 13$

______ 6. $3^{x+1} = 7^x$

Check your answers in the back of your book.

If You Missed Problems:	You Should Review Examples:
1, 2	5–7
3	8
4	10
5	12
6	13

Summary

In this chapter we studied two new functions that are very closely related.

Exponential Function	Logarithmic Function
$y = a^x$	$y = \log_a x$
$(a > 0,\ a \neq 1)$	$(a > 0,\ a \neq 1)$
Domain: $x \in R$	Domain: $x \in R,\ x > 0$
Range: $y \in R,\ y > 0$	Range: $y \in R$

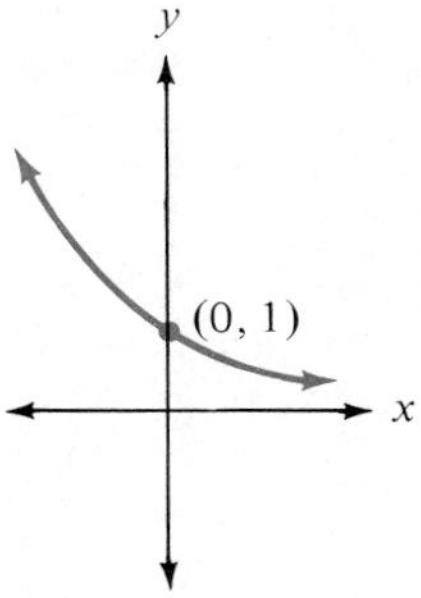

$y = a^x$

$0 < a < 1$

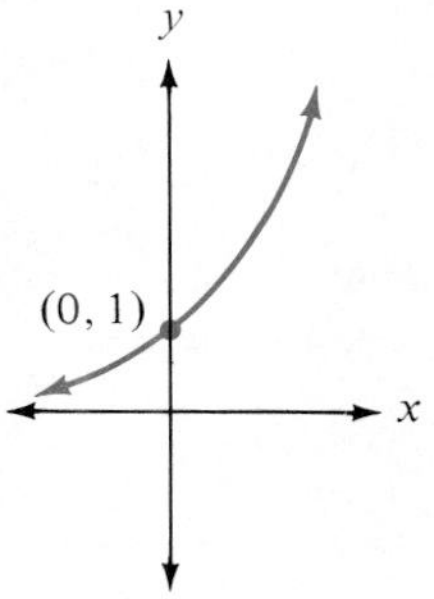

$y = a^x$

$a > 1$

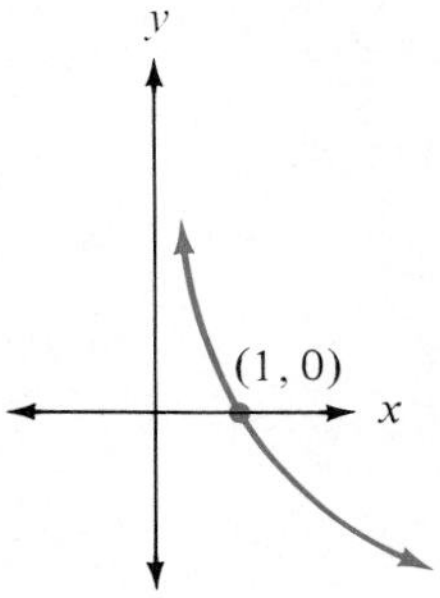

$y = \log_a x$

$0 < a < 1$

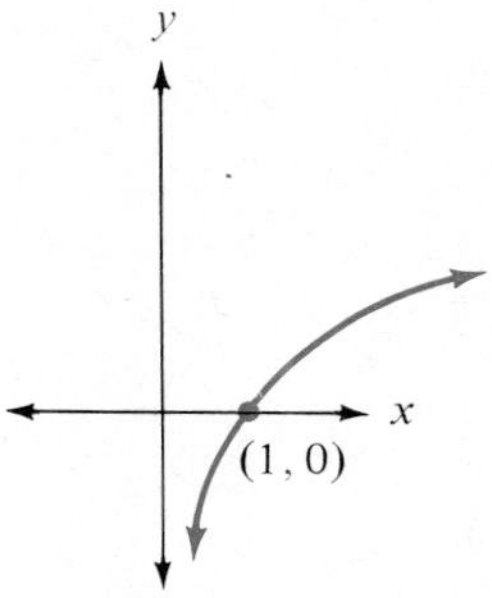

$y = \log_a x$

$a > 1$

We agreed that

$$y = \log_a x \quad \text{means} \quad a^y = x$$

and discussed the three **laws of logarithms** that can be used to rewrite logarithmic expressions.

Name	Symbols	Example
First Law of Logarithms	$\log_a M \cdot N = \log_a M + \log_a N$	$\log_2 x(x + 3) = \log_2 x + \log_2 (x + 3)$
Second Law of Logarithms	$\log_a \dfrac{M}{N} = \log_a M - \log_a N$	$\log_3 \left(\dfrac{x + 5}{8}\right) = \log_3 (x + 5) - \log_3 8$
Third Law of Logarithms	$\log_a M^c = c \log_a M$	$\log_{10} (x + 3)^2 = 2 \log_{10} (x + 3)$

Common logarithms are logarithms for which the base is 10. Because every positive number can be rewritten in scientific notation as a product of some number between 1 and 10, and an appropriate power of 10, we can readily approximate the common logarithms of such numbers using the table inside the back cover.

In solving exponential and logarithmic equations, we used four important facts about exponential and logarithmic statements.

Fact	Example
If $a^m = a^n$ then $m = n$	$2^x = 4^{x+1}$ $2^x = 2^{2(x+1)}$ $x = 2(x+1)$ $x = 2x + 2$ $-2 = x$
If $\log_a M = \log_a N$ then $M = N$	$\log_3(2x - 3) = \log_3(x + 1)$ $2x - 3 = x + 1$ $x = 4$
$y = \log_a x$ means $a^y = x$	$\log_3(5x - 1) = 2$ $3^2 = 5x - 1$ $9 = 5x - 1$ $10 = 5x$ $2 = x$
If $M = N$ then $\log M = \log N$	$3^x = 8$ $\log 3^x = \log 8$ $x \log 3 = \log 8$ $x = \frac{\log 8}{\log 3} \doteq \frac{0.9031}{0.4771}$ $x \doteq 1.89$

When solving logarithmic equations, we agreed to always check our solutions. If the logarithm of a negative number appeared in checking a solution, we agreed to reject that solution because *the logarithm of a negative number or zero does not exist.*

❑ Speaking the Language of Algebra

Complete each sentence with the appropriate word or phrase.

1. The function $y = a^x$ (where $a > 0$, $a \neq 1$) is called a ________ function. Its domain is ________ and its range is ________ , ________ .

2. The function $y = \log_a x$ (where $a > 0$, $a \neq 1$) is called a ________ function. Its domain is ________ , ________ and its range is ________ .

3. The statement $y = \log_a x$ means ________ .

4. A logarithm for which the base is 10 is called a ________ logarithm.

5. To solve an exponential equation, we first try to rewrite both sides as powers of the ________ ________ .

6. To solve an equation such as $\log_2(x + 5) = 4$, we should rewrite it in ________ form.

7. After solving a logarithmic equation, we must check the solutions to be sure that the logarithm of a ________ number or zero does not appear.

Writing About Mathematics

Write your response to each question in complete sentences.

1. If $a > 0$ and $x \in R$, explain why $a^x > 0$.

2. Using the graph of $y = 3^x$, explain why $3^m = 3^n$ implies that $m = n$.

3. Explain why the logarithm of a negative number or zero does not exist.

4. Explain, in your own words, the three laws of logarithms.

5. Explain why you must check every solution for a logarithmic equation.

Name ______________________ Date ______________

REVIEW EXERCISES for Chapter 11

________ 1. Evaluate 5^{2x-1} when $x = 1$, $x = 0$, $x = -1$, $x = \frac{1}{4}$, $x = \frac{1}{2}$.

________ 2. Graph $y = 3^{x+1}$.

________ 3. Solve $\left(\frac{1}{81}\right)^x = 3^{x+5}$.

________ 4. Write $3^{-2} = \frac{1}{9}$ as a logarithmic statement.

________ 5. Write $125^{-\frac{1}{3}} = \frac{1}{5}$ as a logarithmic statement.

________ 6. Write $\log_2 64 = 6$ as an exponential statement.

________ 7. Write $\log_8 \frac{1}{2} = -\frac{1}{3}$ as an exponential statement.

________ 8. Find $\log_3 \frac{1}{27}$.

________ 9. Find $\log_4 \frac{1}{2} - \log_2 1$.

________ 10. Solve $\log_a 216 = 3$.

________ 11. Solve $\log_4 b = -3$.

________ 12. Solve $\log_8 (x + 2) = \frac{1}{3}$.

________ 13. Graph $y = \log_5 x$.

________ 14. Find $\log_3 \sqrt[4]{243}$.

________ 15. Find $\log_2 \frac{\sqrt{16} \cdot 32^2}{128}$.

Given that $\log_{10} 2 \doteq 0.3010$ *and* $\log_{10} 3 \doteq 0.4771$, *find each logarithm in Exercises 16 and 17.*

________ 16. $\log_{10} \frac{32}{27}$

________ 17. $\log_{10} \sqrt{540}$

________ 18. Rewrite $\log_a x^5(x + 5)$ using the laws of logarithms.

________ 19. Rewrite $\log_a \frac{x^3\sqrt{x^2 - 1}}{(x + 1)^5}$ using the laws of logarithms.

________ 20. Rewrite $2 \log_a x + \frac{1}{2} \log_a (x - 1)$ as a single logarithm.

________ 21. Rewrite $3 \log_a (2x + 1) - \frac{4}{5} \log_a x - \log_a (x + 1)$ as a single logarithm.

________ 22. Use scientific notation to rewrite 0.00729.

________ 23. Write 9.34×10^5 in standard form.

________ 24. Use common logarithms to approximate $N = \frac{(34.7)^2(123)}{\sqrt{80.3}}$.

Solve each equation.

________ **25.** $\log_{10}(7x - 2) = \log_{10}(3x + 2)$

________ **26.** $\log_a x + \log_a(x + 14) = \log_a(16x + 63)$

________ **27.** $\log_3(x - 9) - \log_3 5 = 1$

________ **28.** $\log_{10} x + \log_{10}(3x - 5) = 2$

________ **29.** $4^{2x} = 27.3$

________ **30.** $9^{x-1} = 12^{2x}$

________ **31.** Find the total amount in an account after 5 years if $1000 is invested at 2.5 percent per quarter.

________ **32.** If the number of bacteria in a culture triples every hour and there are 500 bacteria present initially, find the number of bacteria present after 30 minutes.

Check your answers in the back of your book.

If You Missed Exercises:	You Should Review	Examples:
1	SECTION 11.1	1, 2
2		6
3		12
4, 5	SECTION 11.2	1
6, 7		2
8, 9		3–7
10–12		10–15
13		16
14, 15	SECTION 11.3	1–6
16, 17		7–10
18, 19		11, 12
20, 21		13, 14
22	SECTION 11.4	4
23		6
24		12–14
25	SECTION 11.5	3
26		5–7
27, 28		8–11
29, 30		12, 13
31		14
32		16

If you have completed the **Review Exercises** and corrected your errors, you are ready to take the **Practice Test** for Chapter 11.

Name ______________________________ Date ______________

PRACTICE TEST for Chapter 11

		Section	Examples
________	1. Evaluate 9^{2-2x} when $x = 0$, $x = 1$, $x = \frac{3}{4}$.	11.1	1
________	2. Write $36^{-\frac{1}{2}} = \frac{1}{6}$ as a logarithmic statement.	11.2	2
________	3. Write $\log_5 125 = 3$ as an exponential statement.	11.2	4
________	4. Find $\log_7 \frac{1}{49}$.	11.3	4, 5
________	5. Find $\log_2 \dfrac{32\sqrt{2}}{(8)^4}$.	11.3	4, 5
________	6. Given that $\log_{10} 3 \doteq 0.4771$ and $\log_{10} 7 \doteq 0.8451$, find $\log_{10} 6300$.	11.3	9, 10
________	7. Rewrite $\log_a \dfrac{x^5(x^2-3)^2}{\sqrt[3]{x+1}}$ using the laws of logarithms.	11.3	11, 12
________	8. Rewrite $\frac{1}{5} \log_a (x+1) - 2 \log_a (x-6) - \frac{2}{3} \log_a x$ as a single logarithm.	11.3	13, 14
________	9. Use scientific notation to rewrite 2,350,000.	11.4	4
________	10. Write 1.72×10^{-3} in standard form.	11.4	6
________	11. Use common logarithms to approximate $N = (68.5)\sqrt[3]{3.24}$	11.4	12–14

12. Graph $y = 3^{x+2}$. — Section 11.1, Examples 6

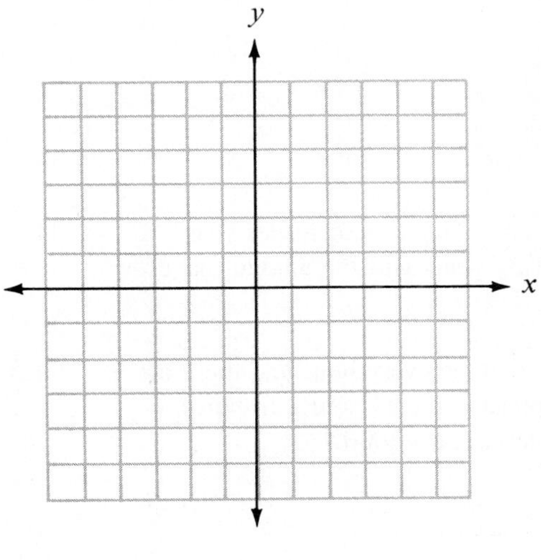

	Section	Examples
13. Graph $y = \log_2 x$.	11.2	16

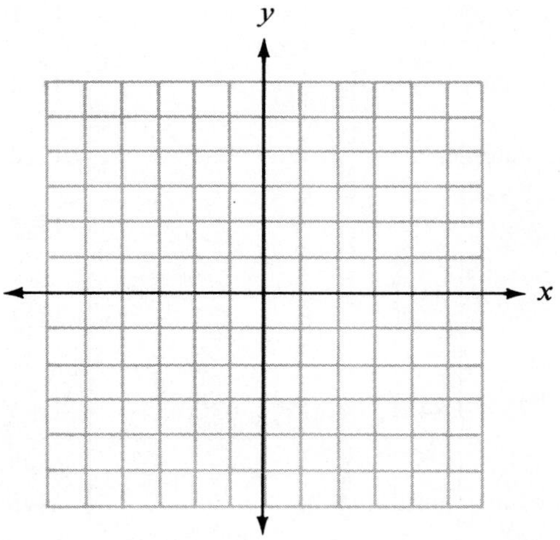

Solve for the variable.

	Section	Examples
________ 14. $8^{2x} = 2^{x-3}$	11.1	11
________ 15. $\log_a \frac{1}{16} = -4$	11.2	11
________ 16. $\log_{25} (2x + 1) = \frac{1}{2}$	11.5	1
________ 17. $\log_4 (2x - 1) = \log_4 (8 - x)$	11.5	3
________ 18. $\log_7 (x + 4) - \log_7 x = \log_7 5$	11.5	8
________ 19. $\log_2 (x + 6) + \log_2 x = 4$	11.5	9
________ 20. $8^x = 3^{x-1}$	11.5	13
________ 21. If \$3000 is invested today in an account paying 8 percent per year compounded semiannually, in how many years will the amount have doubled?	11.5	15
________ 22. If the number of bacteria in a culture doubles every hour and there are 50 present initially, find the number present after 3 hours. In how many hours would the number of bacteria reach 1000?	11.5	16, 17

Name ______________________ Date __________

SHARPENING YOUR SKILLS after Chapters 1–11

		Section
______	1. Evaluate $8(x - 2)^2 - 11x$ when $x = 0$.	1.4
______	2. Evaluate $\lvert(-2)(4)\rvert + \lvert -3 \rvert$.	1.4
______	3. Give the restrictions on the variable and reduce the fraction $\dfrac{9y^2 - 144}{3y^2 - 24y + 48}$.	6.1
______	4. Write $\left(\dfrac{27}{64}\right)^{-\frac{2}{3}}$ in radical form and simplify.	7.1
______	5. Evaluate $-i^{11}$.	7.5
______	6. Find the slope of the line joining $(-\frac{1}{2}, 2)$ and $(\frac{1}{4}, \frac{2}{3})$.	8.3
______	7. Find the equation of the line with x-intercept of $\frac{1}{2}$ and y-intercept of $-\frac{1}{4}$.	8.3
______	8. If $f(x) = 4x - 3$ and $g(x) = x^2 - 5x + 6$, find $g(1) - f(-3)$.	8.5

Perform the indicated operation and simplify.

______	9. $\dfrac{-6(3 - 8) - 5(-8)}{-4[(-12) + 7]}$	1.3
______	10. $x^2y^3(3xy^2)^2 - 2xy^4(xy)^3$	3.1
______	11. $\left(\dfrac{-7x^4y^{-3}}{x^{-2}y^{-2}}\right)^{-2}$	3.2
______	12. $7x^2 - 9[2(3x^2 + x) - 5x] + 2$	3.3
______	13. $(2a - 3)^3$	3.4
______	14. $(3a^3 - 8a^2 - 20a + 16) \div (3a - 2)$	3.5
______	15. $\dfrac{9x^2 - 25}{x^2 - 3x - 40} \cdot \left(\dfrac{x^2 - 64}{3x^2 + x - 10} \div \dfrac{3x^2 + 29x + 40}{2x^2 + 9x - 5}\right)$	6.2
______	16. $\dfrac{2}{x + 5} - \dfrac{3}{x + 3} - \dfrac{x - 3}{x^2 + 8x + 15}$	6.3
______	17. $\dfrac{8 - 10x^{-1} + 3x^{-2}}{2 - 13x^{-1} + 6x^{-2}}$	6.4

SECTION

________	18. $7\sqrt[3]{a^4b^3} - 2b\sqrt[3]{27a^4}$	7.2
________	19. $(\sqrt{3} - 2\sqrt{y})(4\sqrt{3} + \sqrt{y})$	7.3
________	20. $\dfrac{-10}{2 - \sqrt{-36}}$	7.5

Write each expression as a product of factors.

________	21. $(x + 3)^3 + 64$	4.2
________	22. $5h^2 + 22hk + 24k^2$	4.3
________	23. $27 + 42(2x - y) - 5(2x - y)^2$	4.3

Solve each equation or inequality.

________	24. $4[x - (2x + 1)] = -2[3(x - 1) + 7]$	2.1
________	25. $2x - 9 < 5x - 3$ or $x - 1 > 7x + 17$	2.3
________	26. $\lvert 5m - 9\rvert \leq -3$	2.4
________	27. $5x(x - 2) = 10(9 - x)$	5.3
________	28. $(x - 3)(x + 1) < x(5 - x) - 6$	5.4
________	29. $\dfrac{x + 5}{x - 4} = \dfrac{64}{x^2 - 16} + \dfrac{8}{x + 4}$	6.5
________	30. $\dfrac{6}{x(x + 2)} < 0$	6.5
________	31. $\sqrt{x - 4} = 5 + \sqrt{x + 1}$	7.4
________	32. $\dfrac{y}{y - 7} = \dfrac{1}{y}$	7.6
________	33. $\begin{aligned} 2x - 3y &= 7 \\ x + 4y &= -2 \end{aligned}$	10.2
________	34. $\begin{aligned} x &= y^2 - 4 \\ x &= 3y \end{aligned}$	10.4

Graph. SECTION

35. $2x - 3y > -9$ 8.4

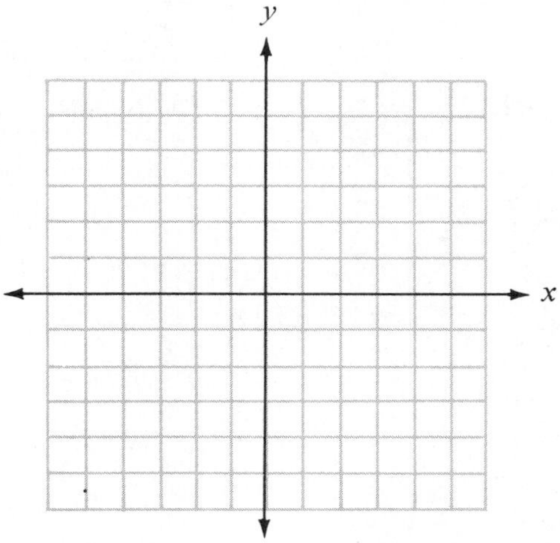

36. $y = x^2 + 3x + 7$ 9.2

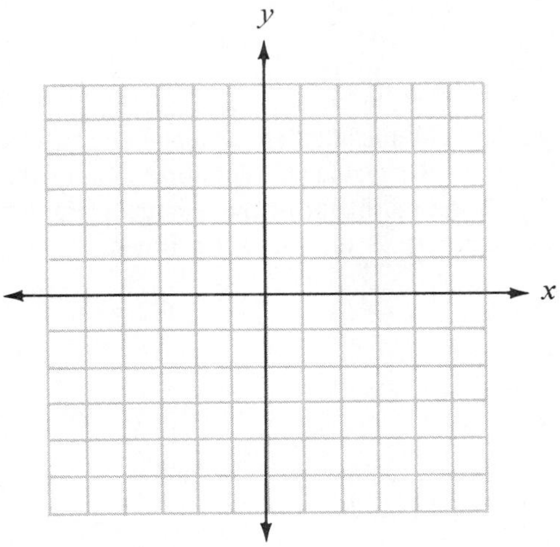

37. $(x - 2)^2 + (y - 2)^2 \leq 81$ 9.4

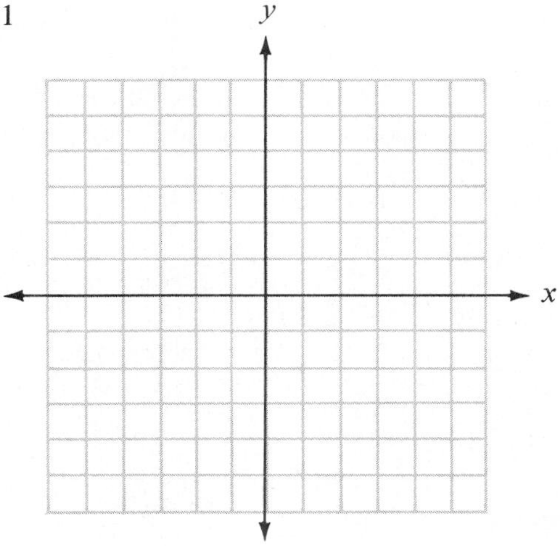

SECTION

38. $x^2 + y^2 \leq 25$ 10.4
$2x - y \geq 1$

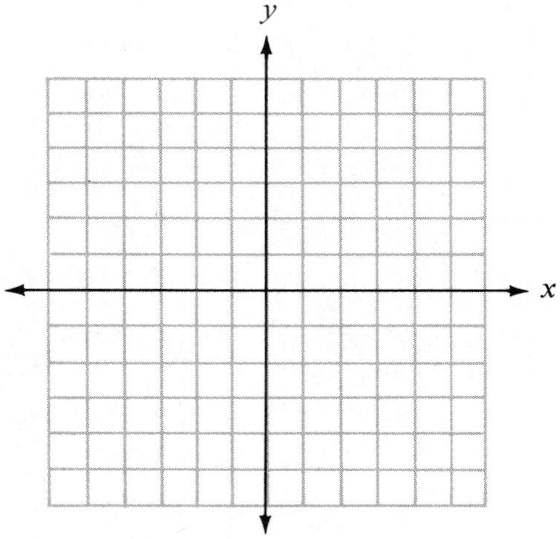

________ 39. Solve $t = 2\pi\sqrt{\frac{l}{g}}$ for g. 7.4

________ 40. Katrina has a rectangular garden and a triangular garden. Each of two sides of the triangle is as long as the length of the rectangle. The third side of the triangle is as long as the width of the rectangle. If the perimeter of the triangular garden is 24 feet and the perimeter of the rectangular garden is 30 feet, find the dimensions of the rectangular garden. 10.5

Answers

This section includes the answers to Odd-Numbered Exercises, Stretching the Topics, Checkups, Speaking the Language of Algebra, Review Exercises, Practice Tests, and Sharpening Your Skills.

Chapter 1

Exercise Set 1.1 (page 10)

1. Q, R **3.** W, J, Q, R **5.** H, R **7.** Q, R
9. 0, 9 **11.** $-8, -5.2, -\frac{1}{2}, 0, \frac{7}{2}, 5.\overline{6}, 9$ **13.** -8
15. True **17.** True **19.** True
21.

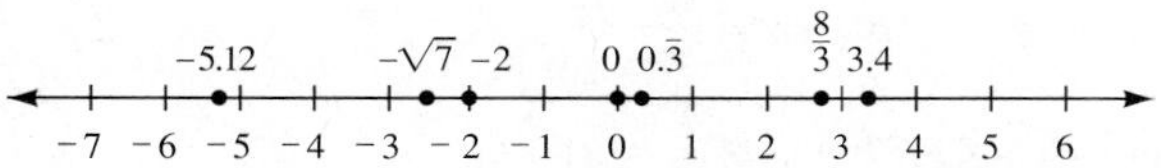

23. $3 > \sqrt{7}$ **25.** $-\frac{9}{2} < -3$ **27.** $\frac{10}{2} = \sqrt{25}$
29. $-1.6 < -0.7$ **31.** $-\frac{8}{3} > -\sqrt{16}$
33. $0.29 > 0$ **35.** $-\sqrt{6} < 0$ **37.** $\frac{3}{4} > 0$
39. $-9 < 0$ **41.** -5 **43.** $\frac{7}{8}$ **45.** 3.2
47. $-\sqrt{5}$ **49.** 6 **51.** 6.8 **53.** $\sqrt{3}$ **55.** 5
57. 8

Stretching the Topics

1. 7 **2.** > **3.** 0

Checkup 1.1 (page 12)

1. J, Q, R **2.** $-3, 0, 4$ **3.** $\frac{8}{2} = \sqrt{16}$
4. $-\frac{7}{2} < -2.5$ **5.** $-\frac{9}{5} < 0$ **6.** $\sqrt{36} > 0$ **7.** $\frac{5}{12}$
8. -9 **9.** 3 **10.** 9

Exercise Set 1.2 (page 22)

1. 14 **3.** -10 **5.** 4 **7.** -9 **9.** $-\frac{7}{9}$
11. -7 **13.** -2 **15.** 1.5 **17.** -40 **19.** 74
21. -4 **23.** 72 **25.** -105 **27.** 0 **29.** 16
31. -4 **33.** 7 **35.** $-\frac{6}{7}$ **37.** 12 **39.** 0
41. 23 **43.** 0 **45.** Not uniquely defined
47. -13 **49.** Undefined

Stretching the Topics

1. -5 **2.** Undefined **3.** 1°, 22°

Checkup 1.2 (page 24)

1. -52 **2.** -38 **3.** 33 **4.** 7 **5.** 7 **6.** 28
7. -27 **8.** -9 **9.** 15 **10.** Undefined

Exercise Set 1.3 (page 34)

1. Associative property for multiplication
3. Closure property for addition
5. Multiplicative identity
7. Distributive property for multiplication over addition
9. Multiplicative inverse **11.** $8 + (-4)$
13. $-9 + (3 + 4)$ **15.** 0 **17.** -19
19. $2(3 - 5)$ **21.** 23 **23.** 290 **25.** -3
27. 168 **29.** 0 **31.** 18 **33.** -87 **35.** -20
37. -65 **39.** -8 **41.** 0 **43.** -13 **45.** -3
47. Undefined **49.** $-\frac{3}{4}$ **51.** 11 **53.** 2
55. -47 **57.** $-\frac{5}{12}$

Stretching the Topics

1. -6.7 **2.** 0 **3.** 496,800; 360 gain

Checkup 1.3 (page 36)

1. Commutative property for multiplication
2. Multiplicative inverse **3.** 17 **4.** 28 **5.** 5
6. -180 **7.** -92 **8.** 15 **9.** $-\frac{1}{2}$ **10.** $\frac{10}{3}$

Exercise Set 1.4 (page 47)

1. 9 **3.** 10 **5.** 0 **7.** 45 **9.** 30 **11.** 38
13. 459.2 **15.** 13 **17.** 100 **19.** $-4a - 7$
21. $2x^2 - 11x$ **23.** $-9y^2 - 1$ **25.** $11x - 6y - 2$
27. $3x^2 - 5xy + y^2$ **29.** $-12a^2 - 11b^2$ **31.** a
33. $-7x$ **35.** $12y^2$ **37.** $6x^2 - 7$
39. $13y^2 - 16y + 14$ **41.** $8x + 15y - 52$
43. $-14x$ **45.** $19m - 4$ **47.** $-7x + 10y$
49. $-67x + 12$ **51.** $-20x + 79$ **53.** $70x + 120$
55. $-2x + 38y$ **57.** $-38a + 6b$ **59.** $9x + 5y$
61. 11 **63.** $3x$ if $3x \geq 0$; $-3x$ if $3x < 0$ **65.** $5 - \pi$
67. $x - 2$ if $x - 2 \geq 0$; $-(x - 2)$ if $x - 2 < 0$
69. $3 - \sqrt{7}$ **71.** -5 **73.** 13

Stretching the Topics

1. $5x - 83y$; 43 **2.** $11x - 5$ **3.** 29

Checkup 1.4 (page 50)

1. 18 **2.** 784 **3.** $-4x^2 + x - 4$ **4.** $3x^2 - 9y^2$

5. $-26a$ **6.** $17y^2 + 6y - 8$ **7.** $-14n + 45$
8. $-28x + 51y$ **9.** $4x$ if $4x \geq 0$; $-4x$ if $4x < 0$
10. $9 - \sqrt{2}$

Exercise Set 1.5 (page 57)

1. $27.50x$ dollars **3.** $(18 + 0.3n)$ dollars
5. $6w$ ft or $3l$ ft **7.** $2x + 7$ **9.** $(940 + 10n)$ dollars
11. $110t$ mi or $55b$ mi **13.** $9x$ dollars **15.** $24s$ or $8w$
17. $(0.8c + 27.8)$ dollars or $(-0.8p + 43.8)$ dollars
19. $(15f + 70)$ dollars or $(15t - 35)$ dollars
21. $(2500 + 3.85x)$ dollars **23.** $(41.3x - 2475)$ dollars
25. $(1.75x - 10{,}250)$ dollars

Stretching the Topics

1. $2n - 4$ **2.** $58h$
3. $(0.0125x + 900)$ dollars or $(-0.0125y + 1050)$ dollars

Checkup 1.5 (page 60)

1. $(420 + 12x)$ dollars **2.** $(4l - 10)$ ft or $(4w + 10)$ ft
3. $(50 + 3n)$ dollars
4. $(-25t + 220)$ mi or $(25h + 120)$ mi
5. $(4x + 50)$ dollars

Speaking the Language of Algebra (page 62)

1. Constants; variables **2.** Real
3. Terms; sum; difference
4. Combine like terms **5.** Factors; product
6. Dividend; divisor; quotient **7.** 0; undefined
8. Negative

Review Exercises (page 65)

1. $-\sqrt{2}, \sqrt{3}$ **2.** $-4, 0, 5$ **3.** $\frac{15}{3} = \sqrt{25}$
4. $-\frac{9}{4} < 13$ **5.** $-19 < -\frac{3}{4}$ **6.** $5 > \sqrt{7}$
7. $-\sqrt{6} < 0$ **8.** $8.75 > 0$ **9.** -55 **10.** -14
11. -4.29 **12.** -4 **13.** 17 **14.** 45
15. -5.6 **16.** -26 **17.** 5 **18.** Undefined
19. 4 **20.** 0 **21.** 840 **22.** 3360 **23.** 48
24. -72 **25.** 96 **26.** $\frac{1}{2}$ **27.** -8 **28.** 0
29. -3 **30.** 14 **31.** \$2720 **32.** 22
33. $-2x^2 + x - 7$ **34.** $3x^2 - 3y^2$ **35.** $-37a$
36. $4y^2 - 21y - 15$ **37.** $2n + 53$ **38.** $3x - 36$
39. $79 - 48x$ **40.** $-2x - 9y$ **41.** $-84x - 8y$
42. $48x + 53y$ **43.** 21 **44.** a if $a \geq 0$; $-a$ if $a < 0$
45. $6 - \sqrt{10}$ **46.** 4 **47.** -4 **48.** 22
49. $6l - 14$
50. $(2.7p - 4.75)$ dollars or $(2.7c + 8.75)$ dollars
51. $6i - 15$ **52.** $(500 - 12x)$ dollars

Practice Test (page 69)

1. $\sqrt{10} > 3$ **2.** $-\sqrt{5} < 0$ **3.** -47 **4.** -3
5. 15 **6.** 44 **7.** -4 **8.** 0 **9.** -32
10. 28 **11.** -3 **12.** 142 **13.** $-3x^2 - 2x - 11$
14. $-ab - b^2$ **15.** $9m$ **16.** $-11x - 20$
17. $12a - 24$ **18.** $16y - 19$ **19.** $-21x + 12$
20. $6x - y$ **21.** -7 **22.** -6 **23.** $3 - \sqrt{5}$
24. $(10i + 135)$ mi or $(-10t + 165)$ mi
25. $(4w + 4)$ ft

CHAPTER 2

Exercise Set 2.1 (page 83)

1. $x = 5$ **3.** $y = -2$ **5.** $m = -6$ **7.** $y = \frac{2}{3}$
9. $a = 0$ **11.** $x = -\frac{5}{9}$ **13.** $x = 12$
15. $x = -26$ **17.** $x = 7$ **19.** $x = -\frac{1}{2}$
21. $y = 3$ **23.** $x = 12$ **25.** $x = -4$
27. $a = 0$ **29.** $x = 4$ **31.** $y = 3$ **33.** $x = -4$
35. $m = \frac{5}{3}$ **37.** $a = 1$ **39.** $x = -10$
41. $x = 0$ **43.** Any real number **45.** $y = 21$
47. No solution **49.** $x = \frac{95}{4}$ **51.** $x = -\frac{8}{3}$
53. $x = -3$ **55.** $x = -\frac{7}{2}$ **57.** $x = 3$
59. $h = \dfrac{2A}{b}$ **61.** $t = \dfrac{d}{r}$ **63.** $t = \dfrac{A - P}{Pr}$
65. $x = \dfrac{2y + 1}{3}$ **67.** $x = \dfrac{5 - by}{a}$ **69.** $x = 6a$
71. $n = \dfrac{l - a + d}{d}$ **73.** $r = \dfrac{V}{2d}$

Stretching the Topics

1. $x = \dfrac{3b}{2a}$ **2.** $a = d + l - nd$ **3.** $x = 16$

Checkup 2.1 (page 85)

1. $x = 27$ **2.** $x = -\frac{36}{11}$ **3.** $a = -24$ **4.** $x = 3$
5. $x = \frac{1}{3}$ **6.** $y = 4$ **7.** No solution **8.** $x = -\frac{16}{11}$
9. $x = \frac{15}{2}$ **10.** $r = \dfrac{A - P}{Pt}$

Exercise Set 2.2 (page 94)

1.

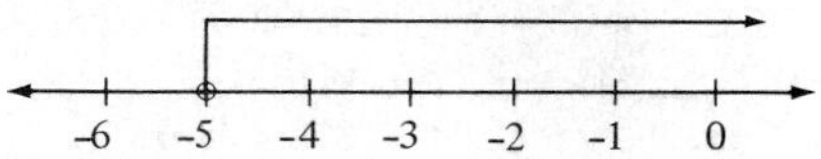

$(-5, \infty)$

3.

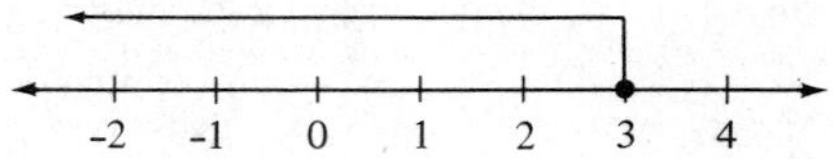

$(-\infty, 3]$

5.

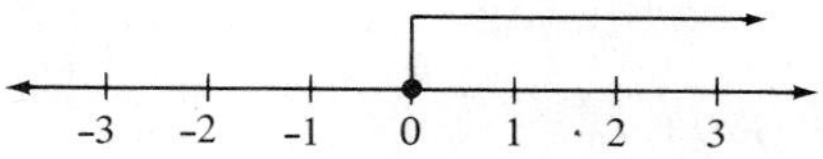

$[0, \infty)$

7.

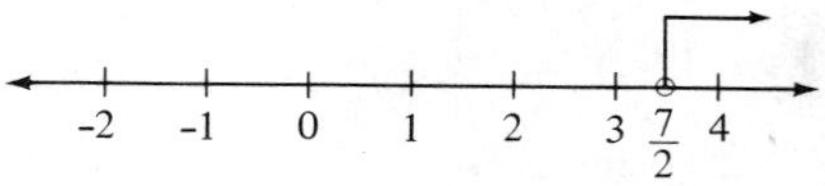

$(\frac{7}{2}, \infty)$

9.

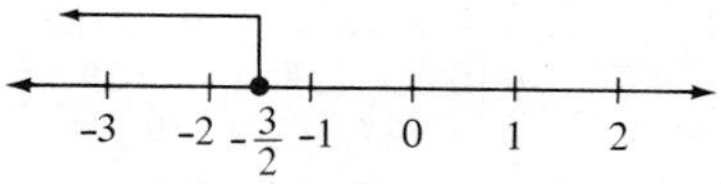

$(-\infty, -\frac{3}{2}]$

11. $x < -3$; $(-\infty, -3)$ **13.** $x \geq 3$; $[3, \infty)$ **15.** $\varnothing$
17. $2 < x \leq 5$; $(2, 5]$ **19.** $-1 \leq x \leq 4$; $[-1, 4]$
21. $\varnothing$ **23.** $x = -3$ **25.** $x \geq 2$; $[2, \infty)$
27. $x < 4$; $(-\infty, 4)$
29. $x \in R$, $x \neq 3$; $(-\infty, 3) \cup (3, \infty)$
31. $x \in R$; $(-\infty, \infty)$
33. $x < -\frac{1}{2}$ or $x \geq 0$; $(-\infty, -\frac{1}{2}) \cup [0, \infty)$

35. $x \in R$; $(-\infty, \infty)$ **37.** $x < -3$; $(-\infty, -3)$
39. $-4 < x < \frac{7}{2}$; $(-4, \frac{7}{2})$
41. $x < -1$ or $x > 0$; $(-\infty, -1) \cup (0, \infty)$
43. $x < 5$; $(-\infty, 5)$ **45.** $x \geq \frac{3}{5}$; $[\frac{3}{5}, \infty)$
47. $x \in R$; $(-\infty, \infty)$ **49.** $\varnothing$

Stretching the Topics

1. $x = 3.5$ **2.** $x \geq -\frac{3}{4}$; $[-\frac{3}{4}, \infty)$
3. $\pi \leq x \leq \frac{22}{7}$; $[\pi, \frac{22}{7}]$

Checkup 2.2 (page 96)

1.
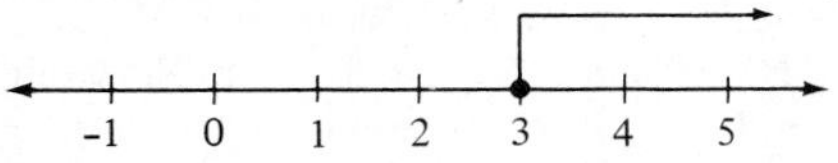

$[3, \infty)$

2.
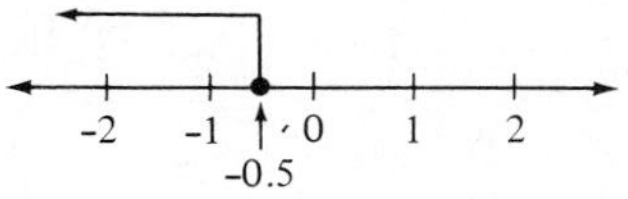

$(-\infty, -0.5]$

3.
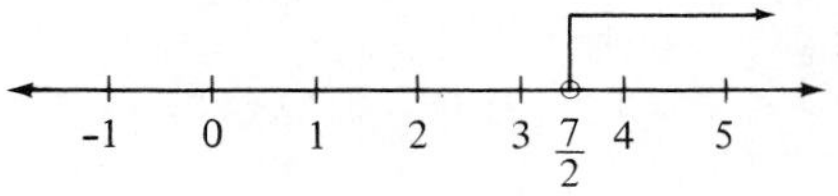

$(\frac{7}{2}, \infty)$

4. $0 < x \leq 2$; $(0, 2]$ **5.** $x < 3$; $(-\infty, 3)$ **6.** $\varnothing$
7. $x \in R$; $(-\infty, \infty)$ **8.** $0 \leq x < 5$; $[0, 5)$
9. $x < -3$ or $x > 2$; $(-\infty, -3) \cup (2, \infty)$
10. $x \in R$, $x \neq \frac{1}{2}$; $(-\infty, \frac{1}{2}) \cup (\frac{1}{2}, \infty)$

Exercise Set 2.3 (page 108)

1. $x \geq 2$; $[2, \infty)$
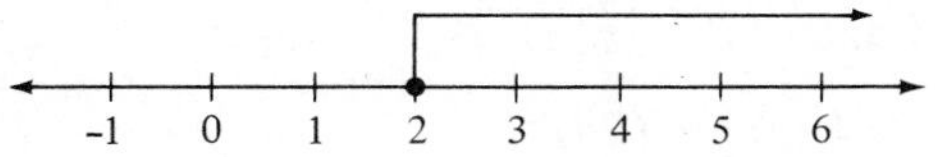

3. $y < 1$; $(-\infty, 1)$
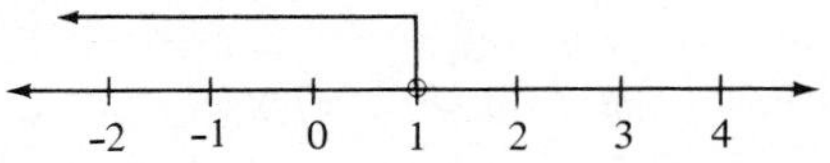

5. $a \leq -\frac{1}{2}$; $(-\infty, -\frac{1}{2}]$
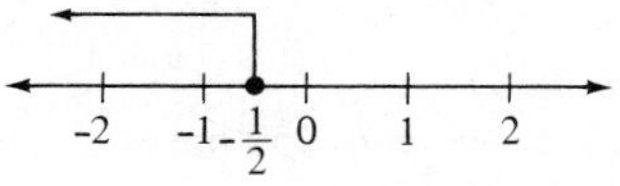

7. $x > -2$; $(-2, \infty)$
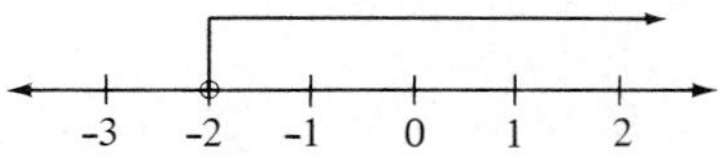

9. $a \geq 2$; $[2, \infty)$
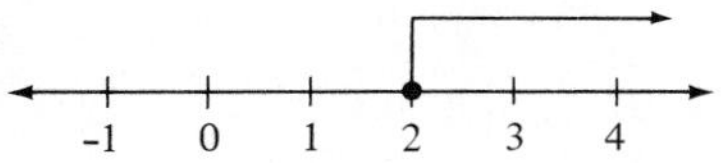

11. $x > \frac{3}{2}$; $(\frac{3}{2}, \infty)$
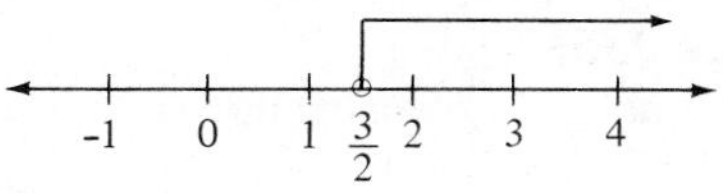

13. $m \leq 4$; $(-\infty, 4]$
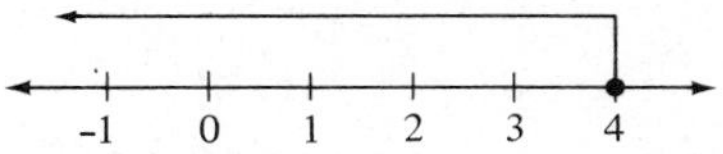

15. $x > 12$; $(12, \infty)$
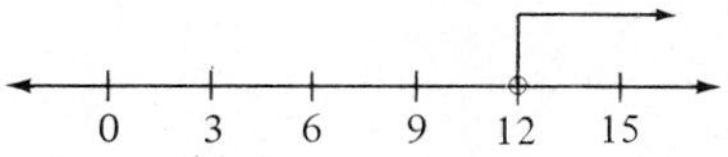

17. $x \leq \frac{7}{3}$; $(-\infty, \frac{7}{3}]$
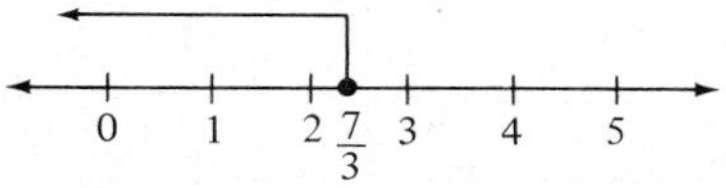

19. $y \geq -4$; $[-4, \infty)$
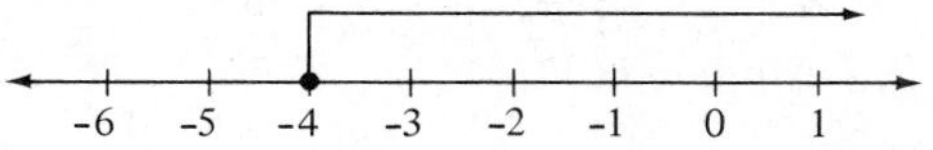

21. $\varnothing$ **23.** $\varnothing$

25. $0 \leq x < 2$; $[0, 2)$
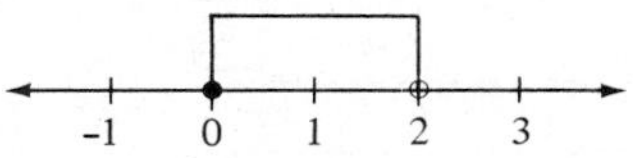

27. $x < 16$; $(-\infty, 16)$
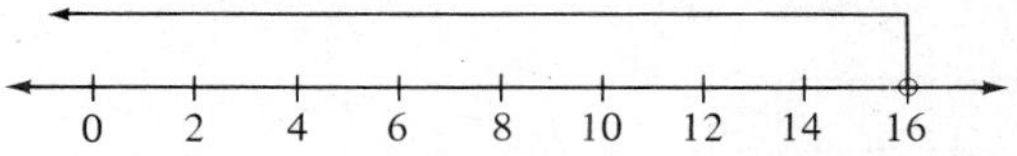

29. $x < 5$ or $x \geq 6$; $(-\infty, 5) \cup [6, \infty)$
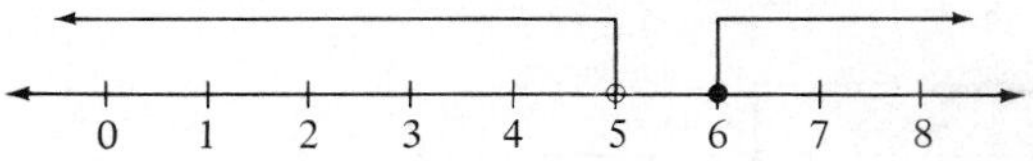

31. $x \in R$; $(-\infty, \infty)$

33. $7 \leq m \leq 12$; $[7, 12]$
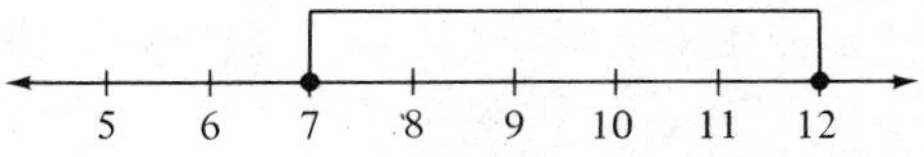

35. $-8 < a \leq -6$; $(-8, -6]$
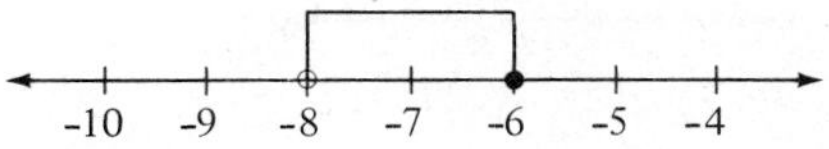

37. $-7 \leq y < 5$; $[-7, 5)$
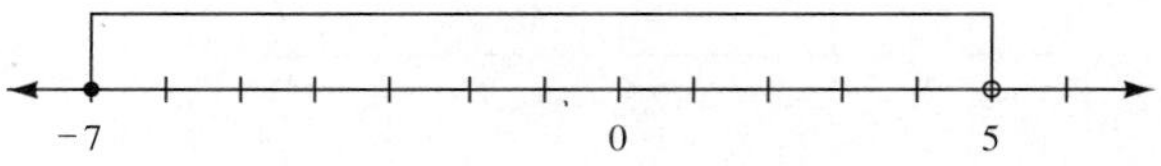

39. $0 < x \leq 2$; $(0, 2]$
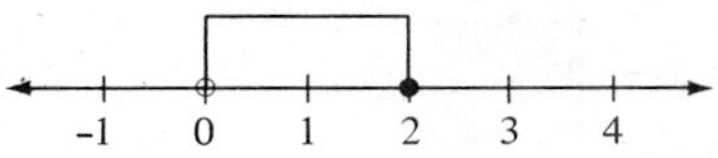

Stretching the Topics

1. $b \geq 13$; $[13, \infty)$

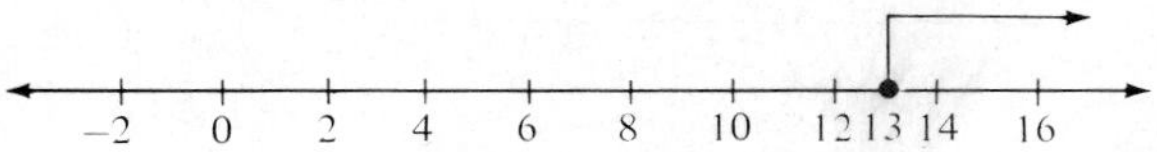

2. $x \geq 4$; $[4, \infty)$

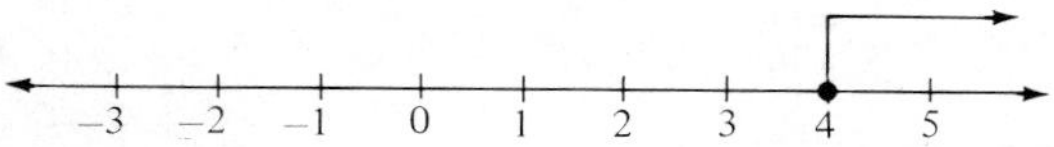

3. $-2 \leq x < 3$; $[-2, 3)$

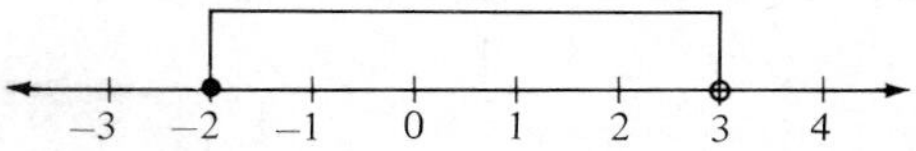

Checkup 2.3 (page 110)

1. $x \geq 3$; $[3, \infty)$

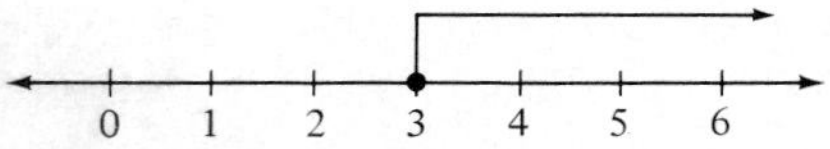

2. $a > 4$; $(4, \infty)$

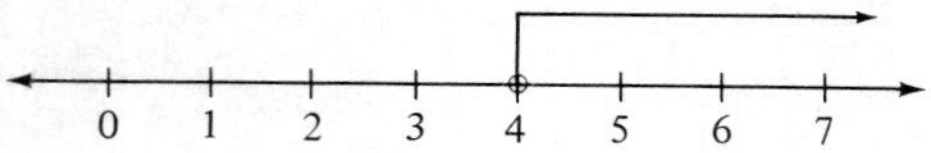

3. $y < -10$; $(-\infty, -10)$

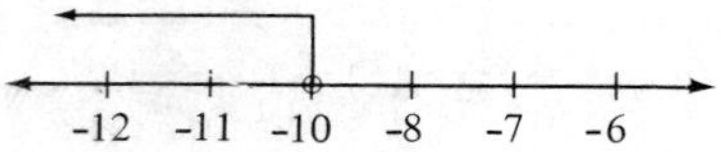

4. $x \leq -2$; $(-\infty, -2]$

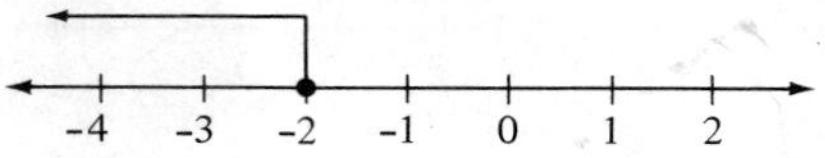

5. $\frac{1}{2} < x < 2$; $(\frac{1}{2}, 2)$

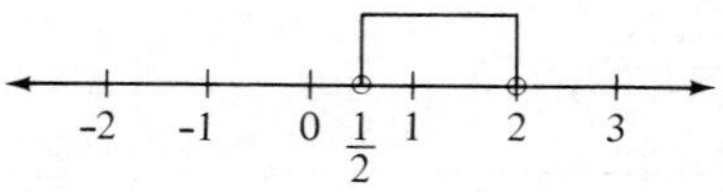

6. $x > 3$; $(3, \infty)$

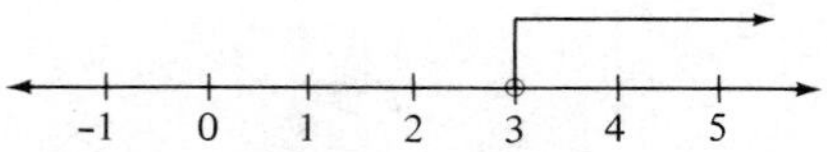

7. $x > -\frac{7}{2}$; $(-\frac{7}{2}, \infty)$

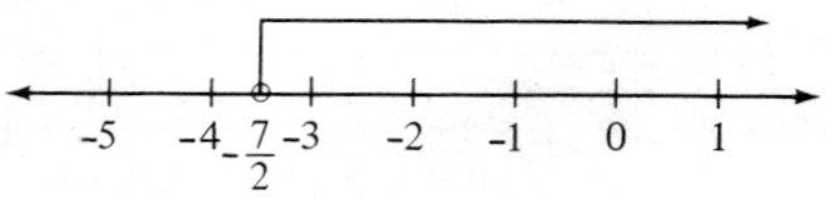

8. $x \leq 2$ or $x > 3$; $(-\infty, 2] \cup (3, \infty)$

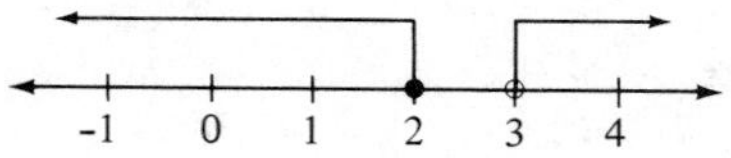

9. $-4 \leq a \leq -2$; $[-4, -2]$

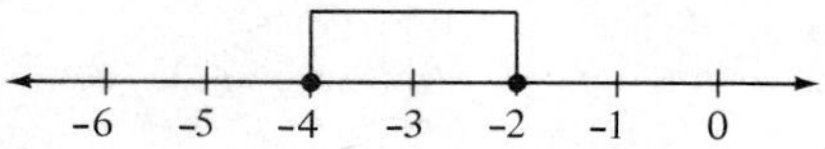

10. $-4 < y \leq 2$; $(-4, 2]$

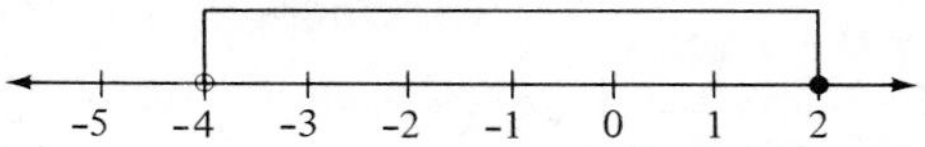

Exercise Set 2.4 (page 122)

1. $x = 9$ or $x = -9$ **3.** $a = 21$ or $a = -21$
5. $x = 0.7$ or $x = -0.7$ **7.** $y = 5$ **9.** No solution
11. $x = 1$ or $x = \frac{4}{3}$ **13.** No solution
15. $a = -24$ or $a = -36$ **17.** $x = \frac{8}{3}$ or $x = 6$
19. $x = -13$ or $x = 27$ **21.** $-\frac{5}{3} < y < \frac{5}{3}$; $(-\frac{5}{3}, \frac{5}{3})$
23. $-35 < a < 35$; $(-35, 35)$
25. $-2 < x < 2$; $(-2, 2)$ **27.** $-\frac{2}{3} \leq x \leq 2$; $[-\frac{2}{3}, 2]$
29. $12 \leq x \leq 18$; $[12, 18]$ **31.** No solution
33. $-9 < x < \frac{29}{3}$; $(-9, \frac{29}{3})$
35. $a > 2.5$ or $a < -2.5$; $(-\infty, -2.5) \cup (2.5, \infty)$
37. All real numbers; $(-\infty, \infty)$
39. $x \leq -2$ or $x \geq 2$; $(-\infty, -2] \cup [2, \infty)$
41. $x > 10$ or $x < -10$; $(-\infty, -10) \cup (10, \infty)$
43. $x \geq 1$ or $x \leq -\frac{3}{5}$; $(-\infty, -\frac{3}{5}) \cup [1, \infty)$
45. $m < -15$ or $m > 21$; $(-\infty, -15) \cup (21, \infty)$
47. All real numbers; $(-\infty, \infty)$
49. $x \geq \frac{13}{7}$ or $x \leq -1$; $(-\infty, -1] \cup [\frac{13}{7}, \infty)$

Stretching the Topics

1. $m = 1$ or $m = \frac{3}{2}$ **2.** $\frac{1}{2} < x < \frac{3}{2}$; $(\frac{1}{2}, \frac{3}{2})$
3. $1 < x < 8$; $(1, 8)$

Checkup 2.4 (page 124)

1. $x = 5.2$ or $x = -5.2$ **2.** $x = -1$ or $x = 17$
3. No solution **4.** $-3 < x < 3$; $(-3, 3)$
5. $4 \leq x \leq 24$; $[4, 24]$ **6.** $-\frac{5}{3} \leq a \leq 3$; $[-\frac{5}{3}, 3]$
7. $x > 2$ or $x < -2$; $(-\infty, -2) \cup (2, \infty)$
8. $m \leq 6$ or $m \geq 12$; $(-\infty, 6] \cup [12, \infty)$
9. $x > 6$ or $x < -5$; $(-\infty, -5) \cup (6, \infty)$
10. $x > 14$ or $x < -2$; $(-\infty, -2) \cup (14, \infty)$

Exercise Set 2.5 (page 131)

1. 6 **3.** 400 boxes of chocolate chip, 250 boxes of sugar
5. 12 lb **7.** 9 quarters, 14 dimes
9. \$7000 at 7%, \$5000 at 9% **11.** \$10,000
13. 12 lb of ground beef, 8 lb of sausage
15. 50 and 55 mph **17.** 247.5 mi
19. 12 books for \$2.50, 18 books for \$4 **21.** 12
23. 51 \$10 bills, 43 \$5 bills, 86 \$1 bills **25.** At least 19
27. No more than 14 m **29.** No more than 60 by 120 ft
31. No more than 100 **33.** Less than 40
35. At least 4000

Stretching the Topics

1. \$12,500 **2.** 7 by 16 ft **3.** No more than 11 cm

Checkup 2.5 (page 135)

1. 3 by 5 ft **2.** $3\frac{11}{24}$ hr **3.** No more than 9
4. At least 35

Speaking the Language of Algebra (page 138)
1. 1 **2.** Isolate the variable
3. Remove the parentheses and combine like terms
4. Open; left **5.** $a < x < b$ **6.** Intersection; union
7. Preserved; reversed
8. $x = a$ or $x = -a$; $-a \le x \le a$; $x \ge a$ or $x \le -a$

Review Exercises (page 139)
1. $x = 7$ **2.** $x = -6$ **3.** $x = -42$
4. $x = -45$ **5.** $x = -3$ **6.** $x = -\frac{1}{2}$
7. $x = \frac{8}{15}$ **8.** $x = 5$ **9.** $x = -\frac{19}{11}$
10. $y = \frac{5}{7}$ **11.** $x = 0$ **12.** $C = \dfrac{5F - 160}{9}$

13.
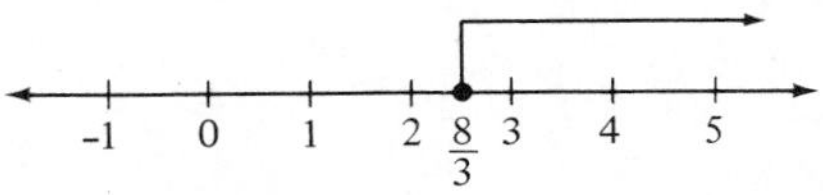

14.
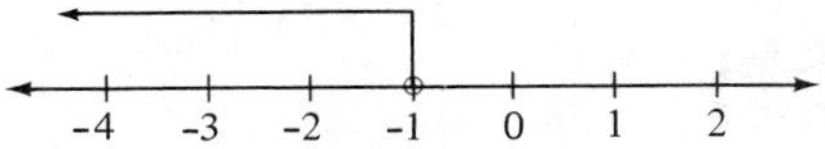

15. $-5 \le x < 2$; $[-5, 2)$ **16.** $\varnothing$
17. $x > -1$; $(-1, \infty)$ **18.** $x \in R$; $(-\infty, \infty)$
19. $x \le -10$; $(-\infty, -10]$

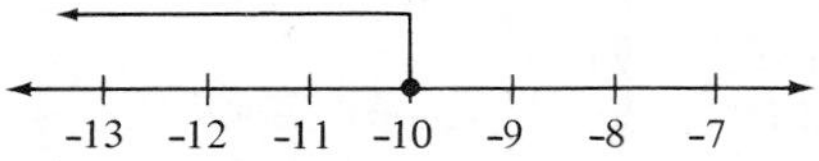

20. $x < 1$; $(-\infty, 1)$

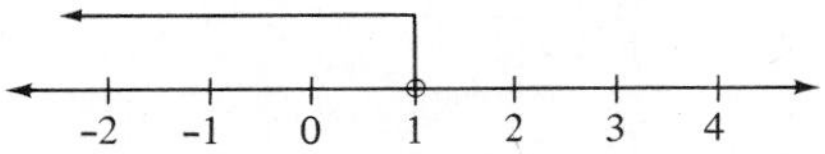

21. $x > -5$; $(-5, \infty)$

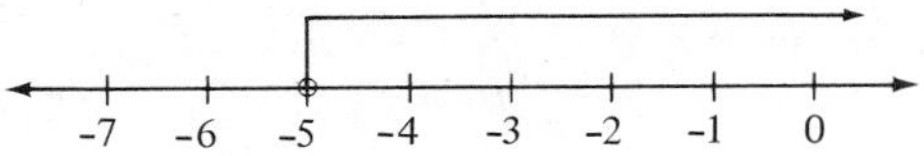

22. $y \ge \frac{1}{2}$; $[\frac{1}{2}, \infty)$

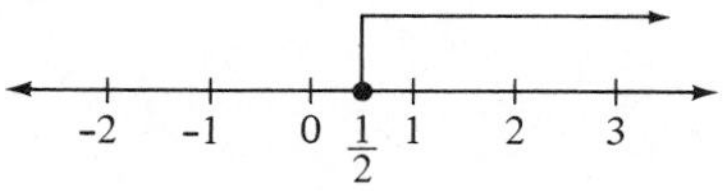

23. $x \le -2$; $(-\infty, -2]$

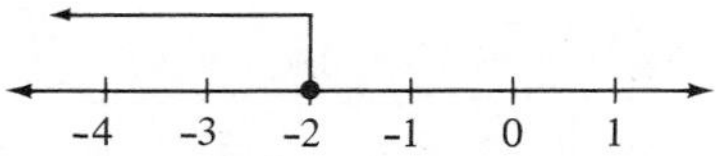

24. $3 < x < 4$; $(3, 4)$

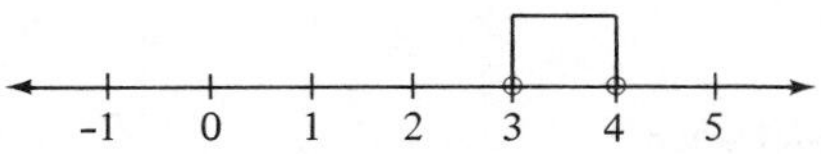

25. $x \ge -5$ or $x < -7$; $(-\infty, -7) \cup [-5, \infty)$

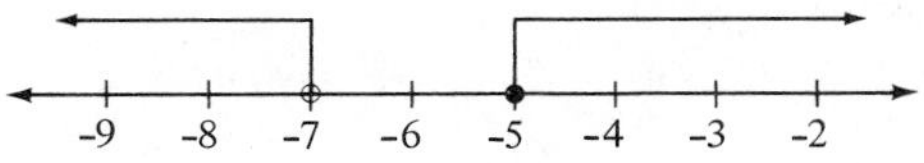

26. $x < 3$; $(-\infty, 3)$

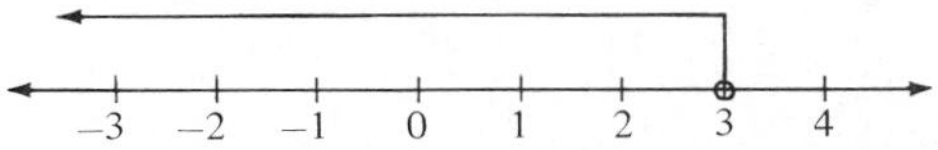

27. $7 \le x \le 13$; $[7, 13]$

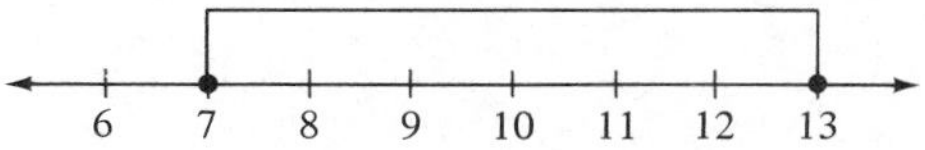

28. $4 < x \le 10$; $(4, 10]$

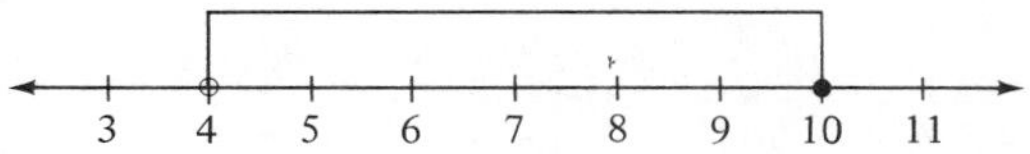

29. $x = 5$ or $x = -2$ **30.** $x = 5$ or $x = 13$
31. No solution **32.** $-\frac{11}{3} \le x \le 5$; $[-\frac{11}{3}, 5]$
33. No solution
34. $x > 7$ or $x < -4$; $(-\infty, -4) \cup (7, \infty)$
35. All real numbers
36. $x \le 2$ or $x \ge 10$; $(-\infty, 2] \cup [10, \infty)$ **37.** At least 5
38. \$12,000 at 14%, \$17,000 at 12%
39. 45 and 55 mph **40.** 15 by 45 ft

Practice Test (page 143)
1. $x = 2$ **2.** $y = 1$ **3.** $x = 4$ **4.** $a = -\frac{2}{5}$
5. $x = \frac{1}{9}$ **6.** $b = \dfrac{2A - ha}{h}$ **7.** $x \ge 2$; $[2, \infty)$
8. $y > 2$; $(2, \infty)$ **9.** $-\frac{1}{8} \le x < 3$; $[-\frac{1}{8}, 3)$
10. $x = \frac{9}{4}$ or $x = \frac{5}{4}$ **11.** $x = -5$ or $x = 10$
12. $-5 < x < 6$; $(-5, 6)$
13. $x \ge 1$ or $x \le -\frac{7}{3}$; $(-\infty, -\frac{7}{3}] \cup [1, \infty)$
14. $-6 < x \le -2$; $(-6, -2]$ **15.** $x < 7$; $(-\infty, 7)$
16. $-3 < x < 4$; $(-3, 4)$
17. $x < -1$ or $x \ge 3$; $(-\infty, -1) \cup [3, \infty)$
18. 8 mph **19.** 435 **20.** $1\frac{1}{2}$ in. or more

CHAPTER 3

Exercise Set 3.1 (page 156)
1. 186 **3.** 2 **5.** 1 **7.** 2 **9.** $-20x^8$
11. $-12x^7y^5$ **13.** $6x^7y^4$ **15.** $42x^4y^4$
17. $35a^3b^4c^5$ **19.** $14x^2y^8z^7$ **21.** $64y^4$
23. $-32x^{10}y^{15}$ **25.** $20x^7$ **27.** $9x$ **29.** $-50x^8y^5$
31. $288a^7b^{11}$ **33.** $12a^{24}b^8$ **35.** $-448x^{17}y^7z^8$
37. $11x^2y^6$ **39.** $-a^6b^{12} + 3a^{12}b^6$ **41.** $\dfrac{1}{9y^2}$
43. $\dfrac{125a^6}{27b^3}$ **45.** $\dfrac{a^4b^{20}}{16c^8}$ **47.** $\dfrac{16y^{12}}{81z^{20}}$ **49.** 1 **51.** x^4
53. $-3a^4$ **55.** x^6y^2 **57.** $-x^3y^3z$ **59.** $\dfrac{a^4}{7}$
61. $\dfrac{3a^6b^2c}{2}$ **63.** $8a^3b^3c^3$ **65.** $49a^5b^5$

Stretching the Topics
1. $-64a^5b^{12}c^5$; 262, 144 **2.** $x^{3n}y^{2n+11}$ **3.** $a = -7$

Checkup 3.1 (page 158)
1. -55 **2.** 13 **3.** $42x^8y^3$ **4.** $-45a^5b^6c^3$

5. $121y^6$ **6.** $-75x^{18}$ **7.** $-\frac{27a^6}{b^9}$ **8.** $\frac{16a^6}{9c^4}$

9. $\frac{27a^5}{8}$ **10.** $-3a^4b^4c$

Exercise Set 3.2 (page 168)

1. $\frac{1}{16}$ **3.** $-\frac{1}{243}$ **5.** $\frac{1}{81}$ **7.** $-\frac{1}{25}$ **9.** $\frac{7}{36}$

11. $\frac{16}{9}$ **13.** $\frac{1}{x^6}$ **15.** $\frac{a^2}{b^6}$ **17.** $\frac{1}{xz^6}$ **19.** $\frac{ac^3}{b^4}$

21. $\frac{1}{x^6}$ **23.** $\frac{1}{32x^5}$ **25.** $\frac{9}{a^2}$ **27.** $x^{15}y^3$

29. $-\frac{x^3}{27y^6}$ **31.** $\frac{49y^4}{x^{10}}$ **33.** a^9b^7 **35.** xy^9z^3

37. $\frac{a^{12}}{125}$ **39.** $\frac{y^3}{x^6z^3}$ **41.** $\frac{y^6}{x^5}$ **43.** $\frac{c^6}{a^9b^{12}}$

45. $\frac{1}{x^{10}y^{12}}$ **47.** $\frac{1}{9x^{16}y^{10}}$ **49.** $-\frac{64c^9}{a^{18}b^{12}}$ **51.** $\frac{4b^{20}}{27a^{16}}$

53. $\frac{2}{b^7}$ **55.** y^{10}

Stretching the Topics

1. $\frac{1}{x^{3b-2a}}$ **2.** $-\frac{32b^{13}e^8}{a^5c^9d^{12}f^2}$ **3.** $x^{4a+6}y^{2b-5}$

Checkup 3.2 (page 170)

1. $\frac{1}{64}$ **2.** $-\frac{1}{64}$ **3.** xz^5 **4.** $\frac{1}{x^9}$ **5.** $\frac{1}{36x^2}$

6. $\frac{x^6}{y^{14}z^6}$ **7.** $\frac{y^{14}}{x^5}$ **8.** $\frac{y^4}{x^5z^3}$ **9.** $\frac{y^{10}}{9x^{18}}$ **10.** $\frac{1}{3x^4y^5}$

Exercise Set 3.3 (page 177)

1. Fifth-degree binomial in x
3. Third-degree binomial in x **5.** Constant monomial
7. Second-degree trinomial in y
9. Fourth-degree, four-term polynomial in x
11. $x^3 - 4x^2 + 3x + 7$ **13.** $x^4 + 8x^2 + 12$
15. $-a^3 + a^2 - a + 1$ **17.** $3x + 5$
19. $x^3 + 4x^2 - 3x + 2$ **21.** $-x^2 + 4x - 6$
23. $3a^2 - 8ab + 16b^2$ **25.** $-a^2 + 10a - 7$
27. $-a^2 - 7a + 6$ **29.** $-3b$ **31.** $-x^2 - y^2$
33. $-3a^2 - 18a - 35$ **35.** $-y^2 + 3$
37. $9x^2 - 2x + 7$ **39.** $-3x^2 - 6x - 4$
41. $7a + 23b + 3$ **43.** $3x^2 + 3x - 4$
45. $2m^3 - 10m^2 + 12m$
47. $-15x^3y^3 + 5x^2y^2 + 20xy$ **49.** $14a^4b^2 - 2a^2b^4$
51. $32m^4n - 8m^3n^2 + 4m^2n^3 - 12mn^4$
53. $7x^2y - 10xy^2$ **55.** $3x^3 - 6x^2 + 9x$
57. $-14a^3 - 6a^2 + 35a$ **59.** $-5x^3y^3 - 13x^3y^7$

Stretching the Topics

1. $9x + 1$ **2.** $-9a^3b - a^2b^2 + 28ab^3; -20$
3. $2y^2 + 2xy - \frac{2x^2}{y^2} - 4x^2$

Checkup 3.3 (page 179)

1. Fourth-degree binomial in x
2. Second-degree trinomial in y
3. $-x^4 - 2x^3 + x^2 + 2$ **4.** $-7a^2 + 8a - 9$
5. $5a^2 + 2ab - 15b^2$ **6.** $6x^2 + 17x - 7$
7. $15a - 65b$ **8.** $5x^3 - 15x^2 + 25x$
9. $13m^2n + 45mn^2$ **10.** $15a^3 - a^2 - 24a$

Exercise Set 3.4 (page 190)

1. $12y^2 - 23y + 5$ **3.** $12x^2 + 8x - 15$
5. $18 - 13y + 2y^2$ **7.** $28x^2 + 29xy + 6y^2$
9. $a^2 + 6ab + 9b^2$ **11.** $4x^2 - 9y^2$
13. $x^2y^2 - 5xy - 50$ **15.** $9x^2y^2 - 12xyz + 4z^2$
17. $0.36a^2 - 25b^2$ **19.** $3x^4 - x^2y^2 - 24y^4$
21. $a^4 + a^2b^2 + \frac{1}{4}b^4$ **23.** $x^3 - 5x^2 + 7x - 3$
25. $-56y^2 - y^3 + y^4$ **27.** $3x^2 + x - 15$
29. $4a^2 + 24a + 13$ **31.** $-4x^2 + 32x - 44$
33. $79x^2 - 70x + 18$ **35.** $a^3 - 11a^2 + 6a$
37. $a^3 - 4a^2b - 7ab^2 + 10b^3$
39. $x^3 - 4x^2 - 3x + 18$ **41.** $a^3 + 6a^2 + 12a + 8$
43. $x^5 - 2x^4 - 3x^2 - x + 5$
45. $3x^4 + 5x^3 - 10x^2 - 9x + 5$
47. $a^4 - 10a^2b^2 + 25b^4$ **49.** $x^4 - 4y^4$
51. $x^2 - 2xy + y^2 - 10x + 10y + 25$
53. $a^2 + 8a + 16 - 9b^2$
55. $16 - 32x + 24x^2 - 8x^3 + x^4$

Stretching the Topics

1. $4a^2 - 4ab + b^2 - 12ac + 6bc + 9c^2$
2. $8x^{6a} - 18x^{3a}y^b + 7y^{2b}$ **3.** $x^{-3} - 8y^{-3}$

Checkup 3.4 (page 192)

1. $2x^2 + 11x - 21$ **2.** $a^2 - 20ab + 100b^2$
3. $9x^2 - 25y^2$ **4.** $x^3 - 8x^2 + 9x + 18$
5. $13x^2 - 2xy - 4y^2$ **6.** $2a^2 - 19a + 9$
7. $x^3 - 2x^2y - xy^2 + 2y^3$
8. $a^3 - 15a^2 + 75a - 125$
9. $x^4 - 4x^3 + 6x^2 - 5x + 2$
10. $a^2 + 2ab + b^2 - 18a - 18b + 81$

Exercise Set 3.5 (page 201)

1. $x^3 - 7x^2 + 3x + 8$ **3.** $3a^2 - 2ab + 5b^2$

5. $x - 2 + \frac{5}{3x}$ **7.** $-1 + 4z - 5xy$

9. $x^3 - 3x^2 + 2x - 1 + \frac{3}{x}$ **11.** $x - 8$

13. $4x + 3$ **15.** $x - 2 - \frac{2}{2x + 5}$

17. $x^2 - 4x + 9 - \frac{2}{x + 1}$ **19.** $a^2 + 3a - 4$

21. $x^2 - xy + 3y^2$ **23.** $x^2 - 4x + 2$

25. $x^2 + 2x + 4$ **27.** $x^3 + 2x^2 + 3x + 6 + \frac{25}{2x - 4}$

29. $2x^2 - x + 4 + \frac{x + 3}{4x^2 - x - 3}$

31. $x^3 - 3x^2 + 2x - 2 + \frac{1}{x + 1}$ **33.** $3x^3 - x^2 + x$

35. $x^3 - 2x^2 + 4x - 8 + \frac{32}{x + 2}$

37. $x^4 - 2x^3 + x - 2 - \frac{2}{x + 5}$

39. $x^5 - x^4 + x^3 - x^2 + x + 1 - \frac{1}{x - 1}$

Stretching the Topics

1. $4x^2 - 5xy + 3xy^2 - y^3 + \frac{y^4}{2x + y}$
2. $x^{2a} + 2x^a - 3$ **3.** No; 205

Checkup 3.5 (page 203)

1. $2x^2 - xy + 3y^2$ **2.** $-\frac{x^2}{2} + x - 2 + \frac{4}{x}$
3. $2x + 3$ **4.** $x^2 - 4x - 5 - \frac{2}{x + 3}$
5. $x^2 + 2x + 1 + \frac{1}{2x^2 - 5x - 3}$
6. $a^4 + 2a^3 + 4a^2 + 8a + 16$
7. $x^2 - 2x - 4 + \frac{2}{x + 3}$ **8.** $2x^3 - x^2 + x - 2$

Exercise Set 3.6 (page 211)

1. $2x^2 + 12x$ **3.** $6x^2 - 35x$ or $6y^2 + 49y + 49$
5. $18x^2 + 18x + 9$ **7.** $7x^3 + 21x^2 + 14x$
9. $2l^2 - 8l$ sq ft **11.** $4l - 16$ or $4w + 16$ in.
13. $20s + 100$ sq cm **15.** $5x^2 + 20x + 25$ sq ft
17. $\frac{x^2 - 7x}{2}$ or $\frac{a^2 + 7a}{2}$ sq in. **19.** $2n^2 - 12n + 9$
21. $w^2 + 50w$ or $l^2 - 50l$ sq ft
23. $\frac{x^2 + 9x}{2}$ or $\frac{y^2 - 9y}{2}$ sq ft **25.** $x^2 + 9x + 20$ sq m
27. $10w + 65$ sq ft **29.** $w^3 + 4w^2$ cu ft

Stretching the Topics

1. $300x - 2x^2$ sq ft **2.** $18w^2 - 288w + 864$ cu cm
3. $\frac{35x - x^2}{2}$ sq in.

Checkup 3.6 (page 214)

1. $4w^2 - 3w$ sq ft **2.** $\frac{2x^2 + 5x}{2}$ sq ft
3. $2x^2 + 12x + 36$ or $2y^2 - 12y + 36$ sq in.
4. $2x^2 + 2x$ **5.** $30s + 225$ sq cm

Speaking the Language of Algebra (page 216)

1. Base; exponent; a; n **2.** Add **3.** Multiply
4. Factor **5.** Numerator; denominator
6. Subtract; denominator; numerator **7.** First; binomial
8. Second; trinomial **9.** Difference of the squares
10. Square of the hypotenuse

Review Exercises (page 219)

1. $-50x^6y^5$ **2.** $49y^{10}$ **3.** $-288x^{17}$ **4.** $-\frac{27x^{12}}{y^6}$
5. $\frac{8a^3}{27c^3}$ **6.** $\frac{14a^3}{3}$ **7.** $\frac{1}{16}$ **8.** $-\frac{1}{8}$ **9.** $\frac{z^3}{x^2}$
10. $\frac{5}{x^3}$ **11.** $\frac{x^6z^4}{y^{10}}$ **12.** $\frac{y^{12}}{x^3}$ **13.** $\frac{y^{12}}{8x^{21}}$ **14.** $\frac{9y^8}{x^4}$
15. Monomial in x and y **16.** Third-degree trinomial in x
17. $2y^4 + y^3 - y^2 + 6$ **18.** $-2x^2 + 3x - 5$
19. $5x^2 + 23$ **20.** $13a - b$ **21.** $3x^3 - 15x^2 + 18x$
22. $17m^2n + mn^2$ **23.** $6a^3 - 2a^2 + 10a$
24. $-2x^2y^3 + 7x^3y^5$ **25.** $3x^2 + 14x - 5$
26. $a^2 - 14ab + 49b^2$ **27.** $16x^2 - 9y^2$
28. $x^3 - 6x^2 + 5x + 6$ **29.** $7x^2 + 2xy - 9y^2$
30. $a^2 - 25a + 27$ **31.** $2a^3 - 3a^2b - 3ab^2 + 2b^3$
32. $a^3 - 6a^2 + 12a - 8$
33. $x^4 + x^3 - 6x^2 + x + 3$
34. $a^2 + 4ab + 4b^2 + 6a + 12b + 9$
35. $a^2 - 3ab + 2b^2$ **36.** $3x - 1 - \frac{3}{x + 6}$
37. $x^2 - 4x + 16$ **38.** $x^2 + 3xy - 10y^2$
39. $x^2 - 3x + 9$ **40.** $2x^2 + 3x + 7 + \frac{12}{x - 2}$
41. $3x^3 - 4x^2 + 2x - 2 + \frac{4}{x + 1}$ **42.** $\frac{5x^2}{2}$
43. $5x^2 + 5x$ **44.** $x^2 + 2x - 15$ sq cm
45. $-6x^2 + 2$

Practice Test (page 223)

1. -96 **2.** $-\frac{1}{32}$ **3.** $8a^{12}b^3$ **4.** $-2x^7y^4$
5. $-125x^9y^3$ **6.** $-a^{12}b^6c^9$ **7.** $\frac{81x^6}{y^4z^6}$ **8.** $-2ab^3$
9. $\frac{x^9z^3}{8}$ **10.** $\frac{8x^2}{y}$ **11.** $\frac{a^3c^6}{b^9}$ **12.** $-x^9y^3z^3$
13. $-\frac{2y^4}{x^6}$ **14.** $-6x^2 + 69x - 6$
15. $12a^2 + 41a - 36$ **16.** $12x^2 - x - 35$
17. $9c^2 - 42cd + 49d^2$ **18.** $81y^2 - 4z^2$
19. $3x^3 - 17x^2 + 9x + 5$
20. $a^3 - 15a^2 + 75a - 125$ **21.** $2x^2 - 5x + 14$
22. $-4x^2y^2 - xy + 2$ **23.** $7x + 24 + \frac{70}{x - 3}$
24. $y^2 - 3y + 9$
25. $2x^2 + 16x + 64$ or $2y^2 - 16y + 64$ sq in.
26. $6x + 3$ **27.** $3x^2 + 2x$ sq ft

Sharpening Your Skills after Chapters 1–3 (page 225)

1. $-\frac{11}{2} < 4.5$ **2.** $8.2 > 0$ **3.** -9 **4.** 7
5. Commutative property for addition **6.** 0 **7.** 18
8. $-x + 9y$ **9.** $-2x^2 - 2$ **10.** $x = \frac{1}{2}$
11. $y = 8$ **12.** $x < -7$; $(-\infty, -7)$
13. $-1 < y \le 5$; $(-1, 5]$ **14.** $x = -4$ or $x = 22$
15. $-\frac{4}{5} \le a \le 2$; $[-\frac{4}{5}, 2]$ **16.** $\varnothing$
17. $-3 \le x < -2$; $[-3, -2)$
18. $x \le 3$ or $x > 6$; $(-\infty, 3] \cup (6, \infty)$
19. $500 - 15n$ dollars **20.** $\frac{1}{2}$ hr

CHAPTER 4

Exercise Set 4.1 (page 233)

1. $10(5x^2 - xy + 2y^2)$ **3.** $2b(b^2 - 3b - 2)$
5. $-y(y^2 - y + 2)$ **7.** $3x(x^2 - 2xy + 4y^2)$
9. $-a^2b(a^2 - 8ab + 5b^2)$ **11.** $5x^2y^2(x - 16y)$
13. $3xy(3x^2 - 4xy + 8y^2)$ **15.** $8a^2(9a^2 - 5a + 1)$
17. $4x^2y^2(3x^2 + 2xy + y^2)$
19. $9xyz(x^3 + 2x^2y - 3xy^2 - 5y^3)$ **21.** $(y - 1)(y - 3)$
23. $(x + 3)(2x - 1)$ **25.** $(3x - y)(2a - b)$
27. $(x - y)(x - y - 3)$ **29.** $(x - y)(a + b - 2c)$
31. $2x(2a + b)$
33. $(x - 2)[5(x - 2)^2 - 4(x - 2) + 1]$

35. $2(x + 2y)[2(x + 2y)^2 - 3(x + 2y) - 1]$
37. $(6 + c)(a + b)$ **39.** $(x - 7)(x^2 + 6)$
41. $(2x + 3y)(7a - 5)$ **43.** $(7a - 4b)(10a + 3)$
45. $3(2x + y)(a^2 + b^2)$ **47.** $(x + 1)(x^2 + 1)$
49. $(b - c)(9a - 7)$ **51.** $(2a - 3b)(x^2 + 1)$

Stretching the Topics

1. $x^{4a}(x^a + 4)$ **2.** $x^2(x^{3a} + 3x^{2a} - 5)$
3. $x^{-3}(1 + 5x + x^2)$

Checkup 4.1 (page 235)

1. $7(5x^2 - 2x + 1)$ **2.** $7b(b^2 - 2b + 4)$
3. $5y(x^2 - 5y^2)$ **4.** $-a^4b^2(a^2 - 3ab^2 - 6b^4)$
5. $9x^3yz(x^2y^2 - 3xy - 2)$ **6.** $(x - 9)(2x - 3)$
7. $(a + b)(a + b - 12)$ **8.** $(x - y)^2(6 - x + y)$
9. $(a + b)(2x - 3y)$ **10.** $(x - 7)(x^2 + 1)$

Exercise Set 4.2 (page 243)

1. $(3y + 1)(3y - 1)$ **3.** $\frac{1}{4}(a + \frac{5}{3})(a - \frac{5}{3})$
5. $(mn + 15)(mn - 15)$ **7.** $(5xy + 4z)(5xy - 4z)$
9. $3(x + 0.5)(x - 0.5)$ **11.** $6m(m + 5)(m - 5)$
13. $2y^2(x + 6y)(x - 6y)$ **15.** $(5x^2 + 11)(5x^2 - 11)$
17. $(m^2 + 4)(m + 2)(m - 2)$
19. $3(m^2 + 9)(m + 3)(m - 3)$
21. $(x + y + 7)(x + y - 7)$
23. $(3a + 2b + 2c)(3a + 2b - 2c)$
25. $2xy(x + 2y)(x - 2y)$
27. $(16 + x + 2y)(16 - x - 2y)$
29. $(x + y + 1)(x - y - 5)$
31. $(x - 9)(x + 3)(x - 3)$ **33.** $(x - 2)(x + 5)(x - 5)$
35. $(x - y)(x^2 + xy + y^2)$ **37.** $(a + 2)(a^2 - 2a + 4)$
39. $(4 - m)(16 + 4m + m^2)$
41. $(xy + 1)(x^2y^2 - xy + 1)$
43. $(5x - 3y)(25x^2 + 15xy + 9y^2)$
45. $2x(x + 2)(x^2 - 2x + 4)$
47. $(ab - 4c)(a^2b^2 + 4abc + 16c^2)$
49. $4x^2y^2(x + 2y)(x^2 - 2xy + 4y^2)$
51. $7xy^2(x - 2y)(x^2 + 2xy + 4y^2)$
53. $(x - y + 5)(x^2 - 2xy + y^2 - 5x + 5y + 25)$
55. $(5 - x)(7 - 4x + x^2)$
57. $3(x + y)(x - y)(x^2 + xy + y^2)(x^2 - xy + y^2)$
59. $(x - 1)(x + 1)(x^2 - x + 1)$

Stretching the Topics

1. $4(x^a + 3y^a)(x^a - 3y^a)$
2. $(x + 2)(x - 2)(x^2 + 4)(x^4 + 16)$
3. $x^a(x^a - 1)(x^{2a} + x^a + 1)$

Checkup 4.2 (page 245)

1. $(5a + 2)(5a - 2)$ **2.** $(10xy + 11z)(10xy - 11z)$
3. $3(x + 12)(x - 12)$ **4.** $5m(2m + 3n)(2m - 3n)$
5. $(x^2 + 16)(x + 4)(x - 4)$
6. $(x + 4y)(x^2 - 4xy + 16y^2)$
7. $3(a - 2b)(a^2 + 2ab + 4b^2)$
8. $(x - 2 + 3y)(x - 2 - 3y)$
9. $(x - y + 6)(x^2 - 2xy + y^2 - 6x + 6y + 36)$
10. $(a + b)(a - b)(a^2 + ab + b^2)(a^2 - ab + b^2)$

Exercise Set 4.3 (page 257)

1. $(x - 12)^2$ **3.** $(y - 6)(y + 5)$
5. $(b + 11)(b - 3)$ **7.** $(3a - 4b)^2$ **9.** $(x - 5y)^2$
11. $(13 - m)(3 - m)$ **13.** $(xy + 5)(xy - 3)$
15. $3(x - 3)(x + 2)$ **17.** $5a(a - 4)^2$
19. $2y(5 - y)(2 + y)$ **21.** $-10a(b^2 - 3)^2$
23. $xy(x - 10y)(x - 8y)$ **25.** $abc(a - 9b)(a + 7b)$
27. $(a - 5)^2$ **29.** $(x + y - 4)(x + y - 3)$
31. $(x + 13)(x + 1)$ **33.** $(2y + 1)(y + 3)$
35. $(5x - 2)(2x - 3)$ **37.** $(3a + 8b)(2a - 7b)$
39. $5(3c - d)(2c - 9d)$ **41.** $-x^2(3x - 7)(3x - 4)$
43. $(4m + 11n)(3m - 5n)$ **45.** $2xy(4x - 5)(3x + 7)$
47. $(5a - 5b + 2)(a - b + 1)$ **49.** $(5x + 5y - 2)^2$
51. $(4 - 13x + 26y)(2 - 3x + 6y)$
53. $(x^2 + 8)(x + 3)(x - 3)$
55. $3(x^2 - 5)(x + 2)(x - 2)$ **57.** $3x^2(x^2 - 5y^2)^2$
59. $2(x^5 + 11)(x^5 - 3)$ **61.** $A = \frac{1}{2}h(a + b)$; 28 sq in.
63. $s = 2(5 + 8t)(3 - t)$; 36.16 ft
65. $C = (2x - 3)(x + 2)$; \$5044

Stretching the Topics

1. $(2x^a + 11)^2$ **2.** $[3(x + y)^m + 5][2(x + y)^m + 3]$
3. $(x^{a+3} + 6y^{a-2})^2$

Checkup 4.3 (page 259)

1. $(x - 6y)(x - 3y)$ **2.** $2(y + 9)(y - 6)$
3. $(5x - 9y)(2x + 3y)$ **4.** $(7a - 3b)^2$
5. $-6y(y - 2)(y + 1)$ **6.** $(x^2 + 7)(x + 3)(x - 3)$
7. $(2x + 1)(2x - 1)(x + 3)(x - 3)$
8. $(x + y + 10)(x + y - 2)$
9. $(2x + 2y - 7)(x + y - 3)$
10. $S = 16t(50 - t)$; 10,000 ft

Exercise Set 4.4 (page 262)

1. $(x - y)(4x - 4y + 1)(x - y - 1)$
3. $2m(m + 5)(m - 5)$ **5.** $6a(a - 8)(a + 3)$
7. $abc(a - 9b)(a + 6b)$
9. $(3x^3 - 1)(x - 2)(x^2 + 2x + 4)$
11. $(7a - 7b - 3)(a - b + 1)$ **13.** $(x - 5)(x^2 + 1)$
15. $3(x - 10)(x + 9)$ **17.** $-2x(x - 4)(x^2 + 4x + 16)$
19. $8x^2(3x^2 - 5xy^3 + 7y^4)$ **21.** $(9x - 2y)^2$
23. $(11x + 11y - 2)(x + y - 3)$
25. $4x^2(x^2 - 5y^2)(x + y)(x - y)$ **27.** $(x - 15y)^2$
29. $3(x^2 - 2x - 90)$ **31.** $(xy + 6)(x^2y^2 - 6xy + 36)$
33. Prime **35.** $4(x - 2y)(a + b)(a - b)$
37. $(x^2 + 1)(3x + 4)(3x - 4)$
39. $3(2x - 5)(4x^2 + 10x + 25)$ **41.** $4ab(3a + 5b)^2$
43. $(0.5xy + 1.1z)(0.5xy - 1.1z)$ **45.** $(2a^3b^2 + 7)^2$
47. $-2(3x + 4)(3x - 4)$
49. $4x^3yz^2(11z + 8y)(z - 3y)$

Stretching the Topics

1. $2x^a(x^{2a-4} + 3x^{a-2} + 9)$
2. $ab(a^x - 9b^x)(a^{2x} + 9a^xb^x + 81b^{2x})$
3. $(-x - 7y)(37x^2 + 14xy + 13y^2)$

Checkup 4.4 (page 264)

1. $(5x - 9y)^2$ **2.** $3y^2(x + 8y)(x - 8y)$
3. $4x^2y^2(x^3 - 3)$ **4.** $(3x - 2y)(5a - 3)$
5. $(3x + 8)^2$ **6.** $(x + 6)(x - 6)(x^2 + 1)$
7. $(7a - 7b - 3)(a - b + 1)$
8. $(x + 2y)(x - 2y)(x^2 - 2xy + 4y^2)(x^2 + 2xy + 4y^2)$
9. $(6x - 7y)(4x - 3y)$
10. $(x + 1)(x - 1)^2(x^2 + x + 1)$

Speaking the Language of Algebra (page 265)

1. Product **2.** Common factor **3.** Grouping
4. Trial and error **5.** Prime

Review Exercises (page 267)

1. $5a(a - 2)(a - 1)$ **2.** $-x^3y^2(x^2 - 5xy + 2y^2)$
3. $3x^2yz(xy - 1)^2$ **4.** $(x - 2)(3x + 4)$
5. $(a - 2b)(a - 2b - 3)$ **6.** $(b + 2)(a - 2)$
7. $(a + b)(5 + c)$ **8.** $(x + y)(3a - 2b)$
9. $(x - 9)(x^2 + 2)$ **10.** $2(y - z)(3x - 5)$
11. $(3a + 4)(3a - 4)$ **12.** $(2ab + 3c)(2ab - 3c)$
13. $3(x + 4)(x - 4)$ **14.** $6m(2m + 5n)(2m - 5n)$
15. $(x^2 + 9)(x + 3)(x - 3)$
16. $(x + 3y)(x^2 - 3xy + 9y^2)$
17. $(3a - 2b)(9a^2 + 6ab + 4b^2)$
18. $3z(xy - 2)(x^2y^2 + 2xy + 4)$
19. $(x + y + 3)(x^2 + 2xy + y^2 - 3x - 3y + 9)$
20. $(a + 4)(a - 1)(a^2 - 4a + 16)(a^2 + a + 1)$
21. $(x - 7y)(x - 5y)$ **22.** $3(a - 8)(a + 7)$
23. $-2y(y + 5)(y - 3)$ **24.** $(x^2 + 1)(x + 4)(x - 4)$
25. $(x + y + 9)(x + y - 7)$ **26.** $(9x - 2)(2x + 3)$
27. $(3x - 5y)^2$ **28.** $3x^2(6x - 7)(2x + 3)$
29. $(3x - 7)(x - 7)$
30. $(2a + 1)(2a - 1)(a + 5)(a - 5)$

Practice Test (page 269)

1. $4w(2w^3 - w + 3)$ **2.** $-5ab(ab^2 - 3a + 5b)$
3. $(3x - 4)(7 - 6x)$ **4.** $(3y - 2z)(5 + 3y - 2z)$
5. $(a + b)(2 + c)$ **6.** $(5 - x)(a - 1)$
7. $(11x + 2y)(11x - 2y)$ **8.** $4(x + 9)(x - 9)$
9. $(a^2 + 1)(a + 1)(a - 1)$
10. $(2a - 5b + 10)(2a - 5b - 10)$
11. $(2x + 1)(x + 5)(x - 5)$
12. $(5x + y)(25x^2 - 5xy + y^2)$
13. $5(a - 2)(a^2 + 2a + 4)$ **14.** $(5x - 3)(6x + 1)$
15. $7(x - 3y)(x + 2y)$ **16.** $(5a - 3b)^2$
17. $x^3(2x + 3)(x - 4)$ **18.** $(10x + 9)(4x - 1)$
19. $(3x + 3y - 4)(2x + 2y + 1)$
20. $(x + 6)(x - 6)(x + 1)(x - 1)$

Sharpening Your Skills after Chapters 1–4 (page 271)

1. $-9, -3.2, \frac{1}{3}, 0, \frac{5}{2}, 4, 5.\overline{13}$
2. $-3x^4 + 3x^3 + 4x^2 - 5x$ **3.** -20 **4.** -21
5. $\pi - 2$ **6.** -20 **7.** $\frac{1}{16}$ **8.** 1 **9.** -1
10. $-15n + 19$ **11.** $14a - 50b$
12. $8x^2 + 22xy - 21y^2$ **13.** $x^3 + 6x^2 + 12x + 8$
14. $a^3 - 6a^2 + 12a - 8$ **15.** $x^2 - 5x + 25$
16. $8x^9y^5$ **17.** $9a^2b^4c^6$ **18.** x^2y^5 **19.** $\dfrac{x^{12}}{8y^{18}z^6}$
20. $\dfrac{7x^{11}}{y^5}$ **21.** $x = -7$ **22.** $x \le \frac{4}{3}; (-\infty, \frac{4}{3}]$
23. $m \le 11$ or $m \ge 15$; $(-\infty, 11] \cup [15, \infty)$
24. $x > -6$; $(-6, \infty)$

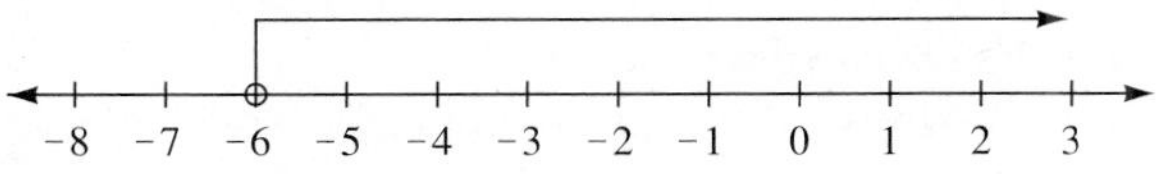

25. $12s + 36$ sq ft

Chapter 5

Exercise Set 5.1 (page 284)

1. $x = 9, x = -11$ **3.** $x = \frac{3}{5}$ **5.** $x = \frac{1}{8}, x = -\frac{5}{3}$
7. $x = \frac{9}{2}, x = -7$ **9.** $x = \frac{1}{3}, x = -\frac{1}{3}$
11. $x = 0, x = 2, x = \frac{7}{3}$ **13.** $x = 6, x = 5$
15. $x = -10, x = 2$ **17.** $x = -5$
19. $m = 4, m = -11$ **21.** $y = -1, y = -\frac{2}{5}$
23. $a = 5, a = \frac{2}{3}$ **25.** $a = 8$ **27.** $x = 3, x = -3$
29. $y = 0, y = 13$ **31.** $m = \frac{3}{4}, m = -\frac{5}{2}$
33. $y = 3, y = \frac{7}{2}$ **35.** $m = \frac{8}{3}, m = -\frac{2}{3}$
37. $x = 5, x = -3$ **39.** $x = 0, x = -9$
41. $x = \frac{23}{14}$ **43.** $x = \frac{9}{2}, x = -1$ **45.** $x = 0, x = 2$
47. $x = -\frac{10}{3}, x = -1$ **49.** $x = 4, x = -2, x = \frac{3}{2}$
51. $x = 0, x = \frac{8}{5}, x = \frac{2}{5}$
53. $x = 1, x = -1, x = 4, x = -4$
55. $x = 3, x = -3, x = 1, x = -1$
57. $x = 0, x = 2, x = -2$
59. $x = 0, x = 8, x = -8$
61. $x = -3, x = 4, x = -4$
63. $x = \frac{1}{2}, x = \frac{3}{2}, x = -\frac{3}{2}$ **65.** $x^2 + 11x + 18 = 0$
67. $x^3 - 9x = 0$ **69.** $3x^2 - 17x + 10 = 0$
71. $2x^4 - 3x^3 - 11x^2 + 6x = 0$

Stretching the Topics

1. $x = 4, x = -\frac{17}{2}$ **2.** $x = 2, x = -\frac{6}{23}$ **3.** $x = 3$

Checkup 5.1 (page 286)

1. $x = \frac{10}{3}, x = -\frac{1}{5}$ **2.** $x = 9, x = -7$
3. $y = \frac{7}{3}, y = -\frac{1}{2}$ **4.** $a = \frac{5}{2}, a = -\frac{2}{3}$
5. $x = 3, x = \frac{1}{3}$ **6.** $x = \frac{5}{4}, x = -3$
7. $x = 0, x = \frac{2}{13}, x = \frac{3}{2}$
8. $x = 6, x = -6, x = 2, x = -2$
9. $x = 0, x = 1, x = 64$ **10.** $x^2 - 3x - 4 = 0$

Exercise Set 5.2 (page 296)

1. $3\sqrt{6}$ **3.** $11\sqrt{2}$ **5.** $6\sqrt{5}$ **7.** $13\sqrt{3}$ **9.** 15
11. $\frac{3}{4}$ **13.** $\frac{11}{2}$ **15.** $\dfrac{\sqrt{7}}{5}$ **17.** $\dfrac{2\sqrt{2}}{3}$ **19.** $\dfrac{7\sqrt{2}}{17}$
21. $x = 2\sqrt{2}, x = -2\sqrt{2}$; irrational
23. No real number solutions
25. $x = 7, x = -7$; rational
27. $y = \frac{1}{3}, y = -\frac{1}{3}$; rational
29. $x = \frac{7}{15}, x = -\frac{7}{15}$; rational
31. $a = \frac{1}{4}, a = -\frac{1}{4}$; rational
33. No real number solutions
35. $x = \sqrt{10}, x = -\sqrt{10}$; irrational
37. $x = \frac{10}{3}, x = -\frac{10}{3}$; rational
39. $x = \sqrt{15}, x = -\sqrt{15}$; irrational
41. $a = \pm\sqrt{\dfrac{x + y}{3}}$ **43.** $s = \pm\sqrt{\dfrac{A}{6}}$
45. $t = \pm\sqrt{\dfrac{2S}{g}}$ **47.** $r = \pm\sqrt{\dfrac{V}{\pi h}}$

Stretching the Topics

1. $x = \dfrac{\sqrt{5}}{2}, x = -\dfrac{\sqrt{5}}{2}, x = \sqrt{3}, x = -\sqrt{3}$
2. $x = 2, x = -2$
3. $x = a, x = -a, x = 3a, x = -3a$

Checkup 5.2 (page 298)

1. $2\sqrt{3}$ **2.** $5\sqrt{7}$ **3.** $\frac{5}{9}$ **4.** $\frac{5\sqrt{5}}{12}$ **5.** $\frac{4\sqrt{2}}{5}$
6. $x = 2\sqrt{6}, x = -2\sqrt{6}$; irrational
7. No real number solutions
8. $x = \frac{9}{5}, x = -\frac{9}{5}$; rational
9. $x = \frac{2\sqrt{5}}{3}, x = -\frac{2\sqrt{5}}{3}$; irrational
10. $b = \pm\sqrt{c^2 - a^2}$

Exercise Set 5.3 (page 311)

1. $x = -5, x = 3$ **3.** $x = 0, x = -4$
5. $x = \frac{-1 + \sqrt{37}}{6}, x = \frac{-1 - \sqrt{37}}{6}$
7. $x = \frac{-9 + \sqrt{17}}{8}, x = \frac{-9 - \sqrt{17}}{8}$
9. No real number solutions **11.** $x = 3, x = 1$
13. $x = 0, x = -8$
15. $x = \frac{-1 + 3\sqrt{5}}{2}, x = \frac{-1 - 3\sqrt{5}}{2}$
17. $x = \frac{-1 + \sqrt{17}}{4}, x = \frac{-1 - \sqrt{17}}{4}$
19. $x = \frac{1 + \sqrt{3}}{5}, x = \frac{1 - \sqrt{3}}{5}$ **21.** $x = 4, x = 5$
23. $x = 3, x = -\frac{1}{2}$ **25.** No real number solutions
27. $x = \frac{-3 + \sqrt{77}}{2}, x = \frac{-3 - \sqrt{77}}{2}$
29. $x = \frac{-4 + \sqrt{10}}{6}, x = \frac{-4 - \sqrt{10}}{6}$ **31.** $x = -3$
33. $x = \sqrt{2}, x = -\sqrt{2}$
35. $x = 1 + \sqrt{5}, x = 1 - \sqrt{5}$
37. $x = \frac{-7 + \sqrt{37}}{6}, x = \frac{-7 - \sqrt{37}}{6}$
39. No real number solutions
41. $x = 2 + \sqrt{2}, x = 2 - \sqrt{2}$
43. $x = \frac{5}{4}, x = -1$ **45.** $x = \frac{3 + \sqrt{2}}{2}, x = \frac{3 - \sqrt{2}}{2}$
47. $x = \frac{-c + \sqrt{c^2 + 4aM}}{2a}, x = \frac{-c - \sqrt{c^2 + 4aM}}{2a}$
49. $r = \frac{-\pi h \pm \sqrt{\pi^2 h^2 + A\pi}}{\pi}$
51. One rational solution
53. Two different rational solutions
55. Two different irrational solutions
57. Two different rational solutions
59. Two different rational solutions

Stretching the Topics

1. $x = 2\sqrt{3}, x = -4\sqrt{3}$ **2.** $x = -3\sqrt{5}$
3. Two different rational solutions

Checkup 5.3 (page 313)

1. $y = 1 + 2\sqrt{3}, y = 1 - 2\sqrt{3}$
2. No real number solutions
3. $x = 1 + \sqrt{5}, x = 1 - \sqrt{5}$
4. $x = \frac{-3 + \sqrt{13}}{2}, x = \frac{-3 - \sqrt{13}}{2}$
5. $a = \frac{4 + \sqrt{6}}{5}, a = \frac{4 - \sqrt{6}}{5}$
6. $x = \frac{-1 + \sqrt{7}}{2}, x = \frac{-1 - \sqrt{7}}{2}$
7. $r = \frac{-\pi h \pm \sqrt{\pi^2 h^2 + 2\pi A}}{2\pi}$
8. Two different rational solutions

Exercise Set 5.4 (page 322)

1. $-7 < x < 7$; $(-7, 7)$
3. $x < -\frac{5}{2}$ or $x > \frac{5}{2}$; $(-\infty, -\frac{5}{2}) \cup (\frac{5}{2}, \infty)$
5. $x \le -3$ or $x \ge 0$; $(-\infty, -3] \cup [0, \infty)$
7. $0 \le x \le \frac{1}{2}$; $[0, \frac{1}{2}]$ **9.** $-2 < x < -1$; $(-2, -1)$
11. $-4 \le x \le 1$; $[-4, 1]$
13. $x < -5$ or $x > 1$; $(-\infty, -5) \cup (1, \infty)$
15. $3 \le x \le 7$; $[3, 7]$ **17.** $-5 < x < 5$; $(-5, 5)$
19. $x < -6$ or $x > \frac{1}{3}$; $(-\infty, -6) \cup (\frac{1}{3}, \infty)$
21. $-\frac{1}{2} \le x \le \frac{8}{3}$; $[-\frac{1}{2}, \frac{8}{3}]$ **23.** $x \in R$; $(-\infty, \infty)$
25. $x \le -5$ or $x \ge \frac{1}{3}$; $(-\infty, -5] \cup [\frac{1}{3}, \infty)$
27. $-3 < x < \frac{2}{7}$; $(-3, \frac{2}{7})$

Stretching the Topics

1. $x \le -\sqrt{2}$ or $x \ge \sqrt{2}$; $(-\infty, -\sqrt{2}] \cup [\sqrt{2}, \infty)$
2. $x < -2$ or $0 < x < 2$; $(-\infty, -2) \cup (0, 2)$
3. $-3 < y < 1$ or $y > 3$; $(-3, 1) \cup (3, \infty)$

Checkup 5.4 (page 323)

1. $x \le -5$ or $x \ge -3$; $(-\infty, -5] \cup [-3, \infty)$
2. $0 < x < 4$; $(0, 4)$
3. $x < -12$ or $x > 3$; $(-\infty, -12) \cup (3, \infty)$
4. $\frac{1}{2} < x < 9$; $(\frac{1}{2}, 9)$
5. $x \le -\frac{3}{4}$ or $x \ge 7$; $(-\infty, -\frac{3}{4}] \cup [7, \infty)$
6. $3 < x < 6$; $(3, 6)$

Exercise Set 5.5 (page 330)

1. 8 by 11 ft **3.** 150 by 200 ft
5. 38 by 50 ft; 300 sq ft
7. 5 by 5 cm and 10 by 10 cm **9.** 12 by 18 ft
11. 12 by 12 by 4 in.; 20 by 20 in. **13.** 10 rods
15. 11 and 12 or -3 and -2 **17.** 8 cm and 15 cm
19. 14 ft **21.** 25 ft **23.** 10 **25.** 35 **27.** 3 sec

Stretching the Topics

1. 18 by 24 ft **2.** 6, 7, 8 or $-8, -7, -6$

Checkup 5.5 (page 333)

1. 12 by 16 ft **2.** $\frac{1}{2}$ ft **3.** 17 cm

Speaking the Language of Algebra (page 336)

1. Quadratic **2.** At least one of the factors must be zero
3. $x - r$; factor **4.** $\frac{-b \pm \sqrt{b^2 - 4ac}}{2a}$
5. Discriminant; no real **6.** Positive; negative

Review Exercises (page 339)

1. $x = -7, x = \frac{5}{3}$ **2.** $x = 8, x = -3$
3. $x = -7, x = \frac{5}{2}$ **4.** $x = -6$ **5.** $x = 3, x = -3$
6. $x = 0, x = 4$ **7.** $a = 4, a = \frac{2}{3}$ **8.** $a = 3, a = \frac{1}{5}$

9. $x = 4, x = -4$ **10.** $y = 0, y = \frac{7}{5}$
11. $x = 7, x = \frac{4}{3}$ **12.** $x = 6, x = \frac{1}{5}$
13. $x = 0, x = -3, x = \frac{5}{12}$
14. $x = 2, x = -2, x = 5, x = -5$
15. $x = 0, x = 7, x = -8$ **16.** $x^2 - 5x - 24 = 0$
17. $5\sqrt{3}$ **18.** $9\sqrt{2}$ **19.** $\frac{\sqrt{5}}{3}$ **20.** $\frac{3\sqrt{2}}{7}$
21. $x = 4\sqrt{2}, x = -4\sqrt{2}$; irrational
22. No real number solutions
23. $x = \frac{5}{3}, x = -\frac{5}{3}$; rational
24. $x = \frac{1}{4}, x = -\frac{1}{4}$; rational **25.** $x = 8, x = -6$
26. $x = \frac{3 + \sqrt{33}}{2}, x = \frac{3 - \sqrt{33}}{2}$
27. $x = \frac{7 + \sqrt{33}}{4}, x = \frac{7 - \sqrt{33}}{4}$
28. $x = -2 + \sqrt{5}, x = -2 - \sqrt{5}$
29. $x = \frac{1 + \sqrt{37}}{18}, x = \frac{1 - \sqrt{37}}{18}$
30. $y = -1, y = \frac{5}{3}$
31. $x = -1 + \sqrt{7}, x = -1 - \sqrt{7}$
32. $x = \frac{3 + \sqrt{41}}{8}, x = \frac{3 - \sqrt{41}}{8}$
33. $n = \frac{3 \pm \sqrt{9 + 8d}}{2}$
34. $t = \frac{-g \pm \sqrt{g^2 + 64S}}{32}$
35. Two different irrational solutions
36. Two different rational solutions
37. No real number solutions **38.** One rational solution
39. $-12 < x < 12$; $(-12, 12)$
40. $x \le 0$ or $x \ge 3$; $(-\infty, 0] \cup [3, \infty)$
41. $-\frac{1}{2} \le x \le 5$; $[-\frac{1}{2}, 5]$
42. $-2 < x < 0$ or $x > 2$; $(-2, 0) \cup (2, \infty)$
43. 60 by 84 in. **44.** 6 by 6 cm, 12 by 12 cm
45. 3 in. **46.** 7 ft

Practice Test (page 341)

1. $x = 4, x = -9$ **2.** $x = \frac{5}{3}$ **3.** $x = 0, x = 3$
4. $x = 0, x = 2, x = -\frac{3}{5}$ **5.** $x = 2, x = -7$
6. $x = 2, x = -2, x = 4, x = -4$
7. $x = \frac{1}{2}, x = 1, x = -1$ **8.** $x^2 - 16 = 0$
9. $a = \frac{2}{9}, a = -\frac{2}{9}$ **10.** $x = \frac{2\sqrt{5}}{3}, x = -\frac{2\sqrt{5}}{3}$
11. $x = \frac{-7 + \sqrt{37}}{2}, x = \frac{-7 - \sqrt{37}}{2}$
12. $x = \frac{1 + \sqrt{7}}{3}, x = \frac{1 - \sqrt{7}}{3}$
13. No real number solutions
14. $y = \frac{1 \pm \sqrt{1 + 12aK}}{6a}$ **15.** One rational solution
16. $0 \le x \le \frac{7}{2}$; $[0, \frac{7}{2}]$
17. $x < -\frac{4}{5}$ or $x > 2$; $(-\infty, -\frac{4}{5}) \cup (2, \infty)$
18. 5 in., 12 in., 13 in. **19.** 4 by 9 cm **20.** 60

Sharpening Your Skills after Chapters 1–5 (page 343)

1. $x - 2 < 0$ **2.** $-\frac{16}{3} > -\sqrt{36}$ **3.** Undefined
4. 2 **5.** $-\frac{1}{16}$ **6.** 1 **7.** $-2x - 2y$
8. $-63x^6y^4z^6$ **9.** $\frac{9x^{10}}{25z^6}$ **10.** $\frac{y^9}{x^{17}}$
11. $-21x^6y^6 - 7x^5y^5$
12. $a^3 - 4a^2b - 31ab^2 + 70b^3$
13. $x^2 + 2xy + y^2 - 12x - 12y + 36$
14. $2a^2 - 4ab + 3b^2$ **15.** $x^2 + 2x - 5$
16. $(a + 9b)(a + 3b)$ **17.** $5(x^2 + 4)(x + 2)(x - 2)$
18. $(3x - 4)(9x^2 + 12x + 16)$ **19.** $(x - 5)(x^2 + 9)$
20. $-9x^2(2x - 3)(x + 1)$ **21.** $(x - 9)^2$ **22.** \$6900
23. $x = 1$ **24.** $x < 3$; $(-\infty, 3)$ **25.** No solution
26. $-1 < x \le 3$; $(-1, 3]$
27. $-\frac{10}{3} < x < 4$; $(-\frac{10}{3}, 4)$ **28.** $x = \frac{ab - by}{a}$
29. At least 10 **30.** $(s + 3)(s - 2) = s^2 + s - 6$

Chapter 6

Exercise Set 6.1 (page 354)

1. $a \ne 0, b \ne 0$; $-\frac{2a^2}{b^3}$ **3.** $a \ne 0, b \ne 0, c \ne 0$; $\frac{c}{3a^2b^2}$
5. $x \ne 0, y \ne 0, z \ne 0$; $\frac{1}{x^2y^2}$ **7.** $x \ne 0$; $2(2y - 1)$
9. $y \ne 0, y \ne \frac{4}{5}$; $\frac{1}{5y - 4}$
11. $a \ne 0, b \ne 0$; $-\frac{4a + 5c}{6}$ **13.** $x \ne -y$; $\frac{1}{2}$
15. $x \ne -y$; $\frac{x - y}{2(x + y)}$ **17.** $x \ne -9$; $\frac{x - 9}{3}$
19. $x \ne 4, x \ne -4$; $\frac{x + 5}{x - 4}$
21. $x \ne -2, x \ne -5$; $\frac{x + 4}{x + 5}$
23. $y \ne 7, y \ne 3$; $\frac{5(y + 3)}{y - 7}$ **25.** $b \ne \frac{1}{2}$; $\frac{1}{2b - 1}$
27. $x \ne -\frac{2}{3}, x \ne \frac{7}{2}$; $\frac{x}{2x - 7}$
29. $a \ne 0, a \ne \frac{1}{2}, a \ne -5$; $\frac{a + 7}{a(2a - 1)}$
31. $x \ne -3, x \ne \frac{5}{4}$; $\frac{x^2 - 3x + 9}{4x - 5}$ **33.** $\frac{3}{a - 5}$
35. $\frac{(2a + 1)(2a - 1)}{5}$ **37.** $\frac{1}{x + 1}$ **39.** $\frac{x + 5y}{x + y}$
41. $-\frac{1}{b - 2a}$ **43.** $-\frac{x + 5}{4 - 3x}$ or $\frac{x + 5}{3x - 4}$
45. $-\frac{x + y}{4y - 5x}$ or $\frac{x + y}{5x - 4y}$ **47.** $\frac{a - b}{3(a + 5)}$
49. $2x - 3$ **51.** $\frac{4(3x - 2)}{(x - 2)^3}$ **53.** $(a + 5)(a - 2)$

Stretching the Topics

1. $-\frac{6(x - 2)}{x^2(x + 6)}$ **2.** $x^n + 2$
3. $x \ne 0, x \ne 5y, x \ne -y$

Checkup 6.1 (page 356)

1. $x \neq 0, y \neq 0, z \neq 0; -\frac{2x}{z^2}$ **2.** $y \neq 0, y \neq \frac{3}{2}; \frac{1}{2y-3}$

3. $x \neq y; \frac{x+y}{2(x-y)}$ **4.** $a \neq 0, b \neq 0; -\frac{2a+3}{4}$

5. $x \neq 9, x \neq -9; \frac{x+6}{x+9}$ **6.** $x \neq 6, x \neq -6; \frac{3x}{x+6}$

7. $\frac{9x+y}{x+y}$ **8.** $-\frac{1}{2b-3a}$ **9.** $a+5$ **10.** 8

Exercise Set 6.2 (page 364)

1. $\frac{6}{x}$ **3.** $\frac{a^2+b^2}{a+b}$ **5.** $\frac{2(x-5)}{3(x+5)}$ **7.** $3(x+9y)$

9. $\frac{5y(2x+5)}{8z^2(2x-5)}$ **11.** 1 **13.** $\frac{x+5y}{5}$ **15.** 3

17. $-\frac{3}{2(3x-8)}$ **19.** $-\frac{x+3}{2x-1}$ **21.** $-\frac{1}{x(1-x)}$

23. $\frac{x-3}{x+4}$ **25.** $3x^3$ **27.** $\frac{y^2}{9(x+9)}$ **29.** $\frac{x+5}{49}$

31. $\frac{x(2x-3)}{3(x-8)}$ **33.** $\frac{3}{5(x-12)}$ **35.** $\frac{1}{x+3}$

37. $\frac{3x}{x-3}$ **39.** 1 **41.** -1 **43.** $-\frac{x+5}{6x+1}$

45. $\frac{5}{2}$ **47.** $\frac{1}{4}$ **49.** $\frac{2}{x}$

Stretching the Topics

1. $\frac{1}{2}$ **2.** $\frac{3x^a+1}{3x^a(x^a+3)}$ **3.** $2x^{2a-2c}$

Checkup 6.2 (page 366)

1. $\frac{12}{x}$ **2.** $\frac{x(x-7)}{8(x-4)}$ **3.** $\frac{5(2x-7)}{6y(2x+7)}$ **4.** $\frac{x+9y}{2y}$

5. $\frac{y^3(x-15)}{3x^5}$ **6.** $\frac{y^3}{4(x+4)}$ **7.** $\frac{1}{5x+4}$

8. $\frac{7x(x-10)}{(x+2)(x-7)}$ **9.** $-\frac{2}{5x}$ **10.** 1

Exercise Set 6.3 (page 377)

1. 2 **3.** $\frac{2}{x}$ **5.** $\frac{y+3}{x}$ **7.** $-\frac{2x}{2x-1}$ **9.** 3

11. $\frac{1}{x-9}$ **13.** 2 **15.** $\frac{1}{x+4}$ **17.** $\frac{2a-3}{a}$

19. $\frac{x+1}{x-4}$ **21.** $x-4$ **23.** $\frac{15x-27}{5x^2}$

25. $\frac{36y-50x}{15xy}$ **27.** $\frac{x+3}{x(x+1)}$ **29.** $\frac{2x^2-5x-15}{2x^2(x-3)}$

31. $\frac{7x+9}{(x+2)^2}$ **33.** $\frac{-2x-32}{(x+4)(x-4)}$ **35.** $\frac{-11}{3(x+1)}$

37. $\frac{3}{2(x-3)}$ **39.** $-\frac{2}{x}$ **41.** $\frac{x^2-x-5}{x(x+5)(x-5)}$

43. $\frac{x+4}{x^4}$ **45.** $\frac{4x}{(x+4)(x-2)}$ **47.** $\frac{2x-5}{x(2x-1)}$

49. $\frac{3x-7}{(x+3)(x-2)}$ **51.** $\frac{1}{x+5}$ **53.** $\frac{2}{x+5}$

55. $\frac{x-2}{x^2+2x+4}$

Stretching the Topics

1. $\frac{6x^{3a}-4x^ay^b+5y^{3b}}{10x^{2a}y^{2b}}$ **2.** $\frac{x^a-1}{x^a}$ **3.** $\frac{x-1}{x+1}$

Checkup 6.3 (page 379)

1. $\frac{1}{x^2}$ **2.** $\frac{10x}{5x-3}$ **3.** $-\frac{1}{x+2}$ **4.** $\frac{x+5}{4x-3}$

5. $\frac{9x+10y}{6x^2y^2}$ **6.** $\frac{11x-21}{x(x-3)}$ **7.** $\frac{7x-2}{(x+2)(x-2)}$

8. $\frac{x-3}{2x^2(x-1)^2}$ **9.** $\frac{17}{(x-9)(x+7)}$

10. $\frac{x+10}{(x-6)(x+2)}$

Exercise Set 6.4 (page 388)

1. $\frac{1}{10}$ **3.** $\frac{x^2}{3y}$ **5.** $\frac{5x^2}{x-y}$ **7.** $\frac{3}{5(x+4)}$

9. $\frac{x^2}{x+2}$ **11.** $\frac{x}{2(x^2+3x+9)}$ **13.** $2x^3(3x+4)$

15. $\frac{28}{5}$ **17.** $\frac{x-3}{x}$ **19.** $\frac{1}{x+1}$ **21.** $\frac{x}{4x+1}$

23. $\frac{4a^2+2a+1}{a^2}$ **25.** $\frac{2}{15}$ **27.** $\frac{9-x}{9x}$

29. $\frac{15(2x+1)}{x^2}$ **31.** $\frac{5(x+6)}{x+5}$ **33.** $\frac{(x+7)^2}{(x+10)^2}$

35. $\frac{5x+6}{9x-2}$ **37.** $a-4$ **39.** $2b$

41. $\frac{(x+1)(x+2)}{(x-2)(x+14)}$ **43.** $\frac{x}{x-4}$

45. $\frac{1-2x+4x^2}{x^2}$ **47.** $\frac{2x-1}{x-2}$ **49.** $\frac{y-2x}{y+2x}$

51. $\frac{4x^2}{1+2x^2y^3}$ **53.** $\frac{x^2}{(x+2)(x-2)}$

Stretching the Topics

1. $\frac{(2x^2-3x+25)(x+3)}{(x-4)(7-x)}$ **2.** $\frac{1}{x(x+d)}$

3. $3(x^n+1)$

Checkup 6.4 (page 391)

1. $\frac{x^7y^3}{5}$ **2.** $\frac{5x^3}{2x+3}$ **3.** $\frac{x^3}{x-11}$ **4.** $\frac{x-9}{x}$

5. $\frac{4a^2+6a+9}{a^2}$ **6.** $\frac{7+x}{7x}$ **7.** $\frac{7x(x-5)}{(x+2)(x-7)}$

8. $\frac{2(a-18)}{7(a-2)}$ **9.** $\frac{x(4x+1)}{x-5}$ **10.** $\frac{xy}{y-3x}$

Exercise Set 6.5 (page 401)

1. $x = -\frac{14}{5}$ **3.** $x = 3$ **5.** $x = 3$ **7.** $x = \frac{44}{9}$

9. $x = \frac{33}{4}$ **11.** $x = -\frac{1}{10}$ **13.** $x = -\frac{2}{5}$

15. $x = 39$ **17.** $x = -\frac{17}{3}$ **19.** $a = \frac{3}{5}$

21. $x = 0$ **23.** No solution **25.** $x = -7$

27. $x = \frac{5}{3}, x = -2$ **29.** $y = 2\sqrt{2}, y = -2\sqrt{2}$

31. $x = -6, x = 5$ **33.** $y = \frac{x}{1-x^2}$

35. $y = \frac{4xz}{2x-z}$ **37.** $t = \frac{A-P}{Pr}$

39. $x < 0$; $(-\infty, 0)$ 41. $x > 3$; $(3, \infty)$
43. $x \le -3$; $(-\infty, -3]$ 45. $x > \frac{5}{2}$; $(\frac{5}{2}, \infty)$
47. $x \le 3$; $(-\infty, 3]$ 49. $x > 5$; $(5, \infty)$
51. $x < 0$ or $x > 4$; $(-\infty, 0) \cup (4, \infty)$
53. $0 < x < 1$; $(0, 1)$
55. $x < 0$ or $x \ge \frac{9}{2}$; $(-\infty, 0) \cup [\frac{9}{2}, \infty)$
57. $x < -1$ or $x > 1$; $(-\infty, -1) \cup (1, \infty)$

Stretching the Topics

1. $x = 0, x = -4$ 2. $x = \dfrac{5 \pm \sqrt{41}}{2}$ 3. $x = -\dfrac{3k}{4}$

Checkup 6.5 (page 404)

1. $x = -23$ 2. $a = \frac{23}{2}$ 3. $b = \frac{1}{13}$
4. $a = -\frac{12}{7}$ 5. $x = -5$ 6. $y = \dfrac{x}{x - 5}$
7. $x > 5$; $(5, \infty)$ 8. $0 \le x < 10$; $[0, 10)$
9. $-3 \le x < 0$; $[-3, 0)$
10. $x < -1$ or $x > 3$; $(-\infty, -1) \cup (3, \infty)$

Exercise Set 6.6 (page 412)

1. $12\frac{1}{3}$ 3. 4 5. Ann, \$680; Meredith, \$850
7. Burt 22, Johnny 33
9. Gretchen, 10 mph; George, 12 mph
11. Michelle, 24 min; Carolyn, 30 min
13. Bus, 50 mph; train, 80 mph 15. $7\frac{1}{2}$ days
17. $17\frac{1}{7}$ hr 19. $17\frac{1}{2}$ min

Stretching the Topics

1. $22\frac{1}{2}$ 2. 14.8 mph

Checkup 6.6 (page 414)

1. 280 2. $26\frac{2}{3}$ hr 3. 55 mph

Speaking the Language of Algebra (page 416)

1. Rational algebraic expression 2. Division by zero
3. Divide; common factor 4. Factor 5. LCD
6. Negative 7. Proportion
8. Rate; time worked

Review Exercises (page 417)

1. $x \ne 0, y \ne 0, z \ne 0$; $-\dfrac{3x^3}{2z^3}$ 2. $x \ne 0, x \ne 2$; $\dfrac{1}{x - 2}$
3. $x \ne y$; $\dfrac{x + y}{4(x - y)}$ 4. $x \ne 7, x \ne -7$; $\dfrac{x - 5}{x - 7}$
5. $x \ne \frac{1}{7}y, x \ne -2y$; $\dfrac{3x - 2y}{x + 2y}$
6. $a \ne \frac{5}{3}b, a \ne 2b$; $\dfrac{a^2 + 2ab + 4b^2}{3a - 5b}$
7. $x \ne \frac{1}{4}y, x \ne y$; 1 8. $x \ne 8, x \ne \frac{3}{2}$; $-\dfrac{x + 8}{2x - 3}$
9. $\dfrac{2}{x}$ 10. $\dfrac{x(x + 6)}{5(x + 2)}$ 11. $\dfrac{3x - 5y}{2x + y}$
12. $\dfrac{y(x + 11y)}{5x^4}$ 13. $-\dfrac{9x - y}{x(3x - y)}$
14. $\dfrac{8}{3}$ 15. $\dfrac{8x}{2x - 3}$ 16. $\dfrac{1}{x - 3}$ 17. $\dfrac{8x + 30}{x(x + 10)}$
18. $\dfrac{3x^2 + 5x}{(x + 5)(x - 5)}$ 19. $\dfrac{1}{x + 7}$ 20. $\dfrac{2}{x + 2}$
21. $\dfrac{x^3}{3y}$ 22. $\dfrac{3x}{3x + 1}$ 23. $\dfrac{x + 8}{x}$ 24. $\dfrac{5 + x}{5x}$
25. $\dfrac{a^2 + 2a + 4}{a^2}$ 26. $\dfrac{2x + 1}{x}$ 27. $x = 29$
28. $a = 7$ 29. $x = -6$ 30. $x > 9$; $(9, \infty)$
31. $x < 0$ or $x \ge 1$; $(-\infty, 0) \cup [1, \infty)$
32. $-1 < x < 5$; $(-1, 5)$ 33. 20 ft 34. $4\frac{2}{7}$ hr
35. Mac, 45 mph; Becky, 55 mph

Practice Test (page 419)

1. $x \ne 0, y \ne 0, c \ne 0$; $\dfrac{3x}{2c^2}$ 2. $x \ne -5, x \ne 2$; $\dfrac{x - 3}{x - 2}$
3. $y \ne 1, y \ne -1$; $\dfrac{y^2 + y + 1}{y + 1}$
4. $m \ne -2n, m \ne 3n$; $\dfrac{m - 1}{m - 3n}$ 5. $\dfrac{2(x + 8)}{x(x + 1)}$ 6. $4x^2$
7. $\dfrac{3b(a + 5b)}{4a^3}$ 8. $\dfrac{7}{x^2}$ 9. $\dfrac{1}{x - 7}$ 10. $\dfrac{2(5x - 1)}{5x(5x + 2)}$
11. $-\dfrac{y^2 + 36}{(y + 6)(6 - y)}$ or $\dfrac{y^2 + 36}{(y + 6)(y - 6)}$ 12. $\dfrac{3}{x + 4}$
13. $\dfrac{-3x^2 + 15x + 62}{(3x - 4)(x + 6)}$ 14. $\dfrac{2}{7x(1 + 2x)}$
15. $\dfrac{5(2x - 3)}{x - 5}$ 16. $\dfrac{2x - 5}{x}$ 17. $x = \dfrac{8}{21}$
18. No solution 19. $-4 < x \le \frac{1}{5}$; $(-4, \frac{1}{5}]$
20. $23\frac{1}{3}$ hr 21. \$821.80
22. Chico, 62 mph; Vagas, 57 mph

Sharpening Your Skills after Chapters 1–6 (page 421)

1. $-\sqrt{13} < 0$ 2. 0 3. 27 4. $-\frac{1}{16}$
5. $16\sqrt{3}$ 6. Two different irrational solutions
7. $x^2 - x - 6 = 0$ 8. $14x + 9y$
9. $a^{15}b^7 + 3a^7b^{15}$ 10. $\dfrac{27b^{24}}{16a^{12}}$ 11. $a^3 - 9a + 6$
12. $-63y^2 + 16y^3 - y^4$ 13. $x^2 - 9x + 5 - \dfrac{11}{x}$
14. $x^2 - 5xy - 6y^2$ 15. $-3x^2y(x^4y^2 - 7x^2y + 9)$
16. $(a - b)(a - b - 5)$ 17. $(3xy + 11)(3xy - 11)$
18. $(5y + 4)(2y - 3)$ 19. $3(8 + y)(5 - y)$
20. $y = -2$ 21. No solution
22. $m < -5$ or $m > 25$; $(-\infty, -5) \cup (25, \infty)$
23. $x = 0, x = -8, x = \frac{5}{9}$ 24. $m = \frac{2}{5}, m = -\frac{1}{6}$
25. $x = \frac{8}{11}, x = -\frac{8}{11}$
26. $x = 1 + \sqrt{5}, x = 1 - \sqrt{5}$
27. $y \ge -\frac{20}{3}$; $[-\frac{20}{3}, \infty)$ 28. $-3 < a \le 6$; $(-3, 6]$
29. $-4 \le x \le \frac{2}{5}$; $[-4, \frac{2}{5}]$ 30. $x = \dfrac{-1 \pm \sqrt{1 + 8aM}}{4a}$

Chapter 7

Exercise Set 7.1 (page 433)

1. x^2 3. $a^{\frac{5}{3}}$ 5. x^4 7. $\dfrac{x}{y^4}$ 9. $\dfrac{b^{\frac{2}{3}}}{a^{\frac{3}{2}}}$ 11. $x^{\frac{16}{15}}$

13. $\frac{x^{\frac{1}{2}}}{y^{\frac{1}{4}}z^{\frac{1}{3}}}$ **15.** $x^{\frac{1}{2}}y^{\frac{1}{3}}$ **17.** $\frac{b^{\frac{5}{2}}}{a^{\frac{3}{8}}}$ **19.** $\frac{x^{\frac{1}{9}}z^{\frac{1}{14}}}{y^{\frac{1}{20}}}$ **21.** -3
23. 3 **25.** Not a real number **27.** $-\frac{4}{5}$ **29.** 2
31. $\frac{5}{9}$ **33.** $(2x)^{\frac{1}{4}}$ **35.** $5a^{\frac{1}{2}}$ **37.** $(x^2 + 16)^{\frac{1}{2}}$
39. $-y^{\frac{1}{5}}$ **41.** $(-a)^{\frac{1}{3}}$ **43.** $x^{\frac{1}{3}} + y^{\frac{1}{3}}$ **45.** $x^{\frac{2}{5}}y^{\frac{1}{5}}$
47. $(x - y)^{\frac{2}{5}}$ **49.** $\frac{1}{x^{\frac{4}{3}}y^{\frac{2}{3}}}$ or $x^{-\frac{4}{3}}y^{-\frac{2}{3}}$ **51.** $\sqrt[5]{3y^3}$
53. $4\sqrt{x}$ **55.** $\sqrt[3]{a + b}$ **57.** $\sqrt[5]{x} + \sqrt[5]{y}$ **59.** 5
61. $\sqrt{13}$ **63.** $\sqrt[3]{y^2}$ **65.** $\sqrt[3]{(x - 3)^2}$
67. $\frac{1}{\sqrt[5]{(a - b)^3}}$ **69.** 4 **71.** $\frac{1}{27}$ **73.** 5 **75.** $\frac{6}{5}$

Stretching the Topics

1. $\frac{1}{250}$ **2.** $\frac{n^4}{36{,}864}$ **3.** $9xy^{\frac{5}{12}}$

Checkup 7.1 (page 435)

1. a **2.** $\frac{y^2}{x^{\frac{1}{2}}}$ **3.** $\frac{x^{\frac{1}{4}}}{y^{\frac{31}{10}}}$ **4.** $\frac{x^{\frac{1}{15}}z^{\frac{1}{2}}}{y^{\frac{9}{4}}}$
5. Not a real number **6.** $-\frac{1}{2}$ **7.** $x^{\frac{2}{7}}y^{\frac{3}{7}}$
8. $(x + y)^{\frac{3}{2}}$ **9.** $\sqrt[3]{(x - y)^2}$ **10.** 32

Exercise Set 7.2 (page 443)

1. $|a|$ **3.** y **5.** $|x - 5|$ **7.** $2b\sqrt{3}$
9. $2x^2y^3\sqrt{6xy}$ **11.** $36a^7$ **13.** $2c^5d\sqrt{6d}$
15. $-2ab\sqrt[3]{5a^2b}$ **17.** $\frac{4a^2\sqrt{3a}}{3b}$ **19.** $\frac{2x\sqrt[3]{2x}}{3y^2}$
21. $\frac{12x^3\sqrt{x}}{5y^2}$ **23.** $2x\sqrt[5]{x^2}$ **25.** $7x^2$ **27.** $2x\sqrt[3]{2x}$
29. $2x^2y^2\sqrt[4]{2xy^2}$ **31.** $(x - 3)\sqrt[5]{x - 3}$ **33.** $x - 3$
35. $2x + 1$ **37.** $7\sqrt{5}$ **39.** $6\sqrt{3} + 6\sqrt{7}$ **41.** 0
43. $-4\sqrt[3]{5}$ **45.** $15 - 7\sqrt{x}$ **47.** $-3\sqrt{5} - 11y$
49. $-2\sqrt{a} - 23\sqrt[3]{b}$ **51.** $-6\sqrt[5]{2} - 3\sqrt[3]{2}$
53. $3\sqrt{3}$ **55.** $\sqrt{5}$ **57.** $4x\sqrt{3}$ **59.** $22\sqrt[3]{2}$
61. 0 **63.** $\frac{4\sqrt{2}}{15}$ **65.** $13\sqrt{2} - 5\sqrt{3}$
67. $11ab\sqrt[3]{3a^2b}$

Stretching the Topics

1. $x^{2m}\sqrt{x}$ **2.** $x^{2m}y^{3n}$ **3.** $\frac{3y^2\sqrt{2y} - 10y^2\sqrt{2}}{z^2}$

Checkup 7.2 (page 445)

1. $2x^2y\sqrt{11x}$ **2.** $-\frac{3a\sqrt[3]{a}}{2b}$ **3.** $(x + 1)\sqrt[4]{x + 1}$
4. $-9x\sqrt{2}$ **5.** $5\sqrt{7} - 21x$ **6.** 0 **7.** $-\sqrt[3]{10}$
8. $4\sqrt[3]{3} + 4\sqrt{3}$ **9.** $\frac{\sqrt{2}}{21}$ **10.** $3x^2y\sqrt[3]{xy}$

Exercise Set 7.3 (page 458)

1. $7\sqrt{2}$ **3.** $3a^2\sqrt{2}$ **5.** $-3x\sqrt[3]{2}$ **7.** $3x^2$
9. $-2x\sqrt[3]{3}$ **11.** $2\sqrt[5]{x^3}$ **13.** $8x\sqrt{xy}$ **15.** $5x$
17. $3x\sqrt{10y}$ **19.** $12x^2y^2z\sqrt{y}$ **21.** $2\sqrt{3} + 2$
23. $x\sqrt{3} - 3\sqrt{x}$ **25.** $3x\sqrt{2y} + 2y\sqrt{3x}$
27. $3\sqrt[3]{a} - a$ **29.** $2 - 3\sqrt{x} + x$
31. $10 - 14\sqrt{2y} - 3y$ **33.** $x - 6\sqrt{x} + 9$
35. $2x - 9$ **37.** $16x - 20$ **39.** $x^2 - 2y$
41. $9x - 30\sqrt{xy} + 25y$ **43.** $x + 4\sqrt{x - 3} + 1$
45. $\frac{2\sqrt{6}}{3}$ **47.** $\frac{\sqrt{2}}{2}$ **49.** $\frac{\sqrt{6}}{3}$ **51.** $\frac{3x\sqrt{10xy}}{5y}$
53. $\frac{3xy\sqrt{15xz}}{5z}$ **55.** $\frac{4\sqrt{a + 5}}{a + 5}$ **57.** $\frac{\sqrt{x^2 - 9}}{x + 3}$
59. $\frac{3\sqrt[3]{25}}{5}$ **61.** $\frac{64\sqrt[4]{27}}{3}$ **63.** $\frac{2\sqrt[3]{9x^2}}{x}$ **65.** $\frac{x\sqrt[4]{9x}}{3}$
67. $\frac{\sqrt[4]{12y^3}}{2y}$ **69.** $\frac{\sqrt[3]{10y}}{2y}$ **71.** $3(2 + \sqrt{3})$
73. $\frac{\sqrt{7} - \sqrt{2}}{5}$ **75.** $\frac{x + 2\sqrt{x}}{x - 4}$ **77.** $\frac{a + 3\sqrt{a}}{a - 9}$
79. $3\sqrt{2} - 2\sqrt{3}$ **81.** $\frac{18 - 7\sqrt{5}}{79}$

Stretching the Topics

1. $x - 2$ **2.** $(3\sqrt{x} + 2\sqrt{y})(3\sqrt{x} - 2\sqrt{y})$
3. $\frac{x + 2 + 5\sqrt{x + 2}}{x - 23}$

Checkup 7.3 (page 460)

1. $2x\sqrt{15x}$ **2.** $4x\sqrt{y}$ **3.** $5\sqrt{2} + 5$ **4.** 10
5. $x - 12\sqrt{x} + 36$ **6.** $\frac{2\sqrt{15}}{5}$ **7.** $\frac{\sqrt{15}}{5}$
8. $-5\sqrt[3]{9x^2}$ **9.** $\frac{3x\sqrt{10xy}}{2y}$ **10.** $-4 + 2\sqrt{6}$

Exercise Set 7.4 (page 469)

1. $x = 25$ **3.** $x = 50$ **5.** $x = 6$ **7.** $x = 3$
9. No solution **11.** $x = 4$ **13.** $x = 6$
15. $x = -7$ **17.** $x = 3$ **19.** $x = 4$
21. $x = 9, x = 2$ **23.** $x = 2$ **25.** $x = 2$
27. No solution **29.** $x = 1$ **31.** $x = 9$
33. $x = \frac{9}{8}$ **35.** No solution **37.** $x = -10, x = 2$
39. $x = 2$ **41.** $y = 3, y = -1$
43. 5 in., 12 in., 13 in. **45.** 28 ft
47. 3 cm, 4 cm, 5 cm **49.** 972π cu in.
51. $w = \sqrt{d^2 - l^2}$ **53.** $s = gt^2 + a + k$

Stretching the Topics

1. $x = 3, x = 2$ **2.** $x = 6$ **3.** $l = \frac{T^2g}{4\pi^2}$

Checkup 7.4 (page 471)

1. $x = 7$ **2.** $x = \frac{4}{3}$ **3.** No solution **4.** $x = 9$
5. No solution **6.** $x = -9$ **7.** $x = 9$
8. $x = 6, x = 3$ **9.** $x = 2$ **10.** $x = 2$

Exercise Set 7.5 (page 485)

1. $0 + 4i$ **3.** $0 + \sqrt{21}i$ **5.** $0 + 5\sqrt{2}i$
7. $0 - 5\sqrt{3}i$ **9.** $4 - 3\sqrt{6}i$ **11.** $18 + 7i$
13. $x = 1, y = -4$ **15.** $x = -4, y = \frac{1}{3}$
17. $x = -\frac{1}{2}, y = -\frac{5}{3}$ **19.** $x = 3, y = \frac{3}{2}$
21. $x = 2, y = 0$ **23.** $x = 4, y = -\frac{4}{3}$ **25.** i
27. 1 **29.** i **31.** i **33.** $5 + 9i$ **35.** $0 - 9i$

37. $9 - 5\sqrt{2}i$ **39.** $5 + 0i$ **41.** $-2 + i$
43. $12 + 2i$ **45.** $-18 - 12i$ **47.** $-3 + 21i$
49. $-10 + 10i$ **51.** $31 + 29i$ **53.** $4 - 6\sqrt{5}i$
55. $-2 - 21\sqrt{6}i$ **57.** $19 + 0i$ **59.** $-26 + 7i$
61. $2 - 5i$ **63.** $\frac{4}{5} - 2i$ **65.** $3 - \frac{\sqrt{3}}{4}i$
67. $0 - 5i$ **69.** $0 + 4\sqrt{5}i$ **71.** $-\frac{1}{2} + \frac{\sqrt{3}}{2}i$
73. $6 + 9i$ **75.** $\frac{7}{25} + \frac{1}{25}i$ **77.** $-\frac{\sqrt{3}}{4} + \frac{7}{4}i$
79. $-\frac{2}{29} - \frac{5}{29}i$ **81.** $\frac{13}{15} - \frac{2}{5}i$ **83.** $-\frac{1}{9} - \frac{4\sqrt{5}}{9}i$

Stretching the Topics
1. -8 **2.** $(x + 3i)(x - 3i)$
3. $x = -4, y = 5; -4 + 5i$

Checkup 7.5 (page 487)
1. $0 + 11i$ **2.** $9 - 2\sqrt{7}i$ **3.** $x = 4, y = -4$
4. -1 **5.** $6 - 5i$ **6.** $3 - 9\sqrt{3}i$ **7.** $17 - 6i$
8. $21 - 20i$ **9.** $-2 - 3i$ **10.** $-1 - i$

Exercise Set 7.6 (page 492)
1. $x = 3i, x = -3i$ **3.** $x = \frac{5}{2}i, x = -\frac{5}{2}i$
5. $x = \sqrt{3}i, x = -\sqrt{3}i$ **7.** $x = -\frac{1}{2} \pm \frac{\sqrt{3}}{2}i$
9. $x = \frac{4}{3} \pm \frac{\sqrt{5}}{3}i$ **11.** $x = \frac{3}{4} \pm \frac{\sqrt{23}}{4}i$
13. $x = -\frac{3}{2} \pm \frac{\sqrt{3}}{2}i$ **15.** $x = -\frac{3}{5} \pm \frac{\sqrt{11}}{5}i$
17. $x = -2, x = 1 \pm \sqrt{3}i$
19. $x = 5, x = -\frac{5}{2} \pm \frac{5\sqrt{3}}{2}i$
21. $x = \frac{1}{2}, x = -\frac{1}{4} \pm \frac{\sqrt{3}}{4}i$
23. $x = 3, x = -3, x = 3i, x = -3i$
25. $x = \sqrt{11}, x = -\sqrt{11}, x = \sqrt{11}i, x = -\sqrt{11}i$
27. $x = 3i, x = -3i, x = 2i, x = -2i$
29. $x = \sqrt{5}i, x = -\sqrt{5}i, x = \sqrt{2}, x = -\sqrt{2}$
31. $x = -\frac{2}{3} \pm \frac{\sqrt{2}}{3}i$ **33.** $x = \frac{1}{3} \pm \frac{\sqrt{2}}{3}i$
35. $x = \frac{1}{6} \pm \frac{\sqrt{23}}{6}i$

Stretching the Topics
1. $x = -8$ **2.** $x = -2i, x = -\frac{1}{2}i$
3. $x^2 + ix + 6 = 0$

Checkup 7.6 (page 493)
1. $x = 5i, x = -5i$ **2.** $x = 2\sqrt{6}i, x = -2\sqrt{6}i$
3. $x = \frac{1}{2} \pm \frac{3\sqrt{11}}{2}i$ **4.** $x = \frac{6}{5} \pm \frac{3}{5}i$
5. $x = -5, x = \frac{5}{2} \pm \frac{5\sqrt{3}}{2}i$
6. $x = \frac{3}{4}, x = -\frac{3}{8} \pm \frac{3\sqrt{3}}{8}i$

7. $x = \sqrt{10}, x = -\sqrt{10}, x = \sqrt{10}i, x = -\sqrt{10}i$
8. $x = 4i, x = -4i, x = 3i, x = -3i$
9. $x = 1 \pm \sqrt{6}i$ **10.** $x = \frac{3}{4} \pm \frac{\sqrt{15}}{4}i$

Speaking the Language of Algebra (page 498)
1. Real **2.** Index; radicand **3.** Like radicals
4. Rationalizing the denominator **5.** Extraneous
6. Complex number; real; real; $\sqrt{-1}$
7. Real; imaginary; imaginary unit
8. Real parts; imaginary parts **9.** Conjugate
10. Conjugate; denominator

Review Exercises (page 499)
1. x **2.** $\frac{x^2}{y^8}$ **3.** $\frac{x^{\frac{3}{4}}y^{\frac{7}{4}}}{z}$ **4.** $x^{\frac{1}{6}}y^{\frac{1}{10}}$ **5.** -5 **6.** $\frac{1}{3}$
7. $a^{\frac{2}{5}}b^{\frac{3}{5}}$ **8.** $(x - y)^{-\frac{2}{5}}$ or $\frac{1}{(x - y)^{\frac{2}{5}}}$ **9.** $(\sqrt[3]{27})^2 = 9$
10. $\sqrt[4]{(a + b)^3}$ **11.** $3xy\sqrt{2x}$ **12.** $-\frac{2x\sqrt[3]{3x^2}}{5y}$
13. $2\sqrt{7} - 11x$ **14.** $x\sqrt{5}$ **15.** $5\sqrt[3]{2} - \sqrt{3}$
16. $-\frac{16\sqrt{2}}{9}$ **17.** $5x^2\sqrt{6}$ **18.** $3a\sqrt{b}$
19. $3\sqrt{2} + 3$ **20.** 78 **21.** $2x - 6\sqrt{2x} + 9$
22. $\frac{2\sqrt{6a}}{3a}$ **23.** $\frac{3\sqrt[3]{4x}}{2x}$ **24.** $\frac{5x\sqrt{6xy}}{3y}$ **25.** $\frac{\sqrt{xy}}{2y^2}$
26. $-(2\sqrt{3} + 3\sqrt{2})$ **27.** $x = 20$ **28.** $v = \pi r^2 h$
29. $x = 3$ **30.** $x = -13$ **31.** $x = 5$ **32.** $x = 2$
33. $0 + 12i$ **34.** $2 + 5i$ **35.** $6 - 2\sqrt{3}i$
36. $0 + 9i$ **37.** $x = 3, y = -4$ **38.** $-i$
39. $1 - 12\sqrt{2}i$ **40.** $23 + 14i$ **41.** $-\frac{2}{3} - \frac{1}{3}i$
42. $-\frac{1}{2} - \frac{5}{2}i$ **43.** $x = 3\sqrt{3}i, x = -3\sqrt{3}i$
44. $x = \frac{1}{3} \pm \frac{\sqrt{2}}{3}i$ **45.** $x = 1 \pm \sqrt{3}i$
46. $x = 4, x = -2 \pm 2\sqrt{3}i$ **47.** $x = -\frac{1}{2} \pm \frac{\sqrt{3}}{2}i$
48. $x = -\frac{1}{5} \pm \frac{3}{5}i$

Practice Test (page 503)
1. $x^{\frac{3}{4}}y^{\frac{1}{4}}$ **2.** $3\sqrt[5]{(x - y)^2}$ **3.** $\frac{1}{\sqrt[3]{a^2 - b^2}}$
4. $0 + 7i$ **5.** $5 - 5\sqrt{3}i$ **6.** 6 **7.** $\frac{4}{9}$ **8.** $x^{\frac{7}{5}}$
9. $\frac{b^{10}}{a^9}$ **10.** $x^{\frac{3}{8}}y^{\frac{1}{5}}$ **11.** $\frac{y^{\frac{1}{2}}}{x^{\frac{1}{4}}z^{\frac{5}{4}}}$ **12.** $\frac{3x\sqrt{7x}}{7y}$
13. $13\sqrt{2x}$ **14.** $-\sqrt[3]{3} + 4\sqrt{3}$ **15.** $\frac{4x\sqrt{5}}{3}$
16. $6x^2\sqrt{6x}$ **17.** $-2 + 5\sqrt{7}$
18. $3x - 10\sqrt{3x} + 25$ **19.** $3x^2$ **20.** $\frac{10x\sqrt{14xy}}{7y}$
21. $\frac{\sqrt[3]{15x}}{5x}$ **22.** $\frac{5(\sqrt{7} - \sqrt{3})}{2}$ **23.** $-18 - 24i$

24. $-\frac{1}{4} - i$ **25.** $\frac{13}{10} + \frac{11}{10}i$ **26.** $x = 2, y = -\frac{3}{7}$

27. $x = 13$ **28.** No solution **29.** $a = \frac{1}{3} \pm \frac{\sqrt{2}}{3}i$

30. $S = 4\pi r^2$

Sharpening Your Skills after Chapters 1–7 (page 505)

1. -24 **2.** $5 - \sqrt{10}$ **3.** 8 **4.** $\frac{1}{64}$ **5.** $\frac{3\sqrt{2}}{5}$

6. 7 **7.** $-118x + 16$ **8.** $5x^5y^{11}z^5$ **9.** $-\frac{54}{y^8}$

10. $-27a^3 - 8a^2$ **11.** $x^3 + 3x^2 - 9x + 5$

12. $7x - 2 + \frac{1}{3x - 2}$ **13.** $\frac{a - 11}{5a - 2}$ **14.** $\frac{x^2(x - 13)}{2y^5}$

15. $\frac{x + 3}{x(x + 6)}$ **16.** $\frac{5(2x - 5)}{x}$

17. $2xyz(3x^3 - 6x^2y - 15xy^2 - 2y^3)$

18. $3x^2y^2(x + 3y)(x^2 - 3xy + 9y^2)$

19. $(3a - 2)(a - 5)$ **20.** $2(x + 3)(x - 3)(x^2 + 2)$

21. $(x - 2)(x + 2)(x^2 - 2x + 4)$ **22.** $x = -1$

23. $x \geq 1$ or $x \leq -\frac{3}{7}$; $(-\infty, -\frac{3}{7}] \cup [1, \infty)$

24. $x = \frac{7}{20} \pm \frac{\sqrt{71}}{20}i$

25. $x = \frac{1 + \sqrt{2}}{3}, x = \frac{1 - \sqrt{2}}{3}$ **26.** $x \in R$

27. $x = 0$ **28.** $x < 0$ or $x > 9$; $(-\infty, 0) \cup (9, \infty)$

29. \$2250 at 12%; \$2750 at 9% **30.** 4 ft

CHAPTER 8

Exercise Set 8.1 (page 527)

1.

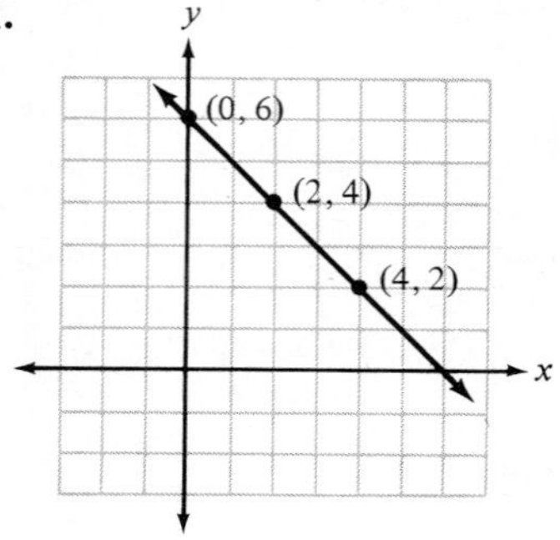

3.

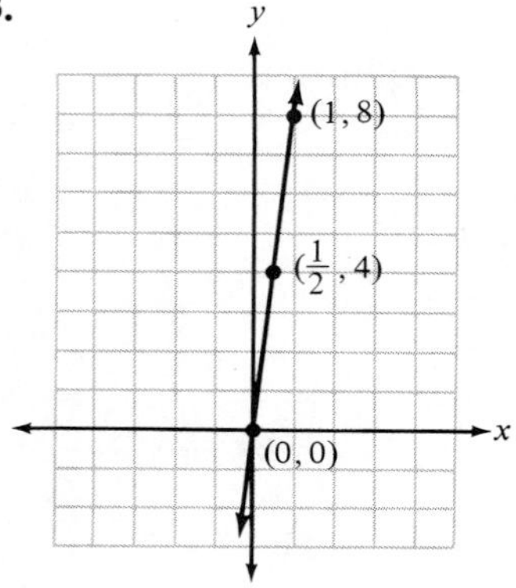

5.

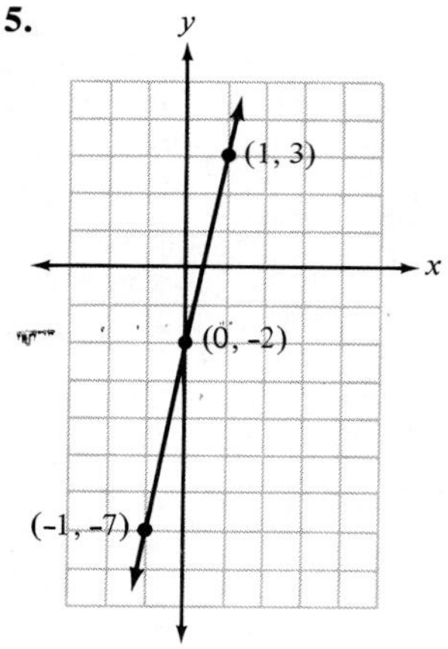

7.

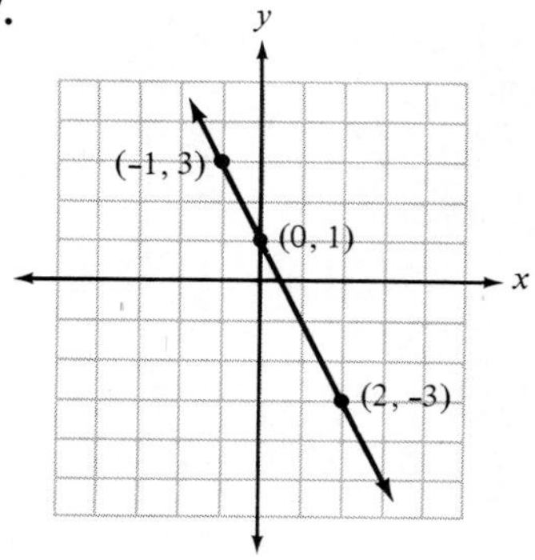

9.

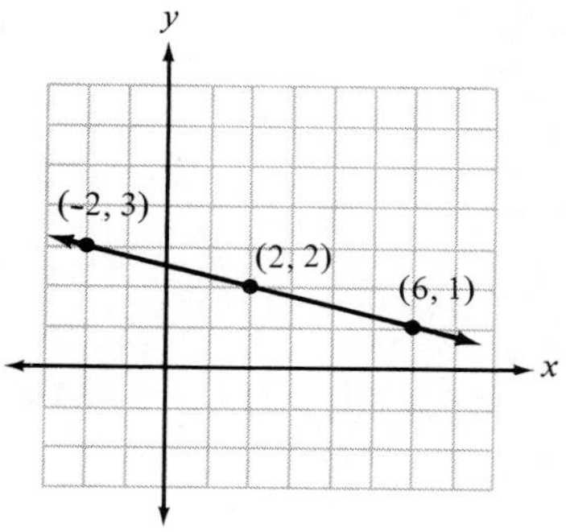

11.

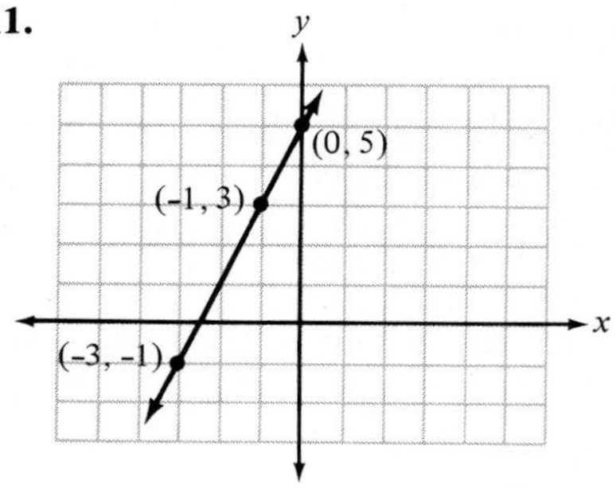

13.

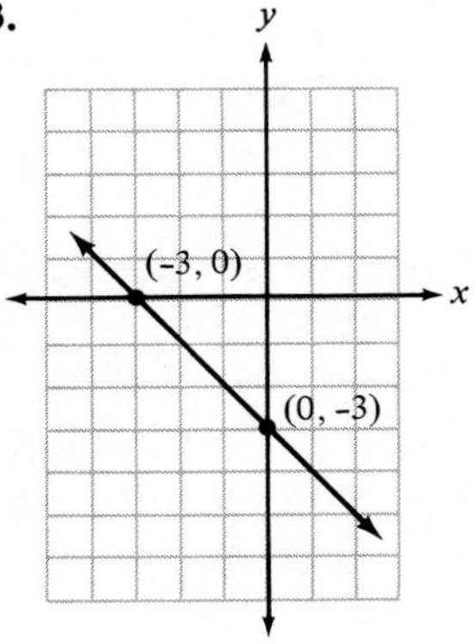

15.

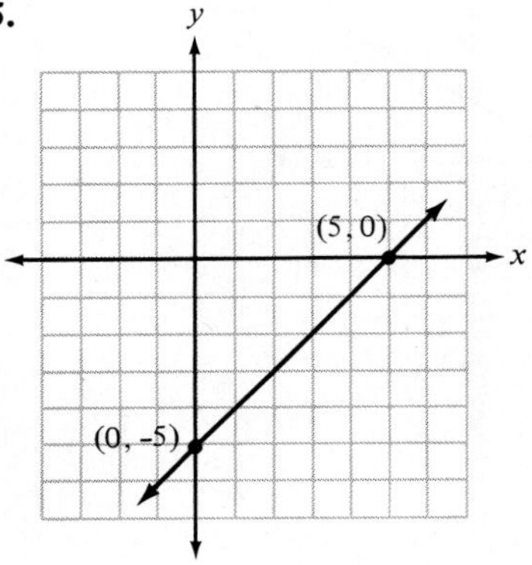

17.

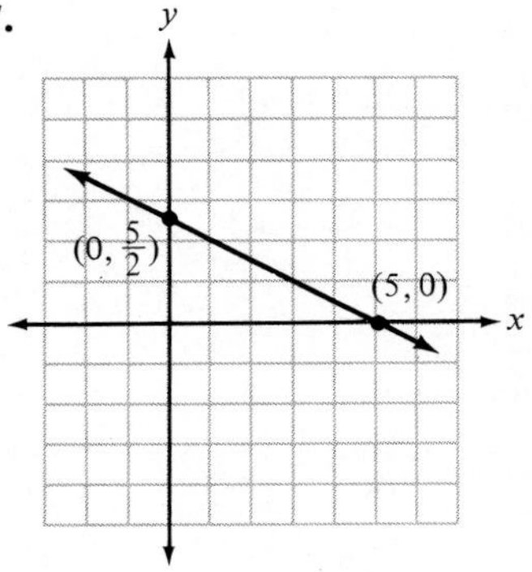

19.

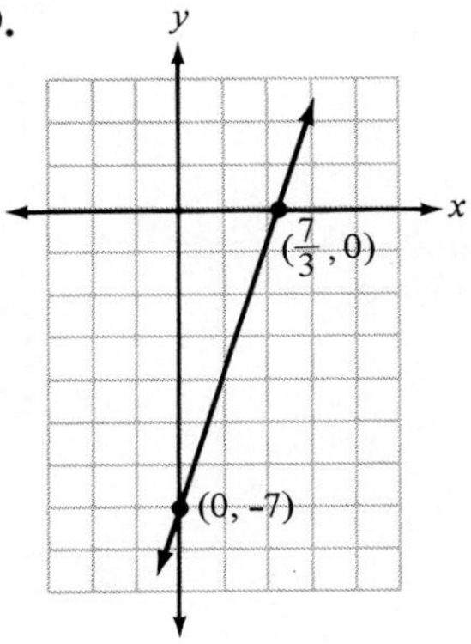

21.

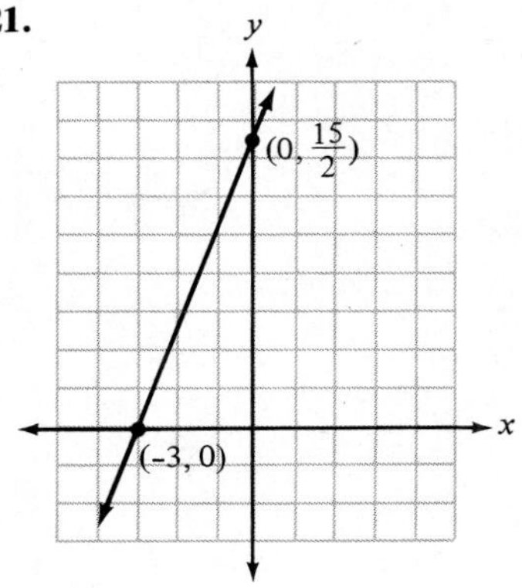

23.

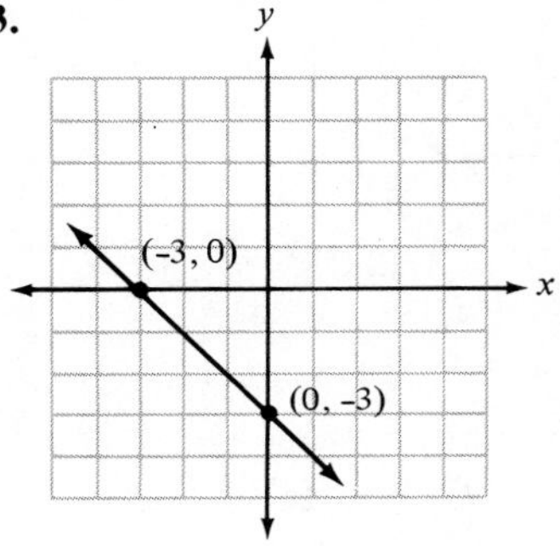

25.

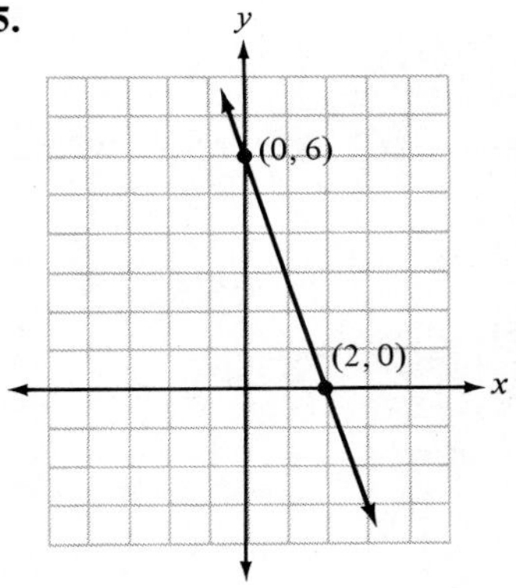

27.

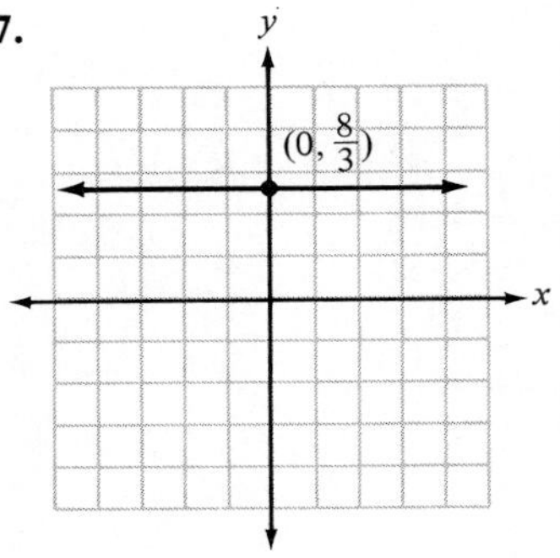

29.

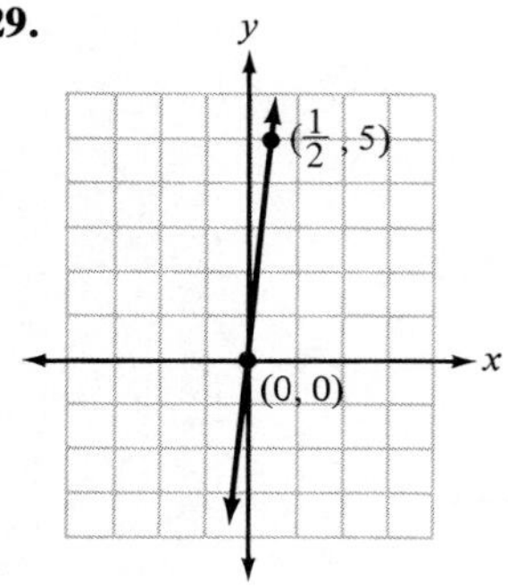

31.

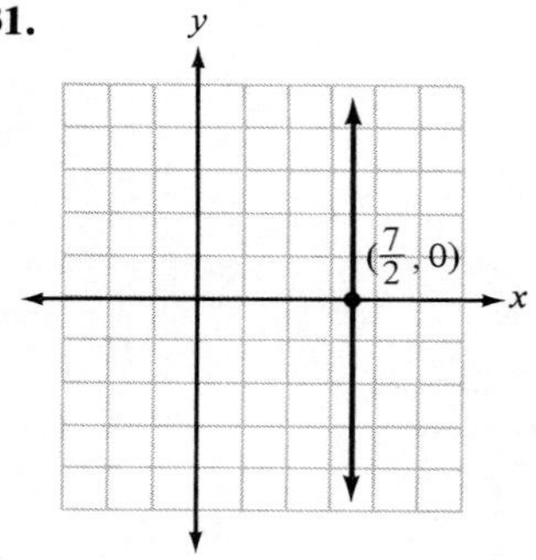

33.

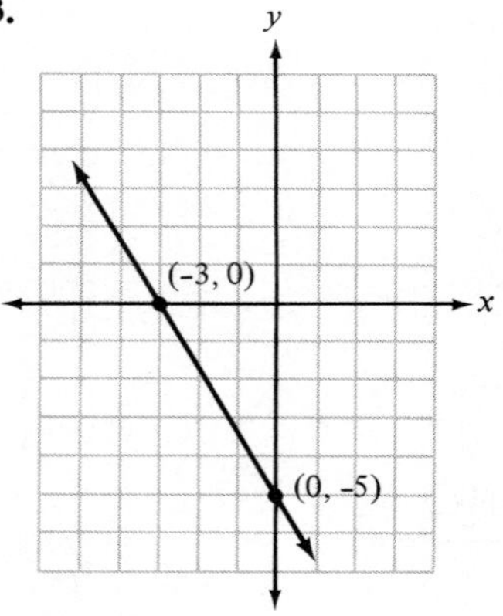

35.

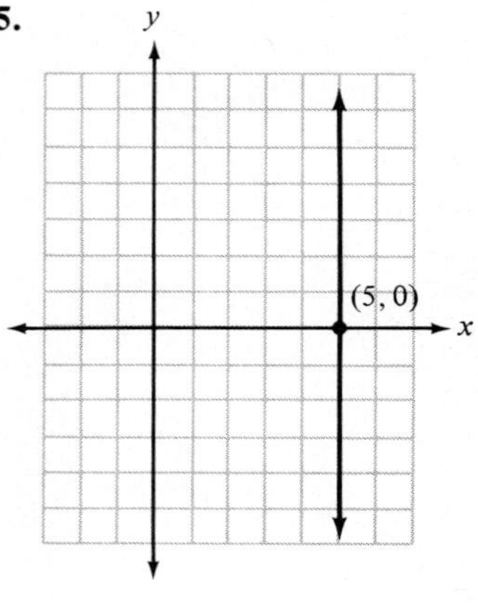

37.

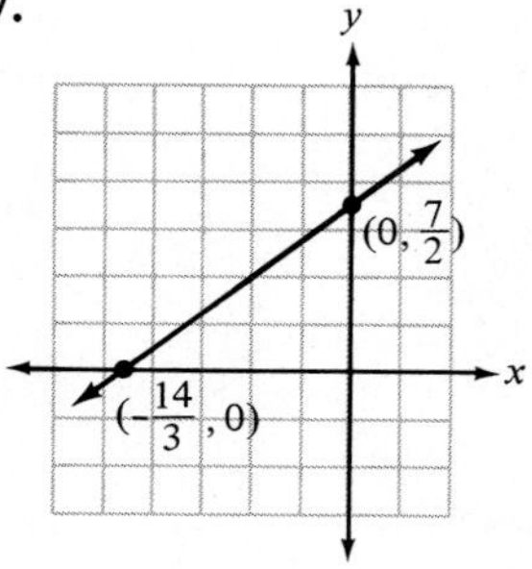

39.

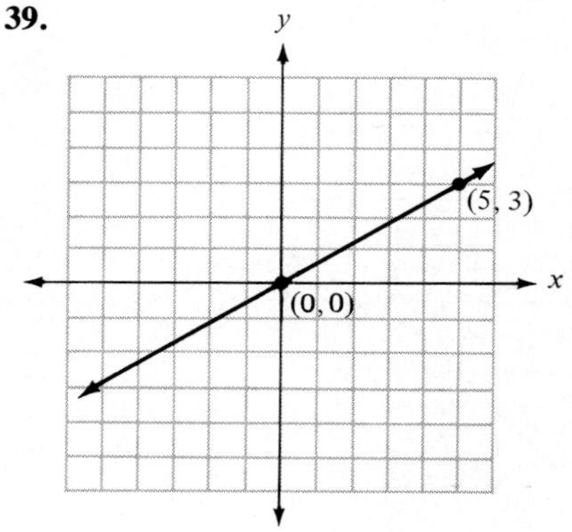

Stretching the Topics

1.

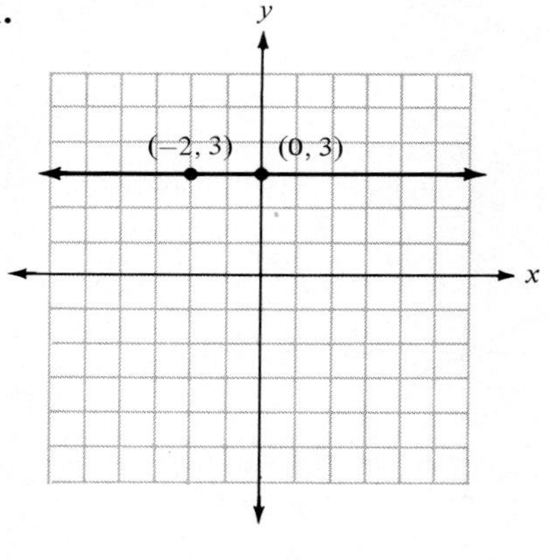

2.

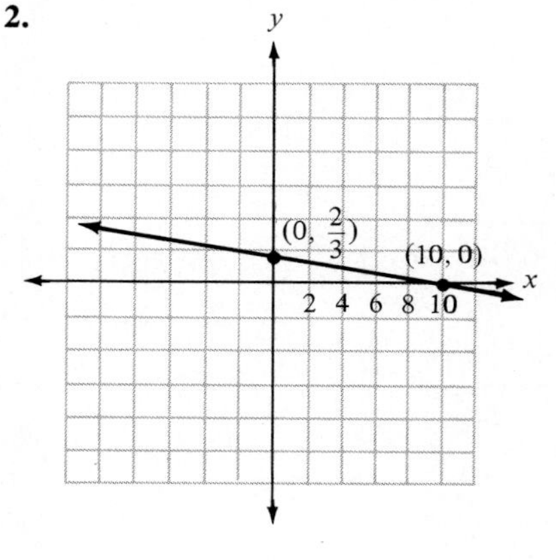

3.

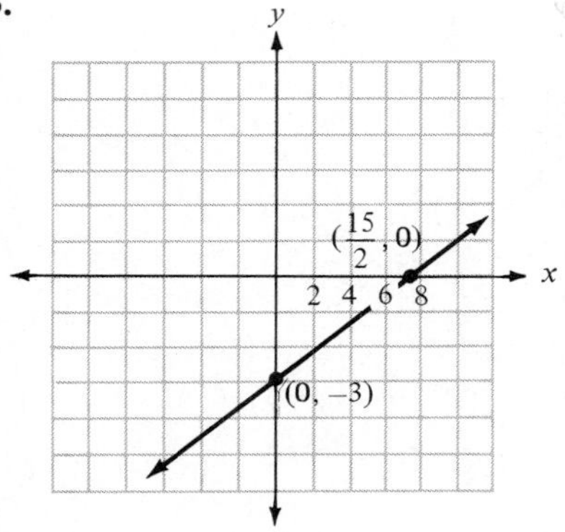

4.

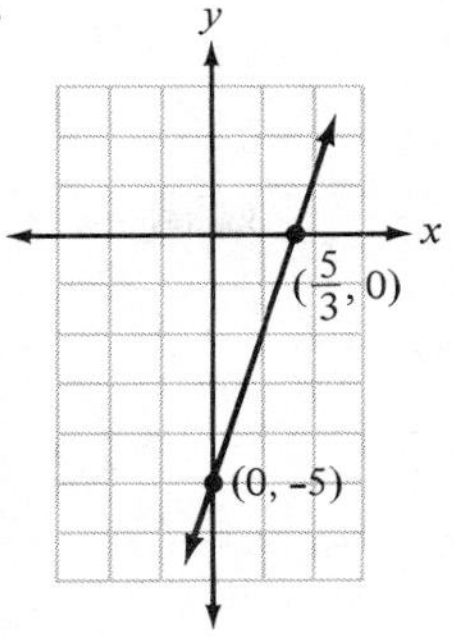

Checkup 8.1 (page 528)

1.

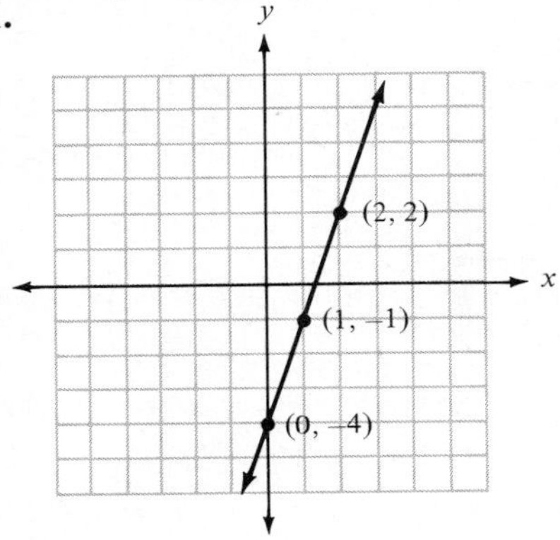

2.

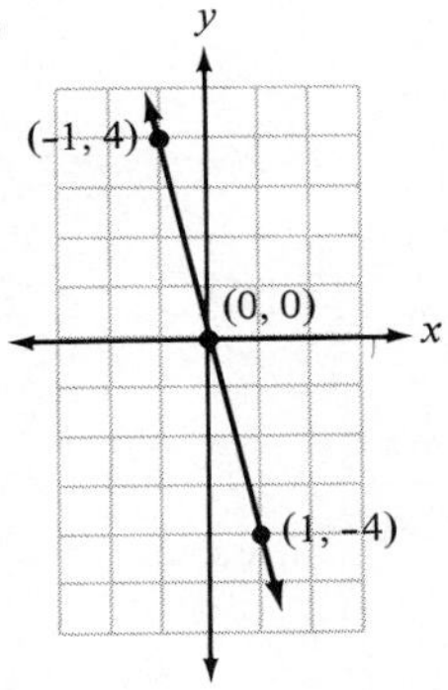

3.

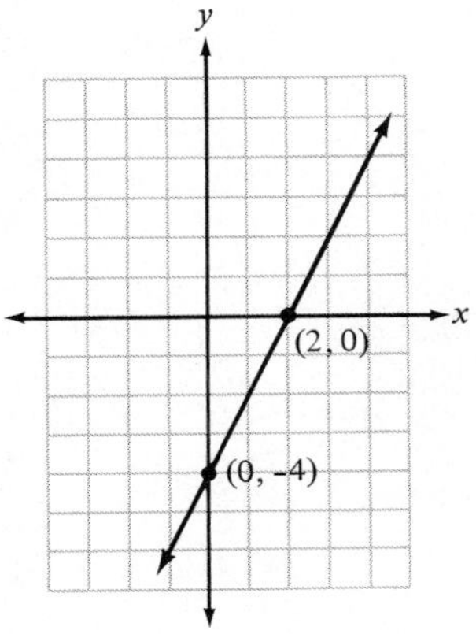

5.

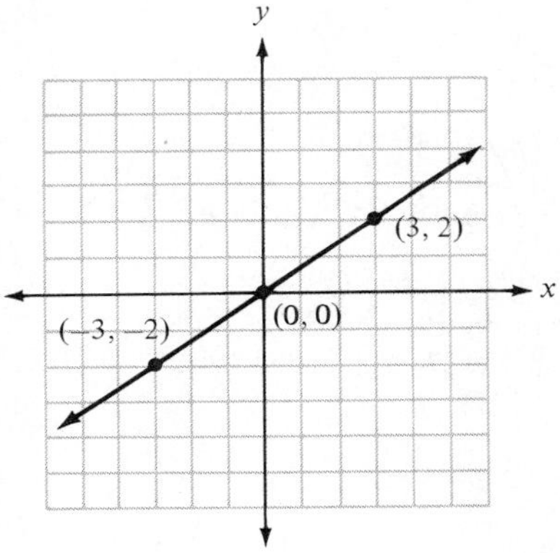

6.

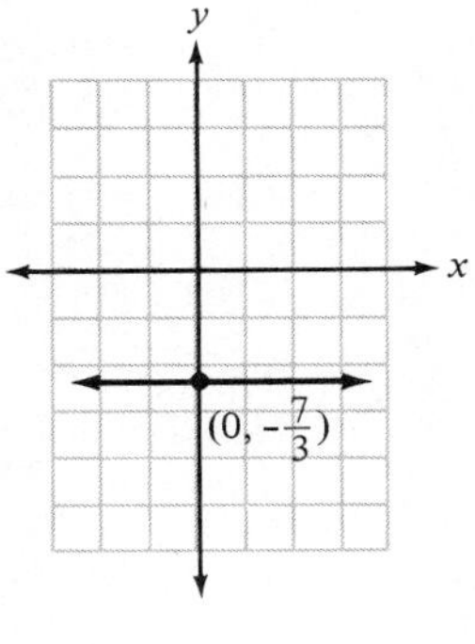

7.

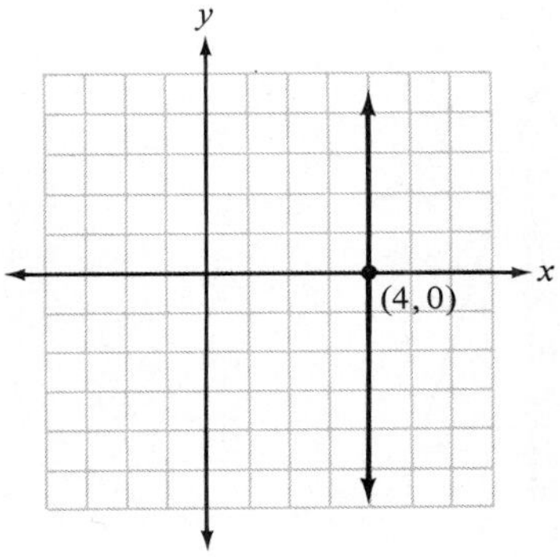

8.

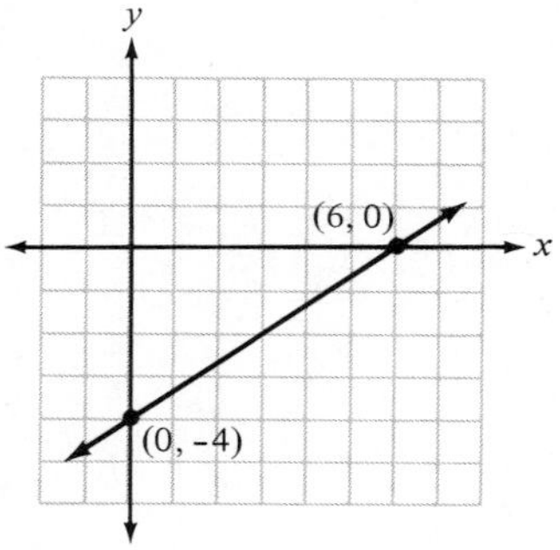

41.

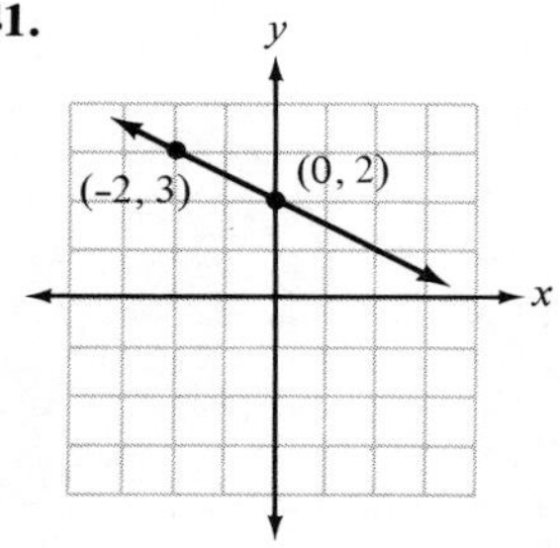

43.

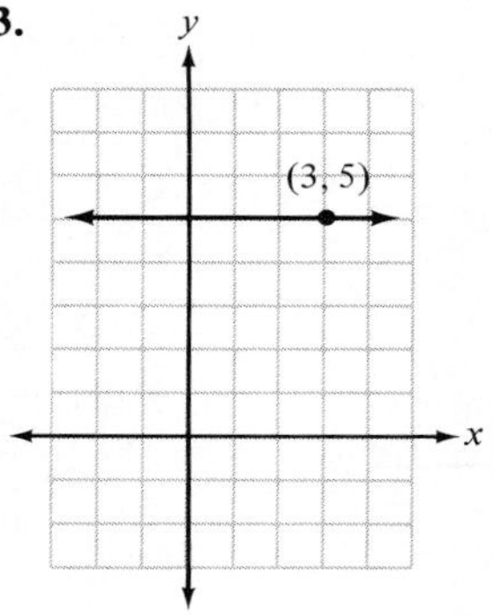

Exercise Set 8.2 (page 547)

1. 5 **3.** $3\sqrt{5}$ **5.** $\sqrt{5}$ **7.** 13 **9.** $\dfrac{\sqrt{5}}{2}$
11. $\sqrt{29}$ **13.** (6, 11) **15.** (1, −2)
17. $(-\frac{5}{2}, 3)$ **19.** $(-6, \frac{11}{2})$ **21.** (4, 0)
23. $(-\frac{1}{4}, \frac{7}{4})$ **25.** $\frac{5}{3}$ **27.** Undefined **29.** −1
31. 0 **33.** $-\frac{39}{2}$ **35.** 3

37.

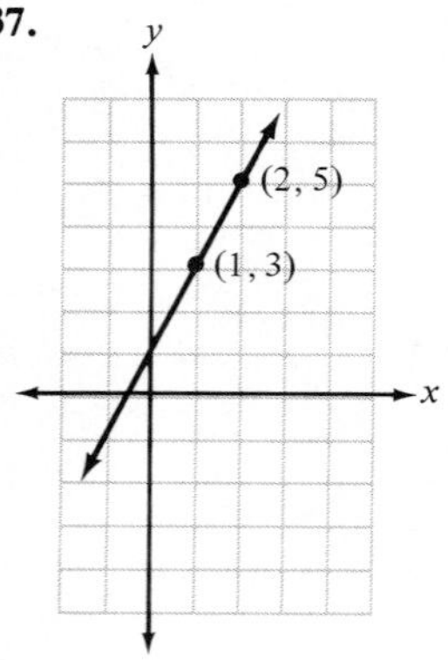

45.

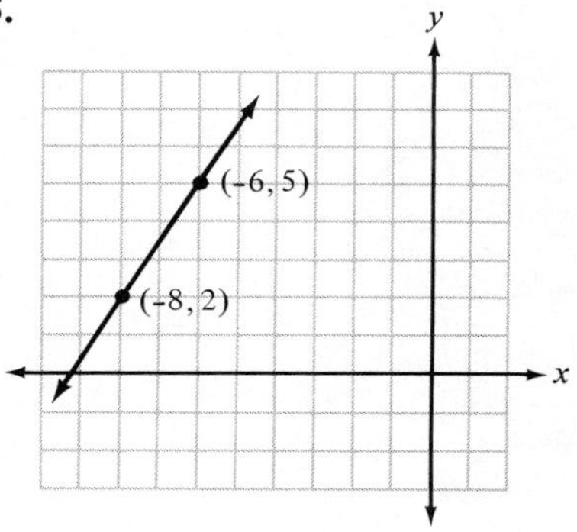

39.

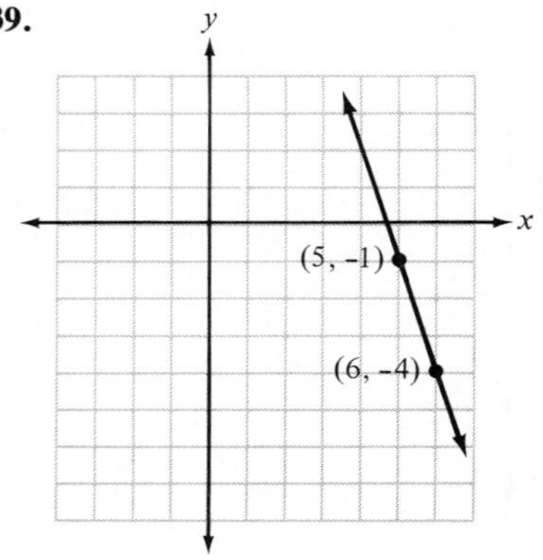

47.

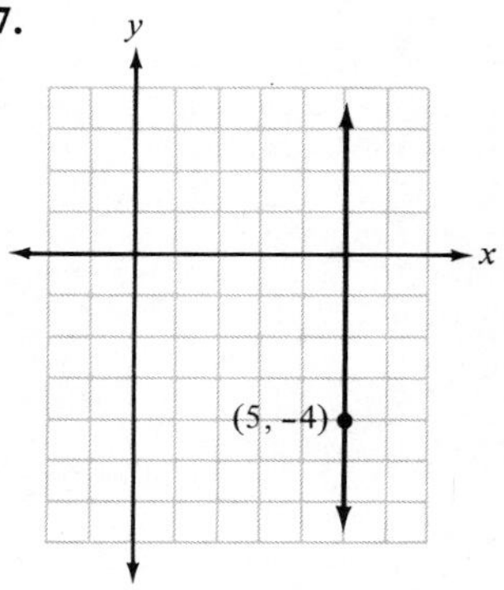

49.

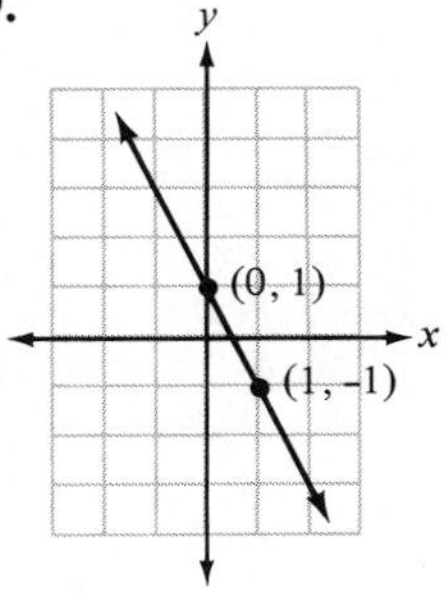

51.

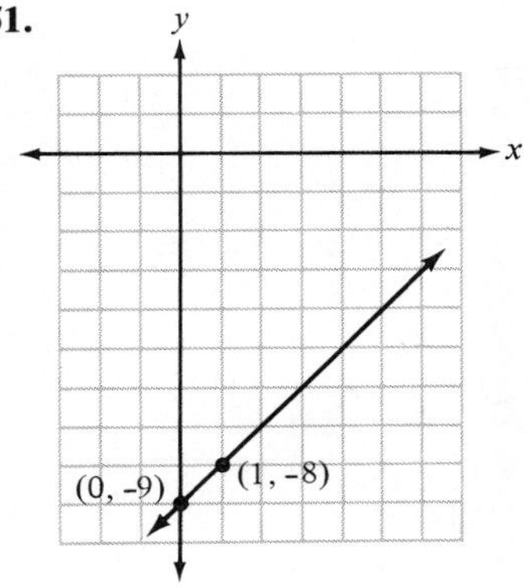

53.

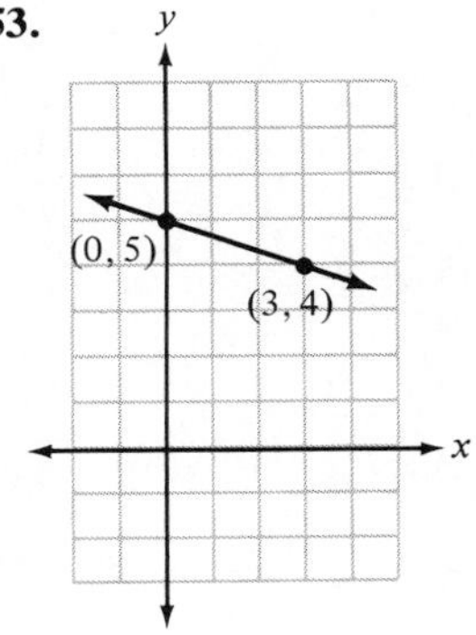

55.

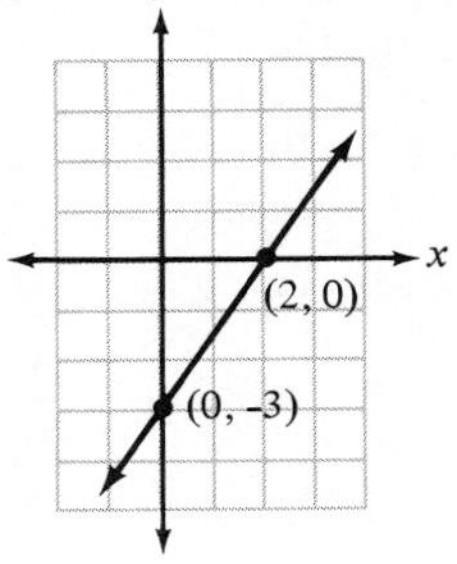

57.

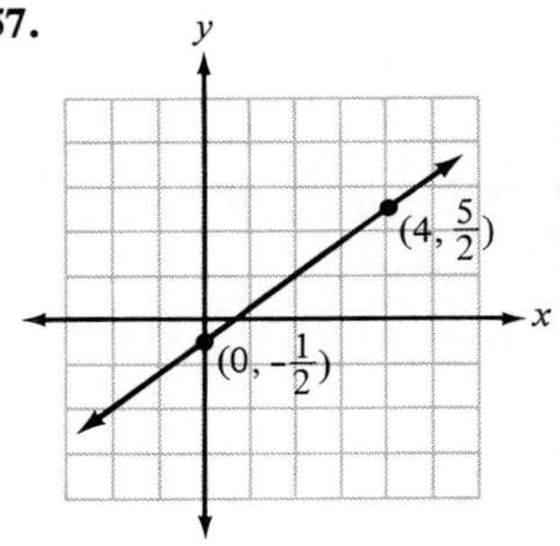

59.

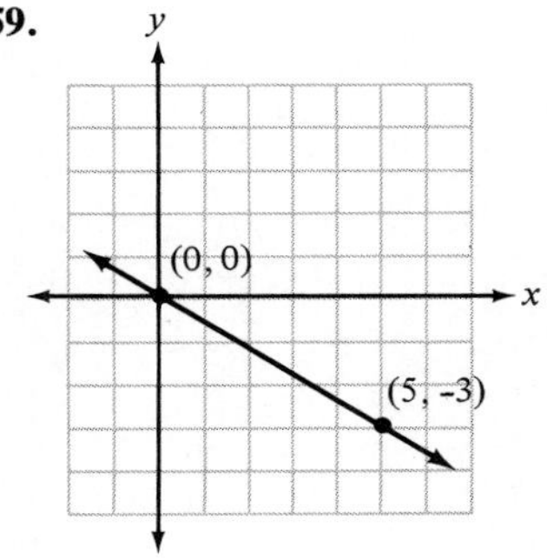

Stretching the Topics

1. 24 **2.** $\left(-\frac{5}{3}, -\frac{19}{6}\right)$ **3.** $m = \frac{13}{42}$, $b = \frac{37}{28}$

Checkup 8.2 (page 549)

1. 5 **2.** $(7, -8)$ **3.** -1 **4.** $-\frac{3}{4}$

5.

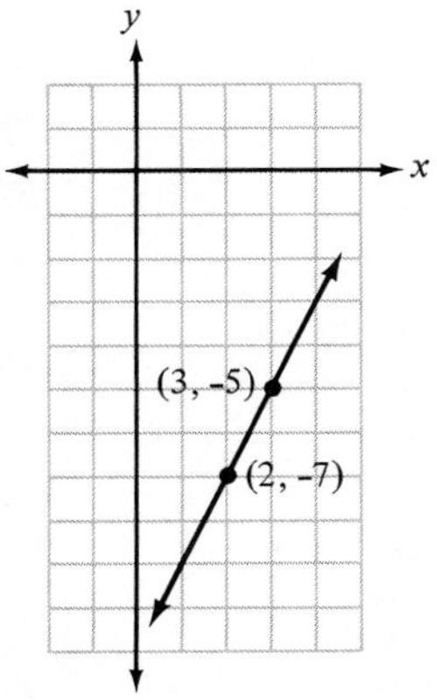

6.

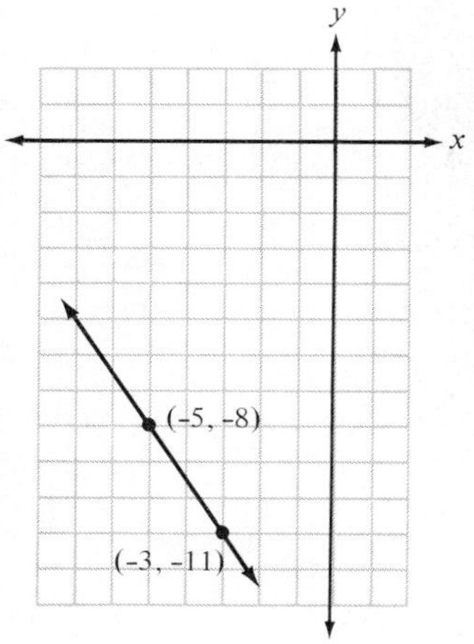

7.

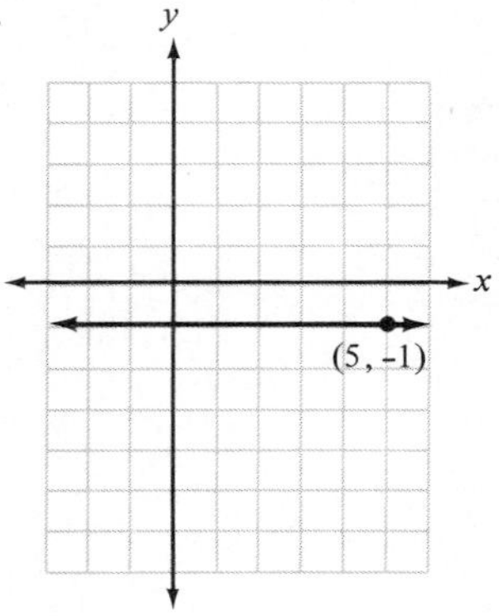

8.

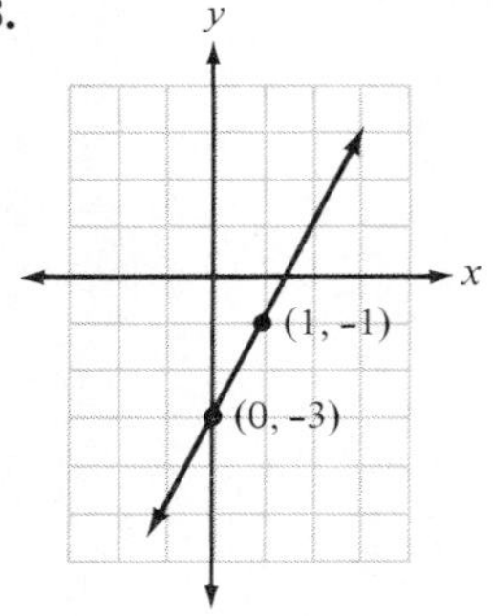

9.

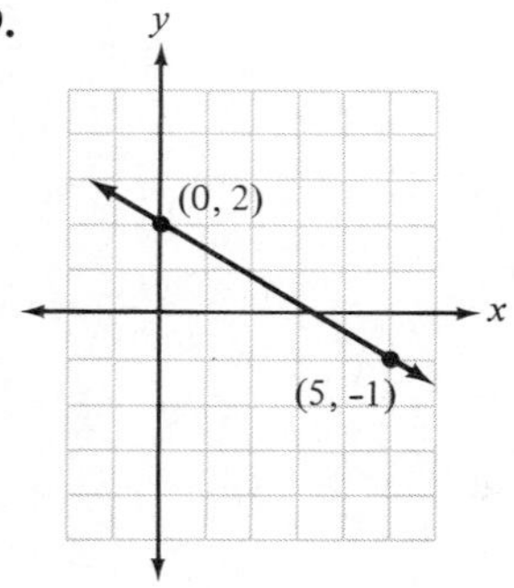

10.

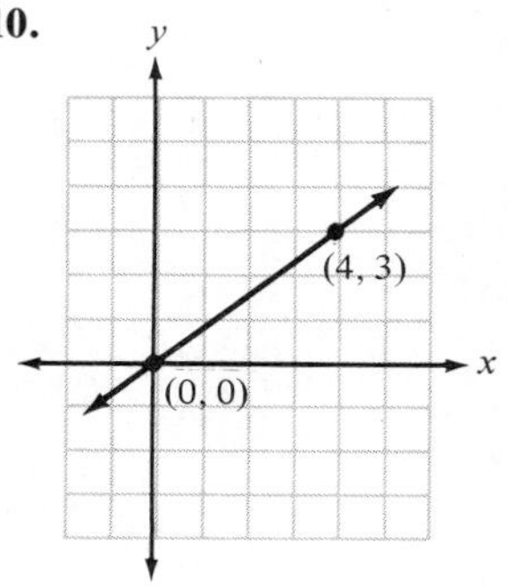

Exercise Set 8.3 (page 561)

1. $y = -5$ **3.** $y = \frac{1}{2}$ **5.** $x = 2$ **7.** $x = -\frac{2}{3}$
9. $y = 4x + 2$ or $4x - y = -2$
11. $y = -2x + 1$ or $2x + y = 1$
13. $y = \frac{1}{3}x - \frac{2}{3}$ or $x - 3y = 2$ **15.** $y = 10$
17. $y = \frac{1}{2}x - \frac{5}{4}$ or $2x - 4y = 5$
19. $y = 7x - 2$ or $7x - y = 2$
21. $y = 8x + 13$ or $8x - y = -13$
23. $y = -x + 9$ or $x + y = 9$ **25.** $y = -3$
27. $y = \frac{3}{4}x + \frac{19}{4}$ or $3x - 4y = -19$ **29.** $x = -2$
31. $y = -\frac{2}{3}x + \frac{7}{12}$ or $8x + 12y = 7$
33. $y = -2x + 2$ or $2x + y = 2$
35. $y = -\frac{5}{3}x + \frac{16}{3}$ or $5x + 3y = 16$ **37.** $y = -8$
39. $y = -\frac{5}{2}x - \frac{1}{2}$ or $5x + 2y = -1$ **41.** $x = 6$
43. $y = x - \frac{11}{10}$ or $10x - 10y = 11$
45. $y = -2x - 4$ or $2x + y = -4$
47. $y = 2x + 3$ or $2x - y = -3$
49. $y = -\frac{1}{3}x + 4$ or $x + 3y = 12$
51. $y = -\frac{1}{3}x + \frac{7}{3}$ or $x + 3y = 7$
53. $y = 2x + 4$ or $2x - y = -4$
55. $y = -\frac{1}{2}x - \frac{1}{2}$ or $x + 2y = -1$
57. $y = \frac{5}{3}x - 5$ or $5x - 3y = 15$
59. $y = -2x + 4$ or $2x + y = 4$

Stretching the Topics

1. $y = \frac{13}{8}x + \frac{33}{2}$ or $13x - 8y = -132$
2. $y = 4x - 4$ or $4x - y = 4$
3. $y = \frac{10}{3}x - \frac{34}{3}$; $y = -\frac{3}{7}x + \frac{26}{7}$; $y = -\frac{13}{4}x - \frac{19}{4}$

Checkup 8.3 (page 563)

1. $y = 7$ **2.** $x = \frac{1}{2}$
3. $y = -4x + 2$ or $4x + y = 2$
4. $y = -\frac{2}{3}x$ or $2x + 3y = 0$
5. $y = -4x - 3$ or $4x + y = -3$ **6.** $x = -3$
7. $y = 5x - 6$ or $5x - y = 6$
8. $y = -x + \frac{3}{10}$ or $10x + 10y = 3$
9. $y = 3x + 5$ or $3x - y = -5$
10. $y = 4x - 5$ or $4x - y = 5$

Exercise Set 8.4 (page 570)

1.

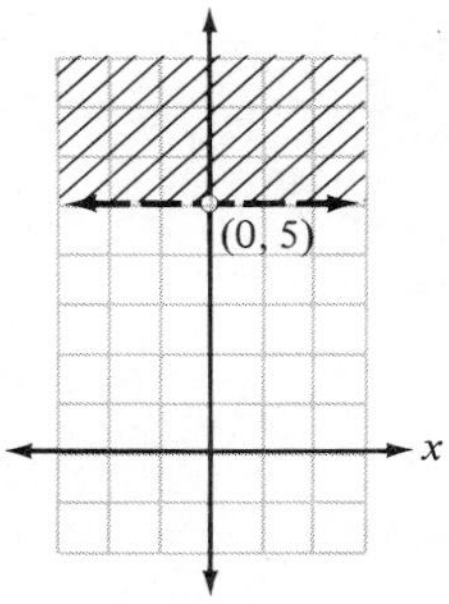

3.

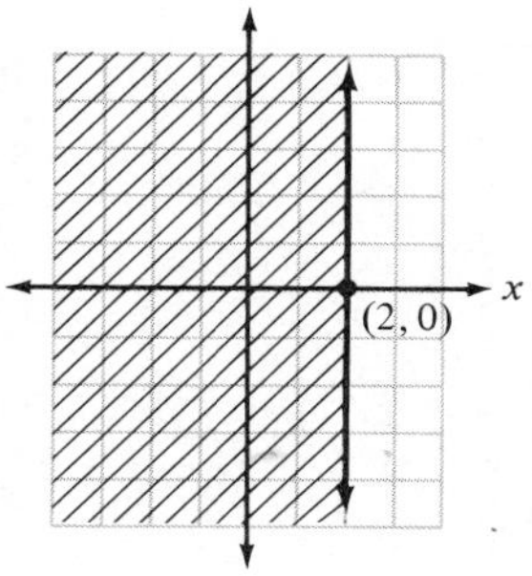

5.

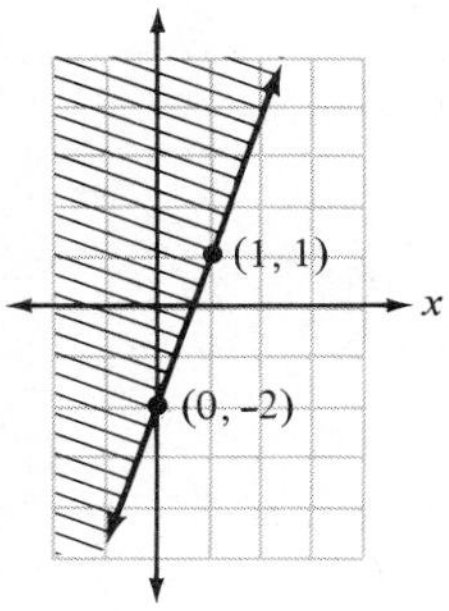

7.

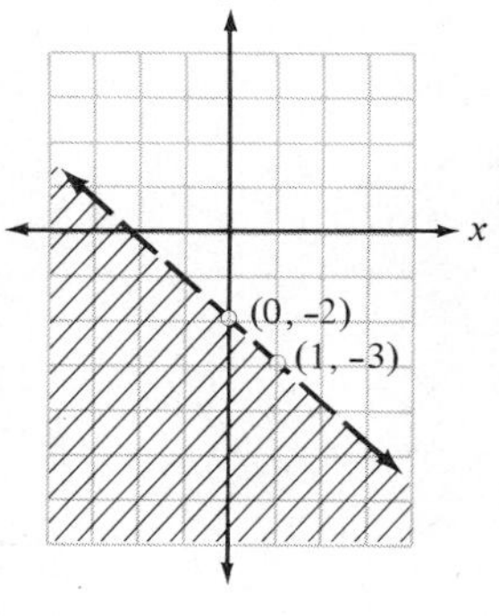

9.

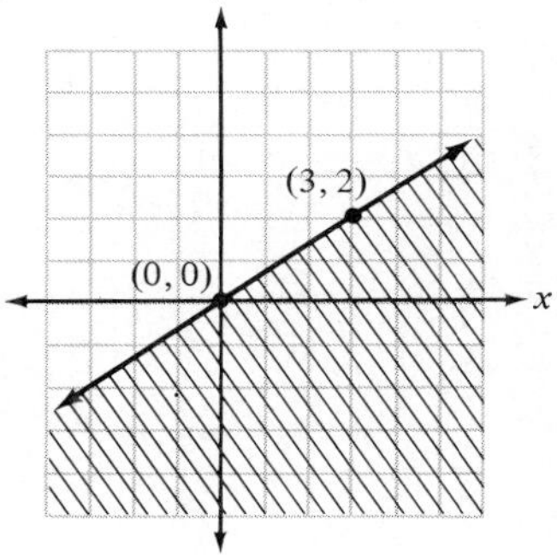

11.

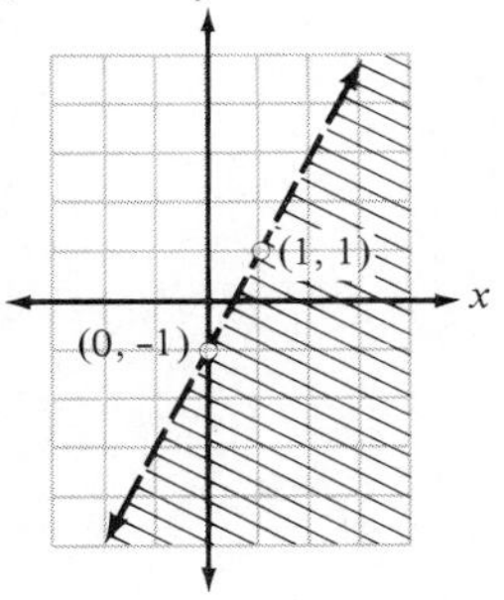

13.

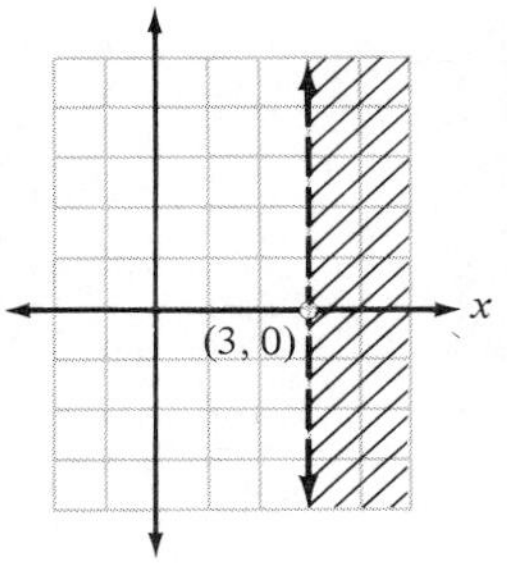

15.

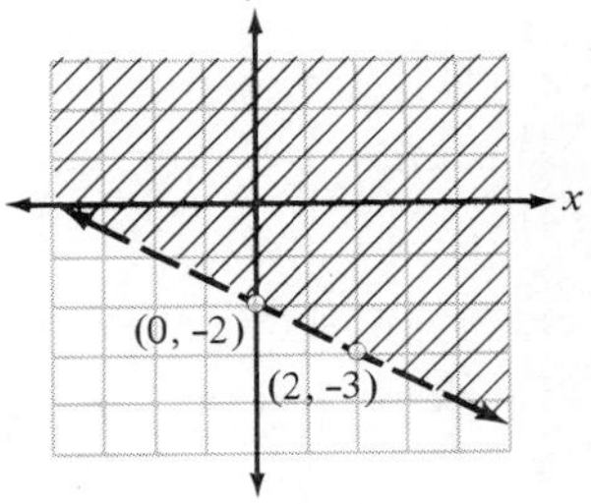

17.

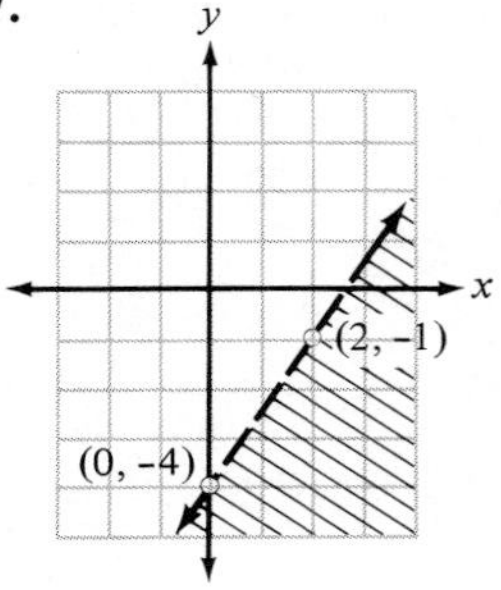

19.

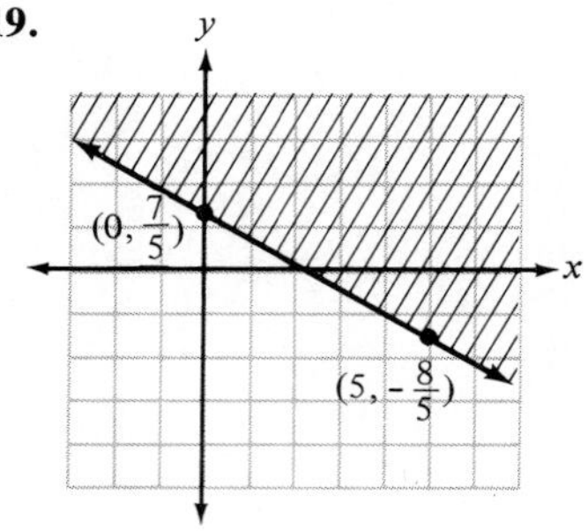

Stretching the Topics

1.

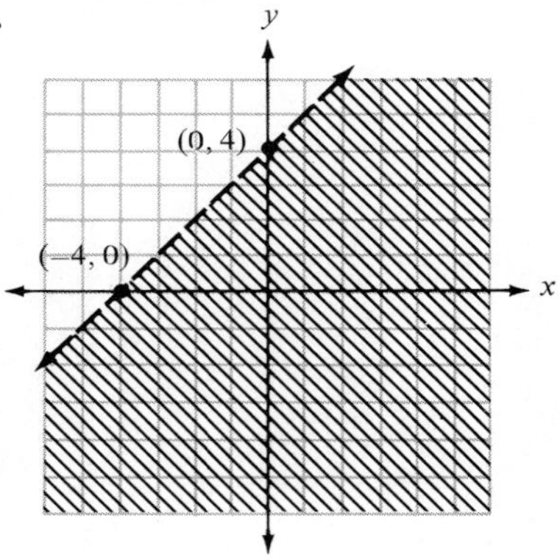

2.

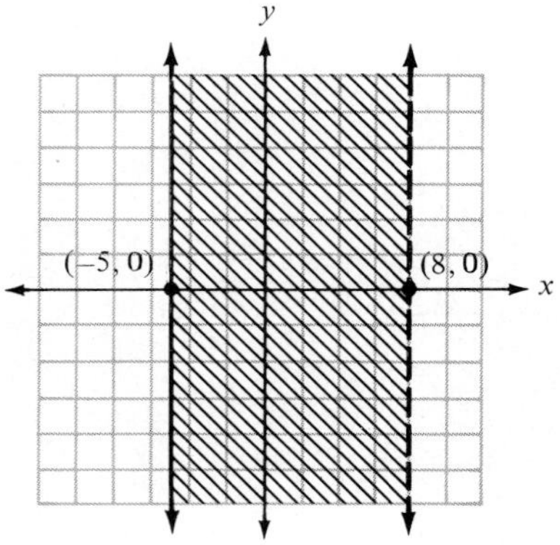

3.

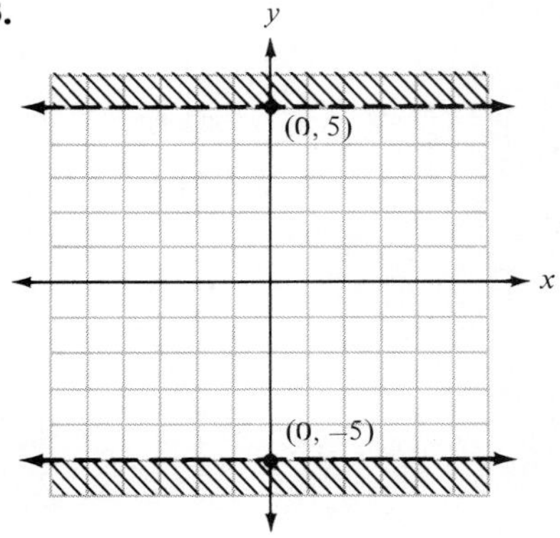

Checkup 8.4 (page 571)

1.

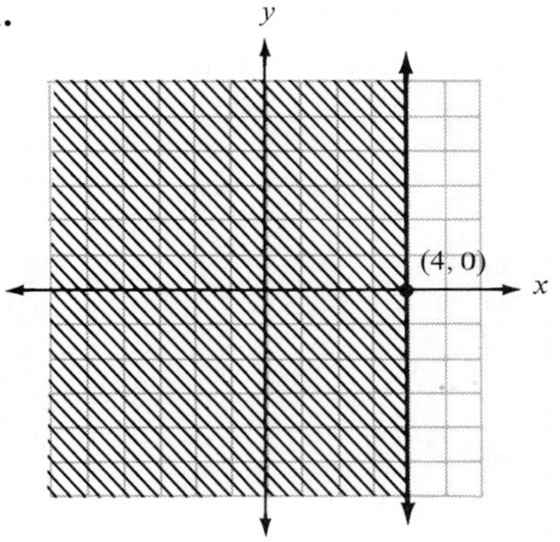

2.

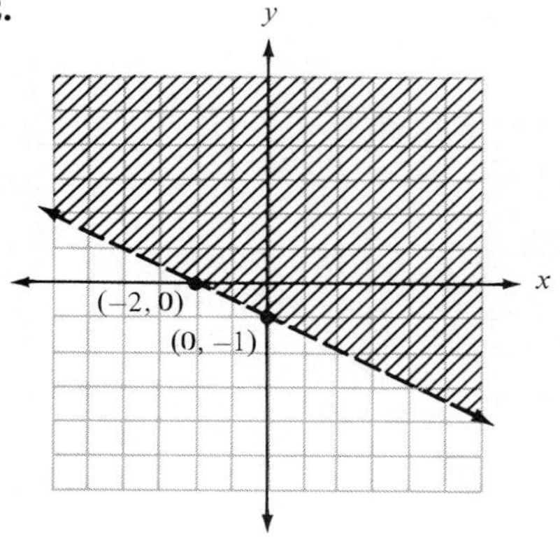

3.

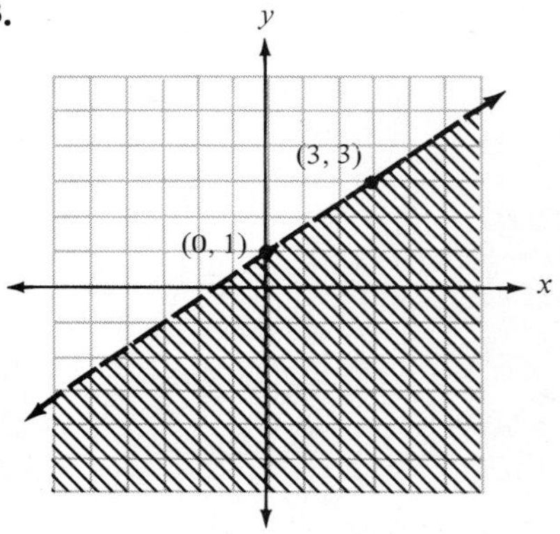

4.

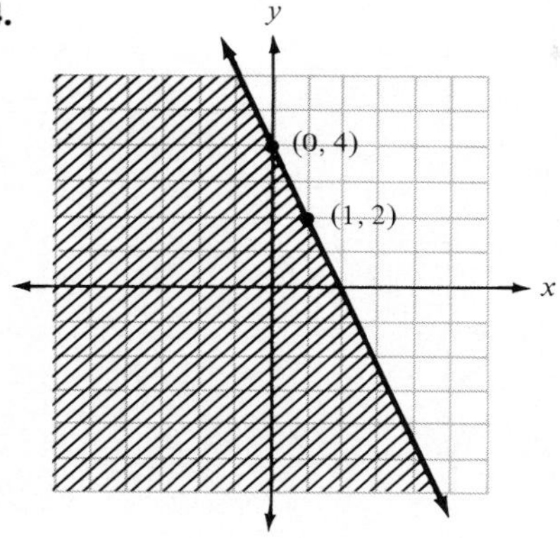

Exercise Set 8.5 (page 580)

1. Domain {1, 2, 3}; range {−3, 0, 3, 5}; no
3. Domain $x \in R$; range $y \in R$; yes
5. Domain {−6, −2, 0, 3}; range {3}; yes
7. Domain {3}; range $y \in R$; no
9. Domain {−2, −1, 1, 2}; range {3, 5}; yes **11.** Yes
13. Yes **15.** No **17.** No **19.** 7 **21.** −9
23. $3x + 3h + 1$ **25.** −16 **27.** $2x - 1$
29. $a^2 - a - 1$ **31.** −2 **33.** $a^2 + a - 1$
35. $x \in R, x \neq -4$ **37.** $x \in R, x \neq \frac{1}{2}$
39. $x \in R, x \geq -1$ **41.** $x \in R, x \neq 3$
43. $x \in R, x \leq 5$ **45.** $x \in R, x \neq 4, x \neq -4$
47. $x \in R, x \geq 3$ **49.** $x \in R, x \neq 1, x \neq \frac{7}{2}$

Stretching the Topics

1. $4a + 2h - 3$ **2.** $\dfrac{a}{a + 1}$
3. $x \in R, x > 0, x \neq 4, x \neq \frac{3}{2}$

Checkup 8.5 (page 583)

1. Domain {5}; range $y \in R$; no **2.** No **3.** Yes
4. 7 **5.** 1 **6.** −10 **7.** $x \in R, x \neq \frac{3}{2}$
8. $x \in R, x \geq 4$ **9.** $x \in R, x \neq 7, x \neq -4$

Exercise Set 8.6 (page 591)

1. $y = 12x$ **3.** $y = \frac{1}{5}x$ **5.** $y = \dfrac{60}{x}$ **7.** $y = \dfrac{4.5}{x}$
9. $z = \frac{3}{4}xy$ **11.** $y = \dfrac{16x}{3z^2}$ **13.** 3200 **15.** 20 min
17. 12 lb **19.** 200 **21.** 5 sec **23.** 225 lb
25. $C = 18 + 0.30m$
27. $s = 100 + 75n$ **29.** $A = 5d$
31. $P = 4w + 24$ **33.** $A = \frac{1}{3}l^2$
35. $A = \frac{1}{2}h^2 + 2h$ **37.** $d = 14j$
39. $a = \dfrac{12 + x}{4}$ **41.** $C = 150 + 55n$
43. $5x + 3y \geq 2000$

Stretching the Topics

1. Multiplied by 9; doubled **2.** 312.5 lb
3. −90; The loss is $90 when 20 items are manufactured.

Checkup 8.6 (page 595)

1. 2 **2.** $5\frac{3}{5}$ hr **3.** $w = xy^2$
4. $F = 6.00 + 1.75n$ **5.** $P = 12w - 10$

Speaking the Language of Algebra (page 598)

1. Ordered pairs **2.** Independent; dependent
3. Domain; range **4.** Straight line; slope; *y*-intercept
5. Horizontal line; vertical line **6.** Dashed; above
7. Slope **8.** Standard; slope-intercept **9.** 17
10. Direct; inverse

Review Exercises (page 601)

1.

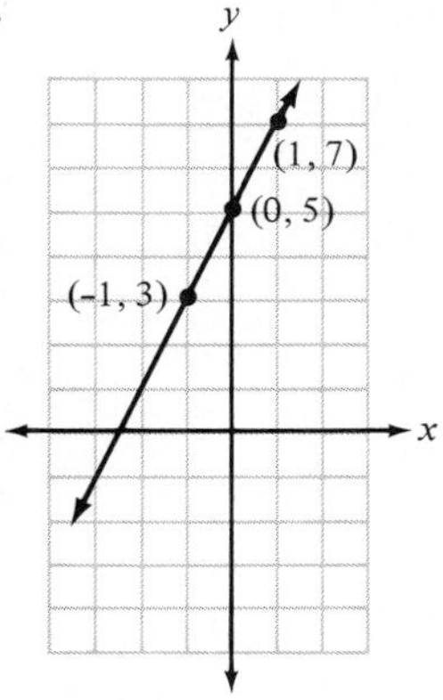

2.

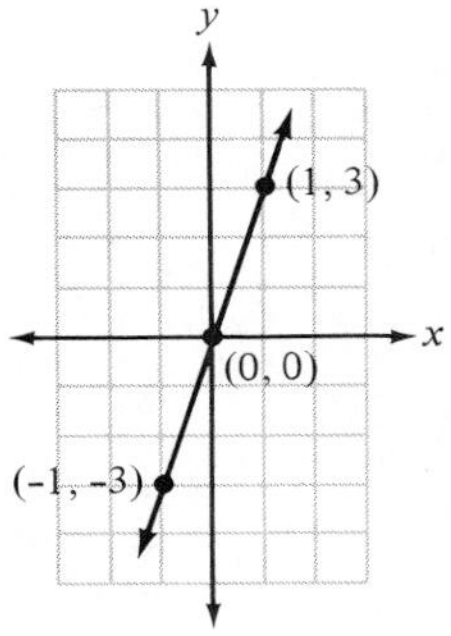

3.

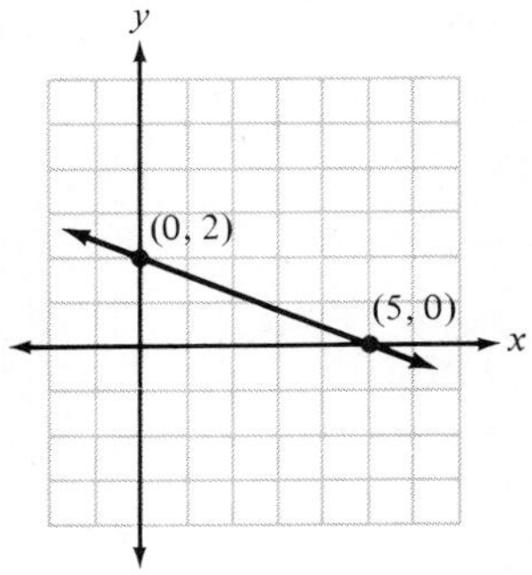

4.

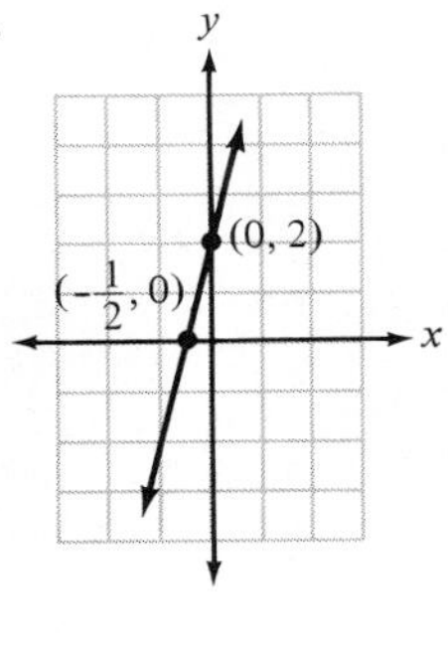

5. 5 **6.** $(2, \frac{3}{2})$ **7.** -2

8.

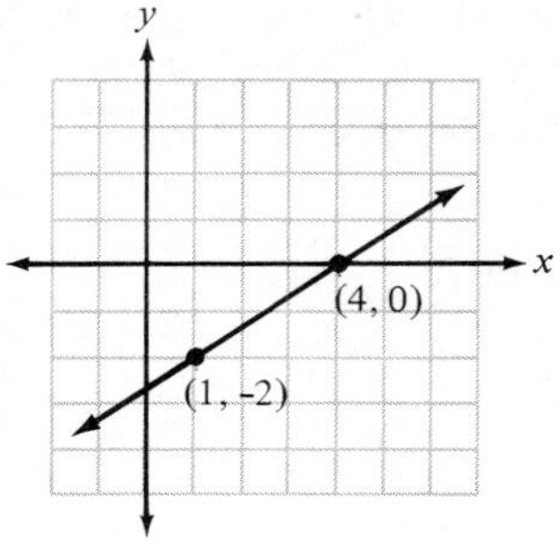

9.

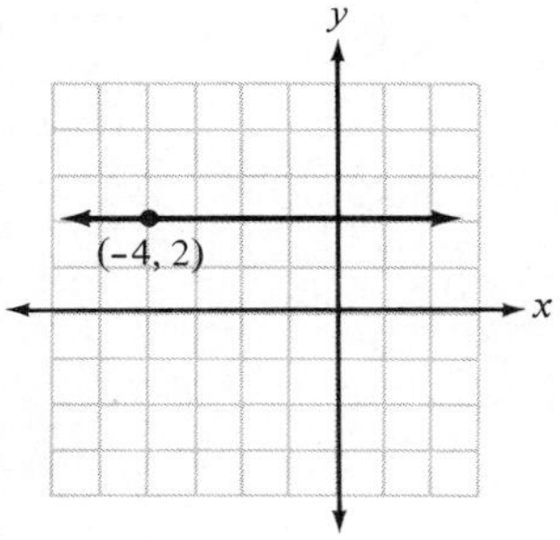

10.

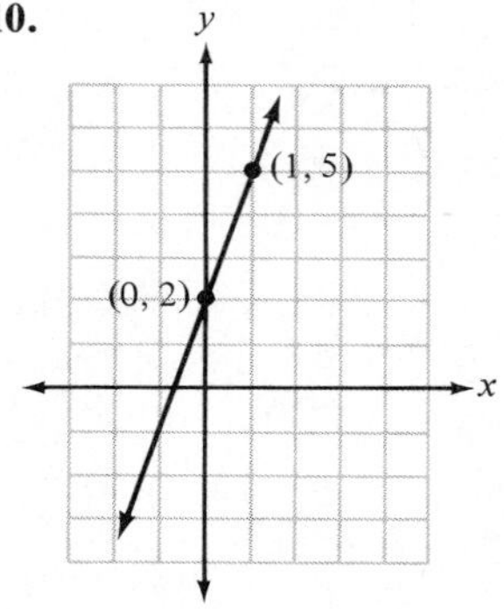

11.

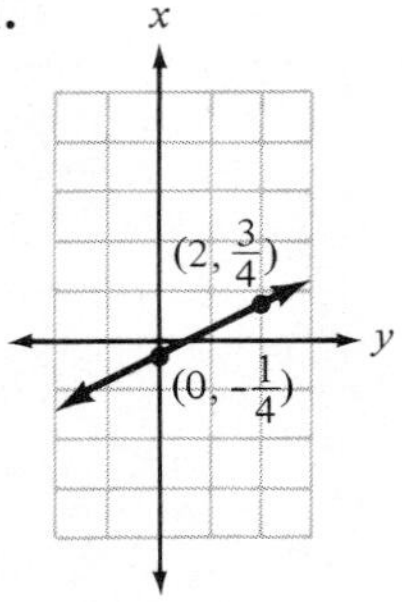

12.

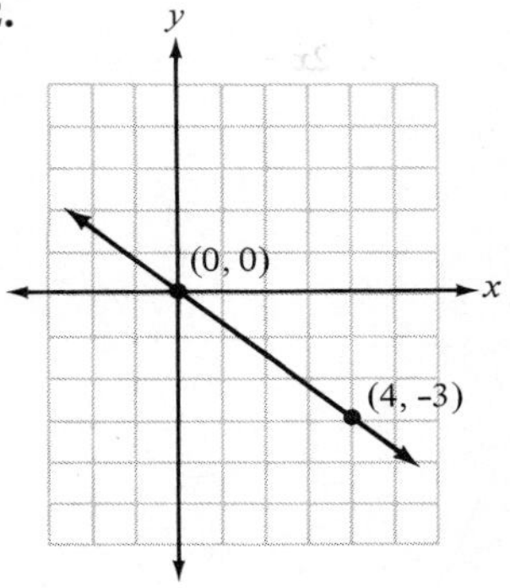

13.

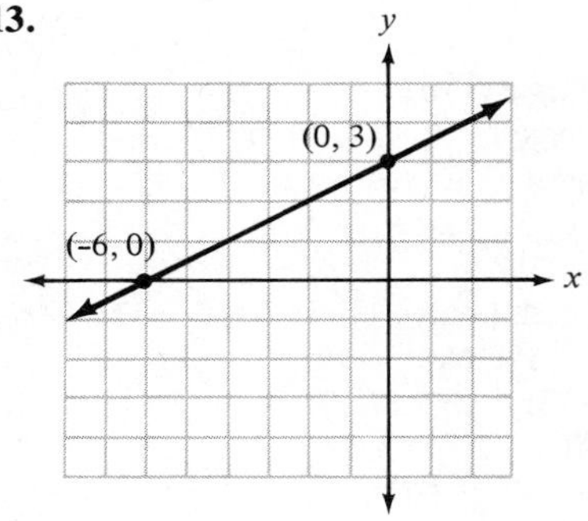

14.

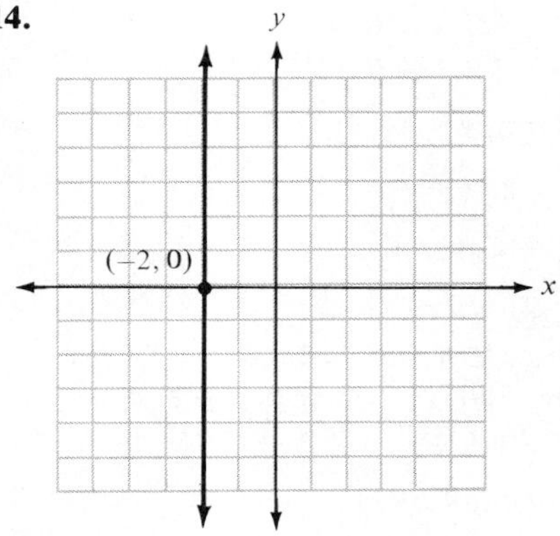

15.

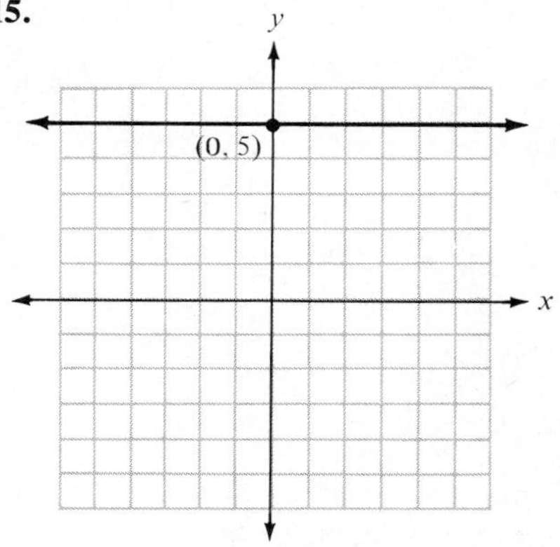

16. $y = -3x - 1$ or $3x + y = -1$
17. $y = \frac{1}{2}x$ or $x - 2y = 0$ **18.** $x = -2$
19. $y = \frac{3}{4}x + \frac{15}{4}$ or $3x - 4y = -15$
20. $y = -x - 5$ or $x + y = -5$

21. $y = 2x - \frac{11}{6}$ or $12x - 6y = 11$ **22.** $x = 3$
23. $y = -5$ **24.** $y = 2x + 5$ or $2x - y = -5$
25. $y = 2x - 12$ or $2x - y = 12$

26.

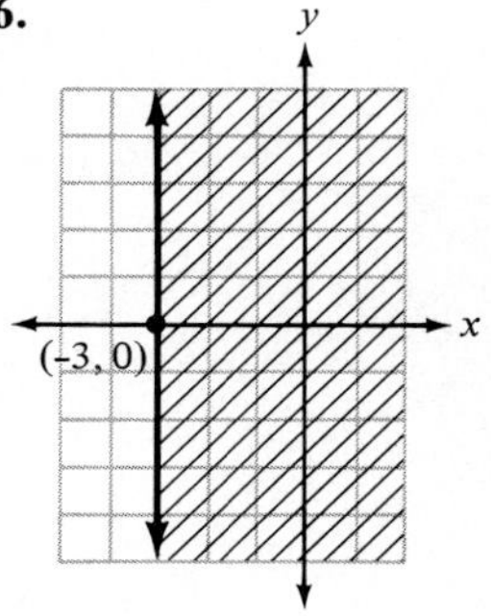

27.

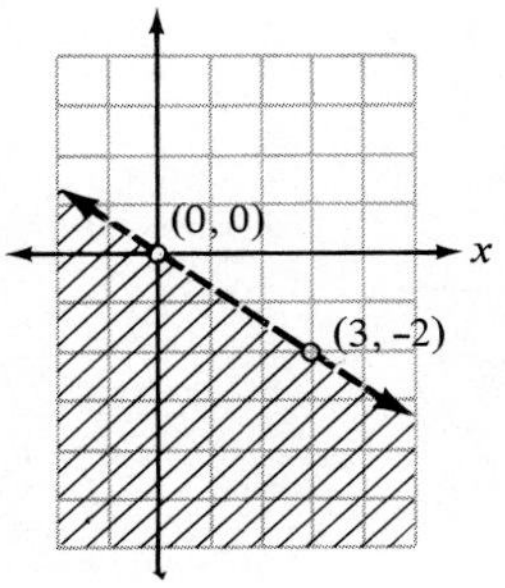

28.

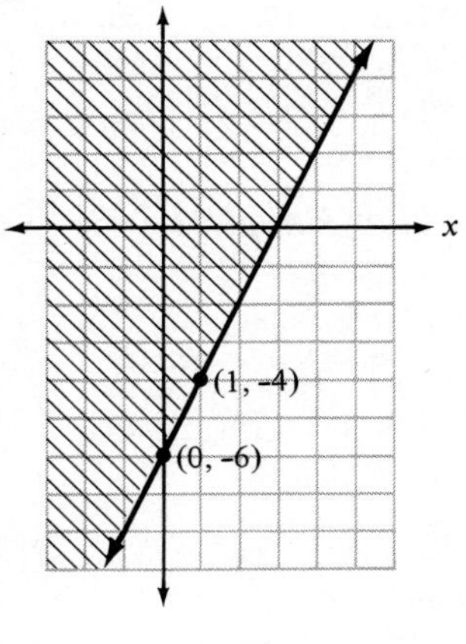

29.

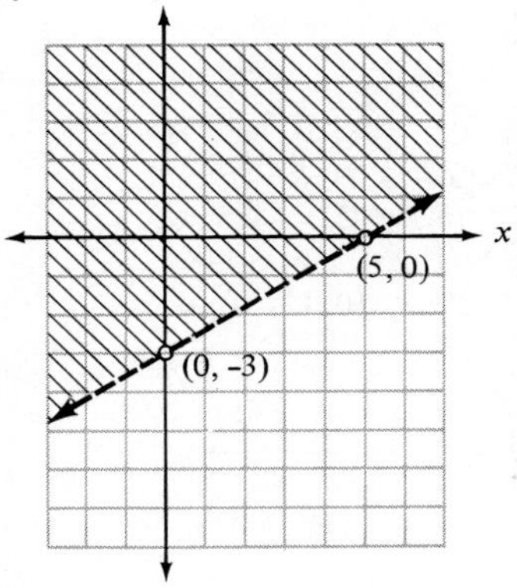

30. Domain $\{-2, -1, 0, 1, 2\}$; range $\{-2, -1, 0, 1, 2\}$
31. Domain $x \in R$; range $\{3\}$ **32.** Yes **33.** No
34. -7 **35.** 11 **36.** $x \in R, x \neq 2, x \neq -2$
37. $x \in R, x \leq 3$ **38.** $y = 7x$ **39.** $y = \frac{3}{x}$
40. $z = \frac{1}{576}xy^2$ **41.** $T = \frac{60D}{R}, T = 400$
42. \$5.60 **43.** 2304 ft **44.** $A = 2w^2 - 3w$

Practice Test (page 605)

1.

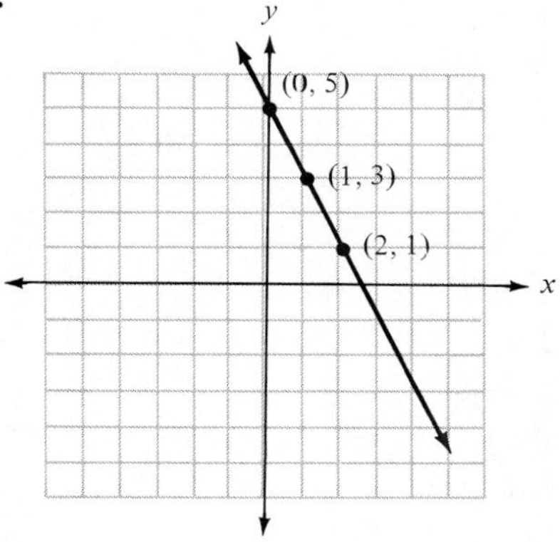

2.

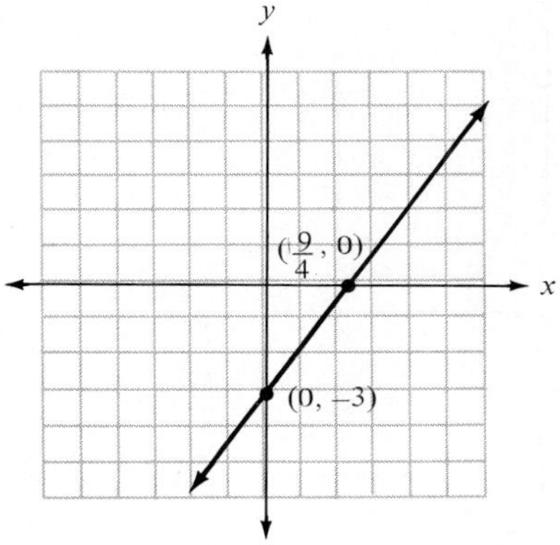

3. $2\sqrt{10}$ **4.** $(-\frac{3}{2}, \frac{7}{2})$ **5.** -1

6.

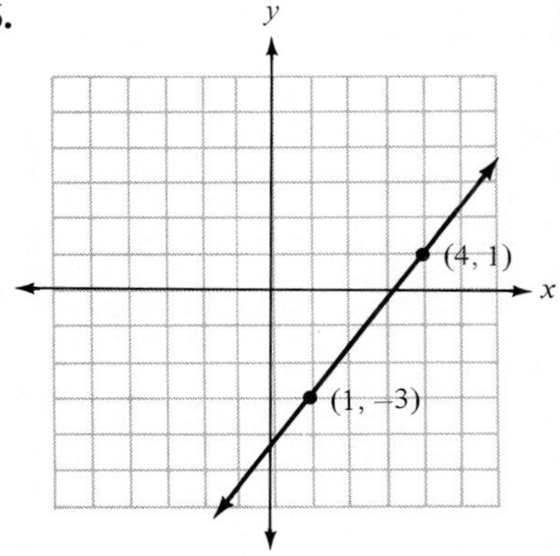

7.

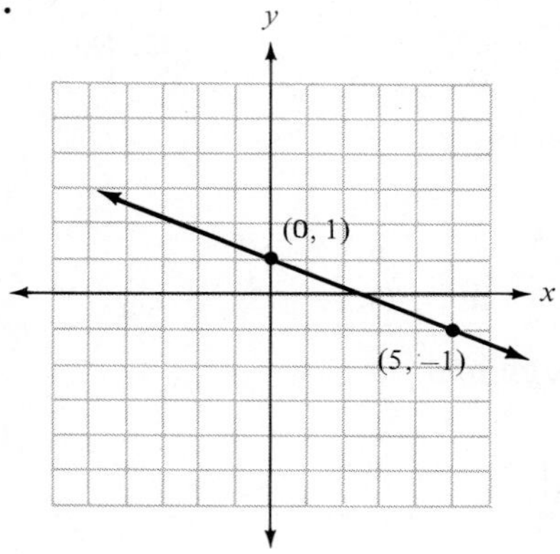

8.

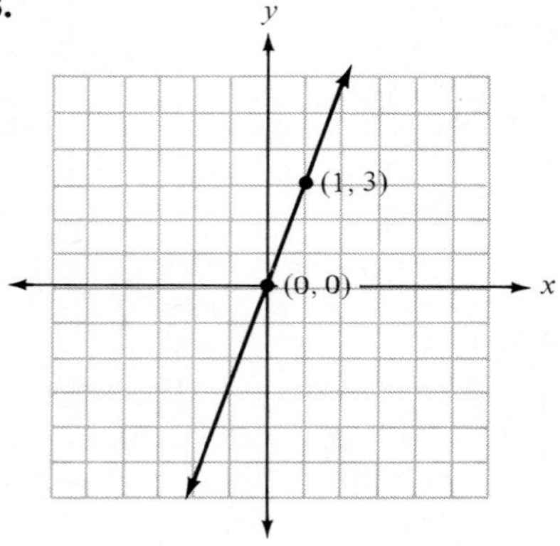

9.

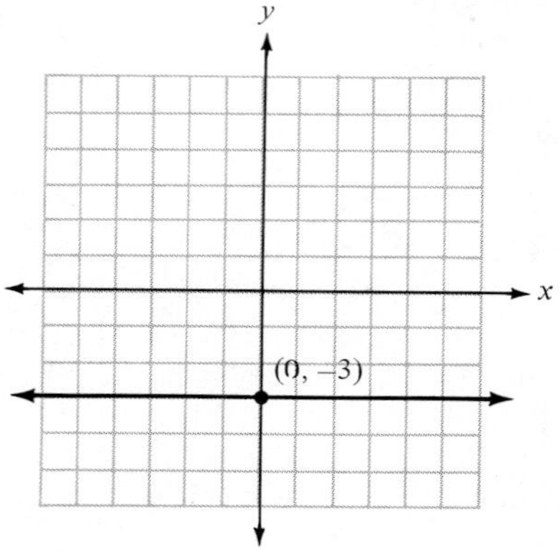

10. $y = -2x$ or $2x + y = 0$
11. $y = -\frac{2}{3}x - 3$ or $2x + 3y = -9$ **12.** $x = 5$
13. $y = \frac{2}{3}x - \frac{16}{3}$ or $2x - 3y = 16$
14. $y = x - 8$ or $x - y = 8$
15. $y = \frac{1}{2}x + \frac{9}{2}$ or $x - 2y = -9$

16.

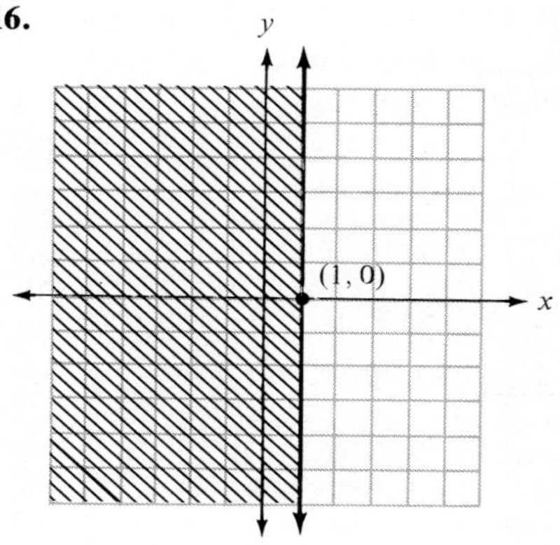

17.

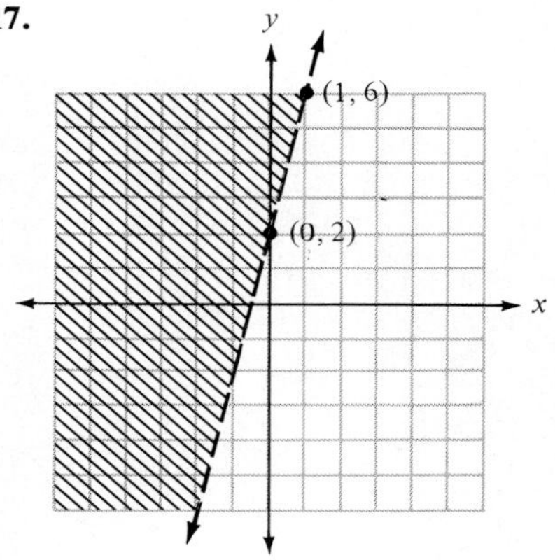

18. Domain $x \in R$; range $\{8\}$; yes **19.** 7 **20.** 0
21. $x \in R,\ x \geq \frac{1}{2}$ **22.** $x \in R,\ x \neq 0,\ x \neq -\frac{3}{4}$
23. $y = 2x^2$ **24.** $A = \frac{1}{2}bh$; $\frac{35}{2}$
25. $A = 700 - 25m$; $400

Sharpening Your Skills after Chapters 1–8 (page 609)

1. $-\frac{1}{27}$ **2.** $7 - 2\sqrt{3}i$ **3.** 1
4. No real number solutions **5.** $8\sqrt[3]{a}$ **6.** $(a - b)^{\frac{4}{3}}$
7. -71 **8.** $118x - 63$ **9.** 1 **10.** $\frac{a^2c}{b^2}$
11. $3b^3 - 15a^3 + 4a^3b^3$
12. $x^4 - 8x^3 + 7x^2 + 72x - 144$ **13.** $\frac{1}{2x - 3}$
14. $\frac{5x}{(x + 5)(x - 2)}$ **15.** $\frac{x}{x - 1}$ **16.** $11\sqrt{3x}$
17. $3x - 25$ **18.** $\frac{23}{41} - \frac{2}{41}i$
19. $(x + 5)(x^2 - 5x + 25)$ **20.** $(3x^2 - 5)^2$
21. $(9a - 2b)(7a - 5)$ **22.** $x = 2$
23. $x \geq 15$ or $x \leq -45$; $(-\infty, -45] \cup [15, \infty)$
24. $x = 0,\ x = \frac{3}{11},\ x = 2$ **25.** $x = 30$
26. No solution **27.** $2 \leq x \leq 10$; $[2, 10]$
28. $\frac{1}{2} \leq a < \frac{3}{2}$; $[\frac{1}{2}, \frac{3}{2})$

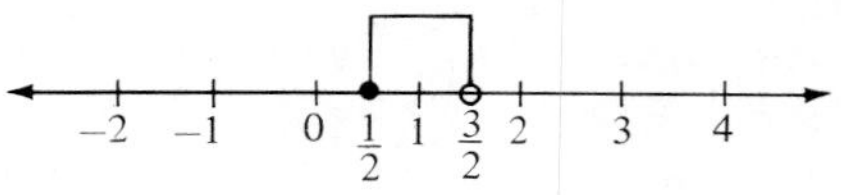

29. 24 by 40 ft

Chapter 9

Exercise Set 9.1 (page 619)

1.

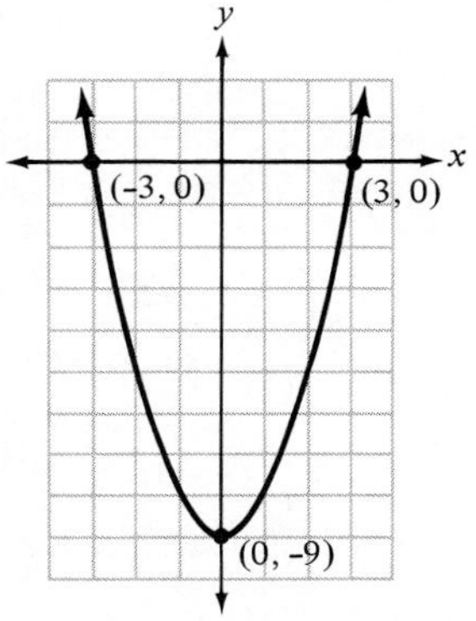

Do: $x \in R$
Ra: $y \in R, y \geqslant -9$

3.

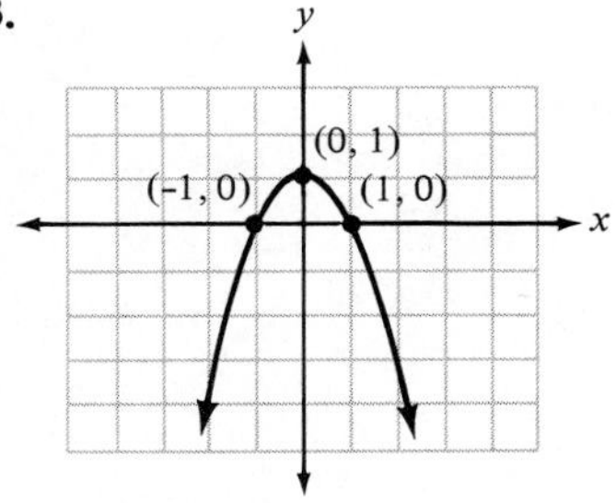

Do: $x \in R$
Ra: $y \in R, y \leqslant 1$

5.

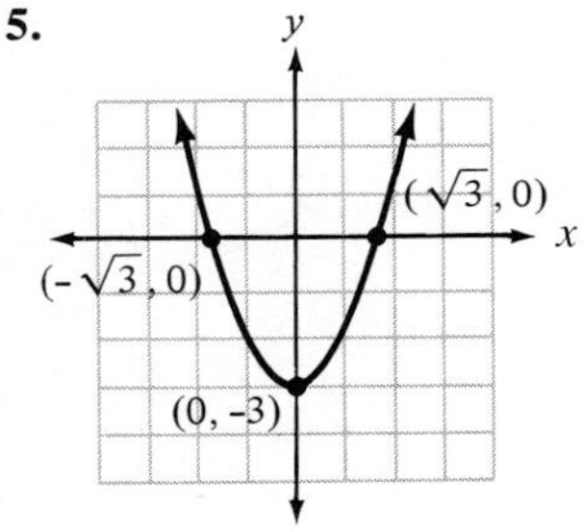

Do: $x \in R$
Ra: $y \in R, y \geqslant -3$

7.

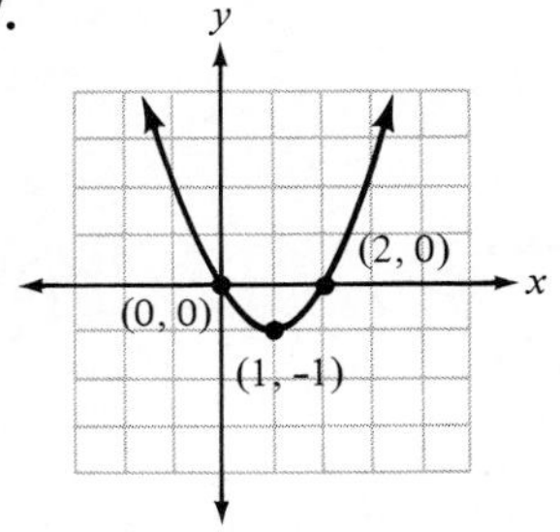

Do: $x \in R$
Ra: $y \in R, y \geqslant -1$

9.

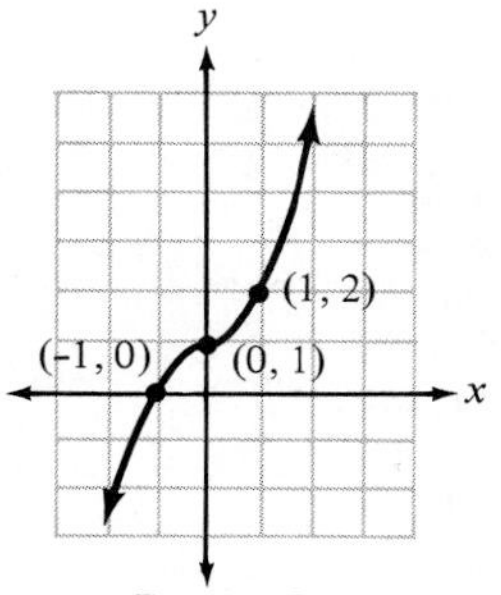

Do: $x \in R$
Ra: $y \in R$

11.

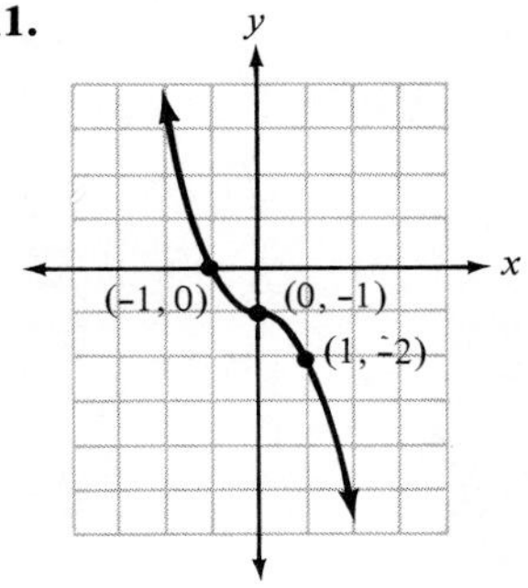

Do: $x \in R$
Ra: $y \in R$

13.

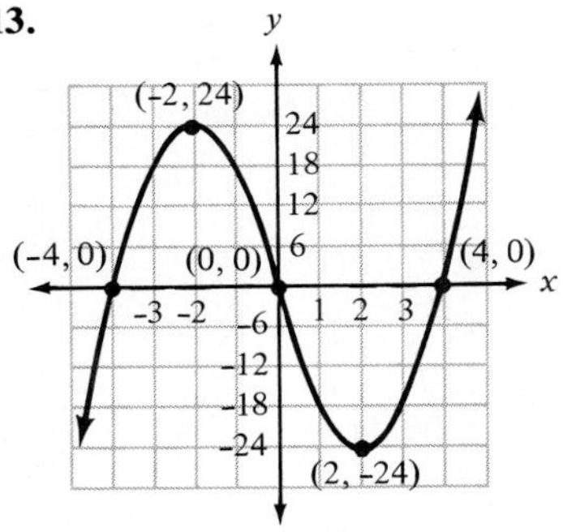

Do: $x \in R$
Ra: $y \in R$

15.

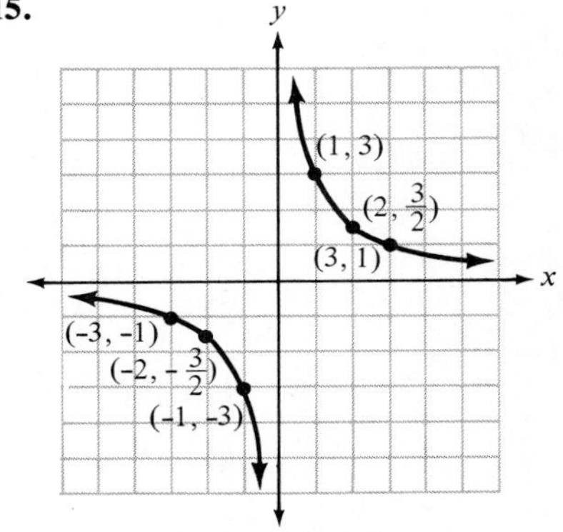

Do: $x \in R, x \neq 0$
Ra: $y \in R, y \neq 0$

17.

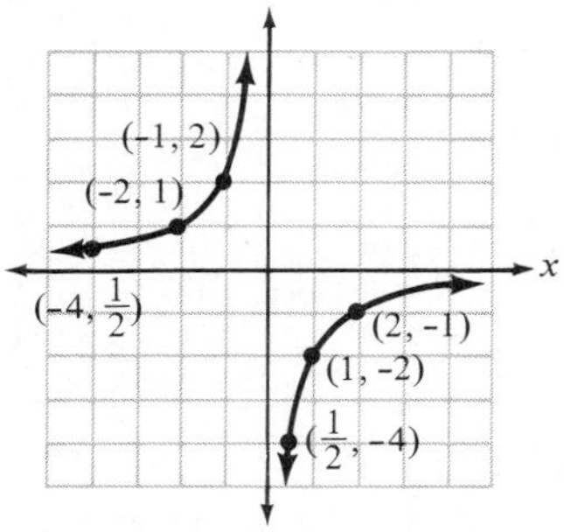

Do: $x \in R, x \neq 0$
Ra: $y \in R, y \neq 0$

Stretching the Topics

1.

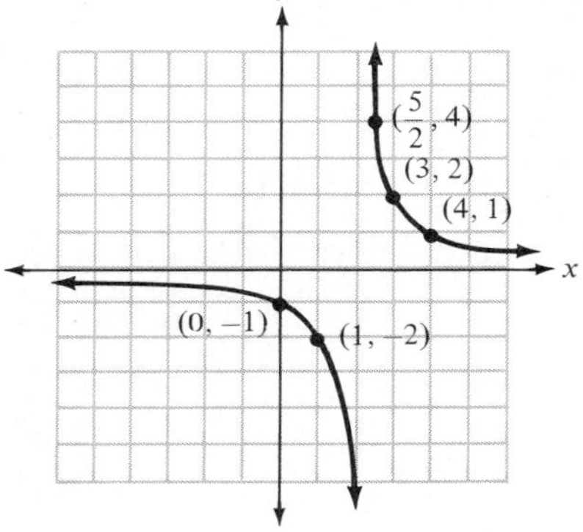

2.

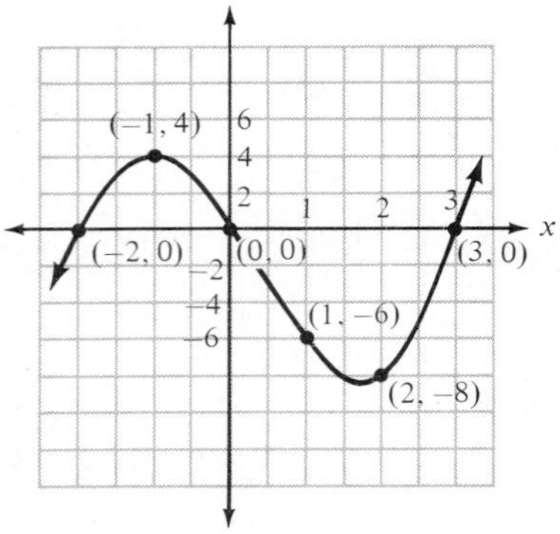

Checkup 9.1 (page 620)

1.

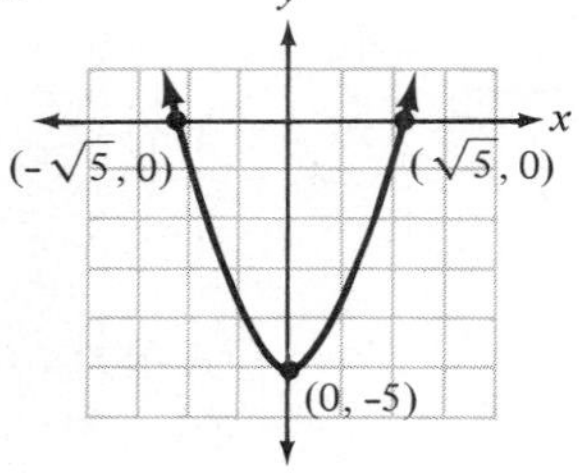

Do: $x \in R$
Ra: $y \in R, \ y \geq -5$

2.

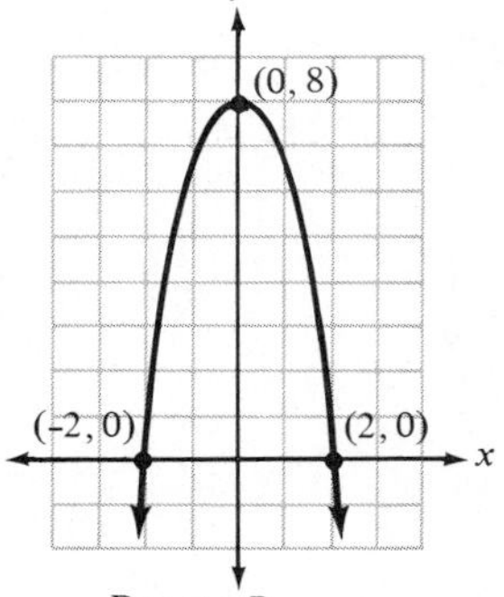

Do: $x \in R$
Ra: $y \in R, y \leq 8$

3.

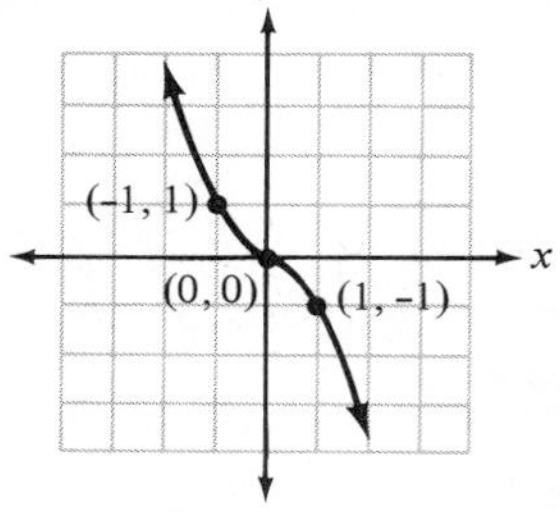

Do: $x \in R$
Ra: $y \in R$

4.

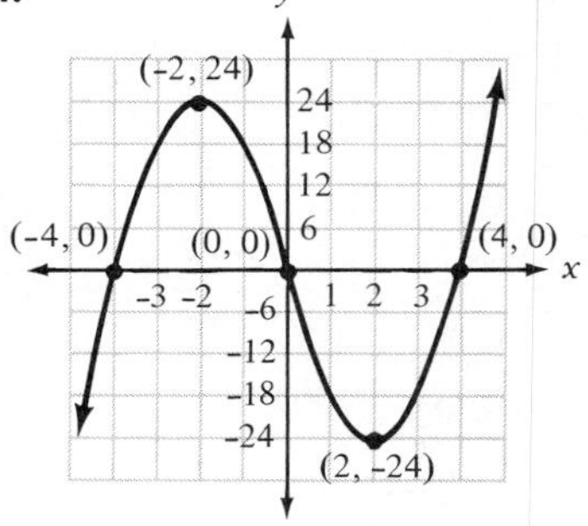

Do: $x \in R$
Ra: $y \in R$

5.

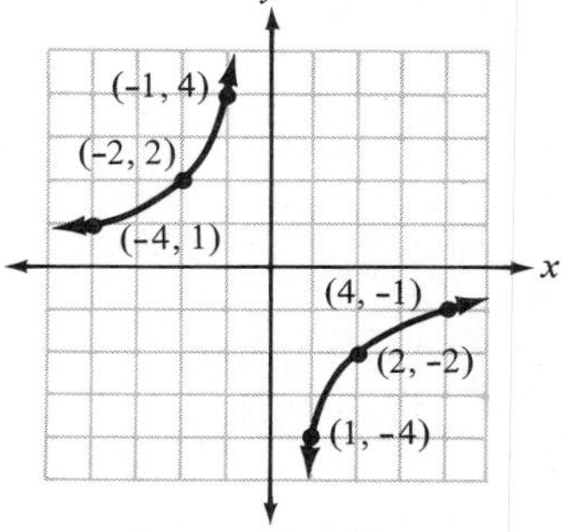

Do: $x \in R, x \neq 0$
Ra: $y \in R, y \neq 0$

Exercise Set 9.2 (page 637)

1.

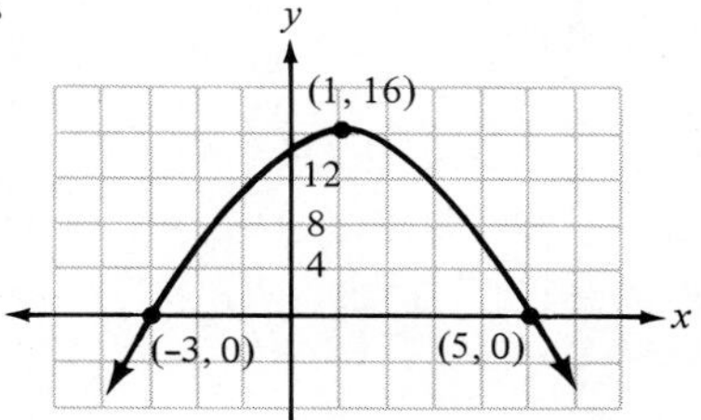

Do: $x \in R$
Ra: $y \in R, y \leqslant 16$

3.

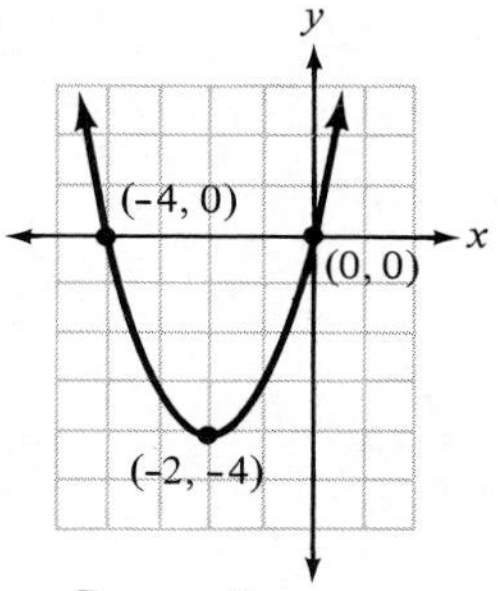

Do: $x \in R$
Ra: $y \in R, y \geqslant -4$

5.

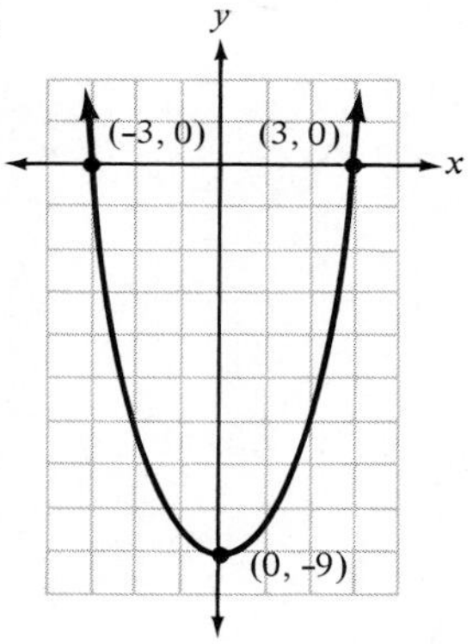

Do: $x \in R$
Ra: $y \in R, y \geqslant -9$

7.

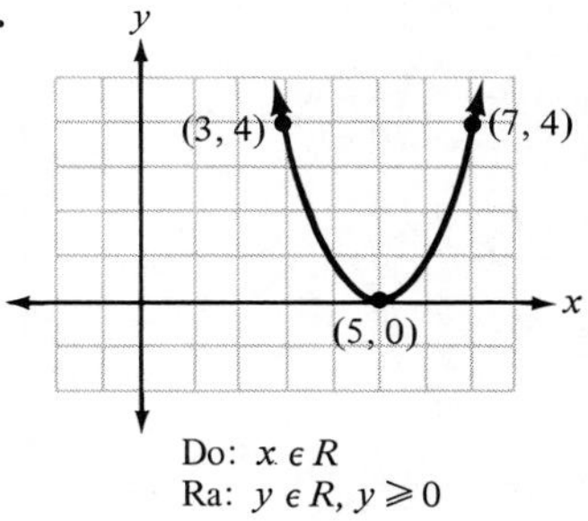

Do: $x \in R$
Ra: $y \in R, y \geqslant 0$

9.

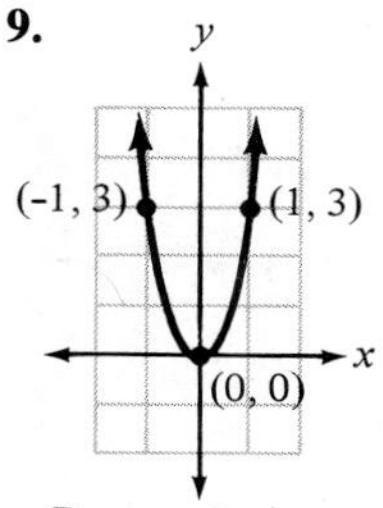

Do: $x \in R$
Ra: $y \in R, y \geqslant 0$

11.

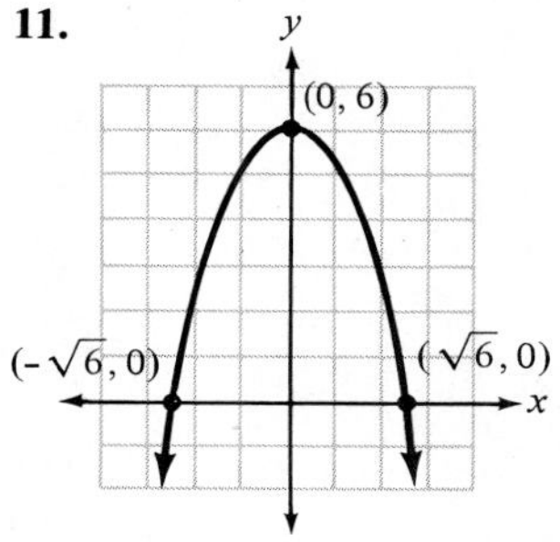

Do: $x \in R$
Ra: $y \in R, y \leqslant 6$

13.

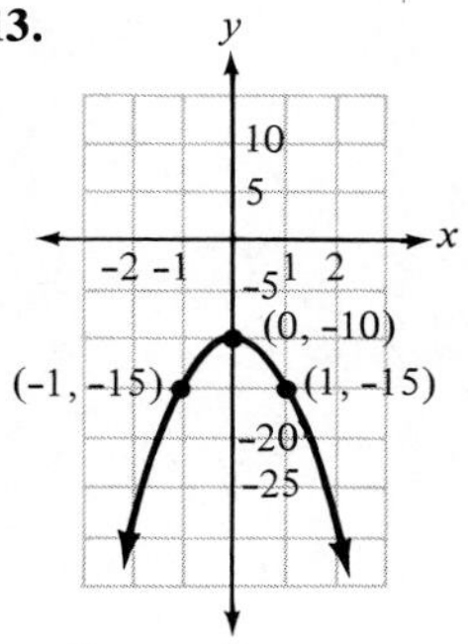

Do: $x \in R$
Ra: $y \in R, y \leqslant -10$

15.

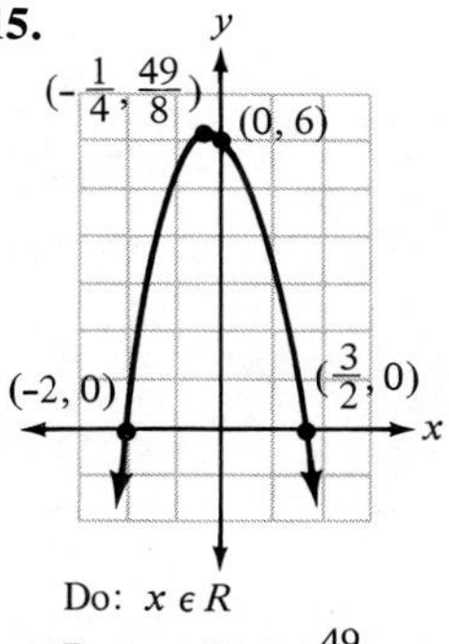

Do: $x \in R$
Ra: $y \in R, y \leqslant \frac{49}{8}$

17.

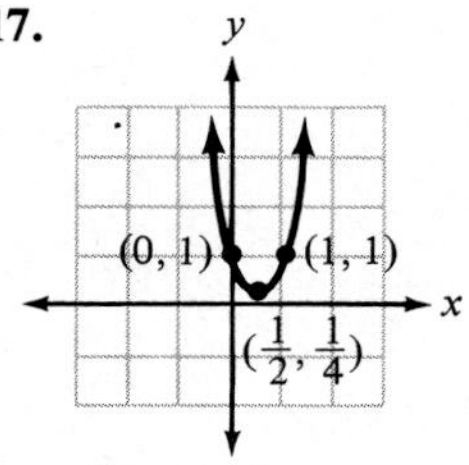

Do: $x \in R$

Ra: $y \in R, y \geqslant \frac{1}{4}$

19.

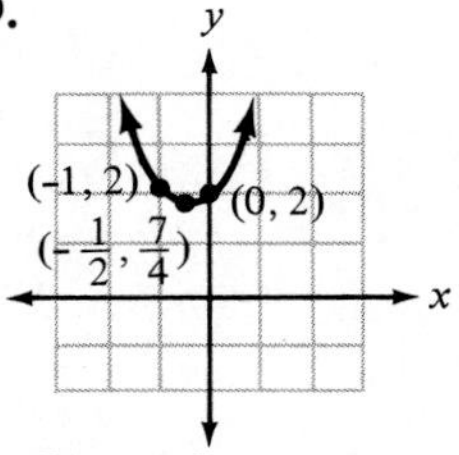

Do: $x \in R$

Ra: $y \in R, y \geqslant \frac{7}{4}$

21.

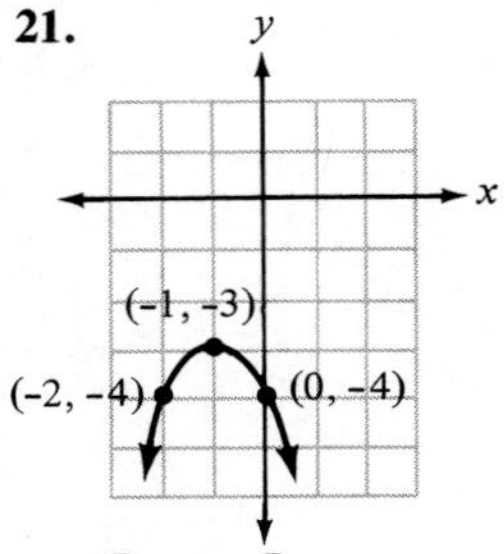

Do: $x \in R$

Ra: $y \in R, y \leqslant -3$

23.

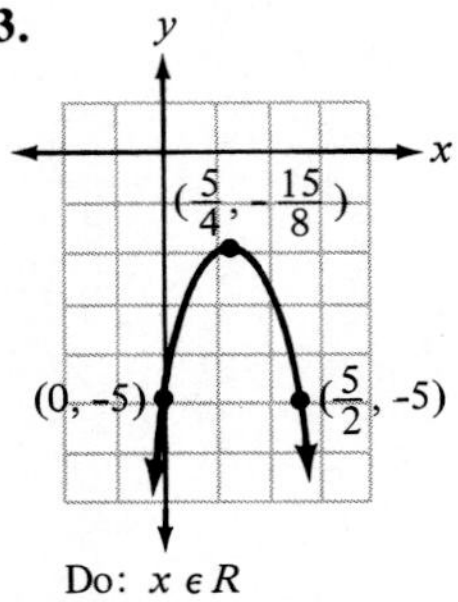

Do: $x \in R$

Ra: $y \in R, y \leqslant -\frac{15}{8}$

25.

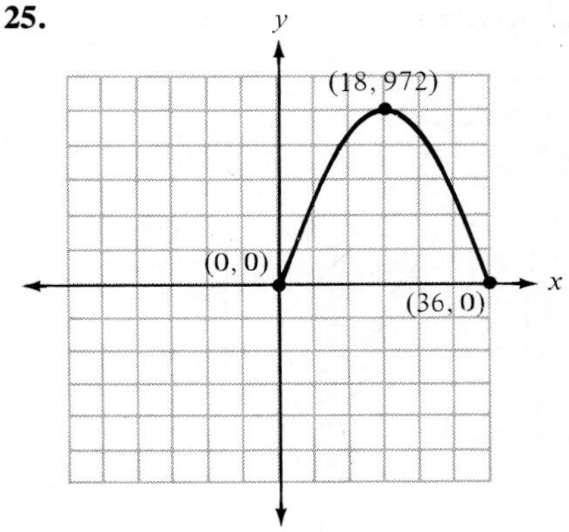

18, \$972

27.

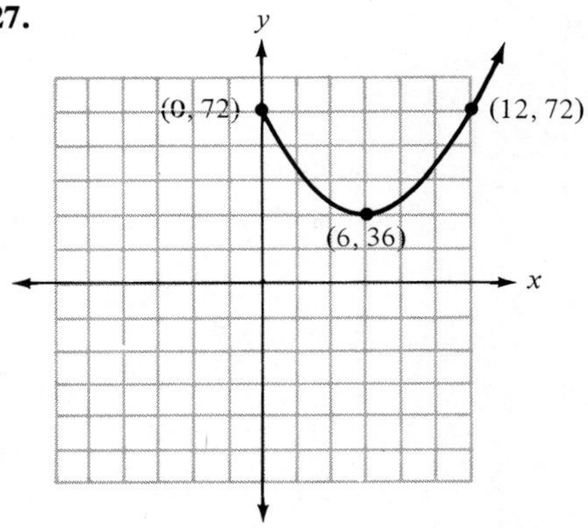

6, \$36

Stretching the Topics

1. $x = 4.2, x = -1.2$

2. $A = 18w - w^2$

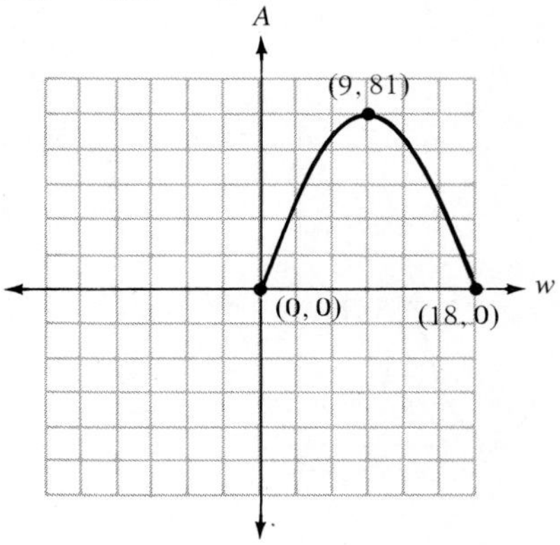

9 ft; 81 sq ft

Checkup 9.2 (page 639)

1.

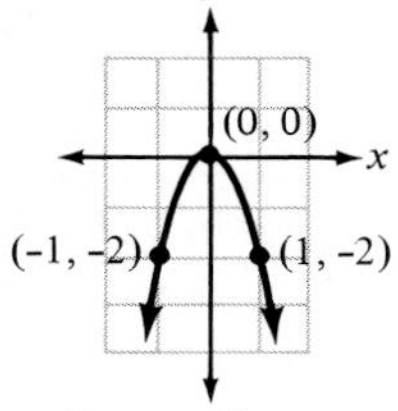

Do: $x \in R$
Ra: $y \in R, y \leqslant 0$

2.

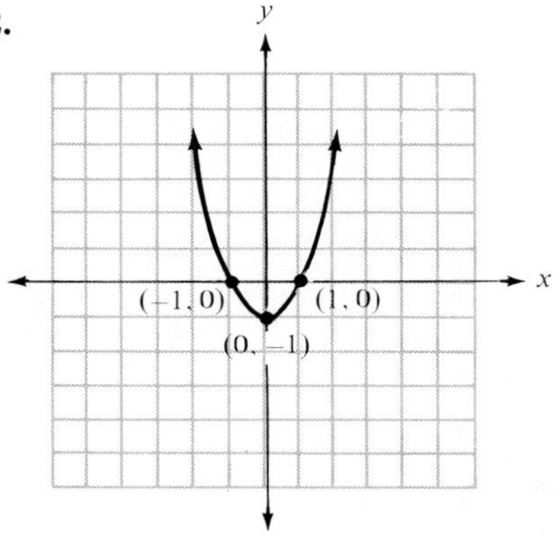

Do: $x \in R$
Ra: $y \in R,\ y \geqslant -1$

3.

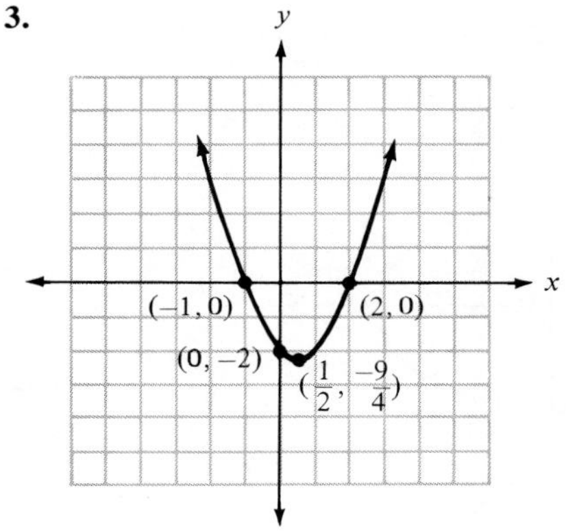

Do: $x \in R$
Ra: $y \in R,\ y \geqslant -\frac{9}{4}$.

4.

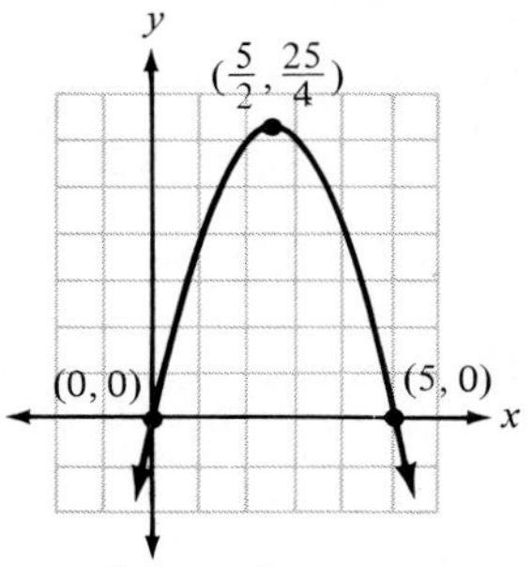

Do: $x \in R$
Ra: $y \in R, y \leqslant \frac{25}{4}$

5.

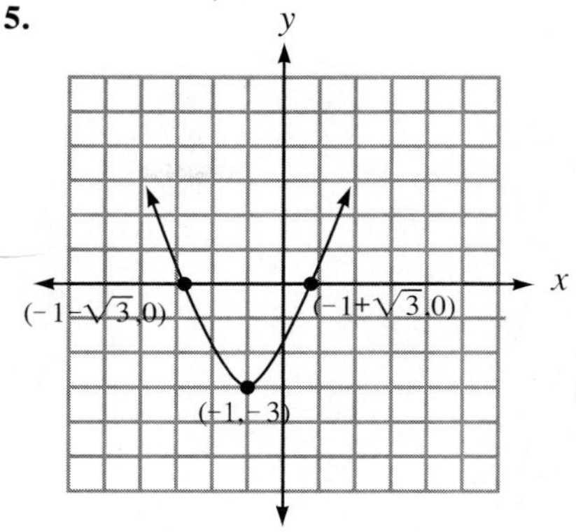

Do: $x \in R$
Ra: $y \in R,\ y \geqslant -3$

6.

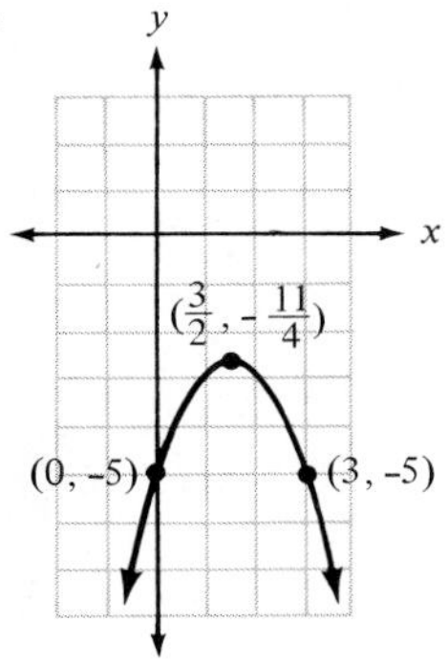

Do: $x \in R$
Ra: $y \in R, y \leqslant -\frac{11}{4}$

Exercise Set 9.3 (page 655)

1.

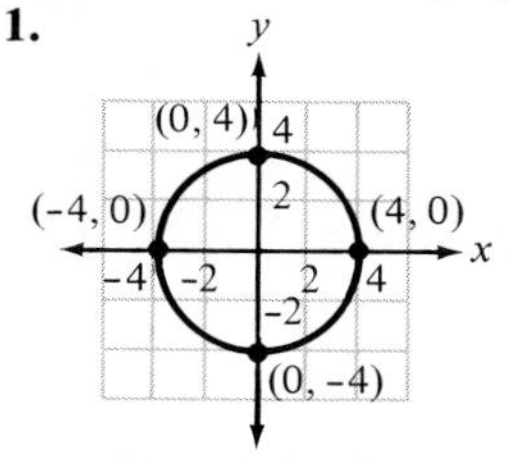

Do: $-4 \leqslant x \leqslant 4$
Ra: $-4 \leqslant y \leqslant 4$

3.

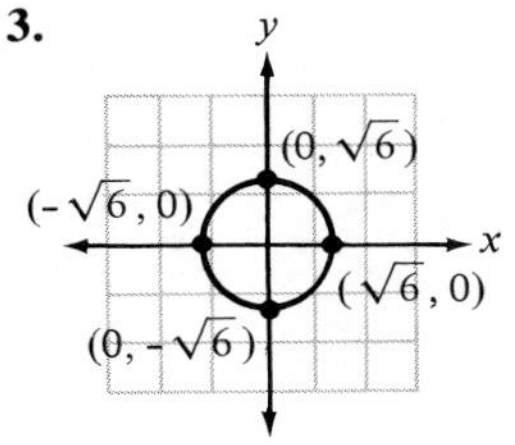

Do: $-\sqrt{6} \leqslant x \leqslant \sqrt{6}$
Ra: $-\sqrt{6} \leqslant y \leqslant \sqrt{6}$

5.

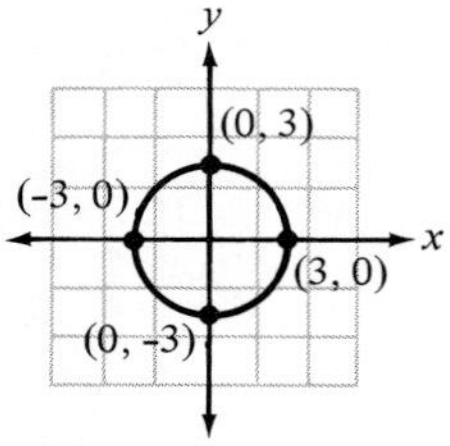

Do: $-3 \leqslant x \leqslant 3$
Ra: $-3 \leqslant y \leqslant 3$

7.

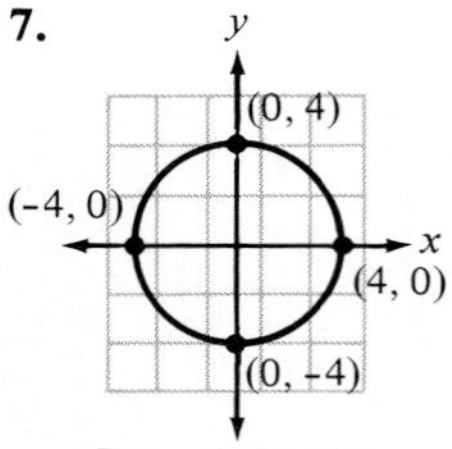

Do: $-4 \leqslant x \leqslant 4$
Ra: $-4 \leqslant y \leqslant 4$

9.

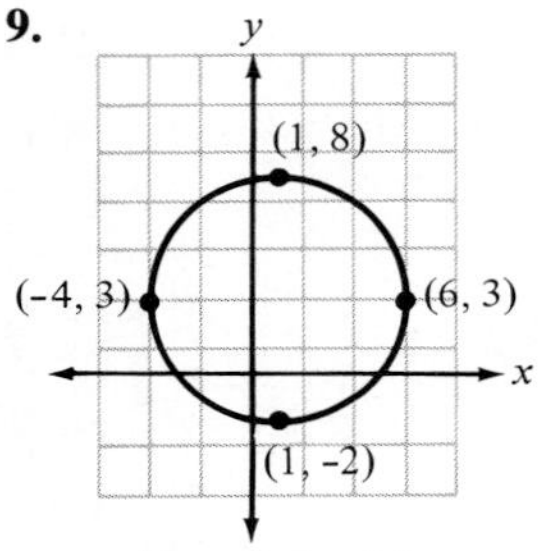

Do: $-4 \leqslant x \leqslant 6$
Ra: $-2 \leqslant y \leqslant 8$

11.

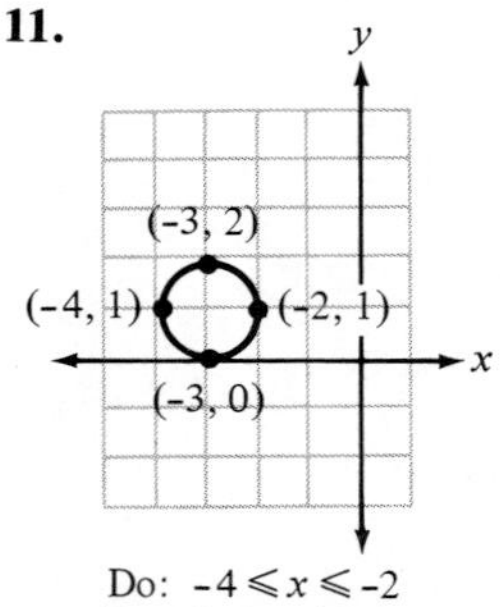

Do: $-4 \leqslant x \leqslant -2$
Ra: $0 \leqslant y \leqslant 2$

13.

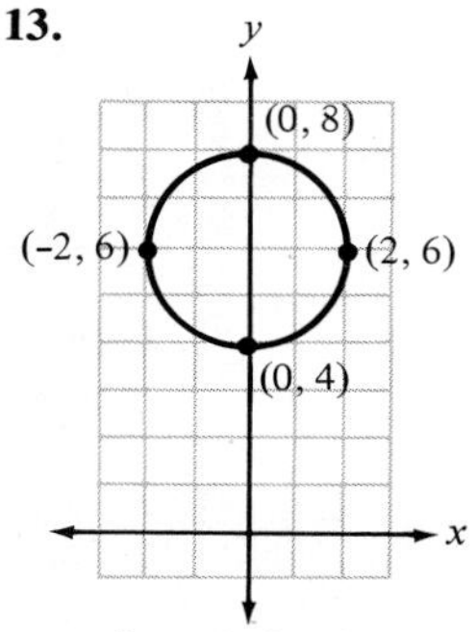

Do: $-2 \leqslant x \leqslant 2$
Ra: $4 \leqslant y \leqslant 8$

15.

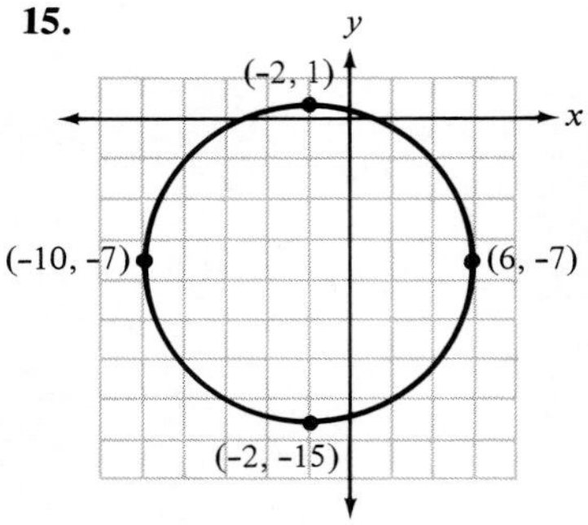

Do: $-10 \leqslant x \leqslant 6$
Ra: $-15 \leqslant y \leqslant 1$

17.

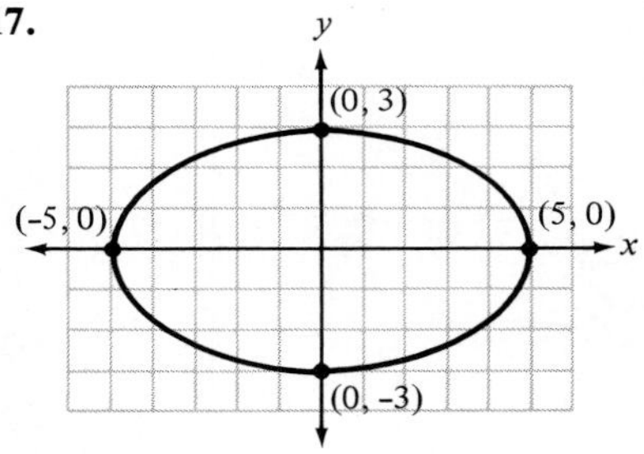

Do: $-5 \leqslant x \leqslant 5$
Ra: $-3 \leqslant y \leqslant 3$

19.

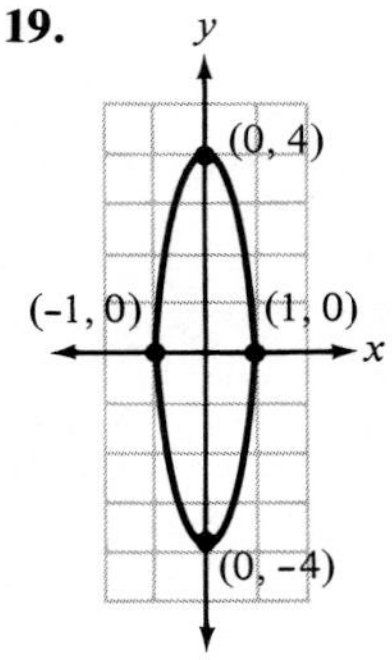

Do: $-1 \leqslant x \leqslant 1$
Ra: $-4 \leqslant y \leqslant 4$

21.

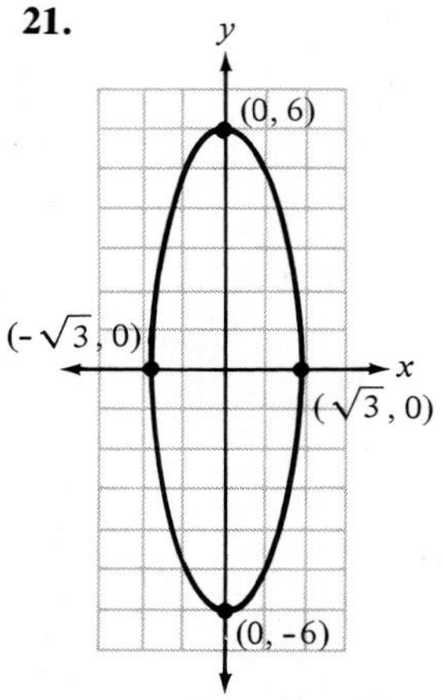

Do: $-\sqrt{3} \leqslant x \leqslant \sqrt{3}$
Ra: $-6 \leqslant y \leqslant 6$

23.

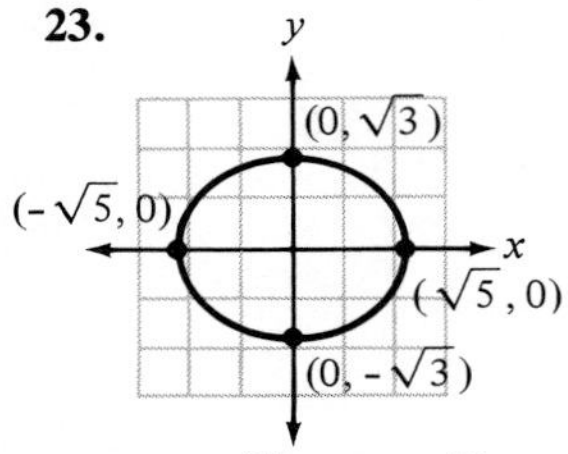

Do: $-\sqrt{5} \leqslant x \leqslant \sqrt{5}$
Ra: $-\sqrt{3} \leqslant y \leqslant \sqrt{3}$

25.

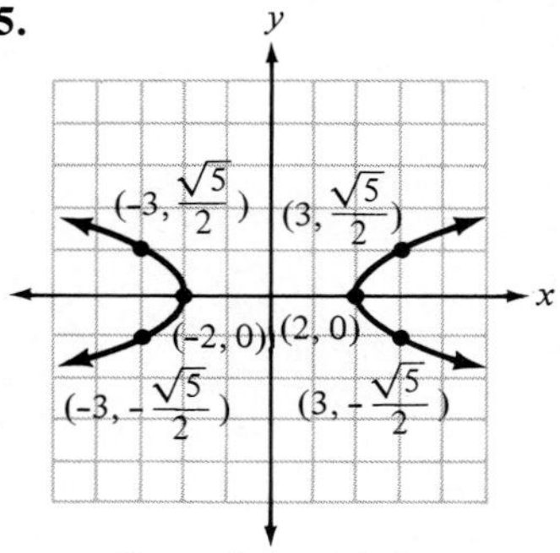

Do: $x \leqslant -2$ or $x \geqslant 2$
Ra: $y \in R$

27.

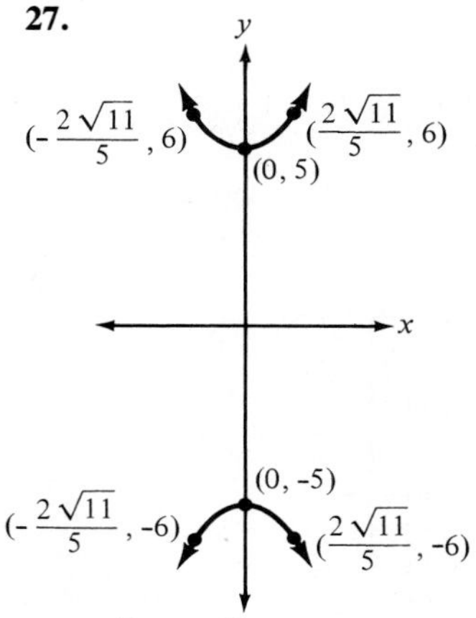

Do: $x \in R$
Ra: $y \geqslant 5$ or $y \leqslant -5$

29.

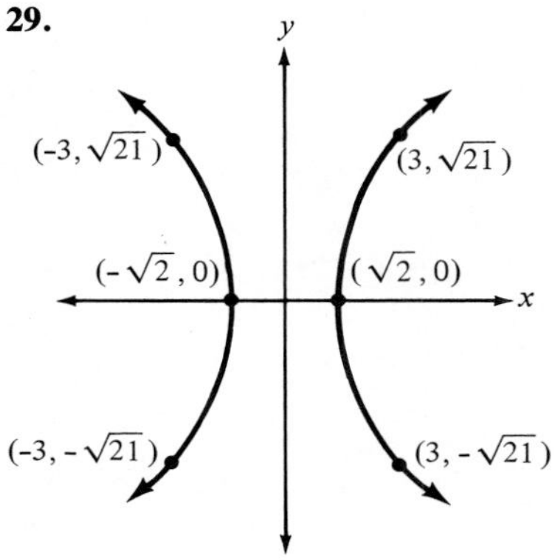

Do: $x \leqslant -\sqrt{2}$ or $x \geqslant \sqrt{2}$
Ra: $y \in R$

31.

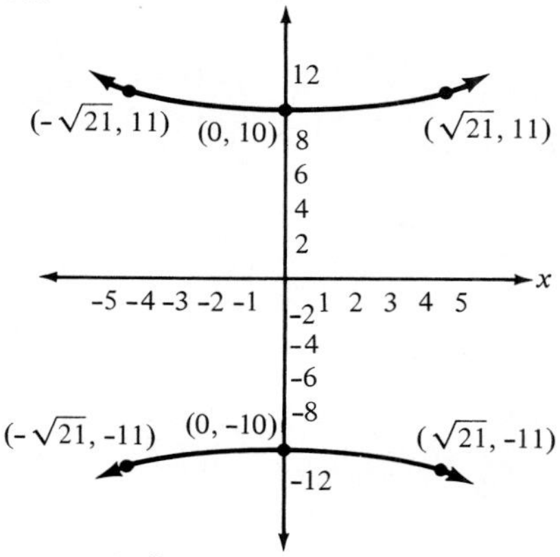

Do: $x \in R$
Ra: $y \leqslant -10$ or $y \geqslant 10$

Stretching the Topics

1. $(x + 3)^2 + (y - 4)^2 = 25$

2. $x = 2.1, x = -2.1; y = 1.7, y = -1.7$

3.

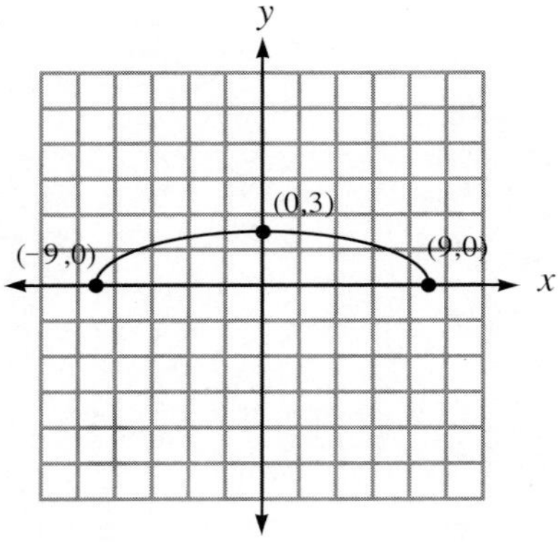

Do: $-9 \leq x \leq 9$
Ra: $0 \leq x \leq 3$

Checkup 9.3 (page 657)

1.

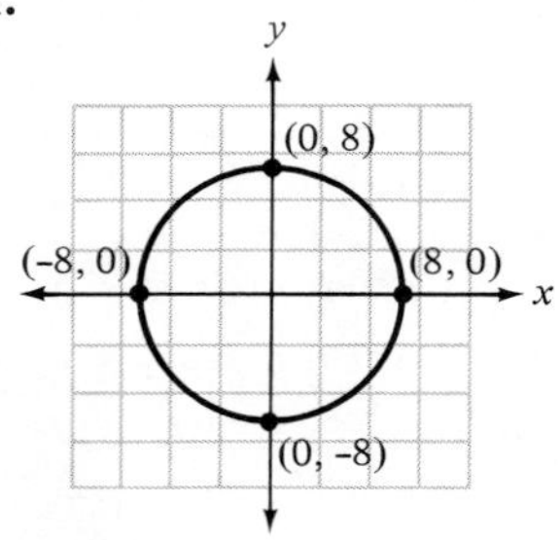

Do: $-8 \leqslant x \leqslant 8$
Ra: $-8 \leqslant y \leqslant 8$

2.

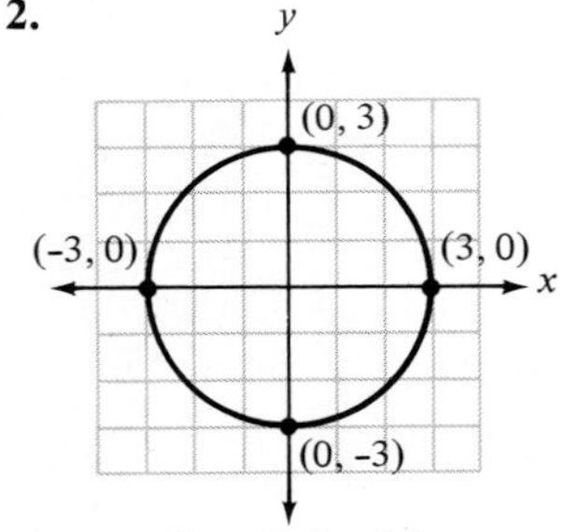

Do: $-3 \leqslant x \leqslant 3$
Ra: $-3 \leqslant y \leqslant 3$

3.

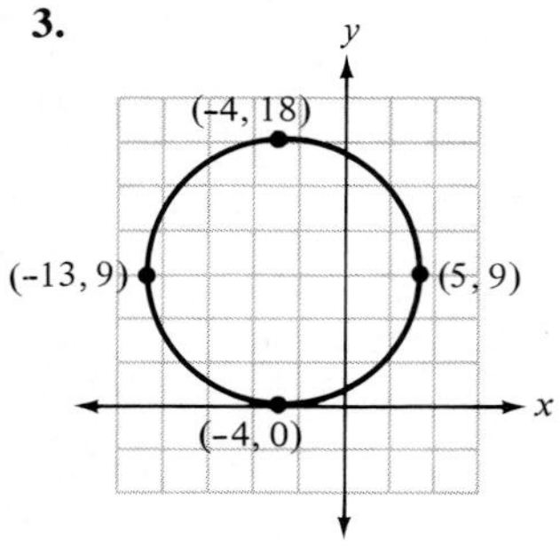

Do: $-13 \leqslant x \leqslant 5$
Ra: $0 \leqslant y \leqslant 18$

4.

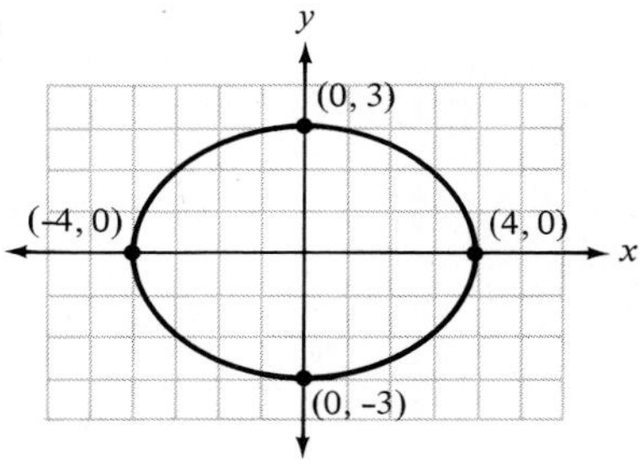

Do: $-4 \leqslant x \leqslant 4$
Ra: $-3 \leqslant y \leqslant 3$

5.

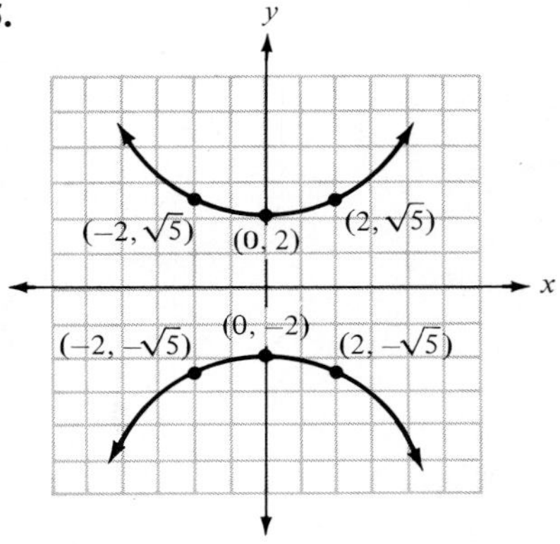

Do: $x \in R$
Ra: $y \geqslant 2$ or $y \leqslant -2$

Exercise Set 9.4 (page 664)

1.

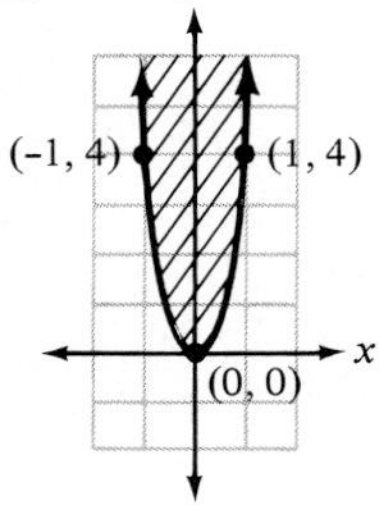

3.

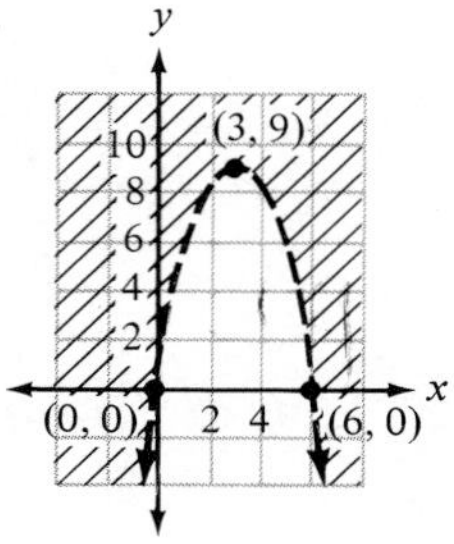

5.

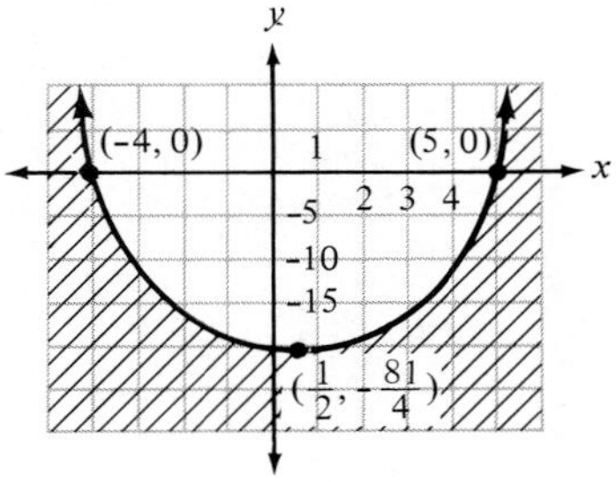

7.

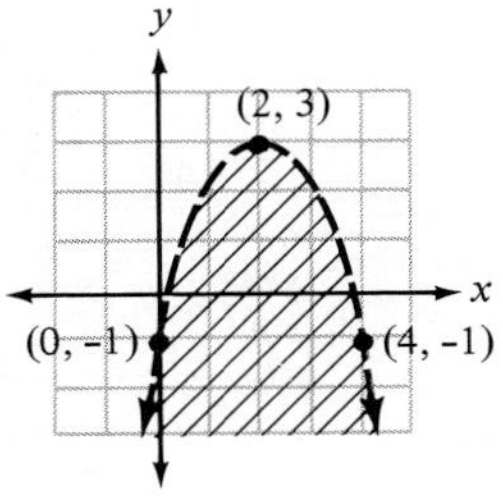

9.

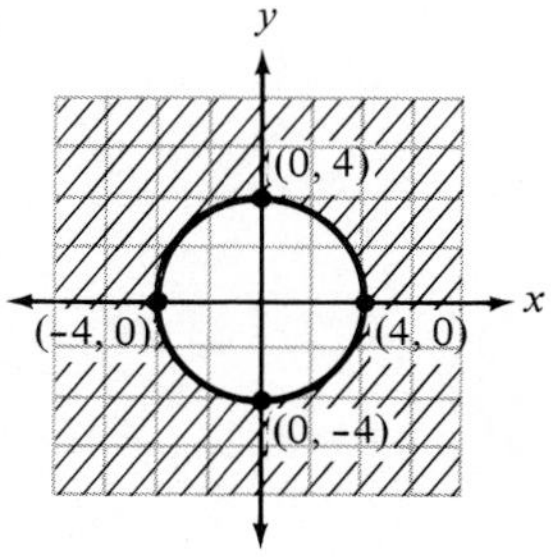

11.

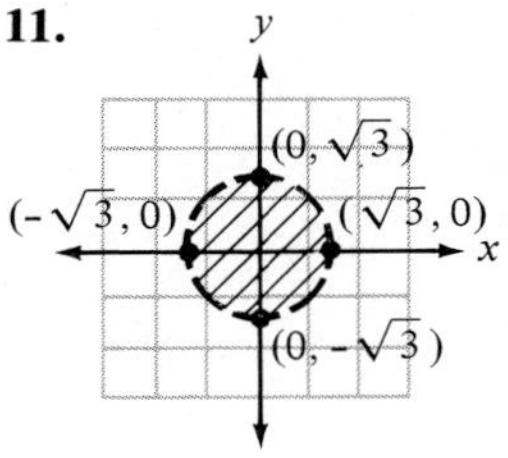

13.

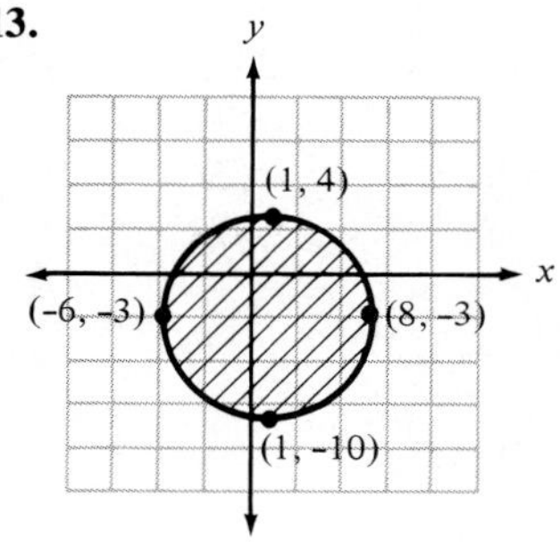

15.

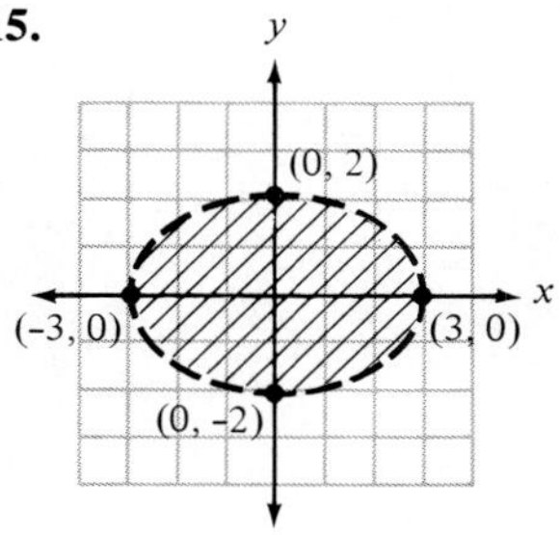

17.

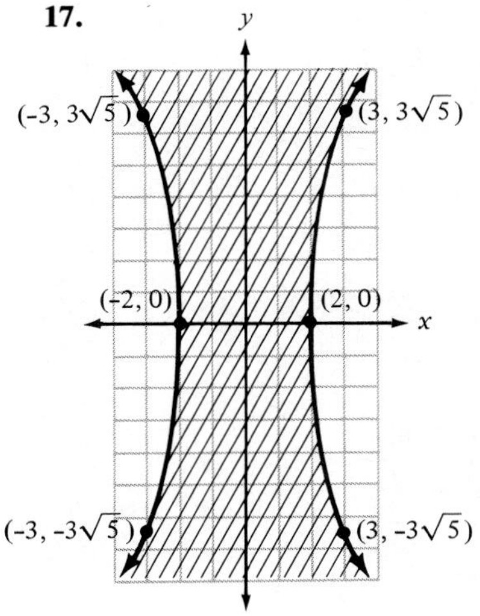

19.

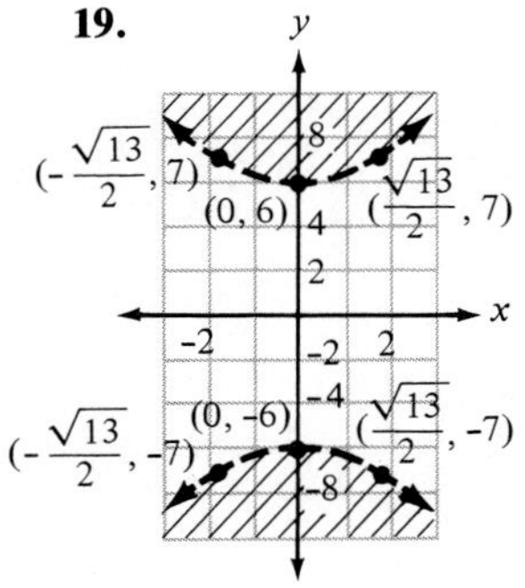

Stretching the Topics

1.

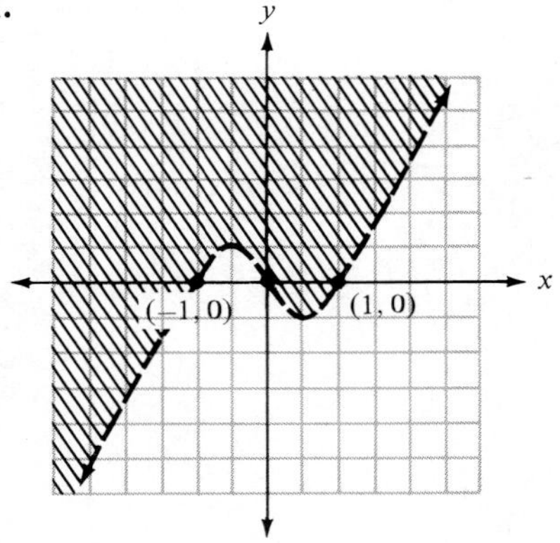

2.

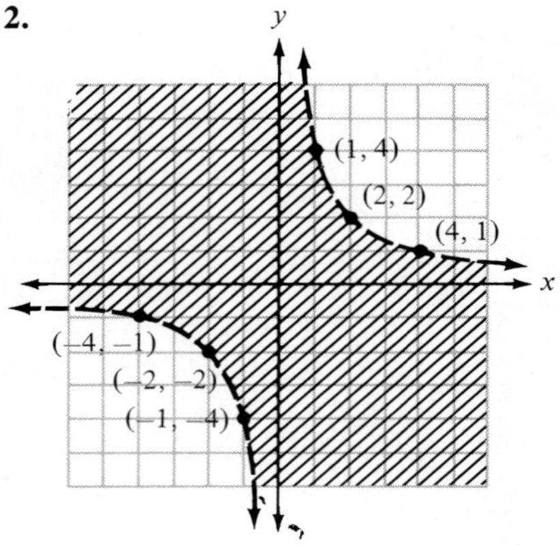

3.

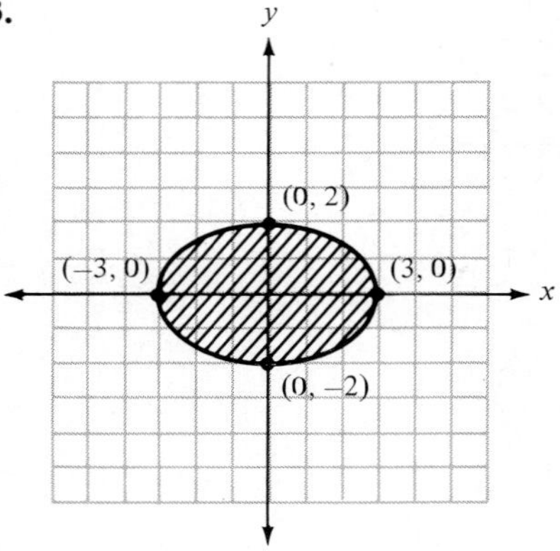

Checkup 9.4 (page 665)

1.

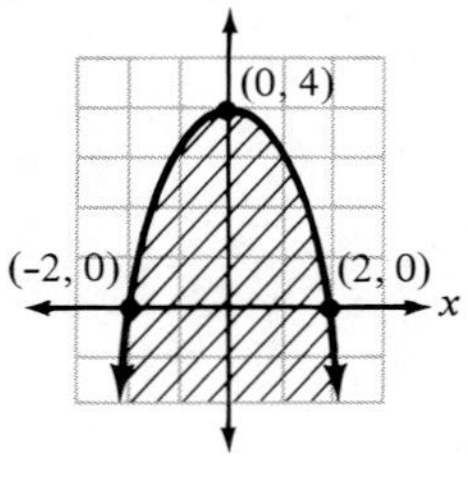

2.

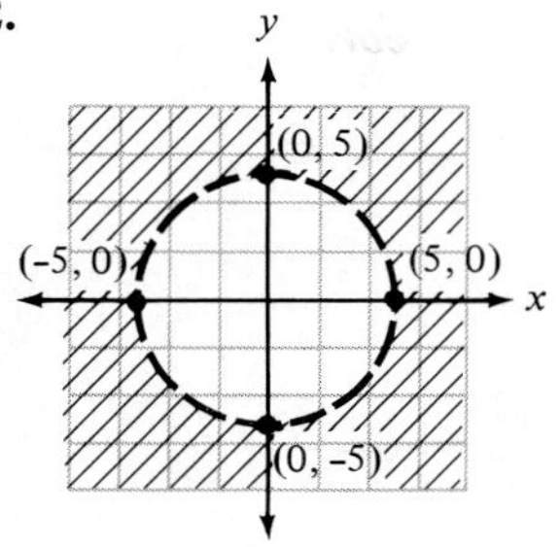

3.

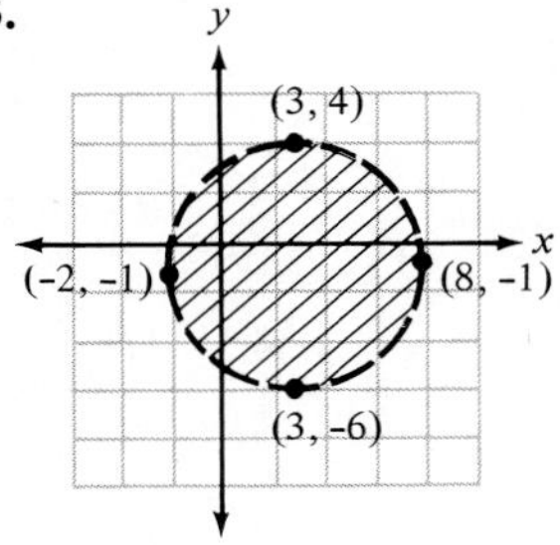

4.

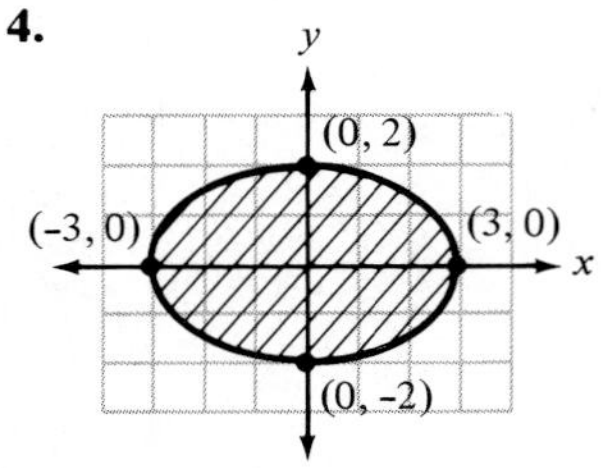

5.

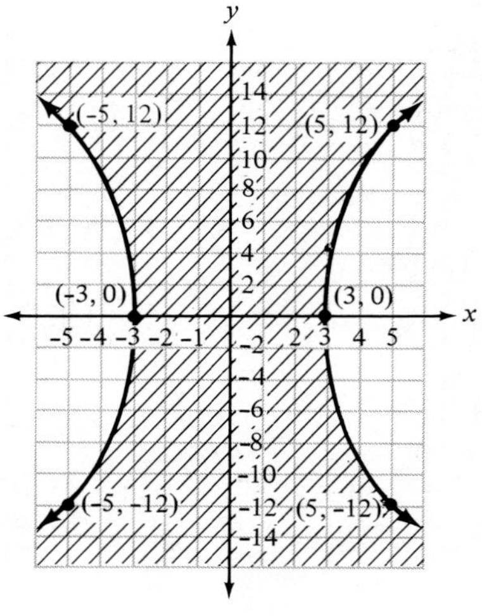

Speaking the Language of Algebra (page 667)

1. Parabola; upward **2.** $-\frac{b}{2a}$; vertex

3. Circle; ellipse; hyperbola

4. Unit; inside; outside

Review Exercises (page 669)

1.

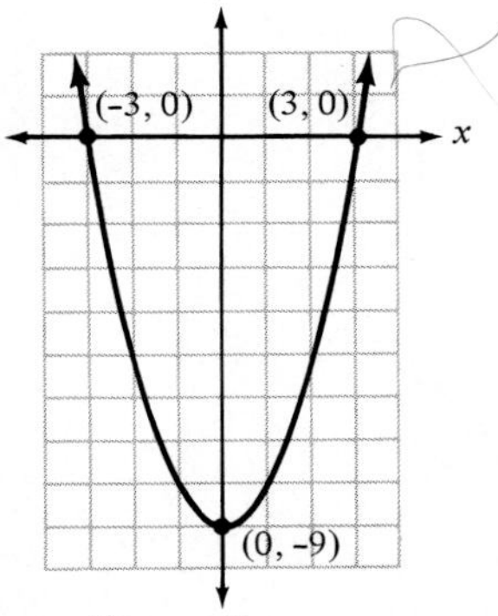

Do: $x \in R$
Ra: $y \in R, y \geqslant -9$

2.

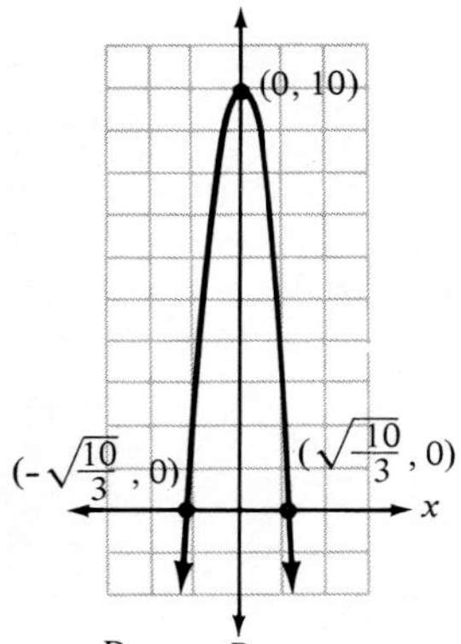

Do: $x \in R$
Ra: $y \in R, y \leqslant 10$

3.

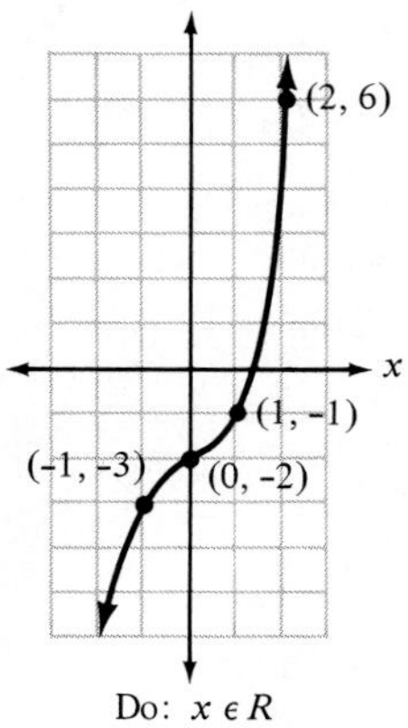

Do: $x \in R$
Ra: $y \in R$

4.

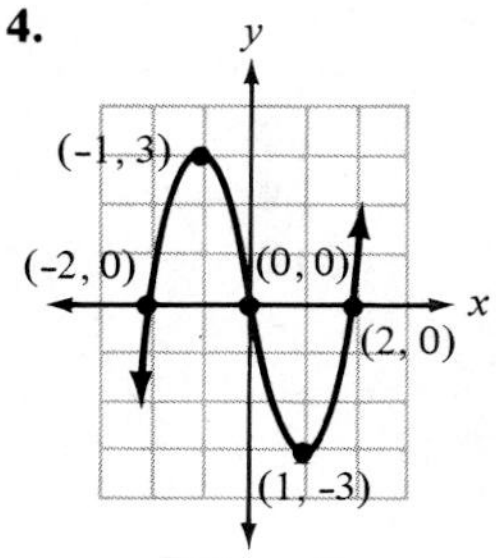

Do: $x \in R$
Ra: $y \in R$

5.

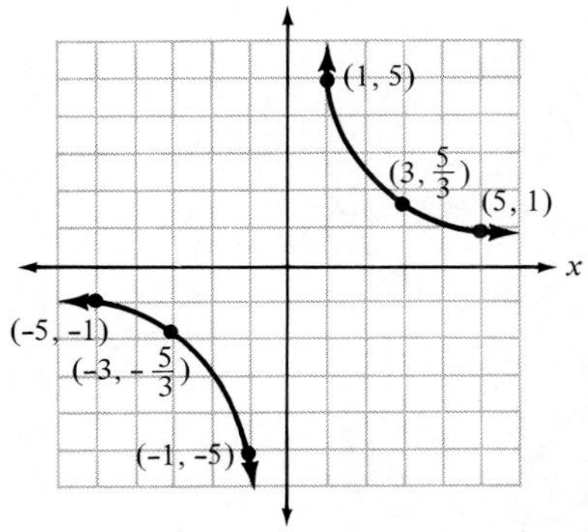

Do: $x \in R, x \neq 0$
Ra: $y \in R, y \neq 0$

6.

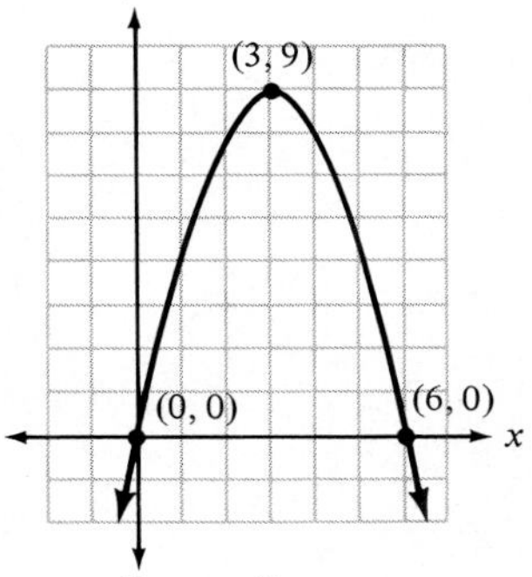

Do: $x \in R$
Ra: $y \in R, y \leqslant 9$

7.

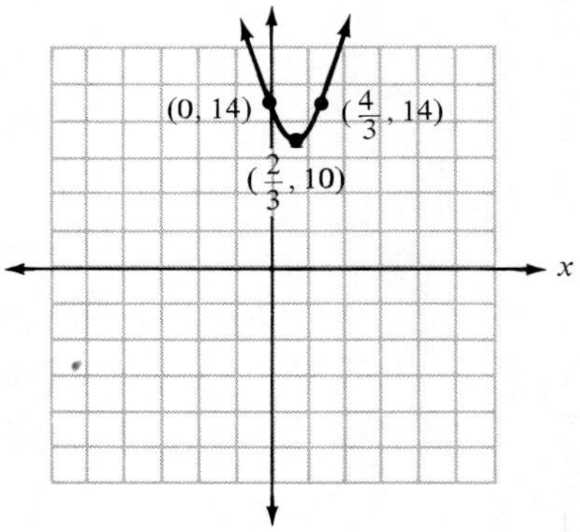

Do : $x \in R$
Ra : $y \in R, y \geqslant 10$

8.

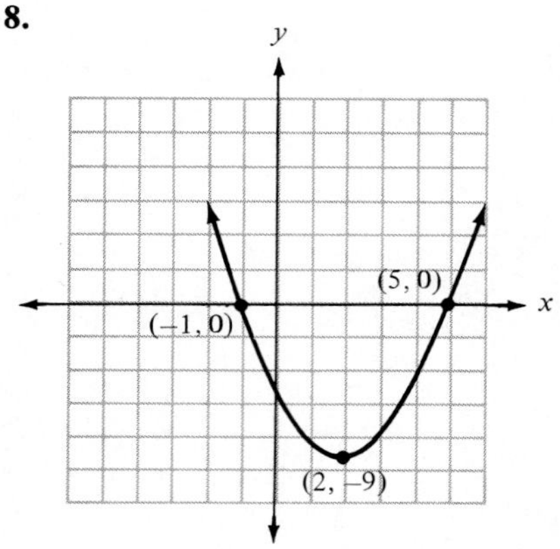

Do : $x \in R$
Ra : $y \in R, y \geqslant -9$

9.

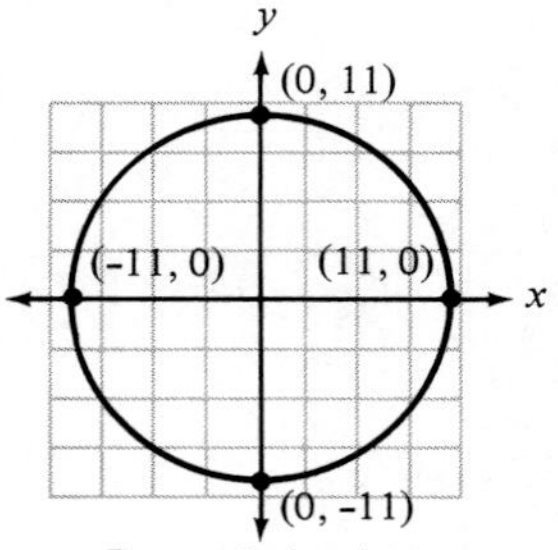

Do: $-11 \leqslant x \leqslant 11$
Ra: $-11 \leqslant y \leqslant 11$

10.

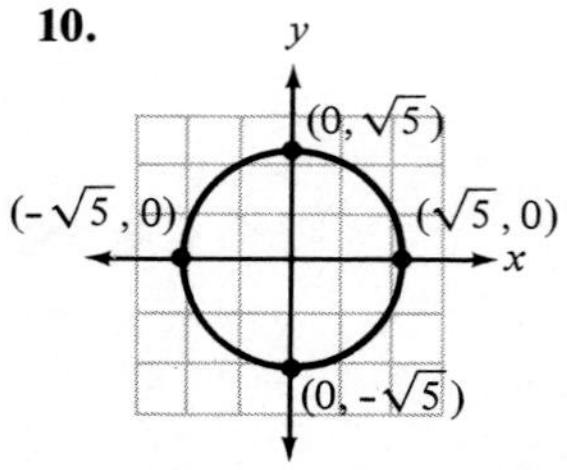

Do: $-\sqrt{5} \leqslant x \leqslant \sqrt{5}$
Ra: $-\sqrt{5} \leqslant y \leqslant \sqrt{5}$

11.

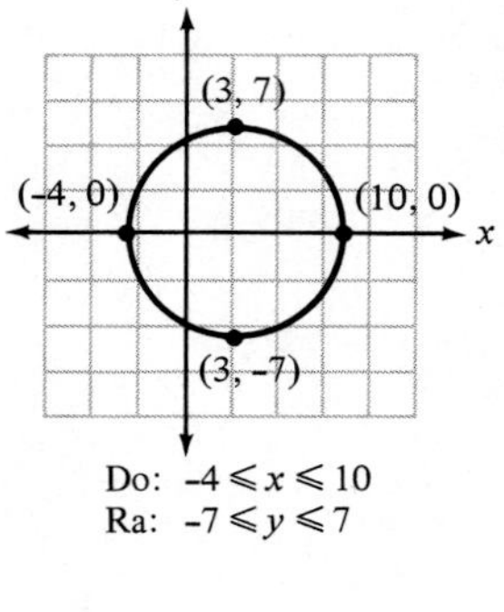

Do: $-4 \leqslant x \leqslant 10$
Ra: $-7 \leqslant y \leqslant 7$

12.

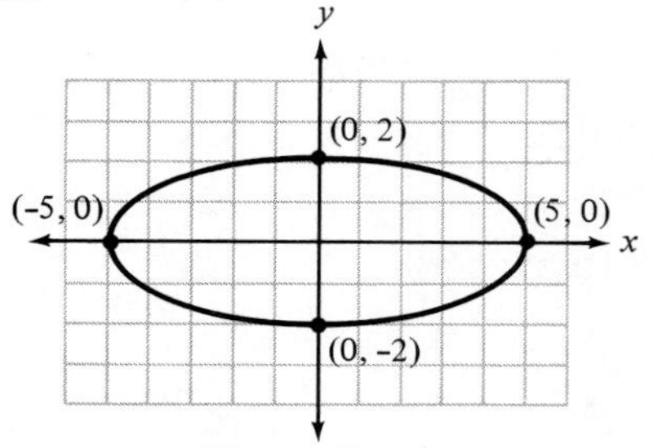

Do: $-5 \leqslant x \leqslant 5$
Ra: $-2 \leqslant y \leqslant 2$

13.

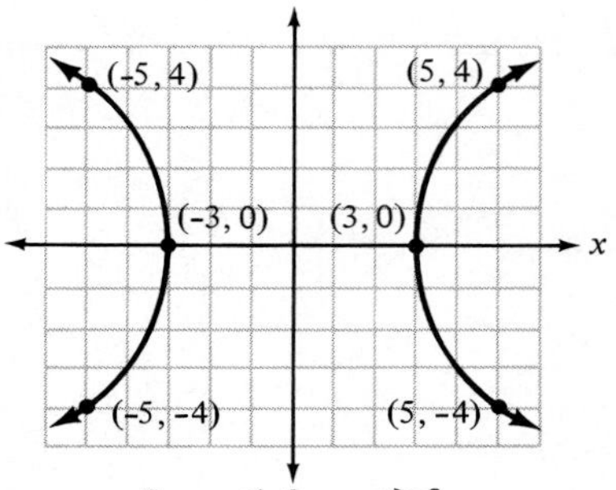

Do: $x \leqslant -3$ or $x \geqslant 3$
Ra: $y \in R$

14.

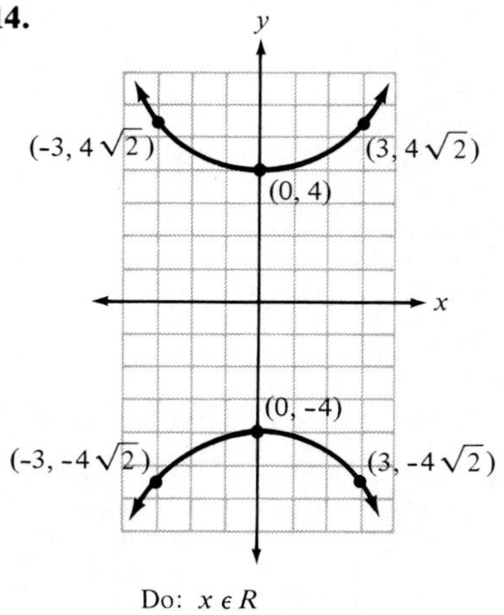

Do: $x \in R$
Ra: $y \leqslant -4$ or $y \geqslant 4$

15.

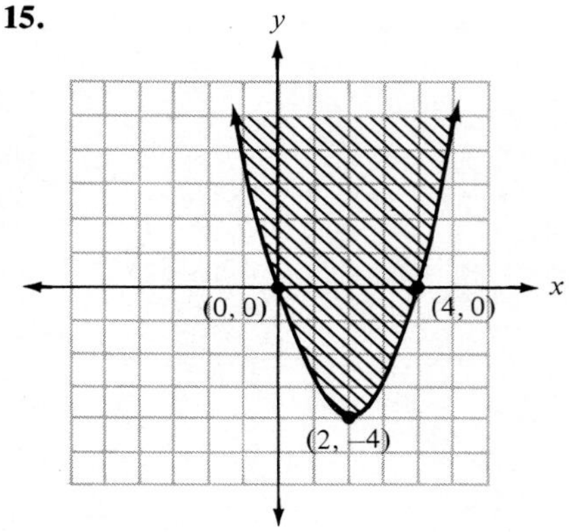

16.

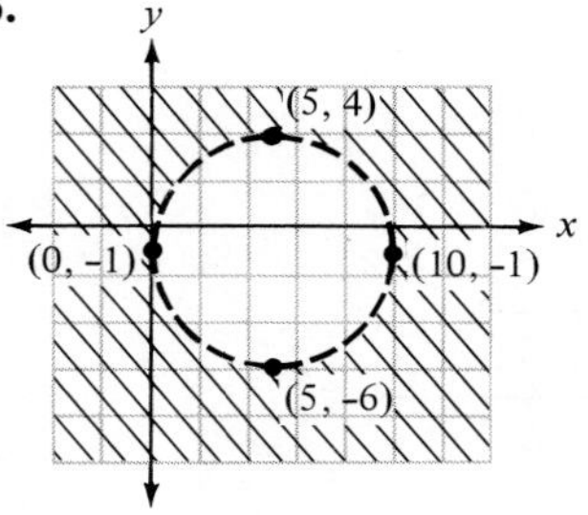

17.

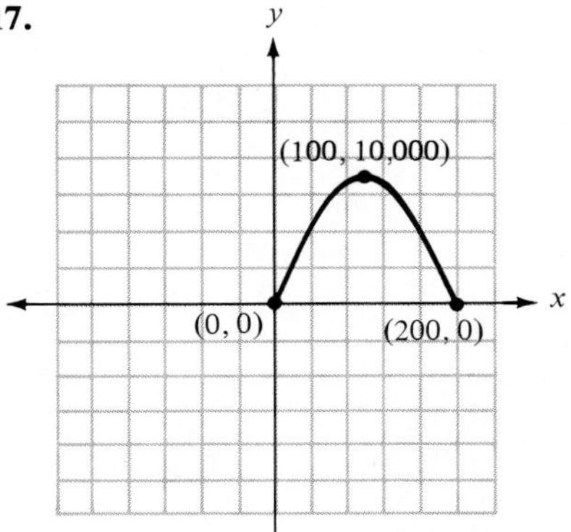

$10,000

Practice Test (page 671)

1.

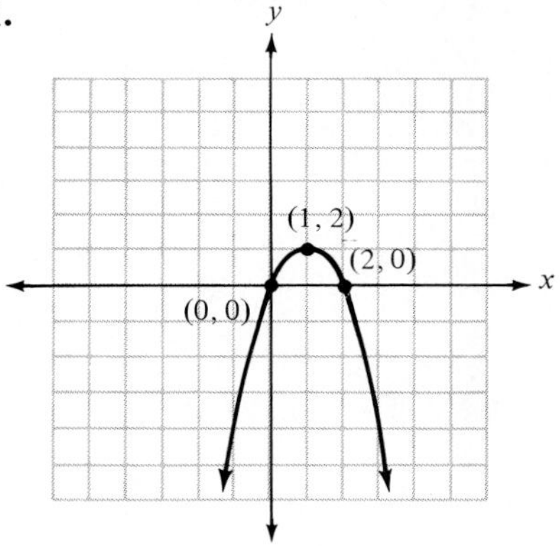

Do : $x \in R$

Ra : $y \in R, y \leqslant 2$

2.

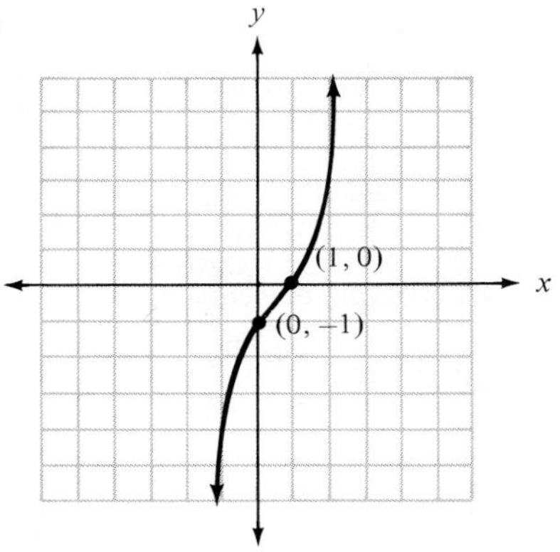

Do : $x \in R$

Ra : $y \in R$

3.

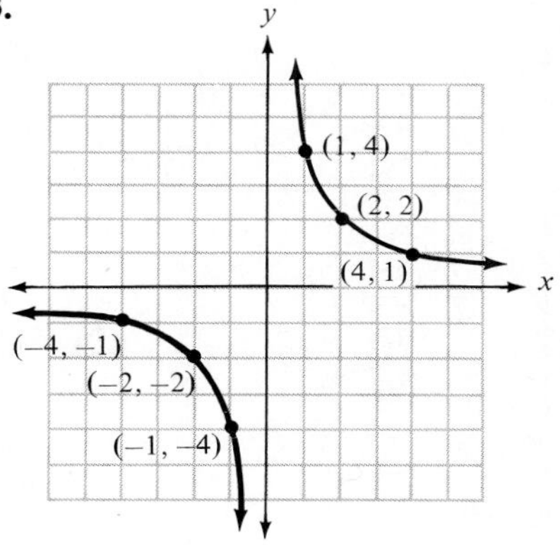

Do : $x \in R, x \neq 0$

Ra : $y \in R, y \neq 0$

4.

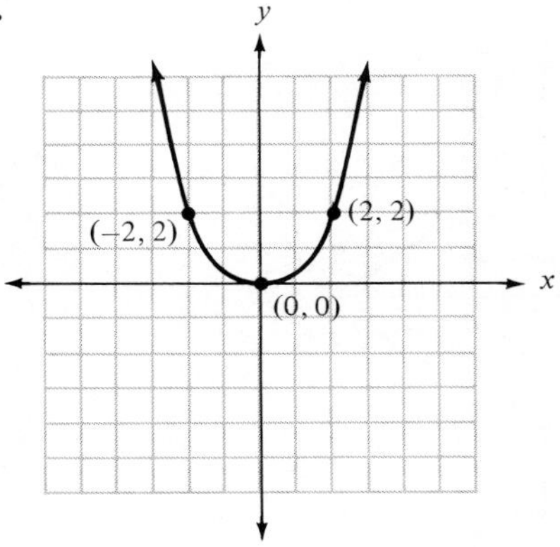

Do : $x \in R$

Ra : $y \in R, y \geqslant 0$

5.

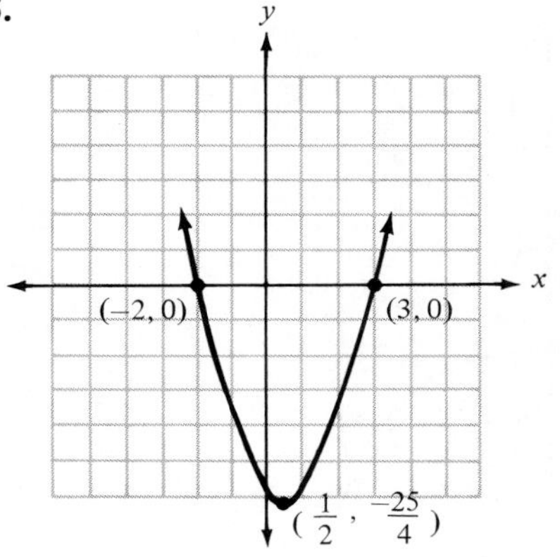

Do : $x \in R$

Ra; $y \in R, y \geqslant -\frac{25}{4}$

6.

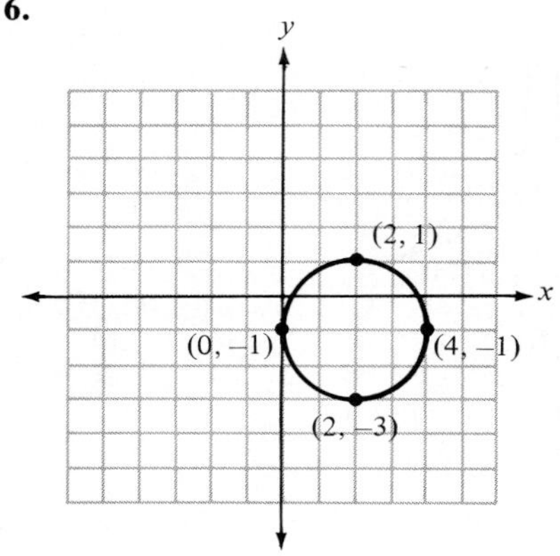

Do : $0 \leqslant x \leqslant 4$
Ra : $-3 \leqslant y \leqslant 1$

7.

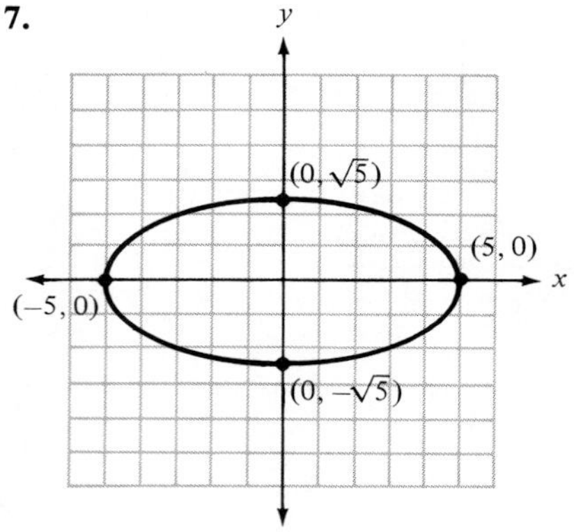

Do : $-5 \leqslant x \leqslant 5$
Ra : $-\sqrt{5} \leqslant y \leqslant \sqrt{5}$

8.

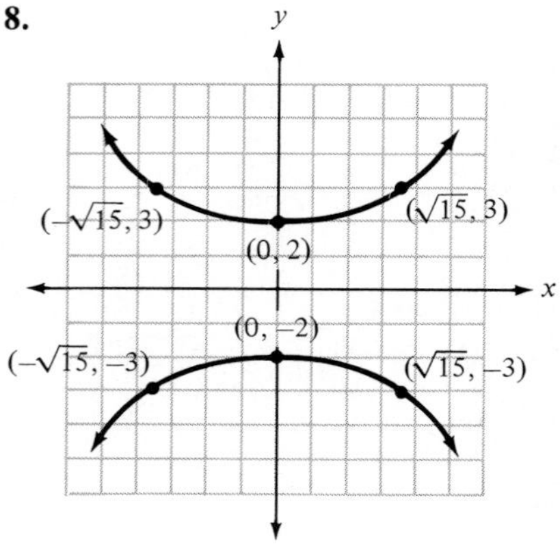

Do : $x \in R$
Ra : $y \in R, y \geqslant 2$ or $y \leqslant -2$

9.

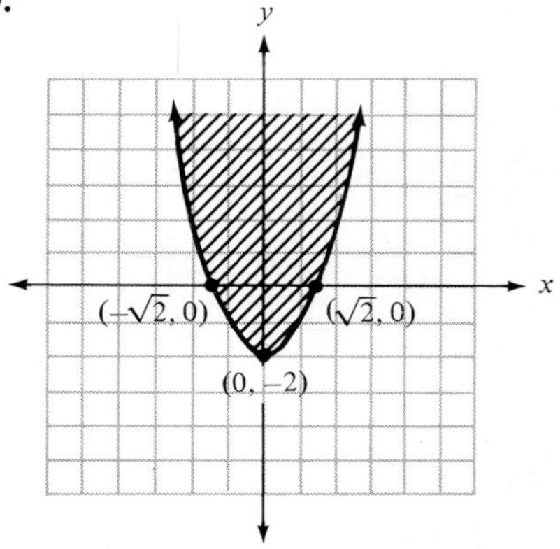

10.

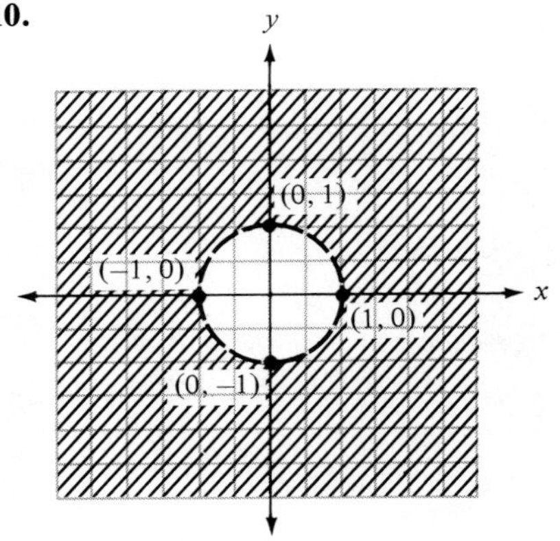

Sharpening Your Skills after Chapters 1–9 (page 677)

1. 33 **2.** $7 - 6\sqrt{2}i$

3. $y = -\frac{2}{5}x + \frac{89}{60}$ or $24x + 60y = 89$ **4.** 17

5. $x^2 - 6x - 7 = 0$ **6.** $-17a$ **7.** $90a^{18}b^{17}$

8. $\frac{z^2}{x^3}$ **9.** $-\frac{7x^6}{y}$ **10.** $a^3 - 12a^2 - 45a$

11. $a^2 - 2a - 8$ **12.** $\frac{4x(x-2)}{(x+4)(x-5)}$ **13.** $\frac{2}{x+3}$

14. $\frac{xy}{y - 2x}$ **15.** $\frac{x^{\frac{1}{3}}}{y^{\frac{5}{27}}z^{\frac{1}{9}}}$ **16.** $2xy^2\sqrt[4]{x}$

17. $-\frac{\sqrt{2} - \sqrt{5}}{3}$ **18.** All real numbers

19. $x = 0, x = \frac{8}{7}$ **20.** $x = \frac{3 \pm \sqrt{19}}{5}$

21. $x \leq 4; (-\infty, 4]$ **22.** $x = -5$

23. $x < \frac{3}{2}$ or $x > 2$; $(-\infty, \frac{3}{2}) \cup (2, \infty)$

24. $x = -4, x = 3$ **25.** $x = \frac{3}{5}, x = -\frac{3}{10} \pm \frac{3\sqrt{3}}{10}i$

26.

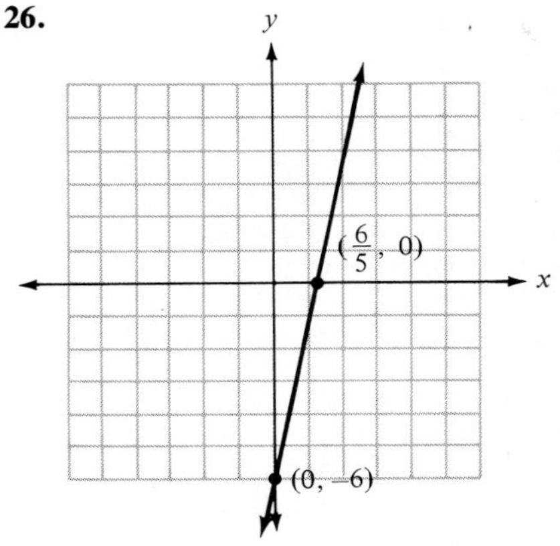

27.

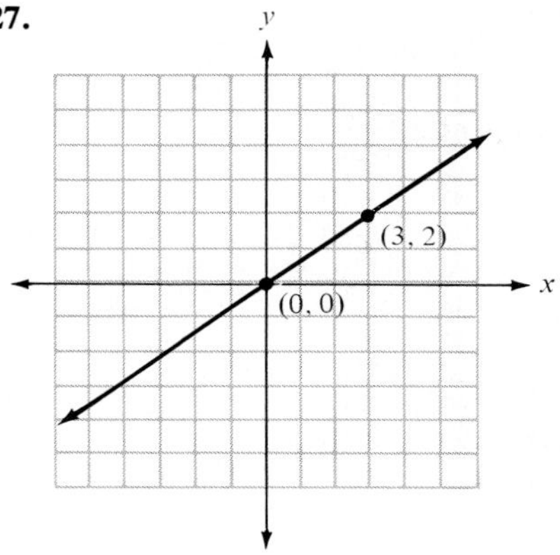

28.

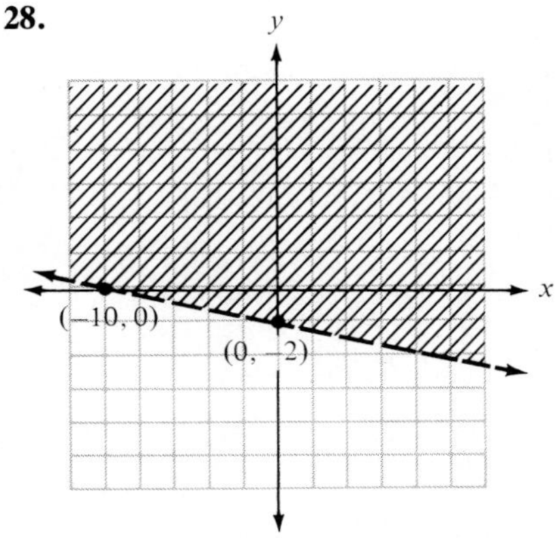

29. $V = 32t$; 256 ft per sec **30.** $V = \pm\frac{\sqrt{2hg}}{2}$

Chapter 10

Exercise Set 10.1 (page 692)

1.

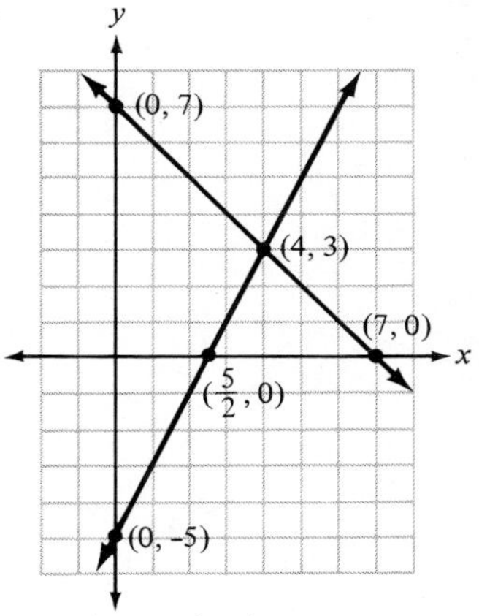

Solution: (4, 3)
Independent and consistent

3.

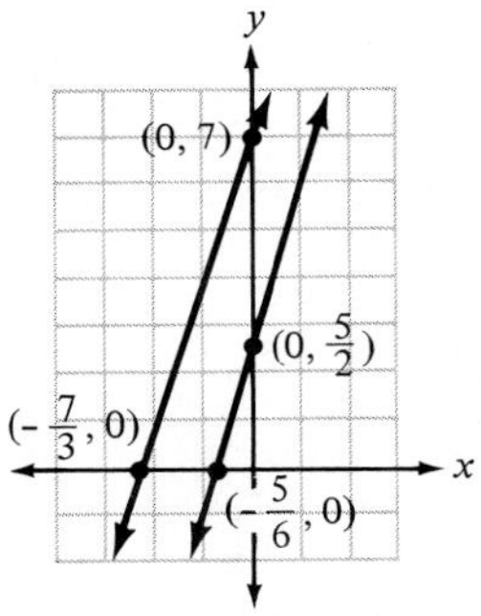

Solution: None
Inconsistent

5.

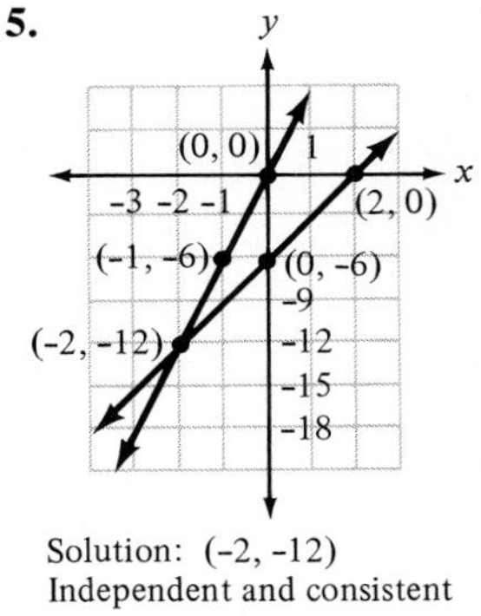

Solution: (-2, -12)
Independent and consistent

7.

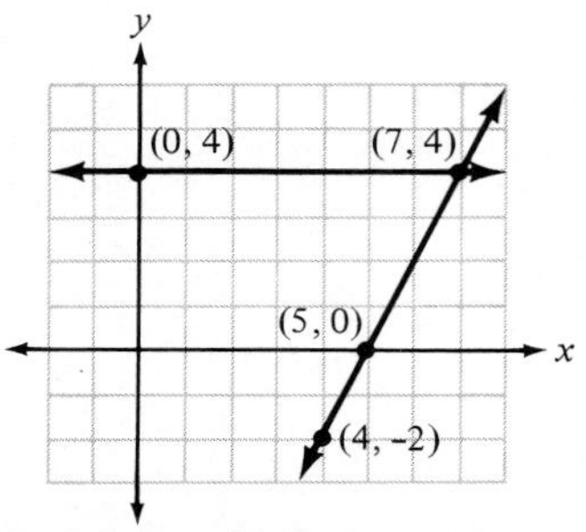

Solution: (7, 4)
Independent and consistent

9.

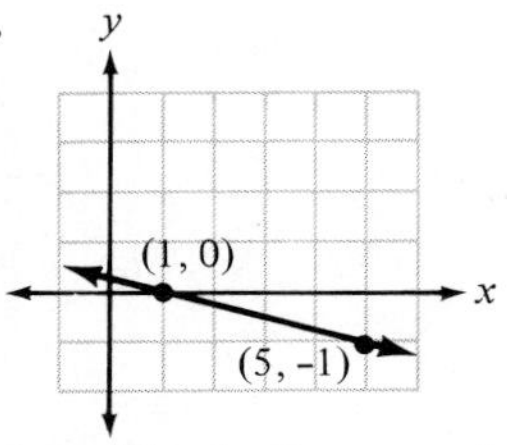

Solution: All ordered pairs satisfying $x + 4y = 1$
Dependent

11.

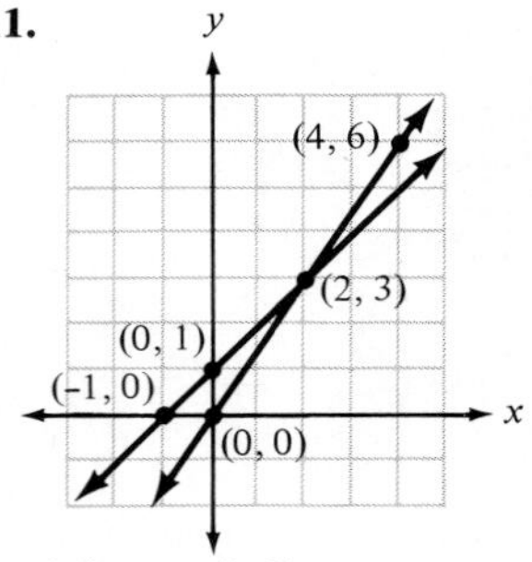

Solution: (2, 3)
Independent and consistent

13.

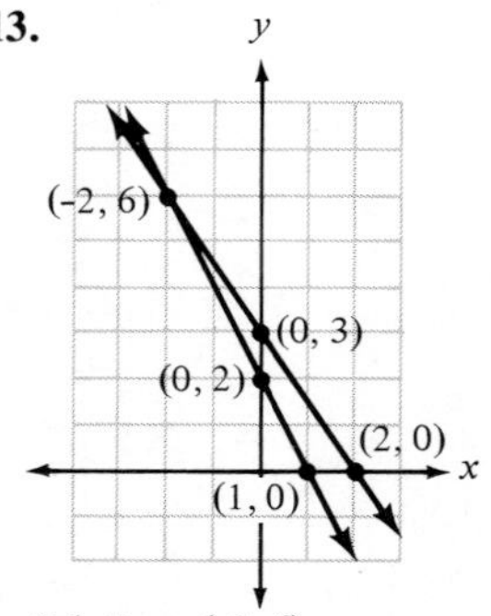

Solution: (-2, 6)
Independent and consistent

15.

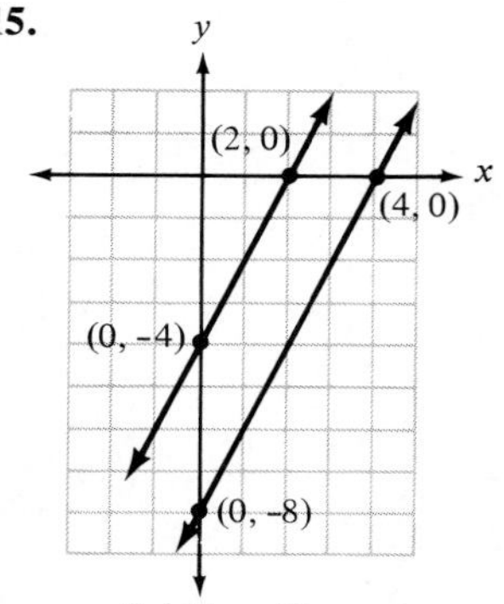

Solution: None
Inconsistent

17. (2, −3) **19.** $(\frac{1}{10}, \frac{17}{10})$ **21.** (2, 3)
23. $(-\frac{13}{21}, -\frac{13}{7})$ **25.** No solution **27.** $(\frac{25}{14}, -\frac{6}{7})$
29. All ordered pairs satisfying $y = 2x - 4$ **31.** (1, 3)

Stretching the Topics

1. (1, −2) **2.** $k = 3$

Checkup 10.1 (page 694)

1.

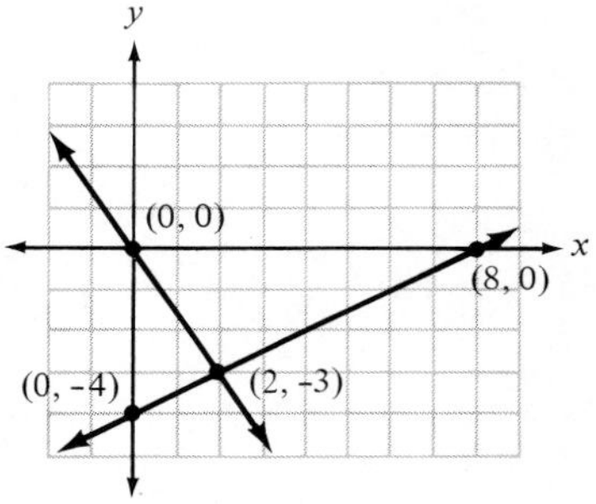

Solution: (2, -3)
Independent and consistent

2.

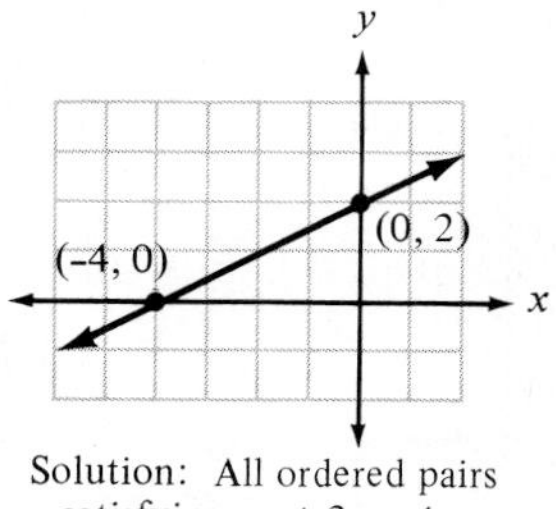

Solution: All ordered pairs satisfying $-x + 2y = 4$
Dependent

3. (2, 2) **4.** No solution **5.** $(\frac{5}{3}, \frac{4}{3})$

Exercise Set 10.2 (page 703)

1. (1, −1) **3.** $(\frac{3}{4}, \frac{9}{8})$ **5.** (2, 1) **7.** No solution
9. (1, 1) **11.** (1, 1)
13. All ordered pairs satisfying $3x - 6y = 9$
15. $(0, -\frac{5}{2})$ **17.** $(-\frac{1}{25}, -\frac{11}{25})$ **19.** $(-\frac{6}{37}, -\frac{27}{37})$
21. $(\frac{18}{7}, \frac{30}{7})$ **23.** (4, 3) **25.** (−170, 70)
27. (0, −1, 1) **29.** (2, −1, 3) **31.** (4, 0, 2)
33. (1, −1, 4) **35.** $(\frac{3}{2}, 3, -3)$

Stretching the Topics

1. $(\frac{31}{9}, \frac{25}{12})$ **2.** (0, 4) **3.** $m = 1, b = -1$

Checkup 10.2 (page 705)

1. $(-1, -1)$ **2.** (1, 1) **3.** (5, 2)
4. All ordered pairs satisfying $3x - 6y = 3$
5. (2, 3, 1)

Exercise Set 10.3 (page 712)

1. 16 **3.** 12 **5.** $\frac{1}{3}$ **7.** $(5, -1)$ **9.** No solution
11. $(-1, 0)$ **13.** $(-\frac{2}{3}, \frac{3}{2})$ **15.** $(\frac{67}{25}, \frac{12}{25})$ **17.** (5, 3)

Stretching the Topics

1. $a = -\frac{38}{7}$ **2.** $\left(\dfrac{3b - 4}{ab + 4}, \dfrac{-2(a + 3)}{ab + 4}\right)$

Checkup 10.3 (page 713)

1. -15 **2.** 3 **3.** $(2, -1)$ **4.** No solution
5. $(3, -2)$ **6.** $(\frac{30}{97}, \frac{19}{97})$

Exercise Set 10.4 (page 724)

1. $(1, -4), (-2, -1)$ **3.** $(-2, -4), (4, 2)$
5. $(4, -3), (-4, -3), (3, 4), (-3, 4)$
7. $(\frac{5}{2}, 1), (-\frac{5}{2}, -1)$
9. $(2, 2), (2, -2), (-2, 2), (-2, -2)$ **11.** No solution
13. $(-5, -2), (2, 5)$ **15.** $(1, 2), (1, -2)$

17.

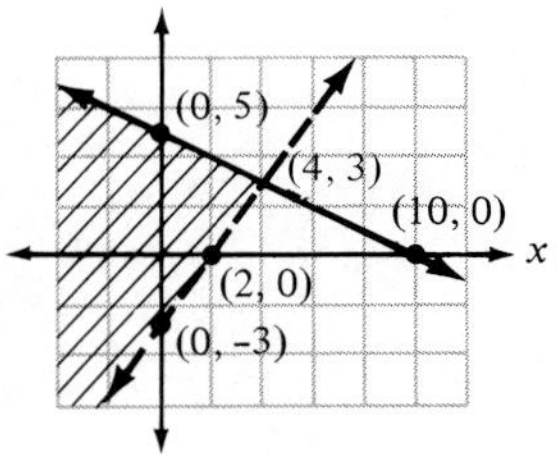

19.

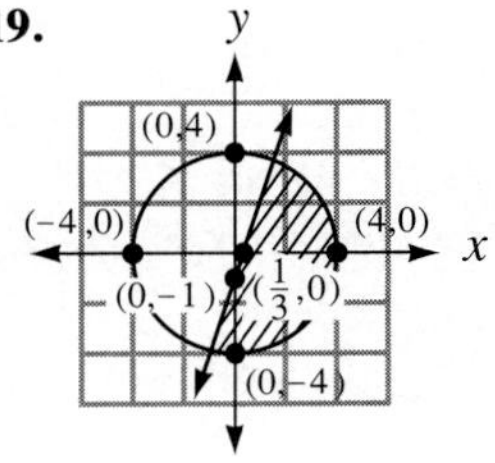

21.

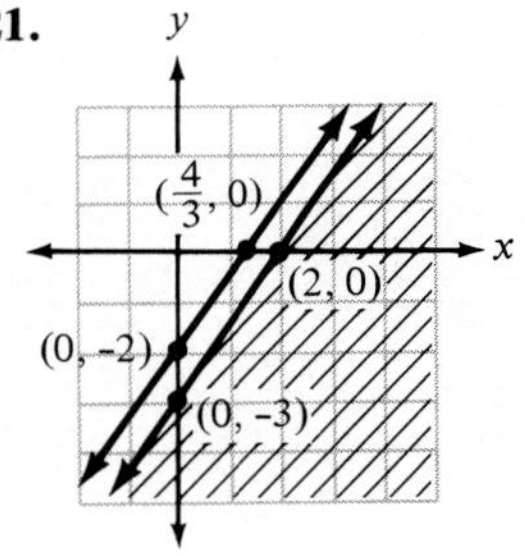

23.

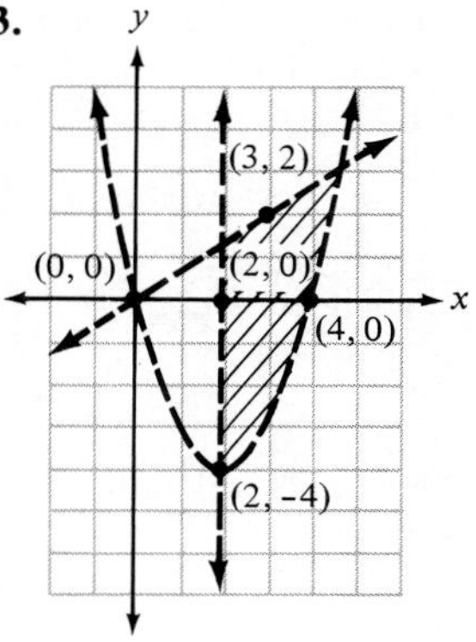

Stretching the Topics

1. $y = \dfrac{a \pm \sqrt{32 - a^2}}{2}$
2. $(-4\sqrt{2}, 4\sqrt{2})$; $(-\infty, -4\sqrt{2}) \cup (4\sqrt{2}, \infty)$; $a = \pm 4\sqrt{2}$

3.

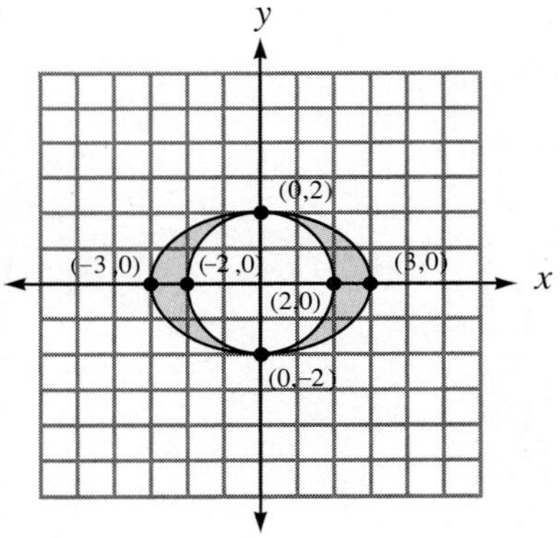

Checkup 10.4 (page 725)

1. $(-4, 8), (1, -7)$
2. $(0, 3), (\sqrt{5}, -2), (-\sqrt{5}, -2)$ **3.** No solution

4.

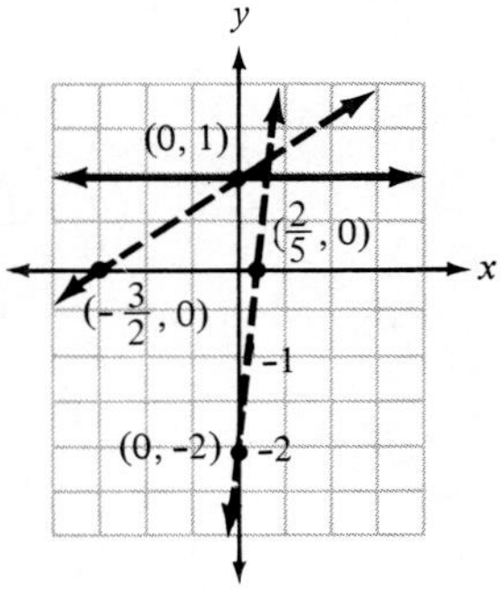

Exercise Set 10.5 (page 731)

1. 193 adults, 52 children **3.** \$5 mowing, \$4 guarding
5. 14 at \$2, 16 at \$1.75
7. \$15,000 at 10%, \$10,000 at 12%

9. Sam, 55 mph; roommate, 65 mph **11.** 12 by 16 ft
13. 175 at \$1.10, 325 at \$0.95 **15.** 100 of each
17. 15 and 20 in. **19.** 8 by 9 ft **21.** 3 by 4 ft
23. 300 pitchers at \$22

Stretching the Topics

1. 4 and 9 in. **2.** 50 bales weighing 60 lb each
3. 5 by 12 ft

Checkup 10.5 (page 734)

1. Lettuce, \$5.40; cabbage, \$4.20 **2.** 9 and 12 ft
3. $3\frac{5}{8}$ hr at 45 mph, $2\frac{3}{8}$ hr at 53 mph
4. 50 by 160 ft or 80 by 100 ft

Speaking the Language of Algebra (page 736)

1. Ordered pair **2.** Substitution **3.** Opposite
4. No; parallel **5.** 4

Review Exercises (page 739)

1.

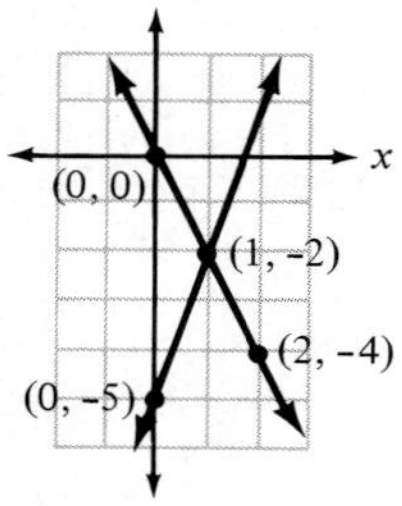

Solution: (1, -2)
Independent and consistent

2.

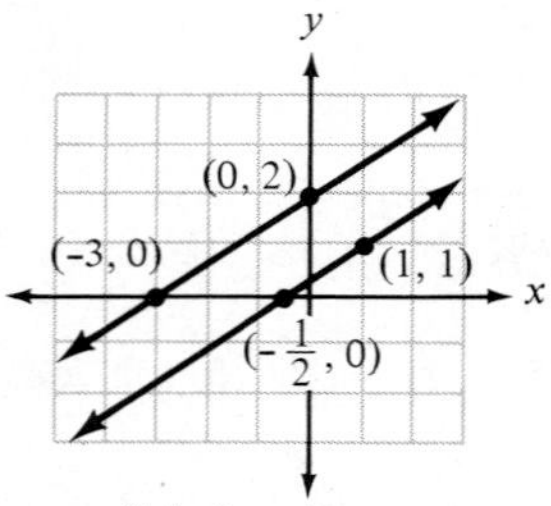

Solution: None
Inconsistent

3. $(-4, 3)$ **4.** $(2, -3)$ **5.** $(-8, 5)$
6. No solution **7.** $(10, -3)$ **8.** $(0, 2, -1)$
9. $(8, 49)$, $(-3, -6)$
10. $(4, 2)$, $(4, -2)$, $(-4, 2)$, $(-4, -2)$

11.

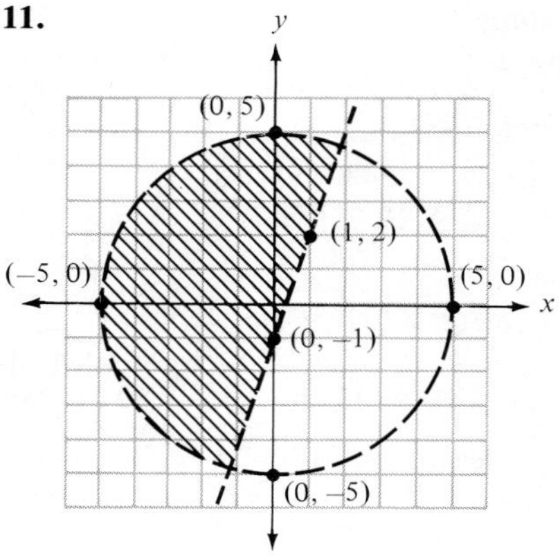

12.

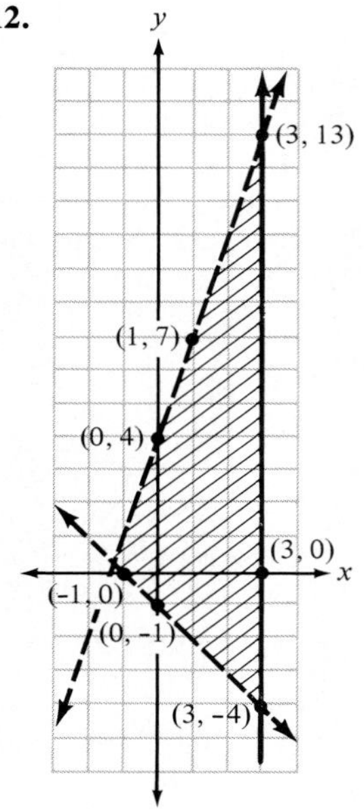

13. 83 speeding, 34 failure to obey traffic signals
14. 21 mi **15.** 5 by 12 m **16.** 1500 lamps at \$39

Practice Test (page 741)

1.

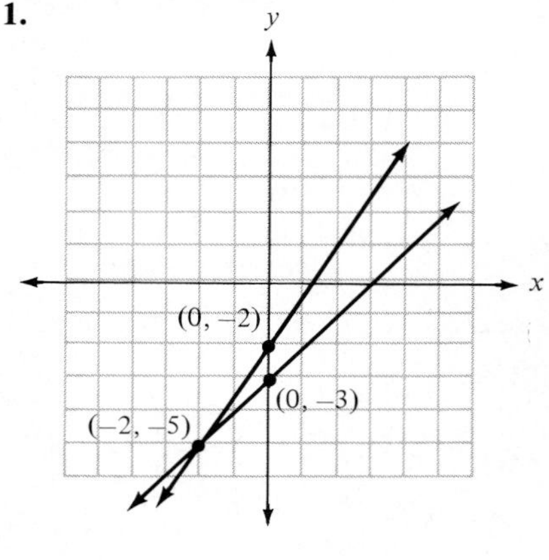

2. $(-\frac{1}{2}, 3)$ **3.** $(-\frac{19}{3}, \frac{10}{3})$
4. All ordered pairs satisfying $6x - 4y = -20$
5. $(-15, -7)$ **6.** No solution **7.** $(1, -2, 0)$
8. $(3, 3), (1, -1)$
9.

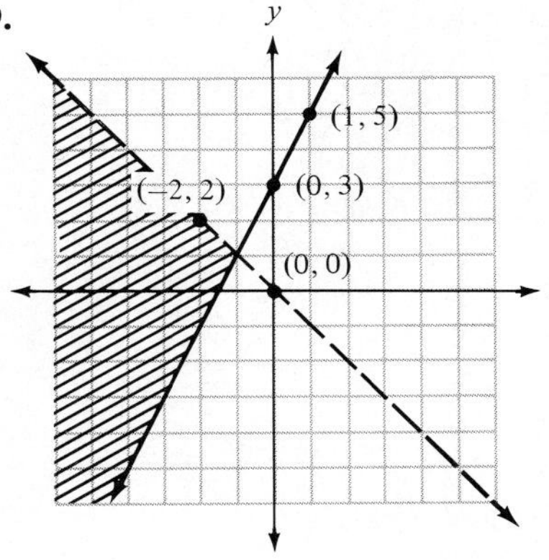

10.

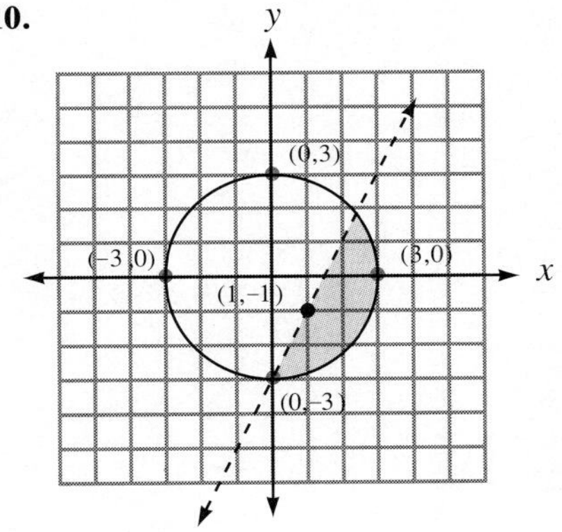

11. Apple, 80 calories; peach, 60 calories
12. 2500 stereos at $80

15. $\frac{(x-3)^2}{(x+2)^2}$ **16.** $\frac{9}{4}$ **17.** $\frac{x^{\frac{1}{20}} z^{\frac{1}{6}}}{y^{\frac{1}{24}}}$ **18.** $-2x\sqrt[3]{2}$

19. $12 - 7\sqrt{3y} - 2y$ **20.** $-\frac{5}{7} - \frac{2\sqrt{6}}{7}i$

21. $x = -\frac{1}{5}, x = -\frac{3}{2}$ **22.** $x = 3, x = \frac{1}{4}, x = -\frac{1}{4}$

23. $x = \frac{-5 + \sqrt{17}}{8}, x = \frac{-5 - \sqrt{17}}{8}$

24. $-12 < x < 12; (-12, 12)$ **25.** No solution
26. $x < 0$ or $x \geq 15; (-\infty, 0) \cup [15, \infty)$ **27.** $x = 36$

28. $x = \frac{4}{7} \pm \frac{\sqrt{5}}{7}i$

29.

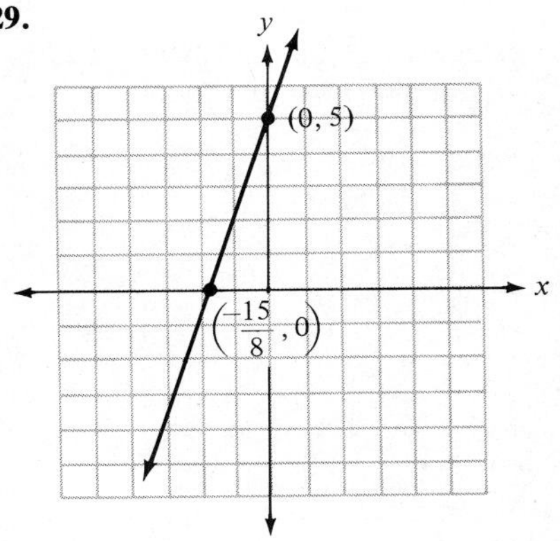

30.

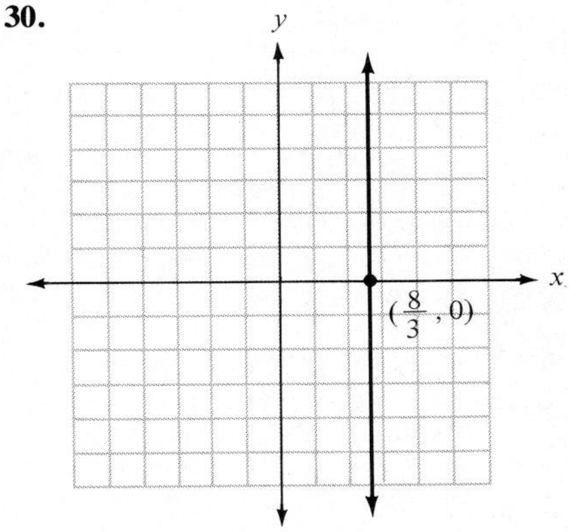

31.

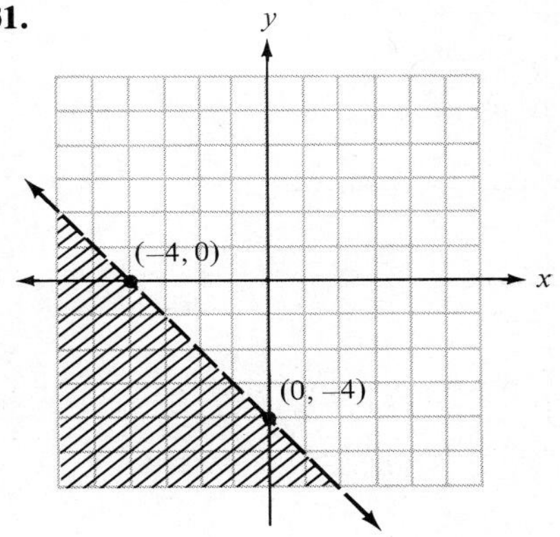

Sharpening Your Skills after Chapters 1–10 (page 743)

1. $a \neq \frac{2}{5}, a \neq -5, \frac{a - 11}{5a - 2}$ **2.** $x = 5, y = -\frac{7}{5}$

3. $y = 1$ **4.** $(4, \frac{7}{2})$ **5.** $y = 2x + 9$ **6.** -9

7. $x \leq \frac{5}{2}; (-\infty, \frac{5}{2}]$ **8.** -4

9. $x < \frac{4}{5}; (-\infty, \frac{4}{5})$

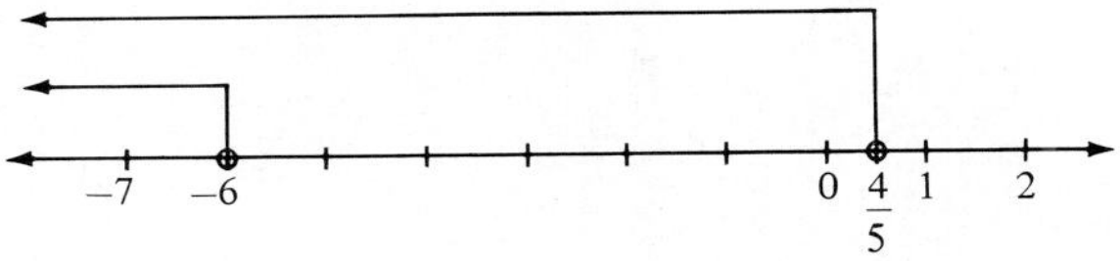

10. $\frac{2a^9}{3b^5}$ **11.** $x^3 + 8x^2 + 7x$

12. $x^4 - 4x^3 + 6x^2 - 4x + 1$

13. $-\frac{x + 4}{5x - 2}$ **14.** $\frac{4x^2 + 15x - 11}{(x - 2)(x - 1)(x + 4)}$

32.

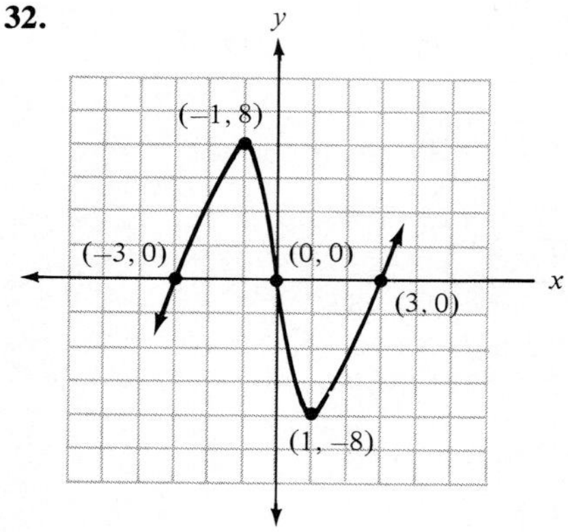

Do: $x \in R$
Ra: $y \in R$

33.

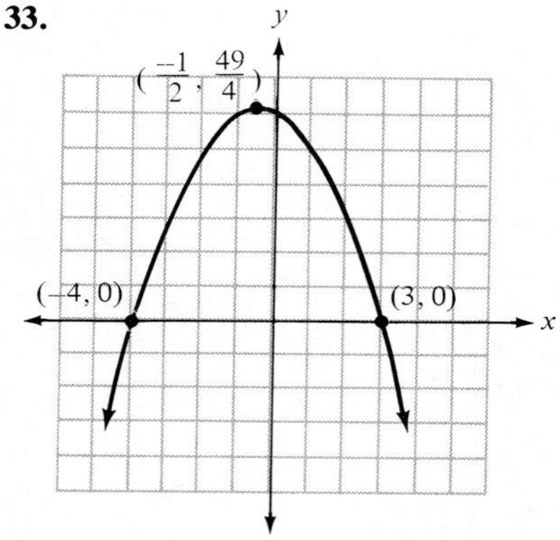

Do: $x \in R$
Ra: $y \in R, y \leq 12\frac{1}{4}$

34.

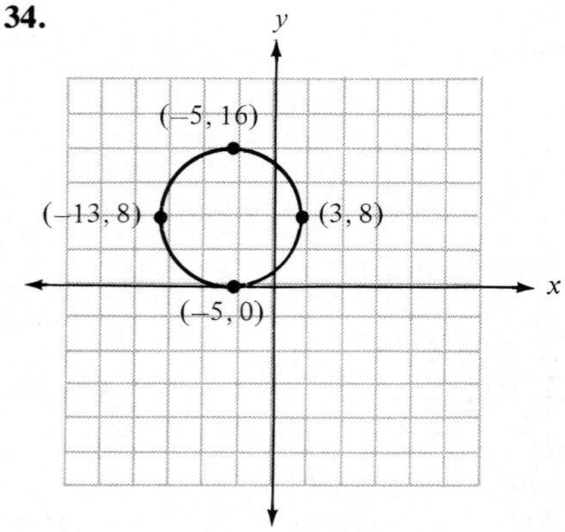

Do: $-13 \leq x \leq 3$
Ra: $0 \leq y \leq 16$

35.

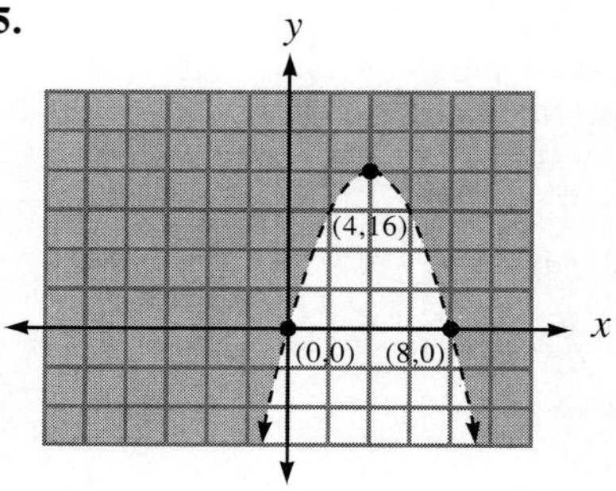

36. $7\frac{1}{2}$ hr

Chapter 11

Exercise Set 11.1 (page 756)

1. $\frac{1}{2}$, 1, 2, 16 **3.** $\frac{1}{25}, \frac{\sqrt{5}}{5}, \sqrt{5}, 25$

5. $-\frac{1}{8}, -\frac{1}{2}, -1, -4$ **7.** $\frac{1}{4}, 1, \sqrt{2}i, 16$

9. $\frac{1}{2}, \frac{1}{4}, \frac{\sqrt{2}}{2}, 4$

11.

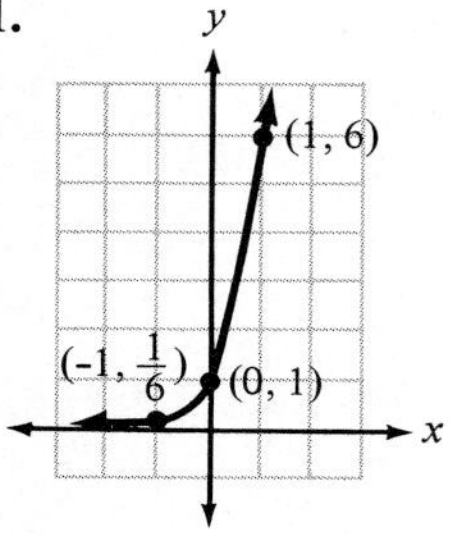

13.

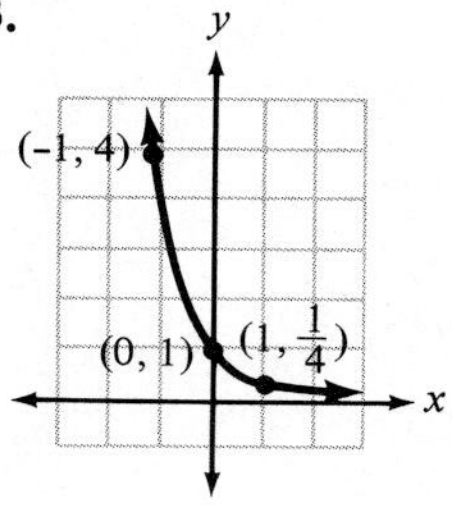

15.

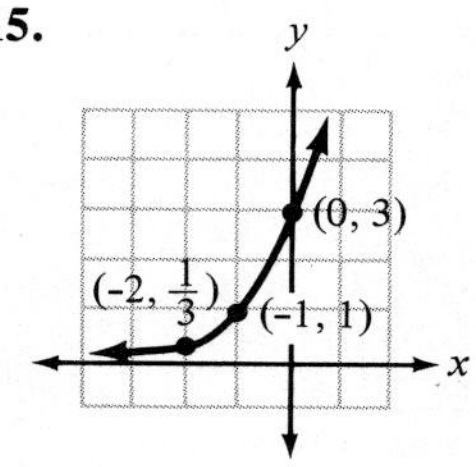

17.

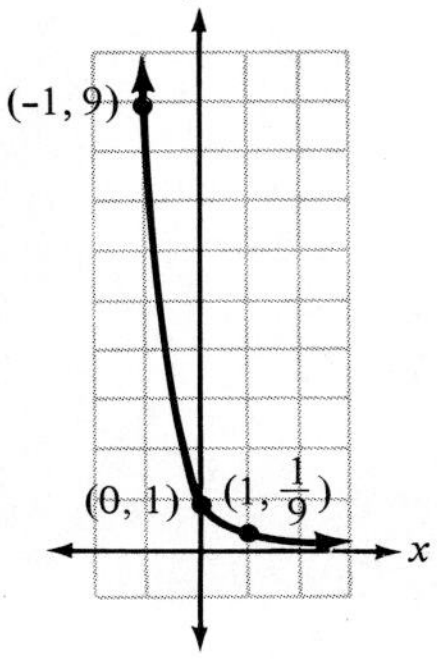

19.

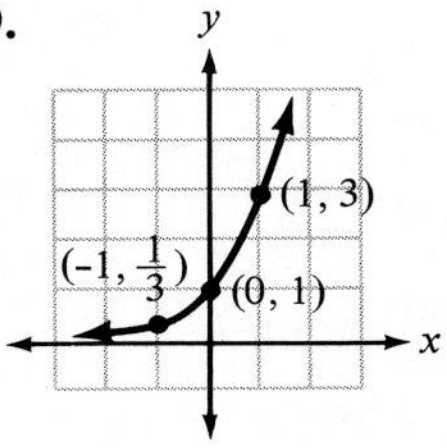

21.

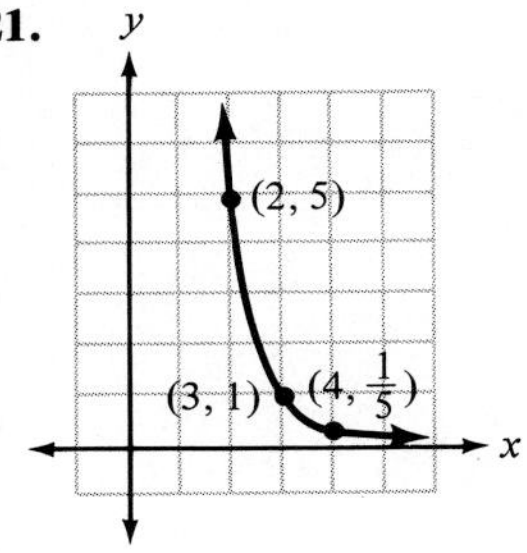

23.

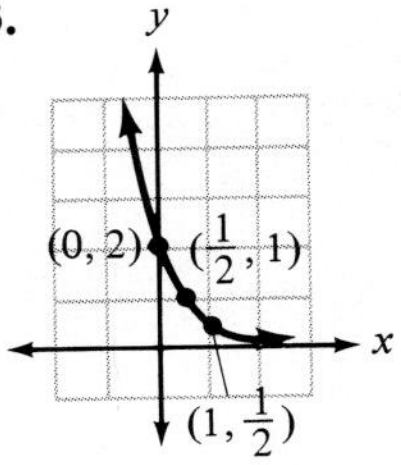

25. \$1104.08 **27.** \$1268.24 **29.** $x = -1$ **31.** $x = 2$ **33.** $x = -4$ **35.** $x = -1$ **37.** $x = -\frac{1}{2}$ **39.** $x = -2$ **41.** $x = -\frac{3}{4}$

Stretching the Topics

1.

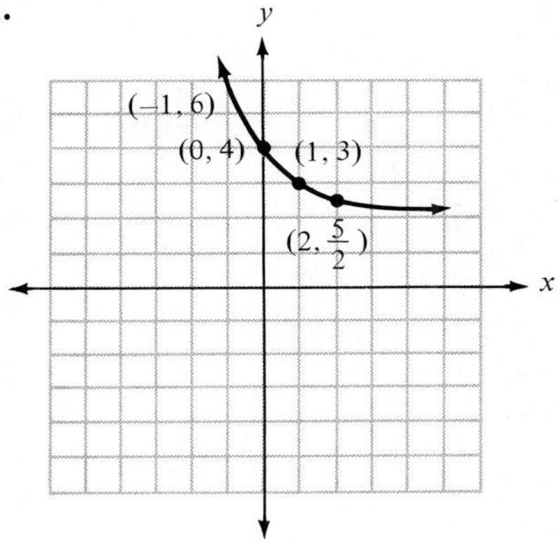

2. $x = 4, x = -1$ **3.** \$8832.16

Checkup 11.1 (page 758)

1. $\frac{1}{9}, \frac{1}{3}, \frac{\sqrt{3}}{3}, 3, \sqrt{3}$

2.

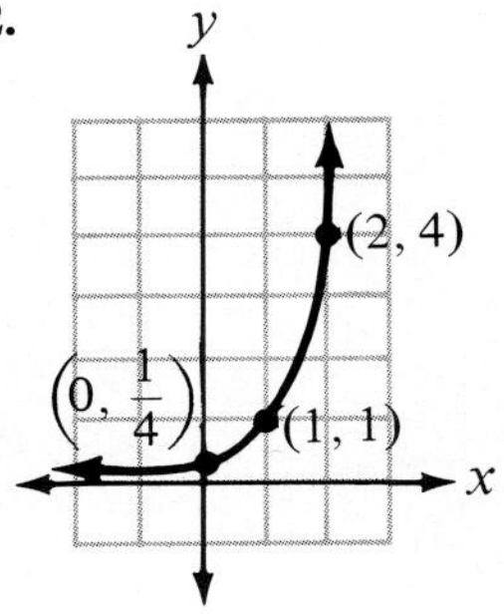

3.

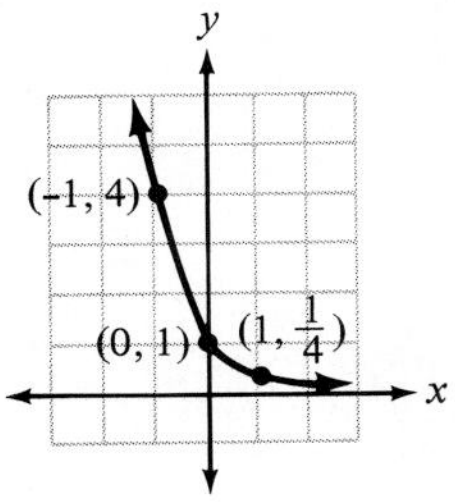

4. $x = -2$ **5.** $x = \frac{2}{7}$

Exercise Set 11.2 (page 767)

1. $\log_3 9 = 2$ **3.** $\log_4 \frac{1}{16} = -2$ **5.** $\log_{16} 4 = \frac{1}{2}$ **7.** $\log_9 \frac{1}{3} = -\frac{1}{2}$ **9.** $5^3 = 125$ **11.** $2^{-5} = \frac{1}{32}$ **13.** $16^{\frac{1}{2}} = 4$ **15.** $36^{-\frac{1}{2}} = \frac{1}{6}$ **17.** -1 **19.** 0 **21.** 3 **23.** 7 **25.** Does not exist **27.** 5 **29.** 0 **31.** $x = -1$ **33.** $a = 7$ **35.** $b = \frac{1}{32}$ **37.** $x = 63$ **39.** $x = -\frac{1}{2}$ **41.** $x = 3$

43.

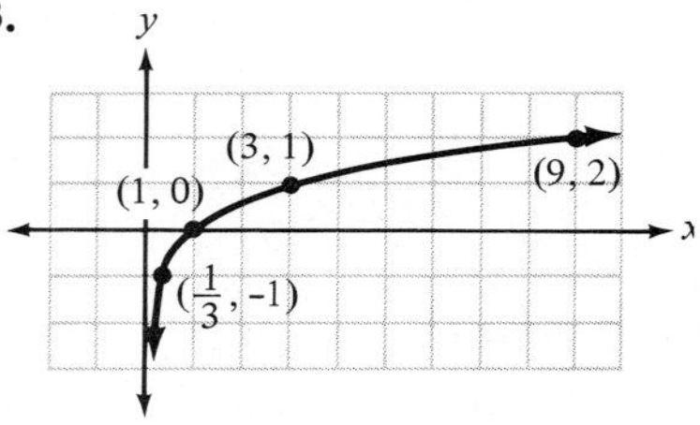

45.

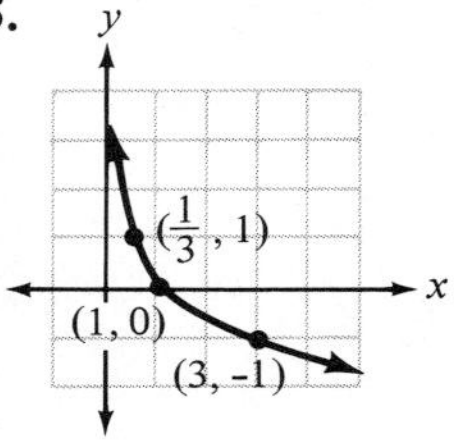

Stretching the Topics

1. $x = \frac{5 \pm \sqrt{41}}{2}$

2.

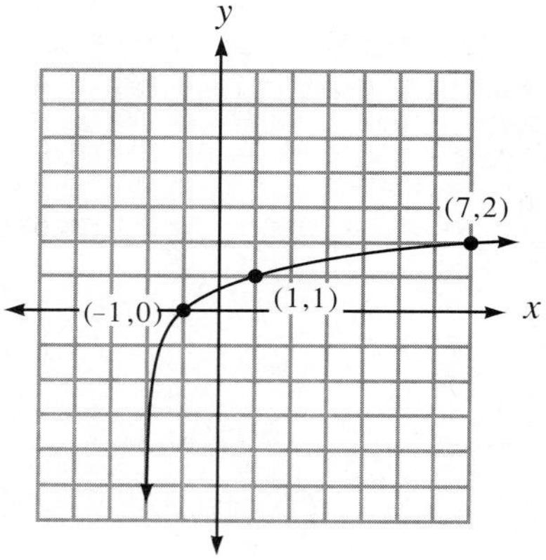

3. $x = -6, x = 1$

Checkup 11.2 (page 769)

1. $\log_5 25 = 2$ **2.** $\log_{64} \frac{1}{8} = -\frac{1}{2}$ **3.** $3^4 = 81$
4. $5^{-2} = \frac{1}{25}$ **5.** -5 **6.** $\frac{1}{3}$ **7.** $a = 5$
8. $x = 5$

9.

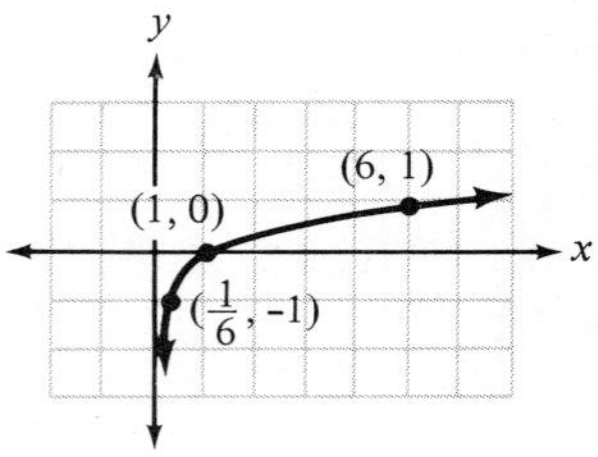

10.

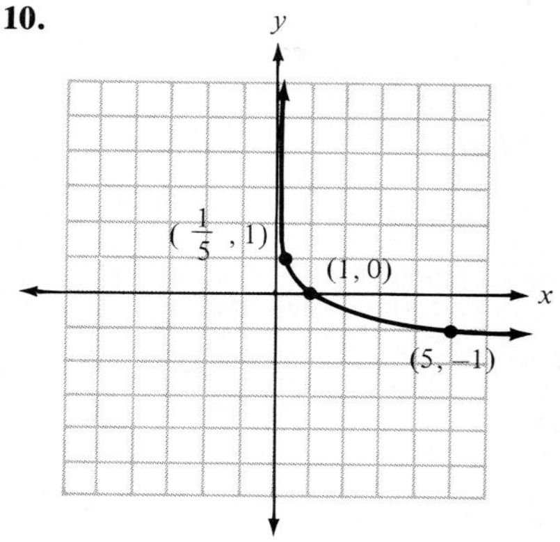

Exercise Set 11.3 (page 778)

1. 7 **3.** -1 **5.** 4 **7.** $\frac{3}{2}$ **9.** 7 **11.** 0
13. -1 **15.** 1.1761 **17.** 1.1505 **19.** 2.5441
21. 0.9121 **23.** 1.6532
25. $3 \log_a x + 2 \log_a (x + 5)$
27. $5 \log_a (x - 3) - \frac{1}{2} \log_a 3 - \frac{1}{2} \log_a x$
29. $-\frac{4}{3} \log_a x - 5 \log_a (x - 9)$
31. $\log_a (x - 3) + \frac{1}{2} \log_a (2 - x) - \frac{1}{2} \log_a x - 5 \log_a (x + 1)$
33. $\log_a \sqrt[3]{x}\,(x - 1)^2$ **35.** $\log_a \frac{\sqrt{x}}{(x + 2)^4}$
37. $\log_a \frac{x^3\sqrt{3x}}{x^2 + 1}$ **39.** $\log_a \frac{\sqrt[3]{(x + 1)^2}\,(x + 1)^3}{x(x^2 + 1)^5}$

Stretching the Topics

1. $\log_a r = \frac{1}{2}(\log_a 3 + \log_a V - \log_a \pi - \log_a h)$
2. $5 \log_a \sqrt[3]{a} - 7 \log_a \sqrt{a} + 10 \log_a \sqrt[3]{a} - \frac{3}{2} \log_a a$

$$= \log_a \frac{a^{\frac{5}{3}} \cdot a^{\frac{10}{3}}}{a^{\frac{7}{2}} \cdot a^{\frac{3}{2}}}$$

$$= \log_a \frac{a^{\frac{15}{3}}}{a^{\frac{10}{2}}}$$

$$= \log_a \frac{a^5}{a^5}$$

$$= \log_a 1$$

$$= 0$$

Checkup 11.3 (page 780)

1. 7 **2.** $\frac{1}{2}$ **3.** 1.8572 **4.** 0.9515 **5.** 2.5562
6. 1.0512 **7.** $2 \log_a x + \frac{1}{2} \log_a (x^2 + 1)$
8. $\frac{2}{3} \log_a x + \log_a (x^2 + 1) - 5 \log_a (x^2 + 4)$
9. $\log_a \frac{(x + 1)^2}{\sqrt[3]{x^2}}$ **10.** $\log_a \frac{\sqrt{3x}\,(x + 5)}{x^4(2x + 1)}$

Exercise Set 11.4 (page 790)

1. 5.42×10^{-4} **3.** 2.35×10^1 **5.** 8.9×10^7
7. 7.1×10^{-4} **9.** 7.83×10^4 **11.** 0.000162
13. 382,000 **15.** 30.1 **17.** 0.0421
19. 15,000,000 **21.** 4.9420 **23.** $-4 + 0.1430$
25. 1.2718 **27.** 7.9294 **29.** $-9 + 0.7931$
31. 0.0757 **33.** 156,000 **35.** 0.0000000444
37. 2.27 **39.** 88,600 **41.** $N \doteq 3.07$

43. $N \doteq 0.575$ **45.** $N \doteq 2.91$ **47.** $N \doteq 15.0$
49. $N \doteq 4.87$ **51.** $N \doteq 5.34$ **53.** $N \doteq 111$

Stretching the Topics

1. $N \doteq 1.80$ **2.** 105 m

Checkup 11.4 (page 792)

1. 2.31×10^{-3} **2.** 12,100 **3.** 2.8609
4. $-3 + 0.9926$ **5.** 0.00123 **6.** 81,400
7. $N \doteq 0.0175$ **8.** $N \doteq 2.09$ **9.** $N \doteq 20.3$
10. $N \doteq 0.0000276$

Exercise Set 11.5 (page 805)

1. $x = 7$ **3.** $x = 4, x = -\frac{4}{3}$ **5.** $x = -8, x = 2$
7. $x = 5$ **9.** No solution **11.** $x = \frac{7}{2}$ **13.** $x = 3$
15. $x = 38$ **17.** $x = 4$ **19.** $x = 11$ **21.** $x = \frac{9}{4}$
23. $x = 12$ **25.** $x = 33$ **27.** $x = 4$ **29.** $x = \frac{7}{3}$
31. No solution **33.** $x = 5$ **35.** $x = 8$

37. $x = \dfrac{\log 25}{\log 4} \doteq 2.322$ **39.** $x = \dfrac{\log 16}{2 \log 3} \doteq 1.262$

41. $x = \dfrac{\log 17 + \log 5}{\log 5} \doteq 2.760$

43. $x = \dfrac{\log 78.3 - \log 9}{2 \log 9} \doteq 0.492$

45. $x = \dfrac{\log 23.4}{\log 1.76} \doteq 5.577$

47. $x = \dfrac{\log 8 + \log 14}{\log 14} \doteq 1.788$

49. $x = \dfrac{\log 5}{\log 5 - \log 6} \doteq -8.826$

51. $x = \dfrac{\log 3 + 2 \log 7}{\log 7 - 2 \log 3} \doteq -19.865$ **53.** $A \doteq \$6090$

55. $A \doteq \$21,800$ **57.** 15 years (14.3)
59. 16,200; 3.56 hr

Stretching the Topics

1. $x = \dfrac{\log 3}{\log 7 - \log 3} \doteq 1.296$

2. $A \doteq \$111,000$ (With calculator: \$116,486.26)
3. 40 min

Checkup 11.5 (page 808)

1. $x = 7$ **2.** No solution **3.** $x = 6$ **4.** $x = 10$

5. $x = \dfrac{\log 13 - \log 5}{\log 5} \doteq 0.594$

6. $x = \dfrac{\log 3}{\log 7 - \log 3} \doteq 1.296$

Speaking the Language of Algebra (page 810)

1. Exponential; $x \in R$; $y \in R$, $y > 0$
2. Logarithmic; $x \in R$, $x > 0$; $y \in R$ **3.** $a^y = x$
4. Common **5.** Same base **6.** Exponential
7. Negative

Review Exercises (page 813)

1. $5, \frac{1}{5}, \frac{1}{125}, \frac{\sqrt{5}}{5}, 1$

2.

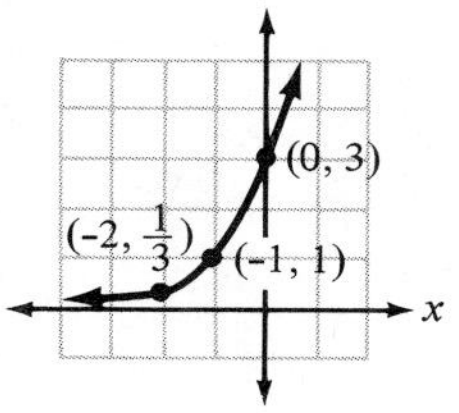

3. $x = -1$ **4.** $\log_3 \frac{1}{9} = -2$ **5.** $\log_{125} \frac{1}{5} = -\frac{1}{3}$
6. $2^6 = 64$ **7.** $8^{-\frac{1}{3}} = \frac{1}{2}$ **8.** -3 **9.** $-\frac{1}{2}$
10. $a = 6$ **11.** $b = \frac{1}{64}$ **12.** $x = 0$
13.

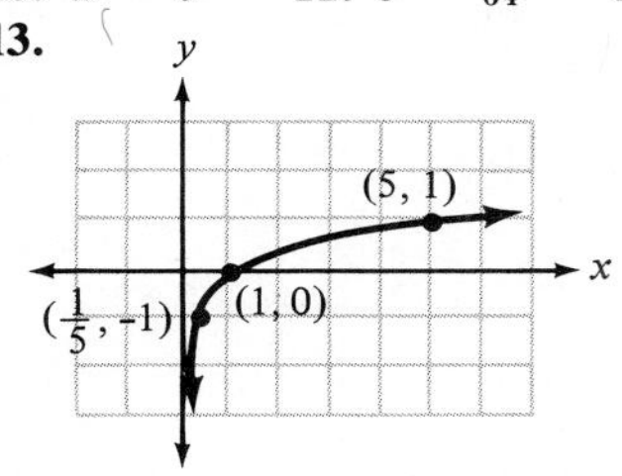

14. $\frac{5}{4}$ **15.** 5 **16.** 0.0737 **17.** 1.3662
18. $5 \log_a x + \log_a (x + 5)$
19. $3 \log_a x + \frac{1}{2} \log_a (x^2 - 1) - 5 \log_a (x + 1)$

20. $\log_a x^2 \sqrt{x - 1}$ **21.** $\log_a \dfrac{(2x + 1)^3}{\sqrt[5]{x^4}\,(x + 1)}$

22. 7.29×10^{-3} **23.** 934,000 **24.** $N \doteq 16,500$
25. $x = 1$ **26.** $x = 9$ **27.** $x = 24$ **28.** $x = \frac{20}{3}$

29. $x = \dfrac{\log 27.3}{2 \log 4} \doteq 1.193$

30. $x = \dfrac{\log 9}{\log 9 - 2 \log 12} \doteq -0.792$ **31.** \$1638.62

32. 866

Practice Test (page 815)

1. 81, 1, 3 **2.** $\log_{36} \frac{1}{6} = -\frac{1}{2}$ **3.** $5^3 = 125$
4. -2 **5.** $-\frac{13}{2}$ **6.** 3.7993
7. $5 \log_a x + 2 \log_a (x^2 - 3) - \frac{1}{3} \log_a (x + 1)$

8. $\log_a \dfrac{\sqrt[5]{x + 1}}{\sqrt[3]{x^2}\,(x - 6)}$ **9.** 2.35×10^6 **10.** 0.00172

11. $N \doteq 101$

12.

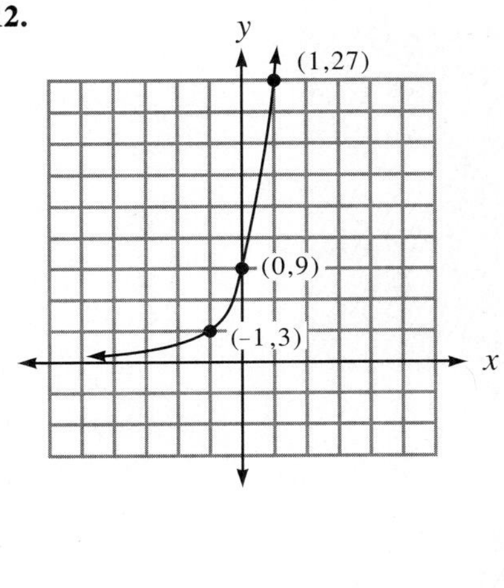

13.

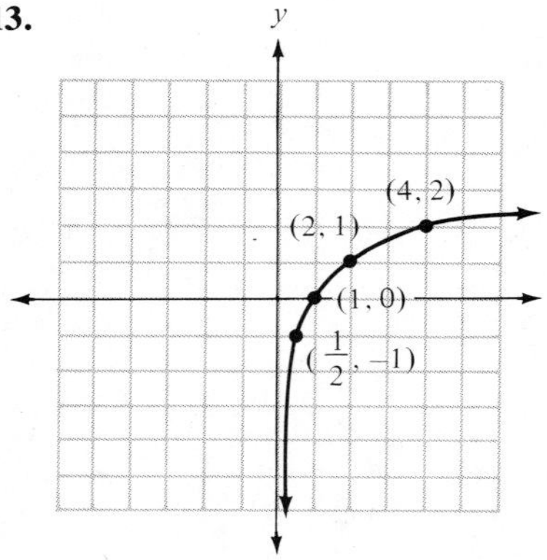

14. $x = -\frac{3}{5}$ **15.** $a = 2$ **16.** $x = 2$ **17.** $x = 3$
18. $x = 1$ **19.** $x = 2$
20. $x = \dfrac{\log 3}{\log 3 - \log 8} \doteq -1.12$ **21.** About 8.85
22. 400; about 4.32

Sharpening Your Skills after Chapters 1–11 (page 817)

1. 32 **2.** 11 **3.** $y \neq 4, \dfrac{3(y+4)}{y-4}$ **4.** $\dfrac{16}{9}$ **5.** i
6. $-\frac{16}{9}$ **7.** $y = \frac{1}{2}x - \frac{1}{4}$ or $2x - 4y = 1$ **8.** 17
9. $\frac{7}{2}$ **10.** $7x^4y^7$ **11.** $\dfrac{y^2}{49x^{12}}$ **12.** $-47x^2 + 27x + 2$
13. $8a^3 - 36a^2 + 54a - 27$ **14.** $a^2 - 2a - 8$
15. $\dfrac{2x-1}{x+2}$ **16.** $-\dfrac{2}{x+5}$ **17.** $\dfrac{4x-3}{x-6}$
18. $ab\sqrt[3]{a}$ **19.** $12 - 7\sqrt{3y} - 2y$ **20.** $-\frac{1}{2} - \frac{3}{2}i$
21. $(x + 7)(x^2 + 2x + 13)$ **22.** $(5h + 12k)(h + 2k)$
23. $(9 - 2x + y)(3 + 10x - 5y)$ **24.** $x = -2$
25. $x > -2$ or $x < -3$; $(-\infty, -3) \cup (-2, \infty)$
26. No solution **27.** $x = 3\sqrt{2}, x = -3\sqrt{2}$
28. $\frac{1}{2} < x < 3$; $(\frac{1}{2}, 3)$ **29.** $x = 3$
30. $-2 < x < 0$; $(-2, 0)$ **31.** No solution
32. $y = \dfrac{1}{2} \pm \dfrac{3\sqrt{3}}{2}i$ **33.** $(2, -1)$
34. $(12, 4), (-3, -1)$

35.

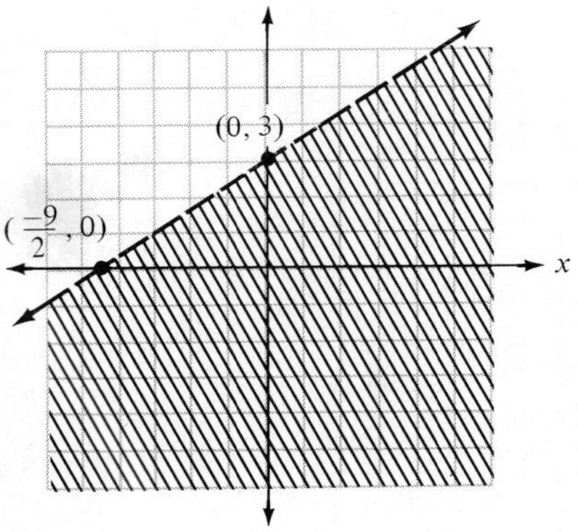

36.

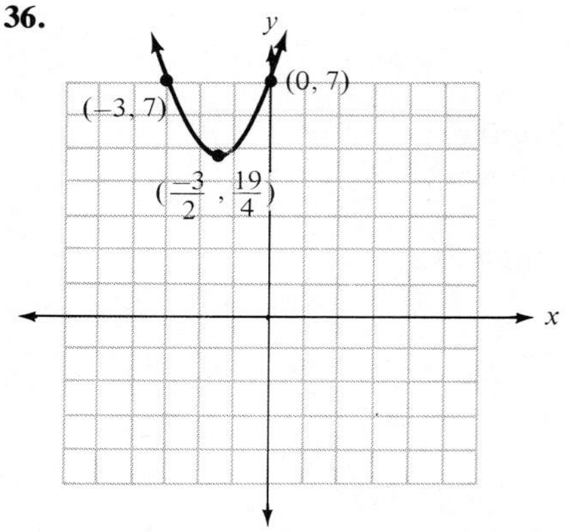

37.

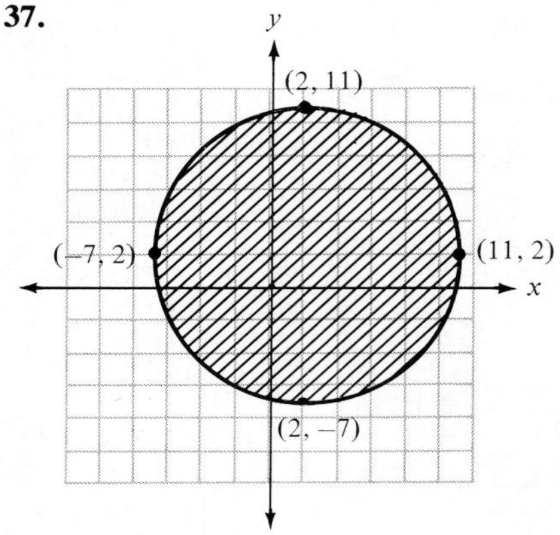

38.

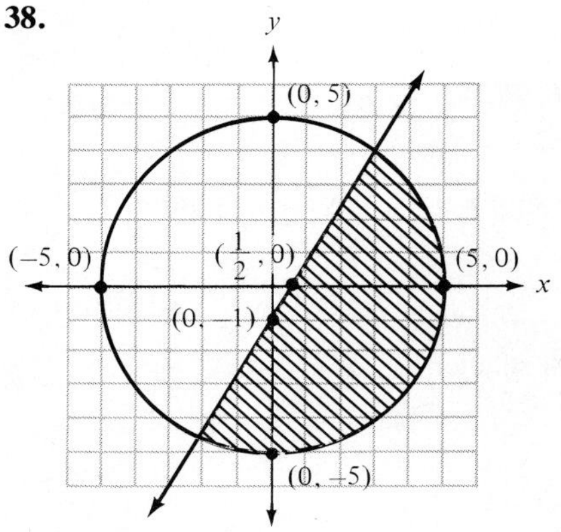

39. $g = \dfrac{4\pi^2 l}{t^2}$ **40.** 6 by 9 ft